MECHANICAL BEHAVIOR
of MATERIALS

This book is in the
**ADDISON-WESLEY SERIES IN
METALLURGY AND MATERIALS**

Morris Cohen, *Consulting Editor*

FRANK A. McCLINTOCK and ALI S. ARGON
Editors
Massachusetts Institute of Technology

MECHANICAL BEHAVIOR of MATERIALS

Ali S. Argon, Stanley Backer,
Frank A. McClintock, George S. Reichenbach,
Egon Orowan, Milton C. Shaw,
Ernest Rabinowicz

ADDISON-WESLEY PUBLISHING COMPANY
READING, MASSACHUSETTS · MENLO PARK, CALIFORNIA
LONDON · AMSTERDAM · DON MILLS, ONTARIO · SYDNEY

Copyright © 1966

ADDISON-WESLEY PUBLISHING COMPANY, INC.

Printed in the United States of America

All rights reserved. This book, or parts thereof, may not be reproduced in any form without written permission of the publisher.

Library of Congress Catalog Card No. 64–16906

ISBN 0-201-04545-1
FGHIJKLMN-MA-79

PREFACE

Although the mechanical behavior of materials is by no means fully understood, the surging activity in this field over the past 15 years makes it possible to present a fairly coherent and quantitative treatment of the subject. Developments have occurred not only in the theory of dislocations and other phenomena on an atomic scale, but also in the methods of continuum mechanics to which one must turn when a myriad of elements are interacting in a way that is simple on a large scale, even if extremely complex on the atomic level.

The interactions between these various scales of study have led us to begin with a review of the various kinds of mechanical behavior and the development of stress, strain, and elasticity. The elastic concepts are then used at nearly the atomic scale to develop the theory of dislocations. After considering polymers, which exhibit all levels of structural perfection and behavior ranging from elastic to viscous to plastic, we return to the macroscopic scale with a general formulation of the constitutive equations. Solutions of these equations, especially around holes and notches, provide important tools for studying the various modes of fracture. Other applications of the physics and mechanics are to damping, creep, friction, wear, and the behavior of composite materials as illustrated by textiles. We conclude with several case studies.

Following an introductory course in materials, of general interest to all engineering students, subsequent courses may well concentrate on specific aspects of materials behavior. This book is intended for such a second course, with emphasis on mechanical behavior from the standpoint of structural, mechanical, marine, and aerospace engineering. Students in these fields will have had mechanics and a general introduction to materials, and some may be taking advanced calculus concurrently. In order to serve the wide range of background and degree of interest found even in one classroom, we have included many topics in depth and breadth, denoted by the symbol ◄. Even though no student could follow all these topics and derivations in a one-semester course, we feel he learns more by delving into those of interest to him and by skimming others than he would from a more uniform, qualitative treatment. He also has more information for further study.

This book has been written not only for those interested in assimilating and using new knowledge, but also to challenge those exceptional students, graduates or undergraduates, who can contribute to the growth of the subject. As an aid in the selective reading that is required with a book of this scope, we have provided a synopsis at the beginning of each chapter, and at the end a quantitative summary.

We hope that the relative completeness of this book will make it possible, although admittedly difficult, for the student to continue directly to most of the current scientific and engineering literature on the subject. For this purpose we have included a liberal number of references. These generally cite the original work, but often later references are given for their clarity and synthesis. Extensive data, including costs, have been provided, and there is an index according to materials as well as the more conventional subject and author indexes.

We have written this book for our colleagues as well as for our students. As each of us in engineering education develops the instruction in his own area, it is easy for him to lose touch with corresponding developments of his colleagues. We hope this book will help bridge that gap by showing the materials scientist and the stress analyst some of the problems we face, and by showing the materials processer and the design engineer some of the possibilities at his disposal.

Finally we have written this book for ourselves, for in so doing we have found how fragmentary is our knowledge in our own fields. The writing has spurred us to try to fill at least some of the gaps, and has disclosed interrelations we had not appreciated.

In our own teaching of the subject, although we generally prefer the order given in the book we have sometimes started with the mechanics section and only referred to the fundamental mechanisms as necessary. Alternatively, we have sometimes presented elasticity and elastic stress concentrations leading to brittle fracture; plasticity and limit analysis leading to ductile fracture; and then polymers and crystalline behavior as needed for transitional modes of fracture. For laboratory work, these latter arrangements, though not as coherent, have the advantage of providing earlier study at the macroscopic scale, on which it is easier to base experiments.

Laboratory work is an essential part of a developing field. In our laboratory, perhaps four different projects selected by the students are studied in a semester by each group consisting of about four students. The reports written are kept on file for reference by others. To offset such concentration on a few kinds of behavior, and to emphasize how much can be learned from simple experiments, we often assign homework problems whose results can be confirmed by "desk-top experiments." A list of the necessary materials is given on p. viii, and such problems are denoted in the book by the letter **E**.

Chapters of this book have been written by various authors, read and criticized by other contributors, and finally revised and often rewritten one or more times. However, the final responsibility for this text lies with the editors, who will appreciate any comments they receive. Chapters were written by the individual authors as follows:

>Ali S. Argon (1, 4, 5, 15, 17, 22)
>Stanley Backer (21)
>Frank A. McClintock (2, 3, 7 through 12, 14, 16, 19)
>Egon Orowan (6)
>Ernest Rabinowicz (20)
>George S. Reichenbach (18)
>Milton C. Shaw (Carnegie Institute of Technology) (13)

Chapters on metallurgical techniques by A. S. Argon, experimental stress analysis by J. Catz, instrumentation by N. H. Cook, and statistics by F. A. McClintock could not be included because of lack of space, but have been made available for reference to our students and are on microfilm. Chapter 14 was originally drafted by A. N. Stroh, who unfortunately had to dissociate himself from the work later due to other commitments. M. E. Shank, through his contributions to the teaching of the mechanical behavior of materials at M. I. T., has also had an important influence on this work.

The editors would like to thank the authors for their participation in this writing project and for their persistence in standing up for what was right when the editors strayed onto unfamiliar ground. We express our appreciation to the Ford Foundation and to Dean Brown for the opportunity to publish this work in a preliminary form about three years ago. The knowledge that this work would soon see the light of day was a strong incentive in undertaking the book.

We are indebted to many individuals, in particular to Professors I. Finnie, D. S. Wood, Morris Cohen, and R. D. Andrews and also to some of our students for having read and criticized all or parts of the book. We are grateful to Miss Karen J. Hall for carrying out the large job of typing the indexes.

Cambridge, Massachusetts F.A.M.
June 1965 A.S.A.

Materials for desk-top experiments for problems denoted by E

Quantity	Item	Approximate cost per pound
12 pieces	chalk	$ .30
¼ lb	modeling clay, very ductile "Plasticine", by Harbutt	.70
	modeling clay, less ductile "Clayrite" by Milton Bradley	.50
1 oz	silicone putty* by Silicone Products Dept., General Electric Co.	7.00 (10 lb lots)
20	paper clips	1.00
10	rubber bands	1.50
2	balloons	6.00
2 ft	steel music wire, 0.026 in. in diameter	2.00
2	wooden stick (tongue depressor)	.50
1	wooden roller	.70
1	plastic screen	2.00
6	polyethylene bags	2.00

* Caution: *Silicone putty will seep through clothing and cannot be removed.*

Forces can be measured with a spring (Problem 9.39) or a number of rubber bands, taking non-linearity into account (Problem 6.34). Most of the material can be assembled locally. The total cost is about $1.00 per kit; costs per pound are included for background information.

CONTENTS

Part I   PHYSICS OF MECHANICAL BEHAVIOR

Chapter 1   **Survey of Structure and Mechanisms of Deformation of Solids**

1.1	Synopsis	3
1.2	Mechanical Phenomena in Solids	3
1.3	Atomic Bonding and Thermal Motion	9
1.4	Structure and Defects in Crystals	17
1.5	Molecular Structure in Polymers	25
1.6	Mechanisms of Deformation	29
1.7	Micro- and Macrostructure of Materials	31
1.8	State of Knowledge	37
1.9	Summary	39

Chapter 2   **Stress and Infinitesimal Strain**

2.1	Synopsis	47
2.2	Definition of Stress	47
2.3	Transformation of Components of Stress	50
2.4	Equations of Equilibrium	52
2.5	Couple Stresses	54
2.6	Definition of Strain	55
2.7	Transformation of Components of Strain	61
2.8	Concept of a Tensor	62
2.9	Measurement of Strain	64
2.10	Work in Terms of Stress and Strain	65
2.11	Summary	66

Chapter 3   **Constitutive Relations for Small Elastic Strain**

3.1	Synopsis	72
3.2	Form of the Elastic Stress-Strain Relations	73
3.3	Effect of Symmetry on Elastic Stress-Strain Relations	76
3.4	Uniqueness of Elastic Stress Distributions	81
3.5	Magnitude of the Elastic Constants	82
3.6	Photoelasticity, Piezoelectricity, and Magnetostriction	89
3.7	Summary	91

Chapter 4   **Dislocation Mechanics**

4.1	Synopsis	96
4.2	The Geometry of Dislocations	97

4.3	Motion of Dislocations in Crystals	102
4.4	Stress Field Around a Dislocation	106
4.5	Elastic Strain Energy, "Line Tension," and "Mass" of a Dislocation	108
4.6	"Force" on a Dislocation	111
4.7	Critical Shear Stress Necessary to Move a Dislocation	114
4.8	Intersection of Dislocations	118
4.9	Dislocation Reactions	121
4.10	Dislocation Multiplication	126
4.11	Tilt and Twist Boundaries and Dislocation Pile-Ups	131
4.12	Modes of Plastic Deformation Other Than Slip	134
4.13	Generation and Mobility of Point Defects in a Lattice	136
4.14	Role of Thermal Activation in Plastic Deformation	139
4.15	Summary	141

Chapter 5 Plastic Deformation in Crystalline Materials

5.1	Synopsis	152
5.2	Structural Hardening of Crystalline Materials	153
5.3	Strain-Hardening in Single Crystals of Single Slip Orientations	160
5.4	Strain-Hardening in Multiple Slip	167
5.5	Thermal Activation in Strain-Hardening	172
5.6	Stored Energy of Strain-Hardening	179
5.7	The Role of Diffusionless Transformations, Surface Films, and Grain Boundaries in Strain-Hardening	180
5.8	Bauschinger Effect	184
5.9	Mechanical Instabilities	186
5.10	Mechanical Equation of State	190
5.11	Data for the Strength of Metals and Alloys	194
5.12	Summary	204

Chapter 6 Deformation in Polymers: Viscoelasticity

6.1	Synopsis	215
6.2	Structure of Polymers	215
6.3	Stages of Deformation in Polymers	230
6.4	Viscosity and Viscoelasticity	233
6.5	Rubber Elasticity	238
6.6	Representation of Small-Strain Viscoelasticity	243
6.7	Crystallinity and the Plastic Behavior of Polymers	253
6.8	Mechanical Behavior of Common Polymers (Data)	257
6.9	Summary	261

Part II MECHANICS OF MATERIALS

Chapter 7 Fundamental Equations of Continuum Mechanics

7.1	Synopsis	273
7.2	Idealization of Mechanical Behavior	274
7.3	Plastic Yielding Under Combined Stresses	276
7.4	Effect of Plastic Deformation on the Equivalent Flow Stress	279
7.5	Relations between Stress and Plastic Strain Increment	283

	7.6	Constitutive Relations for Creep	290
	7.7	Viscoelastic Constitutive Equations	290
	7.8	Dynamic Stress Analysis	295
	7.9	Summary	298

Chapter 8   Tensile and Compressive Deformation

	8.1	Synopsis	309
	8.2	Elastic Tensile Deformation of Anisotropic Materials	309
	8.3	The Beginning of Plastic Flow	310
	8.4	Localization of Plastic Flow	311
	8.5	Finite Strain	316
	8.6	Maximum Load in Tension	319
	8.7	Necking	320
	8.8	Distribution of Strain and Stress in the Neck of a Circular Specimen	322
	8.9	Ductility	325
	8.10	Data	326
	8.11	Instability in Tension and Compression	330
	8.12	Summary	333

Chapter 9   Bending and Torsion

	9.1	Synopsis	339
	9.2	Torsion of Round Cylindrical Bars	339
	9.3	Determining the Stress-Strain Relation from the Torque-Twist Curve	343
	9.4	Bending of Beams of Symmetrical Cross Section	344
	9.5	Bending of Plates	349
	9.6	Thermal Effects in Isotropic Materials	351
	9.7	Summary	352

Chapter 10   Approximate Stress Analysis

	10.1	Synopsis	359
	10.2	Energy Methods in Elasticity	360
	10.3	The Plastic Limit Load	363
	10.4	The Principle of Virtual Work	364
	10.5	The Theorems of Limit Analysis	364
	10.6	Applications of the Theorems of Limit Analysis	368
	10.7	Bounds for Problems in Plane Strain	375
	10.8	Uniqueness of Strain and Stress Distributions	378
	10.9	Less Exact Approximations to the Limit Load	380
	10.10	Summary	384

Chapter 11   Stress and Strain Concentrations

	11.1	Synopsis	393
	11.2	Introduction	393
	11.3	Stress and Strain Distributions Around Cylindrical Holes Under Biaxial Stress	394
	11.4	Stress and Strain Distributions Around Cylindrical Holes Under Uniaxial Stress	398

	11.5	Stress and Strain Concentration Around Spherical Holes	401
	11.6	Stress and Strain Distributions Around Elliptical Holes and Sharp Cracks	403
	11.7	Contact Stress	410
	11.8	Estimating Elastic Stress Concentrations	411
	11.9	Effects of Strain Concentrations	412
	11.10	Amelioration of Strain Concentration	413
	11.11	Summary	413

Chapter 12 Residual Stress

	12.1	Synopsis	420
	12.2	Classification of Residual Stresses	420
	12.3	Mechanical and Thermal Sources of Residual Stress	421
	12.4	Metallurgical and Chemical Sources of Residual Stress	427
	12.5	Composite Sources of Residual Stress	429
	12.6	Measurement of Residual Stress	429
	12.7	Effects of Residual Stress	432
	12.8	Relief of Residual Stress	433
	12.9	Summary	434

Part III APPLICATIONS

Chapter 13 Hardness

	13.1	Synopsis	443
	13.2	Indentation Tests	443
	13.3	Other Measures of Plastic Hardness	450
	13.4	Indentation Tests and Stress-Strain Behavior	453
	13.5	Data and Effects of Variables	458
	13.6	Hardness of Rubber	464
	13.7	Summary	465

Chapter 14 Damping

	14.1	Synopsis	471
	14.2	Models of Materials with Damping	471
	14.3	Sources of Internal Friction	475
	14.4	Structural Damping	483
	14.5	Summary	484

Chapter 15 Brittle Fracture

	15.1	Synopsis	488
	15.2	Brittle Fracture Under Uniaxial Stress	488
	15.3	Brittle Fracture Under Biaxial and Triaxial Stress	490
	15.4	Experimental Verification of the Griffith Theory	495
	15.5	Origin of Strength-Impairing Cracks	497
	15.6	Static Fatigue	499
	15.7	Phenomena of Fracture Propagation	500

15.8	Fracture Surfaces	502
15.9	The Size Effect and Statistical Aspects of Brittle Fracture	504
15.10	Summary	508

Chapter 16 Ductile Fracture

16.1	Synopsis	518
16.2	Rupture	518
16.3	Mechanisms of Ductile Fracture	521
16.4	Ductile Fracture by Growth of Holes	524
16.5	Fracture of Unnotched Specimens	528
16.6	Notch Sensitivity	533
16.7	The Mechanics of Elastic-Plastic Fracture	534
16.8	Summary	540

Chapter 17 Transitional Modes of Fracture

17.1	Synopsis	546
17.2	Transition Between Fracture Modes in Glassy Solids	547
17.3	Transition Between Fracture Modes in Crystalline Materials	548
17.4	Crack Formation by Plastic Deformation	549
17.5	The Process of Crack Propagation	556
17.6	Variables Affecting the Fracture-Mode Transition	557
17.7	The Davidenkov Diagram	564
17.8	Fracture Markings	565
17.9	Common Tests for the Engineering Evaluation of Brittle-Fracture Tendencies in Metals	567
17.10	Summary	570

Chapter 18 Fatigue

18.1	Synopsis	576
18.2	The History and Methods of Study of Fatigue	576
18.3	Mechanisms of Crack Initiation	578
18.4	Mechanisms of Crack Propagation	584
18.5	Fatigue as an Engineering Problem	586
18.6	Effects of Stress	588
18.7	Effects of Material Variables	600
18.8	Effects of Environment	605
18.9	Rolling Contact Fatigue	607
18.10	Engineering Considerations in Design and Service	609
18.11	Summary	611

Chapter 19 Creep

19.1	Synopsis	625
19.2	Mechanisms of Creep in Single Crystals	626
19.3	Mechanisms of Creep in Polycrystals	632
19.4	Presentation and Correlation of Creep Data for Engineering Materials	637

	19.5	Effects of Combined Stress and History	640
	19.6	Creep Fracture	643
	19.7	Choice of Materials	645
	19.8	Summary	648

Chapter 20 Friction and Wear

20.1	Synopsis	657
20.2	Origin of Surface Interaction	657
20.3	The Size of the Real Area of Contact	658
20.4	The Adhesional Theory of Friction	660
20.5	Critique of the Adhesional Theory of Friction	662
20.6	The Laws of Friction	666
20.7	Lubrication	666
20.8	Wear	667
20.9	Adhesion of Solids	671
20.10	Summary	671

Chapter 21 Fibrous Materials

21.1	Synopsis	675
21.2	Introduction	675
21.3	Structure of Polymeric Fibers	677
21.4	Mechanical Behavior of Fibers	680
21.5	Mechanics of Textile Structures	687

Chapter 22 Case Studies

22.1	Synopsis	707
22.2	Determination of Optimum Preload in Connecting Rod Cap Bolts in Internal Combustion Engines	707
22.3	Development of Ultra-Strong Solid-Fuel Rocket Casings	712
22.4	Advisability of "On the Spot" Modification of Pressure Vessels for Gas Storage	717
22.5	Distortion of Springs Due to Stress-Relief	719
22.6	Investigation of the Cause of Service Fracture of a Turbine Wheel	721

Author Index	735
Subject Index	744
Materials Index	761

Part I

PHYSICS OF MECHANICAL BEHAVIOR

CHAPTER 1

SURVEY OF STRUCTURE AND MECHANISMS OF DEFORMATION OF SOLIDS

1.1 SYNOPSIS

The deformation and fracture of materials under applied forces are the principal phenomena associated with the mechanical behavior of materials. Other phenomena that may interact with these are thermal, electrical, magnetic, chemical, and optical effects. These interactions and effects depend on the structure of the material. We shall begin our discussion of the mechanical behavior of materials with a description of the phenomena that are observed on a macroscopic scale. This will be followed by a brief look at the structure of solids, where we shall consider various interatomic bonding forces and compare these with the thermal (vibrational) energy of the atoms. We shall consider some of the conditions which lead alternatively to the formation of crystalline, partially crystalline, or amorphous solids. We shall examine some aspects of crystalline symmetry and introduce the notation for the description of planes and directions in crystals. Then we shall discuss lattice defects in crystals, and, with this background, survey the basic mechanisms of deformation in all solid materials—whether crystalline or amorphous. It will be of interest to establish the effects of stress, temperature, and structural order on these mechanisms of deformation. This will enable us to enumerate some of the deformation-induced structural inhomogeneities, such as various kinds of deformation bands, and anisotropies, such as preferred orientations.

We intend to make this survey descriptive. Detailed discussion of bonding forces and structure of the atom, and proof of statements about lattice symmetry are beyond this level of treatment. For such discussions, we shall refer the inquisitive reader to sources in modern physics, quantum theory, statistical mechanics, and crystallography. The modes of deformation of various structures are of more direct interest to us. These modes will be discussed quantitatively in later chapters, where necessary proof of the statements in this chapter can be found.

1.2 MECHANICAL PHENOMENA IN SOLIDS

Because the deformation of a body resulting from forces applied to it depends so strongly on the size and shape of the body, it is convenient to describe the behavior of a material in terms of force per unit area, or *stress*, and relative displacement per unit distance, or *strain*. At low enough values of applied stress, the

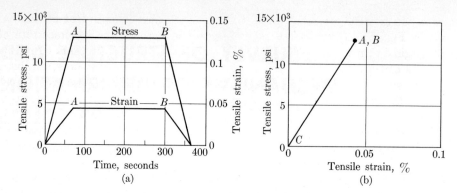

Fig. 1.1. Linear elastic behavior in 1020 steel at room temperature.

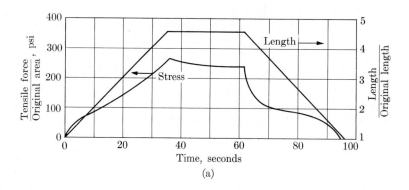

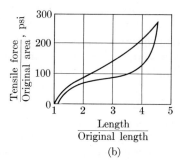

Fig. 1.2. Nonlinear elastic deformation in a rubber band at room temperature.

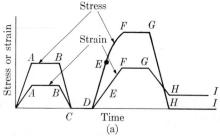
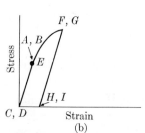

Fig. 1.3. Linear elasticity followed by plasticity.

stress-strain relation of many solids is nearly linear and independent of time, as shown in Fig. 1.1. Furthermore, when the load is released, the specimen returns to its original shape. Time-independent deformation which disappears on release of load is called *elastic* deformation. Some materials, such as rubber, exhibit a nonlinear stress-strain curve, as shown in Fig. 1.2, but the deformation is still elastic in the sense that the material returns to its original shape on release of the load.

All other types of deformation, in which the removal of the applied stress does not result in almost instantaneous (with the speed of sound) corresponding decreases of strain, are called *inelastic* deformation.

When materials are subjected to increasing stress, there may come a point before fracture occurs where, on release of stress, the specimen no longer returns to its original shape. The stress at which this phenomenon begins is called the *elastic limit* of the material. For metals at low temperature, say below half the absolute melting temperature of the metal, the remaining deformation may be

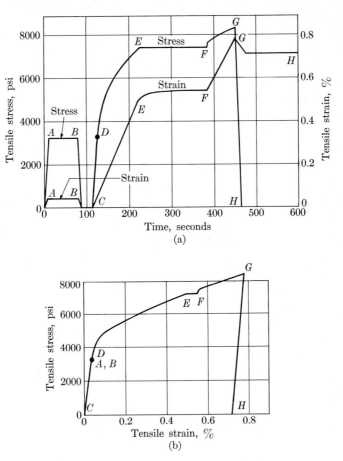

Fig. 1.4. Transient creep at constant stress in annealed copper at room temperature.

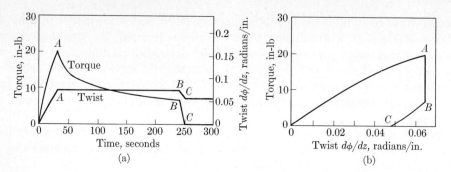

Fig. 1.5. Torque relaxation at constant twist angle in polycrystalline MgO at 1500° C. Dimensions of specimen: cross section, ¼ in²; length, 1 in.

assumed to occur almost immediately and hence be independent of time, as shown in Fig. 1.3. Time-independent deformation which remains on release of load is called *plastic* deformation.

For metals at elevated temperatures, and for polymers under most conditions, deformation continues to increase with time even at constant stress, as illustrated in Fig. 1.4. This phenomenon is termed *creep*. At temperatures and stress levels at which creep occurs, if a deformation is applied and held constant, as indicated in Fig. 1.5, a stress will develop with the application of the deformation and then fall off with increasing time. This decrease is known as *relaxation*. To some extent in metals, and to an even greater extent in polymers, if conditions are such that creep occurs, it is found that on release of the load, some of the strain disappears more slowly than the purely elastic strain, as shown in Fig. 1.6. This effect is known as *delayed elasticity*, or *elastic aftereffect*. We term this inelastic deformation, although some consider it to be another form of elastic deformation. At sufficiently high temperatures, especially in glassy materials, creep will begin when a vanishingly small stress is applied, and will continue at a rate that is nearly proportional to the applied stress, as shown in Fig. 1.7. A material exhibiting this behavior is termed *viscous*.

Under cyclic straining, any kind of inelastic strain leads to a hysteresis loop on a stress-strain plot. This hysteresis loop is an indication of energy dissipation, or *damping*, and is shown in the stress-strain plot for rubber in Fig. 1.2.

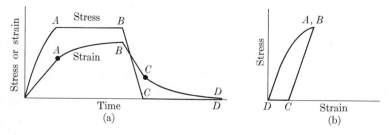

Fig. 1.6. Delayed elasticity, or elastic aftereffect, illustrated by the path CD.

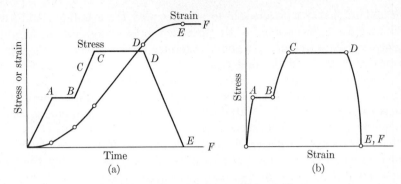

Fig. 1.7. Linearly viscous deformation.

By far the greatest tonnage of metals is used for structural purposes in which resistance to deformation is of primary importance. The plastic resistance of a material is commonly determined from either a tensile test or a hardness test.

Several terms are used to describe the results of a tensile test of a material whose stress-strain behavior is nearly independent of time, as illustrated in Fig. 1.8. The *yield strength* is defined as the stress required to produce some specified plastic deformation, usually of the order of a few tenths of one percent. This figure is useful for design purposes where it is desired to limit the amount of plastic deformation to a value of the order of the elastic deformation.

When it is desired to specify the maximum load-carrying capacity of a part, the test is carried further. As the specimen is extended, the stress required to sustain plastic deformation, called the *flow stress*, increases. At the same time, the cross-sectional area of the specimen decreases. Eventually this decrease in cross-sectional area more than offsets the increase in the flow stress, and the load passes through a maximum, as shown in Fig. 1.8(b). The value of this maximum

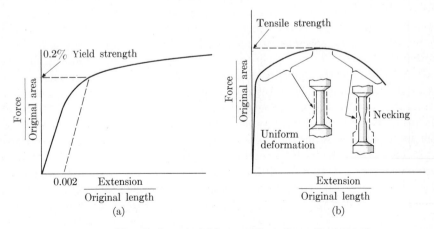

Fig. 1.8. Illustration of yield strength and tensile strength.

load per unit original area is called the *tensile strength*.* If a maximum load is caused by a thinning down, or necking, of the specimen in this way, further deformation usually occurs before fracture. In some materials, the load may still be increasing when fracture occurs. In the first case, the tensile strength is a measure of resistance to deformation, while in the second case, it may be a measure of resistance to fracture (as with glass, where there is no measurable inelastic deformation, or some brasses, where there is). Materials or structures that can undergo considerable amounts of plastic deformation before fracture occurs are called *ductile*, while those that show no plastic deformation prior to fracture are called *brittle*.

A quick measure of resistance to plastic deformation is given by the *hardness* test, in which an indenter is pressed into the surface of the material, as shown in Fig. 1.9. The hardness of the material may be defined as the force per unit area of indentation; it then has the dimensions of stress. For ductile materials, the stress figure obtained by a hardness test will normally be 2.5 to 3 times higher than the tensile strength, because of the confining action of the elastic material surrounding the plastic indentation which prevents free lateral expansion of the indented region. As discussed later in the chapter on hardness, an arbitrary scale of hardness is often used. This scale is determined from the depth of penetration of the indenter under a constant load.

Fig. 1.9. Hardness test. Hardness $= P/A$, where P is the force and A is the area of a shallow indentation.

After a certain history of deformation, a specimen or part will separate into two or more pieces. If there is no appreciable plastic deformation or creep, the process is called *brittle fracture*. At the other extreme is *rupture*, which is the progressive thinning down of the specimen to zero thickness by continued plastic or viscous deformation. *Ductile fracture* is separation under plastic flow before rupture takes place. A measure of the ductility of a material is the percent reduction of area at fracture in a tensile test. For a brittle material, the reduction of area is zero. For rupture, it is 100%. For most metals and alloys, it is of the order of 30 to 70%. Even in ductile materials, fracture can occur under repeated application of loads which in themselves would not produce any large-scale plastic flow. Localized plastic flow in this case leads to cracking and eventually to fracture. This process is termed *fatigue*. Fracture can also occur after creep, as a result of the opening up and joining together of voids in a material, either from grain boundaries or inclusions.

---

* Also called "ultimate strength," or, somewhat redundantly "ultimate tensile strength."

Corrosion is an important factor affecting the mechanical behavior of materials, because stress and corrosion interact to produce a sometimes unexpectedly serious form of fracture known as *stress corrosion*. This can cause cracking under both static and repeated loadings at stresses far below those that otherwise would cause fracture.

Friction and *wear* are also processes which depend to a large extent on the character of the surface of a material, especially its degree of contamination or its lubrication. While failure by wear is not so spectacular as failure by excessive plastic deformation or by fracture, wear and corrosion are the primary sources of failure in most machines and structures.

1.3 ATOMIC BONDING AND THERMAL MOTION

A. Atomic Bonding. The mechanical behavior of materials depends on their structure. The structure, or texture, must be considered at various levels: nuclear, atomic, molecular, crystal, dislocation, slip band, subgrain, deformation band, phase, inclusion, fibril, fiber, fabric, and running up to the beams, columns, stringers, and plates in the structure of a ship or aircraft. In the remainder of this chapter we shall discuss, in turn, the ways these different levels of structure interact with the mechanical behavior of materials.

Changes in the structure of the nucleus affect the mechanical behavior of materials by changing the elements and also by emitting high-energy particles which disarrange the atomic, molecular, or crystal structure. On the other hand, mechanical pressures of millions of atmospheres, sufficient to allow substantial changes in density, allow nuclear reactions to occur which would be impossible at atmospheric pressure. Except for the technology of nuclear explosives, however, the main interaction between nuclear structure and the mechanical behavior of materials is in the effect of nuclear reactions on the atomic, molecular, and crystal structures.

A neutral atom consists of a positively charged nucleus surrounded by electrons whose total charge equals that of the nucleus. The wave-mechanical description of the electrons indicates a series of orbitals in which the electrons are likely to be found. The energy, angular momentum, and spin of the electrons can be thought of as assuming only discrete quantized values. For our purposes here, we shall need only to recognize that the interactions of the electrons with each other and with the nucleus are responsible for the binding of atoms into molecules and crystals. It turns out that the wave-mechanical interactions of electrons can be considered in parts; that is, one can speak of various contributions to the energy of a crystal or of a molecule, such as a repulsive energy and one or more forms of attractive energy. This is represented in Fig. 1.10(a), where the repulsive, attractive, and total energies are plotted as functions of spacing for the uniform dilatation of a sodium chloride crystal. At zero stress, the total energy is nearly equal to the attractive energy and is called the *binding* energy. The derivative of the total-energy curve determines a force-intensity curve, as shown in Fig.

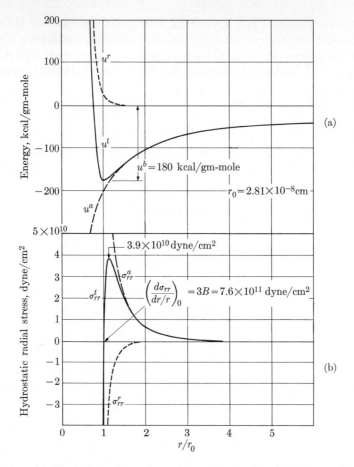

Fig. 1.10. (a) Variation of energy with uniform dilatation in sodium chloride. (b) Derivatives of the energy curves of part (a) (for a spherical shape, referred to the current surface area of the sphere). The initial slope of the stress curve at $r/r_0 = 1$ gives three times the bulk modulus.

1.10(b), and the derivative of that in turn determines the stiffness [in Fig. 1.10(b), the bulk modulus].

The repulsive forces resulting from the interpenetration of tightly bound inner-electron shells dominate at small distances and fall off roughly as the ninth power of the distance, or perhaps more accurately, exponentially (Seitz, 1940, pp. 79–82).

Binding forces may be divided into three different classes, according to their magnitudes:

1. Primary Bonds. (a) *Ionic bond.* The atoms of certain elements, such as the alkali metals, possess easily detachable valence electrons in addition to filled electron shells. The atoms of other elements, such as the halogens, have a tendency to acquire electrons to form a filled electron shell. When neutral alkali atoms are in the presence of halogens, they will lose their valence electrons easily to the latter. The resulting alkali cations and halogen anions will then attract each other electrostatically without any further significant change in their electronic structure. In sodium chloride, for example, the sodium cations will surround themselves with six chlorine anions, and vice versa. The result is the formation of a giant molecule held together by electrostatic forces, without the formation of individual NaCl molecules as such. This type of bond is known as an *ionic* bond.

The electrostatic attractive energy is found by summing the electrostatic interactions between all pairs of positive and negative ions; it varies inversely with the first power of the interionic spacing. The resulting binding energies are of the order of 40,000 to 200,000 calories per gram-mole (Probs. 1.1, 1.2, 1.3).

(b) *Covalent bond.* Atoms of certain elements have a tendency to share some of their valence electrons. The bond resulting from the reduction of energy due to this exchange of valence electrons between atoms is known as a *covalent* bond. The covalent bond is responsible for the formation of most molecules, such as N_2, O_2, CH_4, as well as more complex chemical compounds.

(c) *Metallic bond.* Atoms of elements with easily detachable valence electrons have lower energy when they are close enough together so that the electrons can circulate from one ion site to another throughout the crystal, forming a kind of free-electron gas which holds the positively charged ions together in a close-packed configuration. This bond is characteristic of metals, including the alkalis, and is known as a *metallic* bond. The mobility of the free electrons under an applied electric field accounts for the high electrical conductivity. One of the principal consequences of the metallic bond is the close-packing, or high number of nearest neighbors, of metals.

2. Secondary (*Intermolecular, van der Waals*) Bonds. Even though the valency forces of atoms are saturated within a molecule, molecules still attract each other with much weaker forces. These forces arise from permanent or induced dipoles. Wave mechanics shows that even if the molecules do not contain a permanent dipole, there are electronic charge oscillations representing oscillations of dipole moments which automatically synchronize each other in neighboring molecules and produce attractive forces. These are the forces responsible for the condensation and solidification of substances consisting of molecules (but not, e.g., for the condensation and solidification of metals). They were first considered by van der Waals as the cause of deviations from the ideal gas law.

3. Hydrogen Bond. Hydrogen can exist both as a positively charged and as a negatively charged ion. The positive hydrogen ion, or proton, results from the removal of the only electron. The negative ion, on the other hand, is formed by the imperfect shielding of the positively charged nucleus by the single electron in the neutral atom. This imperfect shielding will result in a constantly shifting dipole which has a weak tendency to acquire another electron by purely ionic attraction. This property of the hydrogen atom enables it to bridge two negative ions, in what is known as a *hydrogen* bond. The hydrogen bond is responsible for the fact that such a light molecule as water boils at so relatively high a temperature and that it expands on crystallizing. The hydrogen bond also plays an important role in strengthening linear polymers.

The binding forces in elements and chemical compounds are usually mixtures of the idealized types discussed above. The total binding energy can be measured either thermochemically or spectroscopically. The energy of the primary bond is usually of the order of 40,000 to 200,000 calories per gram-mole (or 2 to 10 electron volts per atom). The energy of a hydrogen bond is between 5000 and 10,000 cal/gm-mole (or 0.2 to 0.5 ev/atom), while that of a secondary bond is between 500 and 5000 cal/gm-mole (or 0.02 to 0.2 ev/atom). The magnitude of these binding energies is important in determining the elastic constants of crystalline materials and the resistance to deformation in spite of thermal motion of the atoms.

B. Thermal Motion. The atoms in a crystal vibrate about their equilibrium sites with amplitudes which increase with the absolute temperature. The time and temperature dependences of the mechanical behavior depend on this motion. We shall, therefore, review the subject of statistical mechanics briefly by following the treatment given by Joos (1950), to which the reader is referred for a more complete discussion.

◀Statistical mechanics tells us how the total energy of the thermal motion depends on the temperature, and gives the distribution of energies among the various modes of motion, or, what is assumed to be the same thing, gives the fluctuations of energy in a given mode with time. The classical statistics of Boltzmann applies to collections of particles whose energies can be described in terms of position and momentum coordinates. These coordinates provide a multidimensional space, the phase space, through which the particles move. For instance, an undamped spring-mass system follows an elliptical path in a two-dimensional phase space (Prob. 1.4). On the other hand, three-dimensional translation and rotation would require a twelve-dimensional space—six coordinates of position q_i, and six of momentum p_i.

◀The motion of a collection of particles can be thought of as a swarm of points moving through phase space. It can be proven from Hamilton's formulation of the equations of mechanics that for conservative forces, the density of the swarm of points in the neighborhood of any given point remains constant with time. If the swarm is divided into cells of volume Δv, which move with the points, the volumes therefore contain a constant number of points as they move through the phase space. Thus the cells provide a convenient frame of reference for study. Their actual size is determined from wave mechanics.

◀The *macrostate* of a system is described in terms of the numbers, N_i, of points representing particles in each cell. A *microstate* is a particular way of producing a macrostate by assigning specific particles to specific cells. In view of the constraint conditions on the problem, not all cells can be occupied. For example, the motion may be restricted to a plane or a line, and rotation of the particles may be excluded. Among the admissible cells, however, we shall assume all to be equally likely. Then the probability of a macrostate will be proportional to the number of different microstates which have that macrostate.

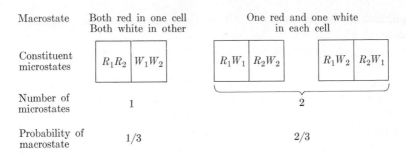

FIG. 1.11. Calculation of probability of a macrostate from numbers of microstates.

◂As an example, consider the experiment of Fig. 1.11, where two individually distinguishable red and two white balls have to be divided in pairs between two boxes, each of which can accommodate only two balls. If the relative positions of the balls in the boxes and the relative positions of the boxes are immaterial, two possible macrostates can be defined for the four balls in two boxes: one, where the two colors are in separate boxes, and the other, where the two colors are mixed in each box. As is shown in Fig. 1.11, there is only one way (microstate) of arranging the balls in the two boxes leading to the separate-color macrostate, while there are two ways (microstates) of arranging the macrostate of mixed colors. Thus, of the three microstates, two give the second macrostate and one gives the first macrostate; hence the probability of occurrence of the first macrostate is $\frac{1}{3}$, while that of the second is $\frac{2}{3}$.

◂In Boltzmann's definition, the entropy of a macrostate is taken to be the product of Boltzmann's constant k and the logarithm of the number of microstates which could give rise to the macrostate; aside from an additive constant, it is proportional to the probability of the macrostate:

$$s = k \ln W. \tag{1.1}$$

For a given total number of particles N and a given total energy of the state, the maximum entropy, or most stable state, will be reached when the distribution of the particles among the cells is given in terms of the total energy u_i of each cell by*

$$\frac{N_i}{\Delta v} = f_i = \frac{N e^{-u_i/kT}}{\sum e^{-u_i/kT} \Delta v}. \tag{1.2}$$

This is the Maxwell-Boltzmann distribution function, where the summation is over all phase space.▸

The above discussion tells us that in the state of stable equilibrium, the particles do not all have the same energy, but rather a few of them have (are in cells

---

*For the derivation of this relation, the reader is referred to Joos (1950, pp. 578–580).

corresponding to) higher energy. The fraction of particles with energy u or greater is given by (Prob. 1.5)

$$p(u_i \geq u) = e^{-u/kT}. \tag{1.3}$$

The average energy associated with each coordinate of the phase space, describing the position and momentum coordinates of all the particles, is differentiated to give the specific heat of the collection. The specific heat can be obtained relatively easily if the energy is expressible as the sum of terms involving the various coordinates. In particular, if the contributions to the energy are found by summing terms each of which is proportional to the square of a momentum coordinate, the average contribution of each coordinate to the total energy is $kT/2$ (Prob. 1.6).

The specific heat at constant volume is the rate of change of energy with temperature, and can be expressed on a mole basis in terms of Avogadro's number N_A. Thus for a monatomic ideal gas, where each atom has only three coordinates of momentum, the specific heat is $3N_A k/2$; for a diatomic gas, where each molecule has two angular momentum coordinates as well as three linear momentum coordinates, the specific heat is $5N_A k/2$. In a crystal, one might consider the atoms to be oscillating about equilibrium sites, and the potential energy function to be parabolic, corresponding to a linear restoring force from small displacements from the equilibrium lattice, as shown in Fig. 1.12. Actually this picture is not exact; due to the motion of the neighboring atoms, the walls of the potential trough for each atom are themselves shifting, and for really violent thermal motion, the restoring force will become nonlinear. Nonetheless, taking it as an approximation, for each atom there are three coordinates of position and three of momentum, each contributing quadratically to the energy. Since the average potential energy of position for a linear oscillator is equal to its average kinetic energy, the specific heat at constant volume of a crystal can be given on a mole basis in terms of Avogadro's number as (Probs. 1.7, 1.8, 1.9)

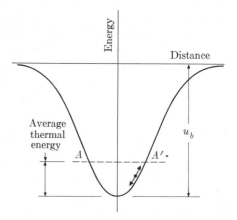

FIG. 1.12. Atom at finite temperature T, vibrating in a potential well.

$$c_v = 3N_A k = 3R \approx 6 \text{ cal/gm-mole·°C}, \tag{1.4}$$

where R is the universal gas constant. The constancy of the molar specific heat for many simple substances had been experimentally determined by Dulong and Petit before the statistical interpretation could be given, and it is known as their law.

At this point we are in a position to return to the question of how to determine the number of coordinates which should enter into the statistical mechanical model. In other words, in what modes is the crystal actually vibrating? In a crystal, should we consider rotation of the atoms as well as their translation? In a crystalline compound, should we consider each atom separately, or do pairs of atoms move together? The answers to these questions are supplied by quantum mechanics, which states that a mode of vibration can be excited only in increments of energy Δu proportional to its frequency of oscillation ν

$$\Delta u = h\nu, \tag{1.5}$$

where h is Planck's constant (6.62×10^{-27} erg·sec). In the simple harmonic oscillator model of a crystal discussed above, the average energy (kinetic plus potential) associated with a single mode of oscillation is kT. Thus, for a mode of vibration of frequency ν to be excited and hence enter into classical statistical calculations, it is necessary that the temperature be high enough that

$$kT \geq h\nu. \tag{1.6}$$

To determine which modes will be excited in a crystal with many atoms, we need to know their respective frequencies. Most of the frequencies are near the highest frequency of the lattice (see for instance, Mott and Jones, 1936, p. 3), which can be roughly estimated from the frequency of each atom oscillating between rigid boundaries, as indicated in Fig. 1.13. In terms of the atomic weight M, the modulus of elasticity E, Avogadro's number N_A, and the interatomic distance b, the highest frequency of vibration is given by (Prob. 1.10)

$$\nu = \frac{1}{2\pi}\sqrt{\frac{k}{m}} \approx \sqrt{\frac{Eb}{M/N_A}}. \tag{1.7}$$

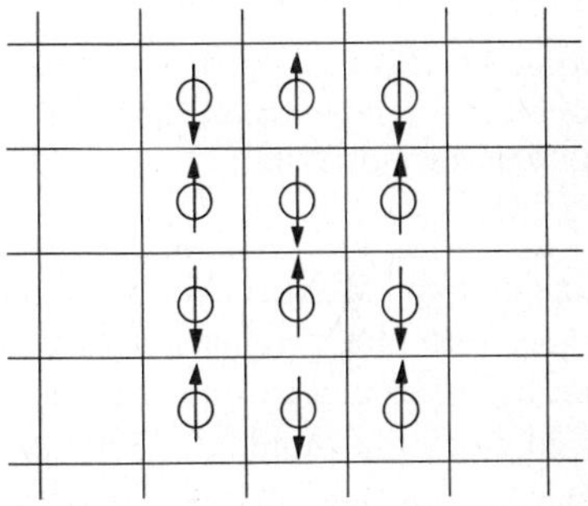

FIG. 1.13. Highest frequency mode of vibration of a crystal lattice.

Equation 1.7 can also be derived by considering the time required for an elastic wave to pass from one atom to its neighbor (Prob. 1.11). The temperature which is just high enough to excite this frequency, according to Eq. 1.6, is a characteristic temperature nearly equal to the Debye temperature. At higher temperatures, one would expect all $3N$ modes of vibration of the lattice to be excited, giving the specific heat of 6 cal/gm-mole·°C, discussed above. At still higher temperatures, some of the modes of electron motion will begin to be excited, and the specific heat will rise further. At temperatures below the Debye temperature, some of the modes of vibration will be "frozen out," and the specific heat will be

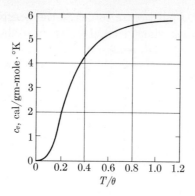

Representative Values of θ

Substance	Temperature, °K	Substance	Temperature, °K
Ag	215	Mg	290
Al	398	Mo	379
Au	180	Na	159
C (diamond)	1860	Ni	370
Cd	160	Pb	88
Cu	315	Ta	245
Fe	420	W	310
		NaCl	281

FIG. 1.14. Specific heat of solids at low temperatures. The values of θ, the Debye characteristic temperature, for some common substances are given in the table. (Kittel, 1953. Courtesy of Wiley.)

less, as shown in Fig. 1.14. Thus we can determine whether or not we have taken into account the proper number of modes of vibration in statistical mechanics either by reference to quantum mechanics (Eq. 1.6) or by comparing the specific heat to 6 cal/gm-mole·°C. This process will be done for polymers in Section 6.4.

Thermal motion allows a gradual change from one state to another. For example, for an atom to evaporate from a solid surface into a vacuum, it and its neighbors must achieve a configuration such that the atom will be thrown off in spite of the binding energy holding it to its neighbors. Roughly speaking, this requires that the atom acquire a kinetic energy equal to the binding energy. The probability of its having at least this kinetic energy is given by Eq. 1.3. The frequency with which the energy states will be shuffled is of the order of the frequency of vibrations in the lattice, given by Eq. 1.7. Thus, the rate at which atoms leave a surface containing N atoms by overcoming the binding energy u_b is

$$dN/dt = N\nu e^{-u_b/kT}. \tag{1.8}$$

Similar equations occur in many other situations, where thermal activation plays a dominant role. Thus we may express the rate $\dot{R}$ at which a process is activated in terms of the number of activation sites N_s, the dominant frequency of the process being activated ν_a, and the activation energy u_a:

$$\dot{R} = N_s \nu_a e^{-u_a/kT}. \tag{1.9}$$

Strictly speaking, the quantity u_a should be the change in the Gibbs free energy required to get the atom out of its trough. See, for example, Barrer (1942); Swalin (1962, p. 44); Glasstone, Laidler, and Eyring (1941, p. 195); or Zener (1952, p. 296). According to quantum mechanics, the entropy will change not only with temperature, as determined from the specific-heat curve, but also with frequency. At room temperature, this contribution to the free energy is only 0.003 ev if the frequency is changed by 10% (Prob. 1.12). Because of this small effect, and because we are already making the greater approximation that equilibrium statistical mechanics applies to these transient states, we shall neglect the difference between the Gibbs free energy and the internal energy of activation, and consider only the latter in the remainder of this book.

As an illustration of rate effects, it is interesting to study some numerical examples of Eq. 1.8. For a material whose atoms or molecules are bound to the interior by bonds having an energy of, say 1 ev, the rate of evaporation at room temperature will be about one molecule out of ten thousand each second. Thus, if there is no redeposition from the gas phase, one cubic centimeter will evaporate in about ten thousand years. If the binding energy is $\frac{1}{5}$ ev, the same amount of evaporation requires only 0.01 second, provided that the vacuum can be maintained. On the other hand, if the binding energy is 2 ev, the process would take 10^{12} times the age of the universe (Prob. 1.13). Evidently, 1 ev is the approximate limit above which thermal movement at room temperature is practically powerless to break bonds.

1.4 STRUCTURE AND DEFECTS IN CRYSTALS

A. Crystal Structure. During cooling at normal rates from the liquid state, most metals and simple compounds can undergo the long-range ordering necessary to crystallize. That is, they form a regular lattice in which a certain arrangement is repeated in space at regular intervals, as shown in Fig. 1.15. The basic arrangement, containing from one to a large number of atoms, is called a *unit cell*. In most cases, a unit cell is chosen which is larger than the *primitive cell** in order to represent the high degree of symmetry of the crystal, which may not be apparent from the primitive cell. Examples are the face-centered cubic, body-centered cubic, sodium chloride, and hexagonal close-packed cells of Fig. 1.16. The arrangement of atoms within a unit cell may itself be more or less symmetrical. Even in the cubic structures, the complexity can vary from the face- and body-centered cubic to diamond cubic and α or β manganese, the former containing 58 atoms in each unit cell. These structures are, however, still simple in comparison with the complex structure of organic compounds. For a discussion of the structure of crystals with various degrees of symmetry, see M. J. Buerger (1956).

---

*A primitive cell contains only one atom per cell.

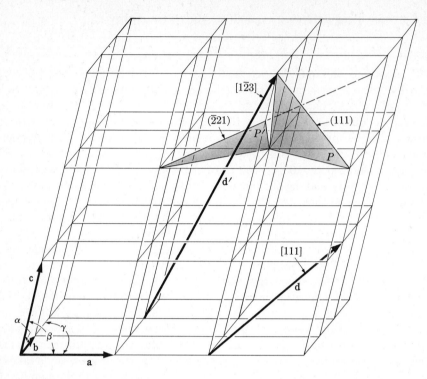

FIG. 1.15. A space lattice characterized by a triad of identity vectors **a**, **b**, and **c**, making angles α, β, and γ.

Most of the familiar engineering metals crystallize in lattices of high symmetry, such as cubic or hexagonal lattices. Tables 1.1 and 1.2 give the types of lattices and their dimensions for a number of common metals and simple inorganic compounds. It is important to note in our study of deformation in crystals that crystal structure appears quite different from different points of view. For instance, both the hexagonal close-packed structure and the face-centered cubic structure allow the densest packing of rigid spheres. This is evident in the hexagonal close-packed structure shown in Fig. 1.17. In the face-centered cubic structure, this is most easily seen by looking at planes formed by three nonadjacent corner atoms of the unit cell, as shown in Fig. 1.18. These planes are like the planes of the hexagonal close-packed structure, but as shown in Fig. 1.18, they are stacked in a sequence of three ($\ldots CABCAB \ldots$), rather than the sequence of two ($\ldots BABA \ldots$) of the hexagonal close-packed structure. If the interatomic forces affected only nearest neighbors, these two configurations would be equally likely. However, the second nearest-neighbor interactions, by shifting the pattern to the face-centered structure from the hexagonal close-packed structure, allow the possibility of four families of close-packed planes of the type shown in Fig. 1.18, rather than just one. This leads to many more possibilities for plastic deformation.

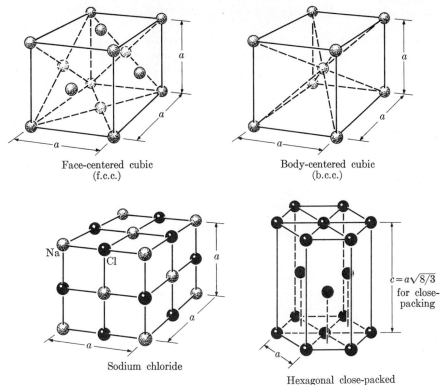

FIG. 1.16. Some of the most common unit cells of crystals.

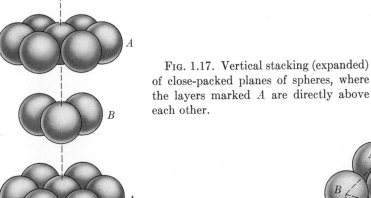

FIG. 1.17. Vertical stacking (expanded) of close-packed planes of spheres, where the layers marked A are directly above each other.

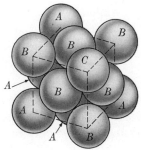

FIG. 1.18. Face-centered cubic structure, showing the $ABCABCA$ sequence of close-packed (111) planes.

TABLE 1.1*

Crystal Structure of Some Common Elements and Lattice Dimensions at Room Temperature (20°C)

Element	Type of structure	Atoms per unit cell	Lattice constants, A†		Distance of closest approach, A
			a	c	
Aluminum	f.c.c.	4	4.0490		2.862
Beryllium	h.c.p	2	2.2854	3.5841	2.225
Cadmium	h.c.p.	2	2.9787	5.617	2.979
Carbon (diamond)	diamond cubic	8	3.568		1.544
Chromium	b.c.c.	2	2.8845		2.498
Cobalt α	h.c.p.	2	2.507	4.069	2.506
Cobalt β	f.c.c.	4	3.552		2.511
Copper	f.c.c.	4	3.6153		2.556
Germanium	diamond cubic	8	5.658		2.450
Gold	f.c.c.	4	4.0783		2.884
Iron α	b.c.c.	2	2.8664		2.481
Iron γ (extrapolated)	f.c.c.	4	3.571		2.525
Lead	f.c.c.	4	4.9495		3.499
Lithium	b.c.c.	2	3.5089		3.039
Magnesium	h.c.p.	2	3.2092	5.2103	3.196
Manganese α	cubic	58	8.912		2.24
Manganese β	cubic	20	6.300		2.373
Molybdenum	b.c.c.	2	3.1466		2.725
Nickel	f.c.c.	4	3.5238		2.491
Platinum	f.c.c.	4	3.9237		2.775
Potassium	b.c.c.	2	5.344		4.627
Silicon	diamond cubic	8	5.4282		3.351
Silver	f.c.c.	4	4.0856		2.888
Sodium	b.c.c.	2	4.2906		3.715
Tantalum	b.c.c.	2	3.3026		2.860
Tin α	diamond cubic	8	6.47		2.81
Titanium α	h.c.p.	2	2.9504	4.6833	2.89
Tungsten α	b.c.c.	2	3.1648		2.739
Zinc	h.c.p.	2	2.664	4.945	2.664

*Compiled from Barrett (1952). Courtesy of McGraw-Hill.
† 1 A = 1 Angstrom = 10^{-8} cm.

B. The Miller Notation. Planes and directions related to a crystal lattice can be described conveniently by a three-index notation, referring them to the triad of identity vectors bounding the unit cell. Because of the repetitive nature of a crystal, we need only consider any one of a family of parallel planes, or lines.

TABLE 1.2*

Crystal Structure of Some Common Inorganic Compounds and Alloys and Their Lattice Dimensions at Room Temperature (20°C)

Compound	Type of structure	Total atoms per unit cell	Lattice constants, A		Distance of closest approach, A
			a	c	
NaCl (rock salt)	rock salt f.c.c.	8	5.627		Na-Na 3.97 Na-Cl 2.81
KCl (sylvine)	rock salt f.c.c.	8	6.28		K-K 4.42 K-Cl 3.14
AgCl	rock salt f.c.c.	8	5.545		Ag-Ag 3.92 Ag-Cl 2.78
CsCl	cesium chloride	2	4.110		Cs-Cs 4.110 Cs-Cl 3.560
CaF_2 (fluorite)	cubic (b.c.c.)	12 (4 moles)	5.451		
$CaCO_3$ (calcite)	rhombohedral	10 (2 moles)	6.361		
$Al_2O_3\beta$	hexagonal		5.56	22.55	
α brass	f.c.c.	4	varies with concentration of Zn		
β brass	b.c.c.	2	varies with concentration of Zn		

*Compiled from *Handbook of Chemistry and Physics*, 42nd ed. Courtesy of Chemical Rubber Co.

To represent a direction, such as d or d' of Fig. 1.15:

1. Determine the vector parallel to the directions as a sum of multiples of the triad of identity vectors, for example,

$$d: 1, 1, 1 \quad \text{or} \quad 2, 2, 2; \quad d': 1, -2, 3 \quad \text{or} \quad 2, -4, 6.$$

A line passing through two lattice points, that is, a crystallographic direction, will always have integral coefficients.

2. The direction of the line is then denoted by the smallest integral set of such coefficients, enclosed in brackets, with a bar over negative coefficients:

$$d: [111]; \quad d': [1\bar{2}3].$$

To represent a crystallographic plane:

1. First determine the intercepts which the plane, or one parallel to it, makes with the three axes corresponding to the identity vectors.

2. Take the ratios of these intercepts to the magnitudes of the corresponding identity vectors and list them as three numbers corresponding to the three axes, in a right-hand screw order. For example, for the planes P and P' of Fig. 1.15, the ratios are

$$P: 1, 1, 1; \qquad P': \bar{1}, 1, \tfrac{1}{2}.$$

3. Take the reciprocals of these ratios and restore them to the smallest integral set by multiplying them by an appropriate factor. These three integers, enclosed in parentheses, are known as the *Miller indices* of the plane. A negative index is again denoted by a bar over it. Thus the planes P and P' are described as

$$P: (111); \qquad P': (\bar{2}21).$$

In crystals with a high degree of symmetry, it is often necessary to label a whole family of equivalent planes or directions of a certain type. This is achieved by writing the Miller indices of positive numbers in braces for planes and in carets for directions. Thus we write, for example, {110} to represent the family of planes (110), (101), (011), and ⟨110⟩ to represent the family of directions [110], [101], [011], [1$\bar{1}$0], etc.

In cubic crystals, the crystallographic system of axes can be considered also as a triad of cartesian coordinate axes. In this case, vector operations can be performed with great ease by using the Miller indices of planes and directions. We note in this case that the vector **n** normal to the plane (h, k, l), and passing through the origin, will have direction cosines $h/\sqrt{h^2 + k^2 + l^2}$, $k/\sqrt{h^2 + k^2 + l^2}$, and $l/\sqrt{h^2 + k^2 + l^2}$ (Prob. 1.14). Thus any crystallographic plane in a cubic crystal may be represented uniquely by its unit normal vector

$$\mathbf{n} = \frac{h\mathbf{i} + k\mathbf{j} + l\mathbf{k}}{\sqrt{h^2 + k^2 + l^2}},$$

where **i**, **j**, **k** are the unit vectors of the cartesian triad. Similarly, we may represent a crystallographic direction by a unit direction vector

$$\mathbf{d} = \frac{u\mathbf{i} + v\mathbf{j} + w\mathbf{k}}{\sqrt{u^2 + v^2 + w^2}}.$$

The following operations are among the more useful.

The cosine of the angle ϕ between two planes $(h_1 k_1 l_1)$ and $(h_2 k_2 l_2)$:

$$\cos \phi = \mathbf{n}_1 \cdot \mathbf{n}_2 = \frac{h_1 h_2 + k_1 k_2 + l_1 l_2}{\sqrt{(h_1^2 + k_1^2 + l_1^2)(h_2^2 + k_2^2 + l_2^2)}}. \tag{1.10}$$

The cosine of the angle λ between two directions $(u_1 v_1 w_1)$ and $(u_2 v_2 w_2)$:

$$\cos \lambda = \mathbf{d}_1 \cdot \mathbf{d}_2 = \frac{u_1 u_2 + v_1 v_2 + w_1 w_2}{\sqrt{(u_1^2 + v_1^2 + w_1^2)(u_2^2 + v_2^2 + w_2^2)}}. \tag{1.11}$$

The direction of the line of intersection of planes $(h_1k_1l_1)$ and $(h_2k_2l_2)$:

$$\mathbf{d} = \frac{u\mathbf{i} + v\mathbf{j} + w\mathbf{k}}{\sqrt{u^2 + v^2 + w^2}} = \mathbf{n}_1 \times \mathbf{n}_2$$

$$= \frac{1}{\sqrt{(h_1^2 + k_1^2 + l_1^2)(h_2^2 + k_2^2 + l_2^2)}} \begin{vmatrix} \mathbf{i} & \mathbf{j} & \mathbf{k} \\ h_1 & k_1 & l_1 \\ h_2 & k_2 & l_2 \end{vmatrix}. \quad (1.12)$$

C. Defects in Crystal Lattices. A perfect crystal lattice is an idealization. Real crystals contain a variety of lattice defects. Thermal motion is a form of defect that has already been mentioned. The next simplest defects in an ideal lattice are: an atom missing from the lattice, generating a *vacancy;* an extra atom of the same kind wedged in the lattice where it cannot be accommodated at a normal lattice site, forming an *interstitial* atom; and a foreign atom, or *impurity* not normally a part of the lattice, usually found in even the purest of materials. Impurity atoms may replace a regular lattice atom and become *substitutional impurities*, or occupy interstices between normal lattice atoms and become *insterstitial impurities*. Such defects involving only one atom are called *point defects*. The next higher level of imperfections in a lattice are *line defects*. The incomplete translation of one part of the lattice relative to the other, as shown in Fig. 1.19, gives rise to a line defect known as a *dislocation*. The plasticity of crystals is the result of the generation and motion of dislocations. Dislocation lines are often interconnected in a network. Such situations are much too complex to be considered atom by atom, and

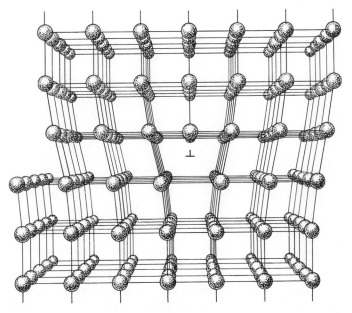

Fig. 1.19. One kind of dislocation. (Guy, 1959, Addison-Wesley.)

instead, we think of the dislocations as entities and speak of the forces on them, their interactions with each other, and their organization into a structure.

So far we have discussed point and line defects in crystals. There may also be surface defects, that is, surfaces across which the identity vectors change direction. For example, there may be a crystallographic plane across which the crystal structure changes to form a mirror image of itself. This surface defect is called a *twin* boundary, and is shown in Fig. 1.20. Crystals of differing orientations join together in grain boundaries, as in Fig. 1.21. Crystals of different compositions

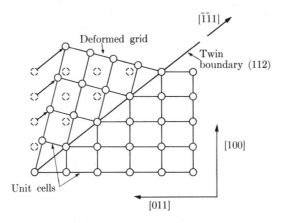

FIG. 1.20. Twinning in a body-centered cubic lattice resulting from shear parallel to (112) planes in the $[\bar{1}\bar{1}1]$ direction.

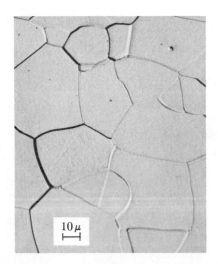

FIG. 1.21. Grain boundaries in α iron. (Van Vlack, 1959, Addison-Wesley. Courtesy of United States Steel Corp.)

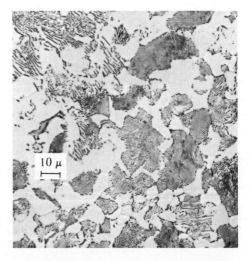

FIG. 1.22. Grains of two distinct phases in annealed 0.4% carbon steel. (Van Vlack, 1959, Addison-Wesley. Courtesy of United States Steel Corp.)

join at phase boundaries, as shown in Fig. 1.22. A free surface could be considered as another kind of surface imperfection, across which an extreme change in crystal structure occurs. Just as we speak of the energy of a dislocation, which is a line imperfection, so we speak of the energy of a twin, grain, or phase boundary or of a free surface. By this we mean, in each case, the difference between the total energy of the crystal containing these defects and the energy of the same crystal without the defects.

At temperatures where point defects become mobile, the excess energy of the crystal due to the point defects which it contains can be reduced if the defects coalesce, forming *vacancy clusters* (incipient holes) or *precipitates*. The structure may transform to an entirely new *phase*. Configurations such as these, having three dimensions, could be called volume defects, but are more often called just by their individual names.

Equilibrium between different phases can be studied from the point of view of thermodynamics, where the atomic details need not be considered. In considering the rate of approach to equilibrium, however, it is often necessary to take into account the atomic and dislocation processes which lead to equilibrium.

1.5 MOLECULAR STRUCTURE IN POLYMERS

A polymer is a material consisting of very large molecules. In each molecule, an atomic group, or *repeat unit*, is repeated at least hundreds or thousands of times. In this sense, any ionic, metallic, or covalent crystal is a polymer, but conventionally the term is applied only to the following materials.

(a) Materials consisting of long-chain, or *linear*, molecules, which may be arranged in a crystalline or noncrystalline manner (Fig. 1.23). In fibers, the molecules are usually aligned substantially parallel to the fiber axis.

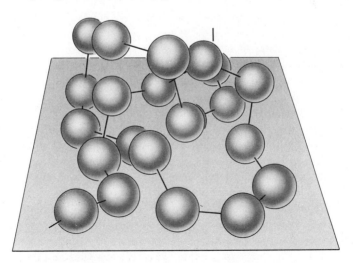

Fig. 1.23. Chain polymer with two bonds per repeat unit.

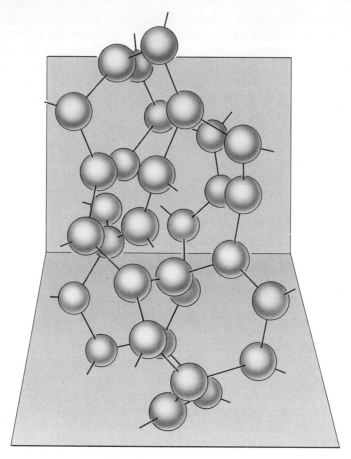

Fig. 1.24. Space network polymer with three bonds per repeat unit.

(b) Materials consisting of a noncrystalline *space* network of atoms joined by primary bonds, as in inorganic glasses or phenol-formaldehyde (Fig. 1.24).

Upon heating, a polymer may soften before decomposing. Such a material is called *thermoplastic*. Or it may decompose before softening. Such a material is called *thermosetting*. This is another way of classifying polymers. Both linear and space polymers are found in each group; for example, in long-chain linear polymers, the bonds between the chains may be weak, of the van der Waals type, so that at a moderate temperature the chains can slide on each other (Prob. 1.15); such polymers are thermoplastic. In cellulose, on the other hand, the chains are connected with numerous hydrogen bonds, and the material decomposes before the chains slide. Cellulose is, therefore, not thermoplastic, even though it consists of chain molecules. Most of the organic space-network polymers, such as the formaldehydes, decompose before viscous flow can occur. By contrast, space-network polymers based on silica or boron oxide (glasses), are thermoplastic. Space-network polymers are amorphous by definition: if they were crystalline, they would be called simply crystals. The best known and simplest of such polymers are the inorganic glasses. In silica glasses, for instance, the quadrivalent silicon ion forms strong single bonds with four divalent oxygen ions, giving a tetrahedron with the silicon ion in the body center and the four oxygen ions at the corners. This tetrahedral nearest-neighbor relation is preserved even in the liquid state. Within normal rates of cooling, the tetrahedra cannot fully arrange themselves in a regular network, such as the one of crystalline quartz shown in Fig. 6.5, but will instead form an irregular space network made up of tetrahedra sharing corners.

The crystallization of polymers is retarded in two ways. First, since the long-chain structure persists even in the liquid state, the molecules may not have time to order completely at normal cooling rates. Second, a chain on which bulky side groups or branches are hanging is much more difficult to pack in a regular fashion. These side groups may be chemical radicals or they may be side branches of the same chemical composition as the main structure.

Crystalline polymers differ from ordinary crystals in two important ways. First, the primary bonding provides a strong anisotropy to the lattice. Second, in many cases, the crystals form tiny lamellae, the thickness of which is generally of the order of 100 angstroms (1 A = 10^{-8} cm).

Electron-diffraction patterns have shown that the molecular chain in a crystalline polymer is perpendicular to the plane of the lamella, yet the length of the chain molecule is at least about 1000 A. Consequently, the molecules must be folded into the crystal lamellae by zig-zagging at least ten times. On a larger scale, the structure of many crystalline polymers consists of spherulites, as shown in Fig. 1.25. A spherulite consists of many crystals growing outward from a center in such a way that a certain crystallographic direction points toward the common center.

FIG. 1.25. Spherulites in polyethylene.

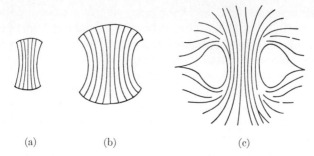

(a)   (b)   (c)

Fig. 1.26. Schematic representation of the growth of a spherulite.

A frequent mechanism of formation of spherulites is shown in Fig. 1.26. The crystal starts to grow in the usual way, but after a while, its ends begin to widen (Fig. 1.26a) and fan out (Fig. 1.26b) to form a wheatsheaf-like shape. The process progresses as shown in Fig. 1.26(c), until a spherulite arises with a wheatsheaf-like core. An electron micrograph of a typical wheatsheaf-shaped nucleus of a spherulite in "Nylon 6" is shown in Fig. 1.27. Thus, some undrawn polymers are made up of spherulites just as a polycrystalline metal is made up of crystalline grains.

Chain molecule polymers can also crystallize by stretching. If ordinary soft rubber is stretched, the x-ray diffraction pattern ceases to be that typical of a liquid, and diffraction spots characteristic of crystalline materials appear.

When a polymer is stretched, the development of crystallinity due to chain alignment may cause a yield phenomenon, which will be discussed in detail in Chapter 6.

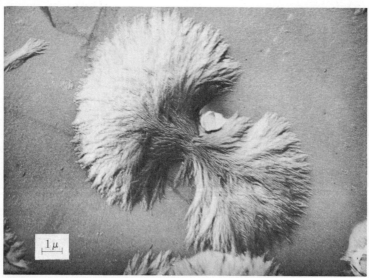

Fig. 1.27. An electron micrograph of an initial "wheatsheaf" in Nylon 6. (Stuart, 1955. Courtesy of Springer.)

1.6 MECHANISMS OF DEFORMATION

The deformation of a solid material is of two kinds: elastic and inelastic. In the usual small-strain kind of elastic deformation, the work done by external forces is stored in the body as a more or less uniform distortion of the interatomic bonds, as shown in Fig. 1.28. If the loads are removed at a rate slower than the natural frequency of vibration of the structure, the bond energy is transferred back to the agency applying the external force. Thus, elastic deformation is reversible and nearly instantaneous.

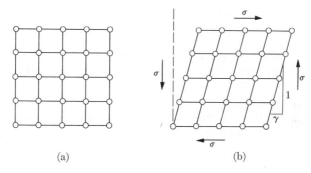

FIG. 1.28. (a) An unstrained simple cubic lattice. (b) A simple cubic lattice strained elastically.

The elasticity of rubber results from a different mechanism. In rubber, the long-chain molecules have many kinks; these are constantly shuffled by random thermal motion. When a stress is applied, the molecule is straightened out against the randomizing effect of the thermal motion. Thus, work is done on rubber to reduce its molecular disorder, that is, to reduce its entropy. When the stress is removed, random thermal motion reintroduces a kinking similar to the original, restoring the specimen to its original shape.

For both rubbery and ordinary elastic materials, the compressibility, or its reciprocal the bulk modulus ($B = -V\,dp/dV$), can be roughly estimated from the binding energy u and the density ρ. For most materials, $B = (1 \text{ to } 5)\rho u$, being higher with metals which have shorter-range forces and hence relatively less energy (Prob. 1.16). Thus for materials with primary bonds, bulk moduli are typically 3×10^6 to 3×10^7 psi, while for materials with secondary bonds, the bulk moduli are lower by about another factor of ten (Prob. 1.17).

Inelastic deformation, as opposed to elastic deformation, involves so much deformation that the bonds between initially neighboring atoms can be considered broken, and a new set of bonds is established which are as stable as the original ones. Since on release of load, the new configuration will be maintained, the deformation is inelastic. However, if new bonds are not established, fracture ensues. This is usually the case when the bond rupture is brought about by increasing the stress throughout the part to the ideal strength, which (from Fig. 1.10) is of the order of $B/10$. Likewise, fracture results if bond rupture is brought about by extremely high stress concentrations due to cracks. On the other hand, in many materials, inelastic deformation occurs at stress levels far lower than the ideal strength. We shall consider two limiting kinds of inelastic deformation here, amorphous and crystalline.

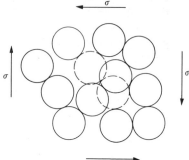

FIG. 1.29. Mechanism of inelastic deformation in an amorphous polymer.

When inelastic deformation occurs at stresses below the ideal strength in an amorphous material, it is because thermal motion is continually breaking the bonds. This is possible because the atoms or molecules are not closely packed, and there are spots where displacement can occur without seriously affecting the neighbors, as shown in Fig. 1.29. The energy required to push the pair of atoms by each other is of the order of the binding energy per atom, Bb^3. If the temperature is high enough that (according to the Boltzmann factor of Eq. 1.3) the probability of providing this energy from lattice vibrations becomes significantly large, such pairs of atoms will switch back and forth between the two positions indicated. The role of the applied stress is not to produce the switch, but rather to bias the energy levels required, so that there is more preference in this case for the dashed position than for the solid position. If the deformation is entirely due to switching of a finite number of such sites, on release of the load the material will eventually return to its initial state. On the other hand, if similar deformation elsewhere tends to further deform the neighborhood of this pair of atoms, the dashed configuration may become stable, and a return to the original state unlikely. This is the mechanism of viscous flow in polymers. If the temperature drops low enough that viscous flow is retarded, then deformation will not occur until the stresses are high enough for the bonds to be ruptured mechanically, in which case brittle fracture usually ensues.

In a crystal, on the other hand, the presence of a dislocation leads to a region in which the order of the surrounding crystal makes the energy required to break one set of bonds available from the healing of the preceding set. Thus, small stress levels are required for deformation. The amount of plastic deformation that can arise if all existing dislocations move clear through the lattice is small for annealed materials. Therefore, some mechanism must exist by which dislocations are continually generated. Such mechanisms require the formation of dislocation loops with diameters equal to hundreds of interatomic distances. As will be shown in Chapter 4, the strain energies associated with a dislocation in a crystal of bulk modulus B and atomic diameter b are of the order of $0.1\ Bb^3$ per atomic length of the dislocation. Thus, even a very tiny loop would require an energy much larger than the binding energy, which itself is almost never supplied by thermal activation at room temperature in crystals with primary bonds. Loops with lengths of hundreds of atomic units, which are actually required, would be produced even less often by thermal activation. A finite, or threshold, applied stress will generate these loops from short segments of dislocations, because as we shall see in Chapter 4, the regular nature of the crystal lattice allows the strain energy from large regions to be drained into the dislocation. The permanent nature of plastic flow arises from the fact that as these loops spread out, they begin to interact with each other, becoming entangled at new locations, so that they do not return to their original positions on release of the load.

In summary, the contrast between plastic and viscous deformation is that plastic deformation results from the propagation of deformed regions through a regular crystal at stresses and temperatures low enough so thermal activation is little help. In an amorphous material, inelastic deformation does not begin until thermal activation can play a part, and in this case, no finite threshold stress is required. Instead, since any stress is enough to produce some bias in the series of equilibrium states, viscous flow results, in which the strain rate is proportional to the applied stress.

1.7 MICRO- AND MACROSTRUCTURE OF MATERIALS

A. Microstructure. Line defects in materials are so small that they can be seen only with the electron microscope or detected through their effects on etching. At the level of magnification of an ordinary light microscope, most of the visible structure results from surface defects, which we now consider.

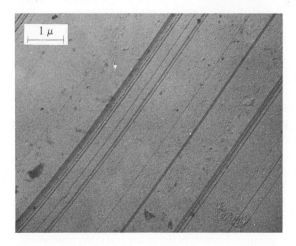

FIG. 1.30. Slip lines on α brass after 8% of shear strain. (Fourie and Wilsdorf, 1959. Courtesy of Pergamon Press.)

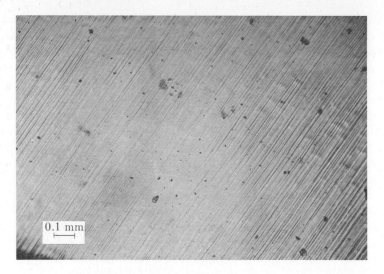

FIG. 1.31. Slip bands on the surface of a single crystal of aluminum.

In the absence of plastic deformation, one can see grain boundaries where crystals of different orientations join. Sometimes, solute atoms aggregate in a regular arrangement to form a new phase, whose boundaries can be revealed by etching. Inclusions are nonmetallic phases which have not gone into solution.

Deformation in materials very often alters the local structure, making further deformation possible in the immediate neighborhood of the original deformation. This confines deformation into bands of various types. The motion of dislocations on a particular slip plane results in a relative translation of the two parts of the crystal separated by the slip plane. The surface step resulting from this operation is a *slip line*. Figure 1.30 shows slip lines on a surface of a crystal of α brass. Since relative translation on a single slip plane rarely exceeds 100 lattice identity vectors,* slip lines are not observable with an optical microscope.

In later stages of plastic deformation, parallel slip lines begin to cluster and produce an optically observable trace on the surface, as shown in Fig. 1.31. These traces are called *slip bands*. With very few exceptions, all optically observable slip traces on metals are slip bands.

In some cases, plastic deformation on two or more intersecting planes may be confined to a broad band, as shown in Fig. 1.32. Such a band might be called a *deformation band*.†

When slip on parallel planes is confined to a band nearly normal to the slip planes, the lattice inside the band will rotate relative to the exterior matrix. Such bands are called *kink bands*. Figure 1.33 shows kink bands in a compressed zinc crystal.

---

*Some alloys, including α brass, are exceptions, as discussed in Section 4.10.

†Slip-band clusters and kink bands have often been called deformation bands. This has resulted in such confusion that the term "deformation band" is almost useless.

1.7 MICRO- AND MACROSTRUCTURE OF MATERIALS 33

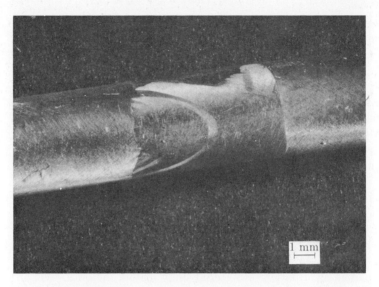

Fig. 1.32. Deformation band in a grain of an aluminum specimen bent back and forth between the fingers.

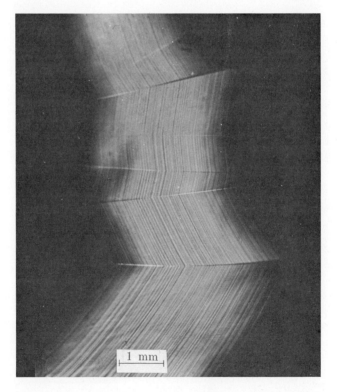

Fig. 1.33. Kink bands in a compressed zinc crystal. (Courtesy of J. J. Gilman.)

Twinning may occur inside a narrow band bound by twin planes, producing a thin lamella with a lattice in twin orientation to the matrix. This is called a *deformation twin band*. Figure 1.34 shows deformation twins in a zinc crystal.

In some polycrystalline metals such as aged low-carbon steel, as well as in many high polymers, the strain concentration of the initial site of deformation will "soften" its immediate vicinity, making further deformation easier at the boundary of the initial band. Deformation bands of this type are often macroscopically visible and are called *Lüders bands*. An example in a strip of steel sheet is shown in Fig. 1.35.

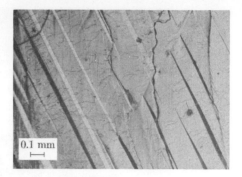

FIG. 1.34. Deformation twins in a single crystal of zinc.

Large plastic deformation of materials may distort the grains out of their original shape, and tend to rotate the operating slip planes toward the direction of deformation, producing preferred orientations and stretching out inclusions into

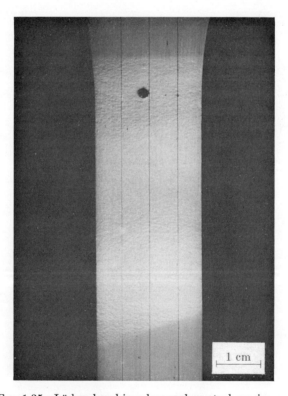

FIG. 1.35. Lüders band in a low-carbon steel specimen.

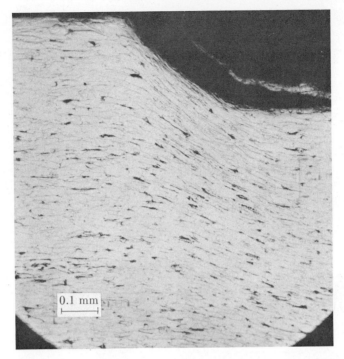

Fig. 1.36. Longitudinal section of dimpled steel wire, showing elongated grains and inclusions. (Courtesy of Pittsburgh Steel Co.)

long lines. Such a structure is commonly produced in forging and rolling and gives rise to the *flow lines* shown in Fig. 1.36.

Among polymers, the microstructure of natural fibers such as cotton, wool, and silk is far more complex than that of crystalline materials. This structure ranges in size from the molecular level up to that of the fiber itself. This structure will be discussed in more detail later, but suffice it to say here that in a cotton fiber with a diameter of 0.1 mm, there are, as shown in Fig. 1.37, three different kinds of structural layers, each consisting of a different arrangement of fibrils. These fibrils, shown in Fig. 1.38, range in size from 50 A to 1 or 2 microns and are themselves not discrete molecular units, but are made up of bundles of the basic cellulose polymer. As will be discussed later, the arrangement of these layers has a profound effect on the mechanical properties and behavior of the fiber as a whole.

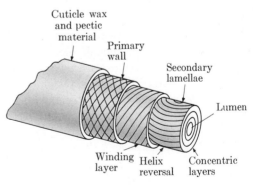

Fig. 1.37. Schematic representation of structural parts of a cotton fiber, before drying and collapsing. (After M. L. Rollins.)

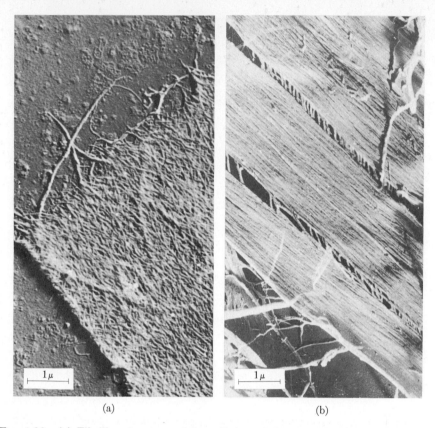

(a) (b)

Fig. 1.38. (a) Fibrillar structure of the cellulose skeleton of the cotton fiber primary wall. (b) Fibrillar structure of first layer of secondary thickening ("winding layer") of cotton fiber. (Courtesy of Mary L. Rollins, Anna T. Moore, and Verne W. Tripp, Southern Regional Laboratory.)

B. Macrostructure. Often, materials are combined into a macroscopic structure, which, on a large enough scale, may again be considered to be a new material. Examples of this are foamed polyesters used in packaging, and sponge rubber. Fiber-reinforced polymers provide another example, which is similar in many ways to reinforced concrete. Cloth is made up of many fibers spun into yarns, which are in turn wound into threads, and then are woven into cloth (Fig. 1.39). The skin of an airplane may be manufactured with integral stiffeners on the back side (Fig. 1.40) or as a honeycomb construction (Fig. 1.41). In each of these materials, the behavior can be considered on two levels: the detailed mechanisms occurring between the elements of the composite, and the description of the composite as a whole. The situation is analogous to the choice between describing a fiber in terms of its fibrils and molecules or in terms of its macroscopic stress-strain behavior. The appropriate scale of study is determined by the problem at hand.

FIG. 1.39. Structure of worsted twill.

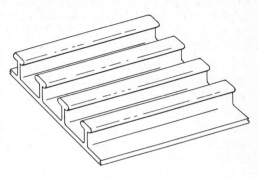

FIG. 1.40. Integral stiffeners on a wing panel of an airplane.

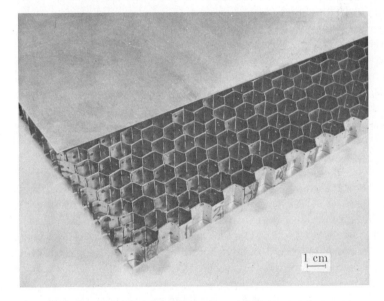

FIG. 1.41. Sandwich construction with honeycomb core.

1.8 STATE OF KNOWLEDGE

We now turn to the question of the extent to which it is possible to predict the mechanical behavior of materials from a knowledge of their structures. On the nuclear level, physicists have gained much understanding, but no comprehensive theory is as yet available. As we have seen, however, nuclear phenomena do not usually play an important role in the mechanical behavior of materials, so the lack of a theory is not serious for this subject.

At the level of atomic structure, the physicists have provided a tool of great power in the form of wave mechanics. By the application of wave mechanics to an atom with a nucleus of known mass and charge, it is possible in principle to derive the structure not only of an individual atom, but of any aggregate of atoms. In this case, however, actual accomplishment is far behind knowledge of the principles. Exact solutions have been obtained only for very simple atoms and molecules. Good approximations have been obtained, however, for certain kinds of crystals and for a variety of chemical compounds, and an understanding has been obtained of the different kinds of bonds. From this kind of work, it has proven possible to calculate to a fair degree of approximation the elastic constants of some crystals.

The next larger scale is concerned with the solubilities of elements in each other and the presence of different phases in adjacent crystals. The factors affecting solubility are reasonably well understood through thermodynamics, although wave-mechanical solutions are not generally available. The thermodynamics of different phases provide important information when the material is in a state of equilibrium. In the solid state, however, it is quite possible for nonequilibrium conditions to exist over long periods of time. The rate of approach to equilibrium is governed by many of the same structural elements, such as vacancies and dislocations, which also affect the mechanical behavior of crystals. Kinetic theory provides a powerful tool for calculating rate phenomena. The diffusion of vacancies is fairly well understood; dislocation interactions can be analyzed in only a relatively few idealized cases.

For plastic deformation, the situation is not so satisfactory as with elasticity, thermodynamics, and statistical mechanics. Plastic deformation depends on the motion of dislocations through a crystal. It has not yet proven possible to solve the equations of wave mechanics to obtain even an approximate solution for the distribution of atoms around the dislocation and the forces required to make dislocations move. All that has been done so far is to extend continuum mechanics down to the atomic level and introduce some simple nonlinear correction in the most highly strained region at the dislocation core, in place of a wave-mechanical analysis. This difficulty in obtaining solutions for even a single dislocation is not too serious, however, in view of the fact that most of the energy associated with the dislocation is not at the very core of the imperfection itself, but is associated with the elastic deformation of a region large enough that the continuum theory of elasticity can be safely applied. Thus it has been possible with relatively few additional assumptions to develop a theory of dislocations that will explain how they move and interact and multiply under the influence of applied stress.

So far, it has been possible to apply dislocation theory only to idealized cases involving the interactions of very few dislocations with each other and with solute atoms and precipitates, but these solutions do give an important insight into the processes of deformation and fracture. Recently, improved metallographic techniques and the use of the electron microscope have made it possible to observe individual dislocations, and at present our experimental knowledge is leaping ahead of our capacity to obtain theoretical solutions. The theoretical difficulty can be seen by recognizing that there may be as many as a million miles of dislocations in a cubic inch of cold-worked metal, and these dislocations are not arranged in an orderly array, but are tangled and interconnected.

On the largest scale, mechanics takes the stress-strain relations as determined empirically, or as suggested by knowledge of the structure of materials, and combines them with the principles of equilibrium and geometric compatibility to predict the behavior of regions large enough to be considered homogeneous. For the elastic stress-strain relations, the theory has been fairly well developed and successfully applied to a variety of cases. For plastic deformation, the situation is more difficult, because the structure of the material is changing. Even when this fact is neglected, the form of the stress-strain relations which seems most appropriate physically leads to a set of equations which can be evaluated only in relatively simple cases. When deformation occurs by creep, the changes in structure cannot usually be neglected, and they depend on the previous history of stress, strain, and temperature. A really suitable and physically sound stress-strain relation has not yet been formulated, much less used to solve practical problems.

In summary, while it may appear that a solution to the equations of wave mechanics would be sufficient to give an understanding of the mechanical behavior of materials, this proves to be quite impracticable (and fortunately not necessary where one can start from an intermediate level). Simplified models of the structure and behavior of the material are required at several scales to obtain the best understanding possible at present. For example, in fracture, one needs to use the macroscopic models of mechanics to find the distributions of stress and strain in regions small enough to contain only a few grains. Very near the tip of a crack, it may be possible to use dislocation theory to determine the plastic flow resulting from local stress. Between the scale of 10^{-2} cm, below which the macroscopic theory of mechanics can no longer be used, and the scale of the order of 10^{-5} cm, above which the dislocation structure is too complicated to follow, there lies a no-man's-land that remains to be mastered.

From this discussion it can be seen that the subject of the mechanical behavior of materials is by no means fully developed. In dealing with it, we shall have to consider phenomena occurring over a wide range of scales; we shall have to use approximate models where exact solutions cannot be obtained; and we shall have to resort to experimental data where the models have not yet been formulated or are too complex to be analyzed.

1.9 SUMMARY

The macroscopic mechanical behavior of materials is conveniently studied as the strain produced by stress, temperature, and time. Various idealized aspects of the behavior are known as elasticity, plasticity, creep, relaxation, elastic aftereffect, and viscosity. Since the mechanical behavior of materials is strongly influenced by structural details, it is essential to study structural forms at nearly all scales, starting from the atomic and going to that of structural parts.

TABLE 1.3
PHYSICAL CONSTANTS AND CONVERSIONS

Name	Symbol	ergs	Energy units calories	electron volts
Planck's constant	h	$= 6.62 \times 10^{-27}$ erg·sec*	$= 1.58 \times 10^{-34}$ cal·sec	$= 4.13 \times 10^{-15}$ ev·sec
Boltzmann's constant	k	$= 1.38 \times 10^{-16}$ erg/°K	$= 3.30 \times 10^{-24}$ cal/°K	$= 8.61 \times 10^{-5}$ ev/°K
for 293°K = 68°F	kT	$= 4.05 \times 10^{-14}$ erg	$= 9.67 \times 10^{-22}$ cal	$= \frac{1}{40}$ ev
Gas constant	$R = kN_A$	$= 8.32 \times 10^{7}$ erg/mole·°K	$= 1.99$ cal/mole·°K	$= 5.18 \times 10^{19}$ ev/mole·°K
for 293°K = 68°F	$RT = kN_AT$	$= 2.43 \times 10^{10}$ erg/mole	$= 584$ cal/mole	$= 1.52 \times 10^{22}$ ev/mole
Erg	1 erg	$= 1$ erg	$= 2.39 \times 10^{-8}$ cal	$= 6.24 \times 10^{11}$ ev
Calorie	1 cal	$= 4.19 \times 10^{7}$ erg	$= 1$ cal	$= 2.61 \times 10^{19}$ ev
Electron volt	1 ev	$= 1.60 \times 10^{-12}$ erg	$= 3.83 \times 10^{-20}$ cal	$= 1$ ev
Electron volt/atom	1 ev/atom	$= 9.64 \times 10^{11}$ erg/mole	$= 23{,}000$ cal/mole	$= 1$ ev/atom

1 atm = 1.03 kg/cm² = 1.01 × 10⁶ dyne/cm² = 1.01 bars = 14.7 psi
1 joule = 1 watt·sec = 10⁷ ergs
Avogadro's No, N_A = 6.02 × 10²³ atom/mole

*Underlined conversions are suggested for use and memory. Moles are gram-moles.

The electronic structure of atoms influences the binding between atoms in solids. Interatomic bonds in solids are of two major types: primary and secondary. The typical range of energy of primary bonds, such as ionic, covalent, or metallic is 2 to 8 ev; that of secondary bonds, such as van der Waals bonds, is 0.02 to 0.2 ev. The hydrogen bond, which is of intermediate type, has an energy of about 0.2 to 0.5 ev. (See Table 1.3.)

A solid has numerous modes of thermal vibration. The highest, corresponding to neighboring atoms in opposing modes, is given roughly in terms of the modulus of elasticity E, Avogadro's number N_A, and the atomic weight M by

$$\nu = \sqrt{\frac{Eb}{M/N_A}}. \tag{1.7}$$

If the temperature is high enough that quantum-mechanical effects do not play a significant role,

$$kT \geq h\nu, \tag{1.6}$$

the thermal motion of atoms in a lattice can be described by the classical statistical mechanics of Boltzmann as a first approximation. In a solid, each atom has three coordinates of position and three coordinates of momentum. Each coordinate has energy $kT/2$, giving a total of $3kT$, or a specific heat of $3N_A k$. The probability of having local energy fluctuations in excess of a magnitude u is given by the Maxwell-Boltzmann factor of

$$P(u_i \geq u) = e^{-u/kT}. \tag{1.3}$$

In a thermally activated process with a single characteristic energy barrier u_a (activation energy), the rate R of the process is determined by the product of the Boltzmann factor, the number of potentially active sites N_s, and the characteristic frequency of vibration ν_a of the process:

$$\dot{R} = N_s \nu_a e^{-u_a/kT}. \tag{1.9}$$

The most stable form of a solid is a crystal. Simple structures such as elemental metals can almost always be found in crystalline form. Structures with complex molecules, on the other hand, cannot go through the ordering process during normal rates of cooling, and will therefore be either partly or fully noncrystalline. Although the structure of crystals can vary greatly in complexity, those of common metals and some inorganic compounds are relatively simple and highly symmetrical.

The crystallographic planes and directions of crystal lattices can be conveniently identified by the three indices of the Miller notation. For a cubic lattice, since the indices are the reciprocals of the direction cosines, the methods of vector algebra can be applied.

Crystals contain a variety of lattice defects. According to their dimension, these defects can be classified as *point*, *line*, or *surface* defects. Vacancies, interstitial atoms, and substitutional foreign atoms are point defects. Lattice dislocations are line defects. Free surfaces and grain, phase, and twin boundaries are surface defects. Precipitates and second phases may be considered as volume defects.

Polymers are molecules consisting of a linear or spatial repetition of smaller groups of atoms. Polymers, depending on their state of crystallinity, can exhibit endless variations of structural forms. Natural fibers made up of polymeric molecules show additional larger-scale structures, such as layers of fibers.

Deformation can produce a certain texture in metals, as well as in polymeric materials, that often has desirable mechanical properties. Deformation of materials under stress is of two types: elastic and inelastic. Elasticity results from the static resistance of bonds when they are stretched or from the tendency to randomness in thermally oscillating links in long molecules, such as in rubber. In both cases, bonds are not severed during the deformation, and the work of deformation is stored, either as potential energy or as free energy associated with a decrease in entropy. Thermal activation plays no role, and the deformation is independent of time.

In inelastic deformation, interatomic bonds are broken and reestablished. In amorphous materials, the disordered structure permits small stable structural rearrangements activated by random thermal oscillations, even at temperatures below the melting temperature. The rate of such deformation is highly temperature sensitive. The stress biases the thermal motion to give a deformation. In crystalline materials, below the temperatures for thermal breaking of bonds, the high degree of order does not permit thermally activated small structural rearrangements. Deformation is caused by the generation and motion of dislocations, principally under an applied shear stress. Since the energy of a stable dislocation configuration is far too large to be provided by thermal fluctuations, thermal activation plays only a minor role in the plasticity of most crystals.

Deformation produces structural inhomogeneities, such as slip, kink, and twin bands, and it can introduce an anisotropy in the form of preferred orientation. Inhomogeneous composite structures are often created artificially to combine certain desirable properties of different materials in one material.

REFERENCES

To see how the binding energy is found from wave mechanics, the interested reader is referred to Hume-Rothery (1955) for a thorough, but descriptive, introduction; to Kittell (1953) for a more analytical introduction; and to Mott and Jones (1936) and Seitz (1940) for more complete treatments. The chemist's point of view, ably presented by Pauling (1948), is based more on the insight obtained from a wide range of data than on the explicit solution of the wave equation for a few relatively simple cases.

For a good description of the classical development of statistical mechanics, see Slater (1939). Hill (1960) develops the subject from quantum mechanics.

In regard to crystallography, Barrett (1953) provides a good introduction and serves as a reference, while Buerger (1956) treats crystal symmetry in more detail.

BARRER, R. M.	1942	"Transition State Theory of Diffusion in Crystals," *Trans. Faraday Soc.* **38**, 78–85.
BARRETT, C. S.	1953	*Structure of Metals*, 2nd ed., McGraw-Hill, New York.
BUERGER, M. J.	1956	*Elementary Crystallography*, Wiley, New York.
FOURIE, J. T. WILSDORF, H.	1959	"A Study of Slip Lines in α-Brass as Revealed by the Electron Microscope," *Acta Met.* **7**, 339–349.
GLASSTONE, S. LAIDLER, K. J. EYRING, H.	1941	*Theory of Rate Processes*, McGraw-Hill, New York.
GUY, A. G.	1959	*Elements of Physical Metallurgy*, Addison-Wesley, Reading, Mass.
HILL, T. L.	1960	*Introduction to Statistical Mechanics*, Addison-Wesley, Reading, Mass.
HUME-ROTHERY, W.	1955	*Atomic Theory for Students of Metallurgy*, Institute of Metals, London.
JOOS, G.	1950	*Theoretical Physics*, 2nd ed., Hafner, New York, pp. 571–580.
KITTELL, C.	1953	*Introduction to Solid State Physics*, Wiley, New York.
MOTT, N. F. JONES, H.	1936	*The Theory of the Properties of Metals and Alloys*, Oxford University Press, London. See also 1958 reprint by Dover Publications, New York.
PAULING, L.	1948	*Nature of the Chemical Bond*, 2nd ed., Cornell University Press, Ithaca, N.Y.
SEITZ, F.	1940	*Modern Theory of Solids*, McGraw-Hill, New York.
SLATER, J. C.	1939	*Introduction to Chemical Physics*, McGraw-Hill, New York.
STUART, H. A. (ed.)	1955	*Die Physik der Hochpolymeren* Vol. 3, Springer, Berlin.
SWALIN, R. A.	1962	*Thermodynamics of Solids*, Wiley, New York.
VAN VLACK, L. H.	1959	*Elements of Materials Science*, Addison-Wesley, Reading, Mass.
ZENER, C.	1952	"Theory of Diffusion," *Imperfections in Nearly Perfect Crystals*, W. Shockley et al., eds., Wiley, New York, pp. 289–316.

PROBLEMS

1.1 Show that if the energy giving attractive forces varies inversely as the first power of the interatomic spacing, whereas the repulsive energy varies inversely as, say, the ninth power, at equilibrium the attractive contribution to the energy will be much greater than the repulsive contribution.

1.2. (a) Because of the very short-range nature of the repulsive energy, the energy of an ionic lattice is largely made up of the electrostatic energies of interaction of any one ion with all the others. Show that the electrostatic energy at equilibrium (approximately the binding energy) per mole of an ionic crystal must depend on the charges on the positive and negative ions e_+ and e_-, on Avogadro's number N_A, and on the cation-anion distance r.

(b) Show that this energy must have the form

$$u = -N_A A_r e_+ e_- / r,$$

where A_r is the so-called Madelung constant, characteristic of the type of crystal structure alone. [Values of the Madelung constant range from 1.74 for the NaCl structure to 5 or even 25 in more complicated ones such as aluminum oxide (Seitz, 1940, p. 78).]

(c) Estimate the binding energy for silver chloride (which has the same structure as NaCl). Is it within a reasonable range for ionic bonds?

1.3. Show that for an ionic lattice, relatively large changes in binding energy, as might be expected on the surface, will lead to relatively small changes in density.

◀ 1.4. Show that in a two-dimensional phase space the path of an undamped spring-mass system of constant energy is an ellipse.

◀ 1.5. If the energies of particles in an assembly are closely spaced to form a continuous distribution in phase space (so that the sum in Eq. 1.2 over all of phase space can be replaced with the integral $\int e^{-u/kT} \, dv$), show that the fraction of particles with energies equal to or greater than u is given by Eq. 1.3.

◀ 1.6. The equation representing the density of particles in phase space (Eq. 1.2) (number dN per volume dv of phase space) can be written more generally as

$$dN/dv = A e^{-u/kT},$$

if the energies form a continuous spectrum from zero to infinity.

(a) Assuming that the phase space is taken to be one dimensional (to represent a one-dimensional ideal gas, where dv stands for the one-dimensional velocity of particles of mass m and their energy is entirely kinetic in nature), show that the constant A can be evaluated in terms of the total number of particles N and that the probability $f(v) \, dv$ of finding any one particle at a velocity between v and $v + dv$, taking signs into account, is then

$$f(v) \, dv = \sqrt{m/2\pi kT} \, e^{-u/kT} \, dv.$$

(b) From the results of part (a), show that the average energy $\bar{u}$ of any one particle would be

$$\bar{u} = \int_0^\infty u f(u) \, du = \int_{-\infty}^\infty u f(v) \, dv = kT/2.$$

(c) Show, by generalizing to three dimensions, that the average energy of an atom of an ideal gas is $\bar{\epsilon} = \tfrac{3}{2}kT$.

1.7. Show that the average potential energy of a simple harmonic oscillator is equal to its average kinetic energy.

◀ 1.8. By generalizing from the steps of Prob. 1.6, show that on a mole basis, the specific heat at constant volume of a simple crystal at high temperatures would be $3R$.

1.9. Verify the law of Dulong and Petit (Eq. 1.4) for some simple elemental solids at elevated temperatures by looking up values from tables.

1.10. Derive the equation for the natural frequency of vibration of an atom in a lattice site (Eq. 1.7), by assuming an atom of mass m to be tied to each surface of its box by restoring forces proportional to Young's modulus.

1.11. Derive Eq. 1.7 by considering the time required for an elastic wave to pass from one atom to its neighbor.

◀1.12. According to quantum mechanics, the mean energy per normal mode of a crystal is (Hill, 1960, p. 90)

$$E(\nu) = \tfrac{1}{2}h\nu + h\nu/(e^{h\nu/kT} - 1).$$

(a) Show that at high temperatures the specific heat obtained from this expression has the classical value.

(b) Show that the entropy per normal mode is

$$s(\nu) = \int_0^T \frac{dE(\nu)}{dT}\frac{dT}{T} = \frac{E(\nu)}{T} + k\ln\left(\frac{e^{-h\nu/2kT}}{1 - e^{-h\nu/kT}}\right) \approx k\left(1 + \ln\frac{kT}{h\nu}\right) \quad \text{for } \frac{h\nu}{kT} \ll 1.$$

(c) Show that for a 10% change in frequency at room temperature, this equation indicates a change in free energy of approximately 0.003 ev per degree of freedom.

1.13. By reproducing the arguments of Section 1.3, calculate the order of magnitude of the time necessary to evaporate a surface layer of 1 cm thickness from a solid in a perfect vacuum at room temperature if the binding energy is $\tfrac{1}{5}$, 1, or 2 ev, respectively.

1.14. Show that the equation of a unit normal vector $\mathbf{n}$ perpendicular to a plane (h, k, l) in a crystal of cubic symmetry is

$$\mathbf{n} = \frac{1}{\sqrt{h^2 + k^2 + l^2}}(h\mathbf{i} + k\mathbf{j} + l\mathbf{k}).$$

1.15. Show that the van der Waals bond can be broken readily at room temperature by thermal vibrations.

1.16. Calculate the binding energy per atom of sodium chloride as a function of the bulk modulus and lattice parameter by referring to the data of Fig. 1.10.

1.17. Using the result of Prob. 1.16, calculate the bulk modulus of a liquid held together by secondary bonds. Compare with handbook values.

1.18. In a cubic crystal, calculate the angle between two crystallographic planes whose Miller indices are (111) and (321).

1.19. Calculate the angle λ between the [100] direction and the [30$\bar{2}$] direction in a cubic crystal.

1.20. Calculate the angle λ between the [110] direction and the line of intersection of the two planes $(2\bar{1}1)$ and (111) in a cubic crystal.

1.21. (a) Sketch the unit lattice cell in a tensile specimen of a cubic crystal with the [101] direction axial and the $(1\bar{1}\bar{1})$ plane lying in the plane of the flat.

(b) If slip occurs on the $(1\bar{1}1)$ plane, what is the angle of the surface slip markings with the axial direction? [*Hint:* What is the direction of the intersection between the slip plane and the surface?]

1.22. Is a kink in a surface marking a sufficient indication of kinking?

CHAPTER 2

STRESS AND INFINITESIMAL STRAIN

2.1 SYNOPSIS

Because of the myriads of atoms involved, we describe the mechanical behavior of a solid, where possible, in terms of average stress and strain, rather than in terms of individual atoms.

On a far larger scale, we find such averages useful in studying the behavior of structures and machines which are large compared with the structure of the materials from which they are made. In a polycrystalline metal, for example, the averages are taken over elements consisting of a number of grains.

We show that a state of stress can usually be defined in terms of six components, based on the force per unit area acting on orthogonal planes through the point in question. Exceptional cases are discussed in which more components must be considered. When stress gradients are present and the three components of force on a small element are in equilibrium, we have three equations of equilibrium. Equilibrium considerations also lead us to the rules for the transformation of components of stress with a change in coordinate axes. Such transformations will be useful later in understanding the effects of structural symmetry on mechanical behavior.

By defining deformation in terms of displacements and subtracting rigid-body motion, we obtain six components of strain, based on relative displacements of points per unit initial distance between them. The fact that these six components are derived from three displacements often limits to three the strain components whose variation may be chosen arbitrarily. This leads us to the so-called compatibility conditions. The transformation law for strain components based on geometrical arguments turns out to be similar to that for stress.

We finally show how the work per unit volume, done when an element under stress deforms by a given strain increment, can be expressed in terms of the stress and the increment of strain.

2.2 DEFINITION OF STRESS

Mechanics would be a very difficult subject if it were necessary to study separately the forces on each atom, and in fact, it does not make sense to talk about forces on individual atoms. But we can speak of average force intensities acting on an element containing many atoms. Sometimes it is convenient to choose an element large enough to contain many dislocations; other times, one large enough to contain many grains of a polycrystal. The element must still be small compared with the dimensions of the test specimen or engineering structure, so that no serious stress gradients occur across it. Such an element is shown in Fig. 2.1. Sometimes the requirements of having an element large compared with an atom but small compared with the rate of change of forces are contradictory; examples are found in the core of a dislocation or at the very tip of a crack. In these cases, the concept of stress is useful in describing the environment of such regions, but not the details in the regions themselves.

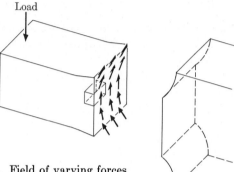

FIG. 2.1. Field of varying forces acting on a section.

In most cases, however, it is possible to choose an element of appropriate size, as shown in Fig. 2.2. The areas of the six different faces of the element are denoted by their outward normals, A_1, A_2, A_{-1}, etc., and on each face there are three components of force. The ratio of one of these components of force to the area on which it acts is called a component of force intensity, or *stress*. Two subscripts are used to denote a component of stress, the first denoting the outward normal of the surface being considered and the second denoting a particular component of the force on that surface. For example,

$$\sigma_{11} = F_1/A_1, \quad \sigma_{12} = F_2/A_1, \quad \text{etc.} \tag{2.1}$$

At first glance, there appear to be 18 different components of stress, but fortunately, two considerations of equilibrium reduce the number of independent components very rapidly. First, the force on one half of a sectioned body is equal and opposite to that on the other; we see that

$$\sigma_{-1-1} = \sigma_{11}, \quad \text{etc.} \tag{2.2}$$

This reduces the number of components of stress to nine. A second reduction is obtained if there are no net moments arising from shear components of stress. As discussed in Section 2.5, this will be the case unless there are significant effects of interatomic moments acting across a section. In the absence of such effects, equilibrium of moments about the x_3 axis, for example, requires that (Prob. 2.1)

$$\sigma_{12} = \sigma_{21}, \quad \text{etc.} \tag{2.3}$$

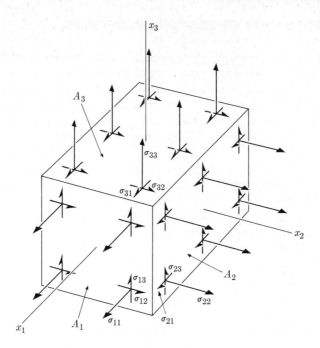

Fig. 2.2. Components of stress on an element.

Thus the order of subscripts is unimportant. The two equilibrium considerations discussed above reduce the number of independent components of stress to six: three components in which the direction of the force is normal to the surface on which it acts, i.e., three *normal* components, and three components in which the direction of the force component is parallel to the surface on which it acts, i.e., three *shear* components.

TABLE 2.1

ALTERNATIVE NOTATIONS FOR COMPONENTS OF STRESS

This book	σ_{11} ... σ_{23} ...
Timoshenko and Goodier (1951), Crandall and Dahl (1959)	σ_x ... τ_{yz} ...
Love (1944)	X_x ... Z_y ...
Bridgman (1952)	$\widehat{xx}$ ... $\widehat{yz}$ ...
Sokolnikoff (1956)	τ_{11} ... τ_{23} ...

◀A number of different notations have been used for the components of stress. The most commonly used of these are summarized in Table 2.1.▶

2.3 TRANSFORMATION OF COMPONENTS OF STRESS

◀Frequently it is convenient to describe the behavior of a material in terms of coordinate axes related to the symmetry of its structure. At the same time, it is convenient to analyze a specimen or machine part in terms of axes related to any symmetry in the shape of the part or the applied loads. If these two requirements conflict, it is necessary to work with two coordinate systems. Equilibrium considerations determine how the components of stress differ from one coordinate system to another.

◀For a transformation consisting of rotation about the x_3 axis, as shown in Fig. 2.3, some of the components of stress acting on the 1' surface can be found by calculating the areas normal to the old coordinate axes, A_1 and A_2, in terms of the area normal to the new; calculating the forces on the various areas in terms of the stress components; and considering equilibrium of forces in the new 1' direction. The result is (Prob. 2.2)

$$\sigma_{1'1'} = \sigma_{11} \cos^2 \theta + \sigma_{12} \sin \theta \cos \theta + \sigma_{21} \cos \theta \sin \theta + \sigma_{22} \sin^2 \theta,$$
$$\sigma_{1'2'} = -\sigma_{11} \sin \theta \cos \theta + \sigma_{12} \cos^2 \theta - \sigma_{21} \sin^2 \theta + \sigma_{22} \sin \theta \cos \theta. \quad (2.4)$$

Similarly, by choosing an element of different shape (Prob. 2.3), we have

$$\sigma_{2'2'} = \sigma_{11} \sin^2 \theta + \sigma_{12} \sin \theta \cos \theta + \sigma_{21} \cos \theta \sin \theta + \sigma_{22} \cos^2 \theta. \quad (2.5)$$

◀The fact that the equations for the transformation of components of stress are more complicated than those for the transformation of a vector should not be surprising, since the stress components involve two directions (the normal to a section and the direction of the force acting on the section), whereas the components of a vector are related to only one direction. Examination of the equations for the transformation of the components of stress, Eqs. 2.4 and 2.5, may

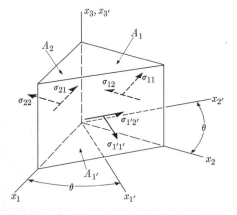

FIG. 2.3. Element for analysis of the transformation of components of stress.

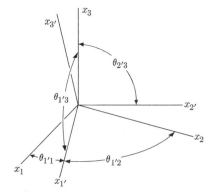

FIG. 2.4. Examples of angles between coordinate axes.

reveal a certain simplifying pattern. This pattern is more clearly revealed if the results are expressed in terms of the cosines of the angles between the various coordinate axes, called the *direction cosines*.

◀ Before further study of the transformation of components of stress, let us develop the concept of direction cosines and apply it to the transformation of components of a vector. (See also Thomas, 1960, p. 615, Problem 15.)

◀ Consider the two sets of coordinate axes shown in Fig. 2.4. The rotated, or "new," axes are denoted by primes. The set of cosines between various coordinate axes can be conveniently presented in an array in which successive columns are associated with the coordinate axes of the original set, and successive rows with axes of the rotated set:

	x_1	x_2	x_3
$x_{1'}$	$\cos \theta_{1'1}$	$\cos \theta_{1'2}$	$\cos \theta_{1'3}$
$x_{2'}$	$\cos \theta_{2'1}$	$\cos \theta_{2'2}$	$\cos \theta_{2'3}$
$x_{3'}$	$\cos \theta_{3'1}$	$\cos \theta_{3'2}$	$\cos \theta_{3'3}$

For brevity, the direction cosines are often denoted as follows:

$$\begin{array}{ccc} l_{1'1} & l_{1'2} & l_{1'3} \\ l_{2'1} & l_{2'2} & l_{2'3} \\ l_{3'1} & l_{3'2} & l_{3'3} \end{array}$$

From Fig. 2.4 and the definition of a direction cosine, we may note that $l_{1'1} = l_{11'}$, $l_{1'2} = l_{21'}$, etc., but that is is not necessarily true that $l_{1'2} = l_{12'}$.

◀ If a vector **a** has components a_1, a_2, a_3 referred to the original set of coordinates, it can be shown from a geometrical study of a vector that the components $a_{1'}$, $a_{2'}$, $a_{3'}$, referred to the transformed set, are given by (Prob. 2.4)

$$\begin{aligned} a_{1'} &= l_{1'1}a_1 + l_{1'2}a_2 + l_{1'3}a_3, \\ a_{2'} &= l_{2'1}a_1 + l_{2'2}a_2 + l_{2'3}a_3, \\ a_{3'} &= l_{3'1}a_1 + l_{3'2}a_2 + l_{2'3}a_3. \end{aligned} \qquad (2.6)$$

A general term in the above equations can be written as $l_{i'j}a_j$, where the subscripts i' and j can take on values from 1 to 3. Equations 2.6 can now be written in a more compact form, which stands for any one of the above equations, depending on whether $i' = 1, 2$, or 3:

$$a_{i'} = \sum_{j=1}^{3} l_{i'j}a_j. \qquad (2.7)$$

Equation 2.7 is sometimes taken as the definition of a vector; that is, a vector is any quantity whose components transform from one set of coordinate axes to another according to Eq. 2.7. Two useful relations between the direction cosines are given in Probs. 2.5 and 2.6.

◀Returning to the transformation of components of stress, we write a general component as σ_{ij}, where i and j may each take on any one of the three subscripts indicating the three coordinate axes. The equations for the components of stress referred to the primed coordinate system, Eqs. 2.4 and 2.5, may now be written as (Prob. 2.7)

$$\sigma_{i'j'} = \sum_{k=1}^{2} \sum_{l=1}^{2} \sigma_{kl} l_{i'k} l_{j'l}. \qquad (2.8)$$

This transformation can readily be extended to cases in which all three coordinate axes are shifted, by considering the tetrahedron of Fig. 2.5 (Prob. 2.8):

$$\sigma_{i'j'} = \sum_{k=1}^{3} \sum_{l=1}^{3} \sigma_{kl} l_{i'k} l_{j'l}. \qquad (2.9)$$

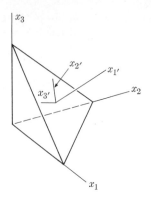

FIG. 2.5. Tetrahedron for deriving the three-dimensional transformation of components of stress.

◀In testing single crystals in the form of long wires, it is possible to apply a single normal component of stress, σ_{33}, as shown in Fig. 2.6. The plastic deformation of these crystals, however, depends on a shear component of stress referred to a particular crystallographic direction on a particular crystallographic plane. Application of the transformation equation (2.9) shows that the shear stress on the 3' plane in the 2' direction, $\sigma_{3'2'}$, can be expressed in terms of the applied stress on the wire, σ_{33}, and the angles ϕ and λ of Fig. 2.6 as (Prob. 2.9)

$$\sigma_{3'2'} = \sigma_{33} \cos \phi \cos \lambda. \qquad (2.10) ▶$$

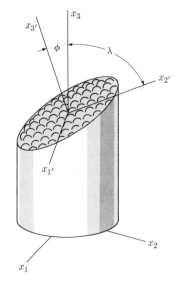

FIG. 2.6. Coordinates referred to crystallographic axes.

2.4 EQUATIONS OF EQUILIBRIUM

If an element is in a varying field of stress, considerations of equilibrium limit the possible distributions of stress. Consider the element in Fig. 2.7, where, for simplicity, only those components of stress resulting from forces in the x_1 direction are shown. If the normal component of stress, σ_{11}, changes from one face to the other, there will be a net force acting in the x_1 direction. This must be offset by changes in the shear-stress components from one A_2 face to the other, or from one A_3 face to the other. By multiplying the stress components by the respective

2.4 EQUATIONS OF EQUILIBRIUM

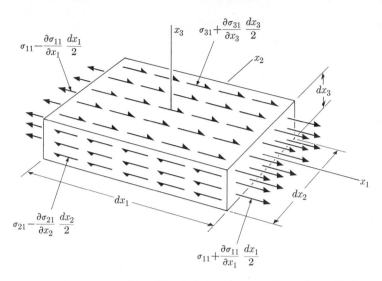

FIG. 2.7. Element considered in deriving the equations of equilibrium.

areas to obtain forces and equating the sum of the forces to zero, we obtain the following equations of equilibrium (Prob. 2.10):

$$\frac{\partial \sigma_{11}}{\partial x_1} + \frac{\partial \sigma_{21}}{\partial x_2} + \frac{\partial \sigma_{31}}{\partial x_3} = 0,$$

$$\frac{\partial \sigma_{12}}{\partial x_1} + \frac{\partial \sigma_{22}}{\partial x_2} + \frac{\partial \sigma_{32}}{\partial x_3} = 0, \quad (2.11)$$

$$\frac{\partial \sigma_{13}}{\partial x_1} + \frac{\partial \sigma_{23}}{\partial x_2} + \frac{\partial \sigma_{33}}{\partial x_3} = 0.$$

The three equations (2.11) can be written more concisely by letting j denote any of the three directions in which force equilibrium is being expressed:

$$\sum_{i=1}^{3} \frac{\partial \sigma_{ij}}{\partial x_i} = 0. \quad (2.12)$$

The three equations of equilibrium (2.11 or 2.12) relate six components of stress. It is therefore not possible to determine the stress distribution from equilibrium considerations alone; the stress-strain relations and the geometrical conditions on the strain must also be considered.

◀ An element may not be in equilibrium, and forces may act not only on its surfaces, but also magnetic, electrostatic, or gravitational forces may act throughout its volume. If the forces are not in equilibrium, there will be an acceleration of the element, $d^2 u_j/dt^2$. A body force may also be considered, where the j component of force per unit volume is denoted by F_j. The resulting equation is a form of the momentum equation, since it gives the rate of change of momentum of a particular element (Prob. 2.11):

$$\sum_{i=1}^{3} \frac{\partial \sigma_{ij}}{\partial x_i} + F_j = \rho \frac{d^2 u_j}{dt^2}. \tag{2.13}$$

◀ If a body has cylindrical symmetry, it may be convenient to refer the equations of equilibrium to cylindrical coordinates (see, for example, Crandall and Dahl, 1959, p. 161)

$$\frac{\partial \sigma_{rr}}{\partial r} + \frac{1}{r} \frac{\partial \sigma_{r\theta}}{\partial \theta} + \frac{\partial \sigma_{zr}}{\partial z} + \frac{\sigma_{rr} - \sigma_{\theta\theta}}{r} = 0,$$

$$\frac{\partial \sigma_{r\theta}}{\partial r} + \frac{1}{r} \frac{\partial \sigma_{\theta\theta}}{\partial \theta} + \frac{\partial \sigma_{z\theta}}{\partial z} + 2 \frac{\sigma_{r\theta}}{r} = 0, \tag{2.14}$$

$$\frac{\partial \sigma_{zr}}{\partial r} + \frac{1}{r} \frac{\partial \sigma_{z\theta}}{\partial \theta} + \frac{\partial \sigma_{zz}}{\partial z} + \frac{\sigma_{zr}}{r} = 0.$$

For applications having spherical symmetry, it is convenient to use spherical coordinates. Considering the special case where there are no corresponding shear components, (i.e., $\sigma_{r\phi} = \sigma_{r\theta} = \sigma_{\theta\phi} = 0$), one can show that the equations of static equilibrium are

$$\frac{\partial \sigma_{rr}}{\partial r} + \frac{1}{r} (2\sigma_{rr} - \sigma_{\theta\theta} - \sigma_{\phi\phi}) = 0,$$

$$\frac{\partial \sigma_{\theta\theta}}{\partial \theta} = 0, \quad \frac{\partial \sigma_{\phi\phi}}{\partial \phi} = 0, \quad \sigma_{\theta\theta} = \sigma_{\phi\phi}. \tag{2.15} ▶$$

2.5 COUPLE STRESSES

◀ So far we have considered the only effect acting across a section to be distributed forces. Distributed couples, as shown in Fig. 2.8, may also exist (Cosserat, 1909). Evidence for the existence of these couples will be given in the chapter on elastic stress-strain relations (Section 3.5). The intensity of these distributed couples per unit area of surface can be denoted by χ_{ij}, where the first subscript denotes the area on which the distributed couple acts and the second denotes the vector direction of the couple. These distributed interface couples (couple stresses) transform as do the components of stress (Prob. 2.12), but it is not necessarily true that $\chi_{ij} = \chi_{ji}$ (Prob. 2.13). When considerations of equilibrium are applied to an element subject to couple stresses, it is no longer necessary that $\sigma_{ij} = \sigma_{ji}$, for (Prob. 2.14)

$$\sigma_{ij} - \sigma_{ji} = \sum_{l=1}^{3} \frac{\partial \chi_{lk}}{\partial x_l}, \tag{2.16}$$

where i, j, and k are all different.

◀ Equation 2.16 indicates that couple stresses may be of interest not only for their own sake, but also, when their gradient is high, for their effect on the

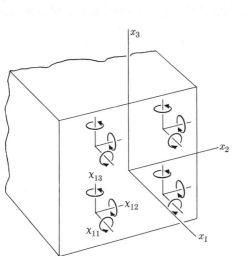

FIG. 2.8. Couple stress on a section. FIG. 2.9. Fiber-reinforced laminate with gradient of distributed couples.

stress distribution. Kröner (1961) has pointed out that the concept may be useful in dealing with elements containing arrays of dislocations. Another possibility is that these couple stresses are of importance in the case of fiber-reinforced laminates, as shown in Fig. 2.9. Under the action of a tensile load in the x_3 direction, the two outer plies will tend to rotate in the $-x_2$ sense, while the middle ply will tend to rotate in the $+x_2$ sense. If this rotation is prevented primarily by regions where the fibers in the different layers cross over each other, then couple stresses will act at each of these crossover points. An element containing a number of fibers may be considered to have couple stresses acting on it. Furthermore, these couples are of the *same* sign acting on *opposite* faces of the element. Therefore, χ_{22} changes sign from one face to the other of the central ply, giving a very high gradient of the couple stresses. According to Eq. 2.16, this high gradient leads to differing shear stresses σ_{13} and σ_{31}. The mechanics of this situation have not been fully developed, but it is known that fracture of laminates sometimes begins with separation at the point where the fibers cross over each other, so that an understanding of the couple stresses might help us achieve a better understanding of the conditions for fracture. ▶

2.6 DEFINITION OF STRAIN

The presence of stress in a body produces deformation. The deformation is described in terms of displacements. However, since even a rigid body undergoes displacements, in describing the deformation of a body we must be concerned with the *relative* displacements of nearby points in the body. Consider, for example, the two points of a body shown in Fig. 2.10. The displacements in the x_1 and x_2 directions are denoted by u_1 and u_2, respectively. The displacement of one point relative to another will be greater, the greater the initial distance between the points. Therefore, in discussing deformation, we will discuss the relative motion of two points per unit initial distance between them. These relative motions will have different components. For the case shown in Fig. 2.10, the 1 and 2 components of relative motion per unit initial distance are

$$\frac{[u_1 + (\partial u_1/\partial x_1)\, dx_1] - u_1}{dx_1} = \frac{\partial u_1}{\partial x_1}$$

and

$$\frac{[u_2 + (\partial u_2/\partial x_1)\, dx_1] - u_2}{dx_1} = \frac{\partial u_2}{\partial x_1}.$$

(2.17)

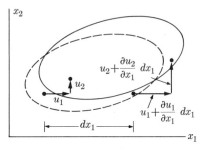

Fig. 2.10. Displacements of nearby points with the same ordinate.

Thus the relative displacements that we wish to study will be related to the various partial derivatives of the displacements in the body. These displacements are functions of initial coordinates:

$$u_i = u_i(x_1, x_2, x_3). \tag{2.18}$$

This definition of relative displacement is meaningful only for elements large enough that the displacements can be averaged so as to vary in a continuous and differentiable way. For example, if an element of a crystal is of the size of the slip-band spacing, an average derivative of the displacements across an element will depend markedly on whether or not the element happened to include a slip band. Just as in the case of stress, it is necessary to find an element of the right size: large enough that an average displacement gradient can reasonably be defined, and yet small enough that the displacement gradients vary only gradually across it.

There are altogether nine partial derivatives of the displacements. These may be presented in an array:

$$\begin{array}{ccc} \dfrac{\partial u_1}{\partial x_1} & \dfrac{\partial u_1}{\partial x_2} & \dfrac{\partial u_1}{\partial x_3}, \\[6pt] \dfrac{\partial u_2}{\partial x_1} & \dfrac{\partial u_2}{\partial x_2} & \dfrac{\partial u_2}{\partial x_3}, \\[6pt] \dfrac{\partial u_3}{\partial x_1} & \dfrac{\partial u_3}{\partial x_2} & \dfrac{\partial u_3}{\partial x_3}. \end{array} \tag{2.19}$$

This array is still not entirely satisfactory for describing stress-induced deformation, because relative motion between the points of a body can be produced by

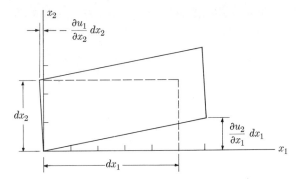

FIG. 2.11. Relative displacements associated with rotation.

rotation without requiring stress. For some problems, such as the development of anisotropy and fracture due to deformation, the rotation may be important, but we shall consider here only the stress-induced deformation. We therefore wish to subtract out the rotations.

◀To eliminate the rotations, we consider the element shown in Fig. 2.11, with the origin of coordinates assumed to translate with the corresponding point of the body. For rotations small compared with unity, the rotation of a line initially in the x_1 direction is

$$\omega = \frac{\partial u_2}{\partial x_1}.$$

The rotation of a line initially in the x_2 direction is

$$\omega = -\frac{\partial u_1}{\partial x_2}.$$

The average rotation of the element in the 1-2 plane can be defined as

$$\omega_{21} = \frac{1}{2}\left(\frac{\partial u_2}{\partial x_1} - \frac{\partial u_1}{\partial x_2}\right). \tag{2.20}$$

The general set of components of rotation can be presented in an array:

$$\begin{bmatrix} 0 & \omega_{12} & \omega_{13} \\ \omega_{21} & 0 & \omega_{23} \\ \omega_{31} & \omega_{32} & 0 \end{bmatrix} = \begin{bmatrix} 0 & \frac{1}{2}\left(\frac{\partial u_1}{\partial x_2} - \frac{\partial u_2}{\partial x_1}\right) & \frac{1}{2}\left(\frac{\partial u_1}{\partial x_3} - \frac{\partial u_3}{\partial x_1}\right) \\ \frac{1}{2}\left(\frac{\partial u_2}{\partial x_1} - \frac{\partial u_1}{\partial x_2}\right) & 0 & \frac{1}{2}\left(\frac{\partial u_2}{\partial x_3} - \frac{\partial u_3}{\partial x_2}\right) \\ \frac{1}{2}\left(\frac{\partial u_3}{\partial x_1} - \frac{\partial u_1}{\partial x_3}\right) & \frac{1}{2}\left(\frac{\partial u_3}{\partial x_2} - \frac{\partial u_2}{\partial x_3}\right) & 0 \end{bmatrix}. \tag{2.21}$$

Subtracting the rotation given by the array 2.21 from the relative displacement gradients given by the array 2.19, we obtain the *strain*, which describes the local changes in shape of the body:*

$$\begin{bmatrix} \epsilon_{11} & \epsilon_{12} & \epsilon_{13} \\ \epsilon_{21} & \epsilon_{22} & \epsilon_{23} \\ \epsilon_{31} & \epsilon_{32} & \epsilon_{33} \end{bmatrix} = \begin{bmatrix} \dfrac{\partial u_1}{\partial x_1} & \dfrac{1}{2}\left(\dfrac{\partial u_1}{\partial x_2} + \dfrac{\partial u_2}{\partial x_1}\right) & \dfrac{1}{2}\left(\dfrac{\partial u_1}{\partial x_3} + \dfrac{\partial u_3}{\partial x_1}\right) \\ \dfrac{1}{2}\left(\dfrac{\partial u_2}{\partial x_1} + \dfrac{\partial u_1}{\partial x_2}\right) & \dfrac{\partial u_2}{\partial x_2} & \cdots \\ \cdots & \cdots & \cdots \end{bmatrix}. \quad (2.22)$$

Individual terms of this array are called *components of strain*, and may be expressed in general form by

$$\epsilon_{ij} = \frac{1}{2}\left(\frac{\partial u_i}{\partial x_j} + \frac{\partial u_j}{\partial x_i}\right). \quad (2.23)$$

◀In problems where only a few components of strain are present, the factor $\frac{1}{2}$ appearing in Eq. 2.22 is often dropped. To distinguish such a component of strain, a different symbol will be used:

$$\gamma_{23} = \frac{\partial u_2}{\partial x_3} + \frac{\partial u_3}{\partial x_2} = 2\epsilon_{23}. \quad (2.24) \blacktriangleright$$

When the rotation has been subtracted from the set of displacement gradients, the components of strain can be written in terms of the displacements as

$$\begin{aligned}
\epsilon_{11} &= \frac{\partial u_1}{\partial x_1}, & \gamma_{23} &= \frac{\partial u_2}{\partial x_3} + \frac{\partial u_3}{\partial x_2}, \\
\epsilon_{22} &= \frac{\partial u_2}{\partial x_2}, & \gamma_{31} &= \frac{\partial u_3}{\partial x_1} + \frac{\partial u_1}{\partial x_3}, \\
\epsilon_{33} &= \frac{\partial u_3}{\partial x_3}, & \gamma_{12} &= \frac{\partial u_1}{\partial x_2} + \frac{\partial u_2}{\partial x_1}.
\end{aligned} \quad (2.25)$$

In either Eqs. 2.22 or Eqs. 2.25, the components with equal subscripts, for example ϵ_{11}, give relative displacements of two points in the same direction as their initial separation, and are called *normal* components of strain. The components with unequal subscripts, for example γ_{12}, give an average transverse displacement per unit separation, and are called *shear* components of strain. Note that by definition, $\gamma_{12} = \gamma_{21}$, and $\epsilon_{ij} = \epsilon_{ji}$, and therefore there are only six different components of strain. For reasons which will become evident in Sections 2.7 and 2.8, the components of strain defined by Eqs. 2.22 or 2.23 will be called *tensor* components of strain. When the components of strain are defined without the factor $\frac{1}{2}$, as in Eqs. 2.24 and 2.25, they will be called *tangential* components of shear strain, since the component γ_{ij} represents the tangent of the angular rotation between two line elements initially parallel to the x_i and the x_j axes, respectively.

---

*Here, as elsewhere in the text, ellipses denote missing terms which may be filled in by the reader.

The volumetric, or dilatational, strain is given by the sum of the normal components, for strains small compared with unity (Prob. 2.15)

$$\epsilon = \Delta V/V = \epsilon_{11} + \epsilon_{22} + \epsilon_{33}. \tag{2.26}$$

Since the expression for volumetric strain is valid only for small strain, and since the expressions for shear strain in terms of displacement were obtained with the aid of an expression for rotation that is valid only for small angles of rotation, it is not surprising that the expressions for strain are physically valid only for small relative displacements. This limitation can be illustrated in the following way. Consider the rotation of a rigid body by an angle θ about the x_3 axis. The displacements of points in the body can be expressed in terms of the angle of rotation θ and the original coordinates, x_1, x_2, and x_3. It can be shown that for θ much less than unity, both Eqs. 2.22 and 2.25 indicate zero strain. However, for large rotations, say $\theta = 180°$, the components of strain given by either of these equations do not turn out to be zero (Prob. 2.16). For large plastic strains, we shall see in Chapter 7 that the physical process is best described by a series of strain increments, each of which is small and relatively independent of the original state, so that Eqs. 2.22 and 2.25 hold when the relative displacements are always referred to the current separation of points in the body. When large strains occur in rubber, as we shall see in Chapter 6, the material does "remember" its original shape, so new definitions of strain must be used.

The six components of strain given by Eqs. 2.22 or 2.25 cannot be chosen arbitrarily, since they are derived from only three components of displacement. Thus one must seek a strain distribution in a solid that can be derived from three components of displacement. As shown below, this requirement turns out to be no restriction in dealing with strain fields which vary linearly through the part, but it is a restriction in more involved cases.

Such a case is the deformation of a plate in its plane, shown in Fig. 2.12. Displacements normal to the plate are assumed zero. For example, we take the normal components of strain to be of the form

$$\epsilon_{11} = -a_1 x_2^2, \qquad \epsilon_{22} = -a_2 x_1^2.$$

Then the shear strain cannot be chosen arbitrarily, for it turns out that the displacements in the plane of the plate must have the form

$$u_1 = -a_1 x_2^2 x_1 + f(x_2),$$
$$u_2 = -a_2 x_1^2 x_2 + g(x_1),$$

and the shear component must be of the form (Prob. 2.17)

$$\gamma_{12} = -2(a_1 + a_2)x_1 x_2 + \frac{\partial f(x_2)}{\partial x_2} + \frac{\partial g(x_1)}{\partial x_1}.$$

◀Instead of actually demonstrating a set of displacements, one can show that if the strain distributions satisfy certain conditions, then it is possible in principle to

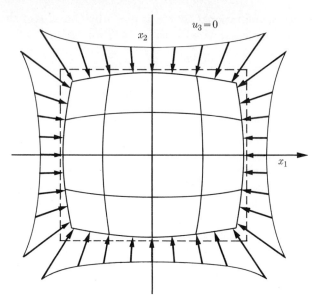

Fig. 2.12. Example of a nonlinear strain field.

integrate the equations to find single-valued displacements. These conditions are called the *compatibility equations*, and are given, for example, by Timoshenko and Goodier (1951, p. 229), or Sokolnikoff (1956, p. 25), as three differential equations of each of the types

$$\frac{\partial^2 \epsilon_{22}}{\partial x_3^2} + \frac{\partial^2 \epsilon_{33}}{\partial x_2^2} = \frac{\partial^2 \gamma_{23}}{\partial x_2 \partial x_3},$$

and (2.27)

$$2 \frac{\partial^2 \epsilon_{11}}{\partial x_2 \partial x_3} = \frac{\partial}{\partial x_1}\left(-\frac{\partial \gamma_{23}}{\partial x_1} + \frac{\partial \gamma_{31}}{\partial x_2} + \frac{\partial \gamma_{12}}{\partial x_3}\right).$$

◄As in the case of stress, for the body with cylindrical symmetry it is often convenient to describe the strain with respect to the corresponding coordinate system. If the displacements in the radial, tangential, and axial directions of cylindrical coordinates are u_r, u_θ, and u_z, respectively, the components of strain are (see, for example, Crandall and Dahl, 1959, p. 166)

$$\epsilon_{zz} = \frac{\partial u_z}{\partial z}, \qquad \gamma_{zr} = \frac{\partial u_r}{\partial z} + \frac{\partial u_z}{\partial r},$$

$$\epsilon_{rr} = \frac{\partial u_r}{\partial r}, \qquad \gamma_{r\theta} = \frac{\partial u_\theta}{\partial r} + \frac{1}{r}\frac{\partial u_r}{\partial \theta} - \frac{u_\theta}{r}, \qquad (2.28)$$

$$\epsilon_{\theta\theta} = \frac{1}{r}\frac{\partial u_\theta}{\partial \theta} + \frac{u_r}{r}, \qquad \gamma_{\theta z} = \frac{1}{r}\frac{\partial u_z}{\partial \theta} + \frac{\partial u_\theta}{\partial z}.$$

For cases having spherical symmetry, it is often convenient to use a spherical coordinate system, denoting the displacements in the r, θ, and ϕ directions as u_r, u_θ, and u_ϕ, respectively. For the special case where displacements occur in the radial direction only ($u_\theta = u_\phi = 0$), it can be shown that the components of strain are

$$\epsilon_{rr} = \frac{\partial u_r}{\partial r}, \qquad \epsilon_{\phi\phi} = \epsilon_{\theta\theta} = \frac{u_r}{r}, \qquad \gamma_{r\theta} = \gamma_{r\phi} = \gamma_{\theta\phi} = 0. \qquad (2.29) \blacktriangleright$$

2.7 TRANSFORMATION OF COMPONENTS OF STRAIN

◀Since the components of strain are defined in terms of the partial derivatives of displacements, the rules governing the transformation of the components of strain due to rotation of coordinate axes are found from a combination of the laws of transformation of displacements and the rules of the transformation of partial derivatives. The displacements, being vectors, transform according to the vector rule (Eq. 2.7)

$$u_{i'} = \sum_{j=1}^{3} l_{i'j} u_j. \qquad (2.30)$$

The partial derivative with respect to a new coordinate axis is found by using the theory of partial differentiation to express the partial derivative with respect to the new coordinate in terms of the partial derivatives with respect to the old coordinates and the rate of change of each of the old coordinates while moving along one of the new axes. The situation is illustrated in Fig. 2.13 and leads to

$$\frac{\partial}{\partial x_{j'}} = \left(\frac{\partial}{\partial x_1}\right)\frac{\partial x_1}{\partial x_{j'}} + \left(\frac{\partial}{\partial x_2}\right)\frac{\partial x_2}{\partial x_{j'}} + \left(\frac{\partial}{\partial x_3}\right)\frac{\partial x_3}{\partial x_{j'}}. \qquad (2.31)$$

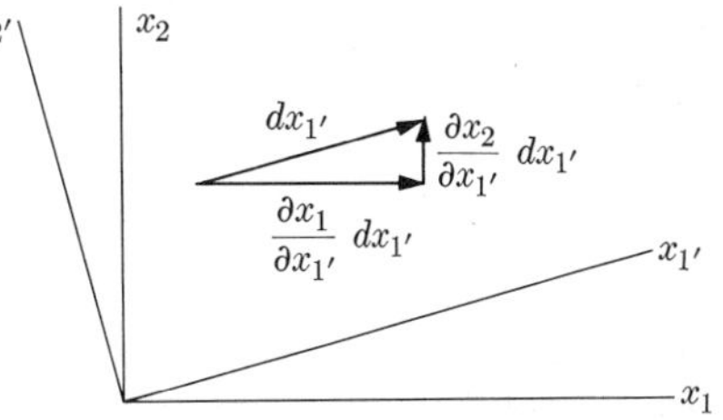

The rate of change of the old coordinate as we proceed along the new one can be expressed in terms of the direction cosines:

$$\partial x_1/\partial x_{j'} = l_{j'1}, \qquad \text{etc.} \qquad (2.32)$$

FIG. 2.13. Partial derivatives between coordinate systems.

The above two equations, along with the tensor definition of strain (Eq. 2.22), can be combined to give the components of strain in the new coordinate system. This is most safely done the first time by writing out all the components. The final result is (Prob. 2.18)

$$\epsilon_{i'j'} = \sum_{k=1}^{3}\sum_{l=1}^{3} \epsilon_{kl} l_{i'k} l_{j'l}. \qquad (2.33) \blacktriangleright$$

2.8 CONCEPT OF A TENSOR

◀It may be noted that the equation for the transformation of components of strain, Eq. 2.33, is identical with the equation for the transformation of components of stress, Eq. 2.9. As a matter of fact, this rule for transformation is encountered in other branches of science as well and has therefore been given a general name: *tensor transformation*. A quantity which transforms according to Eq. 2.9 or Eq. 2.33 is called a second-order tensor, since two direction cosines are involved in the transformation. By this definition, a vector is a first-order tensor. We shall encounter higher-order tensors later in discussing the elastic constants of crystals. Stress and strain are *symmetric tensors* in that components appearing on opposite sides of the diagonal running downward from left to right are equal; for example, $\sigma_{ij} = \sigma_{ji}$. The array representing rotation is *antisymmetric* in that $\omega_{ij} = -\omega_{ji}$. The couple stresses discussed in Section 2.5 are tensor quantities, although they are neither symmetric nor antisymmetric.

◀The reader is probably familiar with the fact that two-dimensional tensors, such as stress and tensor strain in two dimensions, can be transformed with the aid of a graphical construction known as Mohr's circle. This construction is illustrated in Fig. 2.14. The normal components of stress are represented by the abscissa. Shear components are represented by the ordinate with the sign convention that for positive shear the point corresponding to the x_1 component in the usual x_1, x_2 coordinate system is plotted down, and that corresponding to the x_2 component is plotted up. To obtain the components referred to axes rotated by θ in a physical plane, read the coordinates of the ends of the diameter of Mohr's circle that is rotated 2θ, as indicated in Fig. 2.14. If the tangential definition, rather than the tensor definition, of shear strain is used, the shear components must be divided by two before the Mohr's circle construction (or a tensor transformation) can be applied.

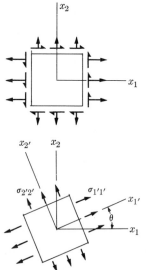

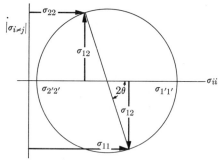

FIG. 2.14. Mohr's circle construction for transformation of tensor components.

◀It is evident from Mohr's circle that in a two-dimensional state of stress or strain, there exist coordinate axes such that shear components referred to them vanish. These are termed *principal* axes. It can be shown (see, for example, Jaeger, 1956) that there are principal axes also in three-dimensional cases. This follows from the fact that three variables are needed to specify one set of coordinate axes relative to another, and it turns out to be possible to choose these so that the three shear components of stress or strain vanish. Thus, in describing the stress-strain behavior of a material, if there is no preferred orientation given by its structure, for convenience the coordinate axes may be chosen to coincide with the principal directions of stress. In solving for the stress or strain distribution of a structure, the directions of the principal axes usually change from point to point. It is then easier to work with a single coordinate system, chosen to fit the shape of the body. Shear components of stress and strain will then usually be present in these coordinates.

◀Equations involving tensors are often written more concisely by using special conventions and symbols. These will not be generally used in this text, but are mentioned so the reader need not immediately abandon papers in which they are used. It is a curious fact that where summation signs are called for, the dummy subscript, which is being summed over, appears just twice in each term of the summation. Because of this fact, the summation sign can be considered to be implied rather than written explicitly whenever a subscript appears twice in a term. This is called the *summation convention*. In terms of the summation convention, the transformation laws for components of vectors and tensors, i.e., Eqs. 2.7 and 2.9, become

$$a_{i'} = l_{i'j}a_j, \tag{2.7}$$

and

$$\sigma_{i'j'} = \sigma_{kl}l_{i'k}l_{j'l}. \tag{2.9}$$

A further step is to abbreviate the notation for partial differentiation with respect to a coordinate variable by using the subscript of that variable, preceded by a comma. For example, the expression for the partial derivative of a stress component becomes

$$\partial \sigma_{31}/\partial x_3 \rightarrow \sigma_{31,3},$$

and the momentum equations can be written as

$$\sigma_{ij,i} + F_j = \rho \, d^2 u_j / dt^2. \tag{2.13}$$

◀A number of expressions turn out to have the value unity or zero, depending on whether the values of the subscripts happen to be the same or different, respectively. To express this result, a unit tensor δ_{ij} is defined, having the value unity when the subscripts are equal, and zero when they are not. This tensor allows us to write the relations between the direction cosines as (Probs. 2.5, 2.6)

$$l_{i'j}l_{i'k} = \delta_{jk}. \tag{2.34}$$

◀Another special variable, less frequently found, is the permutation tensor, ϵ_{ijk}, defined as having the value 1, -1, or 0, depending, respectively, on whether the subscripts have numerical values in cyclic order (123, 231, 312), anticyclic order

(321, 213, or 132), or have the same value at least once (113, 322, 222, etc). Using the permutation tensor, we can write a vector cross product in terms of the unit vectors n_i as

$$\mathbf{a} \times \mathbf{b} = \epsilon_{ijk} \mathbf{n}_i a_j b_k. \tag{2.35}$$

The small rotation of an element can be expressed as a vector:

$$\boldsymbol{\omega} = \omega_k \mathbf{n}_k = \mathbf{n}_k \epsilon_{ijk} u_{i,j}/2. \tag{2.36}$$

The inequality of the shear components of stress can be expressed in terms of the gradients of the couple stresses as

$$\chi_{ik,i} = \epsilon_{ijk} \sigma_{ij}. \tag{2.16}$$

This concise notation is useful for proofs of general theorems, and also in setting up computer solutions to problems. It is less useful in working out solutions to specific problems, which usually have simplified shapes of boundaries and conditions of load, so that a number of the components of stress and strain vanish (beams, plates, bars, plane stress, and plane strain). ▶

2.9 MEASUREMENT OF STRAIN

In many cases it is necessary to determine stress from experimental data in order, for example, to determine the loads a part encounters in service, to check the assumptions used in a theoretical analysis, or to determine the stresses in parts too complicated for theoretical analysis. Except for contact stresses on the surface of a body, it is not possible to measure stress directly. Therefore, strain is measured and the corresponding stress calculated. Various methods of measuring the strain include the following.

(1) Direct mechanical measurement is simple, but usually has a sensitivity not better than 10^{-4} to 10^{-5} (Cook and Rabinowicz, 1963).
(2) Electrical resistance strain-gauge measurement is good to 10^{-6} (or 10^{-8} when semiconducting materials are used) but gives the strain only where they are permanently attached (Murray and Stein, 1961).
(3) Model studies with photoelastic materials require models and are simple to analyze only in the case of plane stress (Frocht, 1948).
(4) Photoelastic coatings require more cumbersome recording equipment than strain gauges and are not suitable for very small parts (Post and Zandman, 1961).
(5) Brittle lacquer coatings are more difficult to apply, relatively insensitive, indicate tension only, but have the advantage of showing the direction of the maximum tensile stress (Durelli, Phillips, and Tsao, 1958).
(6) X-ray analysis has the advantage of giving the elastic part of strain alone, although the equipment is much more expensive and cumbersome than that used with any of the other methods and the sensitivity corresponds to only a few thousand pounds per square inch (see Section 12.6).

As a general reference, Hetenyi (1950) is unexcelled.

2.10 WORK IN TERMS OF STRESS AND STRAIN

The work per unit volume of a deformed solid can be expressed in terms of the stress and the increment of strain using either the tangential or the tensor definition of strain. The expression based on the tangential definition of strain can be derived by considering three different sections through an element, such as the one shown in Fig. 2.15 (Prob. 2.19):

$$dW/V = \sigma_{11}\,d\epsilon_{11} + \sigma_{22}\,d\epsilon_{22} + \cdots + \sigma_{23}\,d\gamma_{23} + \cdots. \tag{2.37}$$

◀With the tensor definition of strain, we can express the work per unit volume more concisely, although the numerical evaluation of the expression may be somewhat longer. The derivation is made by considering the differential work done on a body in equilibrium by displacements of its surfaces. Consider again an orthogonal element having the components of force $\sigma_{ij}\,dA_i$ on the surface normal to the x_i axis. Then under displacement increments du_j, the increment of work is

$$dW = \int_A \sum_{j=1}^{3} \sum_{i=1}^{3} \sigma_{ij}\,dA_i\,du_j. \tag{2.38}$$

◀The surface integral of Eq. 2.38 can be changed into a volume integral with the aid of the Green-Gauss theorem (see, for example, Franklin, 1944, p. 308):

$$\int_A \sum_{i=1}^{3} f_i\,dA_i = \int_V \sum_{i=1}^{3} \frac{\partial f_i}{\partial x_i}\,dV, \tag{2.39}$$

where the function f_i is taken to be

$$f_i = \sum_{j=1}^{3} \sigma_{ij}\,du_j.$$

Expansion of the partial derivative, application of the equations of equilibrium

$$\sum_{i=1}^{3} \frac{\partial \sigma_{ij}}{\partial x_i} = 0,$$

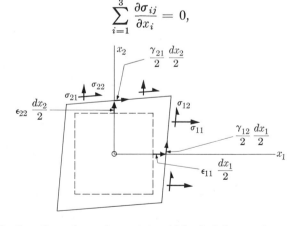

FIG. 2.15 Section through an element, to aid in deriving work per unit volume.

and the definition of strain

$$d\epsilon_{ij} = \frac{1}{2}\left(\frac{\partial\,du_i}{\partial x_j} + \frac{\partial\,du_j}{\partial x_i}\right),$$

show that the differential work done on the body is (Prob. 2.20)

$$dW = \int_V \sum_{i=1}^{3}\sum_{j=1}^{3} \sigma_{ij}\,d\epsilon_{ij}\,dV. \tag{2.40}$$

Thus the differential work per unit volume can be written as

$$dW/V = \sum_{i=1}^{3}\sum_{j=1}^{3} \sigma_{ij}\,d\epsilon_{ij}. \tag{2.41} \blacktriangleright$$

2.11 SUMMARY

The concepts of stress and strain can be applied to elements small enough that the stresses vary only slightly across them and yet large enough compared with the structure of the material that local variations caused by the structure are averaged out.

A component of stress is denoted by two subscripts, the first denoting the area on which the stress acts and the second the component of the force acting on that area:

$$\sigma_{ij} = F_j/A_i. \tag{2.1}$$

Components are called normal or shear, depending on whether the force is normal to the area or parallel to it, respectively. Equilibrium in the absence of couple stress shows that $\sigma_{ij} = \sigma_{ji}$. The six components of stress are related by three equations, each expressing equilibrium in one direction,

$$\frac{\partial \sigma_{11}}{\partial x_1} + \frac{\partial \sigma_{21}}{\partial x_2} + \frac{\partial \sigma_{31}}{\partial x_3} = 0, \quad \text{etc.} \tag{2.11}$$

The *strain* of an element is the set of relative displacements of nearby points per unit separation between them, rearranged to cancel out the relative displacements associated with rotation of the element. For small strains, a definition which satisfies these conditions can be expressed in terms of the displacements u_1, u_2, and u_3 as

$$\begin{aligned}\epsilon_{11} &= \frac{\partial u_1}{\partial x_1}, & \gamma_{23} &= \frac{\partial u_2}{\partial x_3} + \frac{\partial u_3}{\partial x_2}, \\ \epsilon_{22} &= \frac{\partial u_2}{\partial x_2}, & \gamma_{31} &= \frac{\partial u_3}{\partial x_1} + \frac{\partial u_1}{\partial x_3}, \\ \epsilon_{33} &= \frac{\partial u_3}{\partial x_3}, & \gamma_{12} &= \frac{\partial u_1}{\partial x_2} + \frac{\partial u_2}{\partial x_1}.\end{aligned} \tag{2.25}$$

2.11 SUMMARY

The components ϵ and γ are called normal and shear components, respectively. These definitions of strain are not suitable for large strains, except for plastic deformation, where they apply to strain increments. Other definitions of strain must be used in describing the elastic behavior of rubber. We will later find the *mean normal stress*, $\sigma = (\sigma_{11} + \sigma_{22} + \sigma_{33})/3$, and the *dilatational*, or total normal, strain, $\epsilon = \epsilon_{11} + \epsilon_{22} + \epsilon_{33}$, to be useful components of stress and strain.

The increment of work done on an element per unit volume can be expressed in terms of the strain increments:

$$dW/V = \sigma_{11}\, d\epsilon_{11} + \sigma_{22}\, d\epsilon_{22} + \cdots + \sigma_{23}\, d\gamma_{23} + \cdots. \qquad (2.37)$$

◀The transformation of components of strain with change in coordinate axes is shown graphically with the aid of Mohr's circle, Fig. 2.14. An alternative description is to use direction cosines, where $l_{i'j}$ denotes the cosine of the angle between the ith axis in the new coordinate system and the jth axis in the old. If a vector has components a_j referred to the old coordinate system, the components referred to the new coordinate system, $a_{i'}$, are given in terms of the direction cosines by

$$a_{i'} = \sum_{j=1}^{3} l_{i'j} a_j. \qquad (2.7)$$

◀The law for the transformation of the components of stress is similar except that two direction cosines instead of one appear in each term:

$$\sigma_{i'j'} = \sum_{k=1}^{3} \sum_{l=1}^{3} \sigma_{kl} l_{i'k} l_{j'l}. \qquad (2.9)$$

◀The law for the transformation of the components of strain is identical with that for the components of stress, but only if the tensor definition of strain is used:

$$\epsilon_{ij} = \frac{1}{2}\left(\frac{\partial u_i}{\partial x_j} + \frac{\partial u_j}{\partial x_i}\right). \qquad (2.23)$$

Then

$$\epsilon_{i'j'} = \sum_{k=1}^{3} \sum_{l=1}^{3} \epsilon_{kl} l_{i'k} l_{j'l}. \qquad (2.33)$$

When tangential shear-strain components γ_{ij} are employed, they must be divided by 2 prior to the transformation.

◀Coordinate axes such that all three components of shear strain or shear stress vanish are called principal axes.

◀In terms of the tensor definition of strain, the increment of work per unit volume can be expressed as

$$dW/V = \sum_{i=1}^{3} \sum_{j=1}^{3} \sigma_{ij}\, d\epsilon_{ij}. \qquad (2.41)$$

◂The equations of this chapter are often written even more concisely by use of the summation convention discussed in Section 2.8:

Equilibrium: $\sigma_{ij,i} + F_j = \rho(d^2 u_j/dt^2)$.
Transformation: $\sigma_{i'j'} = \sigma_{ij} l_{i'i} l_{j'j}$.
Strain-displacement: $\epsilon_{ij} = (u_{i,j} + u_{j,i})/2$.
Rotation: $\omega_i = \epsilon_{ijk} u_{j,k}/2$.
Work: $dW/V = \sigma_{ij}\, d\epsilon_{ij}$.

◂The equilibrium equations in terms of cylindrical and spherical coordinates are given by Eqs. 2.14 and 2.15; the corresponding definitions of strain by Eqs. 2.28 and 2.29. ▸

REFERENCES

For similar treatments of the definition of stress and strain and the equations of equilibrium, see Crandall and Dahl (1959), Chapter 4; Eirich (1956), Chapter 2; and Timoshenko and Goodier (1951), Chapters 8 and 9. The treatment of infinitesimal strains by Jaeger (1956) is at a comparable level, but Jaeger goes much further in discussing the transformation of components of strain and also in discussing some definitions suitable for large strains. Green and Zerna (1954), Love (1944), Muskhelishvili (1953), and Sokolnikoff (1956) treat finite strains and also present the definition of strain in a general tensor form which is suitable for curvilinear as well as rectangular coordinate systems. These latter books present the theory of elasticity at the graduate level.

BRIDGMAN, P. W.	1952	*Studies in Large Plastic Flow and Fracture*, McGraw-Hill, New York.
COOK, N. H. RABINOWICZ, E.	1963	*Physical Measurement and Analysis*, Addison-Wesley, Reading, Mass.
COSSERAT, E. COSSERAT, F.	1909	*Théorie des corps déformables*, Hermann, Paris.
CRANDALL, S. H. DAHL, N. C. (eds.)	1959	*An Introduction to the Mechanics of Solids*, McGraw-Hill, New York.
DURELLI, A. J. PHILLIPS, E. A. TSAO, C. H.	1958	*Introduction to the Theoretical and Experimental Analysis of Stress and Strain*, McGraw-Hill, New York.
EIRICH, F. R. (ed.)	1956	*Rheology: Theory and Applications*, Vol. I, Academic Press, New York.
FRANKLIN, P.,	1944	*Methods of Advanced Calculus*, McGraw-Hill, New York.
FROCHT, M. M.	1948	*Photoelasticity*, Vols. 1 and 2, Wiley, New York.
GREEN, A. E. ZERNA, W.	1954	*Theoretical Elasticity*, Clarendon Press, Oxford, England.

Hetényi, M. (ed.)	1950	*Handbook of Experimental Stress Analysis*, Wiley, New York.
Jaeger, J. C.	1956	*Elasticity, Fracture, and Flow*, Methuen, London.
Kröner, E.	1961	"Die neuen Konzeptionen der Kontinuumsmechanik der festen Körper," *Physica Status Solidi* **1**, 3–16.
Love, A. E. H.	1944	*Mathematical Theory of Elasticity*, Dover, New York.
Murray, W. M. Stein, P. K.	1961	*Strain Gage Techniques*, Society for Experimental Stress Analysis, 21 Bridge Square, Westport, Conn.
Muskhelishvili, N. I.	1953	*Some Basic Problems of the Mathematical Theory of Elasticity*, Noordhoff, Groningen, Holland.
Post, D. Zandman, F.	1961	"The Accuracy of the Birefringent Coating Method for Coatings of Arbitrary Thickness," *Experimental Mechanics*, **1**, 1–12.
Sokolnikoff, I. S.	1956	*Mathematical Theory of Elasticity*, 2nd ed., McGraw-Hill, New York.
Thomas, G. B.	1960	*Calculus and Analytic Geometry*, Addison-Wesley, Reading, Mass.
Timoshenko, S. Goodier, J. N.	1951	*Theory of Elasticity*, McGraw-Hill, New York.

PROBLEMS

2.1. Show that $\sigma_{21} = \sigma_{12}$ (Eq. 2.3).

◀ 2.2. Derive the equations for the transformation of two stress components, Eqs. 2.4.

◀ 2.3. Derive the equation for the transformation of a third stress component, Eq. 2.5.

◀ 2.4. Derive the equations for the transformation of vector components, Eqs. 2.6.

◀ 2.5. Prove that $\sum_{j=1}^{3} l_{i'j} l_{i'j} = 1$. [*Hint:* Consider the dot product of parallel unit vectors.]

◀ 2.6. Prove that $\sum_{j=1}^{3} l_{i'j} l_{k'j} = 0$ if $i' \neq k'$. [*Hint:* Consider the dot product of two perpendicular unit vectors.]

◀ 2.7. Derive the equation for the plane transformation of stress components, Eq. 2.8.

◀ 2.8. Derive the equation for the general transformation of stress components, Eq. 2.9.

◀ 2.9. Derive the equation for the resolved shear stress under tensile loads, Eq. 2.10.

2.10. Derive the equilibrium equations for statics, Eqs. 2.11.

◀ 2.11. Derive the momentum equations (2.13).

◀ 2.12. Derive the transformation law for couple stresses.

◀ 2.13. Show that it is not necessary that $\chi_{ij} = \chi_{ji}$.

◀ 2.14. Derive the equation for the relation between couple and shear stresses, Eq. 2.16.

2.15. Derive the equation for volumetric strain in terms of normal strain components, Eq. 2.26.

2.16. Prove that Eqs. 2.25 are not valid for large rotations by applying them to a 180° rotation of the element shown in Fig. 2.16.

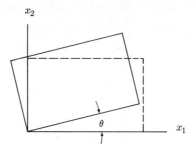

FIGURE 2.16

2.17. Derive the restrictions on the displacements and shear-strain distribution in the plate of Fig. 2.12, if
$$\epsilon_{11} = -a_1 x_2^2 \quad \text{and} \quad \epsilon_{22} = -a_2 x_1^2.$$

◀**2.18.** Derive the equation for the transformation of strain components, Eq. 2.33.

2.19. Derive the equation for the work per unit volume, Eq. 2.37.

◀**2.20.** Carry out the steps in deriving the equation for the work done on a body, Eq. 2.40.

◀**2.21.** Show that the strain distribution of Problem 2.17 satisfies the compatibility equations.

2.22. (a) Write approximate expressions for the components of displacement of the element shown in Fig. 2.11.

(b) Show that the strain of the element is approximately
$$\begin{Bmatrix} \epsilon_{11} & \gamma_{12} \\ \gamma_{21} & \epsilon_{22} \end{Bmatrix} = \begin{Bmatrix} 0.2 & 0.1 \\ 0.1 & 0 \end{Bmatrix}.$$

◀**2.23.** Show that the rotation of the element shown in Fig. 2.11 is
$$\begin{Bmatrix} 0 & \omega_{12} \\ \omega_{21} & 0 \end{Bmatrix} = \begin{Bmatrix} 0 & -0.15 \\ 0.15 & 0 \end{Bmatrix}.$$

◀**2.24.** Sketch an element before and after the strain
$$\begin{Bmatrix} \epsilon_{11} & \epsilon_{12} & \epsilon_{13} \\ \epsilon_{21} & \epsilon_{22} & \epsilon_{23} \\ \epsilon_{31} & \epsilon_{32} & \epsilon_{33} \end{Bmatrix} = \begin{Bmatrix} 0 & 0.1 & 0 \\ 0.1 & 0 & 0 \\ 0 & 0 & 0 \end{Bmatrix}$$

and the rotation
$$\begin{Bmatrix} 0 & \omega_{12} & \omega_{13} \\ \omega_{21} & 0 & \omega_{23} \\ \omega_{31} & \omega_{32} & 0 \end{Bmatrix} = \begin{Bmatrix} 0 & 0.05 & 0 \\ -0.05 & 0 & 0 \\ 0 & 0 & 0 \end{Bmatrix}.$$

Also sketch an element before and after the strain

$$\begin{Bmatrix} \epsilon_{11} & \epsilon_{12} & \epsilon_{13} \\ \epsilon_{21} & \epsilon_{22} & \epsilon_{23} \\ \epsilon_{31} & \epsilon_{23} & \epsilon_{33} \end{Bmatrix} = \begin{Bmatrix} 0 & 0.1 & 0 \\ 0.1 & 0 & 0 \\ 0 & 0 & 0 \end{Bmatrix}$$

and the rotation

$$\begin{Bmatrix} 0 & \omega_{12} & \omega_{13} \\ \omega_{21} & 0 & \omega_{23} \\ \omega_{31} & \omega_{32} & 0 \end{Bmatrix} = \begin{Bmatrix} 0 & -0.05 & 0 \\ 0.05 & 0 & 0 \\ 0 & 0 & 0 \end{Bmatrix}.$$

2.25. Sketch an element before and after the strain

$$\begin{Bmatrix} \epsilon_{11} & \gamma_{12} \\ \gamma_{12} & \epsilon_{22} \end{Bmatrix} = \begin{Bmatrix} 0.2 & -0.1 \\ -0.1 & -0.3 \end{Bmatrix}.$$

Show several sketches, each corresponding to a different rotation occurring in conjunction with the strain.

CHAPTER 3

CONSTITUTIVE RELATIONS FOR SMALL ELASTIC STRAIN

3.1 SYNOPSIS

When the stress-strain relations are extended to include first temperature and later time, they are referred to generally as the *constitutive relations*.

For small elastic strain, it is shown how these relations are derived by linearization of the interatomic force law. In the most general linear relation, each of the six components of strain depends on the six components of stress. It is shown how these relations are simplified, first by considering the reversibility of elastic deformation, and second by referring the stress and strain to crystallographic axes of symmetry. For cubic symmetry, there turn out to be only three independent elastic constants and for isotropy only two; for example, Young's modulus and Poisson's ratio. Even with the symmetry of a cubic crystal, however, if a specimen is not aligned with the axes of symmetry, it is shown how normal components of stress may cause shear components of strain. The anisotropy of iron changes the stiffness by a factor of two between different orientations. Even in many polycrystals there may be enough preferred orientation due to previous mechanical working to cause important elastic anisotropy. In a number of materials, the structure is symmetrical about three mutually perpendicular planes, or *orthotropic*. Wood, cold-rolled steel sheet, and laminate-reinforced plastics are examples.

The derivation of the elastic constants from the basic interatomic force law is beyond the scope of this book, but two results are useful, especially with polycrystals. It is shown how the bulk modulus is approximately related to the binding energy of the lattice, and secondly, that where the interatomic forces are central, Poisson's ratio must be $1/4$.

The linearity of the elastic stress-strain relations leads to the principle of superposition, according to which the stress and strain in a body under combined loads can be found by summing the distributions found for the individual loads, provided that the deformations are small. It also enables us to guess a solution with the confidence that if the guess is correct, it is the only correct solution, according to the uniqueness principle. Values of Young's modulus, Poisson's ratio, and the thermal coefficient of expansion are given for a number of common isotropic materials.

It is pointed out that in certain nonconducting crystals with sufficient asymmetry, the application of electric fields will produce small strains. Likewise, magnetic fields in ferromagnetic materials will produce small but observable strains.

3.2 FORM OF THE ELASTIC STRESS-STRAIN RELATIONS

As we have seen in Chapter 2, equilibrium and strain-displacement equations are not sufficient to determine the distributions of stress and strain in a body under load. A relation between stress and strain is also necessary. Before discussing this relation quantitatively, we need a general form in which to describe it. For normal strain, the energy-strain curve will be similar to that discussed in Chapter 1 and shown again in Fig. 3.1. For shear components of strain, one might assume a sinusoidal stress variation. As a second approximation, one might consider an array of rigid spheres held together by an external pressure. In this case, the energy would vary due to the volume increase required to lift the spheres over one another. Rough interpolation between these assumptions gives the energy curve of Fig. 31. (Prob. 3.1).

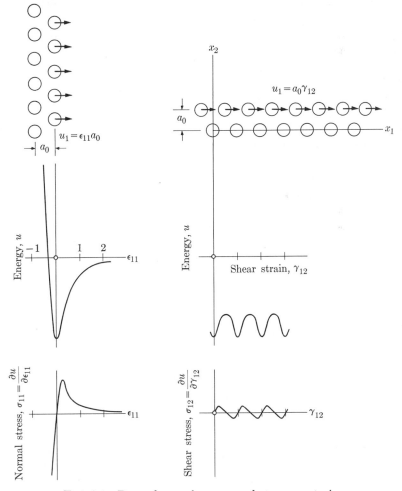

FIG. 3.1. Dependence of energy and stress on strain.

The components of stress can be found from the energy-strain curves, since, for elastic (reversible) deformation, the work done by the applied stress is equal to the increase in internal energy:

$$dW = \sigma_{11}\,d\epsilon_{11} + \sigma_{22}\,d\epsilon_{22} + \cdots + \sigma_{23}\,d\gamma_{23} + \cdots$$

$$= du = \frac{\partial u}{\partial \epsilon_{11}}\,d\epsilon_{11} + \frac{\partial u}{\partial \epsilon_{22}}\,d\epsilon_{22} + \cdots + \frac{\partial u}{\partial \gamma_{23}}\,d\gamma_{23} + \cdots. \quad (3.1)$$

Because this equality must hold for any values of the strain increments, it shows that the components of stress are the derivatives of the strain energy with respect to the corresponding components of strain:

$$\frac{\partial u}{\partial \epsilon_{11}} = \sigma_{11}, \quad \frac{\partial u}{\partial \gamma_{23}} = \sigma_{23}, \quad \text{etc.} \quad (3.2)$$

Two of these derivatives, giving a normal and a shear component of stress, are also shown in Fig. 3.1. As a first approximation, the stress-strain curves are linear near the point of zero strain. The harder the atoms, the smaller is the strain for which the linear approximation is good. An analogous situation arises in the behavior of some composite materials such as solid rocket fuel, which consists of hard particles in a rubbery matrix. In both cases, however, the slope of the stress-strain relation is nearly constant for strains up to 0.01.

◀Once the stress-strain relations are assumed linear, the most general relation between the six components of strain and the six components of stress may be written in terms of constants of proportionality C as

$$\begin{aligned}
\sigma_{11} &= C_{11}\epsilon_{11} + C_{12}\epsilon_{22} + C_{13}\epsilon_{33} + C_{14}\gamma_{23} + C_{15}\gamma_{31} + C_{16}\gamma_{12}, \\
\sigma_{22} &= C_{21}\epsilon_{11} + C_{22}\epsilon_{22} + \cdots \quad\quad + C_{24}\gamma_{23} + \cdots \quad + C_{26}\gamma_{12}, \\
\sigma_{33} &= C_{31}\epsilon_{11} + C_{32}\epsilon_{22} + \cdots \quad\quad\quad\quad\quad\quad\quad\quad + C_{36}\gamma_{12}, \\
\sigma_{23} &= C_{41}\epsilon_{11} + C_{42}\epsilon_{22} + \cdots \quad\quad\quad\quad\quad\quad\quad\quad + C_{46}\gamma_{12}, \\
\sigma_{31} &= C_{51}\epsilon_{11} + C_{52}\epsilon_{22} + \cdots \quad\quad\quad\quad\quad\quad\quad\quad + C_{56}\gamma_{12}, \\
\sigma_{12} &= C_{61}\epsilon_{11} + C_{62}\epsilon_{22} + \cdots \quad\quad\quad\quad\quad\quad\quad\quad + C_{66}\gamma_{12},
\end{aligned} \quad (3.3)$$

or more concisely in terms of the tensor components of strain and different constants C_{ijkl} as

$$\sigma_{ij} = \sum_{k=1}^{3} \sum_{l=1}^{3} C_{ijkl}\epsilon_{kl}. \quad (3.4)$$

The constants C of Eqs. 3.3 and 3.4 are called *components of elastic stiffness*.* The components C_{ijkl}, relating stress to tensor strain, transform from one set of coordinate axes to another according to the rule for fourth-order tensors in terms of the direction cosines:

$$C_{ijkl} = \sum_{i'=1}^{3} \sum_{j'=1}^{3} \sum_{k'=1}^{3} \sum_{l'=1}^{3} C_{i'j'k'l'} l_{i'i} l_{j'j} l_{k'k} l_{l'l}. \quad (3.5)$$

---

*Also called elastic constants or elastic moduli.

This result can be proved from the fact that in view of Eqs. 3.2 and 3.3, the coefficients can be expressed in terms of partial derivatives of the strain energy as

$$C_{ijkl} = \frac{\partial^2 u}{\partial \epsilon_{ij} \partial \epsilon_{kl}}, \tag{3.6}$$

along with the rule for the transformation of partial derivatives and that for components of strain (Prob. 3.2). Alternatively, the transformation rule can be derived from those for stress and strain by expressing the stress and strain in the new coordinate system in terms of those in the old, writing the stress-strain relation in the old coordinate system, and rearranging terms (Prob. 3.3). A general transformation rule for the components of elastic stiffness in terms of the tangential rather than tensor definition of shear strain (Eqs. 3.3) is too lengthy to be of value.
◂Using the tensor notation, there may appear to be 3^4 components of stiffness. The equalities $\sigma_{ij} = \sigma_{ji}$ and $\epsilon_{ij} = \epsilon_{ji}$ reduce the number of coefficients to 36. Not all of these are independent, for the reversibility of elastic deformation leads to the fact that the work done in elastic deformation is a unique function of the strain and independent of the path. Thus a strain-energy function exists, and

$$\frac{\partial^2 u}{\partial \epsilon_{ij} \partial \epsilon_{kl}} = \frac{\partial^2 u}{\partial \epsilon_{kl} \partial \epsilon_{ij}}.$$

From this it can be shown that relations of the type

$$C_{12} = C_{21} \quad \text{or} \quad C_{ijkl} = C_{klij} \tag{3.7}$$

hold between the components, and thus there are only 21 independent components of elastic stiffness (Prob. 3.4). This result can also be proven from the fact that the net work of any cycle involving only elastic deformation must be zero, by considering a cycle in which first one component of strain and then another is applied, and then the first and second are removed in turn (Prob. 3.5).
◂When both elastic and plastic deformations are involved, the strains are additive, and it is therefore easier to work with expressions giving the components of strain in terms of those of stress. The effects of thermal expansion can also be included more easily. A general linear relation can be given in terms of constants S as

$$\begin{aligned}
\epsilon_{11} &= S_{11}\sigma_{11} + S_{12}\sigma_{22} + \cdots + S_{14}\sigma_{23} + \cdots + \alpha_1 \Delta T, \\
\epsilon_{22} &= S_{21}\sigma_{11} + S_{22}\sigma_{22} + \cdots + S_{24}\sigma_{23} + \cdots + \alpha_2 \Delta T, \\
\epsilon_{33} &= S_{31}\sigma_{11} + \cdots \hspace{5em} + \alpha_3 \Delta T, \\
\gamma_{23} &= S_{41}\sigma_{11} + \cdots \hspace{2em} + S_{44}\sigma_{23} + \cdots + \alpha_4 \Delta T, \\
\gamma_{31} &= S_{51}\sigma_{11} + \cdots \hspace{5em} + \alpha_5 \Delta T, \\
\gamma_{12} &= S_{61}\sigma_{11} + \cdots \hspace{5em} + \alpha_6 \Delta T,
\end{aligned} \tag{3.8}$$

or

$$\epsilon_{ij} = \sum_{k=1}^{3} \sum_{l=1}^{3} S_{ijkl}\sigma_{kl} + \alpha_{ij} \Delta T. \tag{3.9}$$

The components of *compliance*, S_{11} or S_{ijkl},* transform in a way similar to the components of stiffness and are subject to the same identities. The components of thermal expansion, α, transform as do the components of strain. The tangential and tensor components of compliance are related by equations of the type (Prob. 3.6)

$$S_{1111} = S_{11}, \quad S_{1122} = S_{12}, \quad S_{1123} = \tfrac{1}{2}S_{14}, \quad S_{2331} = \tfrac{1}{4}S_{45}, \quad \text{etc.} \quad (3.10)$$

◀Thermodynamic reasoning is helpful in determining other relations between the elastic constants. For example, the difference between the isothermal and isentropic compliances is given in terms of the coefficient of thermal expansion and the specific heat at constant stress, C, by

$$\left(\frac{\partial \epsilon}{\partial \sigma}\right)_T = \left(\frac{\partial \epsilon}{\partial \sigma}\right)_s + \frac{T\alpha^2}{\rho C}. \quad (3.11)$$

This can be proven by an adaptation of one of Maxwell's thermodynamic equations to the case of uniaxial stress,

$$\left(\frac{\partial s}{\partial p}\right)_T = -\left(\frac{\partial v}{\partial T}\right)_p \rightarrow \left(\frac{\partial s}{\partial \sigma}\right)_T = \frac{1}{\rho}\left(\frac{\partial \epsilon}{\partial T}\right)_\sigma, \quad (3.12)$$

and the differential expression for strain as a function of stress and entropy (Prob. 3.7). See Mason (1950, p. 31) for a treatment of the relations in three dimensions. A second example arises from the fact that strain depends on stress and temperature, independent of the path. This leads to an effect of stress on the coefficient of expansion, since

$$\frac{\partial \alpha_1}{\partial \sigma_{11}} = \frac{\partial S_{11}}{\partial T}. \quad (3.13)$$

Equation 3.13 may be of importance where there are large enough variations in the temperature in a body to produce significant changes in compliance from point to point.▶

3.3 EFFECT OF SYMMETRY ON ELASTIC STRESS-STRAIN RELATIONS

Since there are only three degrees of freedom in choosing a coordinate system, it is not possible to effect a very great reduction in the number of components of stiffness or compliance by a choice of coordinate system, unless the structure of the material itself has a rather high degree of symmetry. Commonly encountered materials with the least symmetry include wood, rolled metals, and fiber-reinforced laminates, shown in Fig. 3.2. The materials in Fig. 3.2 have three mutually perpendicular axes such that a 180° rotation about any one of them gives an identically appearing structure. Such materials are called *orthotropic*, and de-

---

*Also called elastic moduli, although we shall see that $S_{1111} = 1/E$ for an isotropic material, and E is called the modulus of elasticity.

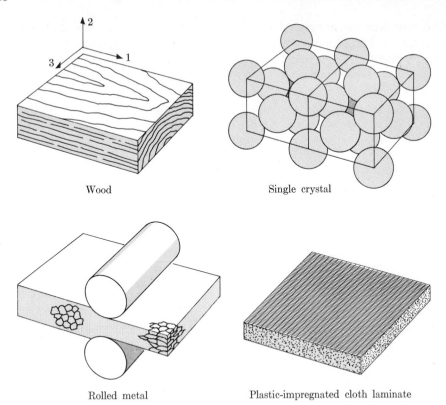

Fig. 3.2. Orthotropic materials. (Crandall and Dahl, 1960. Courtesy of McGraw-Hill.)

scription of their elastic stress compliance referred to these axes of symmetry requires nine components:

Stress Strain	1	2	3	4	5	6
1	$\epsilon_{11} = S_{11}\sigma_{11}$	$+ S_{12}\sigma_{22}$	$+ S_{13}\sigma_{33}$			
2	$\epsilon_{22} = S_{12}\sigma_{11}$	$+ S_{22}\sigma_{22}$	$+ S_{23}\sigma_{33}$			
3	$\epsilon_{33} = S_{13}\sigma_{11}$	$+ S_{23}\sigma_{22}$	$+ S_{33}\sigma_{33}$			
4	$\gamma_{23} = \cdots$			$+ S_{44}\sigma_{23}$		
5	$\gamma_{31} = \cdots$				$+ S_{55}\sigma_{31}$	
6	$\gamma_{12} = \cdots$					$+ S_{66}\sigma_{12}$

(3.14)

If the components of stress and strain are defined in terms of coordinate axes other than those of symmetry, the transformation equations corresponding to Eq. 3.5 will change the form back to that of Eq. 3.8, indicating that a shear stress will produce a normal component of strain, and conversely.

◀The next highest degree of symmetry commonly encountered is hexagonal symmetry. Taking the coordinate system shown in Fig. 3.3, we find that at first glance, it might appear that nine components would again be required, since for example, it would seem that $S_{11} \neq S_{22}$. But application of the transformation equations to the components referred to axes in the basal plane at 60° to the original set, along with the fact that the symmetry of the crystal requires that the components referred to these new axes be identical with those referred to the original axes, shows that only five independent components are required (Prob. 3.8):

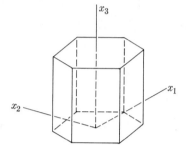

$$\begin{matrix} S_{11} & S_{12} & S_{13} \\ S_{12} & S_{11} & S_{13} \\ S_{13} & S_{13} & S_{33} \\ & & & S_{44} \\ & & & & S_{44} \\ & & & & & 2(S_{11} - S_{12}) \end{matrix} \qquad (3.15)$$

FIG. 3.3. Coordinates for a hexagonal crystal.

◀When a crystal has cubic symmetry, a series of 90° rotations of the coordinate axes shows that only three independent components of compliance, S_{11}, S_{12}, and S_{44}, are required (Prob. 3.9). Even here, a rotation of the axes away from the axes of symmetry introduces all 21 components of the compliance tensor. In cubic crystals, the small number of independent elastic constants makes it possible to write an expression for the component in a new direction $S_{i'j'k'l'}$, in terms of the corresponding components in the old direction, S_{ijkl}, the direction cosines, and the factor $S_{1111} - S_{1122} - 2S_{2323}$ (McClintock, 1950; for special cases see Voigt, 1928, or Boas and Mackenzie, 1950):

$$S_{i'j'k'l'} = S_{ijkl} + (S_{1111} - S_{1122} - 2S_{2323}) \left(\sum_{m=1}^{3} l_{mi'} l_{mj'} l_{mk'} l_{ml'} \right), \qquad (3.16)$$

unless $i' = j' = k' = l'$, in which case

$$S_{i'i'i'i'} = S_{iiii} + (S_{1111} - S_{1122} - 2S_{2323}) \left[\left(\sum_{m=1}^{3} l_{mi'}^{4} \right) - 1 \right]. \qquad (3.16a)$$

Note that if the components are such that

$$(S_{1111} - S_{1122} - 2S_{2323}) = 0, \qquad (3.17)$$

the elastic constants do not change with direction. In other words, the crystal will be elastically isotropic. This happens to be the case for single crystals of tungsten at room temperature.▶

A material is called elastically isotropic if its elastic behavior is independent of orientation. This may arise not only if its crystal structure is elastically isotropic, but also if it consists of a collection of anisotropic crystals arranged at random, so that an element small enough to be of use in stress analysis still can be described in terms of its average behavior. It is important that the crystals be truly randomly oriented. As we shall see in Section 3.4, the crystals in cold-rolled copper may have a variety of orientations, and yet the polycrystalline material can be not at all isotropic.

When a material is isotropic, its elasticity is commonly described in terms of Young's modulus E,* Poisson's ratio ν, and the shear modulus G. That these are not independent can be seen by determining the shear strain per unit shear stress either directly with the shear modulus G, or by rotating the coordinates 45° to get normal stress components, using E and ν to determine strain, and returning to the original coordinates (Prob. 3.10). Alternatively, the relation is found from the condition between the tensor components of compliance for isotropy (Prob. 3.11). The resulting relation is

$$G = \frac{E}{2(1+\nu)}. \tag{3.18}$$

The stress-strain-temperature, or constitutive, relations become:

$$\epsilon_{11} = +\frac{\sigma_{11}}{E} - \frac{\nu\sigma_{22}}{E} - \frac{\nu\sigma_{33}}{E} \qquad\qquad +\alpha\,\Delta T$$

$$\epsilon_{22} = -\frac{\nu\sigma_{11}}{E} + \frac{\sigma_{22}}{E} - \frac{\nu\sigma_{33}}{E} \qquad\qquad +\alpha\,\Delta T$$

$$\epsilon_{33} = -\frac{\nu\sigma_{11}}{E} - \frac{\nu\sigma_{22}}{E} + \frac{\sigma_{33}}{E} \qquad\qquad +\alpha\,\Delta T \tag{3.19}$$

$$\gamma_{23} = \qquad\qquad +\frac{2(1+\nu)}{E}\sigma_{23}$$

$$\gamma_{31} = \qquad\qquad +\frac{2(1+\nu)}{E}\sigma_{31}$$

$$\gamma_{12} = \qquad\qquad +\frac{2(1+\nu)}{E}\sigma_{12}.$$

It is well to bear in mind the actual meaning behind the terms in Eq. 3.19. For example, the terms in the first column indicate that the application of a normal component of stress produces a corresponding component of strain, as well as equal negative components of strain in the two transverse directions.

Physically it is often easiest to consider the compressibility, or its reciprocal, the *bulk modulus*, defined as the pressure change divided by the corresponding fractional volumetric strain, ϵ:

$$B = \frac{-dp}{dv/v} = \frac{-p}{\epsilon}. \tag{3.20}$$

The bulk modulus is related to Young's modulus and Poisson's ratio by (Prob. 3.12)

$$B = \frac{E}{3(1-2\nu)}. \tag{3.21}$$

---

*Also called modulus of elasticity.

TABLE 3.1
Relations Between Isotropic Elastic Constants

In terms of Elastic constants	E, ν	E, G	B, ν	B, G	λ, μ
E	$= E$	$= E$	$= 3(1-2\nu)B$	$= \dfrac{9B}{1+3B/G}$	$= \dfrac{\mu(3+2\mu/\lambda)}{1+\mu/\lambda}$
ν	$= \nu$	$= -1 + \dfrac{E}{2G}$	$= \nu$	$= \dfrac{1-2G/3B}{2+2G/3B}$	$= \dfrac{1}{2(1+\mu/\lambda)}$
G	$= \dfrac{E}{2(1+\nu)}$	$= G$	$= \dfrac{3(1-2\nu)B}{2(1+\nu)}$	$= G$	$= \mu$
B	$= \dfrac{E}{3(1-2\nu)}$	$= \dfrac{E}{9-3E/G}$	$= B$	$= B$	$= \lambda + \dfrac{2\mu}{3}$
λ	$= \dfrac{E\nu}{(1+\nu)(1-2\nu)}$	$= \dfrac{E(1-2G/E)}{3-E/G}$	$= \dfrac{3B\nu}{1+\nu}$	$= B - \dfrac{2G}{3}$	$= \lambda$
μ	$= \dfrac{E}{2(1+\nu)}$	$= G$	$= \dfrac{3(1-2\nu)B}{2(1+\nu)}$	$= G$	$= \mu$

◀ For isotropic materials, the constitutive equations can be simplified with the aid of the bulk modulus by expressing the normal components of stress as differences from their mean normal or hydrostatic value σ; these differences are known as *stress deviators* and are

$$s_{11} = \sigma_{11} - (\sigma_{11} + \sigma_{22} + \sigma_{33})/3 = \sigma_{11} - \sigma,$$
$$s_{22} = \sigma_{22} - \sigma, \quad \text{etc.}, \tag{3.22}$$
$$s_{23} = \sigma_{23}, \quad \text{etc.}$$

Likewise, *strain deviators* are defined as the difference between the three normal components of strain and the mean normal strain $\epsilon/3$:

$$e_{11} = \epsilon_{11} - (\epsilon_{11} + \epsilon_{22} + \epsilon_{33})/3 = \epsilon_{11} - \epsilon/3,$$
$$e_{22} = \epsilon_{22} - \epsilon/3, \quad \text{etc.}, \tag{3.23}$$
$$e_{23} = \epsilon_{23}, \quad \text{etc.}$$

In terms of these quantities, the constitutive equations for small elastic strain become

$$e_{ij} = s_{ij}/2G, \quad \epsilon = \sigma/B + 3\alpha\,\Delta T. \tag{3.24}$$

◀ The constitutive equations (3.19 or 3.24) can also be given as stress in terms of strain. For example (Prob. 3.13),

$$\sigma_{11} = \frac{(1-\nu)E\epsilon_{11}}{(1+\nu)(1-2\nu)} + \frac{\nu E\epsilon_{22}}{(1+\nu)(1-2\nu)}$$
$$+ \frac{\nu E\epsilon_{33}}{(1+\nu)(1-2\nu)} - \frac{\alpha E\,\Delta T}{1-2\nu},$$

and
$$\sigma_{23} = \frac{E}{2(1+\nu)}\gamma_{23}. \tag{3.25}$$

◀ The relations between commonly used elastic constants, including the Lamé constants λ and μ, which are useful in wave propagation, are given in Table 3.1. ▶

3.4 UNIQUENESS OF ELASTIC STRESS DISTRIBUTIONS

◀ The equilibrium equations, strain-displacement equations, and stress-strain relations provide enough equations to solve for the distribution of stress with given boundary loads or displacements. The direct solution of these problems is rarely possible. One indirect method is to guess a solution, which is then modified until it does satisfy the various equations and boundary conditions. The question then arises as to whether this stress distribution is the only one possible, or whether there could be another. In other words, the question arises as to whether a distribution of stress is unique.

◀In proving the uniqueness of elastic stress distributions, we shall find it convenient to use the *principle of superposition,* namely that if one stress distribution satisfies the equilibrium, compatibility, and constitutive equations, then the sum of two such solutions also satisfies all the equations. This can be shown from the linear variation of terms in these equations with stress and strain (Prob. 3.14).

◀The proof of uniqueness proceeds by disproving its converse. Suppose that there are two stress distributions for the same change in external load. Because the equilibrium, stress-strain, and strain-displacement equations are linear, the superposition principle tells us that the stresses corresponding to the difference of these two solutions would also be a solution. The corresponding external load change is zero, and hence the work done is zero. The total strain energy must also be zero. For every element, the strain energy must be zero or positive, or else the material would be unstable. Thus for the total strain energy to vanish, the local strain energy and the stress must vanish everywhere, and the supposed difference between the solutions cannot exist. A more detailed proof is given by Love (1944, p. 170), or by Timoshenko and Goodier (1951, p. 236). The validity of the uniqueness of stress distribution under given loads is restricted to small deformations. For example, a column may buckle to either side. Furthermore, it is implicitly assumed that there are no initial stresses under conditions of zero load. This is not the case in the presence of residual stresses, discussed in Chapter 12. ▶

3.5 MAGNITUDE OF THE ELASTIC CONSTANTS

As discussed in Sections 1.3 and 3.2, the elastic constants depend on the energy of a crystalline lattice. For most crystals, even an approximate quantitative derivation requires the use of wave mechanics, and is beyond the scope of this book. If one neglects structural differences between different kinds of atoms, however,

TABLE 3.2*

Factor Relating Bulk Modulus, Heat of Vaporization, and Density for Liquids, $B/h_v\rho$

Liquid	Compressibility $\chi = 1/B$, cm²/dyne	Heat of vaporization h_v, cal/gm	Density ρ, gm/cm³	Dimensionless factor $B/h_v\rho$
Acetic acid	8.75×10^{-11}	97	1.05	2.7
Benzene	9.37	94.3	0.9	2.4
CS$_2$	9	84	1.3	2.4
CCl$_4$	10	46	1.6	3.2
Glycol	3.7	191	1.1	3.1
Methanol	12.5	262	0.81	0.9
Water	4.5	585	1	0.9

*Data from *Handbook of Chemistry and Physics,* 39th ed.

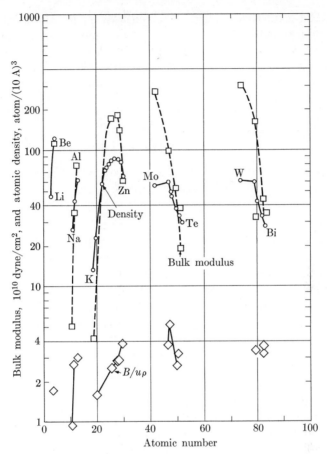

FIG. 3.4. Relation of bulk modulus, density, and binding energy. (Data from Boas and Mackenzie, 1950, Mott and Jones, 1958, and Smith and Arbogast, 1960.)

dimensional analysis leads one to expect that the bulk modulus B will be related to the binding energy u and the atomic density ρ by an equation of the form

$$B/\rho u = C. \tag{3.26}$$

The validity of this equation is illustrated for metals in Fig. 3.4, and for liquids in Table 3.2.

◄Greater insight into the relation between binding energy and bulk modulus can be obtained by considering ionic crystals, in which the attractive energy is known to be electrostatic, and thus varies inversely as the lattice dimension. To a fair approximation, the repulsive energy can be considered to vary inversely as some higher power of the interatomic spacing. This leads to a relation between binding energy and elasticity:

$$u = u_{\text{att}} + u_{\text{rep}} = -C_1/r^m + C_2/r^n, \tag{3.27}$$

where the exponents are left in general form for approximating crystals with other than electrostatic attractive energy. The constant C_1 is obtained by summing the electrostatic attractions and repulsions between any one atom and all of the surrounding ones, and depends on the shape of the lattice. For the special case of pure dilatation (equal normal strain in all directions), this coefficient can be taken to be a constant. The constant C_2 is found from the observed equilibrium spacing, at which the stress is zero:

$$\left(\frac{\partial u}{\partial r}\right)_{r_0} = 0 = + \frac{mC_1}{r_0^{m+1}} - \frac{nC_2}{r_0^{n+1}}. \qquad (3.28)$$

The bulk modulus is found by determining the variation of stress with volume at the equilibrium spacing, obtaining finally (Prob. 3.15)

$$B/\rho u = mn/9. \qquad (3.29)$$

Taking $m = 4$ and $n = 9$ gives $B/\rho u = 4$, in accord with Fig. 3.4. Since the binding energy u (in an ionic crystal relative to the ionized state) can be measured directly, and is roughly known for various types of bonds, as discussed in Section 1.3, Eq. 3.29 provides a useful rule of thumb for estimating the bulk modulus of various materials. Strictly speaking, it is valid only at a temperature of absolute zero, since we have neglected thermal motion. The elastic constants usually do not decrease by more than 10 to 20% from absolute zero to room temperature, so this effect is not important. For further refinement of this relation, the reader is referred to Seitz (1940, p. 379).

◀In liquids and polymers, the atoms in the molecules are held together by primary binding forces, while the molecules themselves are held together by secondary forces. If we estimate the bulk modulus according to Eq. 3.29 from the molecular binding energy and the density of molecules, we are likely to find values too low, because much of the material is essentially rigid, being held by primary bonds. In spite of these differences between liquids and solids, it turns out that the product of the compressibility, the density, and the heat of vaporization is still about 3 to 4 for a number of liquids, although only about unity for water and methanol, as shown in Table 3.2.

◀As noted in Section 3.3, two elastic constants are necessary to describe an isotropic material. Besides the bulk modulus, the easiest one to consider physically is Poisson's ratio. Cauchy found that if central forces were assumed to act between homogeneously deforming lattice points, then relations of the type

$$C_{12} = C_{66}, \quad C_{14} = C_{56}, \quad \text{etc.},$$

would have to exist between the components of stiffness (Love, 1944). For an isotropic material, the Cauchy relations show that Poisson's ratio should be 1/4 (Prob. 3.16). These relations are fairly well satisfied for an ionic crystal, where the electrostatic binding forces are indeed central. They are also well satisfied for glass, in which the crystal structure is so open that little effect can be produced by couples acting between the atoms. For most metals, however, the Cauchy relations are not satisfied, and Poisson's ratio is greater than 1/4. This fact is evidence that couple stresses exist in crystals. Since, however, it is the gradient of couple stress which affects ordinary stress analysis, the existence of couple stress does not usually affect the results.

◀A noncentral force model for a crystal lattice was used by Gibson et al. (1960) in studying radiation damage in a lattice. They assumed the energy to arise from two sources: one, a central repulsive force between pairs of atoms according to the Born-Mayer law, $u_r = Ae^{-b(r_1-r_2)}$, and the second, an attractive energy simply proportional to the volume, $u_a = p_b V$, as indicated in Fig. 3.5. The fact that this is indeed a noncentral attractive force can be recognized from the fact that it does not tend to oppose pure shear. The application of this model to a small "crystal" shows that the ratio of bulk to shear modulus, and hence Poisson's ratio, depends on the constant b and the equilibrium spacing of the atoms, so that Cauchy's relations need not, in fact, be satisfied (Prob. 3.17). ▶

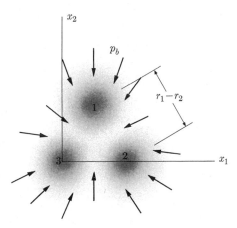

Fig. 3.5. Three-atom element for illustrating noncentral forces.

The approximate relations and ideas discussed above give order of magnitude estimates for the elastic constants of isotropic materials. It should be pointed out that although it has been possible to derive the elastic constants for some of the metals fairly accurately from quantum mechanics, the best values are still obtained from empirical data, such as given in Table 3.3. Crystals are by no means isotropic. Therefore, in dealing with single crystals, it is necessary to consider the components of compliance given in Table 3.4. Even in polycrystalline materials, the arrangement of the crystals may not be random, so that the material is not isotropic. No complete set of data has been compiled, but the variation in Young's modulus with orientation of specimens cut from a sheet is shown in Table 3.5. As indicated in Prob. 3.18, this information is not enough to give us a complete knowledge of the anisotropy. In wood, the material is texturally orthotropic, giving rise to more elastic constants, given for some woods in Table 3.6.

TABLE 3.3*

ELASTIC CONSTANTS FOR ISOTROPIC MATERIALS AT ROOM TEMPERATURE

Material	Composition	Modulus of elasticity E, 10^6 psi	Poisson's ratio ν	Shear modulus G, 10^6 psi	Coefficient of linear expansion α, 10^{-6}/°F	Density, lb/in^3
Aluminum†	Pure and alloy	9.9–11.4	0.32–0.34	3.7–3.85	11.1–13.4	0.096–0.104
Brass†‡	60–70% Cu, 40–30% Zn	14.5–15.9	0.33–0.36	5.3–6.0	11.0–11.6	0.302–0.307
Copper†‡§		17–18	0.33–0.36	5.8–6.7	9.2–9.4	0.323–0.324
Iron, cast‡§	2.7–3.6% C	13–21	0.21–0.30	5.2–8.2	5.8	0.251–0.265
Steel†‡	Carbon and low alloy	28–32	0.26–0.29	11.0–11.9	5.5–7.1	0.279–0.284
Stainless steel†‡§	18% Cr, 8% Ni	28–30	0.30	10.6	8.3–9.4	0.276–0.286
Titanium†§	Pure and alloy	15.4–16.6	0.34	6.0	4.9	0.163
Glass‖	Various	7.2–11.5	0.21–0.27	3.8–4.7	3.3–5.3	0.086–0.14
Methyl methacrylate¶		0.35–0.5			50	0.042
Polyethylene¶		0.02–0.055			100	0.033
Rubber**		0.00011–0.00060	0.50	0.00004–0.00020	70–110	0.036–0.045

*From Crandall and Dahl (1959). (Courtesy of McGraw-Hill.)
†C. J. Smithells, *Metals Reference Book*, Interscience, New York, 1955.
‡L. S. Marks, *Mechanical Engineers Handbook*, McGraw-Hill, New York, 1958.
§ASM, *Metals Handbook*, American Society for Metals, Novelty, Ohio, 1948.
‖G. W. Morey, *Properties of Glass*, Reinhold, New York, 1954, p.16.
¶*Modern Plastics*, Encyclopedia Issue, Vol. 34, 1956.
**U. S. Rubber Co., *Engineering Properties of Rubber*, Fort Wayne, 1950.
††C. L. Mantell, *Engineering Materials Handbook*, McGraw-Hill, New York, 1958.

TABLE 3.4*

Components of Elastic Compliance for Cubic Crystals

Material	S_{11}, 10^{-7} in²/lb	S_{12}, 10^{-7} in²/lb	S_{44}, 10^{-7} in²/lb	$S_{11} - S_{12} - S_{44}/2$ 10^{-7} in²/lb
Al	1.10	−0.40	2.43	0.28
Cu	1.03	−0.43	0.92	1.00
Fe	0.522	−0.195	0.595	0.419
Pb	6.43	−0.29	4.80	4.32
W	0.178	−0.050	0.455	0.000
95% Al, 5% Cu	1.04	−0.48	2.56	0.024
72% Cu, 28% Zn	1.34	−0.58	0.96	0.96

*Compiled from F. Seitz and T. A. Read, "Theory of the Plastic Properties of Solids, I," *J. Appl. Phys.*, **12,** 100–118, 1941; from Crandall and Dahl (1959). (Courtesy of McGraw-Hill.)

TABLE 3.5*

Modulus of Elasticity for Orthotropic Materials in Sheet Form

(For various directions in the plane of the sheet as a function of angle from principal structural direction in plane of sheet)

Material	Principal structural direction	Angle		
		0° E, 10^6 psi	45° E, 10^6 psi	90° E, 10^6 psi
Cold-rolled iron†	Direction of rolling	32.8	29.3	39.1
Cold-rolled copper‡	Direction of rolling	19.8	15.5	20.0
Cold-rolled copper, recrystallized‡	Direction of rolling	10.0	17.5	9.5
Glass-fiber-reinforced polyester§	Direction of warp	2.0–2.7	1.2–1.8	1.7–2.4

*From Crandall and Dahl (1959). (Courtesy of McGraw-Hill.)
†E. Goens and E. Schmid, "On the Elastic Anisotropy of Iron," *Naturwissenschaften*, **19,** 520–524, 1931.
‡J. Weerts, "Elastizität von Kupferblechen," *Z. Metallkunde*, **25,** 101–103, 1933.
§ANC 17 Panel, Civil Aeronautics Authority, "Plastics for Aircraft," part I, *Reinforced Plastics*, U. S. Dept. of Commerce, 1955.

The coefficient of expansion depends on wave mechanics for its solution, but basically arises from the fact that the energy of a lattice does not vary quadratically with the distance from the equilibrium position. Instead, the lattice becomes softer on increased expansion. Therefore, in the presence of thermal vibration, the atoms tend to spend more time in their expanded position than in the contracted position, giving a lower mean density. For a more quantitative study of this phenomenon, see Mott and Jones (1958, p. 17).

TABLE 3.6*
Components of Elastic Compliance for Various Woods
(Axes defined in Fig. 3.2)

Material	Moisture,† %	Density, lb/ft³	S_{11}, 10^{-6} in²/lb	$\dfrac{S_{22}}{S_{11}}$	$\dfrac{S_{33}}{S_{11}}$	$\dfrac{S_{12}‡}{S_{11}}$	$\dfrac{S_{23}}{S_{11}}$	$\dfrac{S_{31}}{S_{11}}$	$\dfrac{S_{66}}{S_{11}}$	$\dfrac{S_{44}}{S_{11}}$	$\dfrac{S_{55}}{S_{11}}$
Balsa	9	4–14	1–8	20	70	−0.3	−15	−0.5	18	200	27
Yellow birch	12	40	0.48–0.50	13	20	−0.5	− 9	−0.5	14	60	15
Douglas fir	12	27–30	0.51–0.69	15	20	−0.4	− 7	−0.5	16	140	13
Sitka spruce	12	22–25	0.59–0.64	13	23	−0.4	− 6	−0.5	16	20	16

*Compiled from *Wood Handbook*, U. S. Department of Agriculture Handbook No. 72, 1955; from Crandall and Dahl (1959). (Courtesy of McGraw-Hill.)
†Thermal coefficients of expansion are not given because moisture effects are much more important than thermal effects under normal conditions.
‡Data were somewhat inconsistent in not indicating $S_{12} = S_{21}$, so average values are given. See Prob. 3.25.

3.6 PHOTOELASTICITY, PIEZOELECTRICITY, AND MAGNETOSTRICTION

◄In some materials, especially polymers, strain affects the dielectric constant, thus producing a velocity of light which, for a given direction of the light ray, varies with the orientation of the transverse field strength vector, or plane of polarization. This leads to the photoelastic effect mentioned in Section 2.9 as a means of experimental stress analysis.

◄In a related effect, an electric field can produce a strain in some solids by polarizing the unit cells and distorting the electron distribution. For this to occur, it is necessary that the crystal be nonconducting, so a significant field can be applied, and that the unit cell not have a center of symmetry, so that the centers of the positive and negative electronic clouds can be acted on separately by a field. Materials having a pronounced piezoelectric effect include rochelle salt, $NaKC_4H_4O_6 \cdot 4H_2O$; quartz, SiO_2; ethylene diamine tartrate (EDT), $(NH_2)_2(CH_2)C_4H_4O_7 \cdot H_2O$; and ammonium dihydrogen phosphate (ADP), $NH_4H_2PO_4$.

◄Piezoelectric crystals are used both to generate and to detect pressure waves in sonar and in ultrasonic detection of flaws. They are used for the generation of waves in ultrasonic cleaning equipment, and as transducers to measure force or to convert mechanical vibrations into electrical signals, as in the crystal of a phonograph pickup. In electrical circuits, they are used to produce stable oscillators. Because the strain is a second-order tensor and the components of the electric field strength constitute a vector, or a first-order tensor, the components of the tensor relating strain to field strength comprise a third-order tensor (Prob. 3.19). As in elasticity, however, the relations are often written in terms of the independent components of strain, and the components of the field vector E:

$$\begin{aligned}
\epsilon_{11} &= d_{11}E_1 + d_{21}E_2 + d_{31}E_3, \\
\epsilon_{22} &= d_{12}E_1 + d_{22}E_2 + d_{32}E_3, \\
\epsilon_{33} &= \cdots \qquad\qquad\quad + d_{33}E_3, \\
\gamma_{23} &= d_{14}E_1 + \cdots \quad + d_{34}E_3, \\
\gamma_{31} &= \cdots \qquad\qquad\quad + d_{35}E_3, \\
\gamma_{12} &= \cdots \qquad\qquad\quad + d_{36}E_3.
\end{aligned} \qquad (3.30)$$

The calculation of strains and displacements in a crystal is a relatively straightforward problem if the field strength is uniform. However, if the strength varies through the crystal, as it does in most arrangements, then the strains calculated from the field alone using Eq. 3.30 no longer can be derived from continuous displacements. Added strain then arises from the resulting stress. Since the crystals exhibiting the piezoelectric effect are quite asymmetrical, the resulting problem in mechanics is a challenging one, especially for cases in which the crystal is not static but oscillating near one of its natural frequencies. For a reference on the piezoelectric effect and its applications, see Mason (1950). For thermoelastic and optical data, see Krishnan (1958).

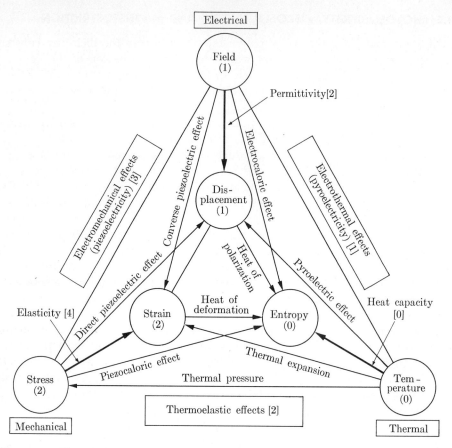

Fig. 3.6. The relations between the thermal, electrical, and mechanical properties of a crystal, showing the order of the tensors describing the variables ( ) and the properties [ ]. (Nye, 1957. Courtesy of Oxford University Press.)

TABLE 3.7

Magnetostriction of Rods

Material	Saturation field, oersteds	Strain, 10^{-6} in/in
Iron	20	4, then decreases through 0 at 200 oersteds
Cobalt		
Cast	100	-10, then rises through 0 at 500 oersteds
Annealed	300	-10
Nickel	100	-35
Invar	60	20

◀Changes in temperature also produce changes in field strength. This leads to a three-way interaction between temperature, stress, and field strength, which is shown schematically in Fig. 3.6. Such a multifaceted interaction occurs between other physical variables and their effects as well. For a more physical explanation of piezoelectricity, see Forsbergh (1956).

◀In ferromagnetic materials, the application of a magnetic field produces a strain. Here again there is a relation between a second-order tensor, the strain, and a first-order tensor, the magnetic field. In this case, however, the relation is so nonlinear that it cannot be stated in an analytical fashion. For a further treatment see Bates (1951). Typical materials, saturation fields, and the corresponding strains are given in Table 3.7.▶

3.7 SUMMARY

Any force-displacement relation which on an atomic scale is reasonably smooth can be closely approximated by a linear relation. The most general linear relation between the six components of stress and those of strain involves 21 independent constants of proportionality, taking into account the reversibility of elastic deformation. The symmetry of the crystal structure often makes possible further reduction of the number of elastic constants. Finally, in an isotropic aggregate of crystals of random orientations, the number of independent elastic constants is reduced to two, such as the modulus of elasticity E, and Poisson's ratio ν. The shear modulus G is also often convenient. In terms of these and the thermal coefficient of expansion α, the constitutive equations for an isotropic material are

$$\epsilon_{11} = \frac{\sigma_{11}}{E} - \frac{\nu \sigma_{22}}{E} - \frac{\nu \sigma_{33}}{E} + \alpha \Delta T, \quad \text{etc.},$$

$$\gamma_{23} = \sigma_{23}/G, \quad \text{etc.,} \tag{3.19}$$

where

$$G = E/2(1 + \nu). \tag{3.18}$$

Typical values of the modulus of elasticity or Young's modulus range from 10^6 to 70×10^6 psi for materials held together by primary bonds, and from 10^5 to 10^6 psi for materials held together by hydrogen bonding or secondary bonds. Poisson's ratio is of the order of 0.25 for materials held together by central forces, but higher in other solids. Coefficients of linear expansion range from 4 to $15 \times 10^{-6}/°F$ for most metals and glasses. Expansion coefficients for borosilicate and silica glass range from 0.3 to $2 \times 10^{-6}/°F$. In ferrous alloys, the presence of magnetic transformations can cancel this expansion and give coefficients which are nearly zero or even slightly negative. In many cases, Young's modulus E can vary by a large factor between different orientations in a single crystal or a cold-rolled material.

◀For a material of the most general anisotropy, or for any anisotropic material at arbitrary directions to the axes of its anisotropy, the stress-strain relations are described by equations of the form

$$\epsilon_{ij} = \sum_{k=1}^{3} \sum_{l=1}^{3} S_{ijkl}\sigma_{kl} + \alpha_{ij}\Delta T. \tag{3.9}$$

If stress is considered a function of strain, the elastic constants can be found by differentiating the internal energy as a function of the components of strain:

$$C_{ijkl} = \partial^2 u/\partial \epsilon_{ij}\,\partial \epsilon_{kl}. \tag{3.6}$$

For an isotropic material, the constitutive equations can also be given in terms of stress and strain deviators

$$s_{ij} = \sigma_{ij} - \delta_{ij}\sigma, \qquad \text{where} \qquad \sigma = \sigma_{kk}/3, \tag{3.22}$$

$$e_{ij} = \epsilon_{ij} - \delta_{ij}\epsilon/3, \qquad \text{where} \qquad \epsilon = \epsilon_{kk} \tag{3.23}$$

giving

$$e_{ij} = s_{ij}/2G, \qquad \epsilon = \sigma/B + 3\alpha\,\Delta T. \tag{3.24} \blacktriangleright$$

REFERENCES

(See also footnotes to the tables for sources of data.)

BATES, L. F.	1951	*Modern Magnetism*, Cambridge University Press, Cambridge, England, p. 401.
BOAS, W. MACKENZIE, J. K.	1950	"Anisotropy in Metals," *Progress in Metal Physics*, Vol. 2, Interscience, New York, pp. 90–120.
CRANDALL, S. H. DAHL, N. C. (eds.)	1959	*An Introduction to the Mechanics of Solids*, McGraw-Hill, New York.
FORSBERGH, P. W., JR.	1956	"Piezoelectricity, Electrostriction, and Ferroelectricity," *Handbuch der Physik*, Vol. 17, Springer, Berlin, pp. 264–392.
GIBSON, J. B. GOLAND, A. M. MILGRAM, M. VINEYARD, G. H.	1960	"The Dynamics of Radiation Damage," *Phys. Rev.*, **120**, 1229–1253.
KRISHNAN, R. S. (ed.)	1958	*Progress in Crystal Physics*, Vol. 1, Interscience, New York.
LOVE, A. E. H.	1944	*Mathematical Theory of Elasticity*, Dover, New York, pp. 618–627.
MCCLINTOCK, F. A.	1950	"A Study of Single and Polycrystalline Ingot Iron under Repeated Stress," Ph.D. Thesis, Calif. Inst. Tech., Pasadena.
MASON, W. P.	1950	*Piezoelectric Crystals and Their Application to Ultrasonics*, Van Nostrand, New York.

Mott, N. F. Jones, H.	1936	*The Theory of the Properties of Metals and Alloys*, Oxford University Press, London. See also 1958 reprint by Dover Publications, New York.
Nye, J. F.	1957	*Physical Properties of Crystals*, Oxford University Press, London.
Petterson, D. R.	1958	"On the Mechanics of Non-Woven Fabrics," Sc. D. Thesis, M.I.T., Cambridge, Mass. See also S. Backer and D. R. Petterson, "Some Principles of Non-Woven Fabrics," *Text. Res. J.* **30,** 704–711 (1960).
Seitz, F.	1940	*Modern Theory of Solids*, McGraw-Hill, New York.
Smith, J. F. Arbogast, C. L.	1960	"Elastic Constants of Single Crystals of Beryllium," *J. Appl. Phys.* **31,** 99.
Timoshenko, S. Goodier, J. N.	1951	*Theory of Elasticity*, McGraw-Hill, New York.
Voigt, W.	1928	*Lehrbuch der Kristallphysik*, 2nd. ed., Teubner, Leipzig.

PROBLEMS

3.1. For both the sinusoidal and hard-sphere models for interatomic forces, sketch the energy and stress as a function of shear strain. Show how the curves of Fig. 3.1 can be obtained by interpolation between these extremes.

◀ 3.2. Derive the transformation rule for the components of elastic stiffness, Eq. 3.5, from Eq. 3.6.

◀ 3.3. Derive the equations for the transformation rule for components of elastic stiffness, Eq. 3.5, from the transformation rules for components of stress and strain.

◀ 3.4. Derive the relations between the components of elastic stiffness, Eq. 3.7, from the strain-energy function. Tabulate the 21 independent components.

◀ 3.5. Derive Eq. 3.7 from the thermodynamic cycle described below Eq. 3.7. Tabulate the 21 independent components.

◀ 3.6. Derive the equations for the relation between tangential and tensor components of elastic compliance, Eq. 3.10.

◀ 3.7. Derive the equation for the relation between isothermal and isentropic compliance, Eq. 3.11.

◀ 3.8. Prove that the only independent components of elastic compliance for hexagonal crystals are those given in Eq. 3.15.

◀ 3.9. Prove that only three independent components of the compliance are required for a crystal with cubic symmetry.

3.10. Derive the relation between G and E by the transformation procedure outlined before Eq. 3.18.

◀3.11. Derive Eq. 3.18 from the compliance relation, Eq. 3.17.

3.12. Derive the equation relating the bulk modulus to Young's modulus, Eq. 3.21.

3.13. Derive Eq. 3.25 for stress in terms of strain.

◀3.14. Prove the principle of superposition by showing that if the two stress distributions $\sigma'_{ij}(x_1, x_2, x_3)$ and $\sigma''_{ij}(x_1, x_2, x_3)$ each satisfy equilibrium, and the strains resulting from them and the stress-strain relations are derivable from a continuous set of displacements, then the stress distribution given by the sum of these two functions $\sigma' + \sigma''$ will satisfy the same equations.

◀3.15. Derive the equation for the relation between bulk modulus and binding energy, Eq. 3.29, using Eqs. 3.27 and 3.28.

◀3.16. Show that the Cauchy relations in Section 3.5 lead to a value of Poisson's ratio equal to 1/4.

◀3.17. Show that the Cauchy relations are not satisfied for the non-central force model of Fig. 3.5.

◀3.18. (a) Why are the data for Young's modulus given in Table 3.5 insufficient to determine the complete anisotropic behavior?

(b) How many constants would be required to determine the complete behavior? How could these be obtained experimentally?

◀3.19. Show that the piezoelectric constants for anisotropic materials are tensor components.

◀3.20. Calculate the coefficient of elastic compliance S_{11} for a $\langle 111 \rangle$ direction in iron and compare it to the value for a $\langle 100 \rangle$ direction given in Table 3.4.

◀3.21. Calculate the components of elastic compliance for an iron crystal referred to coordinate axes rotated 30° about one of the cube axes.

◀3.22. According to Boas and Mackenzie (1950), drawn iron wire tends to have a $\langle 110 \rangle$ crystal direction axial. If all grains had this orientation:

(a) What tensile modulus would you expect?

(b) What equivalent shear modulus would you expect? Since a variety of orientations containing a $\langle 110 \rangle$ direction axial would be present in the wire, a first approximation might be obtained by taking the average of the shear moduli for two such orientations which are perpendicular to each other.

◀3.23. To a first approximation, some woven fabrics may be considered to be infinitely stiff in the directions of the warp and woof, but to have more or less elastic behavior in shear.

(a) What will be the nonvanishing coefficients of elastic compliance necessary to describe such a material?

(b) In terms of the above coefficients, what will be the modulus of elasticity when pulled at a direction θ to the direction of the threads?

3.24. Prove that the coefficient of thermal expansion is isotropic for a cubic crystal.

◀3.25. Nonwoven fabrics are textiles in which fiber-to-fiber bonding replaces twisting and interlacing. The process of manufacture leads to a structure with different properties in the longitudinal and transverse directions. The elastic behavior is often expressed in terms of the following coefficients:

$$\epsilon_L = \frac{1}{E_L} f_L - \frac{\nu_{LT}}{E_L} f_T, \quad \epsilon_T = -\frac{\nu_{TL}}{E_T} f_L + \frac{1}{E_T} f_T, \quad \gamma_{LT} = \frac{1}{G_{LT}} f_{LT},$$

where f denotes stress.

(a) Restate the above equations in terms of the coefficients of compliance, and tell what relations must exist between some or all of five different coefficients used in the above equations.

(b) Petterson (1958) reported the following values for the coefficients with stress in terms of grams per denier (units used in the textile industry and described in Chapter 21). For the present purposes these units may be used without change.

Coefficient	E_L	E_T	ν_{LT}	ν_{TL}	G_{LT}
Value	45	9.0	1.25	0.25	9.0

Are the required relations given by part (a) satisfied?

(c) Calculate Young's modulus for an element cut out of this material at 30° to the longitudinal axis. Petterson measured a value of 35 gm/denier. The agreement was not always so good as you should find in this case.

◀3.26. Predict the stiffness and Poisson's ratio of a piece of screening, pulled in the diagonal direction, from an experiment on the shear stiffness parallel to the wires and from a knowledge of its structure. Compare it with a simple experiment.

CHAPTER 4

DISLOCATION MECHANICS

4.1 SYNOPSIS

As we pointed out in Chapter 1, real crystals differ from ideal crystals by the lattice defects they possess. The importance of such defects in influencing mechanical, electrical, and even chemical behavior is now well recognized. Instead of considering these defects in terms of the motion of the atoms around them, we shall adopt a complementary point of view, which considers the imperfections as separate entities. Thus, we shall consider plastic deformation to be due to the motion and mutual interaction of lattice dislocations in an otherwise ideal crystal. We shall consider the diffusion of atoms in a crystal to be due to the back-diffusion of vacant lattice sites.

In this chapter, we shall first examine the geometrical features of dislocations and see that the motion of such dislocations results in a relative motion, equal to one lattice identity vector, of two parts of the crystal across the plane of motion of the dislocation. We shall then examine the internal stresses associated with a dislocation and consider how these stresses interact with each other and with externally applied stresses or free surfaces. It will be possible to show that the effect of an applied stress on a dislocation can be considered as a distributed force always exerted normal to the dislocation line. We shall show also that the large amount of elastic energy associated with a dislocation line will make it behave like a rubber band that is always under a line tension. From the dynamics of a moving dislocation, we shall learn that the dislocation behaves like a weighted line with a certain mass per unit length. Thus, a dislocation line is a close one-dimensional parallel to a weighted, stretched rubber membrane.

We shall show that a dislocation in a crystal requires a finite applied stress to move it over the potential hills of the lattice, but that the yield strength that is experimentally measured generally results from other considerations. We shall discuss imperfect dislocations in crystals, the motions of which result in planar faults of the regular stacking of the crystal. The intersection of such imperfect dislocations and their combination reactions will be examined.

We shall consider the mechanisms for generation of dislocations, which make large plastic strains possible. Finally, some properties of equilibrium arrangements of dislocations such as low-angle grain boundaries and spatial dislocation networks will be studied.

We shall critically evaluate the role of thermal activation in plastic deformation. Because of its relative simplicity as well as its practical importance, we shall first discuss the concentrations of vacancies in thermal equilibrium, and also the thermally activated diffusion of vacancies and its relation to the diffusion of lattice atoms. Pursuing this line of reasoning further, we find then that thermal activation is insufficient to generate dislocations, but that it can affect the rate at which dislocations can move through the internal stress field in a crystal once a critical shear stress has been applied.

4.2 THE GEOMETRY OF DISLOCATIONS

Plastic deformation and phase transformations are made possible or are enhanced by crystal dislocations, which are a type of line fault in the stacking of the atoms making up the crystal. The dislocation shown in Fig. 4.1, known as an *edge dislocation*, results from the termination of an atomic plane inside the crystal, giving a misfit of atoms between layer I and layer II around the point O.

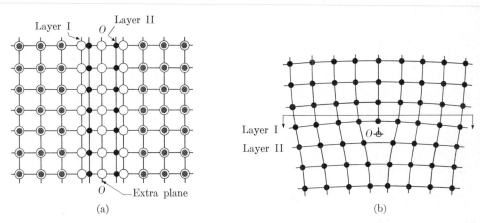

FIG. 4.1. (a) Edge dislocation at point O. (b) Atomic layers I (open circles) and II (smaller) showing the same dislocation as viewed from above.

Figure 4.1 shows that at the dislocation, layer I is in compression, while layer II is in tension. Furthermore, the distortion of the atomic bonds between layers I and II suggests that a shear stress is present between these two layers. An edge dislocation can be thought of as being introduced into a perfect crystal by making a partial cut, the faces of which are subsequently displaced relative to each other by one atomic distance normal to the termination of the cut and then rewelded. This process is equivalent to the introduction of an extra half plane into an alternate cut perpendicular to the first one.

Figure 4.2 shows a *screw dislocation*, which differs fundamentally from an edge dislocation. Here, a partial cut is made into a perfect crystal and the two sides are then subsequently sheared relative to each other in a direction parallel to the termination of the cut by one atomic distance and then welded together. The position of atoms in layers I and II on both sides of the partial cut is illustrated in detail in region A of Fig. 4.3.

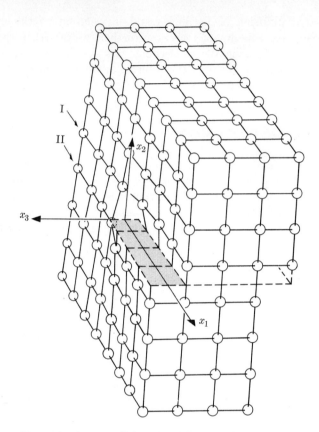

Fig. 4.2. A screw dislocation lying parallel to the x_3 axis.

The basic difference between the two dislocations results from an absence of an extra half plane in a screw dislocation. Dislocations in crystals are very seldom of pure edge or pure screw character, but can blend from one type into another, as shown in Fig. 4.3, forming a dislocation line.

The amount of relative translation of the faces of the cut gives the strength of the dislocation. To evaluate the strength of any dislocation unambiguously in a lattice, we first assign a positive direction to the dislocation line by means of a unit tangent vector. We then compare the dislocated crystal with a perfect crystal. This we do by first drawing in the perfect crystal a large, closed reference circuit in a right-handed screw sense with respect to the previously chosen unit tangent vector, as shown in Fig. 4.4(a). We then attempt to draw an identical circuit around the dislocation in the dislocated crystal, being careful not to bring the circuit too close to the highly strained "bad" material at the core of the dislocation. Because of the relative translation across the faces of the cut of the dislocation, the circuit in the dislocated crystal will not close, as shown in Fig. 4.4(b). The vector from the starting point to the terminus of the circuit in the faulted crystal then represents the strength of the dislocation. This resultant vector is known as the

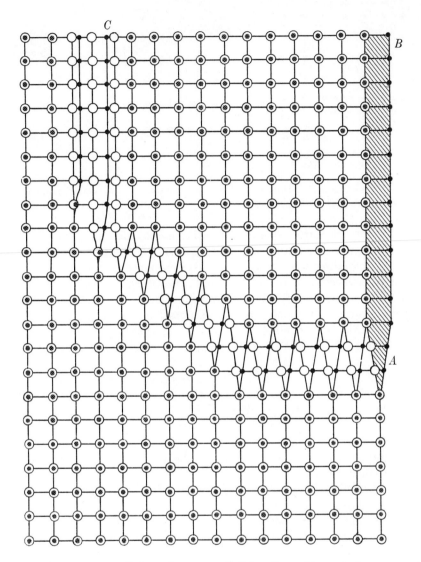

FIG. 4.3. Arrangement of atoms in a curved dislocation. Open circles represent the atomic plane just above the slip plane; smaller circles represent the atoms just below. The segment at C is pure edge dislocation; the segment at A is pure screw dislocation; the region in between is part edge and part screw dislocation.

Burgers vector. The circuit with the right-handed screw sense, advancing in the direction of the unit tangent vector of the dislocation, is known as the *Burgers circuit*.

Although the magnitude of the Burgers vector never changes, its direction is dependent on the initial choice of the unit tangent vector. If the direction of the unit tangent vector is indicated by a small vector attached to the end of the

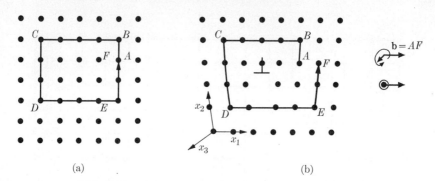

FIG. 4.4. Definition of a Burgers vector with the aid of a perfect crystal. (a) Perfect reference crystal. (b) Dislocated crystal.

Burgers vector, the nature of the dislocation is fully defined (Prob. 4.1). Thus, in Fig. 4.4(b), the circle attached to the tail end of the Burgers vector indicates that the positive direction of the dislocation is assumed to come out of the paper; a cross would indicate that the direction of the unit tangent vector is into the paper.

An edge dislocation of the type shown in Figs. 4.1 and 4.4(b), with an extra half plane containing the positive x_2 axis normal to the x_1 axis, and terminating along the x_3 axis, will have a Burgers vector pointing in the positive x_1 direction if the unit tangent vector is chosen to point in the direction of the positive x_3 axis. Such a dislocation is known as a *positive* edge dislocation.*

The screw dislocation of Fig. 4.2 is of a right-handed screw nature. If the x_3 axis is chosen to coincide with the dislocation line, the Burgers vector will always point in the direction of the positive x_3 axis; such a screw dislocation is a positive screw dislocation. A positive edge dislocation is represented symbolically by an inverted letter ⊥; an upright letter T would stand for a negative edge dislocation. In this notation, the horizontal line represents the plane of the cut, or slip plane, and the vertical line stands for the extra half plane. To represent positive and negative screw dislocations, one might use a letter S and its mirror image Ƨ , respectively. For mixed dislocations, the vector notation is needed.

Dislocations of the type shown in Figs. 4.1 through 4.4, with Burgers vectors equal to one identity vector of the lattice, or an integral multiple of it, are known as *perfect*, or *unit*, *dislocations*.

◂The Burgers vector is the most unchanging property of a dislocation. A dislocation line may alternate between an edge nature and a screw nature, and snake its way through a crystal in an arbitrarily irregular way. In all of this meandering, however, its Burgers vector cannot change. This has the following important consequences.

(a) A dislocation line cannot terminate inside a crystal (Prob. 4.2).
(b) A dislocation can only terminate on a free surface, close on itself creating a loop (Prob. 4.3), or join other dislocations at a node.

---

*The opposite definition is used by some writers.

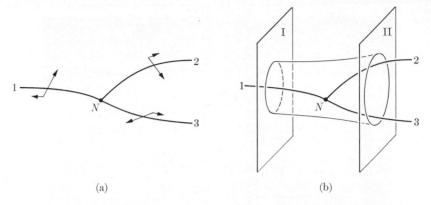

Fig. 4.5. The vector sum of the strengths of dislocations joining at a node is zero.

◀ In the case of several dislocations joining each other at a node, it is necessary that the vector sum of all Burgers vectors of the joining dislocation vanish if the unit tangent vectors of all segments are chosen to point away from the node. This is the dislocation equivalent of the familiar Kirchhoff law in electrical circuit junctions. To demonstrate this in the case of Fig. 4.5(a), we draw two Burgers circuits on plane I and plane II on either side of the node, as shown in Fig. 4.5(b). Since no dislocation threads out of the cylindrical surface connecting the two Burgers circuits, the vector resultant of the circuits on planes I and II must be identical if the dislocations are considered to flow into plane I and out of plane II (all unit tangent vectors point to the right). However, if the dislocations are assumed to emanate from the node N and flow out of the volume through planes I and II, (unit tangent vectors point away from the node as in Fig. 4.5a), the vector resultants of the Burgers circuits on these planes will give equal but opposite Burgers vectors, showing that the vector sum of the Burgers vectors of the two dislocations threading through plane II, when added to the Burgers vector of the dislocation threading through plane I, will close the vector polygon (Prob. 4.4).

Fig. 4.6. Movement of an earthworm (above) and of a snake (below) by dislocations. (After Orowan, 1954.)

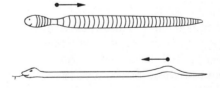

◀ As pointed out by Orowan (1954), dislocations are also of interest in zoology: "Snakes, worms, and mollusks move by developing dislocations [as shown in Fig. 4.6]. The earth-worm starts its movement by developing a tensile dislocation (negative, T) at its neck. The dislocation moves toward the tail of the worm and leaves the body there; after this, or (if the worm is in a hurry) already before, a new tensile dislocation is formed at the neck. In contrast to this, most snakes move by forming compressive (positive, ⊥) dislocations at the tail and moving them toward the head."

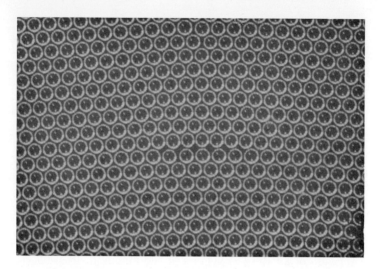

Fig. 4.7. A dislocation in a bubble raft. Diameter of bubbles is 1.2 mm.

◀ A convenient model for the study of edge dislocations was invented by Bragg and associates (Bragg and Nye, 1947; Bragg and Lomer, 1949), who demonstrated that small soap bubbles of uniform size floating on the surface of water interact with forces which resemble those between atoms in close-packed metals. The bubbles are attracted by the long-range surface-tension forces of the water, while their compressibility, arising from the internal pressure of the bubbles, provides the short-range repulsive force. This gives rise to a close-packed layer of compressible spheres floating on a liquid. If the diameter of the bubbles is kept at about 1.2 mm, then the attractive and repulsive forces best simulate those in copper. Figure 4.7 shows an edge dislocation in such a soap-bubble raft (Prob. 4.5). ▶

4.3 MOTION OF DISLOCATIONS IN CRYSTALS

It is possible for dislocations in a crystal to have any arbitrary direction and shape. Very often, however, the anisotropy of the crystal, the presence of preferred growth directions, thermal stresses during solidification, and mutual interactions between dislocations (to be discussed later) will make the dislocations lie on definite crystallographic planes. These planes are usually planes of densest packing with a large interplanar separation.

Consider now an edge dislocation lying in a densely packed plane, as shown in Fig. 4.8(a), where we are viewing it end on as the termination of an extra half plane. If a shear stress is applied to the crystal, as shown in Fig. 4.8(a), the extra half plane can be progressively shifted through the crystal until it reaches the free surface, where it will form a slip step of one atomic distance, as shown in Fig. 4.8(b). A similar motion of a screw dislocation under an applied stress, as seen in Fig. 4.9 (note the spiral nature of a screw dislocation in this figure), will also

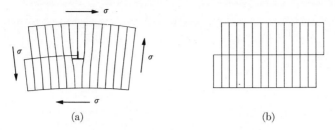

Fig. 4.8. The effect of the motion of an edge dislocation through a crystal.

progressively move the sheared zone through the crystal until the dislocation leaves at the free surface at the far end, creating a slip step of one atomic distance.

Thus we see that the net effect of the motion through the crystal of a dislocation lying on a densely packed plane is a slip displacement of one Burgers vector over the area swept by the dislocation. Inspection of Figs. 4.1 and 4.2 shows that the direction of the slip displacement cannot be arbitrary in a plane, but it is constrained to certain well-defined crystallographic directions by the necessity for crystalline order before and after slip. This picture of a dislocation traveling through a frozen lattice, where every atom is at precisely prescribed lattice sites, is correct only at very low temperatures, where the effect of thermal oscillations can be neglected. Thus, if slip at such temperatures is possible, it will be on densely packed planes with the largest interplanar separation and in directions of densest packing in these planes.

At higher temperatures, the increased level of thermal motion will make slip possible in most crystals also on other less densely packed planes and in less densely packed directions. Table 4.1 shows the slip planes and slip directions of some of the more important ductile materials (Prob. 4.6). The dislocations shown in Figs. 4.8 and 4.9, having their Burgers vectors in the slip plane, are known as *glide dislocations*. As their name indicates, glide dislocations move with relative ease through a crystal and do not require any diffusion of material for this motion;

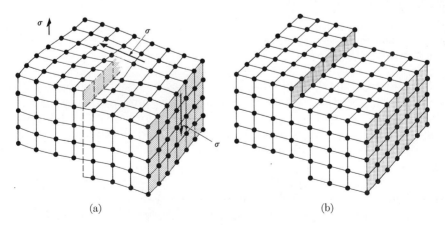

Fig. 4.9. The effect of the motion of a screw dislocation through a crystal.

TABLE 4.1

Slip Planes and Slip Directions of the More Common Ductile Materials

Material	Type of lattice	At low temperature		Additional slip system at high temperature		
		Slip plane	Slip direction	Minimum temp.	Slip plane	Slip direction
Al	f.c.c.	(111)	[10$\bar{1}$]	450°C	(100)	[011]
Cu	f.c.c.	(111)	[10$\bar{1}$]			
Ag	f.c.c.	(111)	[10$\bar{1}$]			
Au	f.c.c.	(111)	[10$\bar{1}$]			
αFe	b.c.c.	(101)(112)(123)*	[11$\bar{1}$]	(At room temperature and higher, slip can take place on many other planes containing the [111] direction.)		
Ta	b.c.c.	(101)(112)	[11$\bar{1}$]			
W	b.c.c.	(101)(112)	[11$\bar{1}$]			
Mg	h.c.p.	(0001)	[11$\bar{2}$0]		(1$\bar{1}$01)	[11$\bar{2}$0]†
Zn	h.c.p.	(0001)	[11$\bar{2}$0]	225°C	(1$\bar{1}$01)	[11$\bar{2}$0]†
Cd	h.c.p.	(0001)	[11$\bar{2}$0]		(1$\bar{1}$01)	[11$\bar{2}$0]†
NaCl	f.c.c. (rock salt)	(110)	[1$\bar{1}$0]	100°C		
LiF	f.c.c. (rock salt)	(110)	[1$\bar{1}$0]	400°C	(001)	[1$\bar{1}$0]
MgO	f.c.c. (rock salt)	(110)	[1$\bar{1}$0]		(111)	[1$\bar{1}$0]

* In iron at room temperature, slip can occur on a wavy surface containing a [111] direction. This is called *pencil glide*.

† The Miller index notation for hexagonal lattices differs somewhat from the three-index notation discussed in Chapter 1. To introduce a uniformity into the description of equivalent planes, the third (redundant) base axis is brought into the picture. The first three terms h, k, i in the notation $hkil$ for planes then represent the reciprocals of the intercepts of the plane with the three base axes. The hexagonal geometry of the base plane demands that the sum of the first two indices equal the negative of the third, i.e., $i = -(h + k)$.

i.e., the extent of the extra half plane is not altered by glide. The motion of glide dislocations is referred to frequently as a *conservative* motion; this is to be understood as the conservation of the extra half plane by an absence of diffusion, and should not be interpreted to mean an absence of a friction process. Because a screw dislocation does not have an extra half plane, its motion is always conservative.

Glide dislocations are thus demarcation lines between slipped and not slipped regions within a crystal. For these dislocations, the Burgers vector indicates the amount of relative slip between parts of a crystal separated by a slip plane.

In contrast, a dislocation loop formed by the accumulation of vacancies on a slip plane and their subsequent collapse, as indicated in Fig. 4.10, has its Burgers vector normal to its plane. Such dislocations, with Burgers vectors normal to their planes, cannot glide in the plane on which they lie, regardless of how they were formed, and are therefore known as *sessile dislocations*.

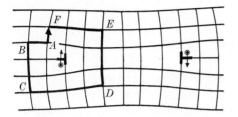

FIG. 4.10. Section through a sessile dislocation loop.

In some cases, where parts of the loop may be on the line of intersection between the plane of the loop and another slip plane containing the Burgers vector of the loop, slip on the intersecting slip plane may become possible for that part of the loop. Barring these instances of slip on other intersecting planes, the loop can move only by diffusion, extending the extra half plane by condensation of interstitial atoms, or alternatively by evaporation of vacancies. Such motion of dislocations normal to their Burgers vector, involving diffusion of atoms or vacancies, is known as *climb*, which is a *nonconservative* motion.

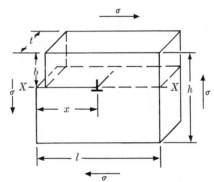

FIG. 4.11. Calculation of plastic shear strain.

When a glide dislocation moves a distance x on a slip plane in a crystal of length l, height h, and thickness t, as shown in Fig. 4.11, the plastic shear strain which results is (Prob. 4.7)

$$\gamma = \frac{xt}{tl} \cdot \frac{b}{h}. \tag{4.1}$$

When there are N dislocations parallel to the t direction of the crystal and all of them move an average distance λ, the total strain is evidently

$$\gamma = \frac{\lambda b}{lh} N. \tag{4.2}$$

With the quantity $N/lh = \Lambda$, the *dislocation density*, Eq. 4.2 can be simplified to

$$\gamma = \Lambda\lambda b. \tag{4.3}$$

The density Λ for well-annealed crystals is usually of the order of 10^4 to 10^6 per square centimeter and rises to about 10^{10} to 10^{12} per square centimeter for highly deformed crystals (Prob. 4.8).

4.4 STRESS FIELD AROUND A DISLOCATION

A dislocation, by its nature, produces a self-balanced stress field in the regions surrounding it. If we disregard the anisotropic nature of a crystal, and consider linear elasticity, it is possible to derive the stress fields of both edge and screw dislocations, and hence by superposition any mixed type. Dislocations in anisotropic media have been considered by Burgers (1939), Eshelby (1949), Eshelby, Read, and Shockley (1953), and Stroh (1958).

Consider the screw dislocation of Fig. 4.2, with the x_3 axis chosen to lie along the dislocation line. We shall proceed by guessing a solution and verifying that it does indeed satisfy the required equations and boundary conditions. If the dislocated block is considered as a large continuum, so that the geometry of the external boundaries does not materially influence the displacements around the center, one might expect from the spiral nature of the dislocation that there would be displacements only in the x_3 direction. Furthermore, in view of the fact that the dislocation is similar to itself at any point along the x_3 axis, these displacements can only be a function of the x_1 and x_2, or r and θ, coordinates. Since the screw dislocation is like a spiral ramp with a pitch equal to the interatomic distance b, the simplest displacement in the x_3 direction that we can imagine would be

$$u_3 = b\frac{\theta}{2\pi} = \frac{b}{2\pi}\tan^{-1}\frac{x_2}{x_1}.$$

The only nonvanishing stress components which result from this displacement are (Prob. 4.9)

$$\sigma_{13} = -\frac{Gb}{2\pi}\frac{x_2}{x_1^2 + x_2^2}, \tag{4.4a}$$

$$\sigma_{23} = \frac{Gb}{2\pi}\frac{x_1}{x_1^2 + x_2^2}, \tag{4.4b}$$

or given more compactly in cylindrical coordinates (Prob. 4.10),

$$\sigma_{\theta z} = \frac{Gb}{2\pi r}. \tag{4.5}$$

This simple stress distribution satisfies equilibrium conditions and since it dies out at large distances, it satisfies the boundary conditions, except possibly at the ends of the dislocation.

◀ The displacements and stresses of an edge dislocation cannot be obtained in so straightforward a manner as those for a screw dislocation. In an edge dislocation of the type shown in Fig. 4.1 (positive, ⊥), with the coordinate axes chosen as indicated in that figure, the dislocation will still be similar at any point on the x_3 axis, so that the condition will be one of plane strain, except near the end at a junction or free surface. For plane strain, there would then be only four stress components: the three normal components σ_{11}, σ_{22}, σ_{33}, and the shear component σ_{12}.

$$\sigma_{11} = -\frac{Gb}{2\pi(1-\nu)r}\sin\theta(2+\cos 2\theta) = -\frac{Gb}{2\pi(1-\nu)}\frac{x_2(3x_1^2+x_2^2)}{(x_1^2+x_2^2)^2}, \quad (4.6a)$$

$$\sigma_{22} = \frac{Gb}{2\pi(1-\nu)r}\sin\theta\cos 2\theta = \frac{Gb}{2\pi(1-\nu)}\frac{x_2(x_1^2-x_2^2)}{(x_1^2+x_2^2)^2}, \quad (4.6b)$$

$$\sigma_{12} = \frac{Gb}{2\pi(1-\nu)r}\cos\theta\cos 2\theta = \frac{Gb}{2\pi(1-\nu)}\frac{x_1(x_1^2-x_2^2)}{(x_1^2+x_2^2)^2}. \quad (4.6c)$$

The third normal component can be found from these (Prob. 4.11). ▶

The nature of the stress components around a positive edge dislocation is shown in Fig. 4.12. Even without the exact solution, one can obtain much of the picture of the stress field from intuition and the equilibrium equations (Prob. 4.12).

All stress components, whether for edge or screw dislocations, have a common factor $Gb/2\pi r$, which describes their dependence on the radial coordinate. This factor is often used in "order of magnitude" calculations to represent the dislocation stress field.

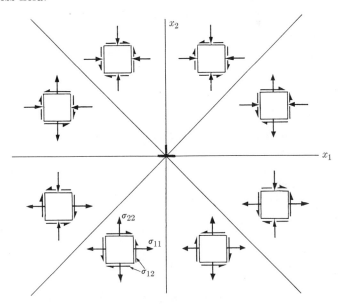

FIG. 4.12. Variation of stress around a positive edge dislocation.

◀The only nonvanishing stress components of a screw dislocation in an isotropic material are shear components. Hence a screw dislocation does not produce any dilatation or volume strain at any point in its surroundings. An edge dislocation, on the other hand, produces local volume changes in its surroundings. The local volume change around an edge dislocation can be readily determined by evaluating Eq. 2.26 with the aid of the isotropic stress-strain relations (Eq. 3.19) and Eqs. 4.6a and 4.6b (Prob. 4.13):

$$\frac{\Delta V}{V} = -\frac{b}{2\pi r}\left(\frac{1-2\nu}{1-\nu}\right)\sin\theta. \tag{4.7}$$

◀Thus, the side of the edge dislocation with the extra half plane has a net decrease in volume, while the opposite side has a net increase in volume. The net volume change around the dislocation, however, is zero (Prob. 4.14). Thus, in the region where linear elasticity applies, outside the very core of the dislocation, neither an edge nor a screw dislocation produces any net volume change. In the core of the dislocation, where linear elasticity no longer applies, there should be a net positive volume change. This is because the high compressive forces in the core will produce a smaller volume decrease than the increase of volume produced by the tensile forces. The net volume increase at the core of the dislocation in simple crystals cannot exceed a fraction of the volume of an atom per atom plane of the dislocation. This would produce a fractional volume change of the order of 10^{-4} in the most heavily deformed materials. The fractional volume changes due to severe plastic deformation in copper reported in the literature are about of this order (2.4×10^{-4}) (Clarebrough, Hargreaves, and West, 1955). The measured values include the effects of point defects generated in the process of plastic deformation (see Section 4.8).▶

Since no volume change will result from the introduction of dislocations into a crystal, save for the small effect due to the dislocation cores, plastic deformation will involve no change in volume. Thus, in isotropic plasticity, the Poisson's ratio must be 1/2.

4.5 ELASTIC STRAIN ENERGY, "LINE TENSION," AND "MASS" OF A DISLOCATION

From the stress field of a dislocation it is possible to calculate the elastic energy stored in the crystal per unit length of dislocation in the x_3 direction by constructing the elastic energy density as a function of r and θ (or x_1 and x_2), and integrating over the volume from r_0 (the radius of the core of the dislocation inside which the strains are too large for linear elasticity to apply) to R (the outside radius of the crystal). The result for a screw dislocation gives, per unit of its length, the strain energy (Prob. 4.15):

$$\mathcal{E}_s = \frac{Gb^2}{4\pi}\ln\left(\frac{R}{r_0}\right). \tag{4.8a}$$

A similar calculation for an edge dislocation gives, per unit length,

$$\mathcal{E}_e = \frac{Gb^2}{4\pi(1-\nu)} \ln\left(\frac{R}{r_0}\right). \tag{4.8b}$$

Equations 4.8a and 4.8b show that the energy of a crystal containing a single dislocation is a function of the dimensions of the crystal. This indicates the long-range nature of the dislocation stress field. We note that the energy of the dislocation is proportional to the square of the magnitude of its Burgers vector. Since a crystal will always seek its lowest energy level, we would expect that super-dislocations with Burgers vectors greater than the smallest identity translation would be unstable, because the elastic strain energy of the crystal would be substantially reduced by the splitting up and separation of such dislocations into unit dislocations. In real crystals containing not one, but many dislocations, the effect of the stress field does not extend over the entire crystal, but only to the neighboring dislocation, which in a random distribution of dislocations, would be one of opposite sign, cancelling the long-range effect of the one considered. Thus, in this case, R may be replaced by $\sqrt{1/\Lambda}$. Problem 4.16 is an example of the application of this technique.

Equations 4.8a and 4.8b represent the major part of the energy of a dislocation. To obtain the total strain energy of a dislocation, however, the energy stored in the core of a dislocation, as well as the effect of the stress-free crystal surfaces, must be considered. These corrections are, in most practical cases, of no importance and will be neglected here for simplicity. It can be shown (Cottrell, 1953) that the configurational entropy contribution to the free energy of a crystal having a single dislocation is also very small compared with the strain energy of the dislocation stress field for any meaningful length of a dislocation. Thus, the free energy of a dislocation is about equal to its strain energy outside the dislocation core.

If the elastic strain energy of a dislocation in a crystal of high perfection is calculated, it is found to be of the order of 5 electron volts per atom slice normal to the dislocation line (Prob. 4.17). This is more than two orders of magnitude larger than the average vibrational energy level per degree of freedom of an atom of a crystal at room temperature. Therefore, it can be concluded that even short segments of dislocations cannot be generated by random thermal motion. We will have occasion to come back to this subject in Section 4.14, where this conclusion will be based on much firmer ground.

If we consider the energy per unit length of a dislocation in a crystal with a dislocation density of $\Lambda = 10^6$ to 10^8 per cm^2, we can simplify Eqs. 4.8a and 4.8b and replace them both with the approximate but more convenient expression

$$\mathcal{E} \approx Gb^2/2. \tag{4.9}$$

Since the dislocation possesses a large elastic energy per unit length, it will always tend to shorten its length between two points inside a crystal by gliding or climbing. Thus, it will behave as if it were subjected to a line tension T all along its length. This behavior is analogous to the surface tension of a thin soap film, which tends to reduce its energy by reducing its surface area. Thus, consider a small portion l of a dislocation line bowed out in its slip plane to a large radius R, as shown in Fig. 4.13. A segment of length dx of the dislocation line would experience a net vertical restoring force equal to (for $dy/dx \ll 1$)

$$T \frac{d^2y}{dx^2} dx = \frac{T}{R} dx.$$

To bow the dislocation segment out of its curved form, an external agency would have to displace an increasing value of this force to the final position y of the dislocation segment. In doing so, a total amount of work will be expended by the external agency equal to

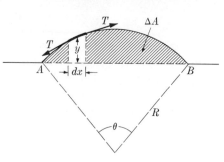

Fig. 4.13. Bowed-out dislocation segment.

$$\frac{T}{R} \int_0^l \frac{1}{2} y(x) \, dx = \frac{T}{2R} \Delta A.$$

This work done on the dislocation line must be stored in it as increased strain energy resulting from an increase in length Δs

$$\frac{T}{2R} \Delta A = \mathcal{E} \, \Delta s.$$

Simplification of the above relation gives (Prob. 4.18)

$$T = \mathcal{E} \approx Gb^2/2. \tag{4.10}$$

If a screw dislocation moves rectilinearly with a constant velocity v in a direction normal to itself, its energy per unit length will increase by a Lorentz factor, very much like the relativistic increase of the energy of a moving particle (Frank, 1949). In an isotropic elastic medium of density ρ, where the velocity of a shear wave is $c = \sqrt{G/\rho}$, the energy of a screw dislocation increases from its rest energy $\mathcal{E}_{s0}$ to

$$\mathcal{E}_{sv} = \mathcal{E}_{s0}/\sqrt{1 - v^2/c^2}. \tag{4.11}$$

Since the stress field of an edge dislocation has both normal stress and shear stress components, its terminal velocity is neither that of a shear wave nor that of a dilatational wave. It rather turns out to be intermediate at the velocity of a surface shear wave (or Rayleigh wave, see Section 7.8). Although the strain energy of an edge dislocation will increase asymptotically at the velocity of a surface shear wave, the relationship is more complicated than Eq. 4.11.

For velocities v of a screw dislocation less than one-tenth the velocity of sound c, the energy per unit length can be written approximately as

$$\mathcal{E}_{sv} \approx \mathcal{E}_{s0} + \frac{1}{2}\left(\frac{\mathcal{E}_{s0}}{c^2}\right)v^2. \tag{4.12}$$

We can interpret these two terms as the elastic self-energy and the kinetic energy of the dislocation. Thus, for small velocities of the dislocation, the inertia of the moving parts of the crystal will produce the same effect as if the dislocation itself had a mass per unit length of

$$\mu = \mathcal{E}_{s0}/c^2 \approx b^2\rho/2. \tag{4.13}$$

4.6 "FORCE" ON A DISLOCATION

A. Glide Force. If a dislocation in a crystal moves under the action of an externally applied stress, the resulting displacements on the boundary will cause the external forces to do work on the crystal. Since the strain energy of the crystal will not be altered by a small displacement of the dislocation in the center of a very large crystal, the work done by the external forces on the free surfaces must be dissipated as a temperature rise within the crystal. We consider this dissipation as having resulted from the displacement of a fictitious force applied normal to the dislocation line by the external stresses. This representation has the advantage that now the dislocation can be considered as a separate entity. Consider a prismatic crystal (Fig. 4.11) containing an edge dislocation on a slip plane X-X. Under the application of a shear stress σ, the dislocation glides a distance x on the slip plane. The work done by the stress σ per unit thickness of the crystal is then (Prob. 4.19)

$$W = \sigma x b.$$

This is taken to be equal to the work done per unit thickness of crystal by the glide component F of a fictitious force acting on the dislocation through the distance x. Hence

$$\sigma x b = F x$$

or

$$F = \sigma b. \tag{4.14}$$

This force acts normal to the dislocation line.

◀When we note that the right-hand side of Eq. 4.14 represents the product of the magnitude of the Burgers vector and the shear stress component on the slip plane in the direction of the Burgers vector, and that the force is always normal to the dislocation line, we can generalize the formula for the glide force as a vector equation. In Fig. 4.14, a dislocation line with a unit tangent vector **s** and a Burgers vector **b** is shown on a slip plane with unit normal vector **n**. If the sum of all stresses other than those due to the dislocation are σ_{ji} at the site of the dislocation,

the net force per unit area acting across
the area will be

$$T_i = \sum_{j=1}^{3} \sigma_{ji} n_j,$$

giving for the product of the magnitudes of the Burgers vector and the component of the shear stress in the direction of the Burgers vector

$$\sigma b = \sum_{i=1}^{3} \sum_{j=1}^{3} b_i \sigma_{ji} n_j.$$

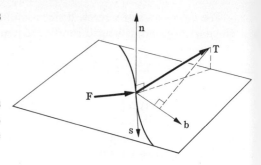

FIG. 4.14. Calculation of the glide force on a dislocation.

◀This is the magnitude of the glide force; its direction is in the slip plane normal to the dislocation line, i.e., parallel to the vector $\mathbf{n} \times \mathbf{s}$, giving for the glide force (Prob. 4.20)

$$\mathbf{F} = (\mathbf{n} \times \mathbf{s}) \sum_{i=1}^{3} \sum_{j=1}^{3} b_i \sigma_{ji} n_j. \tag{4.15}$$

B. Climb Force. For edge dislocations, there may also be a climb force perpendicular to the slip plane. The magnitude of this force is equal to the product σb, where now σ is the normal stress acting across the extra half plane perpendicular to the Burgers vector. Its direction can be obtained by considering whether the normal stress would tend to extrude the extra half plane out of the crystal or pull it in further.

C. Force between Dislocations. Generally the stress fields of two dislocations will interact to produce a repulsion or attraction between them. Such interactions can readily be visualized by utilizing the concept of force introduced in the previous paragraphs. The glide force produced by the stress field of one dislocation on the other can then be calculated according to Eq. 4.14 if the shear stress due to the first dislocation is resolved on the slip plane and parallel to the Burgers vector of the second dislocation, and then multiplied by the magnitude of this Burgers vector. The climb force exerted by the first dislocation on the second can be calculated similarly by evaluating the normal stress due to the first dislocation at the site of the second dislocation acting across its extra half plane and multiplying this by the magnitude of the Burgers vector of the second dislocation. We can then establish by inspection whether the direction of this normal stress would tend to extrude the extra half plane out of the crystal or pull it in.

Figure 4.15 shows the result of such a calculation for the mutual glide force between two edge dislocations lying parallel to the x_3 axis, passing through points $(0, 0)$ and (x_1, x_2), and having their Burgers vectors in the x_1 direction. Here the glide component of the interaction force on the dislocation at (x_1, x_2) is plotted positive when it acts in the positive x_1 direction (repulsion). Hence, curve 1 is for

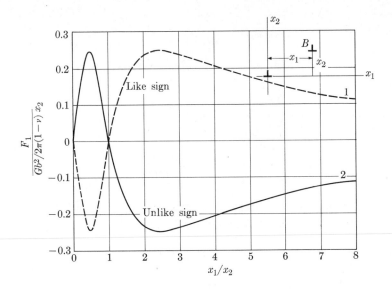

FIG. 4.15. Glide force F_1 on dislocation B due to dislocation A.

two dislocations of like sign, while curve 2 is for two dislocations of unlike sign. Inspection of the diagram shows that two parallel edge dislocations of the same sign are in stable equilibrium when they are above each other and in unstable equilibrium when $x_1 = x_2$, the reverse being true for two parallel edge dislocations of opposite sign. This can also be seen by inspection of Fig. 4.12 in connection with Eq. 4.14. In the special case where the dislocations are on the same slip plane, the force is

$$F_1 = \frac{Gb^2}{2\pi(1-\nu)} \frac{1}{x_1} \qquad (4.16)$$

for dislocations of the same sign, and the negative of Eq. 4.16 for dislocations of mixed sign.

D. Force between Dislocations and Free Surfaces. ◀The strain energy of a crystal will decrease if the dislocation approaches a free surface. This is because the free surface does not offer stress to oppose the displacements of a dislocation. Therefore, the decrease of the strain energy of a dislocation as it approaches a free surface will tend to pull the dislocation to the surface further until the dislocation leaves the crystal by producing a surface step of one interatomic distance. Hence, the reduction of the strain energy of the crystal as a function of its distance from the free surface can be thought of as producing a force pulling the dislocation out of the crystal. It has been shown by Koehler (1941) and Head (1953) that the force tending to pull the dislocation to the surface is of a kind that would be exerted in an infinite crystal on the dislocation by an image dislocation of opposite

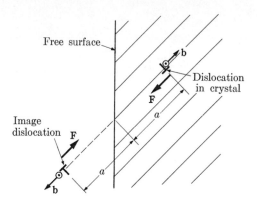

FIG. 4.16. Image dislocation tending to pull a dislocation near a surface out of the crystal by glide.

sign situated with respect to the boundary, as shown in Fig. 4.16. If the surface is covered with a thin impenetrable layer, such as an oxide layer on aluminum, dislocations attracted to the surface may not be able to leave the crystal because of the difficulty of shearing through the layer, and an apparent surface hardening may result. ▶

4.7 CRITICAL SHEAR STRESS NECESSARY TO MOVE A DISLOCATION

◀ In order to understand the kinematics of dislocation motion on the atomic scale, it is necessary to take a closer view of the core of the dislocation. Consider the atoms in the planes on either side of the slip plane of Fig. 4.17(a). The edge of the extra half plane acts like a wedge and displaces pairs of atoms, such as 1–1, relative to each other, thus producing a misfit between layers I and II. The shear resistance of the distorted bonds, such as 1–1, on the right and left of the extra half plane opposes this relative displacement and the extension of the misfit. Thus, the atoms in layer I are compressed, and those in layer II are pulled apart. The dislocation arrangement as it is shown in Fig. 4.17 indicates the equilibrium arrangement. Thus, there are two forces which counterbalance each other; the compressive stress in the upper layer and the tensile stress in the lower layer tend to stretch out the region of misfit, while the shear resistance of the bonds tends to eliminate it.

◀ If the individual atoms have little compressibility, or if the shear resistance of the bonds across the slip plane is small, the misfitting zone of the dislocation will be stretched out. Alternatively, if the atoms have a high compressibility or if the shear strength of the bonds is high, the misfitting area will be confined to a small zone. This effect of variation of the extent of the zone of misfit is shown in the Bragg bubble model. The dislocation in a raft of large, compressible bubbles (Fig. 4.7) shows a very narrow zone of misfit, while the dislocation in a raft of small, incompressible bubbles (Fig. 4.18) is stretched out over a large distance.

4.7 CRITICAL SHEAR STRESS NECESSARY TO MOVE A DISLOCATION

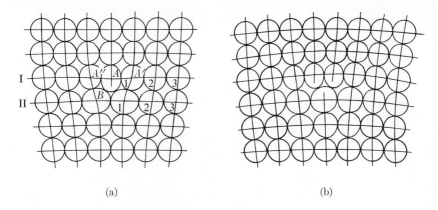

FIG. 4.17. Schematic rendering of two possible distortions of compressible spheres at the core of an edge dislocation in a simple cubic crystal.

More atoms of the dislocation in Fig. 4.7 are nearly in their lowest energy position. In contrast, many atoms of the stretched-out dislocation in Fig. 4.18 are near their position of unstable equilibrium. Under an applied stress, the dislocation with the wide zone of misfit can therefore be displaced more readily than the one with the narrow zone of misfit. Peierls (1940) and Nabarro (1947) have calculated the stress necessary to move an edge dislocation in a simple cubic lattice by considering only pairwise interactions between atoms on opposite sides of the slip plane. Their results show that the stress necessary to move a dislocation in a faultless crystal is very sensitively dependent on the width of the zone of misfit. This stress will be called the *lattice friction stress* (or Peierls-Nabarro stress).

◀Studies of the behavior of individual dislocations suggest that the lattice friction stress is much smaller than the yield strength in pure face-centered cubic and hexagonal close-packed metals and some ionic crystals. In the body-centered

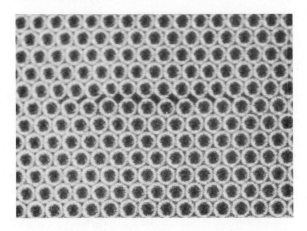

FIG. 4.18. Dislocation in bubble raft with small bubbles (1.1 mm).

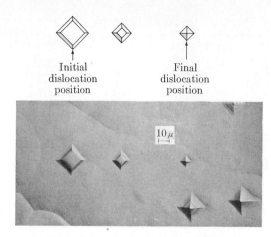

Fig. 4.19. Motion of dislocation in LiF, revealed with etch pits. (Johnston and Gilman, 1959. Courtesy of the American Physical Society.)

cubic transition metals, the lattice friction stress appears to make up a substantial part of the yield strength (Conrad and Hayes, 1963). In crystals having high-energy directional bonds, such as silicon, quartz, and diamond, a large lattice friction stress prevents plastic deformation at room temperature. In the usual f.c.c. and h.c.p. ductile crystals of even so-called "super-purity," the resistance to a directly observable long-range motion of dislocations is higher than the lattice friction stress, because of the additional resistance of minute amounts of impurities or vacant lattice sites. ▶

The shear stress required to move individual dislocations has been measured experimentally for some crystals. In these experiments a crystal with a low initial dislocation density is etched to reveal the sites where dislocations pierce through the external surface, such as in Fig. 4.19 for LiF. The crystal is then subjected to a series of stress pulses of gradually increasing amplitude. Each stress pulse is followed by etching. The stress pulse at which new dislocation pits appear near old ones is then taken as the stress required to move a dislocation. The results are given in Table 4.2. The yield stress at which large amounts of plastic deformation sets in is also entered in Table 4.2. We note that in all cases, dislocations move at stresses lower than the stress necessary for macroscopic yielding. The stress necessary to move a dislocation in a nearly perfect real crystal will be called the *friction stress* of the crystal, to distinguish it from the lattice friction stress, which is assumed to exist in an ideally perfect and pure material.

◀When the stress on the dislocations is increased beyond the friction stress, the dislocations will move through the crystal with increasing velocity, or reversing the statement, to move the dislocations through a crystal with a high velocity, stresses larger than the friction stress are necessary. The relationship between the velocity of individual dislocations and the applied stress in LiF, as measured by the etch pit method by Johnston and Gilman, is shown in Fig. 4.20. ▶

TABLE 4.2

Experimentally Measured Friction Stresses Necessary to Move a Dislocation at Room Temperature

Material	Friction stress, kg/cm²	Macroscopic yield strength, kg/cm²	Reference
LiF	60	90	Johnston and Gilman (1959)
Fe + 3% Si	1100	1500	Stein and Low (1960)
Cu	0.2	4	Young (1962)

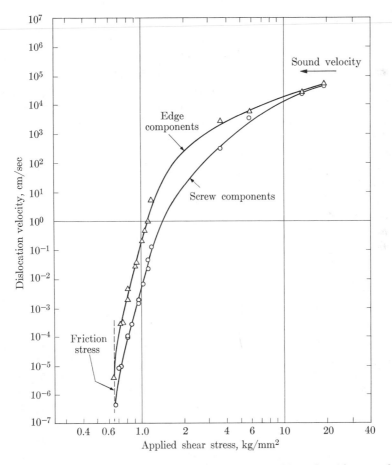

FIG. 4.20. Velocities of edge and screw dislocations in lithium fluoride crystals of high perfection. (Johnston and Gilman, 1959. Courtesy of the American Physical Society.)

It is of interest to compare the experimentally observed friction stress necessary to move a dislocation with the ideal shear strength of a dislocation-free perfect crystal. Consider the two halves of a crystal to be sheared by one interatomic spacing b on a crystallographic plane along a close-packed direction. Because the initial and final positions of the two halves are equilibrium states, the shear stress opposing this motion on planes with separation a_0 must be of a periodic nature with an initial slope equal to the shear modulus G as shown in Fig. 3.1. If for simplicity a sine curve is taken for this stress,

$$\sigma = \sigma_0 \sin \frac{2\pi x}{b}, \tag{4.17}$$

it is then easy to show that the shear strength σ_0 of a perfect crystal must be

$$\sigma_0 = Gb/2\pi a_0. \tag{4.18}$$

This is of the order of $G/10$ across densely packed planes of large separation (higher on less densely packed planes of small separation), and by comparison with the experimentally observed low shear stresses necessary to move dislocations, it is found to be several orders of magnitude higher. Values approaching these high ones ($G/85$) have been observed on dislocation-free parts of large LiF crystals (Gilman, 1959) and on dislocation-free whiskers ($G/20$) of many metals, as well as ionic salts (Gyulai, 1954; Brenner, 1958).

It was this large discrepancy between the ideal shear strength and the experimentally observed critical shear stress which prompted Orowan (1934), Taylor (1934), and Polanyi (1934) to introduce the concept of crystal dislocations into plasticity. The vast amount of experimental work of the last decade (to which references will be made in this and the following chapters), with direct observation of the motion of individual dislocations in a large variety of crystals has put the theory of crystal dislocations on a firm foundation.

4.8 INTERSECTION OF DISLOCATIONS

In the process of slip, dislocations on intersecting slip planes often meet at various angles and cut each other. When two edge dislocations meet and one cuts through the other, as shown in Fig. 4.21, a step is formed in the horizontal dislocation AD equal to the length of the Burgers vector of the intersecting dislocation XY. Naturally, the Burgers vector of the newly created step PP' will be the same as that of dislocation AD (Prob. 4.21).

These steps in dislocations which form in the process of intersection are called *jogs*. Jogs lying in the slip plane, which can be eliminated by glide motion of the dislocation, are called *kinks*. When an edge dislocation intersects a screw dislocation, as shown in Fig. 4.22, the edge dislocation will acquire a jog. The corresponding kink which is formed on the screw dislocation could in principle be easily

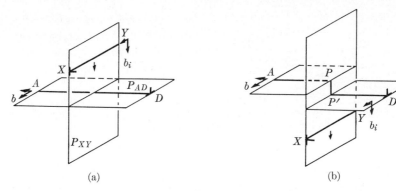

FIG. 4.21. (a) A dislocation XY moving on its slip plane P_{XY} is about to cut the dislocation AD. The Burgers vector of XY is b_i. AD has Burgers vector b and lies on the slip plane P_{AD}. (b) XY has cut through AD. Now AD has a jog PP'. PP' has the direction and magnitude of b_i. (After Read, 1953.)

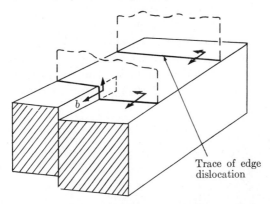

FIG. 4.22. Formation of a jog as a result of an intersection between a screw and an edge dislocation.

eliminated by glide under the action of the line tension of the screw dislocation. In a real crystal this may not be possible, however, if the work that has to be done against the lattice friction is in excess of the energy of the kink.

If the jog is stable, it represents an increase in the length of the dislocation line in the crystal, and hence requires an addition to the work of plastic distortion. The energy of a unit jog would be somewhat less than the energy of a dislocation per atom plane, since its entire length lies in the bad material of the core of the parent dislocation. A frequently used estimate of the energy of a jog of length b_2 in a dislocation of Burgers vector b_1 is (Friedel, 1956)

$$u_j = Gb_1^2 b_2/10, \qquad (4.19)$$

which is about 0.5 to 1.0 ev in metals. Thus, for each jog produced by dislocation intersections, an amount of work equal to the value of the jog energy will have to be provided by the applied stress.

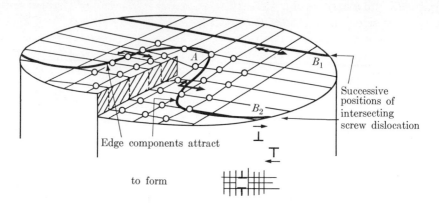

FIG. 4.23. Generation of dipole trails or vacancies by intersecting screw dislocations.

When two screw dislocations intersect, a line defect consisting of a dipole of edge dislocations on neighboring planes is formed between the two intersecting dislocations. This can be visualized by inspection of Fig. 4.23. Here the upper part of the crystal is removed and we are viewing screw dislocation A projecting through the slip plane of screw dislocation B. The initial position of dislocation B is shown as B_1. The intersection can then be considered by first deforming dis-

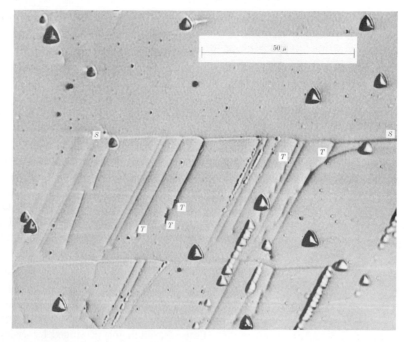

FIG. 4.24. Screw dislocations, S–S, and trails T in silicon. (Dash 1958. Courtesy of the American Physical Society.)

location B into a position B_2. Here the two concave sides of B are edge dislocations of opposite sign, which attract and form a dipole of edge dislocations. Such dipoles, or dislocation trails, have been observed in deformed silicon by Dash (1958) and are shown in Fig. 4.24. In these experiments, silicon crystals were doped with a small amount of copper, which decorated the dislocations by precipitation. The trails are revealed by a process of dissolution normal to the slip plane.

◀ The dipole trails that are produced at screw dislocation intersections can cause a drag on the screw dislocations to which they are attached. Although no detailed calculations are available for the energy of dipole trails, one may take this to be somewhat larger than the core energy of a single dislocation.

◀ Since the dipole trails have no net Burgers vector, they can be sheared over into a line of vacancies or interstitials. These can then break up into individual vacancies or interstitials, which can then diffuse away. The energy of a vacancy in a close-packed metal such as copper is of the order of 1 ev* while that of an interstitial is between 4 and 5 ev. Because the energy of a dipole trail per atom plane would appear to be somewhat less than 1 ev, the interstitial type of dipole trail will require a much higher temperature than the vacancy type for shearing over. As we shall see in Section 4.14, both types will require less than 1 ev for subsequent diffusing. Both types of dipole trails will, however, be fairly stable at room temperature in hard metals.

◀ Jogs have an important role in the climb of dislocations, since they are convenient sites for evaporation of vacancies (or interstitials) into the crystal lattice, making possible the extension or dissolution of the extra half plane. ▶

4.9 DISLOCATION REACTIONS

◀ We have seen that the energy of dislocations, when they are not too close together, is proportional to the square of the magnitude of their Burgers vectors. When two parallel dislocations are brought together, the Burgers vector of the resultant dislocation will be the vector sum of the Burgers vectors of the initial dislocations. If the energy of the resultant dislocation is smaller than the sum of the energies of the two reacting dislocations, these two dislocations will be attracted toward each other to complete the reaction. Alternatively, if the energy of the resultant dislocation is greater than the sum of the energies of the reacting dislocations, there will be a repulsion between the reacting dislocations tending to oppose the reaction. If it is crystallographically possible, a dislocation may dissociate into two component dislocations, the Burgers vectors of which must vectorially add up to the Burgers vector of the parent dislocation. Such a dissociation into two smaller dislocations could actually occur, provided that the sum of the energies of the component dislocations is smaller than the energy of the

---

* The energy of formation of a vacancy must be calculated by considering specific quantum-mechanical details of binding of atoms in the lattice. It is from such calculations that the value of 1 ev has been obtained.

initial dissociating dislocation. It is possible to summarize these rules as follows.
(a) If the angle between the Burgers vectors of two reacting dislocations is *obtuse*, combination will be possible because it would decrease the energy of the crystal.
(b) If the Burgers vector of a dislocation can dissociate into two crystallographically meaningful Burgers vectors making an *acute* angle with each other, the dislocation itself will break up into two smaller dislocations, reducing the energy of the crystal.

◂A dislocation reaction can be written as a vector sum relating the Burgers vectors of the reacting and the resultant dislocations, with an arrow indicating the energetically favored direction. As a simple example we may write the possible reaction in the b.c.c. lattice with a lattice parameter of a (Prob. 4.22):

$$\frac{a}{2}[111] + \frac{a}{2}[\bar{1}\bar{1}1] \rightarrow a[001]. \tag{4.20}$$

◂One immediate conclusion that can be drawn from the above statements is that dislocations with Burgers vectors greater than lattice identity vectors are unstable and will break up into unit dislocations.

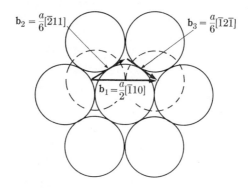

Fig. 4.25. Splitting up of dislocation b_1 into b_2 and b_3 in the (111) plane of an f.c.c. lattice.

◂In certain close-packed lattices, it becomes possible for unit dislocations to split into component dislocations with Burgers vectors smaller than the lattice identity spacing. Such dislocations, when moved through a crystal, will not translate the two parts of the crystal across the slip plane by an identity spacing, but by a smaller amount. Hence, a dislocation with Burgers vector smaller than an identity spacing will leave a faulted layer behind it. For this reason, such dislocations are known as *imperfect*, or *partial*, *dislocations*. Since the faulted layer left behind by a moving partial dislocation raises the energy of the crystal, partial dislocations are encountered in complementary pairs bracketing the faulted layer. A good example of partial dislocations is found in crystals with face-centered cubic (f. c. c.) or hexagonal close-packed (h.c.p.) lattices. Figure 4.25 shows the atoms on a {111} close-packed slip plane of a f.c.c. crystal. The Burgers vector (smallest identity vector) of the perfect dislocation is denoted by b_1. Inspection of the figure will show that slip of the top atomic layer, represented by the dashed circles, over the lower atomic layer, represented by the solid circles, in the direction $\mathbf{b}_1$ can be effected in two stages by first slipping by an amount $\mathbf{b}_2$ and then by an amount $\mathbf{b}_3$. In the language of dislocations, this means the splitting of a dislocation $\mathbf{b}_1$ into two dislocations $\mathbf{b}_2$ and $\mathbf{b}_3$. This splitting reaction

$$\frac{a}{2}[\bar{1}10] \to \frac{a}{6}[\bar{2}11] + \frac{a}{6}[\bar{1}2\bar{1}] \tag{4.21a}$$

should readily take place because it is energetically favored, since (Prob. 4.23)

$$b_2^2 + b_3^2 < b_1^2. \tag{4.21b}$$

Following the splitting of $\mathbf{b}_1$ into $\mathbf{b}_2$ and $\mathbf{b}_3$, the energy of the crystal is reduced by the repulsion of the two dislocations $\mathbf{b}_2$ and $\mathbf{b}_3$, since these are largely of the same sign. As these dislocations are spread apart, they delineate three regions in the slip plane, as shown in Fig. 4.26. Region A has slipped by one identity vector $\mathbf{b}_1$, while region C has not slipped at all. Hence, in both regions, the atomic packing is f.c.c. with a characteristic packing sequence of $ABCABC$, etc., as discussed in Section 1.4. In region B, however, only dislocation $\mathbf{b}_2$ has passed through the crystal, displacing the atoms by less than an identity translation and upsetting the regular $BCABC$ sequence to a sequence of $ABCA|CABC$, as shown in Fig. 4.26. The latter sequence is hexagonal close-packed across the slip plane represented by the vertical line in the sequence. This local change in stacking from face-centered cubic to hexagonal close-packed and back represents a *stacking fault* in a face-centered cubic lattice, which raises the energy of the crystal. Thus spreading of the dislocations $\mathbf{b}_2$ and $\mathbf{b}_3$ reduces the elastic energy, but it also extends the stacking fault and thereby tends to raise the energy of the crystal. At the separation d between $\mathbf{b}_2$ and $\mathbf{b}_3$, where the total energy is minimized, the mutual repulsive force between the dislocations is almost (depending on the amount of screw component)

$$F = \frac{G(\mathbf{b}_2 \cdot \mathbf{b}_3)}{2\pi(1-\nu)d}. \tag{4.22}$$

This must be balanced by the specific stacking-fault energy α_s, from which the equilibrium separation d can be determined as (Prob. 4.24)

$$d = \frac{G(\mathbf{b}_2 \cdot \mathbf{b}_3)}{2\pi(1-\nu)\alpha_s}. \tag{4.23}$$

The partial dislocations $\mathbf{b}_2$ and $\mathbf{b}_3$, together with the ribbon of stacking fault separating them, are called an *extended dislocation*.

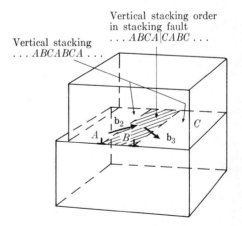

FIG. 4.26. Partial dislocations and stacking fault in a crystal.

◀ Extended dislocations have been observed in f.c.c. and h.c.p. structures as well as in other close-packed layerlike crystals, such as graphite and talc. Figure 4.27 is an electron micrograph of a very thin slab of graphite crystal showing extended dislocations which are visible by virtue of their varying electron diffraction effects.

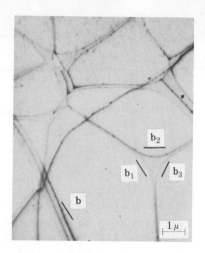

FIG. 4.27. Extended dislocations and nodes in electron-transparent slab of graphite. (Amelinckx and Delavignette, 1960. Courtesy of the American Physical Society.)

◀ A good example of a fusion reaction in an f.c.c. lattice is the Cottrell reaction between two partial slip dislocations of the type discussed above, leading to a sessile partial dislocation. Consider the two extended dislocations on {111}-type slip planes shown in Fig. 4.28. If these extended dislocations encounter each other along the line of intersection of their respective slip planes, the leading partial dislocations will fuse to give another partial dislocation according to the reaction

$$\frac{a}{6}[\bar{1}\bar{2}\bar{1}] + \frac{a}{6}[211] \to \frac{a}{6}[1\bar{1}0]. \tag{4.24}$$

The resulting dislocation at the line of intersection has a Burgers vector that does not lie in either slip plane and therefore cannot glide. Such sessile dislocations, which hold the two remaining partials of the intersecting extended dislocations with ribbons of stacking fault in between, are thought to form barriers to other dislocations on the same slip planes, and hence play a role in strain-hardening of f.c.c. crystals.

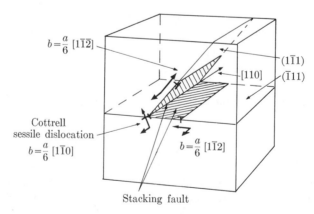

FIG. 4.28. Lomer-Cottrell sessile dislocation in an f.c.c. lattice.

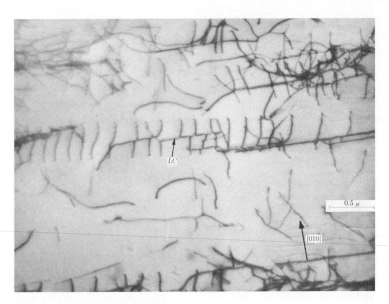

Fig. 4.29. Dislocations in ultra-thin stainless-steel foil viewed in transmission in an electron microscope. (Whelan, 1959. Courtesy of the Royal Society of London.)

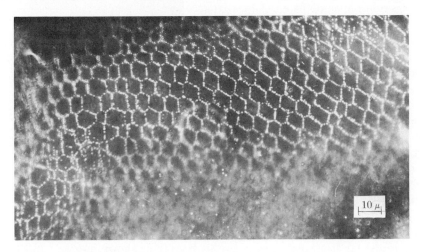

Fig. 4.30. A hexagonal dislocation net in a KCl crystal. (Amelinckx, 1958. Courtesy of Pergamon Press.)

◀ The coalescence and splitting of dislocations are very common phenomena in crystals. Segments of two approaching dislocations on intersecting planes will attract each other and combine over substantial lengths along the line of intersection of their slip planes. In cases like this, the new combined dislocation forms a node with the remaining parts of the two parent dislocations, as seen in Fig. 4.29, which is a transmission electron micrograph of an ultra-thin section of stainless-steel foil (1000 to 3000 A). Part of the straight segment indicated by LC in Fig. 4.29 is an arrangement of the type shown in Fig. 4.28, constrained to be along the line of intersection of two slip planes by the Cottrell sessile dislocation* (Prob. 4.25). Such occurrences of dislocation nodes are easily observable in moderately deformed or recrystallized grains, resulting frequently in stable three-dimensional dislocation nets. Part of such a stable net in a KCl crystal observed by Amelinckx (1958) is shown in Fig. 4.30. It is now almost certain that all but the most perfect crystals have such a net resulting from growth or solidification. ▶

4.10 DISLOCATION MULTIPLICATION

Although annealed crystals, depending on their perfection, will usually contain about 10^4 to 10^6 dislocations per square centimeter, it is not too difficult to see that the magnitude of observed plastic strains could hardly be attributed solely to their motion and "exhaustion" (Prob. 4.26). It is known, on the contrary, from x-ray evidence and direct observation, that plastic deformation raises the density of dislocations by orders of magnitude. As we shall see in Section 4.14, dislocations cannot be generated by thermal fluctuations at the low levels of shear stress sufficient for slip. Therefore, other geometrical mechanisms are necessary for their multiplication. The extreme localization of slip in single crystals like the one in Fig. 4.31 indicates further that dislocations must be generated in large numbers in some slip planes. Such a mechanism, first proposed on theoretical grounds by Frank and Read (1950), is explained below.

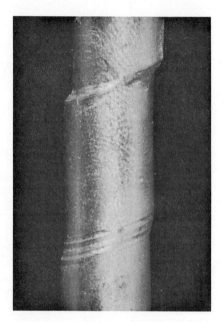

FIG. 4.31. Localized slip in a single crystal of zinc. (Courtesy of E. R. Parker.)

Consider a crystal that has a stepped dislocation $ADD'B$, as illustrated in Fig. 4.32(a). If a shear stress σ is applied to this crystal, as shown in Fig. 4.32(b), only the segment DD' experiences a force, while the segments AD and $D'B$ remain immobile. Thus, under increasing stress σ, the segment DD' will bow out as shown in Fig. 4.32(b) until it becomes unstable in a semicircular configuration. The shaded portion of the slip plane indicates the extent of slip on the plane. From this point on, the loop expands at a lower applied stress and continues to sweep over the slip plane. Figure 4.32(c) shows the loop breaking through the surface where it has almost completed the formation of a surface step of one interatomic

---

* Also called a Lomer-Cottrell lock.

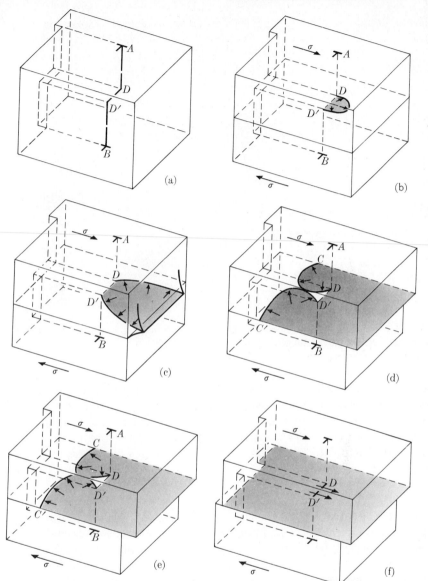

Fig. 4.32. Stages in one cycle of the operation of a Frank-Read dislocation mill.

distance. In Fig. 4.32(d), the two separate legs DC and $D'C'$ have swept out the right-hand side of the slip plane and have met. The bowed-out segments will run together, cancel each other, and form two segments CC' and DD' as shown in Fig. 4.32(e). Segment CC' will continue to sweep out toward the left of the crystal until it produces a slip step of one interatomic spacing on the free surface at the left of the crystal, as in Fig. 4.32(f), while segment DD' will straighten out and be ready to undergo another cycle. The salient point here is that the segment DD'

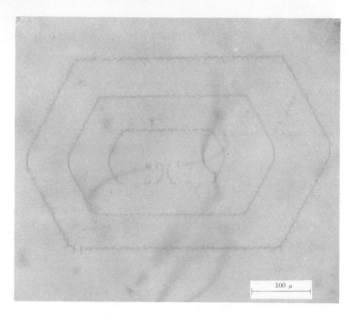

FIG. 4.33. A Frank-Read dislocation mill seen in plastically deformed silicon. Copper precipitates reveal the dislocation lines. The dislocations lie on the (111) plane of silicon. (Dash, 1957. Courtesy of General Electric Company.)

has generated slip of one unit step without disappearing from the crystal and can repeat its operation over and over, giving large amounts of slip on one slip plane.

The critical stress σ required to bow out the loop to a semicircular shape of diameter l can be readily calculated by considering the balance of the line-tension forces acting at points D and D' in the slip plane and the stress force acting on the half loop; it is (Prob. 4.27)

$$\sigma = Gb/l. \tag{4.25}$$

Segments of a dislocation net, such as the one of Fig. 4.30, formed in the process of crystal growth can serve as Frank-Read dislocation mills in crystals. A Frank-Read mill of this type, observed by Dash (1957) in silicon, is shown in Fig. 4.33.

In most pure ductile crystals, the height of individual slip steps is so small that the slip lines can be revealed only under extreme magnification in the electron miscroscope. This indicates that for various reasons a Frank-Read dislocation mill will not continue to operate for many cycles. It then becomes necessary to generate new Frank-Read mills.

A mechanism of generation of Frank-Read mills in the process of deformation is illustrated in Fig. 4.34. Here a screw dislocation moving toward the left on its slip plane P either encounters an obstacle at A, or by virtue of a thermal fluctuation goes into an intersecting slip plane parallel to the Burgers vector of the dislocation. After gliding in this plane for a certain distance, the dislocation may go back into a plane P' parallel to its original plane of glide, as shown in Fig. 4.34(b).

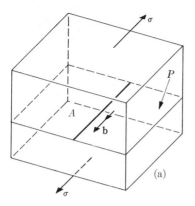 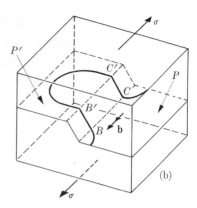

Fig. 4.34. Cross glide and double cross glide of screw dislocation generating Frank-Read mills.

This process of changing of slip planes by a screw dislocation is known as *cross glide*.* Often the intersecting slip plane that the screw dislocation uses as a stepping stone to go on to the parallel plane P' may be one in which dislocations are difficult to move. This may result in the anchoring of the edge segments BB' and CC' on the bridging plane. Thus, when the screw segment $B'C'$ expands on the parallel plane P', it is of the nature of a Frank-Read mill, and can operate to create a number of new dislocations on this plane.

The process of generating Frank-Read mills by double cross glide was first observed in ionic crystals, where it is the chief mechanism of slip-band broadening (Johnston and Gilman, 1960). There is now overwhelming evidence to indicate that the process of double cross glide operates in nearly all ductile crystals.

When dislocations are extended, in the manner discussed in Section 4.9, their screw components cannot cross glide in the extended configurations, since neither of the two partials will be in pure screw orientation. Thus, prior to cross gliding, an extended screw dislocation will have to form a constriction by recombining over a short length. In crystals such as alpha brass, where the stacking fault energy is low, the formation of such a constriction will require a considerable amount of energy fluctuation. Therefore, in alpha brass at low temperatures, slip lines will be unusually deep as shown in Fig. 1.30, and slip bands will not easily widen.

At a certain critical shear stress, many screw segments of dislocations will double cross glide and will generate as many Frank-Read mills as are required to furnish the large numbers of glide dislocations necessary for stable plastic deformation. This critical shear stress will be the yield stress of the crystal in the

---

* In the literature, this process has been named cross slip. In this book, the term *slip* is used in connection with microscopically observed surface traces, while the term *glide* is used to refer to the conservative motion of dislocations themselves. Hence, we will use the term *cross slip* to signify a sharp step in a microscopically observable surface trace and the term *cross glide* to express the submicroscopic motion of a screw dislocation from one slip plane into an intersecting slip plane.

TABLE 4.3*

Critical Resolved Shear Stress for Single Crystals of Metals and Ionic Salts

Metal	Purity of initial material, %	Method of production of crystal	Glide elements		Critical shear stress at yield point, kg/mm^2	Shear modulus, kg/mm^2	Elastic shear at yield point
Copper	>99.9	Solidified in vacuo	(111)	[10$\bar{1}$]	0.10	—	—
Silver	99.99				0.060		
Gold	99.99				0.092		
Nickel	99.8				0.58	—	—
Magnesium	99.95	Recrystallization	(0001)	[11$\bar{2}$0]	0.083	1700	4.95×10^{-5}
Zinc	99.96		—	—	0.094	4080	2.3
Cadmium	99.996	Drawn from melt	—	—	0.058	1730	3.35
β-Tin	99.99		(100)	[001]	0.189	1790	10.6
			(110)		0.133	1790	7.43
Bismuth	99.9		(111)	[10$\bar{1}$]	0.221	970	22.8

Salt	Total of impurities, wt %	Critical shear stress of the {110} ⟨110⟩ glide system, kg/mm^2
NaCl	0.030	0.075
KCl	0.016	0.050
KBr	0.030	0.080
KI	0.020	0.070

* Compiled from Schmid and Boas, 1935. (Courtesy of Springer.)

absence of a large lattice friction stress or other internal stresses resulting from impurities, precipitates, grain boundaries, and the like. In close-packed metals this stress is generally greater than the critical shear stress of Eq. 4.25 (Johnston and Gilman, 1960; Young, 1962) necessary to operate a Frank-Read dislocation mill, and much greater than the lattice friction stress.

Experiments on single crystals have shown that for yielding it is necessary in general that the shear stress on a slip plane in the slip direction reach a critical value. The critical resolved shear stress for a variety of single crystals of some common materials are shown in Table 4.3.

◀ If the direction cosines of the unit normal vector of the slip plane and the unit tangent vector in the slip direction are

$$l_{1p}, \quad l_{2p}, \quad l_{3p},$$
$$l_{1q}, \quad l_{2q}, \quad l_{3q},$$

slip will take place on this plane from motion of dislocations with Burgers vectors

$$\mathbf{b} = b(l_{1q}\mathbf{i}_1 + l_{2q}\mathbf{i}_2 + l_{3q}\mathbf{i}_3) \tag{4.26}$$

under an applied set of stresses σ_{ij}, when the resolved shear stress σ_{pq} reaches a critical value σ_y, that is,

$$\sigma_y = \sigma_{pq} = \sum_{i=1}^{3} \sum_{j=1}^{3} l_{ip}\sigma_{ij}l_{jq}. \tag{4.27}$$

For a long single crystal under a uniaxial stress σ_{33}, Eq. 4.27 reduces to the simple relation

$$\sigma_y = \sigma_{33} l_{3p} l_{3q}. \tag{4.28}$$

◀ It should be noted here that although in general it is necessary that a critical shear stress be reached on a slip plane for slip to develop on that plane, it is not always true that slip will develop first on the most highly stressed plane of a family of slip planes. An example of this has been encountered in tungsten and molybdenum where the behavior appears to be related to a unidirectional mobility of screw dislocations on certain slip planes resulting from a three-dimensional splitting of screw dislocations (Maloof and Argon, 1962). ▶

4.11 TILT AND TWIST BOUNDARIES AND DISLOCATION PILE-UPS

As was shown earlier (Fig. 4.15), the component of the mutual force along the slip plane between two like parallel edge dislocations on parallel planes vanishes when these dislocations arrange themselves above each other in a position of stable equilibrium. This situation holds true also for a set of like dislocations on parallel planes, such as the ones shown in Fig. 4.35. Such a dislocation wall, with spacing h between dislocations, forms a type of low-angle grain boundary known as a *tilt boundary*, which produces a tilt between the two neighboring halves of the crystal about an axis parallel to the dislocations, at an angle θ which must be

$$\theta = b/h. \tag{4.29}$$

Tilt boundaries are very common, since they represent positions of stable equilibrium for dislocations in crystals, and thereby minimize the

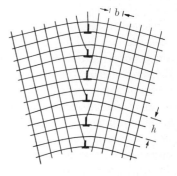

FIG. 4.35. Tilt boundary.

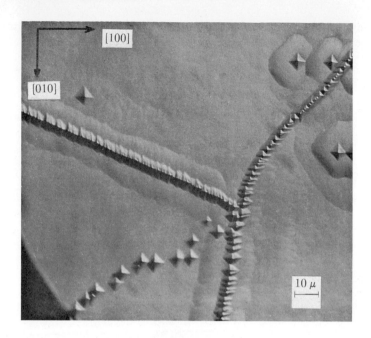

Fig. 4.36. Low-angle boundaries revealed on lithium fluoride crystal by an etching process which preferentially attacks surface terminations of dislocations. (Gilman and Johnston, 1957. Courtesy of General Electric Company.)

total internal strain energy by eliminating the long-range stress fields of the dislocations making up the wall (see Cottrell, 1953). Figure 4.36 shows such a tilt boundary made visible in a LiF crystal by a selective etching technique which preferentially attacks the site of dislocations.

It can be visualized that a parallel array of screw dislocations on one plane will produce a twist between the two halves of the crystal that the plane separates. It can be shown, however, that such an array will have to be held in equilibrium with a large long-range torque or friction stress (Prob. 4.28). When an identical net of the same dislocation density exists on the same or on a nearby plane forming a crossed rectangular grid with the first and having a negative long-range torque, then the two nets could hold each other in equilibrium, as seen in Fig. 4.37. Such a net, observed in a KCl crystal by Amelinckx (1958), is shown in Fig. 4.38. Here the screw dislocations lie on a {001} plane, and the nodes are made visible under the light microscope in thin sections of the crystal by previously introducing free silver atoms into the lattice, which, in a special heat-treatment process, are made to precipitate at the dislocation nodes. (Screw dislocations produce no local volume change; hence, decoration of dislocations by precipitation of silver atoms takes place only at the nodes.) Crossed screw dislocation nets of this form, producing a rotation of the lattice across the plane in which they lie, are known as *twist boundaries*.

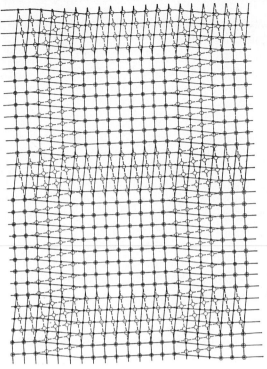

Fig. 4.37. A pure twist boundary. The boundary is parallel to the plane of the figure, and the two grains have a small relative rotation about their cube axis which is normal to the boundary. The open circles represent the atoms just below. The grains join together continuously except along the two sets of screw dislocations which form a crossed grid. (Read, 1953. Courtesy of McGraw-Hill.)

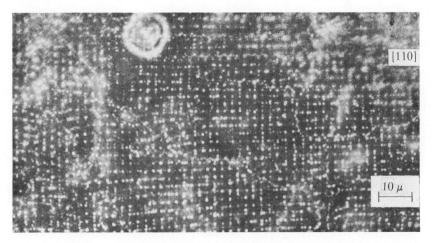

Fig. 4.38. Almost ideal square net in a {100}-type plane of a KCl crystal. Note preferential decoration at the node points. (Amelinckx, 1958. Courtesy of Pergamon Press.)

When the dislocations released by a Frank-Read dislocation mill are held up at a strong obstacle, such as a grain boundary, a *dislocation pile-up* is formed, as shown in Fig. 4.39. A dislocation pile-up has the property of concentrating the applied shear stress at the obstacle. The mutual repulsion of the dislocations in the slip plane tends to spread out the arrangement, whereas the applied stress

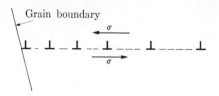

FIG. 4.39. An edge dislocation pile-up at a grain boundary.

squeezes the dislocations closer together against the obstacle. Thus, the obstacle supports not only the applied shear stress, but also all the shear stresses of the piled-up dislocations. If there are n dislocations in the pile-up pressed against the obstacle by the applied stress σ, then the shear stress exerted on the obstacle will be $n\sigma$. This fundamental result can be obtained by giving the whole dislocation arrangement a small virtual displacement to the left; then the work done by the applied stress σ must be expended to displace the force acting between the obstacle and the first dislocation (Prob. 4.29).

4.12 MODES OF PLASTIC DEFORMATION OTHER THAN SLIP

Although slip, as described above, is the predominant mechanism of plastic deformation in crystalline materials, other modes, such as twinning and kinking, also play important roles. When slip is blocked, these mechanisms are the only possible alternatives.

Deformation *twinning* results when a part of the crystal becomes, on the atomic scale, the mirror image of the matrix with respect to a plane by undergoing a uniform twinning shear. Hence twinning differs from slip principally by its uniform translation in every atomic layer by an amount less than the identity distance. The nature of the displacements of twinning is shown in Fig. 1.20.

Twins frequently occur in b.c.c. crystals. Here the mirror plane is the (112) plane, and the shear direction is $[\bar{1}\bar{1}1]$, as shown in Fig. 1.20. Twins grow in the form of flat disks with large diameter-to-thickness ratios, as seen in Fig. 1.34. These thin disks, which have been observed in many b.c.c. materials, are also known as *Neumann bands*.

Twins are very common in hexagonal close-packed materials such as zinc, cadmium, and magnesium. Mechanically formed twins are far less frequent in f.c.c. lattices than in b.c.c. or h.c.p. crystals. They have been observed in single crystals of copper (Blewitt et al., 1957) at 4.2°K. Here the mirror plane is the (111) plane, while the shear direction is $[11\bar{2}]$.

A dislocation model producing the twinning shear on each successive plane parallel to the (112) mirror plane in a b.c.c. lattice has been proposed by Cottrell and Bilby (1951). This model correctly accounts for the geometry of the displacements, but it fails for instance to explain the relatively lower temperature dependence of the twinning stress compared with that of the critical stress required for slip.

When crystals with only one set of easy slip planes, such as hexagonal close-packed crystals with their slip planes nearly normal to the tensile axis, are extended, or crystals with their slip planes nearly parallel to the axis are compressed, slip may be confined entirely to a narrow band, as shown in Fig. 4.40. This produces a condition of crystallographic buckling, which is called *kinking*.

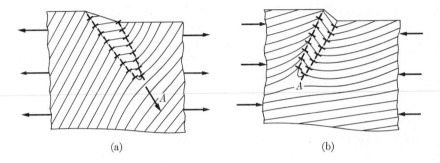

Fig. 4.40. Schematic drawing of kinking in tension (a) and compression (b) of a crystal with only one set of good slip planes.

Kinking is accomplished by the production and separation of dislocation pairs on parallel planes, as shown in Fig. 4.40. The result of kinking then is the formation of a band inside of which the lattice has rotated relative to the outside by an angle whose tangent equals the shear strain γ which the interior of the band has undergone. It has been shown by Frank and Stroh (1952) that the shear stress produced by the kink boundaries at point A can be of the order of the ideal shear strength of the perfect lattice given by Eq. 4.18. Thus, beyond a critical size a kink band is self-propagating; it increases in length by generating dislocation pairs at its tip by overcoming the ideal shear strength of the lattice. Figure 4.41 shows a kinked zinc crystal. Kinking has been observed in h.c.p., f.c.c., b.c.c., and also ionic crystals.

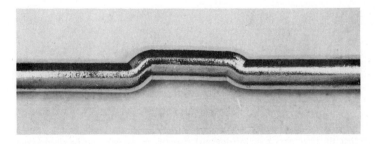

Fig. 4.41. Kinked zinc crystal, actual size. (Courtesy of J. J. Gilman.)

4.13 GENERATION AND MOBILITY OF POINT DEFECTS IN A LATTICE

Thermal motion produces temperature and rate effects in plasticity through its effects on vacancies and dislocations. We first consider the simpler case, the effects on vacancies. Vacancies produce the climbing motion of dislocations by evaporating from or condensing at the extra half plane of edge dislocations. Since the energy of formation of a vacancy is much less than that of an interstitial, vacancies will exist in greater abundance than interstitials and play a more prominent role (see Section 4.8).

In the process of creation of a vacancy, the enthalpy* of the lattice is raised. However, the production of a vacant lattice site will also introduce a disorder into the lattice and therefore cause a rise in its entropy. Hence, if the rise in entropy is substantial, i.e., if its product with absolute temperature exceeds the change of enthalpy of the lattice, there will be a net reduction of the Gibbs free energy $(g = h - Ts)$ of the lattice, so that thermodynamic equilibrium will require the presence of some vacancies. Their equilibrium concentration can be found by minimizing the Gibbs free energy of a crystal.

◀At the atomic level, entropy changes are primarily due to changes in probability of a configuration, with changes in vibration playing only a minor role, which we shall neglect (Prob. 1.12). In the statistical interpretation of Boltzmann, the configurational entropy, aside from an additive constant, is given in terms of Boltzmann's constant k, and is a measure of the degree of disorder, which in statistical thermodynamics is taken simply as the logarithm of the number of possible distinguishable configurations, i.e. (see Joos, 1950, p. 571),

$$s = k \ln W. \tag{4.30}$$

In a crystal having N atoms and n vacant sites (i.e., $N + n$ total sites), the number of possible distinguishable ways of stacking n vacancies (which is proportional to the probability as discussed in Section 1.3) is given by the theory of combinations as

$$W = \frac{(N+n)!}{n!N!}. \tag{4.31}$$

In terms of the enthalpy of formation of a vacancy at constant pressure, h_f, the excess Gibbs free energy of the crystal (i.e., over and above the energy of the atoms of the crystal) is given by

$$g = nh_f - kT \ln \frac{(N+n)!}{n!N!}. \tag{4.32}$$

Using Stirling's formula for the expansion of a factorial with a large argument,

$$N! \approx \sqrt{2\pi}\, N^{N+1/2} e^{-N} \quad \text{or} \quad \ln N! \approx N \ln N, \tag{4.33}$$

---

* We have to consider here the change in enthalpy and not the change in internal energy, because if the lattice is under substantial pressure, additional work has to be done in extracting the atom out of the lattice. This additional work is the difference between the enthalpy and the internal energy of the vacancy configuration.

in Eq. 4.32, followed by the minimization of the free energy as a function of n will lead, for $N \gg n$, to (Prob. 4.30)

$$c_v = n/N = e^{-h_f/kT}, \qquad (4.34)$$

which is, by definition, equal to the probability p_v of finding a vacancy at a lattice site, or alternatively to the fractional concentration of vacancies, c_v.

◀ The fractional concentration of interstitial atoms in thermal equilibrium can be calculated similarly, leading to the same relation as Eq. 4.34, where h_f would then represent the enthalpy of formation of an interstitial.

◀ Equation 4.34 gives the equilibrium concentration of vacancies without regard to the kinetics of the establishment of such a concentration. But vacancies can form easily only at free surfaces, grain boundaries, from a vacancy dipole trail or at a jog on the extra half plane of an edge dislocation, where it is possible to accommodate the atom creating the vacancy. Inside a perfect crystal, such accommodation is very difficult, and requires the formation of vacancies and interstitials only in pairs. Such a process requires an activation energy equal to the sum of the enthalpies of formation of a vacancy and interstitial, and will therefore be quite rare. Thus, the establishment of an equilibrium concentration of vacancies requires diffusion of the vacancies produced at the surface into the interior of the crystal. We shall consider now the problem of diffusion of vacancies from regions of high concentration to regions of low concentration.

◀ For a vacancy to jump into an adjacent lattice site, it must exchange its position with the atom occupying that site. In this exchange, the atom will have to overcome an enthalpy barrier h_j and will have to produce a state of lattice disorder; i.e., it will have to overcome a Gibbs free energy change $h_j - Ts_j$ (see Zener, 1952). The probability p_j that such a free energy change can be provided from the random vibrational energy fluctuations is given by the Boltzmann relation (see Section 1.3 and Eq. 1.3)

$$p_j = e^{-(h_j - Ts_j)/kT} = e^{s_j/k} e^{-h_j/kT}. \qquad (4.35)$$

We shall call the enthalpy barrier h_j the *activation enthalpy* and s_j the *activation entropy*.

FIG. 4.42. Vacancy gradient between adjacent planes. ▶

◀ It is now possible to determine the net diffusion rate of vacancies from one part of the crystal to another. If the vacancy concentration were uniform throughout, the probability of diffusion in all directions would be the same, and no net diffusion would result. If, however, a change in vacancy concentration exists, such as it must around free surfaces, dislocations, etc., vacancies will jump more frequently in the direction of the low vacancy concentration. Consider two adjacent atom layers, as shown in Fig. 4.42. At a given site in layer x, the average rate at which a vacancy will diffuse to layer $x + a$ by an atom coming from layer $x + a$ is

TABLE 4.4

ACTIVATION ENTHALPIES FOR FORMATION OF VACANCIES h_f, MOTION OF VACANCIES h_j, AND OF SELF-DIFFUSION h_D*

Metal	h_f, ev	h_j, ev	h_D, ev
Copper	0.9	0.7	2.0
Gold	1.0	0.7	1.85
Platinum	1.4	1.1	2.96
Aluminum	0.75	0.45	1.4
Nickel	1.4	—	2.7
Iron	—	—	2.6
Tungsten	—	—	5.2

* All values except that for tungsten have been taken from Cottrell, (1958). Tungsten is from Langmuir (1934).

given by the product of the fractional concentration of vacancies in layer x, $c_v(x)$, by the frequency of attempts of the atom to jump, ν, by the probability that an atom exists in an energetically favorable site in layer $x + a$, times the number of these sites, say three, times the probability that the atom which will jump into the vacancy will have the required jump enthalpy and jump entropy $e^{s_j/k}e^{-h_j/kT}$:

Vacancy diffusion rate per site from 1 to 2

$$= 3c_v(x)\nu[1 - c_v(x + A)]e^{s_j/k}e^{-h_j/kT}. \qquad (4.36)$$

Combining Eq. 4.36 with the corresponding rate in the reverse direction and expressing the result as a vacancy flux f_v per unit time per unit area and the gradient of the vacancy concentration, in vacancies per cubic centimeter ($C_v = c_v/a^3$), gives (Prob. 4.31)

$$f_v = -3\nu a^2 e^{s_j/k}e^{-h_j/kT} \text{ grad } C_v. \qquad (4.37)$$

Dropping the activation entropy factor, which is usually of order unity (Prob. 1.12), gives

$$f_v = -D_v \text{ grad } C_v, \qquad (4.38)$$

where $D_v = 3\nu a^2 \exp(-h_j/kT)$ is the diffusion constant of vacancies.

◄ *Self-diffusion* is the diffusion of similar atoms, such as isotopes, through each other. In this case, the vacancies play the role of carriers of isotopes, while the rate of diffusion is governed by the concentration gradient of the isotopes C_I. If the vacancies are in thermal equilibrium (i.e., there are no significant vacancy concentration gradients), the flux of isotopes, f_I, is given by (Prob. 4.32)

$$f_I = -D_I \text{ grad } C_I = -3\nu a^2 e^{-(h_f+h_j)/kT} \text{ grad } C_I. \qquad (4.39)$$

◀ By considering the random walk of a vacancy, we can calculate the time necessary for a vacancy to diffuse a distance of the order of l. From dimensional analysis, this ought to be

$$t = l^2/D_v. \tag{4.40}$$

(The actual coefficient is indeed unity. See, for example, Schoeck, 1961, p. 69.)
◀ The measured enthalpies of formation and of motion of vacancies, and the activation enthalpy of self-diffusion are given for various pure metals in Table 4.4.
◀ The enthalpies of formation of interstitials have not been measured. For copper, the enthalpy of formation of an interstitial has been estimated by Huntington (1953) to be between 4 and 5 ev. The enthalpy of motion of an interstitial is much less than that of a vacancy; experimental evidence as well as theoretical estimates suggest that it is of the order of 0.25 ev for copper (Huntington, 1953). ▶

4.14 ROLE OF THERMAL ACTIVATION IN PLASTIC DEFORMATION

In Section 1.6 it is stated that plastic deformation is temperature insensitive because the motion of dislocations in a crystal is governed primarily by a threshold stress—the critical resolved shear stress for yielding—rather than by temperature. We are now in a position to consider this statement more carefully.

Just as in the case of vacancies, one might attempt to estimate an equilibrium concentration of dislocations. As indicated in Section 4.5, since the strain energy of line dislocations is almost equal to the free energy, they are always unstable, and no concentration can exist in thermodynamic equilibrium. So we turn to the question of an equilibrium concentration of small dislocation loops. Since the total strain energy of a dislocation loop decreases with decreasing size, and since the number of ways of arranging a loop in a crystal rapidly increases with decreasing size, an equilibrium concentration would appear to be a possibility.

In a crystal with no lattice friction stress, all dislocation loops are unstable and tend to collapse in the absence of an external stress, so we consider equilibrium under an applied shear stress σ. The minimum radius R of a dislocation loop which does not collapse is (Prob. 4.33)

$$R = T/\sigma b = Gb/2\sigma. \tag{4.41}$$

Since the loop is generated under this stress, the stress will do an amount of work $\sigma b \pi R^2$ during formation of the loop, reducing the activation energy by that amount to

$$u_a \approx \pi R G b^2 - \pi \sigma b R^2 = \pi R G b^2 / 2. \tag{4.42}$$

The entropy due to the dislocation itself is negligible (Cottrell, 1953, p. 39); the only important contribution, as mentioned above, comes from the number of ways W the loop can be arranged in the crystal. To have at least one stable dislocation loop of a radius given by Eq. 4.41, a decrease of free energy requires that

$$u_a - kT \ln W \leq 0.$$

With an applied stress of even $G/100$, this turns out to be possible only if the diameter of the crystal is 10^{3000} A (or about 10^{3000} diameters of the solar system—the conversion factor is negligible) (Prob. 4.34).

One might inquire whether such loops would be occasionally generated by chance, even though they are not thermally stable. But this can be seen to be extremely unlikely by comparing the energy given by Eq. 4.42 to the available thermal energy kT at room temperature. There is one exceptional case in which nucleation of dislocations is observed, but only at the feather edge of an electron-transparent stainless-steel foil, where the thickness of a few angstroms limits the stress field and hence the required energy of the dislocations sufficiently to allow thermally activated dislocation generation (Whelan et al., 1957).

In normal circumstances, however, it should now be clear that dislocations cannot be thermally generated from a perfect lattice but must be initially present in crystals as growth accidents in geometrical configurations which are stable, such as the networks of Figs. 4.30 and 4.38.

◀Next we may inquire about the role of thermal activation in the motion of dislocations in crystals with no internal stress, i.e., in crystals having only very few dislocations, where the dislocation has to overcome only the lattice friction stress. If we consider, for instance, the measured friction stress for copper, $\sigma_0 = 0.2$ kg/cm^2 (from Table 4.2), we can compute the length of dislocation line which can advance one interatomic spacing under a negligibly small stress as a result of thermal fluctuation of magnitude equal to 0.75 electron volt. This turns out to be a length of about 0.1 mm (Prob. 4.35), leading to the conclusion that if lattice friction were the only resistance to the motion of dislocations, then thermal activation would play a significant role in crystal plasticity. Such idealized plastic deformation involving the unobstructed motion of straight dislocations, however, cannot produce the observed large plastic strains (Prob. 4.26).▶

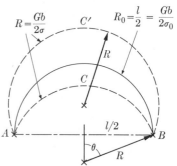

Fig. 4.43. Two positions of equilibrium of the characteristic segment of a Frank-Read dislocation mill under an applied shear stress $\sigma < \sigma_0 = Gb/l$.

We finally inquire about the possible role of thermal activation in the generation of dislocations at a steady rate to replace those which have either left the crystal or have been blocked at regions of high internal stress. The Frank-Read mill of Fig. 4.43 is taken as an example of such generation mechanisms. To bring the dislocation to its unstable semicircular threshold configuration requires a critical shear stress σ_0 of (Prob. 4.27)

$$\sigma_0 = Gb/l. \quad (4.25)$$

If a larger shear stress σ is applied, the dislocation is unstable and will expand. If a smaller stress is applied, the dislocation will bow out only to a shorter length ACB shown by the dashed curve with a larger radius of curvature R. If the dislocation is then to expand, it must be taken through its semicircular configuration into a configuration $AC'B$ of the same radius of curvature R; beyond this the applied stress σ can expand the loop freely. The activation energy which has to be supplied to take the dislocation from the shape ACB to the shape $AC'B$ is the difference between the line energies of these two configurations, minus the work done by the applied stress in sweeping out this area (Prob. 4.36):

$$u_a = \left(\frac{G}{\sigma}\right)\left(\frac{Gb^3}{2}\right)\left(\frac{\pi}{2} - \sin^{-1}\frac{\sigma}{\sigma_0} - \frac{\sigma}{\sigma_0}\sqrt{1 - \frac{\sigma}{\sigma_0}}\right). \quad (4.43)$$

If thermal activation is to play a significant role, this activation energy cannot exceed about one electron volt. If we equate u_a to one electron volt and solve for σ/σ_0, we find that this ratio cannot differ from unity by more than a few parts in 10^4 (Prob. 4.37). Hence, we conclude that thermal activation cannot play a role in the regeneration of dislocations by the Frank-Read mechanism, since the applied stress must reach almost fully the critical stress given by Eq. 4.25.

Once the critical shear stress has been applied to a crystal, deformation will commence at a finite rate. If the temperature of the test is increased under these conditions, the rate of deformation will normally increase in proportion to a Boltzmann factor. The same higher rate of deformation could be obtained, on the other hand, at the low temperature by increasing the applied shear stress a small amount over the required critical threshold value. Thus, although thermal activation does not play a significant role in the generation of dislocations or in the initiation of plastic deformation, it does, nevertheless, govern its rate when the necessary critical stress is applied. More complete discussions of thermal activation in dislocation processes have been given by Nabarro (1952) and also by Friedel (1959).

4.15 SUMMARY

Plastic deformation in crystalline materials is accomplished by the generation and motion of dislocations. A dislocation is a line imperfection delineating a surface inside the crystal across which there has been a relative translation. Because of this, a dislocation cannot end inside the crystal, but must end on a free surface, another dislocation, or on itself. Dislocations may thread through the crystal in an irregular way, or they may lie on definite crystallographic planes.

The strength of a dislocation is measured by the resultant of a circuit around the dislocation. This vector resultant, along with the unit tangent vector indicating the direction of the circuit, is known as the Burgers vector. It is the most unchanging property of a dislocation. When the dislocation lies normal to its Burgers vector, it is called an edge dislocation. Edge dislocations are characterized by an extra half plane terminating on the dislocation line. When the dislocation line lies parallel to its Burgers vector, it is known as a screw dislocation. Screw dislocations do not possess an extra half plane. If the Burgers vector of a dislocation is equal to an identity vector of the lattice, it is a perfect dislocation. The passage of perfect dislocations through a crystal restores full order.

◀ At a node, the sum of the Burgers vectors of all dislocations is zero if the directions of all dislocations are defined as positive, pointing away from the node. Every point on a dislocation is a degenerate node. ▶

The motion of an edge dislocation in the direction of its Burgers vector is called *glide*. Such motion merely translates the edge of the extra half plane parallel to itself on the slip plane. Screw dislocations are not confined to any particular plane, and all motions of a screw dislocation are glide. The displacement of an edge dislocation normal to its Burgers vector, out of its slip plane, is called *climb*. This type of motion requires diffusion of material, and therefore cannot take place at temperatures much lower than half the absolute melting temperature.

When Λ dislocations per unit area all move an average distance λ through the crystal by glide, the resulting plastic strain is

$$\gamma = b\lambda\Lambda. \qquad (4.3)$$

Dislocations, being lines separating slipped from not slipped regions, are associated with large localized stresses of order

$$\sigma \approx Gb/2\pi r. \qquad (4.5)$$

The stress field of an edge dislocation has components that produce local volume changes, while the stress field of a screw dislocation does not produce any local volume change, to a first approximation. When the local volume changes of an edge dislocation are integrated over the whole crystal, however, no net volume change arises outside of the dislocation core. The total volume change due to core effects in a heavily deformed crystal is no more than one part in 10^4. Therefore, plastic deformation is not accompanied by a volume change. The elastic strain energy stored around a dislocation,

$$\mathcal{E} \approx Gb^2/2, \qquad (4.9)$$

is of the order of 4 to 5 ev for a dislocation length equal to one lattice spacing. The introduction of a dislocation of any meaningful length into a crystal will always raise its free energy. This is because the entropy contribution will always be smaller than the internal energy contribution to the free energy. Hence, dislocations cannot be produced by random thermal motion.

The strain energy of a crystal will be reduced if the dislocations it contains assume the shortest possible length between two points, compatible with constraints imposed by the specific crystal structure and other dislocations. This makes the dislocations behave as if they were under a line tension,

$$T = \mathcal{E} \approx Gb^2/2, \qquad (4.10)$$

which is equal to the strain energy per unit length of a dislocation.

4.15 SUMMARY

When a dislocation moves with a constant velocity, the total energy of the crystal increases by a Lorentz factor. If the increase in energy is considered as a kinetic energy of the dislocation, it is possible to ascribe a fictitious distributed mass per unit length to the dislocation:

$$\mu \approx \rho b^2/2. \tag{4.13}$$

If the applied external forces on a crystal result in the glide motion of dislocations, it is convenient to consider the work done by the external forces as the displacement of a fictitious distributed force acting normal to the dislocation line throughout its motion. This line pressure is equal to the product of the magnitude of the Burgers vector and the resolved shear stress on the slip plane in the direction of the Burgers vector, or

$$F = \sigma b. \tag{4.14}$$

◀The distributed climb force on an edge dislocation is of a form identical with that of Eq. 4.14, where the stress component, however, is equal to the normal stress acting across the extra half plane.▶

The stress fields of dislocations usually interact mutually to produce attractive or repulsive forces between dislocations.

◀A surface free of stress will attract a dislocation to itself as if an image dislocation of opposite sign to that inside the crystal existed in space. Dislocations which attract may eventually come together to fuse and form a new dislocation of lower energy. Alternatively, unit dislocations may break up into imperfect partial dislocations to reduce their energy. These partial dislocations are held together at a distance from each other by a *stacking fault*. Such ribbonlike dislocations are encountered in close-packed structures such as face-centered cubic and hexagonal close-packed.

◀The periodic nature of the crystalline lattice will result in the existence of a lattice friction stress opposing the motion of a dislocation. The true lattice friction stress is usually negligibly low for f.c.c. and h.c.p. metal crystals; it is, however, high in b.c.c. transition metals and in covalent crystals such as silicon, alumina, and diamond.▶

In a real crystal, the velocity of a dislocation depends very sensitively on the net stress in excess of the friction stress. The friction stress necessary to move a dislocation will usually be higher than the theoretical lattice friction stress, because of impurities or other minor obstacles. During plastic deformation, dislocations have to intersect each other. This produces either short steps on a dislocation, which are called *jogs* or *kinks*, or a series of edge dislocation dipole trails.

When, in plastic deformation, dislocations run into each other and become entangled, it is necessary that new dislocations be formed. Dislocations are generated by dislocation mills composed of finite segments of dislocations which can expand as a double spiral. Such dislocation mills can produce many dislocations in one slip plane when a critical shear stress

$$\sigma = Gb/l \tag{4.25}$$

is present. Screw dislocation segments can glide out of their slip plane along any other plane that contains the Burgers vector of the dislocation. When such segments of screw dislocations cross glide once again back into a slip plane parallel to the initial plane, they can act as dislocation mills and infect neighboring slip planes with dislocations. This process is known as dislocation multiplication. Observable plastic deformation in a crystal is produced at a critical level of resolved shear stress on a slip plane in a slip direction where a certain rate of dislocation multiplication can be maintained. This stress is usually the yield stress of a crystal in the absence of a high lattice friction stress.

When a sequence of dislocations released by a dislocation mill is intercepted by an obstacle such as a grain boundary, a dislocation pile-up is formed. A dislocation pile-up concentrates the applied shear stress at the obstacle by a factor equal to the number of dislocations in the pile-up.

When edge dislocations are arranged on parallel slip planes to produce an array in a plane normal to the slip planes, they result in a lattice tilt of the two parts of the crystal they separate, with the rotation axis being parallel to the dislocations. If two planar sets of parallel screw dislocations are arranged at right angles to each other, they result in a lattice twist across the plane about an axis to the plane.

In some cases, slip takes place inside a band nearly normal to the boundaries of the band. This process, known as kinking, produces a lattice rotation of the inside relative to the outside of the band.

Certain crystals can undergo a homogeneous constant shear strain parallel to certain crystallographic planes in certain crystallographic directions, producing a mirror image of the lattice across a plane. Such an operation is known as deformation twinning. The twinning shear strain is a constant and is determined by the symmetry properties of the lattice.

◀The most convenient analog for the study of edge dislocations is a raft of small soap bubbles floating on a soap solution. Such a raft of soap bubbles of 1.2-mm diameter will have dislocations in them which move and interact in a two-dimensional analog of a copper crystal.▶

The proper understanding of the effects of temperature and stress on plastic deformation requires an understanding of the nature of processes which can be thermally activated.

Vacancies, which are the simplest defects that play a role in deformation, raise the enthalpy of the crystal, but also produce a rise in entropy, resulting in a reduction of free energy for an equilibrium configuration. The equilibrium concentration of vacancies, c_v, is given by

◀ $$c_v = n/N = e^{-h_f/kT}. \qquad (4.34)$$ ▶

The diffusion of vacancies through the lattice makes the climb of edge dislocations possible. Such diffusion of vacancies requires a certain activation enthalpy which is normally about equal to the enthalpy of formation of a vacancy.

Since the line energy of dislocations is almost exclusively a free energy and has a magnitude of several electron volts per atom length, thermal activation cannot play a role in the generation of dislocations. Therefore, dislocations must be generated by dislocation mills under an applied shear stress. Once a critical shear stress for slip is applied, however, the rate at which dislocations move through the crystal is governed by the level of thermal agitation.

REFERENCES

The classical treatise of the geometrical aspects of plastic deformation in crystalline materials, summarizing the pioneering work, is Schmid and Boas (1935). A lucid and partly qualitative presentation of the effects of dislocations in crystals, written especially for the beginner, is given by Orowan in a book edited by Cohen (1954). The book by Cottrell (1953) is sufficiently detailed to be used by both the beginner and the advanced student. It discusses the effect of dislocations on mechanical behavior rather more than does Read (1953), whose well-illustrated book places more emphasis on theoretical fundamentals, the geometry of dislocations in face-centered cubic crystals, and the application to grain boundaries.

An excellent general quantitative treatment for the beginner is provided by Friedel (1956), while a more advanced summary is given by Nabarro (1952).

AMELINCKX, S.	1958	"Dislocation Patterns in Potassium Chloride," *Acta Met.* **6,** 34–58.
AMELINCKX, S. DELAVIGNETTE, P.	1960	"Electron Optical Study of Basal Dislocations in Graphite," *J. Appl. Phys.* **31,** 2126–2135.
BLEWITT, T. H. COLTMANN, R. R. REDMAN, J. K.	1957	"Low Temperature Deformation of Copper Single Crystals," *Dislocations and Mechanical Properties*, Wiley, New York, p. 179.
BRAGG, W. L. LOMER, W. M.	1949	"A Dynamical Model of a Crystal Structure, II," *Proc. Roy. Soc. (London)* **A196,** 171–181.
BRAGG, W. L. NYE, J. F.	1947	"A Dynamical Model of a Crystal Structure, I," *Proc. Roy. Soc. (London)* **A190,** 474–481.
BRENNER, S. S.	1958	"Properties of Whiskers," *Growth and Perfection of Crystals*, Wiley, New York, p. 157.
BURGERS, J. M.	1939	"Some Considerations in the Field of Stress Connected with Dislocations in a Regular Crystal Lattice" II (Solutions of the Equations of Elasticity for a Nonisotropic Substance of Regular Crystalline Symmetry) *Proc. Kon. Nederlansche Akad. van Wettenschappen* **42,** 378–399.
CLAREBROUGH, L. M. HARGREAVES, M. E. WEST, G. W.	1955	"The Release of Energy During Annealing of Deformed Metals," *Proc. Roy. Soc. (London)* **A232,** 252–270.
COHEN, M. (ed.)	1954	*Dislocations in Metals*, American Institute of Mining and Metallurgical Engineers, New York.
CONRAD, H. HAYES, W.	1963	"Thermally Activated Deformation of the BCC Metals at Low Temperatures," *Trans. ASM* **56,** 249–262.

COTTRELL, A. H.	1953	*Dislocations and Plastic Flow in Crystals*, Oxford University Press, London.
COTTRELL, A. H.	1958	"Point Defects and the Mechanical Properties of Metals and Alloys at Low Temperatures," *Vacancies and Other Point Defects in Metals and Alloys*, Institute of Metals, London, pp. 1–40.
COTTRELL, A. H. BILBY, B. A.	1951	"A Mechanism for the Growth of Deformation Twins in Crystals," *Phil. Mag.* **42**, 573–581.
DASH, W. C.	1957	"The Observation of Dislocations in Silicon," *Dislocations and Mechanical Properties of Crystals*, Fisher et al., eds., Wiley, New York, pp. 57–68.
DASH, W. C.	1958	"Evidence of Dislocation Jogs in Deformed Silicon," *J. Appl. Phys.* **29**, 705–709.
ESHELBY, J. D.	1949	"Edge Dislocations in Anisotropic Materials," *Phil. Mag.* **40**, 903–912.
ESHELBY, J. D. READ, W. T. SHOCKLEY, W.	1953	"Anisotropic Elasticity with Applications to Dislocation Theory," *Acta Met.* **1**, 251–259.
FRANK, F. C.	1949	"On the Equations of Motion of Crystal Dislocations," *Proc. Phys. Soc.* **A62**, 131–134.
FRANK, F. C. READ, W. T.	1950	"Multiplication Processes for Slow Moving Dislocations," *Phys. Rev.* **79**, 772.
FRANK, F. C. STROH, A. N.	1952	"On the Theory of Kinking," *Proc. Phys. Soc.* **B65**, 811–821.
FRIEDEL, J.	1956	*Les Dislocations*, Gauthier-Villars, Paris. English trans., Addison-Wesley, Reading, Mass., 1964.
FRIEDEL, J.	1959	"Dislocation Interactions and Internal Strains," *Internal Stresses and Fatigue in Metals*, Rassweiler and Grube, eds., Elsevier, Amsterdam, pp. 220–263.
GILMAN, J. J.	1959	"Dislocation Sources in Crystals," *J. Appl. Phys.* **30**, 1584–1594.
GILMAN, J. J. JOHNSTON, W. G.	1957	"The Origin and Growth of Glide Bands in Lithium Fluoride Crystals," *Dislocations and Mechanical Properties of Crystals*, Fisher et al., eds., Wiley, New York, pp. 116–163.
GYULAI, Z. Z.	1954	"Strength and Plasticity Properties of NaCl Whiskers," *Z. Physik* **138**, 317–321.
HEAD, A. K.	1953	"Edge Dislocations in Homogeneous Media," *Proc. Phys. Soc.* **B66**, 793–801.
HUNTINGTON, H. B.	1953	"Mobility of Interstitial Atoms in a Face-Centered Cubic Metal," *Phys. Rev.* **91**, 1092–1098.
JOHNSTON, W. G. GILMAN, J. J.	1959	"Dislocation Velocities, Dislocation Densities, and Plastic Flow in Lithium Fluoride Crystals," *J. Appl. Phys.* **30**, 129–144.
JOHNSTON, W. G. GILMAN, J. J.	1960	"Dislocation Multiplication in Lithium Fluoride Crystals," *J. Appl. Phys.* **31**, 632–643.

Joos, G.	1950	*Theoretical Physics*, 2nd ed., Hafner, New York, pp. 571–580.
Koehler, J. S.	1941	"On the Dislocation Theory of Plastic Deformation," *Phys. Rev.* **60**, 397–410.
Langmuir, I.	1934	"Thoriated Tungsten Filaments," *J. Franklin Inst.* **217**, 543.
Maloof, S. R. Argon, A. S.	1962	"Deformation and Fracture of Tungsten Single Crystals at Low Temperatures," *AVCO Rept. No. TR 62-60*.
Nabarro, F. R. N.	1947	"Dislocations in a Simple Cubic Lattice," *Proc. Phys. Soc.* **59**, 256–272.
Nabarro, F. R. N.	1952	"The Mathematical Theory of Stationary Dislocations," *Advances in Physics* **1**, 269–394.
Orowan, E.	1934	"Crystal Plasticity III, On the Mechanism of the Glide Process," *Z. Physik* **89**, 634–659.
Orowan, E.	1942	"A Type of Plastic Deformation New in Metals," *Nature* **149**, 643–644.
Orowan, E.	1954	"Dislocations and Mechanical Properties," *Dislocations in Metals*, **103**, AIME Monograph, New York.
Peierls, R.	1940	"On the Size of a Dislocation," *Proc. Phys. Soc.* **52**, 34–37.
Polanyi, M.	1934	"On a Kind of Glide Disturbance that Could Make a Crystal Plastic," *Z. Physik* **89**, 660–664.
Read, W. T.	1953	*Dislocations in Crystals*, McGraw-Hill, New York.
Schmid, E. Boas, W.	1935	*Kristallplastizität*, Springer, Berlin. English trans., *Plasticity of Crystals*, Hughes, London (1950).
Schoeck, G.	1961	"Theories of Creep," *Mechanical Behavior of Materials at Elevated Temperatures*, J. E. Dorn, ed., McGraw-Hill, New York, pp. 79–107.
Silcox, J. Whelan, M. J.	1960	"Direct Observations of the Annealing of Prismatic Dislocation Loops and Climb of Dislocations in Quenched Aluminum," *Phil. Mag.* **5**, 1–23.
Stein, D. L. Low, J. R.	1960	"Mobility of Edge Dislocations in Silicon Iron Crystals," *J. Appl. Phys.* **31**, 362–369.
Stroh, A. N.	1958	"Dislocations and Cracks in Anisotropic Elasticity, *Phil. Mag.* **3**, 625–646.
Taylor, G. I.	1934	"The Mechanism of Plastic Deformation of Crystals, Part I, Theoretical," *Proc. Roy. Soc. (London)* **A145**, 362–387.
Whelan, M. J.	1959	"Dislocation Interactions in Face-Centered Cubic Metals, with Particular Reference to Stainless Steel," *Proc. Roy. Soc. (London)* **A249**, 114–137.
Whelan, M. J. Hirsch, P. B. Horne, R. W. Bollmann, W.	1957	"Dislocations and Stacking Faults in Stainless Steel," *Proc. Roy. Soc. (London)* **A240**, 524–538.

YOUNG, F. W., JR.   1962   "Elastic Plastic Transition in Copper Crystals as Determined by an Etch-Pit Technique," *J. Appl. Phys.* **32,** 1815–1820.

ZENER, C.   1952   "Theory of Diffusion," *Imperfections in Nearly Perfect Crystals*, W. Shockley et al., eds., Wiley, New York, pp. 289–316.

Note: Librarians list some books under *International, Conference, Congress, Symposium*, the city, or the sponsoring organization rather than under the name of the editor. They do not use *Proceedings* or *Transactions*.

PROBLEMS

4.1. With respect to the dislocation in Fig. 4.4(b), show that so long as the x_3 axis is chosen to coincide with the dislocation line, and the positive x_2 axis is contained in the extra half plane, the resultant of the Burgers circuit will always point in the positive x_1 direction.

◀ 4.2. Show that a dislocation cannot terminate inside a crystal. [*Hint:* Assume the dislocation to terminate inside a crystal. Construct a rectangular prism at the termination of the dislocation. The rectangular circuits of the edges of the prism will close on all surfaces through which the dislocation does *not* thread; the circuit on the surface through which the dislocation enters the prism will, however, not close. Show that this is inconsistent.]

◀ 4.3. The Burgers vector does not change along a closed dislocation loop. Does this mean that the nature of the dislocation is the same everywhere?

◀ 4.4. Show that any point on a dislocation line may be considered as a two-element node.

4.5. For the dislocation in the soap-bubble raft of Fig. 4.7, draw a Burgers circuit and evaluate its Burgers vector.

4.6. Sketch the slip planes on the four unit cells of Fig. 1.16 and label each plane with the proper Miller indices.

4.7. Derive Eq. 4.1 with the aid of Fig. 4.11.

4.8. Estimate the total length of dislocations per cubic centimeter for annealed and also for heavily cold-worked crystals.

4.9. Derive the shear stress components σ_{13} and σ_{23} of a screw dislocation from the assumed displacement u_3. Show that these stresses satisfy the equations of equilibrium.

4.10. Derive Eq. 4.5 from Eqs. 4.4a and 4.4b.

4.11. What is the expression for the stress σ_{33} in an edge dislocation?

4.12. The sign of some of the stress components in Fig. 4.12 can be obtained by reasoning from the nature of the edge dislocation. Obtain the sign of the other components by considering the allowable changes of the stress components given by the equations of equilibrium.

◀ 4.13. Derive the equation for the local volume change around an edge dislocation, Eq. 4.7.

◀4.14. Show that Eq. 4.7 does not lead to any net volume change around an edge dislocation.

4.15. Derive Eq. 4.8a by direct integration of the strain-energy density over the whole volume of a cylindrical crystal with the dislocation at its center.

4.16. Show by a qualitative argument that the energy of a dipole of dislocations made up of two parallel edge dislocations of opposite Burgers vectors, at a distance δ apart, is

$$\mathcal{E} = \frac{Gb^2}{2\pi(1-\nu)} \ln \frac{\delta}{r_0}.$$

See also Prob. 4.45.

4.17. Calculate the strain energy of a length b of a dislocation in electron volts.

4.18. Derive the equation for the line energy of a dislocation from the equation preceding Eq. 4.10.

4.19. Show that if a dislocation as shown in Fig. 4.11 is displaced in its slip plane by an amount x, the externally applied stress will do an amount of work on the crystal equal to σbx per unit thickness.

4.20. What difficulties may arise in using Eq. 4.14 or 4.15 for a screw dislocation due to the possibility of slip on several planes?

4.21. Why does the intersecting X–Y dislocation in Fig. 4.21 not develop a jog?

◀4.22. Visualize the dislocation reaction of Eq. 4.20 by drawing the Burgers vectors on a b.c.c. unit cell, and show that the reaction is energetically favored.

◀4.23. Show that the square of the magnitude of the Burgers vector on the left-hand side of the reaction in Eq. 4.21a is greater than the sum of the squares of the Burgers vectors on the right-hand side.

◀4.24. Derive the repulsive force between two partial dislocations, Eq. 4.22, and the equilibrium spacing, Eq. 4.23.

◀4.25. Assuming that some parts of the long straight lines in Fig. 4.29 are Cottrell sessile dislocations along $\langle 110 \rangle$-type directions, with the aid of the [010] direction given in the figure, determine the type of crystallographic plane of the foil.

4.26. Calculate the amount of plastic strain obtainable by exhausting the dislocations resulting from growth in a crystal of reasonable size.

4.27. Derive Eq. 4.25 and show that the stress necessary to hold the dislocation segment in equilibrium in its bowed-out position will pass through a maximum when the segment is semicircular in shape.

◀4.28. Show that an array of parallel screw dislocations in a plane is unstable.

◀4.29. By giving a dislocation pile-up such as shown in Fig. 4.39 a small virtual displacement, as discussed in the text, show that the concentrated shear stress at the obstacle due to n dislocations in a pile-up is $n\sigma$.

◀4.30. By following the steps from Eq. 4.30 to Eq. 4.33, show that the fractional concentration of vacancies in a lattice at temperature T is that given by Eq. 4.34.

◀4.31. By generalization from Eq. 4.36, show that the vacancy flux across any area is that given by Eq. 4.37.

◀4.32. Show that in self-diffusion, the flux of isotopes across any area is the relation given by Eq. 4.39.

4.33. Show that the critical radius of a dislocation loop under a shear stress σ is that given by Eq. 4.41.

◀4.34. To show the thermal instability of dislocation loops, estimate the size of the crystal needed for one loop to cause a decrease of free energy as follows:

(a) Show that the number of ways (including both translational and rotational coordinates) that a loop of radius R can be arranged in a crystal of linear dimensions L is of the order of $(L/b)^3 (R/b)^2$.

(b) Show that the free energy will decrease for one dislocation loop when $(L/b)^3 = (b/R)^2 \exp(\pi G b^2 R / 2kT)$.

(c) Show that for $G/\sigma = 100$, a crystal dimension of the order of 10^{3000} diameters of the solar system is required.

(d) For what stress could loops be thermodynamically stable in a crystal of reasonable size?

4.35. Show that the length of a straight dislocation which can be advanced by one interatomic spacing in copper against the friction stress, as a result of a thermal fluctuation of 0.75 electron volt, is about 0.1 mm (assume the applied stress is negligible).

◀4.36. Show that the activation energy required to move a dislocation loop from its stable equilibrium shape of ACB in Fig. 4.43 under an applied stress $\sigma < \sigma_0 = Gb/l$ to its unstable equilibrium shape of $AC'B$ is the relation given by Eq. 4.43.

◀4.37. Show that if thermal activation is to play a significant role, the stress ratio σ/σ_0 in Eq. 4.43 must differ from unity by no more than a few parts in 10^4.

4.38. Considering that a dislocation behaves as if it were under a line tension T and has a mass per unit length μ, show by direct comparison that a dislocation line of a given length l, pinned at its ends, will vibrate freely like a pretensioned, uniformly weighted string, with a natural frequency

$$\nu = (1/2l) \sqrt{T/\mu}.$$

Calculate the natural frequency of a dislocation segment of one-micron length in a copper crystal.

◀4.39. In Section 4.7 it is stated that the hardness of silicon, quartz, and diamond likely results from their high lattice friction stress, which would be due to the high energy and strongly directed nature of interatomic bonds of such crystals. Find a number of other crystals that fall into this category.

◀4.40. Estimate the change in vacancy concentration with applied hydrostatic pressure.

4.41. Clarebrough et al. (1955) have obtained calorimetrically a relationship between the stored energy of plastic deformation and compressive plastic strain in pure copper, shown in Fig. 4.44. Assuming that all the stored energy can be accounted for by the total line energy of dislocations in the crystal, estimate the dislocation density after 80% reduction in length.

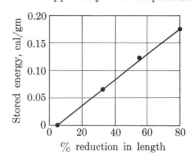

FIGURE 4.44

4.42. A long crystal of square cross section (1 × 1 mm) is plastically bent to a radius of curvature of $R = 20$ cm. Assuming all the bending to result from introduction of edge dislocations, calculate the final dislocation density if the Burgers vector is of length 3×10^{-8} cm.

4.43. Show that introducing some foreign atom which is attracted to the vacancies would or would not affect the rate of diffusion of vacancies in a lattice.

4.44. Silcox and Whelan (1960) observed about 10^{15} dislocation loops per cubic centimeter having a mean diameter of 138 A in aluminum foil quenched from 600°C. What is the activation energy for the formation of vacancies, assuming that all dislocation loops came from the collapse of vacancies?

4.45. Derive the expression in Prob. 4.16 for the energy of an edge dislocation dipole by considering the force exerted by a stationary dislocation on another as it is brought in from large distances.

CHAPTER 5

PLASTIC DEFORMATION IN CRYSTALLINE MATERIALS

5.1 SYNOPSIS

In Chapter 4 it is shown that the presence of dislocations in crystals can explain why the flow stress of ductile crystalline materials is normally far below the ideal shear strength calculated from the atomic bond strength. On the other hand, the stress necessary to move a dislocation in a ductile crystal is usually much lower than either the normally measured or practically desired yield strength. We shall, therefore, consider various ways in which the motion of dislocations can be impeded by obstacles, leading to an elevation of the yield strength.

Obstacles to slip that are effective from the very beginning of deformation will be termed sources of structural hardening. Examples are the introduction and critical dispersal of solute atoms, precipitates, and hard phases, and lattice damage resulting from irradiation or thermal treatment. These processes will distort the lattice and obstruct the motion of dislocations. Simple calculations will give some insight into the manner in which these imperfections can produce hardening.

Hardening also results from the deformation process itself. The quantitative explanation of strain-hardening is still incomplete, and one of the most challenging problems of dislocation theory. We shall qualitatively describe the most plausible picture of dislocation phenomena governing strain-hardening. Although thermal activation by itself cannot produce dislocations, it does control the rate of deformation when the stress is in the neighborhood of a critical value at a given temperature. We shall present an example based on the cutting of a dislocation by another, and show how such a mechanism can give rise to transient creep in metals. Various other rate-controlling mechanisms will also be briefly presented. Furthermore, we shall discuss a mechanism by which most of the work of deformation is dissipated into heat. We shall point out how the flow stress of polycrystalline materials can differ from that of single crystals, and discuss other distinguishing features of grain boundaries.

The anisotropy of strain-hardening, which manifests itself at reversals of deformation, will be dealt with briefly.

Under certain conditions, materials may become softer rather than harder in plastic deformation; this produces an instability under fixed load. Aging associated with impurity atoms, twinning, kinking, and thermal instabilities at very low temperatures can all lead to mechanical instabilities.

The possible existence of a simple equation of state to relate stress, strain, strain rate, and temperature will be considered. While an exact formulation of such a complex problem is clearly not possible, it will be shown that in many ranges a useful relation can be obtained, allowing the convenient correlation and extrapolation of data with a minimum of information.

Finally, representative data concerning the plastic behavior of various metals of engineering interest will be given, to show how the principles discussed in this chapter can be applied to obtain the high-strength alloys that are currently available.

5.2 STRUCTURAL HARDENING OF CRYSTALLINE MATERIALS

The presence of dislocations in a crystal makes plastic deformation possible at stress levels far below those required for the deformation of a perfect lattice. In fact, judging from studies of individual dislocations, it appears that dislocation motion in f.c.c., h.c.p., and some ionic crystals is possible at stress levels as low as a few ten-millionths of the shear modulus (Young, 1961).

These low values of flow stress now raise the inverse question: Why is it that some materials are so strong relative to the stresses required to move dislocations in a pure and nearly perfect crystal? In this section we consider three effects that make major contributions and are under the control of the metallurgist: the effects of solute atoms, precipitated phases, and sessile dislocation loops introduced by quenching or irradiation. When these mechanisms are effectively introduced, hardening arises from the defect structure before much dislocation motion occurs, and so they are termed *structural hardening*.

A. Solution Hardening. Hardening of a metal by alloying it with small amounts of soluble additions was one of man's earliest discoveries. Thus, for instance, the production of bronze, an alloy of tin in copper, ushered in the bronze age, while the alloying of small amounts of copper and antimony into easily fusible tin resulted in strong yet easily workable pewter, previously used as an inexpensive substitute for silver.

Such soluble additions appear as separate foreign atoms in a crystal and produce a misfit by dilatation or by distortion of the lattice around the addition. As we have seen in Section 4.6, the resulting stress fields will exert forces on any nearby dislocations. We first consider the simplest defect, one causing a dilatational misfit, namely a large impurity atom substituted for a regular atom of the lattice.

◀ *1. Dilatational Misfit.* If the parent lattice is considered as an isotropic elastic continuum, the principal effect of the substitutional introduction of an impurity atom with a radial misfit $\epsilon = \Delta r_0/r_0$ is a radial displacement (here taken as positive) decreasing sharply with increasing radius r (see Section 11.5):

$$u_r = \epsilon r_0 (r_0/r)^2. \tag{5.1}$$

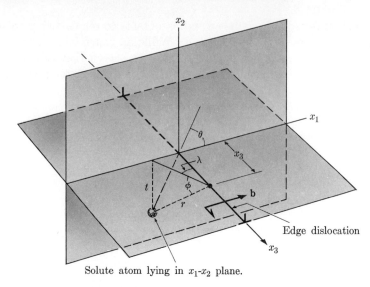

Solute atom lying in x_1-x_2 plane.

Fig. 5.1. Solute atom producing a force on a dislocation in a slip plane at a distance t from the atom.

This displacement produces only a compressive radial strain component and a tensile tangential component:

$$-\epsilon_{rr}/2 = \epsilon_{\theta\theta} = \epsilon_{\phi\phi} = \epsilon(r_0/r)^3. \qquad (5.2)$$

Transforming the strain components to the x_1, x_2, x_3 coordinates and applying the stress-strain relations, we can show that the maximum shear stress on a plane a distance t from the misfitting solute atom is (Prob. 5.1)

$$\sigma_{12\mathrm{max}} = \frac{96}{25\sqrt{5}} \epsilon G \left(\frac{r_0}{t}\right)^3. \qquad (5.3)$$

◀When a straight positive edge dislocation is parallel to the x_3 axis, it experiences an attractive glide force exerted by the misfitting solute atom. Different segments of the dislocation experience different forces. The total force on the dislocation due to the solute atom can be calculated by first obtaining the shear stress on the slip plane, normal to the dislocation line, then multiplying it by the magnitude of the Burgers vector, b, integrating over the whole length of the dislocation, and maximizing with respect to θ (see Fig. 5.1). This leads to (Prob. 5.2)

$$F_T = -\frac{3\sqrt{3}}{2} Gb\epsilon t \left(\frac{r_0}{t}\right)^3. \qquad (5.4)$$

When the solute atom is immediately under the slip plane ($t = r_0 = b/2$), the maximum total force is given by

$$F_{T\mathrm{max}} \approx \epsilon b^2 G. \qquad (5.4\mathrm{a})$$

The maximum total force can be taken as a measure of the solute dislocation interaction.

◀ It is not difficult to see that a misfitting solute cannot produce a net force on a straight screw dislocation similarly situated in the slip plane. For this reason, point defects causing only a spherically symmetrical volume misfit block dislocations only by blocking their edge segments (Prob. 5.3).

◀ **2. Distortional Misfit.** Few solute atoms cause only a spherically symmetrical dilatation. To some extent, substitutional solute atoms, and to a large extent, interstitial solute atoms, will give strongly directional lattice distortions. A good example of such lattice distortions is the interstitial solute carbon atom in body-centered cubic iron. In body-centered cubic metals, the interstitial sites are of tetragonal shape. When carbon atoms situate themselves in these interstices, the resulting strong distortions transform the lattice locally from cubic to tetragonal symmetry, as shown in Fig. 5.2. The resulting displacements are not spherically symmetrical, and strong shear stresses occur in the neighborhood of the distorted lattice. Such distortions in general interact not only with the normal and shear stress components of edge dislocations, but also with the shear stress of screw dislocations. Therefore, the introduction of interstitial solute atoms into a lattice will result in greater hardening than that due to solutes which produce only spherically symmetrical, or dilatational, misfit.

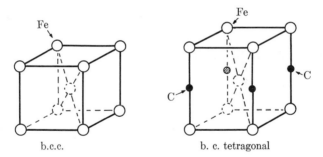

Fig. 5.2. Distortion of body-centered cubic lattice of α-iron by insertion of carbon atoms in interstices, resulting in a body-centered tetragonal lattice.

◀ The maximum total force between a dislocation and a solute atom causing directional distortions is of a form very similar to that of a solute producing only volume misfit, given by Eq. 5.4a. In this case, ϵ would be interpreted as the difference between the local maximum and minimum strain components at the boundary of the solute (Fleischer, 1962).

◀ **3. Stiffness Misfit.** Hardening due to a difference in stiffness between the solute atoms and the matrix can be visualized best by considering the limiting case of a solute of no stiffness, namely a free surface. In Section 4.6, it is shown that a free surface will exert a force on a dislocation through the effective image force due to an imaginary dislocation beyond the surface so situated as to leave zero stress at the free surface. The analysis of a small curved surface, such as a vacancy (Bullough and Newman, 1962), or a partial surface, such as an atom of different modulus, is more difficult to evaluate. An estimate by Fleischer (1961) finds the interaction due to stiffness change between solute and matrix atoms to be of the same order of magnitude as the interaction due to volume misfit in many prominent cases.*

◂The interaction forces arising from dilatational, distortional, or stiffness misfit produce hardening only if they exert a net retarding force on the dislocation. If the dislocation could adjust itself so that a maximum force occurred at intervals l along its length, the contribution to the yield strength in shear would be, from Eq. 4.14,

$$k = F_{T\,\text{max}}/bl.$$

Actually the random distribution of solutes and the finite line tension of a dislocation prevent it from accommodating to the steepest part of the potential hill due to each solute atom, and thus, experiencing $F_{T\,\text{max}}$. Indeed if the line tension were infinitely great, so that the dislocation remained perfectly straight, it would encounter as many positive as negative forces, and the hardening would be zero. Mott and Nabarro have taken these factors into account (see Cottrell, 1953, p. 125).

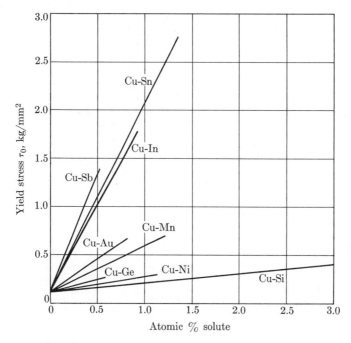

FIG. 5.3. Effect of solute concentration on solution hardening of copper. (Data from Linde, Lindell, and Stade, 1950.)

---

* The separation of the dilatational misfit, distortional misfit, stiffness misfit, and surface tension effects (partial collapse of a vacancy) is rarely possible. They have been discussed here separately for reasons of clarity.

5.2 STRUCTURAL HARDENING OF CRYSTALLINE MATERIALS

In terms of the approximate line tension Gb^2, they find for a fractional concentration of solutes $c < 10^{-2}$ a contribution to the yield strength in shear of

$$\frac{k}{G} = \frac{1}{2}\left(\frac{F_{T\,\max}}{Gb^2}\right)^{4/3} c[2c^{2/3}\ln^4 c]^{1/3}. \tag{5.5}$$

The term in brackets in Eq. 5.5 will be approximately equal to 4 and will not vary much for concentrations between 10^{-3} and 10^{-2}. Thus, substituting the value for $F_{T\,\max}$ from Eq. 5.4a gives

$$k/G \approx 2\epsilon^{4/3} c. \tag{5.6}$$

◀ The dependence of the yield stress on the first power of solute concentration c is shown for copper in Fig. 5.3, while Fig. 5.4 shows that the dependence of the solution hardening rate on the misfit parameter ϵ compares quite well with the $\frac{4}{3}$ power predicted by the rigid solute theory (Linde, et al., 1950; Linde and Edwardson, 1954).

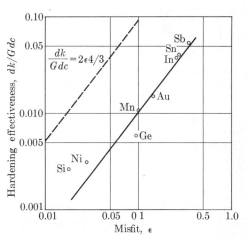

FIG. 5.4. Effect of atomic misfit of solute on the solution hardening rate of copper. The misfit parameter ϵ is measured as the fractional change in lattice parameter of the solid solution as a function of solute concentration. (Data from Linde and Edwardson, 1954.)

◀ As the concentration of solute is increased, precipitation may occur in the form of platelets one or several atomic layers in thickness. These platelets, called Guinier-Preston (G-P) zones (see for example, Preston, 1938) are coherently attached to the solvent lattice. The elastic coherency strains of such relatively widely separated platelets will interact strongly with dislocations and will produce a stronger hardening effect than mere solution-hardening. We shall not discuss these forms here any further, but turn instead to the extreme case in which precipitation has occurred as a distinct and identifiable particle of a second phase with arbitrary shape. ▶

B. Precipitation Hardening. Besides solute hardening, another, and industrially equally important, method of producing hardening is by precipitation from a supersaturated solid solution, obtained by quenching from an elevated temperature. Aging at intermediate temperatures allows the solute to precipitate. If the aging treatment is controlled to give a large number of nuclei before much growth has occurred, a uniformly dispersed collection of closely spaced small precipitates is obtained. Such precipitates may produce hardening in the same way that solute atoms do (Fig. 5.5a). In addition, they may prevent dislocations from passing through them either by a high hardness of their own or because their lattices are not coherent with the matrix (Fig. 5.5b or c).

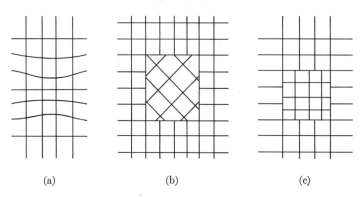

(a)   (b)   (c)

Fig. 5.5. Coherent (a) and incoherent (b and c) precipitates in a lattice.

The impeded dislocations in the matrix will be forced to extrude through the free spaces between the precipitates if the precipitates are impervious to dislocations. Thus, the hardening effect in this mechanism is due to the stresses required to extrude dislocations through the gaps between precipitates. The shear strength would then be identical with the critical shear stress for the operation of a Frank-Read dislocation mill with the critical length being the spacing l between precipitate particles, i.e., from Eq. 4.25,

$$k = Gb/l. \tag{5.7}$$

To obtain maximum hardening, the critical distance l must be made as small as possible. In aluminum alloys, for example, the maximum obtainable b/l ratio is in the range of 1/200 to 1/100. If the spacing between precipitates is decreased too much without increasing the mass fraction of precipitates, the accompanying decrease in size of the precipitates will cause a reversion to the solution hardening described above. The mass fraction in turn is limited by the high-temperature solubility of the hardening constituent. That is, if the spacing l is decreased without decreasing the size of the precipitates, it is necessary to increase the mass fraction of the precipitating phase, and the point will be reached where it can no longer be put in solution without melting the alloy. If the precipitate does not go into solution during heat treatment, it will tend to agglomerate into large and widely separated clusters, which again fail to harden the material. Thus, for ideal precipitation hardening, a solubility is needed which varies markedly within the range of temperatures in which the alloy can be handled, and such that the uniformly dispersed precipitates will be as resistant as possible to the passage of dislocations, through lack of coherence with the matrix, through a high hardness of their own, or through their high misfit stresses.

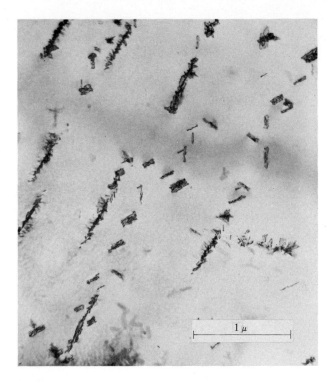

Fig. 5.6. Carbide precipitates in an electron-transparent iron foil. (Leslie, 1961. Courtesy of Pergamon Press.)

If the aging treatment is carried too far, the precipitate clusters will grow larger in size and their separation will increase with an accompanying drop in hardening. Figure 5.6 shows carbide precipitates in an electron-transparent iron foil (Leslie, 1961).

C. Quench-Hardening, Radiation Hardening. If a metal is raised to a temperature close to its melting point, a relatively large vacancy concentration can be held in thermal equilibrium, as discussed in Section 4.13. After rapid quenching, the vacancies in supersaturation will condense into vacancy clusters or disks inside the material instead of escaping to the free surfaces. Such vacancy disks, when large enough (Prob. 5.4), collapse to form sessile dislocation rings, as shown for quenched aluminum in Fig. 5.7.

Irradiation of metals with charged or uncharged heavy particles produces very similar effects. In the latter case, when the radiation is made up of alpha particles, appreciable quantities of helium gas can also be entrapped in the metal (Barnes and Mazey, 1960). Nuclear reactions may produce similar gas entrapment.

The sessile dislocation loops which form from the vacancies produced by quenching or irradiation are strong obstacles to dislocation motion. They produce directional lattice distortions which result in internal stresses of an effective range

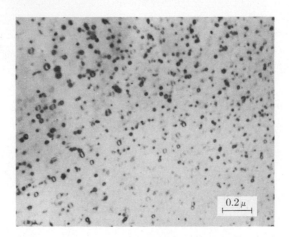

Fig. 5.7. Sessile dislocation loops formed by vacancy condensation in quench-hardened aluminum. (Silcox and Whelan, 1960. Courtesy of Taylor-Francis.)

equal to the size of the loops. In addition, they produce jogs in glide dislocations which cut through them (see Section 4.8). As will be seen in Section 5.4, this can contribute to hardening.

5.3 STRAIN-HARDENING IN SINGLE CRYSTALS OF SINGLE SLIP ORIENTATIONS

The most challenging problem of dislocation theory has been, and to a large extent, still is, the quantitative explanation of strain-hardening (work-hardening). Although no quantitative theory can yet be given for strain-hardening, most of its features can be explained qualitatively.

In high-purity crystals of close-packed metals the study of the behavior of individual dislocations under very low amplitudes of alternating stress suggests that neither the initial yield strength nor the flow stress is much influenced by a lattice friction stress (Young, 1961). Hence, the cause of the flow stress in close-packed crystals must be sought in various interactions between dislocations. In body-centered cubic crystals (Conrad and Hayes, 1963) and in ionic salts (Johnston and Gilman, 1959), however, the lattice friction stress may be substantial.

The simplest case is the hardening of pure ductile face-centered cubic crystals under conditions of single slip, where the resolved shear stress on a good slip system is maximized, as shown in Fig. 5.8*. Even in the pre-yield region, dislocations begin to move, as was observed by Young (1962).

---

* Directions in a crystal are normally presented as a stereographic projection of the points of intersection of these directions with a large reference sphere having the crystal at its origin. For crystals of cubic symmetry $\frac{1}{24}$ of the stereographic projection circle, the "triangle" shown in Fig. 5.8, is sufficient to represent all crystallographic directions (Barrett, 1953).

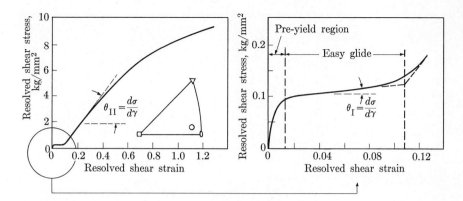

FIG. 5.8. A typical stress-strain curve for a single crystal of copper, oriented for single slip, and deformed at room temperature. (Data from Diehl, 1956.)

The screw segments of these dislocations will occasionally be subject to double cross glide, as outlined in Section 4.10. This should tend to introduce jogs into screw dislocations, thus slowing them down, or stopping them altogether and thereby inactivating the most favorable dislocation mills. Under normal conditions of testing, where the ends of the crystal are displaced at a steady rate, a specimen will continue to extend elastically and the stress will rise. With increasing stress, less favorable dislocation mills then begin to become active, and the rate of double cross glide increases until the rate of production of new dislocation mills is sufficient to counterbalance the rate of inactivation of dislocation mills. At this stage, the number of available mobile slip dislocations will be large enough to furnish a rate of plastic strain which is demanded by the imposed velocity of the ends of the crystal without any further elastic extension. This will be the yield point of the crystal where the stress will appear to level off. In copper crystals, about 10^6 mobile dislocations/cm^2 are required for plastic deformation to commence. If the crystal is of high perfection and has a much lower dislocation density, there will have to be rapid dislocation multiplication in the pre-yield region to make up for the deficiency (Young, 1962).

In iron crystals which exhibit a yield phenomenon, a front of very light plastic deformation travels through a substantial part of the crystal prior to general yielding (Paxton and Bear, 1955). In a crystal of LiF, several slip bands will develop in the pre-yield region (Gilman and Johnston, 1957). Once the requisite number of mobile dislocations is present and the rate of production of effective dislocation mills is sufficient to keep the number of mobile dislocations constant, slip will develop without much increase in the applied stress, as shown in Fig. 5.8. This is the "easy-glide" phenomenon occurring in favorably oriented f.c.c. crystals, h.c.p. crystals, and some ionic salts.

Beyond this stage, the development of slip will show two limiting cases. In hard ionic crystals of LiF, or in particular, MgO, a few slip bands will be nucleated which will then grow laterally at their margins by double cross glide—constantly

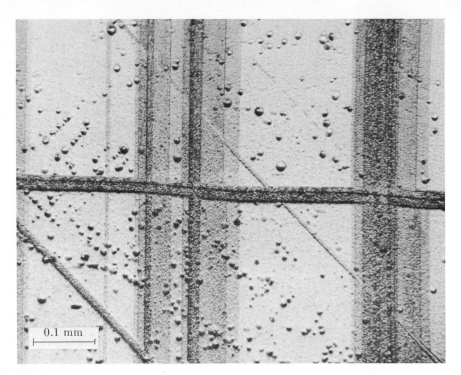

Fig. 5.9. Several stages of widening of slip bands in a compressed MgO crystal. The dislocations inside the bands have been revealed by etching.

infecting the undeformed parts of the crystal with dislocations. In this case, all the slip activity is within a very narrow region of the margins of the broadening band where the density of dislocations rises from a low value of about 10^4 to about 10^8 per square centimeter. Once the boundary of a slip band has moved through a region and it has become part of the interior of a band, it does not undergo much further shear so long as the band can continue to expand into undeformed crystal (Argon and Orowan, 1964) (Fig. 5.9). This type of deformation continues until no undeformed region remains (Johnston and Gilman, 1959).

In metal crystals the "easy-glide" deformation shows somewhat different features. Straining is more homogeneous, and the depth of individual slip lines steadily increases (Seeger et al., 1961). In both of these apparently different modes of deformation the density of dislocations rises linearly with plastic strain (Johnston and Gilman, 1959; Young, 1962; Keh and Weissmann, 1963), and according to Eq. 4.3 the dislocations have a constant mean free path throughout this type of deformation. As tabulated in Table 5.1, these mean free paths are substantial and correspond closely to the best measurements of the length of individual slip lines on face-centered cubic crystals (Seeger et al., 1961).

The only hardening in such crystals which are oriented for single slip is due to dislocation dipoles and other similar defects which result from the double cross-

TABLE 5.1

Mean Free Paths of Easy-Glide Dislocations in Various Crystals

Material	Mean free path, mm	Reference
LiF	0.18	Johnston and Gilman (1959)
MgO	0.12	Argon and Orowan (1964)
NaCl	1.0	Padawer (1963)
Cu	1.2	Young (1962)
Fe*	0.8	Keh and Weissmann (1963)

* Iron does not exhibit easy glide, and has a high initial yield strength; it does, however, exhibit a constant mean free path of dislocations in the early portion of plastic deformation.

glide operation in a manner illustrated in Fig. 5.10. Such dipoles, as shown in Fig. 5.11, are very similar to the sessile dislocation loops produced by radiation damage. They have no long-range stress field, but set up strong directional distortions around them which effectively block the motion of other dislocations (Gilman, 1962). The rate of hardening, $\theta_I = d\sigma/d\gamma$, in this range may be as low as $10^{-4} G$, but is normally somewhat higher. The dislocation dipoles are also the chief cause of the irreversibility of slip on individual slip planes in single-slip situations. When the stress is removed after some plastic deformation has occurred, dislocations cannot retrace their steps backward, and the maze of the three-dimensional dislocation entanglement with its multitude of bridging dislocation dipoles remains essentially intact. There may, however, be some denser dislocation groups pressed against strong obstacles that could produce a backward plastic strain upon removal of the applied stress and make reverse deformation initially

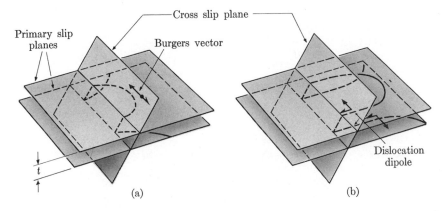

Fig. 5.10. Formation of dislocation dipoles by double cross glide of screw dislocations. Distance t between the two primary slip planes is too small for the members of the dipole to glide past each other under the applied stress.

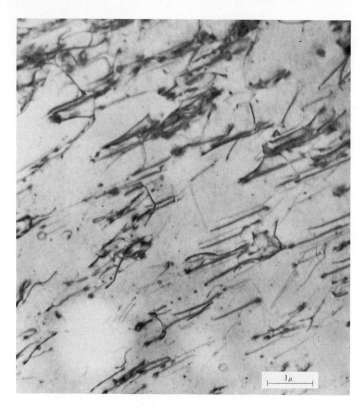

FIG. 5.11 Dislocation dipoles in easy glide of magnesium. (Lally, 1963. Courtesy of Wiley.)

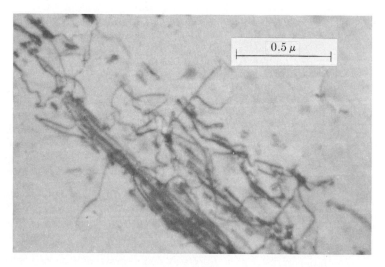

FIG. 5.12. Dislocation entanglement in easy glide. (Kuhlmann-Wilsdorf, and Wilsdorf, 1963. Courtesy of Wiley.)

somewhat easier. This would then produce a Bauschinger effect, which will be discussed in somewhat greater detail in Section 5.8.

The dipoles of Fig. 5.11 appear with both banded and homogeneous easy-glide, but in the homogeneous case, which is usually characterized by a very low friction stress, the arrangement may be less regular, as shown in Fig. 5.12.

Easy-glide ends when a critical dislocation density is reached on the primary slip system. The critical density is established gradually in the homogeneous case; in the banded case it is the density in the spreading bands. In either case, when the critical dislocation density is reached throughout, some slip will start to take place on other intersecting slip systems, and the rate of hardening will, within a small interval of strain, rise to a significantly higher value.

The process of single slip is strongly influenced by many factors. One of the strongest of these is *orientation*. When the orientation of the crystal is such that several slip systems are equally highly stressed, slip will start on all such systems simultaneously. The dislocations of intersecting slip systems are then forced to cut through each other from the very start and produce sessile dislocations and line defects of the type discussed in Sections 4.8 and 4.9. The result will be a rapid hardening. This difference of hardening between nonintersecting slip and intersecting slip is shown for two aluminum crystals in Fig. 5.13.

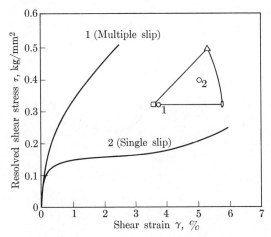

FIG. 5.13. Slow versus rapid hardening in two aluminum crystals at room temperature. (Data from Lücke and Lange, 1952.)

When the *size* of the crystal is comparable to the length of the mean free path of dislocations, a large fraction of dislocations can always escape through the surface, producing plastic strain without hardening the crystal. Surface films inhibit the normal escape of dislocations. The resulting accumulation and pile-ups all but wipe out easy-glide behavior, and affect subsequent strain-hardening, as discussed in Section 5.7B.

The introduction of *alloying elements* which form a solid solution will normally raise the initial shear strength and produce more extensive easy glide. This is often due to a decrease in stacking-fault energy, which makes the dislocations more extended, as shown in Fig. 5.14, and therefore makes double cross glide more difficult—this is in conformity with the deep slip lines that are observed in such alloys (see Section 4.10).

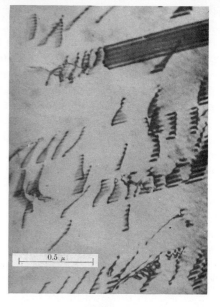

Fig. 5.14. Extended dislocations in easy glide of a copper + 7% aluminum crystal. (Howie, 1960. Courtesy of Wiley.)

In single slip, dislocations of opposite sign may accumulate in long configurations normal to their slip planes and form bands, as shown in Fig. 5.15. These bands strongly resemble the kink bands discussed in Section 4.12. Since the spacing of these bands is always slightly greater than the mean free path of dislocations, they produce little hardening (Mader and Seeger, 1960).

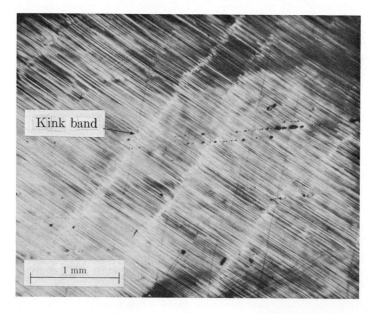

Fig. 5.15. Kink bands formed in easy glide in an aluminum single crystal.

5.4 STRAIN-HARDENING IN MULTIPLE SLIP

The "easy-glide" range appears to be terminated by filling the crystal with dipole-like defects and associated three-dimensional entanglements, which aid slip on intersecting slip systems, on a fine scale at least. Rapid hardening then sets in, as shown in Fig. 5.8. In b.c.c. crystals, where an "easy-glide" stage is absent, and in f.c.c. crystals in multiple-slip orientations, deformation usually starts with this stage. Experiments with periodic twists superimposed on the extension have shown that rapid hardening is characterized by a small amount of intersecting slip even in single-slip orientations, where most of the deformation is still on the primary slip system (Seeger et al., 1957). It appears that this critical amount of secondary slip starts in attempting to relieve some of the internal stress concentrations of the primary dislocation entanglements formed in earlier stages of deformation. As such secondary slip sets in, however, it will provide dislocation intersections with the primary system, which then on the one hand should produce the line defects or trails characteristic of dislocation intersections, and on the other, result in the formation of many segments of sessile dislocations at slip-plane intersections in a manner discussed in Section 4.9. Such sessile dislocation segments will then cement together firmly the densening dislocation entanglements into stable networks and will also anchor these networks to the lattice. This might be expected to produce irregular configurations similar to the one of a deformed silicon crystal in Fig. 5.16.

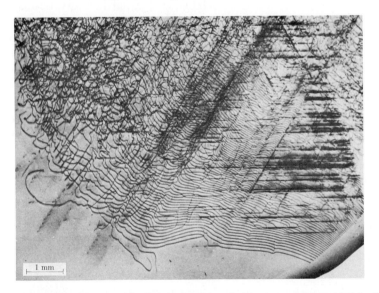

Fig. 5.16. Dislocations in a thin flake of deformed silicon crystal revealed in the image of an x-ray diffraction topograph in transmission, obtained by recording on film the image of a part of a crystal as the x-rays are diffracted by certain crystallographic planes. The dislocations become visible because of the different scattering characteristics of the strained lattice around them. (Lang, 1959. Courtesy of the American Physical Society.)

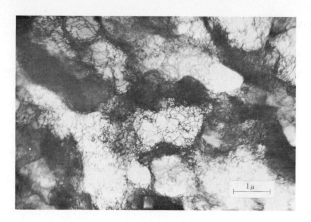

Fig. 5.17. Cell structure in aluminum after 10% reduction in area at 77°K. (Swann, 1963. Courtesy of Wiley.)

Another frequently observed dislocation configuration is a cellular structure, an example of which is shown in Fig. 5.17. It has been observed that in iron, such a structure preferentially develops under conditions of high strain and temperature (Keh and Weissmann, 1963), and that its formation at lower temperatures (below room temperature) is deferred to higher strains. This suggests that the formation of the cellular configuration is associated with rapid recovery effects frequently observed at room temperature and below, at higher strains.

Measurements of dislocation density by surface etch-pit counts show that in the rapid-hardening range, the increase in dislocation density is more than a linear function of the plastic strain, suggesting a steadily decreasing mean free path for glide dislocations (Young, 1962). This is in accord with the observed decrease in slip-line lengths in this range of deformation (Seeger, 1957).

As the dislocation entanglements become more dense, interactions between dislocations will become stronger, and it will become increasingly more difficult to pack more dislocations into the entanglement.

Thus, there are several effects that could result in the rise of the flow-stress curve with increasing plastic deformation. (1) The hardening of individual slip planes in the early stages of deformation due to the lattice distortions set up by the *dislocation dipoles produced primarily by double cross glide* of screw dislocations was already mentioned. This hardening effect, although very strong locally, will not be felt on the flow stress until the crystal is entirely occupied by an initial entanglement. (2) A second effect is due to the line defects, or *dipole trails, produced at screw dislocation intersections*. If such defect configurations cannot be broken up and dissipated in the form of point defects as rapidly as they form (dissipation requires diffusion and can occur only above about one-half the absolute melting temperature) the defects will raise the free energy of the crystal and therefore require additional work by the external forces. It is not too difficult to show that the contribution of this effect to the shear strength (flow stress) of the crystal would be substantial (Prob. 5.5). On the other hand, direct observations of all dipolar line defects* show that they make up only a fraction of the observed dislocation density. The observed dislocation density accounts for all the stored energy of cold-work (Clarebrough, et. al., 1961). The stored energy of cold-work in turn makes up only about 10% of the work of plastic deformation. Thus the actual contribution to the plastic work from the cutting of screw dislocations must be small. Therefore the dipole trail extension mechanism proposed above does not occur with any significant frequency. This is supported by observations of the slip behavior. In crystals extended in high-symmetry orientations, it is often observed that the crystal breaks up into domains in which only one of the possible slip systems develops, while the other slip systems are developed in other domains, as shown for a copper crystal extended in the [110] direction in Fig. 5.18. (3) A third and one of the strong strain-hardening effects is the mutual blocking action of *dislocation stress fields*. This may be estimated from the mutual blocking action of dislocations meeting each other on parallel slip planes. The magnitude of this interaction is given by the curves of Fig. 4.15. From the maximum force between two dislocations on parallel slip planes a distance t apart (Prob. 5.6),

FIG. 5.18. Formation of slip domains in a copper crystal extended about 10% in the [110] direction. (Saimoto, 1963.)

$$\sigma = \frac{Gb}{8\pi(1-\nu)t}, \quad (5.8)$$

it can be concluded that if the spacing between dislocations of opposite signs is a constant fraction of the mean dislocation spacing, this effect should be proportional to the dislocation density Λ, that is,

$$\sigma \propto \sqrt{\Lambda}. \quad (5.9)$$

This type of hardening was suggested first by Taylor (1934) in his pioneering work on dislocations. (4) An extreme form of tangling of dislocations results in the establishment of dislocation-free cells, bounded by cell walls with a high dislocation density, similar to those shown in Fig. 5.17. It may then be necessary for *dislocations to bulge through openings in the cell walls*. If the openings in the cell walls are of size l, the stress required to extrude dislocations through such openings will be given by Eq. 5.7. If the dislocations in the cell walls are the primary contributors to the dislocation density, and the walls are thicker than $\Lambda^{-1/2}$, then the flow strength in shear will again be of the form of Eq. 5.9. This type of hardening was suggested by Kuhlmann-Wilsdorf (1962).

The last two mechanisms appear to be the ones governing the rapid linear hardening of face-centered cubic crystals with a 50-fold rise in the flow stress as

---

* The line defects with atomic separation are not resolvable in the electron microscope.

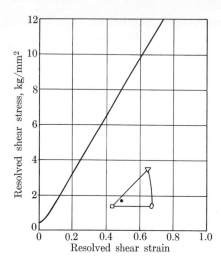

FIG. 5.19. Low yield stress and rapid linear hardening of a single crystal of copper strained at the boiling point of helium. (Data from Blewitt, Coltman, and Redman, 1955.)

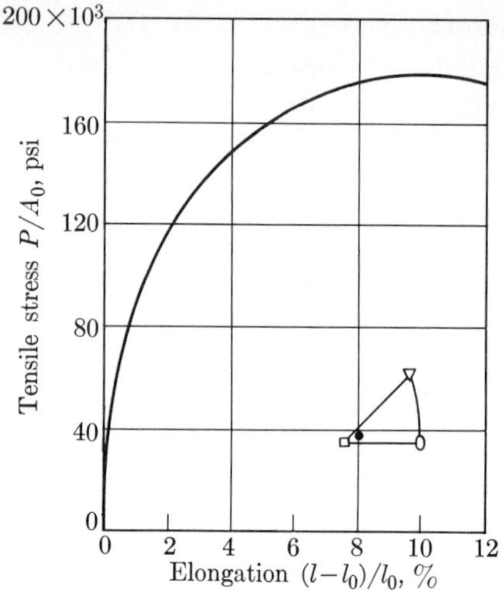

FIG. 5.20. Very rapid hardening in a single crystal of tungsten strained at −74°C. (Argon and Maloof, 1965.)

the dislocation density is increased with increasing strain, as shown in Fig. 5.19 (Blewitt, et al., 1955). These are also the dominant mechanisms of strain-hardening in b.c.c. metals beyond strains of say 10%. Measurements of dislocation density in electron-transparent metal foils made from previously deformed specimens has confirmed a dependence of the flow stress on the square root of the dislocation density in this stage of rapid hardening (Bailey and Hirsch, 1960; Keh and Weissmann, 1963).

The dependence of the dislocation density on strain can differ from one crystal structure to another, and can be affected, for example, by recovery effects, by temperature, and by stacking-fault energy (governing stacking-fault width in f.c.c. and h.c.p. crystals). When the mean free path of dislocations is approximately constant, such as in iron* (Keh and Weissmann, 1963), Eq. 4.3, giving the dependence of shear strain on dislocation motion and density, and Eq. 5.9 give a parabolic hardening curve with a gradually decreasing slope:

$$\theta = d\sigma/d\gamma = \beta_1 G \gamma^{-1/2}, \qquad (5.10)$$

where β_1 is a constant. This would approximate the behavior of the tungsten crystal of Fig. 5.20. When the mean free path of the dislocations is a constant multiple α of the mean dislocation spacing of an entanglement, Eq. 4.3 can be written as

$$\gamma = \alpha b \sqrt{\Lambda}. \qquad (5.11)$$

---

* We are now excluding the easy-glide phenomenon in which all hardening is likely due to dislocation dipoles.

This together with Eqs. 5.8 and 5.9 leads to a linear rate of hardening:

$$\theta_{II} = d\sigma/d\gamma \cong \beta_2 G, \tag{5.12}$$

where β_2 is a constant of proportionality. This approximates the behavior of face-centered cubic crystals in the post "easy-glide" region at low temperature where recovery effects are negligible (Figs. 5.8 and 5.19).

It has been shown by Kuhlmann-Wilsdorf (1962) that the hardening rates of many f.c.c. and b.c.c. metals fall normally within a range of $G/150$ to $G/700$ for cases where the above picture holds true. Occasionally, much higher rates, approaching $G/50$, can be found in the initial stages of hardening of b.c.c. crystals or f.c.c. polycrystals as shown, for example, in Fig. 5.20 for a tungsten single crystal (Prob. 5.7). The very high initial rate of hardening in tungsten appears to be governed by the exhaustion of mobile dislocations, and is therefore distinct from the strain-hardening discussed above (Argon and Maloof, 1965).

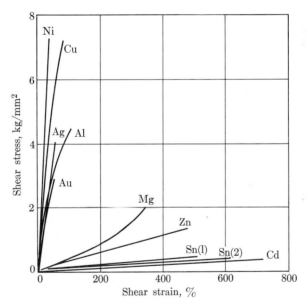

FIG. 5.21. Flow-stress curves of metal crystals. (Data from Schmid and Boas, 1935.)

In lattices where only one good slip system exists, such as h.c.p., dislocation intersections are rare, and sessile dislocation segments are rarely produced. Thus, the dislocation entanglements cannot be stabilized and do not become denser with any appreciable rate. Easy glide then continues to very large strains (as much as 500%). The production of dislocation dipoles does, however, continue and leads to the characteristic easy-glide hardening. This fundamental difference between the hardening rates of the cubic crystals and h.c.p. crystals is dramatically illustrated by the stress-strain curves of Fig. 5.21 (Schmid and Boas, 1935).

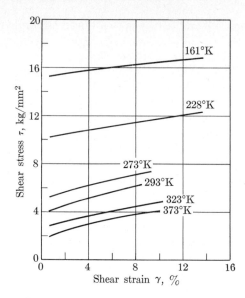

FIG. 5.22. Change in flow stress of single crystals of tantalum with temperature (insensitive to orientation) extended at a shear strain rate of $8 \times 10^{-5}/\text{sec}$. (Data from Mordike, 1962.)

◀ In high-purity b.c.c. transition metals and in ionic salts the flow stress is initially high and increases strongly with decreasing temperature (Fig. 5.22) and increasing strain rate. This behavior appears to be due to a high lattice friction stress, distinguishing these materials from the close-packed metals (Conrad and Hayes, 1963). Although plastically deformed b.c.c. transition metals and ionic salts show internal damping at strain amplitudes of 10^{-7}, and an accompanying reduction in Young's modulus (Chambers and Schultz, 1962; Taylor, 1962), this is compatible with a high lattice friction stress as will be discussed in Section 14.3. ▶

The three-dimensional entanglement of dislocations, with its sessile dislocation segments and dipoles, is essentially irreversible, forming a stable configuration at low temperatures where diffusion phenomena are negligibly slow. Thus, when the externally applied stress is removed, only a very small amount of plastic strain is recovered by the back motion of dislocations (see also Section 5.8). Thus, plastic deformation is an irreversible process and can be undone only by annealing.

5.5 THERMAL ACTIVATION IN STRAIN-HARDENING

As was mentioned repeatedly, plastic deformation is basically temperature insensitive. The picture of strain-hardening sketched in the two preceding sections is in accord with this conclusion. In actual practice, however, the flow stress of ductile metals does show some temperature sensitivity in various ranges of temperature. The stress-strain curves for tantalum single crystals at various

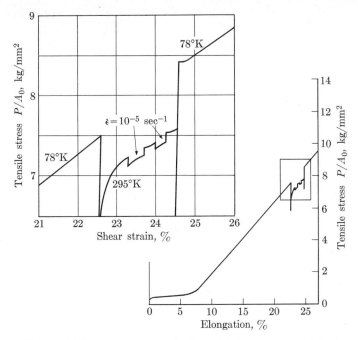

FIG. 5.23. Change in flow stress of a single crystal of copper with change in temperature and strain rate during extension at a strain rate of 1.1×10^{-4}/sec. (Data from Basinski, 1959.)

temperatures (Mordike, 1962) in Fig. 5.22 and the change in flow stress of copper with changes in strain rate (Basinski, 1959) shown in Fig. 5.23 are typical examples. As was mentioned in Section 4.14, in general, a reduction in temperature and a significant increase in strain rate produce a rise in the flow stress, although the effect is much weaker than that for a viscous liquid. Such strain-rate sensitivity or, alternatively, temperature sensitivity of the flow stress indicates that certain stages of the plastic deformation process are in part thermally activated. Examples will now be discussed.

A. Temperature and Strain-Rate Dependence of the Flow Stress. *1. Dislocation Intersections.* Following Cottrell (1953), consider as an example the effect of thermal activation on the motion of dislocations on intersecting slip systems which have to cut through each other. In Fig. 5.24(a), a dislocation is shown hung up at the "trees" of a forest of dislocations threading through the slip plane. If the intersection process involves the formation of only simple jogs on the forest dislocations, (see Section 4.8) the energy of the crystal will be raised by an amount equal to u_j (given by Eq. 4.19) between the solid and dotted positions of the glide dislocation. In going between these two positions separated by a distance δ, however, the glide dislocation may in addition have to extend somewhat and bulge out or pinch the forest dislocation, as shown in Fig. 5.24(b). That is, it may have to go through

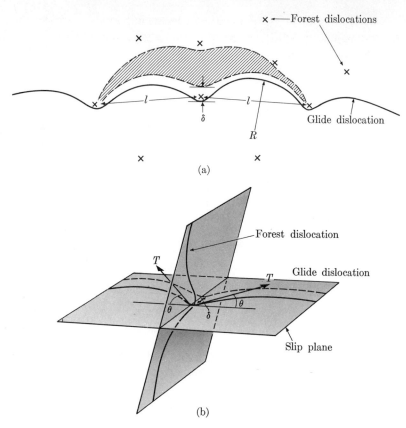

Fig. 5.24. (a) Intersection of a glide dislocation with a forest dislocation in the slip plane. (b) High-energy configuration of dislocations just prior to cutting through each other.

configurations of higher energy differing from the mean energy of the configuration by an amount u_i. The activation energy is less than u_i, however, as will now be shown. The bulge in the glide dislocation of Fig. 5.24(b) will effectively exert a force

$$F_T = 2T \sin \theta = Gb^2 \sin \theta \qquad (5.13)$$

on the forest dislocation, and work will then be done during the intersection process by an amount $F_T \delta$, reducing the activation energy. For small angles, θ can be expressed in terms of the bowing to a radius of curvature R of the dislocations between trees, leading to a net activation energy of

$$u_a = u_i - Tl\delta/R. \qquad (5.14)$$

The radius of curvature R of a dislocation can be related to the net shear stress, $\sigma - \sigma_0$, on the slip plane by Eq. 4.42. The stress σ_0 includes not only the lattice friction stress, but also any structural strengthening due to solution and precipitation hardening, as well as some of the strengthening resulting from strain-hardening. The activation energy is then

$$u_a = u_i - (\sigma - \sigma_0)bl\delta = u_i - (\sigma - \sigma_0)v. \quad (5.15)$$

The quantity $v = bl\delta$ has been called the *activation volume*.

If the dislocations can sweep out an area A (shaded in Fig. 5.24a) after cutting through the forest dislocation, the shear strain rate $\dot{\gamma}$ in a crystal with N active intersections per cubic volume would be*

$$\dot{\gamma} = bAN\nu \exp\left[-\left(\frac{u_i - (\sigma - \sigma_0)v}{kT}\right)\right], \quad (5.16)$$

where ν is a characteristic frequency of vibration of the glide dislocation (Prob. 4.38). When this expression is solved for the shear stress on the slip plane,

$$\sigma = \sigma(0) - \frac{kT}{v}\ln\left(\frac{bAN\nu}{\dot{\gamma}}\right), \quad (5.17)$$

where

$$\sigma(0) = \sigma_0(\gamma) + u_i/v$$

would be the flow stress at absolute zero of temperature. Equation 5.17 shows that the flow stress will rise with a decrease in temperature and, much less rapidly, with an increase in strain rate.

The salient point here is that although thermal activation cannot nucleate dislocations, it can aid them in overcoming barriers. Thus, as Eq. 5.17 shows, a rise in the strain rate can be achieved either by an increase in the shear stress on the slip plane or by an increase in the temperature. Various other dislocation models can lead to thermally activated plastic strain rates with relations very similar to Eq. 5.16. We shall now discuss a few of the more important cases.

2. Overcoming of Point Obstacles by a Dislocation. The most common process that can be thermally activated in commercial-purity metals is the overcoming of the short-range internal stresses of solute atoms by a dislocation. This process appears also to be the governing one in many supposedly pure materials. For example, see Johnston (1962).

◂ **3. Cross Glide of Screw Dislocations.** The cross gliding of screw dislocations past internal stress maxima can make further slip on the principal slip planes possible, and can therefore govern the rate of deformation. An extended screw dislocation

---

* More precisely, the number N of active dislocation intersections per unit volume as well as the frequency of vibration ν of a dislocation segment will increase with increasing strain. The quantities N and ν are also altered by temperature through its effect on the dislocation structure. These effects must be kept in mind when comparisons are made between specimens of widely different strain or thermal histories.

TABLE 5.2
Order of Magnitude Estimates of Terms in Idealized Strain Rate Equation

$$\dot{\epsilon} = \nu_m b A_m n_m e^{-u(\sigma)/kT}$$

Mechanisms	Frequency, ν_m	Area swept out, A_m	Sources per cm^3, n_m	Activation energy at low stress, u_0	Typical order of magnitude, ev	Local stress for zero activation energy, σ_0	References
Intersections forming jogs	$\nu_D b \Lambda^{1/2}$	$1/\Lambda$	$\Lambda^{3/2}$	$Gb^3/10$	1	$Gb\Lambda^{1/2}/10$	
Solute pinning							
Stress effect	$\nu_D b \Lambda^{1/2}$	$(b/c_s^{1/3})^2$	$\ll c_s/b^3$	$<Gb^3 \Delta b/b$	1	$G \Delta b/b$	Cottrell and Bilby (1949)
Modulus effect	$\nu_D b \Lambda^{1/2}$	$(b/c_s^{1/3})^2$	$\ll c_s/b^3$	$<b^3 \Delta G/10$	1	$\Delta G/10$	Fleischer (1961)
Cross glide of screw dislocations							
b.c.c.	$\gg \nu_D b \Lambda^{1/2}$	$1/\Lambda$ to d^2	$\Lambda^{3/2}$	$<(Gb^3/10)\sqrt{G/\sigma}$	1	—	Friedel (1959)
f.c.c.	$\gg \nu_D b \Lambda^{1/2}$	$1/\Lambda$ to d^2	$\Lambda^{3/2}$	$Gb^2 d_s/10$	1–10	—	
Vacancy migration in aging, recovery, and recrystalization	ν_D	b^2	$\dfrac{\exp(-h_f/kT)}{b^3}$	$Gb^3/10$	1		Broom and Ham (1958)

Note: Definitions: b = atomic diameter or Burgers vector, c_s = atomic concentration of solutes, d = grain size, d_s = stacking-fault width $= Gb^2/10\alpha_s$ at zero stress, G = shear modulus, Λ = dislocation density, ν_D = Debye, or highest atomic, frequency $\approx \sqrt{E/\rho}/b$, σ = applied shear stress, h_f = enthalpy of formation of vacancy.

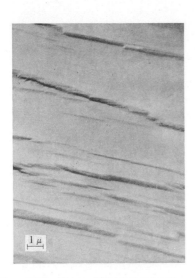
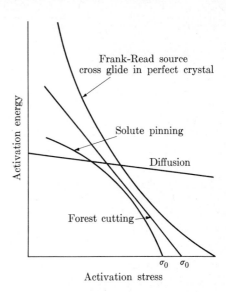

FIG. 5.25. Cross-slip traces on the surface of a copper crystal deformed 65%. (Mader and Seeger, 1960, from van Bueren, 1960. Courtesy of North Holland.)

FIG. 5.26. Types of dependence of activation energy on stress for various mechanisms.

in close-packed metals must first recombine over a small segment and then re-extend in the cross-slip plane. The activation energy required would then be basically due to the constriction. In advanced stages of deformation of f.c.c. crystals, the rate of strain-hardening steadily decreases, giving a roughly parabolic stress-strain curve. This appears to be due to a thermally activated large-scale cross slip, bridging together many individual slip traces, as shown in Fig. 5.25. The resulting deep slip traces are aggregates of slip traces known as *slip bands*. ▶

4. Recovery Effects. In the process of strain-hardening, some of the more unstable dislocation configurations may disintegrate under the action of their mutual interaction stresses and thermal agitation. This process involves diffusion of vacancies over short ranges and has an activation energy of the order of that of self-diffusion.

5. Recrystallization. Simultaneous recrystallization during straining could provide a steady strain rate under a constant stress. This process, which is one of several leading to secondary creep at elevated temperatures, is diffusion controlled and is usually governed by the rate of climb of edge dislocations (see Section 19.2).

In these cases, the activation energy is usually reduced by the shear or normal component of the applied stress. This dependence is schematically summarized in Fig. 5.26. Table 5.2 summarizes some characteristic values of activation energy and other pertinent factors of these mechanisms.

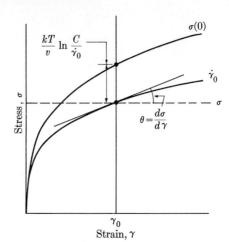

FIG. 5.27. Schematic representation of the flow-stress curves of a metal at absolute zero and at a temperature T.

◀ **B. Transient Creep at Very Low Temperatures.** One manifestation of the partly thermally activated motion of dislocations is the transient creep of metals, an example of which is given in Fig. 1.4. In a strain-hardening crystal at low temperatures, where the stress σ_0 steadily rises with increasing plastic strain, the net stress on the slip plane $\sigma - \sigma_0$, must be maintained constant to maintain a constant strain rate. If a constant strain-rate experiment is suddenly turned into a constant-stress experiment, the strain rate will steadily decrease with time as the internal stress, and consequently the activation energy for the intersection process, rises with additional plastic strain $\Delta\gamma$.* The rate of creep strain is then given by

$$\dot{\gamma} = C \exp\left\{-\frac{u_i - [\sigma - \sigma_0(\gamma_0 + \Delta\gamma)]v}{kT}\right\}, \tag{5.18}$$

where $C = bANv$.

◀ This is graphically illustrated in Fig. 5.27, where the upper curve is the shear stress-shear strain curve at absolute zero, while the lower curve is that at temperature T. If the stress is held constant at σ when the strain is γ_0, then the additional creep strain $\Delta\gamma$ will continue to strain-harden the specimen by an amount $\theta \Delta\gamma$. With the aid of Fig. 5.27, Eq. 5.18 can be written as (Prob. 5.8)

$$\dot{\gamma} = C \exp\left[-\frac{\theta v \Delta\gamma}{kT} - \ln\frac{C}{\dot{\gamma}_0}\right], \tag{5.19}$$

which, upon simplification and integration gives for the creep strain $\Delta\gamma$ (Prob. 5.9) (Mott, 1953)

$$\Delta\gamma = \frac{kT}{\theta v} \ln\left(1 + \frac{t}{kT/\theta v C}\right) = \alpha(T, \theta) \ln\left(1 + \frac{t}{\tau(T, \theta)}\right). \tag{5.20}$$

---

* The dislocation intersection mechanism is not an essential part of transient creep; the other rate-controlling mechanisms mentioned above can serve equally well.

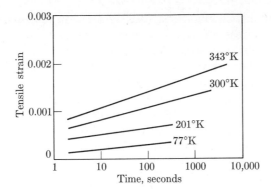

FIG. 5.28. Transient creep strain in copper at a stress of 6 kg/mm². (Data from Wyatt, 1953.)

Such logarithmic transient creep has been observed in many ductile metals at very low temperatures (a small fraction of the absolute melting temperature) (Phillips, 1905). An example is shown in Fig. 5.28 (Wyatt, 1953). ▶

5.6 STORED ENERGY OF STRAIN-HARDENING

◀When a crystal deforms plastically, the temperature is raised and most of the work done during deformation is dissipated as heat. Measurements of heat evolution during plastic deformation (Farren and Taylor, 1925; Taylor and Quinney, 1934) have shown that for strains over 10% only about 10% of the work of deformation is stored in a metal, while about 90% of the work leaves it as heat. The energy stored inside a crystal is almost completely made up of the strain energy of the total length of dislocations introduced by strain-hardening (Clarebrough, et al., 1961). The energy storage due to point defects produced by dislocation intersections makes up a small fraction of the total.

◀The question of the identification of the mechanism of this energy loss is a fundamental one, since it must be an essential part of any successful quantitative theory of strain-hardening. A plausible explanation of this process is to attribute it to the elastic energy radiated by oscillations of dislocations as they move through a fluctuating internal stress field caused by other dislocations. When a dislocation overcomes a region of high internal stress as it moves through the crystal and goes into a trough of an internal stress mimimum, it may violently oscillate back and forth. In crystals with a low friction stress such as f.c.c. metals, theoretical estimates of the elastic energy radiated by an oscillating dislocation (Nabarro, 1951) show that very little energy would be lost this way if the dislocation merely oscillated without interacting with the internal stress trough. In reality, however, the internal stress trough is made up of other dislocations and these will strongly interact with the oscillating dislocation (being set into oscillation themselves) to effectively scatter its kinetic energy throughout the lattice. Thus the kinetic energy of the initially oscillating dislocation will be dissipated around the slip plane into increased lattice vibrations, causing a rise in temperature. In crystals with a high friction stress such as b.c.c. metals, the mere displacement of a dislocation in a perfect lattice is a dissipative process.

◀In many metals, the rate of energy storage decreases with increasing plastic strain and often ceases altogether at high plastic strain values, creating a saturation effect. This saturation corresponds to the leveling off of the flow stress. It results likely from the attainment of a steady-state density of dislocations, where dislocation multiplication is balanced with dislocation annihilation by mutual encounters. Under these conditions, work is also converted into heat when dislocations mutually annihilate each other or combine to form low-energy dipoles. ▶

5.7 THE ROLE OF DIFFUSIONLESS TRANSFORMATIONS, SURFACE FILMS, AND GRAIN BOUNDARIES IN STRAIN-HARDENING

Some obstacles to dislocation motion can raise the dislocation density and therefore affect the rate of hardening. Among such obstacles are grain boundaries and bands or platelets of diffusionless transformations such as twinning and various martensite transformations.

A. Diffusionless Transformations. A strong hardening effect should be present if the distance between the platelets of the diffusionless transformations is of the order of or less than the distance between more common slip obstacles such as subgrain boundaries, forest dislocations, sessile dislocation segments, solute atoms, etc. Otherwise, little hardening is likely to result from this mechanism. In hexagonal close-packed crystals, where there is normally no intersecting slip, the mean free path of a dislocation may easily be of the order of the crystal dimensions. In this case, the formation of twins on the pyramid planes would result in appreciable hardening. In martensite, on the other hand, the spectacular hardening does not result from the dimensions of the martensite platelets, but rather, as Winchell and Cohen (1962) have shown from the softness of iron-nickel martensite, it results from a combination of solution hardening and high lattice friction stress of the body-centered tetragonal martensite lattice itself.

◀**B. Surface Films.** In single crystals of hexagonal close-packed metals and in single crystals of face-centered cubic metals oriented for single slip, the mean free path of dislocations can often be of the order of the cross-sectional dimensions of the crystal, i.e., about one millimeter (see Section 5.3). In such a case, the presence of a hard surface layer, such as an oxide layer, can prevent the escape of dislocations from the surface and will raise the dislocation density somewhat more rapidly than would have otherwise been the case. This would result in an increased rate of hardening. This effect has been demonstrated by many investigators. Garstone et al. (1956) have plated one of two identical crystals of copper with a layer of chromium of 4×10^{-4} cm thickness. The difference in the hardening behavior is shown in Fig. 5.29. Figure 5.30 shows piling-up of edge dislocations beneath a surface coating in a lithium fluoride crystal. In order to

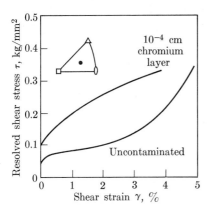

FIG. 5.29. Influence of a thin chromium layer on the hardening of a copper crystal. (Data from Garstone, Honeycombe, and Greetham, 1956.)

FIG. 5.30. Piling-up of dislocations under a surface layer in an LiF crystal. (Westwood, 1960. Courtesy of Taylor-Francis.)

produce appreciable mechanical effects, this pile-up would have to produce a much more widespread increase in dislocation density. Such surface effects are examples of the breakdown of the concept of a stress-strain relation, since the strain in a small region is no longer governed solely by the stress in that region, but by the stress in the neighboring region as well.

◀ In later stages of deformation, when the dislocation density due to dislocation intersections rises and the mean free path of dislocations becomes very much shorter than the cross-sectional dimensions of the crystal, the effect of surface layers will become much less, but will never quite vanish. ▶

C. Grain Boundaries. In a polycrystal, the individual grains are separated from one another by grain boundaries. The structure of a general grain boundary in a real polycrystal is far more complex than the two specialized cases of a tilt and a twist boundary, discussed in the previous chapter. When a boundary separates two crystals of large lattice misorientation, it is, in general, not meaningful to represent it by a dislocation model because the dislocations would have to be so close together that their individual character would be lost. This fact is very clearly illustrated in the bubble-raft analogy of a close-packed crystal lattice, as shown in Fig. 5.31 by arrows a and b. While it is possible to distinguish dislocations in the small-angle boundary marked a, this is not so in the large-angle boundary marked b. For this reason, large-angle grain boundaries have often been considered as a thin transition layer of high disorder with properties resembling those of a liquid or glass.

One of the most important characteristics of a large-angle boundary is its property of obstructing the slip process in individual grains. The spatial misorientation of slip planes across a large angle grain boundary will hinder dislocations from breaking through the boundary. Hence, the slip process in metals with well-

Fig. 5.31. Small-angle a and large-angle b grain boundaries in a raft of soap bubbles.

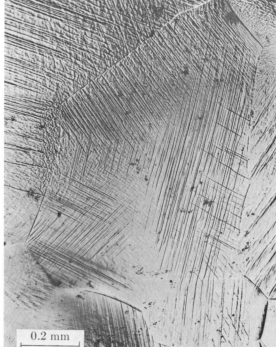

Fig. 5.32. Four sets of slip traces in a grain of deformed aluminum. (Boas and Ogilvie, 1954. Courtesy of Pergamon Press.)

defined slip systems must necessarily be discontinuous on the scale of individual grains. In such a case, it is natural to expect that in a polycrystal subjected to increasing stress, slip will start in grains which contain favorably oriented slip systems. This slip will, however, be blocked, in general, at the boundaries of such (soft) grains. The piling-up of slip on the boundaries will then set up stress concentrations around the grain boundary, strongly aiding the external stress on the less favorably oriented slip systems. In many instances, the constraints imposed on slip in a grain by the surrounding grains lead to multiple slip in the original grain, as shown in Fig. 5.32, where four independent slip systems can be counted. See also Section 7.5. This would produce profuse intersecting slip, which leads to rapid hardening. Therefore, the strain-hardening of polycrystals cannot be obtained simply by considering a weighted average of the strain-hardening characteristics of single grains where single slip systems are active, but consideration must be given to possibilities of multiple slip in the grains themselves. Figure 5.33 shows that the flow-stress curve for aluminum polycrystals falls somewhere between the curves of the $\langle 100 \rangle$ and $\langle 111 \rangle$ oriented single crystals, where many slip systems are simultaneously active. The polycrystalline curve differs greatly from the flow-stress curves of crystals where only one slip system is active. A further point is the limitation which the grain size imposes on the length of slip lines or bands inside it. Thus, in early stages of plastic deformation, where the slip bands would be longer in a single crystal, one would expect the yield strength to be higher in a polycrystal than in a single crystal with the same active slip systems.

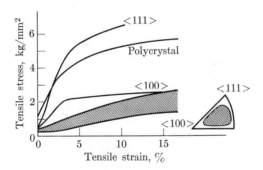

Fig. 5.33. Stress-strain curves of single crystals with orientations $\langle 100 \rangle$ and $\langle 111 \rangle$, and polycrystals of aluminum. (Data from Kocks, 1960.)

The concept of considering a grain boundary as a thin amorphous layer furnishes at once explanations for a large number of grain-boundary phenomena. As we shall see in Chapter 19, beyond a critical combination of high temperature and low stress, grain boundary sliding and fracture can occur. Furthermore, it is known that grain boundaries are likely sites for the accumulation of impurity atoms in the lattice and are most frequently nucleation sites of second phases in two-phase alloys. This results clearly from the large disorder in the grain boundary, which makes it a sink for solute atoms (impurities) where the lattice distortion due to the misfit of the impurities can be largely relieved. The disorder along the grain boundary also enables it to become a preferential path for diffusion of solute atoms and thereby aid in the nucleation of second phases. Due to this tendency of a grain boundary to become a sink for solute impurity atoms, the composition of the grain boundaries and their immediate surroundings may differ from the composition of the grains. In many cases, the chemical affinity of the grain-boundary material to corrosive media, coupled with the higher diffusion rates of such corrosives along the boundary, make the boundary especially susceptible to corrosion. This leads to phenomena such as grain-boundary embrittlement or stress corrosion. Although it will not be possible to discuss fully the significance of such chemo-thermodynamical grain-boundary phenomena here, it has to be kept in mind that they have strong influences on mechanical properties and frequently form the original defects leading to eventual service failure. For a review and further information, see Uhlig (1963).

5.8 BAUSCHINGER EFFECT

Although plastic deformation cannot be undone by removing the externally applied stress, deformation in the reverse direction will normally start at negligibly low stress. Often, a slight amount of reverse deformation may take place even during unloading, as is shown in Fig. 5.34. This anisotropy of work-hardening was observed first by Bauschinger (1886) on wrought iron and is known as the *Bauschinger effect*.

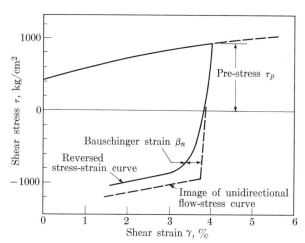

FIG. 5.34. Bauschinger effect on decarburized tubular steel specimen in torsion. (Deak, 1961.)

In polycrystalline materials, it has been observed by Woolley (1953) and by Deak (1961) that the difference between the actual strain and the linear elastic strain, known as the Bauschinger strain, is usually a linear function of the previous flow stress or the *pre-stress*, but may also be a function of *pre-strain*, especially when the stress-strain curve is very high and flat as in alloy steels. The stress-strain curve in reverse never reaches the image of the monotonic flow-stress curve, but remains below it, maintaining, however, the same rate of hardening (see Fig. 5.34). This means that the Bauschinger effect involves not only a premature yielding, but also a certain amount of permanent softening or nonhardening strain.

The premature yielding can be removed by a stress-relieving treatment that produces some readjustments in the dislocation structure but little recovery (Deak, 1961). With this treatment, the material is almost isotropic. Without this thermal treatment, a certain amount of strain is required to bring the flow stress up to its value before reversal. Thus a certain amount of plastic strain is reversible in the sense that it does not contribute to the strain-hardening.

The Bauschinger effect is often ignored in the process of idealization in macroscopic plasticity. In reality, this nonhardening plastic strain is a fundamental part of damping, fatigue, and creep under a varying stress history. Early explanations of the Bauschinger effect relying on macroscopic residual stresses developed due to nonhomogeneous deformation in individual grains, were shown not to be correct because the Bauschinger strain may be many times the yield strain, as is shown in Fig. 5.34. The experiments of Sachs and Shoji (1927) on single crystals of brass, and those of later investigators on single crystals of iron, zinc, cadmium, copper, and aluminum, all disclosed a pronounced Bauschinger effect in pure tension-compression experiments, also indicating that the Bauschinger effect cannot be entirely due to inhomogeneous yielding at the level of individual grains.

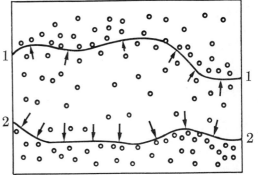

FIG. 5.35. Anisotropy of deformation obstacles causing the Bauschinger effect. Dislocation 1 stopped at a dense row of obstacles; on stress reversal, it moves back and is stopped in position 2 by the nearest row of equal density. (After Orowan, 1959.)

◄Specific mechanisms which have been proposed for the Bauschinger effect include the pile-up of dislocations against obstacles and the stopping of the glide dislocations at especially dense spots in the forest of intersecting dislocations. Thus it can be imagined, for instance (Orowan, 1959), that "forest" dislocations threading the slip planes of the gliding dislocations will accumulate in front of moving dislocations, as shown in Fig. 5.35, perhaps by a mechanism such as the one sketched in Fig. 5.24. Thus, upon reversal of deformation, the dislocations moving backward find fewer obstacles in their paths in the early stages. As the reverse deformation progresses, a new anisotropy of slip obstacles will be produced, now in the reverse direction. The strain expected from the blocking of glide dislocations at dense spots in the dislocation structure, such as subgrain boundaries, can be estimated to be of the order of a few percent (Prob. 5.14). ▶

5.9 MECHANICAL INSTABILITIES

The plastic deformation of crystalline materials is not always stable. Not counting the necking instabilities (to be discussed in Chapter 8), there are many instances where plastic deformation may be accompanied by a drop in flow stress. The most familiar of these is the yield phenomenon in mild steel, which will be discussed first.

A. Impurity Yield Phenomena. *1. Misfit Interactions.* In polycrystalline annealed or normalized mild steel and in a single crystal of iron containing small amounts of carbon or nitrogen, plastic deformation commences by undergoing a yield phenomenon, as shown in Fig. 5.36. The highest stress σ_u before the drop is known as the *upper yield point*, while the constant stress after the drop is called the *lower yield point*. The magnitude of the upper yield point increases with decreasing

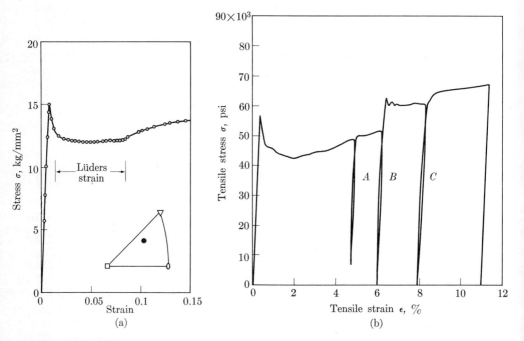

FIG. 5.36. (a) Yield phenomenon in a single crystal of iron containing 0.003% carbon extended at 195°K at a strain rate of 10^{-5}/sec. (Data from Paxton and Bear, 1955.) (b) Yield phenomenon in a normalized specimen of 1020 steel strained at a rate of 10^{-5}/sec. Path A represents unloading and immediate reloading. Unloading with a 10-minute treatment at 100°C has produced a new yield phenomenon (path B). Even a 1-minute treatment at 100°C brings back part of the yield phenomenon (path C).

temperature and increasing strain rate. At room temperature, the yield phenomenon is more pronounced in fine-grained than in coarse-grained specimens. The lower yield point is similarly affected by temperature, strain rate, and grain size.

The extension at the lower yield point is an inhomogeneous process. It starts at a point of stress concentration and spreads out first across the specimen and then as a band along the specimen. Such a band, known as a Lüders band, can be made visible by polishing a specimen prior to extension and then viewing its surface during the extension. The deformed portions inside the band will give a rough appearance, as shown in Fig. 1.35. All current deformation in such a process is concentrated in a thin region at the boundary of the Lüders band, while the region inside the band and the region outside it do not undergo any deformation (see also Section 8.4). Paxton and Bear (1955) have observed formation of slip-band clusters in single crystals of iron during the deformation preceding the upper yield point. The drop in load and the lower yield extension then result from the spread of a Lüders band of very dense slip, which was initiated at the upper yield point. Their observations also indicate that the amount of the lower yield extension increases with a decrease in temperature, going from 1% at room temperature to 5 to 15% at the sublimation temperature of carbon dioxide (195°K).

When a test at room temperature is interrupted in the rising part of the stress-strain curve beyond the lower yield extension, the load is removed and immediately reapplied as shown in Fig. 5.36(b) as path A, yielding will start very nearly at the point where the original straining was interrupted without exhibiting a new yield phenomenon. If the specimen is given a mild heat treatment between the removal and reapplication of the stress, such as an immersion in boiling water for 10 minutes (path B), the yield phenomenon will reappear as illustrated in Fig. 5.36(b). Even a one-minute treatment at 100°C (path C) will bring back part of the yield phenomenon. The mild aging heat treatment does not affect the general flow-stress curve appreciably, but merely reproduces a yield phenomenon and a new Lüders band. The same aging process can be accomplished by prolonged retention of the unloaded specimen at room temperature. At temperatures above 100°C the aging is much more rapid and will at a high enough temperature occur concurrently with straining. Such aging phenomena are not characteristic of iron alone, but have also been observed in nonferrous b.c.c. and in f.c.c. metals.

It has been established that impurity atoms play an essential role in most aging phenomena, since their removal completely eliminates the instability phenomena.

According to one theory (Cottrell and Bilby, 1949, and later refinements) the yield phenomenon of this type caused by interstitial impurities results from the migration of these impurities to the dislocations. As discussed in Section 5.2, an interstitial impurity in the b.c.c. lattice causes a large volume misfit and directional distortions, giving strong interactions with both edge and screw dislocations. At temperatures where the interstitials are mobile* they will migrate toward the

---

* The activation energy of motion of most interstitials is low (e.g., less than 1 ev for iron) whether they are matrix atoms or impurities.

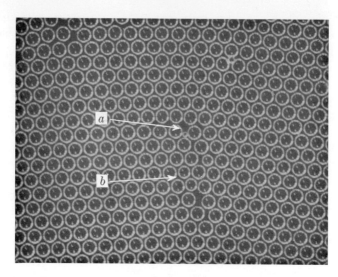

Fig. 5.37. Segregation of impurities to dislocations in a bubble raft. (a) Interstitial impurity atom segregated at the tension side of an edge dislocation. (b) Two substitutional impurity atoms smaller than matrix atoms segregated at the compression side of an edge dislocation.

dislocations where they can reduce the total strain energy of the crystal (see Fig. 5.37). This would anchor dislocations to the lattice and harden the crystal. To initiate plastic deformation, dislocations would have to be torn away from their impurity atmospheres or precipitates at a stress higher than the stress necessary to move them through the solution-hardened lattice. According to another theory (Hahn, 1962), the instability is due to the strong stress dependence of the velocities of dislocations. Thus, when a small number of dislocations are torn away from their impurity atmospheres they are required to move through the crystal at a high velocity to provide the externally imposed rate of grip displacement. The high velocity of the small number of moving dislocations requires a high stress. A high stress, however, is conducive to rapid multiplication of dislocation mills by double cross-glide. As the number of mobile dislocations increases, their velocity decreases and the stress drops. Although these views are generally in accord with observations that large yield drops are associated with high strain rates, Sylvestrowicz and Hall (1951) have observed the largest yield drop with the lowest strain rate.

When the impurity mobility is rapid enough that their mean drift velocity (see Section 4.13) becomes of the order of the mean velocity of dislocations, the latter drag their impurity atmospheres along with them at a high stress. A higher strain rate at the same temperature, however, would free the dislocations from their atmospheres, and deformation would proceed at a lower stress. This results in the unusual negative strain-rate sensitivity of mild steel above 100°C, as shown in Fig. 5.38; this is known as *strain aging*.

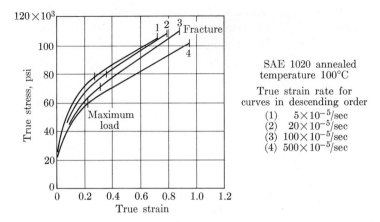

FIG. 5.38. True stress-strain curves for annealed SAE 1020 steel at 100°C for various constant true strain rates, showing reverse strain-rate sensitivity. (MacGregor and Fisher, 1946. Courtesy of ASME.)

In spite of their relative success, the two theories discussed above are in some respects not fully satisfactory. They fail to explain adequately, for instance, the effect of grain size.

◂ **2. The Portevin-LeChatelier Effect.** Yielding is often not a smooth phenomenon, but instead occurs discontinuously in jerks, as observed first by Portevin and LeChatelier (1923) on Duraluminum and soon after by Ehrenfest and Joffe (1924) and Classen-Nekludowa (1929) on rock salt. In the plastic range, when the crystal deforms under a constant stress, the transient creep strain is in the form of jerks. The effect is most prominent at room temperature in dilute solid solutions of zinc, aluminum, and the like, and occurs after some amount of preparatory homogeneous plastic deformation. It has, however, also been observed in very pure zinc crystals at liquid nitrogen temperatures (Bullen, 1962).

The effect has been attributed by Cottrell (1958) to an impurity or vacancy interaction with dislocations. The clocklike regularity of the jerks is presumably due to the thermally activated release of dislocations locked by impurities or point defects, with the released dislocations becoming locked again by impurities or point defects. Here the critical amount of plastic deformation would seem to provide the vacancies which affect the locking either directly or as carriers of impurity atoms. In polycrystals, an avalanche through many grains would be required to explain the observed effect.

◂ **3. Chemical Interactions.** Another effect of impurity interaction has been proposed by Suzuki (1952) for f.c.c. and h.c.p. metals, where the partials of the extended dislocations are separated by a stacking fault. This gives a thin layer of different chemical properties which may cause chemical binding of existing impurities to the stacking fault, reducing the stacking-fault energy. The average chemical binding energy is estimated by Suzuki to be about 10% of the binding energy of a misfitting interstitial solute atom to a dislocation. For the dislocation to become mobile, it again has to be torn away from these impurities at a stress higher than the flow stress, giving a yield phenomenon.

◀ **B. Load Drops in Twinning and Kinking.** The formation of a twin or a kink band will frequently lead to a sharp drop in the driving stress in a stiff testing machine. This results from the fact that in the formation of a twin or a kink, there exists a critical size above which an expansion under a favorable stress will lead to a net reduction of the potential energy of the crystal and the system which is deforming it. Hence, the extension of the twin or kink will be accompanied by a drop in the driving stress. This gives rise to the well-known "cry" of tin as it twins in bending. Many investigators have observed sharp serrations in tension experiments, which could be directly traced to the formation of twin bands (Churchman and Cottrell, 1951), while Holden and Kunz (1953) observed an instability of this type in a single crystal of iron where the instability could be related to the formation of a kink band.

◀ **C. Low-Temperature Thermomechanical Instability.** Another, and quite extraordinary, instability has been observed by Basinski (1957) in single and polycrystals of aluminum extended at the boiling point of helium (4.2°K). This was due to the adiabatic heating of the regions around the deformation bands as a result of the plastic deformation itself. This effect, which was primarily due to the very low specific heat of solids near absolute zero, disappeared in tests at 20°K.

◀ **D. Instability in Prestrained Specimens.** When a face-centered cubic metal is first deformed at low temperature in the linear hardening range and the deformation is continued at higher temperature where the normal hardening is parabolic hardening with a lower flow stress, a yield phenomenon will occur as the denser low-temperature dislocation structure breaks up during deformation (Cottrell and Stokes, 1955).

◀ **E. Order-Disorder Instabilities.** In certain cases of two metals with complete solid solubility of one in the other, a long-range order may be established at certain stoichiometric concentrations, forming a super lattice. When dislocations pass through such long-range ordered lattices, they create disorder, and thereby raise the free energy, requiring a momentary rise in the flow stress, and causing a yield phenomenon (see for example, Marcinkowski, 1963).

5.10 MECHANICAL EQUATION OF STATE

The equation of state of a perfect gas, relating the pressure, volume, and temperature, is familiar to every student of thermodynamics. It has been proposed by Ludwig (1909) and more recently by Hollomon and Zener (1946) that a functional relationship of this sort may apply to the plastic deformation of metals. As shown in Eq. 5.16, such an equation of state must involve the strain rate as well as the stress, temperature, and strain and can be valid only in certain restricted ranges where metallurgical-structural changes are kept to a minimum. Although such changes make a unique functional relationship between stress, strain, and strain rate impossible (Orowan, 1947), we shall explore approximate functional relationships which may be of use in design, where performance may have to be predicted from incomplete data.

The concept of the mechanical equation of state stems largely from a relation such as Eq. 5.16, relating the strain rate to the net shear stress on the slip plane. Recognizing that Eq. 5.16 was developed for a special model, and that its region of applicability may be limited, one may adopt only the notion that part of the resistance to plastic deformation can be overcome by thermal activation, and that the applied stress can strongly influence the activation energy. Hence, for generality we may speak only of a stress-dependent Gibbs free energy of activation $g(\sigma(\epsilon))$. Here we are considering the uniaxial normal components of stress and strain. Furthermore, if the factor $bANv$ is replaced by a strain-rate factor $C\dot{\epsilon}_0$,* Eq. 5.16 then simplifies to

$$\dot{\epsilon} = C\dot{\epsilon}_0 \exp\left[-\frac{g(\sigma(\epsilon))}{kT}\right]. \tag{5.21}$$

Solving this relation for the Gibbs free energy and setting $\ln C = 1/\alpha$, one can obtain the form

$$\frac{\alpha}{k} g(\sigma(\epsilon)) = T\left(1 - \alpha \ln \frac{\dot{\epsilon}}{\dot{\epsilon}_0}\right).$$

An alternative statement of this equation would be that the flow stress is a function of the absolute temperature modified by the strain rate, i.e.,

$$\sigma(\epsilon) = f\left[T\left(1 - \alpha \ln \frac{\dot{\epsilon}}{\dot{\epsilon}_0}\right)\right]. \tag{5.22}$$

This states that the flow stress $\sigma(\epsilon)$ is a function only of the term

$$T_m = T\left(1 - \alpha \ln \frac{\dot{\epsilon}}{\dot{\epsilon}_0}\right), \tag{5.23}$$

which has been called the "velocity-modified temperature" by MacGregor and Fisher (1946), who have also experimentally shown the validity of Eq. 5.23 for various metals. If stress-strain experiments are run at a certain reference rate of $\dot{\epsilon}_0$ at various temperatures, the variation of the flow stress with temperature for any constant strain ϵ gives the functional relationship between the flow stress and the "velocity-modified temperature" for the special case where the latter is equal to the absolute temperature. If the same experiments were repeated at various other strain rates $\dot{\epsilon} \neq \dot{\epsilon}_0$, then it is implied that a correct choice of the

---

* This is the critical assumption, and is only an approximation due to the dependence of N and v on structure as discussed in the footnote to Eq. 5.16. The validity of the approximation is verified by the experiments cited below.

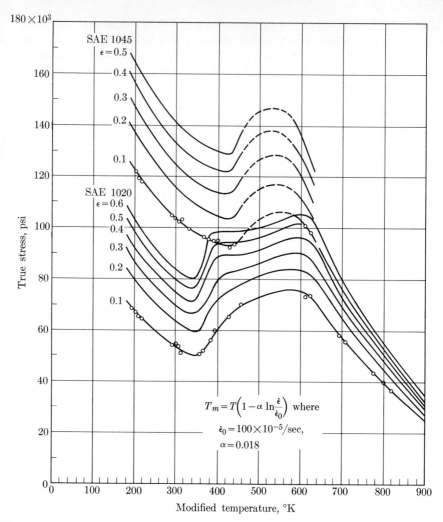

Fig. 5.39. Stress as a function of true strain and velocity-modified temperature for annealed 1020 and 1045 steels. Strain rates from 5 to 500 × 10^{-5}/sec are correlated on each curve. (Data from MacGregor and Fisher, 1946.)

constant α will make all flow-stress versus temperature curves at a constant strain coincident. The validity of this assertion is shown by the data of Figs. 5.39 and 5.40 for 1020 steel, 1045 steel, and annealed brass. The clusters of points in various regions of the curve refer to tests at various strain rates at a constant temperature. The usefulness of the formulation of Eq. 5.23 is apparent from these figures. The hump in the first two curves between 360° and 680°K on the scale of the "velocity-modified temperature" shows the rise in the flow stress due to strain-aging. Thus, it is possible to bring into the picture some diffusional phenomena also.

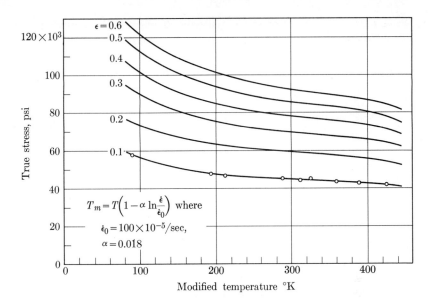

Fig. 5.40. Stress as a function of true strain and velocity-modified temperature for annealed brass. (MacGregor and Fisher, 1946. Courtesy of ASME.)

The set of curves of the type shown in Figs. 5.39 and 5.40 shows that in uniaxial monotonic straining (constantly increasing strain) a relation of the type

$$f(\sigma, \epsilon, \dot{\epsilon}, T) = 0$$

holds and is qualified to be called a mechanical equation of state, although as will be discussed further below, in a much more restricted sense than the analogous equation of state for an ideal gas.

Curves such as those in Figs. 5.39 and 5.40 are particularly attractive to a designer requiring data for creep applications. With such curves it should be possible to obtain information for creep rates at elevated temperatures under constant stress, as discussed in Section 19.4.

In certain cases, the dependence of the Gibbs free energy on stress can be expressed as a power function in the pre-exponential factor of Eq. 5.21, giving a modified relation

$$\dot{\epsilon} = \dot{\epsilon}_0 \epsilon^n (\sigma/\sigma_0)^m e^{-u/kT}, \tag{5.24}$$

for an equation of state where $\dot{\epsilon}_0$, a frequency factor, σ_0 a stress normalization factor, and the exponents n and m are constants which have to be determined experimentally from the strain-hardening, strain rate, and temperature sensitivity of a material at various temperatures. Lubahn and Felgar (1961) have shown various applications of an equation of the type given to some face-centered cubic metals.

Only cases of monotonic loading were discussed above. An equation of state cannot hold if other types of loading are also considered. The simplest departure from monotonic loading, namely, unloading with intervening rest periods, will produce recovery effects which can be very substantial in the range of temperatures where creep occurs. This produces large deviations from the equation of state. Reversal of the applied stress or change of its direction brings equally large deviations. Thus, any application of an equation of state must be restricted to systems of essentially static or monotonically increasing loading.

The reason for the physical nonexistence of a mechanical equation of state is that the flow stress is structure sensitive and is governed by the dislocation configurations which are not in thermodynamic equilibrium. The salient point is that at a given temperature, the dislocation structure of a deformed metal will depend strongly on the temperature and strain rate of the prior deformation.

Below about $\frac{1}{2}$ the absolute melting temperature, the dislocation substructures will be relatively stable and will be determined by the prior deformation history at various temperatures; i.e., the substructure influencing the flow stress at a given plastic strain is a path function of the previous strain-temperature history. In contrast, the molecular arrangements of a gas at a certain temperature are independent of the previous volume-temperature history of the gas and will, for a gas, produce the same pressure regardless of whether the gas has reached its final volume by expansion or contraction from a lower or higher temperature. In essence, while the gas is in thermodynamic equilibrium, the substructure of a deformed metal is not, and its departure from equilibrium is dependent on the previous strain-temperature history. Thus below $\frac{1}{2}T_m$ the equation of state is a useful tool only for moderate changes in history.

At temperatures above $\frac{1}{2}$ the absolute melting temperature, where dislocation substructures anneal out as rapidly as they form, the equation of state will not be a function of strain and will therefore hold more exactly. In this range, however, the material itself becomes less useful for structural applications.

5.11 DATA FOR THE STRENGTH OF METALS AND ALLOYS

In Sections 5.2 through 5.4, various idealized mechanisms of structural hardening and strain-hardening are discussed. Most often these hardening mechanisms do not exist alone, but are found in combinations. The variety of combinations that can occur can provide strengths greater than any one mechanism alone can give. For instance, it is often possible to work-harden a previously solution- or precipitation-hardened sample. On the other hand, aging in previously deformed samples is often found to be more effective than in undeformed samples. In some instances, the precipitation or aging heat treatments not only provide the precipitates, but also dislocate the lattice around the precipitates, as shown in Fig. 5.41.

The structural types of hardening discussed in Section 5.2 have the advantage that they can be applied to finished parts by techniques of heat treatment (including diffusion) without serious accompanying distortions.

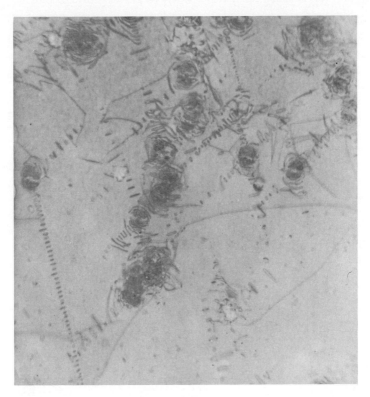

FIG. 5.41. Dislocation loops around precipitates in a quenched magnesium foil. (Lally, 1963. Courtesy of Wiley.)

Solution-hardened alloys find widespread use in parts that have to be fabricated by welding. Since the hardening is due to a constituent in solid solution, no adverse softening will result in the welding process. Precipitation-hardened alloys, on the other hand, do not lend themselves to welding or to use at elevated temperatures, because such alloys will overage and lose their high strength. This is also true of martensite, where the hardness is due to an unstable phase.

Radiation and quench-hardening, although effective, are either commercially unattractive or impractical to apply on a large scale.

In some materials of great commercial importance, it is often difficult to identify uniquely the dominant mechanism of hardening.

The properties of some common metals in various states of thermal or mechanical treatment are given in Table 5.3. Information on other metals and alloys can be found in manufacturers' handbooks such as the *Alcoa Aluminum Handbook* (Alcoa, 1962), general reference handbooks such the *Metals Handbook* (ASM, 1961), *The Metals Reference Book* (Smithells, ed., 1962), *Engineering Alloys* (Woldman, ed., 1962), *Engineering Materials Handbook* (Mantell, ed., 1958), *Materials Handbook* (Brady, ed., 1963), and the *Handbook of Engineering Materials* (Miner and Seastone, eds., 1955).

TABLE 5.3

MECHANICAL PROPERTIES OF SOME COMMON METALS AND ALLOYS

A. CARBON STEELS

Material designation or composition	Condition heat treatment	Cost[a] $/lb	Tensile strength, 1000 psi	Yield strength, 1000 psi	Elongation, %	Red. of area, %	Usage and reference
Armco iron	Normalized (heated to 1700°F and air-cooled to room temperature)	0.038	44	24	45	69	Piping for fresh and salt water, chemical equipment (Smithells, 1962, p. 771)
AISI 1020	Hot-rolled		65	43	36	59	General machine parts (Merriam and Lubars, 1957)
	Cold-worked		78	66	20	55	
AISI 1040	Hot-rolled		91	58	27	50	High-strength machine parts (Merriam and Lubars, 1957)
	Cold-worked		100	88	17	42	
	Hardened (water-quench from 1525°F, tempered at 1000°F)		113	86	23	62	
AISI 1095	Annealed at 1450°F (cool 20°F/hr to 1200°F then air-cool)		100	53	21	42	Heavy machine parts, hand tools, agricultural machinery parts, etc. (Merriam and Lubars, 1957)
	Hot-rolled		142	84	9	18	
	Hardened (oil-quench from 1475°F, tempered at 700°F)		180	118	11	30	

B. Low-Alloy Steels

		0.04	97	83	21	57	
AISI 1330 Manganese steel	Annealed		97	83	21	57	Automotive parts, axles, drive shafts (Merriam and Lubars, 1957)
	Cold-drawn		113	93	15	50	
	Hardened (water-quench from 1525°F, tempered at 1000°F)		122	100	19	52	
AISI 4130 Chromium molybdenum steel	Annealed (heated to 1500°F, cooled 20°F/hr to 1230°F, then air-cool)		81	52	28	56	High-strength aircraft structures (Bethlehem Steel Co., 1955)
	Hardened (water-quench from 1575°F, tempered at 900°F)		161	137	15	54	
AISI 4340 Nickel chromium molybdenum steel	Annealed		119	99	17	43	Large-scale, heavy-duty, high-strength structures (Merriam and Lubars, 1957)
	As-rolled		192	147	9	19	
	Hardened (water- or oil-quench from 1500°F, tempered at 800°F)		220	200	12	48	

(*cont.*)

TABLE 5.3—*Continued*

C. AUSTENITIC STAINLESS STEELS

Material designation or composition	Condition heat treatment	Cost[a] $/lb	Tensile strength, 1000 psi	Yield strength, 1000 psi	Elongation, %	Red. of area, %	Usage and reference
AISI 303	Annealed (rapid cool from 1950°F)	0.39–0.58	90	35	50	55	Atmosphere resistant, good machinability (Merriam and Lubars, 1957)
	Cold-worked	0.39–0.58	110	75	30	50	
AISI 304	Annealed (rapid cool from 1950°F)	0.39–0.58	85	30	60	70	Atmosphere resistant, good weldability (Merriam and Lubars, 1957)
	Cold-worked	0.39–0.58	110	75	60	—	
AISI 316	Annealed (rapid cool from 2000°F)	0.39–0.58	85	35	55	70	Outstanding corrosion resistance (Merriam and Lubars, 1957)

D. MARAGING NICKEL-IRON ALLOYS

Vasco Max 300[b] 18.5% Ni 9.0% Co 4.8% Mo 0.6% Ti	Annealed		150	110	18	72	Aircraft structural parts, dies and tooling, bolts (Yates and Hamaker, 1962)
	Maraged by (a) austenitizing at 1500°F and air-cooling to room temperature followed by (b) aging at 900°F and air-cooling		300	290	10	55	

E. Copper and Its Alloys

Material	Condition						Notes
Copper (commercial purity)	Annealed (400°C 1 hr, furnace cool)		32	10	45	—	Electrical wire, tubing, etc. (International Nickel Co.)
	Cold-drawn	0.31	45	40	15	—	
Cartridge brass 70% Cu 30% Zn	Cold-rolled (annealing 400°C 1 hr, furnace cool)	0.48	76	63	8	—	Deep drawing (from annealed condition) (International Nickel Co.)
Copper-beryllium 1.9% Be 0.25% Co	Annealed (1450°F, water-quench)	1.98	70	—	45	—	(International Nickel Co.)
	Cold-rolled (wire-rod)		200	—	2	—	
	Hardened (after annealing, 2 hr at 600°F)		200	150	2	—	

(cont.)

TABLE 5.3—Continued
F. Aluminum and Its Alloys

Material designation or composition	Condition heat treatment	Cost[a] $/lb	Tensile strength, 1000 psi	Yield strength, 1000 psi	Elongation, %	Red. of area, %	Usage and reference
1100-F (commercial purity)	Annealed (heat to 650°F, equalize, air-cool)	0.24	13	5	40	—	Corrosion resistant, high conductivity (Alcoa, 1962)
2011-T8	Precipitation hardened	0.28	59	45	12	—	Excellent machinability, not weldable, and corrosion sensitive (Alcoa, 1962)
6062-T6	Solution hardened	0.25	45	40	17	—	Corrosion resistant and weldable (Alcoa, 1962)
7178-T6	Fully precipitation hardened	0.30	88	78	10	—	High-strength, not weldable, and corrosion sensitive (Alcoa, 1962)

G. Nickel and Its Alloys

Nickel (pure)	Annealed (heat to 1400°F, hold 2 hr, air cool)	0.84	46	8.5	30	—	Magnetic, high corrosion resistance (International Nickel Co.)
Inconel X Type 550	Annealed (at 2050°F, hold 2 hrs)		125	75	50	—	Excellent high-temperature properties (International Nickel Co.)
6.8% Fe 15.2% Cr 1.1% Al 2.3% Ti	Annealed and age-hardened (2150°F, hold 4 hr, air cool; 1550°F, 24 hr, air cool; 1300°F, 20 hr, air cool)		175	110	20	—	
K-Monel 0.6% Mn 1.0% Fe	Annealed (heat to 1600°F, hold 2 hr, quench)		100	45	40	—	Excellent high-temperature properties and corrosion resistance (International Nickel Co.)
29.6% Cu 2.8% Al	Age-hardened, spring stock		185	160	10	—	

(*cont.*)

TABLE 5.3—Continued
H. Selection of Other Metals

Material designation or composition	Condition heat treatment	Cost[a] $/lb	Tensile strength, 1000 psi	Yield strength, 1000 psi	Elongation, %	Red. of area, %	Usage and reference
Lead (commercial purity)	Rolled	0.095	2.5	1.9	50	—	Pipes, tubes, etc. (International Nickel Co.)
Soft solder 50% Pb 50% Sn	Cast		6.8	—	50	—	Soldering (International Nickel Co.)
Invar 36% Ni 64% Fe	Annealed (heat to 800°C, hold 1 hr, water-quench)		71	40	43	—	Very low coefficient of expansion (0.9×10^6/°C) (International Nickel Co.)
Gold (commercial purity)	Annealed	510.0	17.5	—	40	—	Jewelry (International Nickel Co.)
Platinum (commercial purity)	Annealed	1240.0	27	—	28	—	Outstanding corrosion resistance, chemical ware, jewelry (International Nickel Co.)
Silver (commercial purity)	Annealed (at 300°C)	13.25	23	12	45	—	Jewelry, decorative silverware (International Nickel Co.)
Zinc (commercial purity)	Hot-rolled	0.11–0.12	23	—	50	—	Die casting (International Nickel Co.)

5.11 DATA FOR THE STRENGTH OF METALS AND ALLOYS

I. Refractory Metals (at Room Temperature)

Material	Condition					
Molybdenum (commercial purity)	As-rolled	8	100	75	30	—
Tantalum (commercial purity)	Annealed (1050°C in vacuum)	35	60	45	37	—
Titanium (commercial purity)	As-rolled	1.50–1.60	110	100	3	—
Titanium (commercial purity)	Annealed (1200°F)		90	70	23	—
Titanium 6% Al, 4% V	Annealed (1400°F, 1 hr, air-cool)		135	130	15	—
Titanium 6% Al, 4% V	Heat treated (1650°F, ½ hr, water-quench; 950°F 6 hr, air-cool)		155	145	12	—
Tungsten (commercial purity)	Hard (wire)	2.75–4.00	600	540	0–8	Good high-temperature properties (International Nickel Co.)

[a] These are ingot prices, 1962. In one-half in. diameter bar, for example, steel prices are:

Quantity	10,000 lb	50 lb
1020 hot-rolled	0.10	0.20
Drill rod (high-carbon steel)	0.44	0.80
4340 hot-rolled	0.25	0.33
316 stainless	0.87	1.21

In manufactured parts, prices may be much higher. For instance, for a ⅛-in. wide rack and pinion, the cost is $4.16 per lb in steel and $4.36 per lb in brass.

In 30,000-lb lots, aluminum wire and rod prices are:

1100-F	0.44–0.72
2011-T8	0.52–0.77
6062-T6	0.57–0.77
7178-T6	0.62–1.92

[b] Trade name of International Nickel Co.

5.12 SUMMARY

Crystalline materials can be hardened in various ways. One of the oldest and most effective methods of hardening a ductile metal is by alloying it with small amounts of additives which can form a solid solution substitutionally or interstitially with the parent metal. To produce effective hardening, the solute atom has to produce a local dilatation (or contraction) or directional distortion in the parent lattice due to its misfit. The stresses resulting from such misfit will effectively hinder the motion of dislocations by interacting with them. In the solid solution range, this hardening effect will increase initially linearly with the concentration $c(c < 10^{-2})$ of solute and more than linearly with the fractional difference ϵ of size between the solute and solvent atoms, giving a shear strength

$$k \approx 2G\epsilon^{4/3}c. \tag{5.6}$$

Quenching a metal from high temperatures or irradiating it with high-energy particles will produce small sessile dislocation loops which will produce strong directional distortions normal to their plane, also resulting in very effective hardening.

Precipitation of a solute out of solution in a supersaturated solid solution can give rise to effective hardening if the dispersion of particles is controlled such that their spacing is kept uniform and at a magnitude of about a hundred interatomic spacings. Such precipitates then form impenetrable obstacles for a dislocation. The dislocations can move through the crystal only by extruding through the spaces between precipitates; this leads to a shear strength inversely related to the distance between precipitates as given by Eq. 5.7:

$$k = Gb/l. \tag{5.7}$$

The methods of structural hardening (solid solution, precipitation, transformation to a lattice with high friction stress, irradiation, and quenching of vacancies) lend themselves to hardening finished parts by thermal treatment.

During plastic straining of a crystal, slip obstacles are produced which make the motion of dislocations increasingly more difficult. This is especially the case when dislocations on intersecting slip systems have to cut through each other. The resulting slip defects, such as dislocation dipoles, small segments of sessile dislocations, or vacancy or interstitial type dipole trails will serve to hold together a dislocation entanglement, enabling it to get increasingly denser. A higher and higher flow stress will be required to push new dislocations through such entanglements. The most prominent effect in strain-hardening is the mutual blocking action of dislocations. This gives a shear strength for the crystal which is proportional to the square root of the dislocation density.

The overcoming of some short-range slip obstacles can be aided by thermal activation. This results in a temperature dependence of the flow stress and gives rise to a thermally activated transient creep at a constant stress.

5.12 SUMMARY

◀ Most of the work of plastic deformation is lost as heat. The amount which accounts for the energy of the dislocation entanglements rarely exceeds 10% of the total work of plastic deformation.

◀ Hardening may also result by diffusionless shear transformations, such as twinning and martensite formation, if the spacing of the transformation lamellae is smaller than the mean free path of a dislocation for the same plastic strain level. Hard surface films can produce a hardening effect in cases where the prevention of escape of dislocations from a free surface can cause a significant rise of dislocation density and reduce the mean free path of dislocations. ▶

Grain boundaries in polycrystals are very effective obstacles to dislocation motion. Thus, in plastic deformation, each grain is required to deform compatibly with its neighbors. This requires intersecting slip inside individual grains almost from the very beginning, giving rise to a rate of hardening in a polycrystal equal to that of a single crystal undergoing multiple slip.

Plastic deformation is irreversible because of the irregular dislocation entanglements which form during its course. The resulting work-hardening is not completely isotropic. Initiation of reverse deformation in a previously cold-worked material usually requires a negligible stress, and a small amount of reverse plastic deformation can be obtained under applied stresses much less than the flow stress immediately preceding reverse deformation. Such an anisotropy of work-hardening, resulting partly from the anisotropy of the internal stresses in slip bands and partly from the anisotropy of slip obstacles, is known as the *Bauschinger effect*.

During certain phases of plastic deformation, an increase of strain may be accompanied by a decrease in flow stress, creating a potential mechanical instability. The yield phenomenon in low-carbon steel is a well-known example. In this case, the interstitial carbon or nitrogen atoms will migrate to the dislocations and anchor them to the lattice by relieving part of the high stresses around the dislocation core. The generation of fresh slip dislocations then will require a stress higher than the stress necessary to move them through the crystal, giving rise to the mechanical instability of the yield phenomenon.

◀ Formation of twin bands and kink bands requires the generation of a critical nucleus for which a high shear stress is necessary. The growth of such bands, on the other hand, can be accomplished under a lower stress. Thus, twins and kinks can also cause mechanical instabilities. ▶

In certain ranges of plastic deformation, where diffusion-controlled structural changes are negligibly slow, a relation between flow stress, strain rate, temperature, and plastic strain may exist in restricted ranges for monotonic deformation. Two such forms requiring empirical evaluation are given by

$$\sigma(\epsilon) = f\left[T\left(1 - \alpha \ln \frac{\dot{\epsilon}}{\dot{\epsilon}_0}\right)\right], \tag{5.22}$$

$$\dot{\epsilon} = \dot{\epsilon}_0 \epsilon^n \left(\frac{\sigma}{\sigma_0}\right)^m e^{-u/kT}. \tag{5.24}$$

Such relationships, or equations of state, can be very useful in predicting creep behavior for long loading times from short-term tests run at high temperatures.

Table 5.3 gives some room-temperature strength and ductility data of a selection of some common metals and alloys.

REFERENCES

ALCOA	1962	*Alcoa Aluminum Handbook*, Aluminum Company of America, Pittsburgh.
ARGON, A. S. OROWAN, E.	1964	"Plastic Deformation in MgO Single Crystals," *Phil. Mag.* **9**, 1003–1021.
ARGON, A. S. MALOOF, S. R.	1965	"Plastic Deformation of Tungsten Single Crystals," submitted for publication.
ASM	1961	*Metals Handbook*, 8th ed., T. Lyman, ed., American Society for Metals, Novelty, Ohio.
BAILEY, J. E. HIRSCH, P. B.	1960	"The Dislocation Distribution, Flow Stress, and Stored Energy in Cold-Worked Polycrystalline Silver," *Phil. Mag.* **5**, 485–498.
BARNES, R. S. MAZEY, D. J.	1960	"The Nature of Radiation-Induced Point Defect Clusters," *Phil. Mag.* **5**, 1247–1254.
BARRETT, C. S.	1953	*Structure of Metals*, 2nd ed., McGraw-Hill, New York.
BASINSKI, Z. S.	1957	"The Instability of Plastic Flow of Metals at Very Low Temperatures," *Proc. Roy. Soc. (London)* **A240**, 229–242.
BASINSKI, Z. S.	1959	"Thermally Activated Glide in Face-Centered Cubic Metals and Its Application to the Theory of Strain-Hardening," *Phil. Mag.* **4**, 393–432.
BAUSCHINGER, J.	1886	"On the Change of the Elastic Limit and Hardness of Iron and Steels through Extension and Compression, through Heating and Cooling, and through Cycling," *Mitteilung aus dem Mechanisch, Technischen Laboratorium der K. Technische Hochschule in München* **13**, Part 5, 31.
BECKER, R. OROWAN, E.	1932	"Über Sprunghafte Dehnung von Zinkkristallen," *Z. Physik* **79**, 566–572.
BETHLEHEM STEEL CO.	1955	*Modern Steels and Their Properties*, 3rd ed., Bethlehem, Pa.
BLEWITT, T. H. COLTMAN, R. R. REDMAN, J. K.	1955	"Work-Hardening in Copper Crystals," *Defects in Crystalline Solids*, Physical Society, London, pp. 369–382.
BOAS, W. OGILVIE, J.	1954	"The Plastic Deformation of a Crystal in Polycrystalline Aggregate," *Acta Met.* **2**, 655–659.

BRADY, G. S. (ed.)	1963	*Materials Handbook.* 9th ed., McGraw-Hill, New York.
BROOM, T. HAM, R. K.	1958	"The Effects of Lattice Defects on Some Physical Properties of Metals," *Vacancies and Other Point Defects in Metals and Alloys*, Institute of Metals, London, pp. 41–78.
BULLEN, F. P.	1962	"Discontinuous Yielding in Zinc Single Crystals," *Phil. Mag.* **7,** 133–140.
BULLOUGH, R. NEWMAN, R. C.	1962	"The Interactions of Vacancies with Dislocations," *Phil. Mag.* **7,** 529–531.
CHAMBERS, R. H. SCHULTZ, J.	1962	"Dislocation Relaxation Spectra in Plastically Deformed B.C.C. Metals," *Acta Met.* **10,** 467–485.
CHURCHMAN, A. T. COTTRELL, A. H.	1951	"Yield Phenomenon and Twinning in α-Iron," *Nature* **167,** 943–945.
CLAREBROUGH, L. M. HARGRAVES, M. E. HEAD, A. K. LORETTO, M. H.	1961	"Stored Energy and Flow Stress in Deformed Metals," *Phil. Mag.* **6,** 819–822.
CLASSEN-NEKLUDOWA, M.	1929	"Über die Sprungartige Deformation," *Z. Physik* **55,** 555–568.
CONRAD, H. HAYES, W.	1963	"Thermally Activated Deformation of the BCC Metals at Low Temperatures," *Trans. ASM* **56,** 249–262.
COTTRELL, A. H.	1953	*Dislocations and Plastic Flow in Crystals*, Oxford University Press, London.
COTTRELL, A. H.	1958	"Point Defects and the Mechanical Properties of Metals and Alloys at Low Temperatures," *Vacancies and Other Point Defects in Metals and Alloys*, Institute of Metals, London, pp. 1–40.
COTTRELL, A. H. BILBY, B. A.	1949	"Dislocation Theory of Yielding and Strain Aging of Iron," *Proc. Phys. Soc.* **A62,** 49–62.
COTTRELL, A. H. STOKES, R. J.	1955	"Effect of Temperature on the Plastic Properties of Aluminum Crystals," *Proc. Roy. Soc. (London)* **A233,** 17–34.
DEAK, G.	1961	*A Study of the Causes of the Bauschinger Effect*, Sc. D. Thesis, M.I.T., Cambridge, Mass.
DIEHL, J.	1956	"Zugverformung von Kupfer-Einkristallen I. Verfestigungskurven und Oberflächenerscheinungen," *Z. Metallkunde* **47,** 331–343.
EHRENFEST, P. JOFFE, A.	1924	"On the Jerky Extension Phenomenon," quoted by Classen-Nekludowa (1929).
FARREN, W. S. TAYLOR, G. I.	1925	"The Heat Developed During Plastic Extension of Metals," *Proc. Roy. Soc. (London)* **A107,** 422–451.
FLEISCHER, R. L.	1961	"Solution Hardening," *Acta Met.* **9,** 996–1000.

FLEISCHER, R. L.	1962	"Solution Hardening by Tetragonal Distortions: Application to Irradiation Hardening in F.C.C. Crystals," *Acta Met.* **10**, 835–842.
FRIEDEL, J.	1959	"Dislocation Interactions and Internal Strains," *Internal Stresses and Fatigue in Metals*, Rassweiler and Grube, eds., Elsevier, Amsterdam, pp. 220–263.
GARSTONE, J. HONEYCOMBE, R. W. K. GREETHAM, G.	1956	"Easy Glide of Cubic Metal Crystals," *Acta Met.* **4**, 485–494.
GILMAN, J. J.	1962	"Debris Mechanism of Strain Hardening," *J. Appl. Phys.* **33**, 2703–2710.
GILMAN, J. J. JOHNSTON, W. G.	1957	"The Origin and Growth of Glide Bands in Lithium Fluoride Crystals," *Dislocations and Mechanical Properties of Crystals*, Fischer et al., eds., Wiley, New York, pp. 116–163.
HAHN, G. T.	1962	"A Model for Yielding with Special Reference to the Yield Point Phenomena of Iron and Related B.C.C. Metals," *Acta Met.* **10**, 727–738.
HOLDEN, A. N. KUNZ, F. W.	1953	"Dimension and Orientation Effects in the Yielding of Carburized Iron Sheet Crystals," *Acta Met.* **1**, 495–502.
HOLLOMON, J. H. ZENER, C.	1946	"Problems in Non-Elastic Deformation of Metals," *J. Appl. Phys.* **17**, 69–82.
HOWIE, A.	1960	"Dislocation Arrangements in Deformed F.C.C. Single Crystals," *Proc. European Regional Conf. on Electron Microscopy* (Delft), Vol. 1, pp. 383–386.
INTERNATIONAL NICKEL CO.		*Properties of Some Metals and Alloys*, The International Nickel Co., New York.
JOHNSTON, W. G.	1962	"Effect of Impurities on the Flow Stress of LiF Crystals," *J. Appl. Phys.* **33**, 2050–2058.
JOHNSTON, W. G. GILMAN, J. J.	1959	"Dislocation Velocities, Dislocation Densities, and Plastic Flow in Lithium Flouride Crystals," *J. Appl. Phys.* **30**, 129–144.
KEH, A. S. WEISSMANN, S.	1963	"Deformation Substructure in Body-Centered Cubic Metals," *Electron Microscopy and Strength of Crystals*, Thomas and Washburn, eds., Wiley, New York, pp. 301–332.
KOCKS, U. F.	1960	"Polyslip in Single Crystals," *Acta Met.* **8**, 345–352.
KUHLMANN-WILSDORF, D.	1962	"A New Theory of Work Hardening," *Trans. AIME* **224**, 1047–1061.
KUHLMANN-WILSDORF, D. WILSDORF, H. G. F.	1963	"Origin of Dislocation Tangles and Loops in Deformed Crystals," *Electron Microscopy and Strength of Crystals*, Thomas and Washburn, eds., Wiley, New York, pp. 575–604.

LALLY, J. S.	1963	Quoted by P. B. Price "Direct Observations of Glide, Climb and Twinning in Hexagonal Metal Crystals," *Electron Microscopy and Strength of Crystals*, Thomas and Washburn, eds., Wiley, New York.
LANG, A. R.	1959	"Studies of Individual Dislocations in Crystals by X-Ray Diffraction Radiography," *J. Appl. Phys.* **30**, 1748–1755.
LESLIE, W. C.	1961	"The Quench Aging of Low-Carbon Iron and Iron-Manganese Alloys, An Electron Transmission Study," *Acta Met.* **9**, 1004–1022.
LINDE, J. O. EDWARDSON, S.	1954	"Investigation of the Critical Shear Stress for Single Crystals of Metallic Solid Solutions," *Arkiv för Fysik* **8**, 511–519.
LINDE, J. O. LINDELL, B. O. STADE, C. H.	1950	"Investigation of the Critical Shear Stress for Single Crystals of Metallic Solid Solutions. I," *Arkiv för Fysik* **2**, 89–97
LUBAHN, J. D. FELGAR, R. P.	1961	*Plasticity and Creep of Metals*, Wiley, New York.
LÜCKE, K. LANGE, H.	1952	"Über die Form der Verfestigungskurve von Deformationsbändern," *Z. Metallkunde* **43**, 55–66.
LUDWIG, P.	1909	*Elemente der Technologischen Mechanik*, Leipzig.
MACGREGOR, C. W. FISHER, J. C.	1946	"A Velocity Modified Temperature for the Plastic Flow of Metals," *J. Appl. Mech.* **13**, A11-A16.
MADER, S. SEEGER, A.	1960	"Untersuchung des Gleitlinienbildes Kubisch-Flachenzentrierter Einkristalle," *Acta Met.* **8**, 513–522.
MANTELL, C. L. (ed.)	1958	*Engineering Materials Handbook*, McGraw-Hill, New York.
MARCINKOWSKI, M. J.	1963	"Theory and Direct Observation of Antiphase Boundaries and Dislocations in Superlattices," *Electron Microscopy and Strength of Crystals*, Thomas and Washburn, eds., Wiley, New York, pp. 333–440.
MERRIAM, J. C. LUBARS, W. (eds.)	1957	*Materials Selector*, Special Issue of *Materials In Design Engineering*, 46, No. 4.
MINER, D. F. SEASTONE, J. B. (eds.)	1955	*Handbook of Engineering Materials*, Wiley, New York.
MORDIKE, B.	1962	"Plastic Deformation of Zone Refined Tantalum Single Crystals," *Z. Metallkunde* **53**, 586–592.
MOTT, N. F.	1953	"A Theory of Work-Hardening of Metals. II: Flow Without Slip Lines, Recovery and Creep." *Phil. Mag.* **44**, 742–765.
NABARRO, F. R. N.	1951	"The Interaction of Screw Dislocations and Sound Waves," *Proc. Roy. Soc. (London)* **A209**, 278–290.

Orowan, E.	1947	"The Creep of Metals" *West of Scotland Iron and Steel Institute* **54**, 45–96.
Orowan, E.	1959	"Causes and Effects of Internal Stresses," *Internal Stresses and Fatigue in Metals*, General Motors Symposium, Elsevier, Amsterdam, 59–80.
Padawer, G.	1963	*Dislocation Multiplication as a Rate Controlling Mechanism*, S. M. Thesis, M.I.T., Cambridge, Mass.
Paxton, H. W. Bear, I. J.	1955	"Further Observations on Yield in Single Crystals of Iron," *J. Metals Trans.* **7**, 989–994.
Phillips, P.	1905	"The Slow Stretch in India Rubber, Glass, and Metal Wires when Subjected to a Constant Pull," *Proc. Phys. Soc. (London)* **19**, 491–511.
Portevin, A. LeChatelier, F.	1923	"Sur un Phénomène Observé lors de l'essai de traction d'alliages en cours de Transformation," *Acad. Sci. Compt. Rend.* **176**, 507–510.
Preston, G. D.	1938	"The Diffraction of X-Rays by an Age-Hardening Alloy of Aluminum and Copper. The Structure of an Intermediate Phase," *Phil. Mag.* **26**, 855–871.
Sachs, G. Shoji, H.	1927	"Bauschinger Effect in Brass," *Z. Physik* **45**, 776–796.
Saimoto, S.	1963	*Low Temperature Tensile Deformation of Copper Single Crystals Oriented for Multiple Slip*, Ph.D. Thesis, M.I.T., Cambridge, Mass.
Schmid, E. Boas, W.	1935	*Kristallplastizität*, Springer, Berlin. English trans., *Plasticity of Crystals*, Hughes, London (1950).
Seeger, A.	1957	"The Mechanism of Glide and Work Hardening in F.C.C. and H.C.P. Metals," *Dislocations and Mechanical Properties of Crystals*, Fisher et al., eds., Wiley, New York, pp. 243–329.
Seeger, A. Diehl, J. Mader, S. Rebstock, H.	1957	"Work Hardening and Work Softening of Face-Centered Cubic Metal Crystals," *Phil. Mag.* **2**, 323–350.
Seeger, A. Kronmüller, H. Mader, S. Träuble, H.	1961	"Work Hardening of Hexagonal Close-Packed Crystals and in the Easy Glide Region of Face-Centered Cubic Crystals," *Phil. Mag.* **6**, 639–656.
Silcox, J. Whelan, M. J.	1960	"Direct Observations of the Annealing of Prismatic Dislocation Loops and Climb of Dislocations in Quenched Aluminum," *Phil. Mag.* **5**, 1–23.
Smithells, C. J. (ed.)	1962	*Metals Reference Book*, 3rd ed., Interscience, New York.

Suzuki, H.	1952	"Chemical Interaction of Solute Atoms with Dislocations," *Sci. Rept. Res. Inst., Tohoku University* **A4**, 455–463.
Swann, P. R.	1963	"Dislocation Arrangements in F.C.C. Metals and Alloys," *Electron Microscopy and Strengths of Crystals*, Thomas and Washburn, eds., Wiley, New York, pp. 131–182.
Sylvestrowicz, W. Hall, E. O.	1951	"The Deformation and Aging of Mild Steel," *Proc. Phys. Soc.* **B64**, 495–502.
Taylor, A.	1962	"Low Temperature Internal Friction in Monocrystalline Lithium Fluoride," *Acta Met.* **10**, 490–495.
Taylor, G. I.	1934	"The Mechanism of Plastic Deformation of Crystals, Part I, Theoretical," *Proc. Roy. Soc. (London)* **A145**, 362–387.
Taylor, G. I. Quinney, H.	1934	"The Latent Energy Remaining in a Metal After Cold Working," *Proc. Roy. Soc. (London)* **A143**, 307–326.
Uhlig, H. H.	1963	*Corrosion and Corrosion Control*, Wiley, New York
van Bueren, H. G.	1960	*Imperfections in Crystals*, North Holland, Amsterdam.
Westwood, A. R. C.	1960	"The Effect of Surface Conditions on the Mechanical Properties of Lithium Fluoride Crystals," *Phil. Mag.* **5**, 981–990.
Winchell, P. G. Cohen, M.	1962	"The Strength of Martensite," *Trans. ASM* **55**, 347–361.
Woldman, N. E. (ed.)	1962	*Engineering Alloys*, Reinhold, New York.
Woolley, R. L.	1953	"The Bauschinger Effect in some Face-Centered Cubic and Body-Centered Cubic Metals," *Phil. Mag.*, Series 7, **44**, 597–618.
Wyatt, O. H.	1953	"Transient Creep in Pure Metals," *Proc. Phys. Soc.* **B66**, 459–480.
Yates, D. H. Hamaker, J. C., Jr.	1962	"New Ultrahigh Strength Steels, the Maraging Grades," *Metal Prog.* **82**, 97–100.
Young, F. W., Jr.	1961	"Elastic-Plastic Transition in Copper Crystals as Determined by an Etch-Pit Technique," *J. Appl. Phys.* **32**, 1815–1820.
Young, F. W., Jr.	1962	"On the Yield Stress of Copper Crystals," *J. Appl. Phys.* **33**, 963–970.

PROBLEMS

◀ 5.1. Show that the maximum shear stress that a misfitting solute atom can produce on a slip plane at a distance t from it is given by Eq. 5.3.

◀ 5.2. Show that the total force that a misfitting solute atom can exert on an edge dislocation in a plane a distance t from it is given by Eq. 5.4, when $t = b/2$ and $\tan\theta = 2$.

5.3. Show that a misfitting solute atom cannot exert a net force on a screw dislocation in a plane a distance t from it.

◀ 5.4. Considering the energy of a circular vacancy disk to be made up entirely of the energy of its surfaces, find the minimum radius at which it would be energetically favorable for it to collapse into a sessile dislocation loop in an α-iron crystal. [For α-iron, $b = 2.47 \times 10^{-8}$ cm, shear modulus $G = 7.75 \times 10^{11}$ dyne/cm^2, surface energy $\gamma \approx 300$ erg/cm^2, close to the melting point. Assume the vacancy disk to be on the (112) plane.]

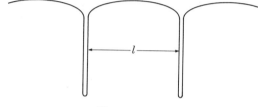

FIGURE 5.42

◀ 5.5.(a) Consider the motion of a screw dislocation held back by dipole trails, as shown in Fig. 5.42. If the segment of length l moves forward by an interatomic distance b, and remains geometrically similar to itself, show that the required applied stress is $\sigma = Gb/20l$, if the energy per unit length of a dipole trail is $Gb^2/20$.

(b) If the density of screw dislocations is $\Lambda/2$, and two slip systems are active to the same extent, show that $l = \sqrt{\Lambda}$, approximately.

(c) Considering the case of a linear rise in dislocation density from 10^6 to 10^{11} per cm^2 for a shear strain of unity, calculate the rate of hardening $d\sigma/d\gamma$ when the dislocation density is 10^8 per cm^2.

◀ 5.6. Show that the maximum shear stress that an edge dislocation can exert on another parallel edge dislocation with parallel Burgers vector is that given by Eq. 5.8.

5.7. Calculate the hardening rate of the single crystal of tungsten of Fig. 5.20 at a strain of 1%.

◀ 5.8. Show that incorporation of the slope of the flow-stress curve as shown in Fig. 5.27 into Eq. 5.18 will lead to Eq. 5.19.

◀ 5.9. Integrate Eq. 5.19 and show that the transient creep strain is correctly given by Eq. 5.20.

5.10. Sketch in detail how a single expanding dislocation loop can double cross glide and produce dislocation dipoles and other similar defects.

5.11. A solute atom producing only volume misfit can interact effectively only with edge dislocations. Consider how this interaction can retard screw dislocations also, by blocking the motion of edge components of dislocation loops.

5.12. For a reasonable dislocation density, how many dislocation segments would be required per unit volume to account for a reduction in Young's modulus from 31×10^6 psi to 29×10^6 psi in steel as a result of increasing cold-work?

◀5.13. What is the amplitude of thermal motion of a dislocation line of length l pinned at the ends?

5.14. Calculate the order of magnitude of the Bauschinger strain from Eq. 4.3, assuming a dislocation density of $\Lambda = 10^{10}$ per cm^2 and a mean free path of dislocations corresponding to a cell size of 10^{-4} cm.

5.15. When a metal is irradiated with high-energy particles, considerable lattice damage results. Apart from some very transient effects, the main effect of radiation is the production of vacant lattice sites and interstitial atoms. Under suitable conditions these point defects can congregate into vacancy and interstitial disks which may then transform into sessile dislocation loops. Such dislocation loops are known to harden the metal. Suppose that the radiation produces an equal concentration of vacancies and interstitials of 10^{19} per cm^3 each, and these coalesce into vacancy and interstitial disks of 10^{-6} cm diameter with 100% efficiency (i.e., with no loss at the surfaces). Calculate the yield strength of the material as a fraction of its shear modulus.

5.16. The introduction of some vacancies into a perfect crystal reduces the free energy of the crystal, so that a certain atomic concentration c of vacancies

$$c = e^{-u_v/kT}$$

can exist in a crystal at thermal equilibrium, where u_v is the internal energy rise of the crystal due to the introduction of one vacancy into a perfect crystal. u_v is about 0.8 ev for aluminum.

It is proposed that aluminum be hardened by heating it to an elevated temperature followed by a subsequent quench. At the elevated temperature, a high concentration of vacancies is established. When quenched, the excess vacancy concentration in the aluminum is eliminated by the formation of sessile dislocation loops. (a) If the average diameter of the sessile dislocation loops which form is 10^{-6} cm, what is the temperature from which aluminum must be quenched to produce a shear strength equal to 10^{-3} times the shear modulus? (Neglect the equilibrium vacancy concentration at room temperature.) (b) Is this a feasible operation?

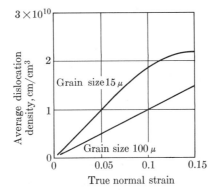

FIGURE 5.43

5.17. Keh and Weissmann (1963) have measured the dislocation density in iron as a function of true strain for two samples with different grain sizes. Their curves are redrawn in Fig. 5.43. (a) From this information obtain the possible stress-strain curves for these two samples. Consider carefully all assumptions leading to the answer. (b) Why is the rate of dislocation multiplication per unit strain more rapid in the fine-grained sample?

5.18. If an edge dislocation has to glide through a forest of screw dislocations threading through the slip plane of the edge dislocation at right angles, jogs must be formed at each intersection. Calculate the stress necessary to drive the edge dislocation through such a forest if the density of the forest is 10^9 lines/cm^2.

5.19. It is desired to develop a strong alloy with a shear strength of the order of $G/100$ by precipitation hardening. Calculate the necessary precipitate particle spacing, and estimate the required percentage of the second phase. Does it make any difference if the precipitates are coherently or incoherently attached to the lattice of the matrix material?

CHAPTER 6

DEFORMATION IN POLYMERS: VISCOELASTICITY

6.1 SYNOPSIS

After a review of the chemical composition and molecular structure of polymers, typical kinds of behavior are discussed—the glassy behavior at low temperatures, the increasingly noticeable viscous effects at higher temperatures, the transition to a viscous liquid in the case of glass or to a rubbery material and then a liquid in the case of a long-chain polymer, the prevention of flow in cross-linked polymers giving a stable rubber, the plastic flow at high stress, and the irreversible chemical decomposition at high temperatures.

The viscoelastic and viscous behavior of a polymer are derived from a simple molecular model. Some of the effects of strain-induced anisotropy on flow phenomena in viscous liquids are shown. When the viscosity controlling the motion of segments of rubber molecules becomes low enough to be neglected, and yet either time or cross-linking does not allow the long-chain molecules to slide over each other, a theory of rubber elasticity, based on the thermal motion of the molecular segments, gives the observed nonlinear stress-strain relations.

The linearity of the viscoelastic stress-strain relations for small strains allows us to represent them by spring and dashpot models, or by a single creep or relaxation curve. The Boltzmann superposition principle allows combining these data to predict the behavior under a complex stress history.

Polymers may exhibit crystallinity, which leads to a behavior more dependent on stress and less on time than encountered in the viscoelastic regime. Drawing of polymers into fine filaments with high degrees of orientation also enhances crystallization, and provides higher strength.

Finally, representative data are given for a variety of polymeric materials of utility for their mechanical behavior.

6.2 STRUCTURE OF POLYMERS

As pointed out in Section 1.5, a polymer is a material in which one or more atomic groups, or *repeat units*, occur hundreds or thousands of times in chains or networks. A variety of common polymers are listed in Table 6.1, classified according to the types of connections between the repeat units. The letter abbreviations given in the table are to be preferred over the trade names, since the trade names, being the property of the manufacturer, can be applied to different materials.

The term "Bakelite," for example, at one time denoted a phenol formaldehyde resin of the original producer, but now denotes their entire line of products.

The chemical terms alone are not always accurate either, for a particular product may be importantly modified by the addition of other polymers or plasticizing agents which make its behavior different and more useful than that of the pure form. We shall use the chemical name as giving the basic substance.

As a matter of general interest and background, the chemical structure and modes of polymerization of some common polymers will be discussed, after which we shall turn to the next larger scale of structure, namely, the arrangements of repeat units relative to each other.

TABLE 6.1

Monomers of Common Polymers Classified According to Mode of Polymerization Including Some Abbreviations and Trade Names

TABLE 6.1A. Vinyl Polymers

Generic reaction

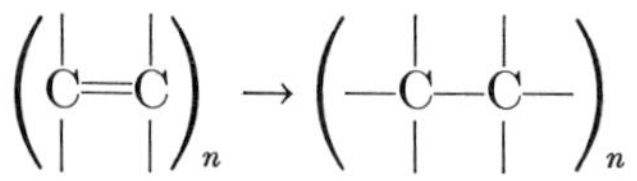

$$\left(\begin{array}{c} | \; | \\ C=C \\ | \; | \end{array} \right)_n \rightarrow \left(\begin{array}{c} | \; | \\ -C-C- \\ | \; | \end{array} \right)_n$$

Ethylene
PE

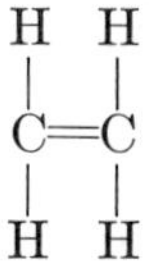

$$\begin{array}{c} H \; H \\ | \; | \\ C=C \\ | \; | \\ H \; H \end{array}$$

Styrene
PS

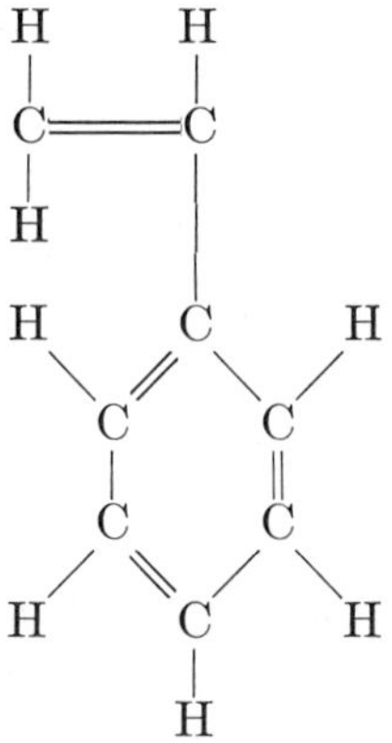

$$\begin{array}{c} H \quad\quad H \\ | \quad\quad | \\ C=\!=\!=C \\ | \\ H \end{array}$$

Vinyl chloride
PVC

$$\begin{array}{c} H \; Cl \\ | \; | \\ C=C \\ | \; | \\ H \; H \end{array}$$

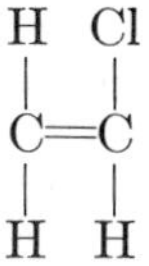

Vinylidene chloride
(Saran)

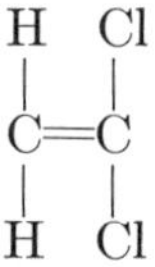

$$\begin{array}{c} H \; Cl \\ | \; | \\ C=C \\ | \; | \\ H \; Cl \end{array}$$

Propylene (Propene)

$$\begin{array}{c} H \quad\quad H \\ | \quad\quad | \\ C=\!=\!=C \\ | \quad\quad | \\ H \quad H-C-H \\ \quad\quad\quad | \\ \quad\quad\quad H \end{array}$$

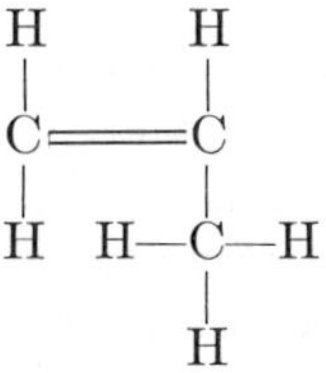

TABLE 6.1A. (Cont.)

Acrylonitrile PAN (Orlon)

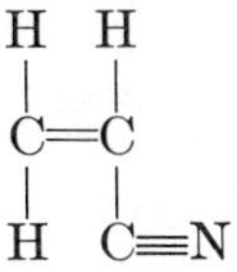

$$\begin{array}{c} H \quad H \\ | \quad | \\ C = C \\ | \quad | \\ H \quad C \equiv N \end{array}$$

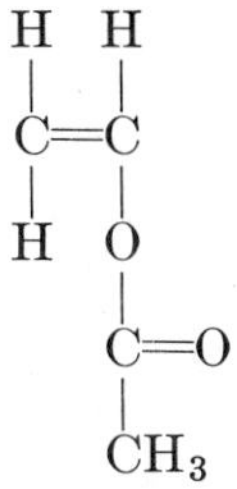

Isobutylene (Propene, 2 methyl) PIB (Vistanex)

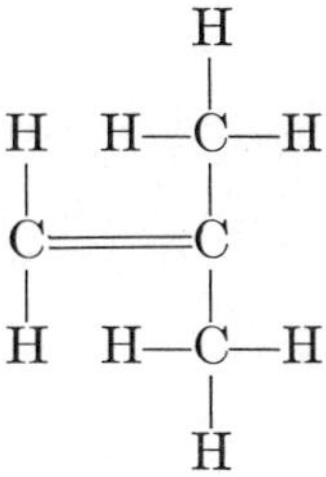

$$\begin{array}{c} \qquad\quad H \\ \qquad\quad | \\ H \quad H-C-H \\ | \qquad | \\ C = C \\ | \qquad | \\ H \quad H-C-H \\ \qquad\quad | \\ \qquad\quad H \end{array}$$

Vinyl alcohol PVA

$$\begin{array}{c} H \quad H \\ | \quad | \\ C = C \\ | \quad | \\ H \quad OH \end{array}$$

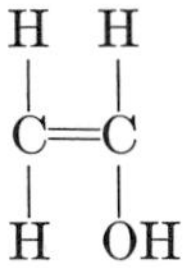

Tetrafluoroethylene PTFE (Teflon)

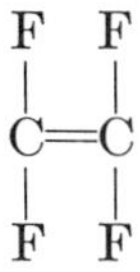

$$\begin{array}{c} F \quad F \\ | \quad | \\ C = C \\ | \quad | \\ F \quad F \end{array}$$

Methyl methacrylate PMMA (Lucite, Perspex, Plexiglas)

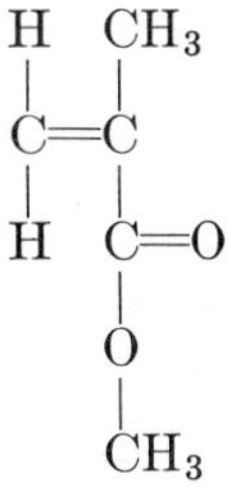

$$\begin{array}{c} H \quad CH_3 \\ | \qquad | \\ C = C \\ | \qquad | \\ H \quad C = O \\ \qquad | \\ \qquad O \\ \qquad | \\ \qquad CH_3 \end{array}$$

Vinyl acetate PVAc

$$\begin{array}{c} H \quad H \\ | \quad | \\ C = C \\ | \quad | \\ H \quad O \\ \qquad | \\ \qquad C = O \\ \qquad | \\ \qquad CH_3 \end{array}$$

◀**A. Vinyl Polymerization.** In the *vinyl* type of polymerization, of which ethylene ($H_2C=CH_2$) is an example, the double bond is changed to a single bond, and one valency electron on each end of the molecule attaches the monomer to other units. The reaction requires an initiator which attaches to one monomer, activating that so it attacks the next, and so on. In polyethylene, the initiator can be oxygen at a high pressure or organo-metallic catalysts at low pressures (Ziegler process). In the first case, a polyethylene with side branches results. In the second case, the result is simple chain, or *linear*, polyethylene, which because of its greater regularity can be more completely crystallized. In general, the less initiator used, the fewer and longer will be the individual chains, although the time required for polymerization will be longer.

TABLE 6.1B. DIENE POLYMERS

Generic reaction

$$\left(\begin{array}{c}|\\ C=C-C=C\\ |\end{array}\right)_n \rightarrow \left(\begin{array}{c}|\\ -C-C=C-C-\\ |\end{array}\right)_n$$

Butadiene
 PB (ingredient of GRS rubber)

H H H H
| | | |
C=C—C=C
| |
H   H

also by partial vinyl addition

    H   H H H H
    | | | | |
 —C—C ——— C—C=C—C—
    | | |
    H C—H   H   H
       ‖
       HCH

Chloroprene (neoprene)

H H Cl H
| | | |
C=C—C=C
| |
H   H

Isoprene (natural rubber, gutta percha)

          H
H H   HCH   H
| | | |
C=C —— C=C
| |
H   H

◀ **B. Diene Polymerization.** The *diene* type of polymerization is illustrated by polybutadiene. The suffix *ene* stands for double bonds. The word *diene* then indicates that there are two double bonds in the backbone chain of the monomer.

These double bonds open to attach repeat units together. As indicated in Table 6.1, butadiene can also polymerize in the vinyl mode. In reality, the two types of addition occur along the same chain. The result, polybutadiene, is Buna rubber ("Bu" from butadiene, "Na" from the sodium catalyst). If one of the central hydrogen atoms in butadiene is replaced by a chlorine atom, the polymerization product is neoprene, an oil- and gasoline-resistant synthetic rubber developed by Wallace H. Carothers, the creator of nylon. If the same hydrogen is replaced by a methyl group, the result is isoprene. The polymer of isoprene is natural rubber or gutta percha, depending on the structure, as discussed below.

◀ **C. Ester Polymerization.** In *ester* polymerization, di-basic or di-acidic monomers combine in a reaction which gives off water. Polymerization in which another compound is given off, in this case water, is called a *condensation* reaction, in contrast to vinyl and diene polymerization, which are simply *addition* reactions. Polyethylene terephthalate is actually obtained by a several-stage process in which methanol and later ethylene glycol are condensed. In making glyptal, a useful laboratory resin for sealing joints, the liberation of water is avoided by starting with the anhydride of the acid, phthalic anhydride. Thus, glyptal can harden between two impermeable surfaces. Furthermore, the presence of *three* hydroxyl ions in glycerol allows the formation of primary bonds between as well as along chains, and thus gives a space network.

TABLE 6.1C. POLYESTERS

Generic reaction

$$\text{HO-R-OH} + \text{H-P-H} \rightarrow \text{-R-P-} + 2\,\text{H}_2\text{O}$$
$$\text{alcohol} + \text{acid} \quad \text{alkyl ester} + \text{water}$$

Ethylene terephthalate
(Dacron, Cronar, Mylar, Terylene)

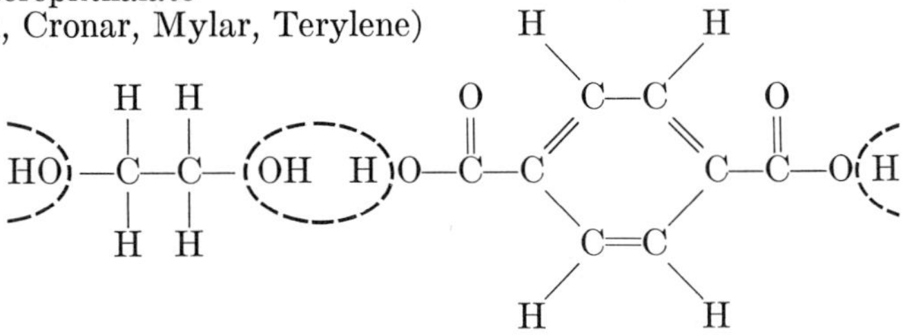

ethylene glycol — terephthalic acid

Urethane

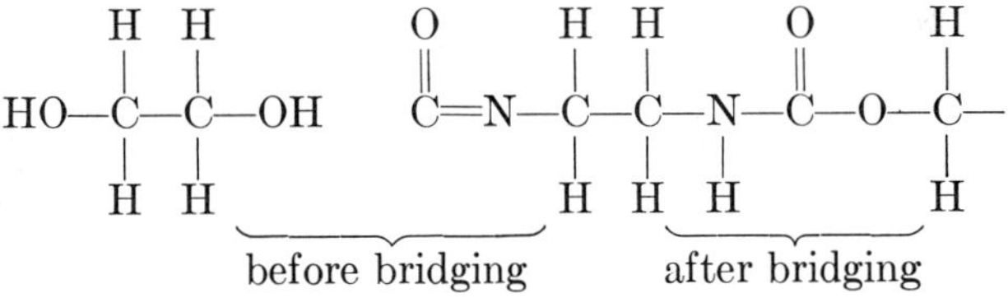

before bridging — after bridging

ethylene glycol — ethylene diisocyanate

Glyptal

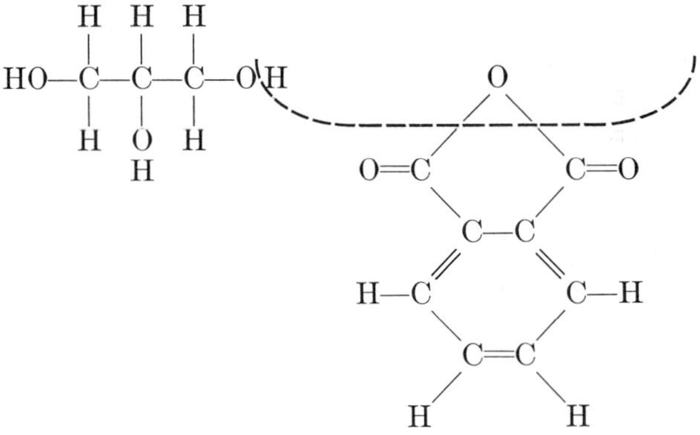

glycerol — phthalic anhydride

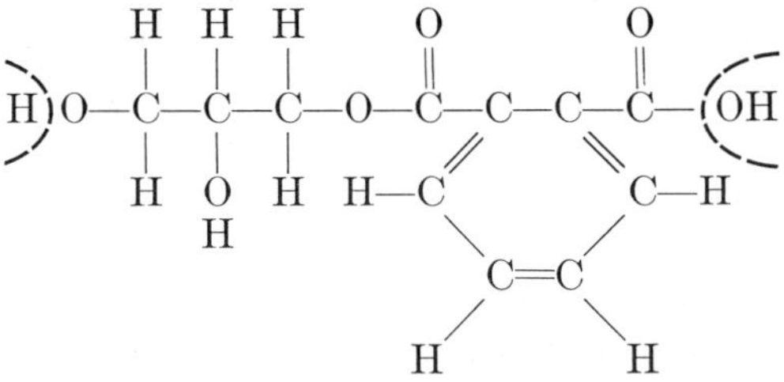

TABLE 6.1D. POLYAMIDES

Generic reaction

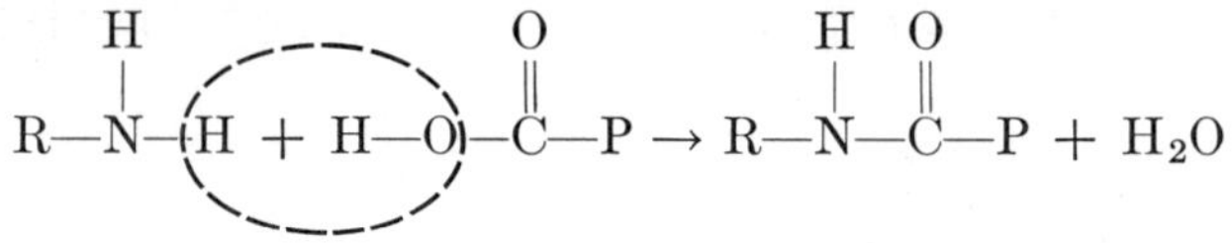

$$R-N(H + H-O)-\overset{O}{\underset{\|}{C}}-P \rightarrow R-N-\overset{O}{\underset{\|}{C}}-P + H_2O$$

Peptides

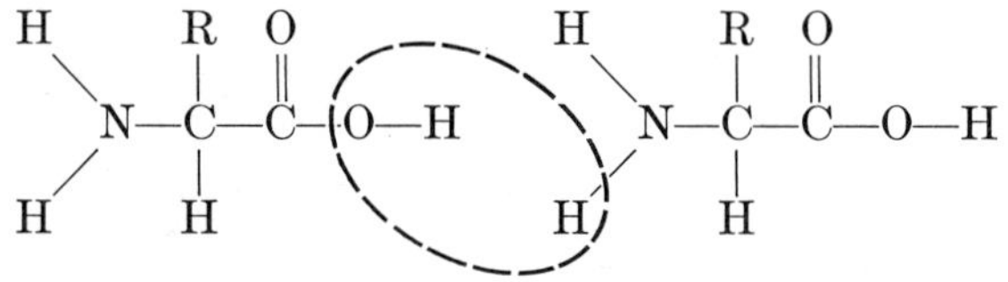

If R is H, peptide is aminoacetic acid (glycine)
If R is CH_3, peptide is D alanine
If R is CH_2OH, peptide is D serine

Nylon 6

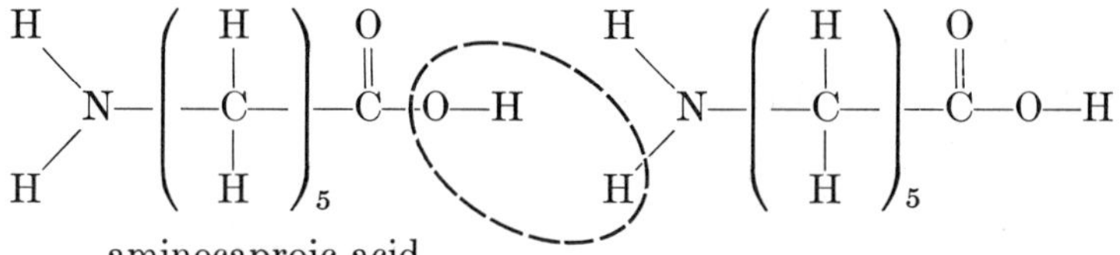

aminocaproic acid

Nylon 6,6

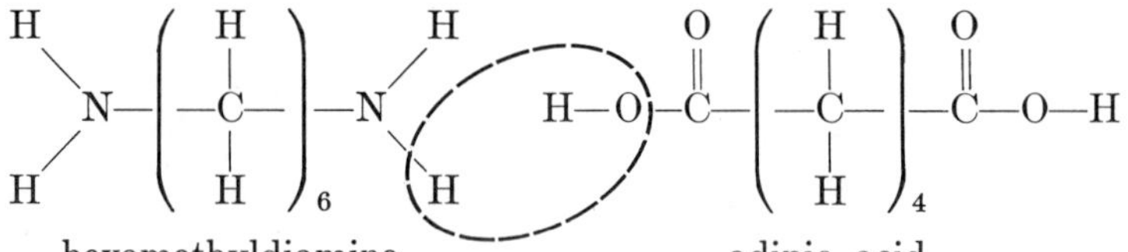

hexamethyldiamine   adipic acid

◂**D. Amine Polymerization.** The *amides* polymerize with the elimination of water from an amine and a carboxylic acid group. There are dozens of aminoacetic acids common in the proteins of living organisms. Hydrogen bonding between oxygen and nitrogen atoms of the same chain makes it curl up into a helix,

TABLE 6.1E. POLYSACCHARIDES

Cellulose

[glucose ring structures shown]

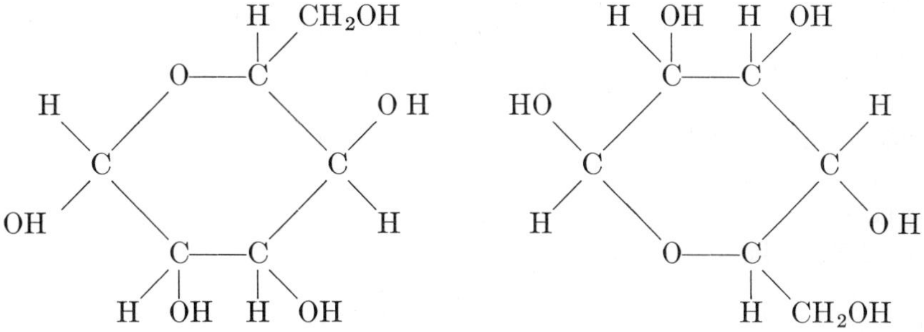

glucose

—OH + HNO$_3$ → —NO$_3$ + H$_2$O gives cellulose nitrate

—OH + HCH$_2$COOH → —CH$_2$COOH + H$_2$O gives cellulose acetate (CA)

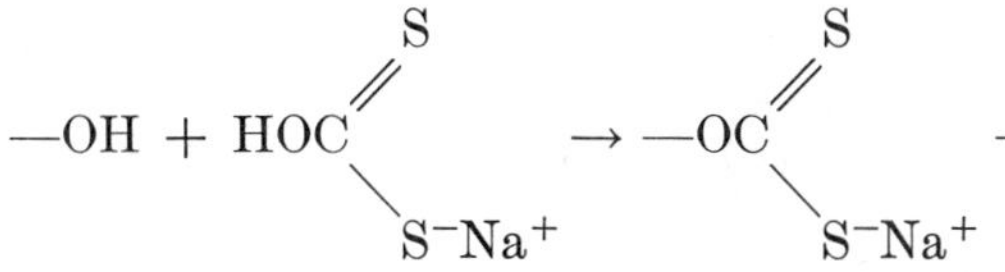

—OH + HOC(=S)(S$^-$Na$^+$) → —OC(=S)(S$^-$Na$^+$) + H$_2$O gives cellulose xanthate (viscose rayon)

---

which may in turn be folded into a complicated knot or arranged in a fibrous or planar structure, forming the basis of a crystallized protein. *Peptides* are polymerization products of substituted aminoacetic acids. The three given in Table 6.1 comprise 85% of a silk fiber. The most important man-made polymer of the amine group is nylon. One variety is formed by the self-polymerization of aminocaproic acid and another by the polymerization of hexamethylenediamine with adipic acid. The numerical designation of nylon indicates the number of carbon atoms in the monomers.

◀ **E. Polysaccharides.** In the group of *polysaccharides*, cellulose can be regarded as a polymer whose repeat units are rings formed from the simple sugar glucose. The large number of hydroxyl groups gives rise to strong hydrogen bonding between the chains. This bonding is responsible for the insolubility and the lack of heat softening. Replacement of one, two, or three of these hydroxyl groups by treatment with nitric, acetic, or xanthic acid gives a polymer that can be softened in a solvent, worked, and then dried to give a regenerated cellulose such as rayon.

TABLE 6.1F. FORMALDEHYDE COPOLYMERS

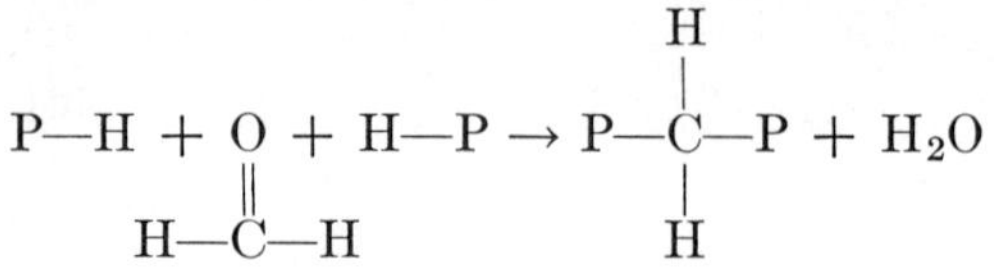

$$\text{P—H} + \underset{\underset{\text{H—C—H}}{\|}}{\text{O}} + \text{H—P} \rightarrow \text{P—}\underset{\underset{\text{H}}{|}}{\overset{\overset{\text{H}}{|}}{\text{C}}}\text{—P} + \text{H}_2\text{O}$$

where the group P may be one of the following:

Phenol formaldehyde
PF, (Bakelite)

P = phenol =

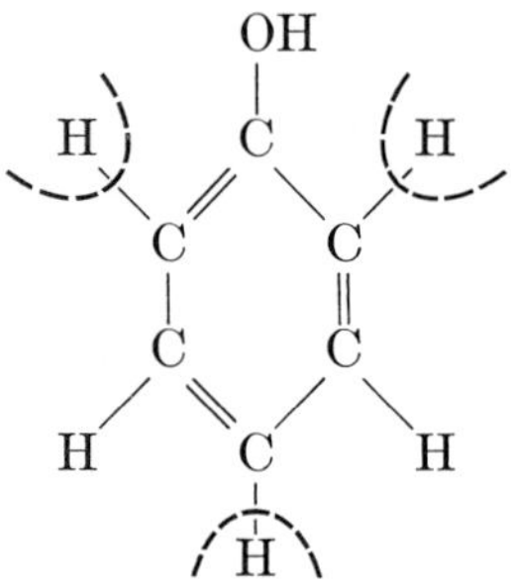

Urea formaldehyde
UF

P = urea =

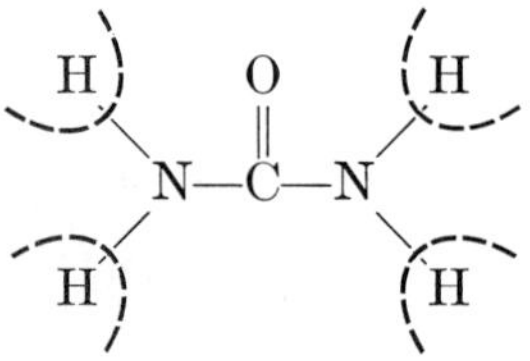

Melamine formaldehyde
MF

P = melamine =

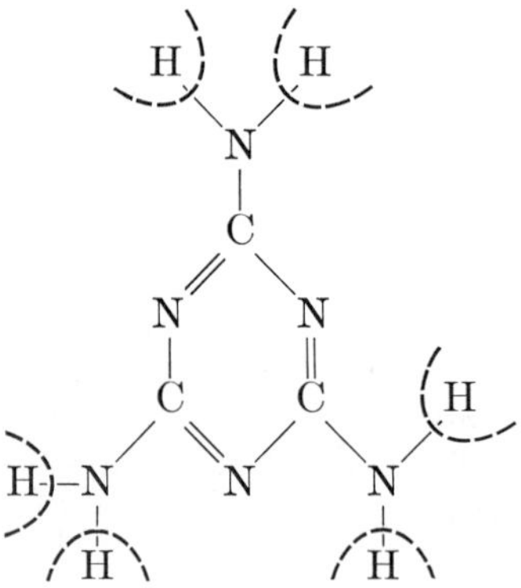

◀ **F. Formaldehyde Resins.** The *formaldehydes* are typically formed by a condensation reaction involving formaldehyde and two other molecules, leaving a methylene (CH_2) group as a bridge between the two larger monomers. Usually, the molecules used have three or more hydrogen atoms available, so that space networks can be formed, as discussed above in connection with glyptal.

TABLE 6.1G. OTHER POLYMERS

Silicones

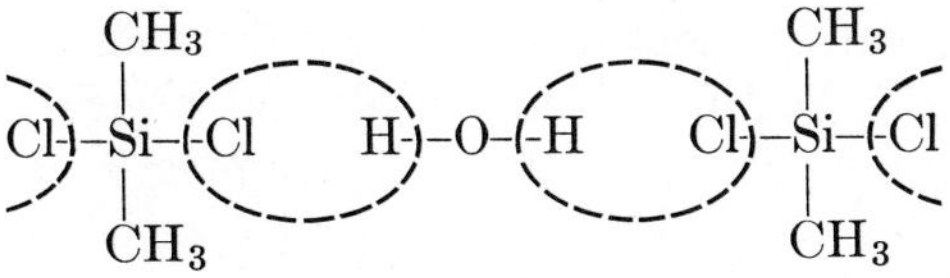

$$\text{Cl}\text{-}\underset{\underset{\text{CH}_3}{|}}{\overset{\overset{\text{CH}_3}{|}}{\text{Si}}}\text{-Cl} \quad \text{H-O-H} \quad \text{Cl-}\underset{\underset{\text{CH}_3}{|}}{\overset{\overset{\text{CH}_3}{|}}{\text{Si}}}\text{-Cl}$$

Silica (quartz)

(Si–O network structure)

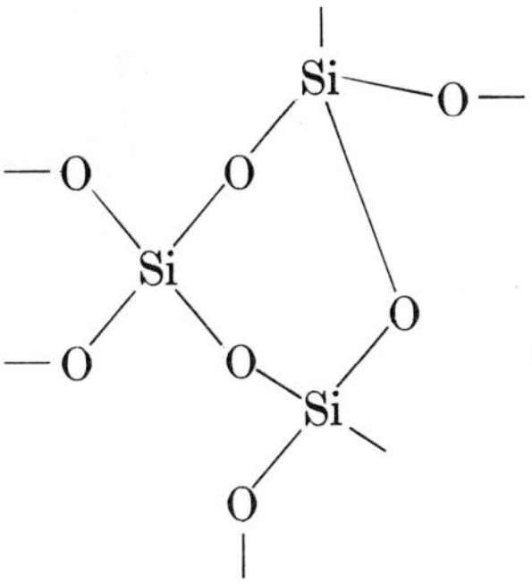

Polysulfides (Thiokol A)

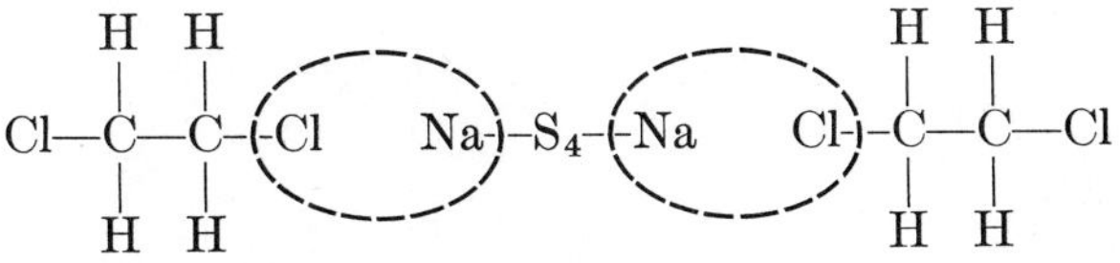

$$\text{Cl}\text{-}\underset{\underset{\text{H}}{|}}{\overset{\overset{\text{H}}{|}}{\text{C}}}\text{-}\underset{\underset{\text{H}}{|}}{\overset{\overset{\text{H}}{|}}{\text{C}}}\text{-Cl} \quad \text{Na-S}_4\text{-Na} \quad \text{Cl-}\underset{\underset{\text{H}}{|}}{\overset{\overset{\text{H}}{|}}{\text{C}}}\text{-}\underset{\underset{\text{H}}{|}}{\overset{\overset{\text{H}}{|}}{\text{C}}}\text{-Cl}$$

ethylene chloride   sodium tetrasulfide

Polyacetal (Delrin)

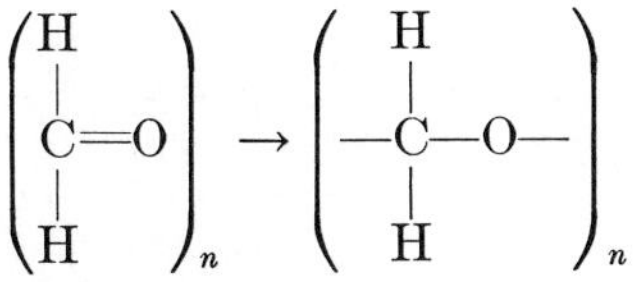

$$\left(\underset{\underset{\text{H}}{|}}{\overset{\overset{\text{H}}{|}}{\text{C}}}\text{=O} \right)_n \rightarrow \left(-\underset{\underset{\text{H}}{|}}{\overset{\overset{\text{H}}{|}}{\text{C}}}\text{-O-} \right)_n$$

formaldehyde

◀ **G. Other Classes.** Many polymers fit into classes other than those explicitly given in Table 6.1. Since silicon has four bonds, as does carbon, it can form polymers, but of fewer types than carbon because double bonds are not known. The *silicones* usually polymerize with the elimination of hydrochloric acid. *Glass* is the term traditionally applied to inorganic polymers with oxygen bonding and no hydrogen atoms. There are a wide variety of glasses, although those based on silicon are most common. *Thiokol* is a polysulfide whose polymerization involves the elimination of sodium chloride. The *epoxy resins* are formed by a series of preliminary reactions forming polymers of low molecular weight. The hardening is then effected by a curing agent which attacks the ends of the chain, but not the elements (Noller, 1957, p. 750). Thus the molecular weight, and hence hardness, of the final structure is determined by the preliminary processing, and the user can get a reproducible material with relatively wide variations in curing conditions. Epoxies are excellent adhesives, and since the final reaction does not involve condensation, they can be used between impermeable surfaces. ▶

Much of the mechanical behavior of polymers depends on the ease with which chains can slide over each other in amorphous polymers, and also on their tendency to crystallize. These factors are markedly affected by the detailed shape of the polymers, which is in turn determined by the chemistry of the monomers and the pattern in which the monomers are put together. For example, the tendency of

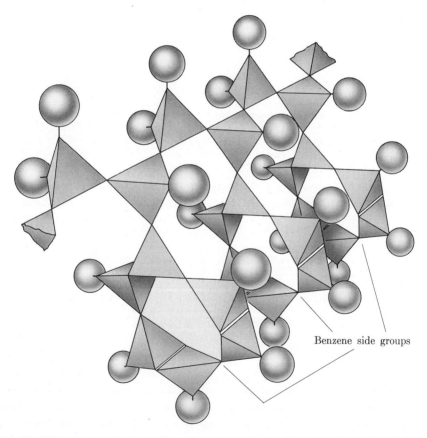

Fig. 6.1. Polystyrene chain, with carbon atoms denoted by tetrahedra bonded at the corners. (Note that this model fails to represent the continual shifting in the double bonds of the benzene ring.)

polyethylene to crystallize is much greater than that of polymethyl methacrylate, with its bulky side groups. To see some of the geometrical effects, consider the polystyrene chain shown in Fig. 6.1, in which the carbon atoms are represented by tetrahedra with bonds at the corners. This representation makes clear how bulky the benzene rings are. Note that a ring could be attached to either side of the group of carbon atoms which forms the backbone of the chain polymer. If all the side groups are on the same side of the main carbon chain, the polymer is called *isotactic;* if on alternate sides, the polymer is called *syndiotactic;* and if irregular, the polymer is called *atactic*. In addition reactions, monomers are almost always joined in a head to tail fashion, but could be also head to head. Examples of these arrangements are indicated schematically in Fig. 6.2.

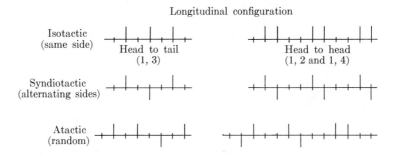

FIG. 6.2. Alternate arrangements of monomers in a polymeric chain. Numbers indicate carbon atoms at which side group is attached, starting from an arbitrary atom with side group.

In the diene polymers, an important factor is whether the chain runs off from the same (*cis*) or opposite (*trans*) sides of the plane through a double bond and its adjacent nuclei. This is shown in Fig. 6.3 for natural rubber and gutta percha, both composed of isoprene. Different structures with an identical chemical composition are called *isomers*. They are determined by x-ray observation of the polymer, as crystallized either by slow cooling, or, in rubber, by stretching. As shown in Fig. 6.3, the different arrangement leads to a much neater chain in gutta percha, increasing its tendency to crystallize and preventing it from giving a satisfactory "rubbery" behavior.

In polymers consisting of two or more different kinds of monomers, called *copolymers*, further differences in structure are possible. As indicated in Fig. 6.4, the different monomers can be arranged in an alternating, random, or block fashion. An example of a copolymer is butadiene copolymerized with styrene, which gives the synthetic rubber GR-S (Government Rubber with Styrene), still the quantitatively most important synthetic rubber used in automobile tires. The rubber is made with approximately 75% butadiene and 25% styrene by weight. Polystyrene by itself is very rigid and brittle, but small additions of butadiene increase its toughness. Acrylonitrile-butadiene-styrene (ABS) terpolymers are of growing importance.

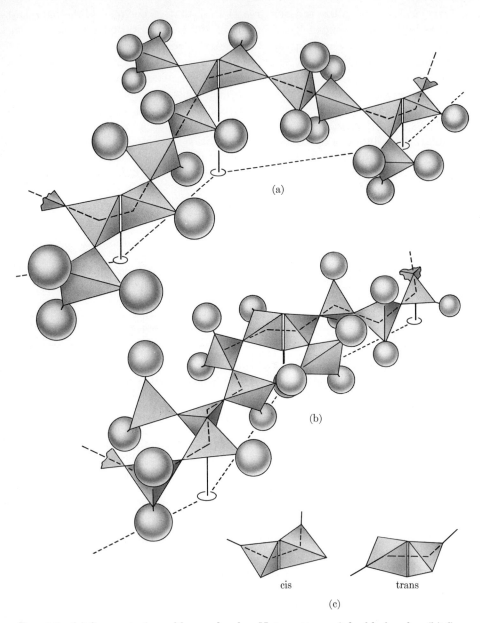

FIG. 6.3. (a) Segment of a rubber molecule. Note pattern of double bonds. (b) Segment of a gutta percha molecule. (c) Comparison of *cis*- and *trans*-bonding.

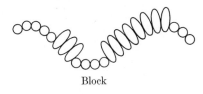
Block

Alternate copolymer

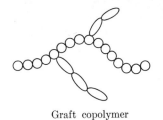
Graft copolymer

Random copolymer

FIG. 6.4. Configurations of copolymers.

An especially important characteristic of the monomer is the number of points of attachment it provides, or its *functionality*. The vinyl and diene polymers discussed above are all bifunctional, and in that form can combine only as long chains. At temperatures high enough to allow sliding between the chains, unlimited flow can occur. This is prevented by linking the chains together, which requires at least occasional trifunctional or polyfunctional connection. This connection must first lead to branching, and later to connections to another chain if extended viscous flow is to be prevented. In the original process of vulcanization of rubber, sulfur provided this function, in some way reacting with the double bonds:

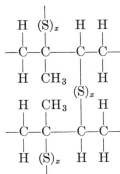

If the network-forming monomers are close together, the whole structure will be linked in a rigid fashion. Since primary bonds are usually involved in network formation, the network can be broken up only at temperatures which also break up the repeat units. In most hydrocarbons, this process is irreversible. Such polymers are called *thermosetting*. Glyptal resins and the formaldehyde copolymers are examples. With glasses, the polymer network reforms on cooling; glasses are therefore *thermoplastic*.

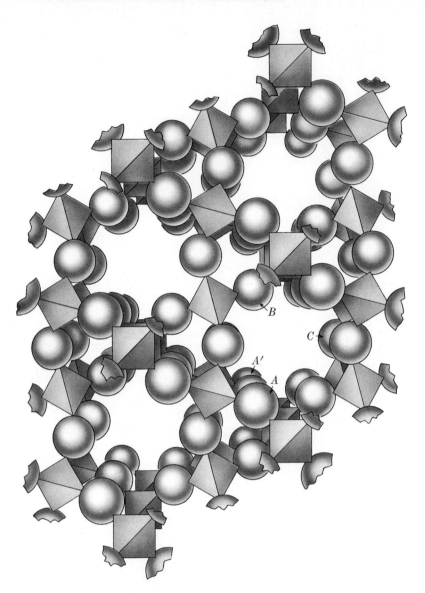

Fig. 6.5. Crystal structure of high quartz with silicon atoms denoted by tetrahedra and oxygen atoms by spheres. Note descending helices $ABCA'$...

Pure vitreous, or fused, silica has a completely three-dimensional network structure. One of the crystalline forms of quartz is shown in Fig. 6.5. Its open, lacelike structure shows its relation to polymers. The addition of sodium oxide tends to break up the network by providing chain terminals. These terminals are not definite, but rather each sodium atom is in a pocket surrounded by about six oxygen atoms, one of which is bound to only one silicon atom. A three-dimensional

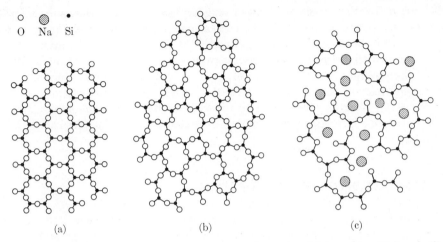

FIG. 6.6. Configurations of a hypothetical two-dimensional (a) quartz, (b) fused silica, and (c) soda glass. [(a) and (b) from Zachariasen, 1932. Courtesy of American Chemical Society. (c) from Warren and Biscoe, 1938. Courtesy of American Ceramic Society.]

representation of this arrangement is rather difficult to present, so recourse is made to the analogous two-dimensional compound shown in Fig. 6.6, along with the crystalline form which in three dimensions would correspond to quartz. The addition of more sodium and oxygen tends to break up the network further. With the composition $SiO_2 \cdot Na_2O$, assuming uniform composition, the silica tetrahedra must be joined in endless chains, as shown in Fig. 6.7. In this case, the glass is water-soluble and becomes the "water glass" useful in laboratory work. The assumption of uniformity of composition is by no means necessarily true, however. Nonuniformity is utilized in the manufacture of Vycor, in which 20% boric oxide is used to make the silica glass workable, by lowering its melting temperature. After working to shape, heat treatment by annealing at 500° to 650°C produces two separate phases. Solution of the boron-rich phase in acid, followed by a further annealing to shrink the structure, produces a transparent, nonporous glass of 96% silica.

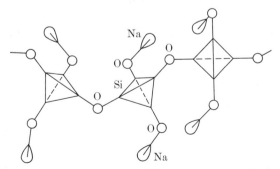

FIG. 6.7. Chain structure of water glass, $SiO_2 \cdot Na_2O$.

The crystallization or devitrification of glass can lead to undesirable contractions, residual stresses, formation of cracks, and loss of strength, but with appropriate additives, a uniform, fine-grained (5×10^{-5} cm) crystallization gives a glass-ceramic with a strength and stiffness twice that of 96% silica (Stookey, 1960). Many other compositions and forms of glass are used for various purposes. For further information the reader is referred to such books as Stanworth (1953), Morey (1954), Kingery (1960), and to the current literature, such as the *Journal of the American Ceramic Society* and the *Transactions of the Society of Glass Technology*.

6.3 STAGES OF DEFORMATION IN POLYMERS

Even at low stress levels, the mechanical behavior of polymers is often time dependent. Therefore, in summarizing behavior over a wide range of temperatures, it is convenient to use a standardized test and to report the behavior at a definite time. For instance, one can apply a fixed strain and measure stress as a function of time in the relaxation test. For strains less than 1 or 2%, the stress at a given time usually turns out to be directly proportional to the applied strain (Schwarzl and Staverman, 1956; Findley, 1954). The ratio of the stress at a certain time to the applied strain is then independent of strain, and is called the *relaxation modulus*, $E_r(t)$. In Fig. 6.8 the relaxation modulus at 10 seconds is plotted as a function of temperature for several forms of polystyrene. At room temperature, all forms are glasslike materials, hard and brittle. Above about 100°C, however, they become much softer. There may then follow a temperature range in which the modulus becomes constant, beyond which the polymers melt to a viscous liquid. This typical sequence is observed with a large number of polymers, not only those which are always amorphous, but also with crystalline polymers. At low temperatures, the chain molecules form a frozen mass and

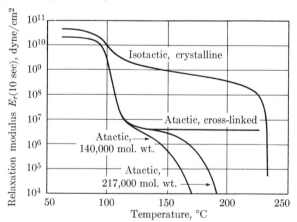

Fig. 6.8. Ten-second relaxation modulus for various forms of polystyrene. Cross-linking by 0.25 mole % divinyl benzene. (Tobolsky, 1960, 1963. Courtesy of Wiley.)

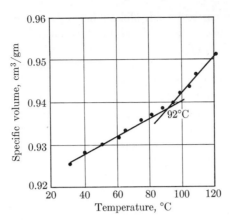

FIG. 6.9. Determination of glass transition temperature T_g from volume data for annealed isotactic polystyrene. (Tobolsky, 1963.)

deformation is possible only by stretching of inter- or intramolecular bonds; the material is a brittle glassy solid. With increasing temperature, sliding between relatively short segments of the chains becomes possible. One might expect this local increase in mobility to be gradual and to extend over a wide temperature range. In fact, the process is self-accelerating; if it has started at a few places, the resulting volume increase makes it easier for the same process to occur in the neighborhood. Consequently, the loosening of the structure with local sliding takes place within a relatively narrow temperature range in which the coefficient of thermal expansion changes rapidly. The center of this range, shown in Fig. 6.9, is called a *second-order transition* temperature. (First-order transitions are discontinuous phase transformations connected with the liberation or absorption of heat, like melting.) At this second-order transition, called the *glass transition temperature* T_g, the change from glassy solid to a rubbery material begins. For mechanical purposes, one often uses an inflection transition temperature T_i, taken to be the nearby point of inflection in a curve such as Fig. 6.8. Because of the difficulty of determining a precise inflection temperature, it is often taken to be the temperature T_m, at which the modulus reaches a root mean square value between the glassy and the rubbery moduli. A value of about 105°C is found for the polystyrene of Fig. 6.8, which is somewhat above the 92°C glass transition temperature of Fig. 6.9. The behavior of polymers in this region is often strongly affected by humidity or various chemical plasticizers which tend to open up the structure or to saturate intermolecular bonds, leaving the chains free to slide over each other.

In the rubbery state, sliding is possible between short parts of the molecules, but the chains are either occasionally bonded together or are so entangled with each other that they cannot easily slide as a whole. In a sense, a rubber is like a tangled yarn, which can be stretched but cannot be pulled apart without tearing the yarn.

As the temperature is increased, sliding becomes easier and the material behaves more and more like a dish of spaghetti. Viscous flow appears in addition to rubber elasticity, and finally becomes a dominating characteristic, so that the polymer becomes a viscous fluid. This final stage can be deferred by increasing the chain length (molecular weight), as shown in Fig. 6.8. Even with a longer chain length, however, flow eventually occurs and the material is useless from a structural point of view. The viscous flow can be much more thoroughly suppressed by cross-linking to form a network between the chains, with one pinning for each 50 to 100 monomer units. Polymers in this state are often called *elastomers*, since the word "rubber" often denotes isoprene, the original rubber compound.

The polystyrenes discussed so far have little tendency to crystallize because they are atactic, with the phenyl groups at random on either side of the polymer chain. In isotactic polystyrene, the phenyl groups are all on one side, and crystallization is easily brought about. As shown in Fig. 6.8, this crystallization impedes the rubbery behavior, so there is a transition from the melt directly to a much more rigid state, and the rigidity is appreciably higher (note the logarithmic scale).

Crystallization has another effect which should be evident from Chapter 4, namely that in a metallic crystalline material, the deformation mechanisms will tend to be stress-activated. Doubling the stress in a creep experiment, for example, will then much more than double the strain rate. In Fig. 6.10, the effect of crystallinity is shown for a series of polyethylene polymers, whose crystallinity is defined by their specific volumes:

$$\text{Weight fraction crystallinity} = \frac{v_{\text{amorph}} - v}{v_{\text{amorph}} - v_{\text{cryst}}}. \tag{6.1}$$

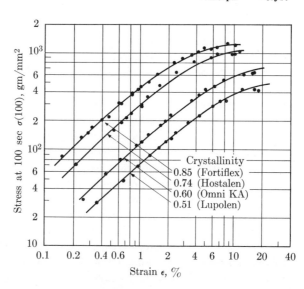

FIG. 6.10. Relaxation data for different polyethylenes at 40.7°C and $t = 100$ sec. (Becker and Rademacher, 1962. Courtesy of Interscience.)

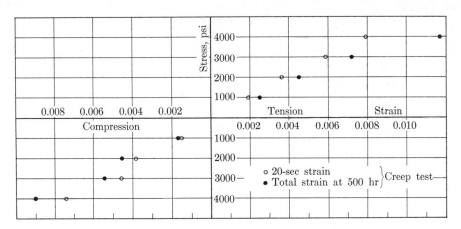

FIG. 6.11. Creep of polyvinyl chloride, showing nonlinear behavior beginning at about 1% strain. (O'Connor and Findley, 1962. Courtesy of ASME.)

Note that the higher density, more crystalline polymer is more rigid and has a higher proportional limit, but that the strain at which nonlinear flow begins is less. This nonlinear behavior can also occur in amorphous polymers, such as the polyvinyl chloride of Fig. 6.11. For most service applications, the allowable strains are such that the linear behavior still holds, and we therefore shall study the mechanics of linear viscoelasticity in more detail in Sections 6.4, 6.6, and 7.7.

At sufficiently high temperatures, the polymer begins to decompose by disruption of the primary bonds. Except in glasses, the structure is not re-established on cooling and the material is permanently altered. This decomposition sometimes occurs without melting, as in the highly cross-linked formaldehyde copolymers and natural cellulose (wood), with its hydrogen bonding.

We now turn to a more detailed but at the same time more idealized discussion of some aspects of the elastic, viscous, viscoelastic, rubbery, and plastic behavior of polymers.

6.4 VISCOSITY AND VISCOELASTICITY

◀ In an amorphous material, there are likely to be a number of spots where thermal activation can switch structural elements back and forth between two similar configurations, as indicated in Fig. 1.29. In soda glass, the structural elements may be network modifiers, such as sodium. In vitreous silica, complete silicon-oxygen tetrahedra may switch; in chain polymers, perhaps side groups or chain segments of molecules constitute the moving elements. Orowan (see 1951) has suggested that an applied stress gives a preference to one or the other of the two configurations. When such a switch occurs, it can be thought of as producing a shear strain of the order of unity over a volume of the order of the cube of the diameter of the switching element, b (in vitreous silica, the diameter of a silicon-

TABLE 6.2

Specific Heats of Various Materials
(Room temperature except as noted)

Substance	Formula	Mass specific heat, cal/gm · °C	Atomic specific heat, cal/gm-atom · °C	
			Based on all atoms	Based on all atoms except H
Polyethylene	C_2H_4	0.55^a	0.26	7.7
Polytetra-fluorethylene	C_2F_4	0.25^a	4.2	6.2*
Polymethyl-methacrylate	$C_4H_8O_2$	0.33^a	2.2	5.2
Polystyrene	C_8H_8	0.33^a	2.1	4.1
Rubber	C_5H_8	0.42^c	2.2	5.7
Benzene	C_6H_6	0.41^b	3.5	7.0
Water, ice	H_2O	0.50^b	3.0	9.0
liquid		1.00^b	6.0	18.0
Carbon tetrachloride	CCl_4	0.20^b	6.2	7.8*
Silica (quartz)	SiO_2	0.19^b	3.8	3.8
Boron oxide	B_2O_3	0.20^c	2.8	4.6*
Lithium nitrate	$LiNO_3$	0.39^b	5.4	6.8*
Lithium hydride	LiH	0.98^b	4.0	8.0
Lithium hydroxide	$LiOH$	0.34^b	2.7	4.0
Sodium chloride	$NaCl$	0.20^b	6	6
Aluminum	Al	0.214^b	6	6
Iron	Fe	0.107^b	6	6
Liquid metals (Li, Na, K, their alloys; Pb, Hg at various temperatures 30 to 600°C)		$6.7-7.3^b$ cal/gm-atom · °C	6	6

* Not counting atoms of lightest element.
Sources: ^{a} Manufacturing Chemists' Association (1957). ^{b} Handbook of Chemistry and Physics (1958). ^{c} International Critical Tables (1930).

oxygen tetrahedron, etc.). Thus a net number of switches in one direction, n_s, in a volume V will produce an average shear strain

$$\gamma = n_s b^3 / V. \qquad (6.2)$$

At temperatures below the glass transition temperature T_g, a new equilibrium state will be reached, after which no further net rearrangements will occur. At higher temperatures, thermal activation will keep generating new sites available for switching, leading to continued viscous flow. To develop this model further, we need to determine whether classical statistical mechanics is applicable, and if so, what modes of vibration will be present.

◀ As indicated in Section 1.3, the suitability of classical statistical mechanics can be determined in either of two ways: by verifying that the thermal activation can supply the quantum of energy required to excite the modes being considered (Prob. 6.1), or by comparing the specific heats with those estimated by assuming different modes to be present. The specific heats of a few polymers and liquids at room temperature are given in Table 6.2. If each atom could vibrate individually, as in the case of a crystalline solid above its Debye temperature, the specific heat would be 6 cal/gm-atom · °C, based on all atoms. Since the specific heats are lower, it is likely that the hydrogen atoms, which are the lightest and have the highest natural frequency, have energy quanta too great to be excited by kT. The last column of Table 6.2 would be 6 cal/mole · °C if this assumption were correct. On the other hand, assuming that the carbon atoms have only four coordinates in phase space, corresponding to the two lateral position and momentum coordinates normal to a chain axis, gives a value of 4 cal/gm-atom · °C, which is generally too low. The assumption that the potential energy effects are negligible, as in the case of a gas, so that the atoms could slide freely past each other, gives an even lower value and is quite unacceptable. We therefore conclude that the chain segments vibrate approximately as units consisting of carbon atoms and the associated hydrogen atoms, and that this configuration can be considered in a classically statistical fashion. A better approximation requires a wave-mechanical treatment, such as that undertaken by Born and Green (1949). The interested reader can get an insight into the nature of the wave-mechanical problem by reading some of the introductory paragraphs, pp. 2 and 64, of that book. An intermediate treatment is given by Glasstone, Laidler, and Eyring (1941).

◀ We now return to the model for viscoelastic behavior and inquire about the rate of switching of the molecules or chain segments from positions unfavorable to the applied stress to favorable positions, as indicated in Fig. 1.29. The stress does not itself force the switch. Rather, its result is to bias the activation energies of the switch in the favorable and unfavorable directions, as indicated in Fig. 6.12.

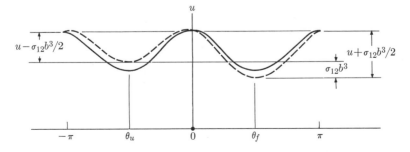

FIG. 6.12. Energy of mobile configuration as a function of pair orientation, showing bias due to shear stress.

◀ The magnitude of the change in activation energy is very difficult to calculate exactly, in view of the dynamics of the process and the deformations of the molecules and their neighbors. However, a rough approximation can be obtained from the change in strain energy in a large volume V containing one switching pair. Assuming the stress in the volume to relax while the boundary of the volume remains fixed, the resulting change in strain energy between the two equilibrium positions is (Prob. 6.2)

$$\Delta u = V \Delta(\sigma_{12}^2/2G) = \sigma_{12} b^3. \tag{6.3}$$

The change in activation energy for either the favorable or unfavorable switches will be half this value, as indicated in Fig. 6.12. The rate of switching to favorable configurations from unfavorable ones will be proportional to the frequency of vibration of the pairs, to the number of pairs in an unfavorable configuration, and to the exponential of the activation energy relative to kT. The net rate of switching is obtained from the difference of the favorable and unfavorable rates:

$$dn_s/dt = \nu n_u e^{-(u-\sigma_{12}b^3/2)/kT} - \nu n_f e^{-(u+\sigma_{12}b^3/2)/kT}. \tag{6.4}$$

The number of pairs in unfavored configurations will be half the total number of sites, $n/2$, less the net number of favorably switched pairs. With this and the corresponding relation for favorable pairs, n_u and n_f can be eliminated from Eq. 6.3, converting it to a differential equation in the number of switched pairs, n_s. The equation can be further simplified if it is noted that the change in activation energy due to stress is small compared with kT. Expressed in another way, the activation volume (the coefficient of the stress term in the exponential) is small. This is an essential feature of linear viscoelasticity. It allows factoring the exponential and expanding the stress-dependent factor in terms of the first two terms of its Taylor series, giving rise immediately to a linear dependence of strain rate on stress. Carrying out the operations, rearranging, integrating, and substituting Eq. 6.2 for the shear strain in terms of the net number of switched sites gives (Prob. 6.3)

$$\gamma_{12} = \left(\frac{nb^3}{V}\right)\left(\frac{\sigma_{12}b^3}{4kT}\right)[1 - \exp(-2\nu t e^{-u/kT})]. \tag{6.5}$$

The first factor of Eq. 6.5 represents the volume fraction of material that is in the form of pairs able to switch back and forth. The second factor indicates that the deformation is linear in the sense that the strain is a linear function of stress, at any given time. The third factor indicates an exponential approach of the strain to a steady-state value, and indicates that the approach of this steady-state value will depend very strongly on the temperature through the second exponential. The reciprocal of the factor multiplying the time t in the exponential is called the time constant of the process,

$$\tau = \frac{1}{2\nu} e^{u/kT}.$$

◀ If the thermal activation is high enough that spots available for switching are continually being generated, then the strain rate will continue at its initial value rather than decaying to zero as the available spots are brought into equilibrium. From the initial strain rate, we can get the viscosity of a liquid (Prob. 6.4):

$$\mu = \sigma_{12}/(d\gamma_{12}/dt) = (V/nb^3)(2kT/vb^3)e^{u/kT} = \mu_0 e^{u/kT}. \tag{6.6}$$

This equation for the coefficient of viscosity can also be derived directly, assuming that the number of sites available remains constant at $n/2$ due to the generation of new sites by thermal activation (Prob. 6.5). The order of magnitude of the viscosity of liquids can be estimated from Eq. 6.5 by taking the fraction of material available for switching to be, say 10^{-1}, the atomic or molecular diameter to be 3 A, and the activation energy to be calculable from the elastic strain energy required to compress the molecules enough to switch freely past a point of minimum spacing into the new position (Prob. 6.6).

◀ The simple picture given above is highly idealized and does not strictly apply to any real glass or liquid. The chief difference results from the fact that in a glassy structure, the sites which switch have a wide spectrum of activation energies because of the various states of extension of bonds. In this case, the viscoelastic strain is not a simple exponential but initially rises much more rapidly than an exponential as sites of low activation energy are used up, and does not approach equilibrium as rapidly as an exponential in the final stages when sites of high activation energy can continue to play a role. As the temperature is increased, the delayed elasticity never disappears and its "time constant" is not much affected. The reason for this is that the sites of low activation energy are used up during the loading process and become part of the elastic strain (resulting in an additional drop of Young's modulus). On the other side of the scale, more and more sites become available as the temperature is increased. Hence, the delayed elasticity process appears to have a time constant equal to the time available to the experimenter for measurement. As the temperature increases further and self-diffusion becomes more rapid, new favorable sites with activation energies on the high side of the scale will be generated at an increasing rate and will result in the establishment of a steady (viscous) strain rate, which will eventually, at high temperatures, mask the delayed elasticity completely. The effect of the specific volume on the number of available sites has been studied by Eyring (see Glasstone et al., 1941, p. 477).

◀ In polymers of increasing molecular weight, mentioned earlier, the delayed elastic strain may result either from the switching of independent bulky side groups or, more likely, from the propagation of kinks or folds along the length of a molecule, aided by the stress concentrations which they produce. In either event, the product of the activation volume and the applied stress in Eq. 6.4 may then become of the order of kT. In such a case, an expansion of the exponential as described above for simple glasses can no longer be allowed. There is then a much stronger dependence of the strain rate on stress than indicated in Eq. 6.5, giving a hyperbolic sine law for amorphous materials, although experiments indicate a power law at low stresses in crystalline materials.

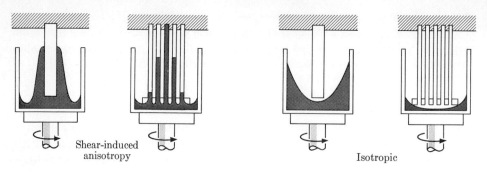

FIG. 6.13. Effect of shear-induced anisotropy on fluid in a rotating cup. (After Weissenberg, 1949. Courtesy North Holland.)

◀In polymers of high molecular weight, shear deformation will tend to stretch the molecular chains out in a certain direction if there is time for lateral viscous flow between chains, but not time for the chains to return longitudinally to their normal random configuration. In this case, the material becomes anisotropic as a result of deformation, and the usual equations of viscosity based on an isotropic material are no longer applicable. This effect may be noticed in silicone putty at room temperature. If a specimen is twisted at a shear strain rate of one per second, the material tends to pinch off and get longer during the test. If, on the other hand, the shear strain rate is dropped to the order of one per minute, inelastic strains of the order of 20 can be introduced without producing appreciable thinning down and lengthening of the specimen. This pinching off is a special example of the Weissenberg effect shown in Fig. 6.13, in which stirring of a viscous liquid produces deformation of a type unexpected from an isotropic material. As we shall see in the next section, this strain-induced anisotropy plays an important role in the elasticity of rubber as well. ▶

6.5 RUBBER ELASTICITY

The modern picture of rubber elasticity is due to Meyer et al. (1932). It arose from the observation that on stretching rubber isothermally, the heat given off is equal to the work done, so the internal energy is unchanged (Prob. 6.7). Since the strained rubber could do work, however, its thermodynamic free energy must have increased. The isothermal change in free energy can be described in terms of the changes in internal energy and entropy:

$$W = \Delta f = \Delta u - T \Delta s. \tag{6.7}$$

The absence of changes in internal energy means that the mechanical behavior must be associated with changes in entropy. Furthermore, the free energy change, and hence the stress for a given strain, must be proportional to the absolute temperature if the entropy is determined by shape and is itself independent of temperature.

◀The change in entropy was calculated from statistical mechanics by Kuhn (1936). The development given here follows the same lines and includes the three-dimensional case, as given for example by Treloar (1958). In statistical mechanics, the entropy is given by the Boltzmann constant k multiplied by the logarithm of the probability (denoted by W for the German word for probability, *Wahrscheinlichkeit*):

$$s = k \ln W. \tag{6.8}$$

We picture the rubber to be a network of chains, whose junction points undergo the same strain as does the volume element as a whole. The problem is to determine the probability of different arrangements of chain segments. For convenience, since translation does not affect the strain, we consider all chains to start at the origin of coordinates. The chains will then terminate in various boxes, one of which is shown in Fig. 6.14. The probability of the entire geometrical state will be the probability of one chain ending in a given box, raised to a power equal to the number of chains which end in that box, multiplied by the probability of a chain ending in the next box, raised to a power equal to the number of chains in the next box, and so on, multiplied by a factor C giving the number of different ways in which the chains can be arranged in the different boxes:

$$s = k \ln C \text{ (prob. box 1)}^{\text{no. in box 1}} \text{ (prob. box 2)}^{\text{no. in box 2}} \ldots \tag{6.9}$$

If we denote the probability of a chain terminating in box i in the unstrained condition by p_{io}, the total number of chains by N_c, and split the logarithm of the product into the sum of logarithms, we obtain the following expression for the entropy:

$$s_o = k \sum_{i=1}^{\infty} \ln p_{io}^{N_c p_{io}} + k \ln C = k \sum_i N_c p_{io} \ln p_{io} + k \ln C. \tag{6.10}$$

◀If the coordinate axes are chosen to be those of principal strain, the deformation can be described in terms of *stretch ratios*, giving the ratio of final to initial

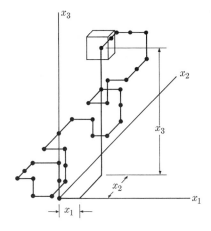

FIG. 6.14. Box containing end of chain composed of random orthogonal elements.

coordinate, with one end fixed at the origin:

$$\lambda_1 = x_1/x_{1o}, \qquad \lambda_2 = x_2/x_{2o}, \qquad \lambda_3 = x_3/x_{3o}. \tag{6.11}$$

Since the chain ends are assumed to move in the same way as the average deformation in the body, we can assume that the number of chains ending in the ith box remains constant, but that the box moves to a new coordinate point whose probability in terms of the old coordinates may be expressed as $p_i(\lambda_1 x_{1o}, \lambda_2 x_{2o}, \lambda_3 x_{3o})$. Thus the difference in entropy between the original and the deformed coordinates is given by

$$s - s_o = k \sum_i N_c p_{io} (\ln p_i - \ln p_{io}). \tag{6.12}$$

◀ We now need to determine the probability of a chain ending in a box having coordinates x_1, x_2, and x_3. The position vector of the end of the chain will be the sum of all the increments in coordinates as we proceed from one end of the chain to the other. In a chain with N_m units, the x_1 coordinate will be a linear sum of the incremental steps, e.g., Δx_{1q}:

$$x_1 = \sum_{q=1}^{N_m} \Delta x_{1q}.$$

The linearity means that if we determine the standard deviation of the distribution of the individual steps, we can obtain the standard deviation of the distribution of final coordinates by the usual equation for the propagation of error.* For simplicity, we shall assume that the chain links are of constant length l, and can lie in either of two directions along any one of the three axes. The frequency distribution of increments per link along the x_1 axis has two vanishingly narrow peaks each with area $1/6$ at $x = \pm l$ and a peak of area $2/3$ at $x_1 = 0$, as shown in Fig. 6.15. The standard deviation is $l/\sqrt{3}$. The distribution of the coordinate x_1 for N_m links therefore has a mean value zero, and, from the propagation of error equation, a standard deviation $\sigma = \sqrt{N_m} l/\sqrt{3}$ (Prob. 6.8). According to the central limit theorem of mathematical statistics, a linear function of a large number of variables will tend to the normal distribution function if all variables contribute more or less equally. We can now express the probability that a chain

---

* If a result R is a linear function of n independent variables x_i, such that

$$R = R_o + \sum_{i=1}^{n} \left(\frac{\partial R}{\partial x_i}\right)_o (x_i - x_{io}),$$

then the standard deviation of the result, σ_R, is given in terms of the standard deviations of the variables, σ_i, by

$$\sigma_R = \left[\sum_{i=1}^{n} \left(\frac{\partial R}{\partial x_i}\right)_o^2 \sigma_i^2\right]^{1/2}.$$

will lie in a particular box as the product of the probability that it will lie in the corresponding intervals of the three coordinate axes:

$$p_i = \left(\frac{1}{\sqrt{2\pi}\,\sigma}\right)^3 e^{-\lambda_1^2 x_{10}^2/2\sigma^2} e^{-\lambda_2^2 x_{20}^2/2\sigma^2} e^{-\lambda_3^2 x_{30}^2/2\sigma^2} \, dx_1 \, dx_2 \, dx_3. \quad (6.13)$$

Substitution of Eq. 6.13 into 6.10, replacement of the summation by an integration, evaluation of the resulting definite integrals, and determination of the change in free energy from the change in entropy, gives the result (Prob. 6.9)

$$\Delta f = -T\,\Delta s = \frac{kTN_c}{2}(\lambda_1^2 + \lambda_2^2 + \lambda_3^2 - 3). \quad (6.14a)$$

It is convenient to express the number of chains per unit volume, N_c, in terms of the density ρ, the universal gas constant R, and the total molecular weight of each network chain, M_c:

$$\Delta f = \frac{\rho RT}{2M_c}(\lambda_1^2 + \lambda_2^2 + \lambda_3^2 - 3). \quad (6.14b)$$

Note that because the density change is small compared with the distortional strain (strain deviator), the product of the three stretch ratios, $\lambda_1\lambda_2\lambda_3$, must equal unity.

FIG. 6.15. Distribution of coordinate increments for links in a chain.

◀ One of the remarkable results of Eq. 6.14b is that it indicates that the stiffness of a rubber is independent of the chemical composition or kind of molecular binding. It depends only on the density, the temperature, and the chain length as expressed in terms of the total molecular weight of the chains. Note also that the particular standard deviation of the distribution of increments in coordinates shown in Fig. 6.15 does not enter. Actually, a more detailed analysis does indicate some effect of the structure of the chain (see Treloar, 1958), but these effects are small compared with the usual uncertainties in the determination of the molecular weight, so that Eq. 6.14 is quite satisfactory in most cases.

◀ To determine the constitutive equations for rubber, the differential of the expression for free energy is equated to the differential of the work done, expressed in terms of the current stresses and the stretch ratios. Since these expressions must be equal for arbitrary increments in the stretch ratios, individual terms can be equated. Adding a term c to allow for superimposed hydrostatic pressure gives the following constitutive relation (Prob. 6.10):

$$\sigma_{11} = \frac{\rho RT}{M_c}[\lambda_1^2 + c(\lambda_1, \lambda_2, \lambda_3)],$$

$$\sigma_{22} = \frac{\rho RT}{M_c}[\lambda_2^2 + c(\lambda_1, \lambda_2, \lambda_3)], \tag{6.15}$$

$$\sigma_{33} = \frac{\rho RT}{M_c}[\lambda_3^2 + c(\lambda_1, \lambda_2, \lambda_3)].$$

◀ From Eqs. 6.15 the stress-strain behavior can be worked out for a number of simple cases of practical interest, choosing c to match the boundary conditions (Prob. 6.11).

◀ For uniaxial tension (current or true stress),

$$\sigma_{11} = \frac{\rho RT}{M_c}\left(\lambda_1^2 - \frac{1}{\lambda_1}\right), \quad \sigma_{22} = \sigma_{33} = 0. \tag{6.16}$$

◀ For biaxial tension (current or true stress),

$$\sigma_{11} = \sigma_{22} = \frac{\rho RT}{M_c}\left(\lambda_1^2 - \frac{1}{\lambda_1^4}\right), \quad \sigma_{33} = 0. \tag{6.17}$$

The case of simple shear is more difficult, because of the problem of determining the principal stretch ratios and their directions for a given tangential displacement, as indicated in Fig. 6.16. However, when the analytic geometry is worked out and expressed in terms of the shear strain given as the tangent of the angle, it turns out that the shear stress σ_{12} is (Prob. 6.12)

$$\sigma_{12} = \frac{\rho RT}{M_c}\gamma_{12} = \frac{\rho RT}{M_c}\frac{\partial u_1}{\partial x_2}. \tag{6.18}$$

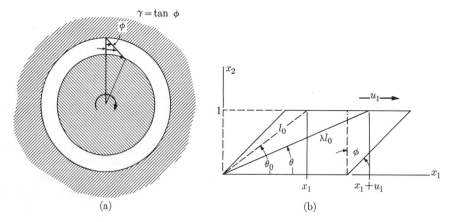

FIG. 6.16. (a) Simple shear in a torsion spring. (b) Notation used in determining line element l_0 undergoing principal stretch in simple shear.

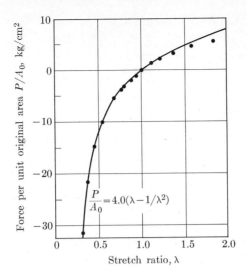

FIG. 6.17. Comparison of theoretical and experimental stress-stretch curves for uniaxial stressing of rubber. (Treloar, 1958. Courtesy of Oxford University Press.)

In other words, the stress-strain relation is linear for simple shear of rubber. It is also interesting to note that the production of simple shear requires at the same time a compressive stress (Prob. 6.13). Under torsion in the absence of this compressive stress, a round bar will lengthen and become thinner, as noted for silicone putty in Section 6.4 (Prob. 6.14). As shown in Fig. 6.17, the above theory is valid for strains in which the length changes by a factor of two. At larger strains, such factors as crystallization and the approach of chains to their fully stretched configuration begin to cause deviations. ▶

6.6 REPRESENTATION OF SMALL-STRAIN VISCOELASTICITY

A. Use of Spring-Dashpot Models. The tangled-chain structure of a polymer is indicated schematically in Fig. 6.18. The transitions from glassy to rubbery to viscous behavior, discussed in Section 6.3 and developed more quantitatively in Sections 6.4 and 6.5, suggest that to obtain the simplest possible model of this behavior, we consider a system of springs of stiffnesses corresponding to the glassy or to the rubbery modes of behavior, and dashpots governing the transition between modes.

Before developing such a model, we should consider some of the simplest kinds of behavior expected from a spring and dashpot model. We define the spring in terms of a modulus of elasticity, $E = \sigma/\epsilon$, and the dashpot in terms of a normal component of viscosity, $\eta = \sigma/(d\epsilon/dt)$. The two basic spring and dashpot elements consist of a spring and dashpot first in series, called a *Maxwell element*, or second in parallel, called a *Voigt element*, as shown in Table 6.3. The Maxwell

TABLE 6.3
Characteristics of Simple Spring and Dashpot Elements

Name	Maxwell	Voigt
Model	(spring E in series with dashpot η)	(spring E in parallel with dashpot η)
Differential equation	$\dfrac{d\epsilon}{dt} = \dfrac{1}{E}\dfrac{d\sigma}{dt} + \dfrac{\sigma}{\eta}$	$\eta\dfrac{d\epsilon}{dt} + E\epsilon = \sigma$
Creep Diagram	(creep curve, linear rise)	(creep curve, exponential approach)
Equation for $t < t_1$	$\epsilon = \dfrac{\sigma}{E} + \dfrac{\sigma t}{\eta}$	$\epsilon = \dfrac{\sigma}{E}(1 - e^{-Et/\eta})$
for $t > t_1$	$\epsilon = \dfrac{\sigma t_1}{\eta}$	$\epsilon = \epsilon_1 e^{-Et/\eta}$
Relaxation Diagram	(stress decays exponentially)	(stress constant)
Equation	$\sigma = E\epsilon e^{-Et/\eta}$	$\sigma = E\epsilon$

element is the simplest element which exhibits the viscous behavior of a material, while the Voigt element is the simplest one which exhibits delayed elasticity, both on loading and unloading.

The differential equation governing the Maxwell element is derived from the behavior of its parts, both of which are subject to the same stress, and the sum of whose strains gives the total strain of the element:

$$\epsilon^e = \sigma/E, \qquad d\epsilon^v/dt = \sigma/\eta, \qquad \epsilon = \epsilon^e + \epsilon^v.$$

Differentiating the first and last equations and combining gives

$$d\epsilon/dt = (1/E)\,d\sigma/dt + \sigma/\eta.$$

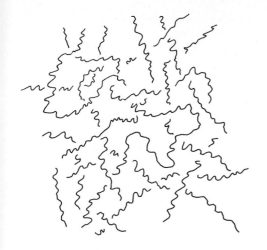

Fig. 6.18. Idealization of tangle of chain polymers.

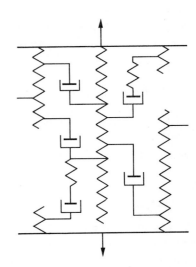

Fig. 6.19. Spring and dashpot model of a chain polymer.

The differential equation for the Voigt element can be derived in a similar fashion, except that the stress, rather than the strain, is additive. Table 6.3 also presents sketches of the behavior and the corresponding equations under step loadings. Note that the Maxwell element exhibits elastic deformation, steady-state creep, and relaxation under a constant applied strain.

The Voigt element exhibits an initial or transient creep and recovery. That is, it has an elastic aftereffect, or delayed elasticity. Together, these elements show the different kinds of time-dependent behavior illustrated in Section 1.2. The factor η/E appearing in the exponentials gives the time at which $1/e$ of the change to the steady-state value has been accomplished. It is called the *relaxation time* or *time constant*:

$$\tau = \eta/E. \tag{6.19}$$

With this brief introduction to spring and dashpot elements, we return to the representation of the viscoelastic behavior of a polymer by a collection of such elements. Figure 6.19 shows a spring and dashpot model of the uniaxial behavior of the idealized polymer of Fig. 6.18. Even this idealization is complicated for analysis, so we first simplify it further to the pair of Maxwell elements shown in

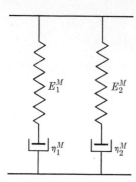

Fig. 6.20. Simplified spring and dashpot model of a chain polymer consisting of two Maxwell elements in parallel.

Fig. 6.20. At low temperatures or after very short times, there will be no flow in the dashpots, so the behavior would be purely elastic, or glassy. As the temperature is raised and the viscosity decreases, flow might begin first in one of the dashpots while the other remained essentially rigid. A further increase in temperature or time would increase the flow in the first element so much that the force in the first spring was practically zero. Under such conditions, the material would again be almost perfectly elastic, but this time with a much lower stiffness because only one of the springs is carrying a load. This stage would correspond to rubber elasticity if E_2 represents the rubbery modulus. Finally, if the temperature were increased still further, or the time were increased, flow would begin in the second dashpot. Over long periods of time this flow would be the major part of the deformation, and ordinary viscosity would result.

B. Equivalent Models. A simple alternative to representing a polymer by a number of Maxwell elements in parallel is to represent it by a number of Voigt elements in series, as shown in Fig. 6.21. It turns out that by appropriate choice of the magnitudes of the elastic and viscous constants, these models can be made equivalent to each other. There is, in fact, an infinite number of different models that can describe a single kind of viscoelastic behavior at one temperature. As an example, the pair of Maxwell elements of Fig. 6.20 is equivalent to the series of three Voigt elements shown in Fig. 6.21, if the viscosity of the first dashpot and the stiffness of the last spring are zero. This might be expected from an examination of the limiting cases of behavior for very long and very short times, and the fact that if the two viscosities within each model differ enough, an intermediate rubbery region can be found for both models (Prob. 6.15). The exact proof of the correspondence can be made by finding the stress-strain-time or constitutive relation for the two different models. In principle, this relation could be found in

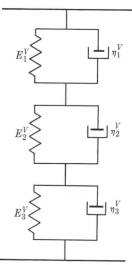

Fig. 6.21. Simplified spring and dashpot model of a polymer, consisting of three Voigt elements in series.

just the same way as the equations were found for the single Maxwell and Voigt models, i.e., by writing down the stress-strain relations for the elements and the conditions for equilibrium and compatibility at the joints, followed by appropriate differentiation and rearranging. Solution is easier if a differential operator notation or the Laplace transform is used, so that the rearranging of the equations can be done algebraically. As in the case of the equations given for the Maxwell and Voigt models in Table 6.3, the constitutive equation turns out to be a linear differential equation with constant coefficients. For both models they are of the same form, of second order with operations on both the stress and strain (Prob. 6.16):

$$p_2 \frac{d^2\sigma}{dt^2} + p_1 \frac{d\sigma}{dt} + p_0\sigma = q_2 \frac{d^2\epsilon}{dt^2} + q_1 \frac{d\epsilon}{dt}. \tag{6.20}$$

Dividing through by one of the coefficients gives four independent coefficients. Any values of these could be matched by appropriate choice of the two viscous and two elastic elements in either the Maxwell or the Voigt type of model; so these models provide equally good descriptions of a polymer. Conversely, if one of these models is a suitable representation of a material, the governing differential equation is of the form of Eq. 6.20. We now consider some important consequences of this fact.

C. Linearity. The constitutive equation (6.20) is *linear* in the sense that if the strain is found as a function of time for one stress history, then for another stress history with double the stress at each particular time, the strain at that time will also be doubled (Prob. 6.17). Likewise, if two solutions $\epsilon^a(t)$ and $\epsilon^b(t)$ are known for two different stress histories, $\sigma^a(t)$ and $\sigma^b(t)$, then for a stress history which is the sum of these two different stress histories, the resulting strain is the sum of the two corresponding strain functions (Prob. 6.18). This is the *Boltzmann superposition principle*. If we wish to predict the strain under a complex loading history such as is shown in Fig. 6.22, this principle allows us to analyze the load as a sum of step loads of various intensities and times of application, determine the strain due to each step from creep data, and sum the resulting strains to find the total strain under a complex service load. Since the strain at any given time is proportional to the applied stress in a creep test, the ratio of strain to stress, called the *creep compliance*, is a function only of time:

$$J(t) = \epsilon(t)/\sigma. \tag{6.21}$$

Similarly, in a mode of service where the history of deformation is known and the stress is to be found, it is easier to work from test data in which a given deformation is applied and the decay of stress with time is noted. As mentioned in Section 6.3, for linear behavior the ratio of stress at a given time to the constant applied strain is a function only of time and is known as the *relaxation modulus*:

$$E_r(t) = \sigma(t)/\epsilon. \tag{6.22}$$

Relaxation data are useful in determining the behavior in an ordinary tensile test, where the strain rate is held more or less constant.

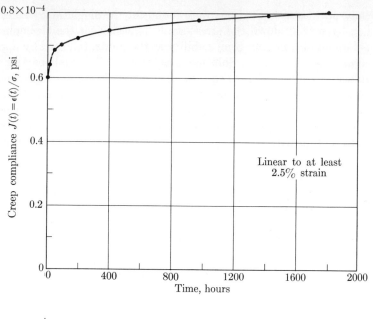

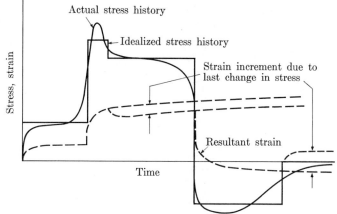

FIG. 6.22. Strain under complex stress history obtained by superposition of creep data. (Data for polyethylene at 75°F and 50% relative humidity from O'Connor and Findley, 1962.)

Note that while the differential constitutive equation (6.20) is linear, the stress-strain curve at constant strain rate is not (Prob. 6.19). While in principle it is possible to predict the behavior under other stress histories from a tensile test at a constant strain rate, (Prob. 6.20), analytical procedures are difficult (Ferry, 1961, p. 74). For these reasons, creep or relaxation tests have been preferred in working with viscoelastic materials, although digital computers may make constant strain-rate data more practical in the future.

◀A more concrete proof of the equivalence of the pair of Maxwell elements of Fig. 6.20 to the series of Voigt elements of Fig. 6.21 (with the modulus of one Voigt element and the viscosity of another taken to be zero) can be given in terms of the creep compliance or of the relaxation modulus. This is done by expressing the creep compliance, for example, in terms of the elastic and viscous characteristics of the elements in each model. Comparison of the expressions shows the relations needed for equivalence (Prob. 6.21, 6.22). For example, the constants for the elements of the Voigt model in terms of the coefficients for the Maxwell model are (Alfrey, 1948)

$$E_1^V = E_1^M + E_2^M,$$
$$E_2^V = E_1^M E_2^M (\eta_1^M + \eta_2^M)^2 (E_1^M + E_2^M)/(\eta_1^M E_2^M - \eta_2^M E_1^M)^2,$$
$$E_3^V = 0, \qquad (6.23)$$
$$\eta_1^V = 0,$$
$$\eta_2^V = \eta_1^M \eta_2^M (E_1^M + E_2^M)^2 (\eta_1^M + \eta_2^M)/(\eta_1^M E_2^M - \eta_2^M E_1^M)^2,$$
$$\eta_3^V = \eta_1^M + \eta_2^M.$$

More general models and transformations between various kinds of test data will be discussed further in Section 7.7, on constitutive relations for materials. ▶

D. Example of Representation at Different Temperatures. As an example of the representation of the time and temperature dependence of the mechanical behavior of a real polymer in terms of the spring and dashpot models discussed above, consider the data plotted in Fig. 6.23.

Let us first consider the temperature of 115°C, which, with a little extrapolation, includes the end of the transition from glassy to rubbery behavior, as well as the beginning of the transition to flow behavior. If we consider a single Maxwell element, we can obtain the relaxation modulus (from its definition, Eq. 6.22) and the stress-strain behavior (Table 6.3), introducing, if desired, the time constant (Eq. 6.19). The result is plotted in Fig. 6.24 on log-log coordinates with a scale similar to that of the data of Fig. 6.23. Clearly at least two Maxwell elements are required to represent the two drops in the curve. Fitting the data is simplified by noting that the relaxation modulus drops off so quickly that for the polymethyl methacrylate at 115°C, at times approaching the transition from rubbery to flow behavior, there will be virtually no contribution at all from the glassy behavior. In fitting a double Maxwell model, as shown in Fig. 6.20, to the data of Fig. 6.23, we therefore choose one modulus equal to the modulus in the glassy state, a second equal to that in the rubbery state, and relaxation times equal to those at which the modulus has dropped to $1/e$ of the value before the transition. A plot of this model is shown in Fig. 6.25. It is apparent that a qualitative description is obtained, but that a good fit in the glass transition region would require a model with a whole series of different relaxation times. Rather than develop the model further, we turn to the effect of temperature.

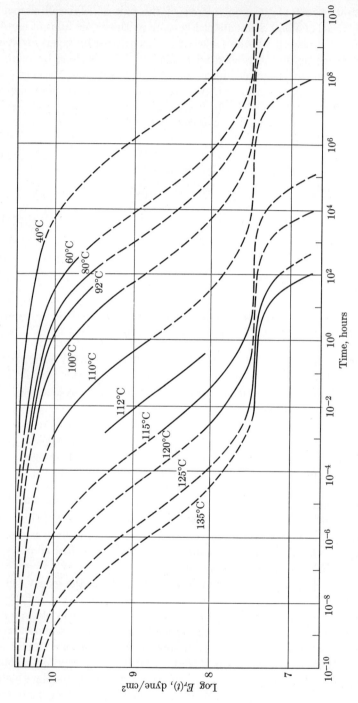

FIG. 6.23. Stress relaxation of polymethyl methacrylate ($M_v = 3{,}600{,}000$) between 40° and 155°C. The dashed lines indicate the extrapolated portion of each curve. (After McLoughlin and Tobolsky, 1952.)

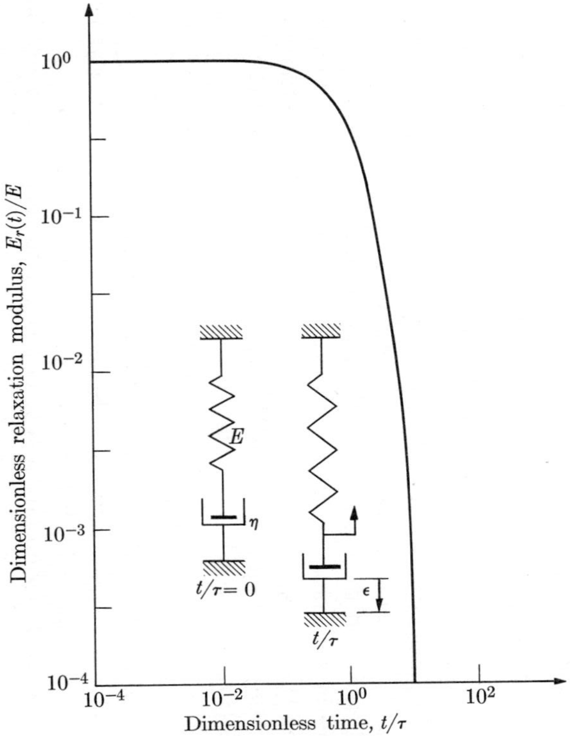

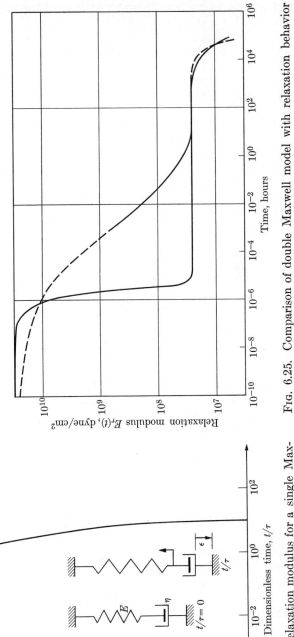

Fig. 6.24. Relaxation modulus for a single Maxwell element, $E_r(t) = Ee^{-t/\tau}$, where $\tau = \eta/E$.

Fig. 6.25. Comparison of double Maxwell model with relaxation behavior of PMMA at 115°C, from Fig. 6.24.

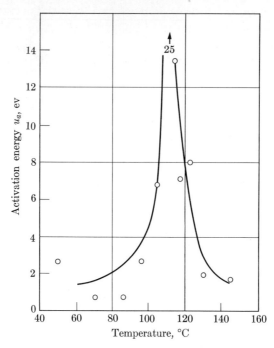

Fig. 6.26. Apparent activation energy determined from shift of isothermal curves of Fig. 6.23 for PMMA. $u_a = k \ln(t_2/t_1)/(1/T_2 - 1/T_1)$.

The effect of changes in temperature is similar to that of changes in time, for in Fig. 6.23 a horizontal translation of the curves for the different temperatures will bring them very nearly into coincidence. This is also to be expected from the exponential variation of the viscosity η with temperature T as derived in Section 6.5, giving an equation of the form

$$\eta = \eta_0 e^{u_a/kT}, \tag{6.24}$$

where k is Boltzmann's constant and u_a is an activation energy, independent of temperature. For the same relaxation modulus at different times and temperatures, Eq. 6.24 indicates that the logarithm of the time for a given stress should vary as the difference in reciprocal temperatures, with a constant of proportionality corresponding to the activation energy. While this idea seems qualitatively correct, quantitative application of it to the PMMA of Fig. 6.23 gives the results shown in Fig. 6.26 (Prob. 6.23), showing apparent activation energies which not only change markedly, but are also higher than expected from the energies of the primary and secondary bonds. The reason is that the apparent activation energy is calculated as if all the temperature effect were caused by thermal motion, whereas actually increases in volume also play a strong role by decreasing the activation energy. Similarly, high apparent activation energies would be found for the temperature variation in viscosity of steam near its critical point, because increases in specific volume with temperature would reduce the activation energy at the same time that the thermal motion was increasing. In such cases, it is important to consider the specific volume of the material relative to that of a close-packed structure ("free volume," see Ferry, 1961, p. 224 and also Glasstone et al., 1941, p. 477). On the other hand, the apparent activation energies found above are useful for extrapolation over limited ranges, bearing in mind that the activation energies so obtained are not physically realistic. The disadvantage of this procedure, which is discussed in more detail by Tobolsky (1960), is that one is tempted to use it beyond the range of experimental data. One may thus be misled in predicting the behavior of materials under the long times needed for structural integrity.

F. Inequivalence of Models at Different Temperatures. When different temperatures are being considered, we lose the equivalence of various spring and dashpot models for representing viscoelastic behavior. For suppose that the behavior of a polymer could be represented by a pair of Maxwell elements in parallel, and that on physical grounds the spring constants did not vary significantly with temperature, whereas the coefficients of viscosity varied exponentially as given by Eq. 6.24. A representation in terms of Voigt elements would require that elastic moduli and coefficients of viscosity vary according to Eq. 6.23. Since the coefficients of viscosity of the Maxwell model affect the spring constants of the Voigt model, these spring constants would now have to depend on temperature, in contrast to our physical assumption. In other words, spring and dashpot models which are equivalent at one temperature will not be equivalent at other temperatures, and apparently only one spring and dashpot model could be correct. As we have seen from Figs. 6.18 and 6.25, this is likely to be a very complicated model. Thus, while the spring and dashpot model is useful at one temperature, to obtain the constitutive relations for viscoelastic materials at various temperatures, we must either make some simplifying approximations or use data directly.

6.7 CRYSTALLINITY AND THE PLASTIC BEHAVIOR OF POLYMERS

Some polymers show evidence of crystallization. They exhibit x-ray diffraction patterns, although the spots are broadened so we know that amorphous material or numerous defects are present. In addition, the specific gravity of the polymer is lower than would be expected from the atomic spacing revealed by the x-ray diffraction patterns. This is conventionally expressed by saying that the polymer has only a partial crystallinity which is estimated, for example, from the specific volume relative to the values for amorphous and crystalline material (Eq. 6.1).

With increasing crystallinity, there are increases in the softening temperature, the specific gravity, and the elastic modulus. As a rule, a crystalline polymer is opaque or at least cloudy or milky. In the amorphous state, the same polymer can be glassy clear. In this case, the opacity is a consequence of scattering of light by "grain" boundaries.

Fig. 6.27. Electron micrograph of the cleavage surface of a singly oriented bristle of nylon 6,10 containing kink bands. The micron marker is in the direction of compression and the axis of the bristle. (Zaukelies, 1962. Courtesy of American Institute of Physics.)

Some polymers, like polyethylene, have a strong tendency to crystallize, while others, like polymethyl methacrylate, are practically impossible to crystallize. One difference between these polymers is in the regularity, or smoothness, of the chain. Evidently a chain on which bulky groups are hanging is much more difficult to pack in a regular fashion. The groups hanging from the chain may be dissimilar, such as the methyl and methyl ester groups, or they may be "side branches," like short polyethylene chains substituted on the main chain in replacement of hydrogen atoms. The oldest type of polyethylene, polymerized at a very high pressure, is strongly branched; the "low-pressure" polyethylene, polymerized in solution with the Ziegler process, is relatively free from side branches. Accordingly, the high-pressure material is less crystalline, much softer (Fig. 6.10), and has a lower melting point and higher solubility. Low-pressure polyethylene behaves oppositely in these respects.

Single crystals which are practically 100% crystalline can be grown from solutions. On the other hand, polyvinyl chloride-polyvinyl acetate copolymers have only a few percent crystalline content. The ordinary crystalline polymers have usually between 30 and 80% crystallinity. As mentioned earlier, this depends very much on how fast they have solidified from the molten state. Thus, nylon can be obtained 100% crystalline when it is crystallized from a solution; on the other hand, rapid quenching of the melt produces completely amorphous nylon.

As was shown in Fig. 1.26, polymers may crystallize in spherulites* with a radial crystallographic orientation. Recent studies (Niegisch, 1959) of the growth of these spherulites during precipitation from solution show that in many cases they start out as relatively flat crystals, with the polymer chain axis normal to the plane. Later an overgrowth of fine fibrils begins to grow out in a progressively less regular orientation, finally giving the radial orientation of the spherulite. Similar observations were made much earlier on inorganic crystals (Lehmann,

---

* A more correct but rare spelling, keeping the Greek roots, is "spherolites."

1888). What little work has been done on melt-grown crystallites indicates that some have a helical structure (Fujiwara, 1960), analogous to the structures reported by Bernauer (1930) for inorganic crystals. Other work using x-ray microbeam techniques shows a variety of structures (Mann and Roldan-Gonzales, 1962). Considering how recent such work on polymers is, as well as the recent electron microscopic studies (Reding and Walter, 1959), considering how many defects there are in the structure, and considering the asymmetry of the crystals due to the different binding forces along and between chains, it is not surprising that relatively little is known about the actual mechanisms of deformation in crystalline polymers. At their most regular, one might expect some of the mechanisms observed in metallic crystals, such as the kink bands in nylon shown in Fig. 6.27. One would expect that the modulus and proportional limit would be higher (although the strain at the proportional limit might be less), as shown in Fig. 6.10. Beyond the proportional limit, one might expect the deformation to be more stress dependent and less time dependent, but this expectation is not fulfilled, at least for the polyethylenes of Fig. 6.10. Perhaps the reason is that the crystallinity is still low in comparison with metallic crystals.

Crystalline polymers, in common with other crystals, exhibit a distinct melting temperature. The glass transition temperature is one-half to two-thirds the absolute melting temperature (see, for example, Tobolsky, 1960, p. 69). This result is very similar to that for metals, in which creep effects become important at about one-half the melting point temperature.

A drop in the load may be observed on extending a crystalline polymer at a constant rate. At a constant loading rate, a much higher strain rate would develop. An example of a load drop is shown in Fig. 6.28. Larger drops may occur. As we

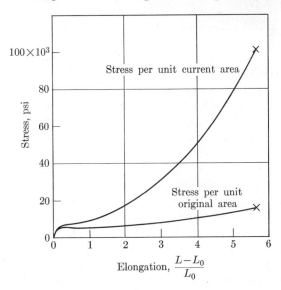

Fig. 6.28. Stress-elongation curve of undrawn nylon at an elongation rate of 6.7 per minute.

shall see in Section 8.4, the presence of a maximum in the load-elongation curve leads to deformation bands in which the plastic flow is concentrated, as shown in Fig. 6.29. The occasional milky appearance of deformation bands arises from scattering or refraction of light in regions of inhomogeneous deformation. X-ray diffraction studies show that the initial crystallinity of the polymer is being destroyed, as if the polymer chains, or larger elements (fibrils) were being unfolded from their crystalline array. Fracture in the bands is at first prevented by an enormous hardening effect as the chains or fibrils are realigned parallel to the specimen axis. X-rays show that the crystallinity is now increasing. A spontaneous rise in crystallinity sometimes causes further extension of the fiber with no applied load, especially if heated (Nebel, 1959). With less crystalline fibers, heating causes shrinkage. Commercial fibers may be "heat-set" to reduce these effects.

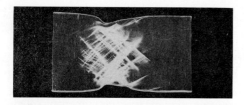
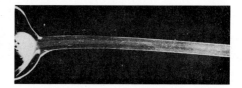

Fig. 6.29. Stages in drawing of low-pressure polyethylene. (Richard and Graube, 1956. Courtesy of Carl Hauser Zeitschriften Verlag.)

As the fiber is further extended, the presence of oriented, strong material shifts the deformation elsewhere, and the zone in which deformation is occurring sweeps along the specimen until it is all highly strained. The deformation becomes homogeneous as the load first begins to rise. This entire process is called *drawing*, or *cold-draw* (although if done rapidly the specimen may become quite hot).

Drawing leads to very high strength. The modulus of elasticity may increase by a factor of as much as five in the axial direction, and three in the transverse direction, although it may drop by a factor of two at 45°, indicating highly anisotropic material (Raumann and Saunders, 1960). The high modulus and strength are due to the replacement of the weak van der Waals and hydrogen bonding by primary bonds in the fiber direction, and closer packing in the transverse direction. The increased crystallinity leads to a higher softening temperature and a lower solubility in solvents. The process of cold-drawing occurs in a number of crystallizable polymers, such as nylon, polyethylene, polyethylene terephthalate, and polyvinylidene chloride, and is of enormous practical value in the production of commercial fibers.

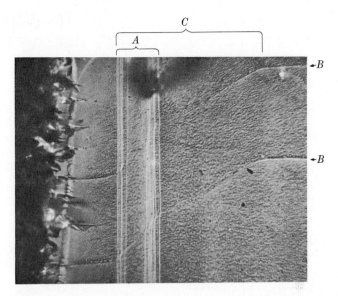

Fig. 6.30. Deformation bands on cleaved surface of preoriented atactic polystyrene. 150% preorientation. The vertical band A appears to have formed first, followed by the horizontal bands B, which were later deformed by the broad band of homogeneous shear in C. (Whitney, 1963. Courtesy of American Institute of Physics.)

The process is not, however, entirely confined to crystalline polymers. Non-crystalline polystyrene will cold-draw when preoriented. It also exhibits deformation bands (when compressed) remarkably similar to those in metal crystals, as shown in Fig. 6.30. This shows that the localized deformation associated with the bands may be explainable generally in terms of a large-scale mechanical instability rather than from a number of different atomic mechanisms which give rise to the stress-strain curve; just as the formation of a kink band in a crystal is a mechanical instability at a scale much larger than that of single dislocations.

6.8 MECHANICAL BEHAVIOR OF COMMON POLYMERS

The choice of appropriate parameters to use in describing the mechanical behavior of polymers is a triangular one among those parameters fundamental to understanding the behavior, such as crystal structure or activation energy, those easy to determine in tests, such as yield point or tensile strength, and those useful to the designer, such as cost, limit of linear viscoelasticity, or allowable stress for 1% strain in five years. Table 6.4 lists our selection of available and relevant parameters. For further information the reader is referred to the Manufacturing Chemists Association (1957), such trade periodicals as *Modern Plastics, Encyclopedia Issue* or *Materials in Design Engineering*, or the technical literature such as the *Journal of Applied Polymer Science*.

TABLE 6.4 Properties

Polymer	Bulk price 1962[9], $/lb[f]	Specific gravity	Room temperature	
			Elastic modulus[e] E, 10^6 psi	Coefficient of linear expansion, $10^{-5}/°C$
Vinyl				
Polyethylene, low density	0.23–0.24	0.92	0.014–0.038	15–30
high density	0.28	0.95	0.035–0.090	15–30
high-density film		0.94–0.95[9]		
Polyvinylchloride,				
nonrigid, general purpose,	} 0.24–0.46	1.4	0.001–0.002	7–25[14]
rigid, normal impact		1.5	0.2–0.6	5–6
Polyvinylidene chloride,				
unoriented	} 0.39–0.54[15]	1.7	0.05–0.08	19
oriented monofilaments		1.7	0.07–0.20	
Polystyrene, general purpose	0.17–0.18	1.06	0.4–0.5	6–8
Polypropylene	0.38–0.58	0.90[8]	0.17[8]	6.2[8]
Polyisobutylene, butyl rubber	0.23–0.30	0.91[7]	0.0002–0.002[11]	19[11]–31[7]
Polytetrafluoroethylene	3.25–5.00	2.2	0.033–0.065	5.5
Polymethylmethacrylate,				
general purpose	0.46–0.55	1.2	0.35–0.45	9
Diene				
Polybutadiene, 25% styrene	0.14–0.35	0.93[7]	0.0002–0.006[11]	22[7]
Polychloroprene	0.35–0.75	1.23[7]	0.00025–0.003[11]	20[7]
Polyisoprene, rubber	0.30	0.91[7]	0.00015–0.006[11]	22[7]
Ester				
Polyethylene terephthalate film		1.4[9]		
Polyurethane	1.15–1.65	1.2–1.3[9]	0.0004–0.0026[7]	10–20[9]
Amide				
Nylon 6, bulk		1.14[8]	0.16–0.28	4.8–5.3[8]
film	} 0.98–2.18	1.12[9]		
Nylon 6, 6		1.14	0.21–0.41	10
Formaldehyde				
Phenol formaldehyde,				
general purpose	0.20–0.35	1.43	0.8–1.2	3.0–4.5
Urea formaldehyde,				
alpha cellulose filled	1.19–0.34	1.5	1.3–1.4	2.2–3.6
Melamine formaldehyde,				
alpha cellulose filled	0.42–0.45	1.5	1.3	2.0–5.7
Polysaccharide				
Cellulose nitrate,				
general purpose	0.45–0.50[15]	1.38	0.19–0.22	8–12
Cellulose acetate, H6-1	0.40–0.52	1.31	0.26–0.4	8–16
Other				
Soda lime glass	0.01	2.40[13]	9.8[13]	0.82[13]
Borosilicate glass		2.23[13]	9.8[13]	0.32[13]
96% silica		2.18[13]	9.7[13]	0.08[13]
Silicone, mineral filled	2.40–3.55	1.9	1.0–1.3	1–10
unfilled	2.50–4.00	1.1–1.69	0.00025–0.002[11]	27[9]–40[11]
Polysulfide	0.60–1.25	1.3–1.4[7]	0.00025–0.003[11]	21[14]
Polyacetal	0.65	1.42[8]	0.41	4.5[8]
Epoxy, general purpose	0.62	1.12	0.3–1.0	3–9

6.8 MECHANICAL BEHAVIOR OF COMMON POLYMERS

of Common Polymers

behavior	Temperature sensitivity					
Tensile strength[a] (Compressive strength, C, if fract.), 10^3 psi	Fracture elongation $\Delta l/l_0$, %	Density trans. T_g, °C	Inflection trans. T_i, 1/sec, °C	RMS mod. trans. T_m, 1/sec, °C	Heat distortion ASTM D648, __ °C at __ psi	Continuous heat resistance, °C
1.5–2.4 2.5–5.0 3.5–8.0[9]	400–700 100–300 50–400[9]		-3^1		40–50, 66 60–82, 66 —	60–80 93 121[9]
1.4–3.0 6–9,12C	200–400 5–25[9]	74^2–82^5	90^3		— 60–80, 264	54–80 71–74
3–5 20–50 5.5–8 5^8 0.8–2.5[11] 1.5–3.3	160–240 20–30 1–3 >220[8] 650–850[9] 120–135	100^3 -20^5 -70^5 29^5	80^1 116^3 -20^1–(-18^5) -66^5 125^1		54–66, 264 — 80–91, 264 105, 66[8] — 60, 264	70–93 100[9] 66–74 149[9] 260
6–8	2–7	90^4–105^5	109^5–120^3		66–77, 264	60–70
0.8–4.0[11] 3.46[6] 0.8–4.5[11]	500^7–820^6 500^6 480–510[6]	-85^5 -46^{11} -72^5	-68^5–(-43^3)	-52^5	— — —	82[9] 116[9] 82[9]
17–28[9] >5[9]	70–130[9] 540–750[9]	-35^2	97^1		— —	121[9] 85^{12}–116[9]
12 13.8–17[9] 8.1–11	50–300[8] >200[9] 60–300	≈60[1] ≈60[1]			63–66, 264[8] — 66, 264	107–121[9] 193[9] 121
6.5–8.5 24–36C 5.5–7.0 30–38C 7–8 40–45C	0.4–0.8 0.6 0.6				143–171, 264 130, 264 204, 264	150–177 77 99
7–8 22–25C 6–8.5 25–36C	40–45 6–31				60–71, 264 80–113, 264	50–60
10^{13d} 5–10[10d] 5–10[10d] 3.5–4.5 16–20C 0.4–1.0[11] 0.6–1.5[11] 10^8 2–12	elastic elastic elastic 500–600[7] 16–75[8] 2–6	-123^5 -123^5		470^{13c} 515^{13c} 820^{13c} -52^5	>260, 264 — — 100, 264[8] >120, 264	288 249[9] 121[9] 85[9] 80

Notes for Table 6.4

Data from Manufacturing Chemists Association (1957) unless otherwise noted.

[1] Ferry (1961, p. 332)
[2] Ferry (1961, p. 251)
[3] Ferry (1961, p. 254)
[4] Flory (1953, p. 52)
[5] Tobolsky (1960, pp. 70, 76, 320)
[6] Born (1961)
[7] Whitby (1954)
[8] Simonds (1959)
[9] Materials in Design Engineering, Clauser (1962)
[10] Stookey (1960)
[11] Payne and Scott (1960)
[12] Modern Plastics, McCann (1961)
[13] Corning Glass Works (1961)
[14] Handbook Chem. Phys. (1958)
[15] Personal communication.

[a] Engineering tensile strength (load over original area) unless otherwise indicated. Compressive strength given where fracture occurs, except for glass for which compressive strength is approximately 8 times tensile strength.

[b] Depends strongly on degree of vulcanization and amount of filler (e.g., carbon black) for rubbers.

[c] Strain point is temperature from which there are negligible residual stresses on cooling. Viscosity at strain point = $10^{-14.6}$ poise.

[d] Abraded.

[e] Modulus quoted in literature for rubbers is frequently stress at a given elongation; this is initial tangent modulus.

[f] Prices per pound are markedly higher for small amounts or for bars or sheets in quantities of a few pounds:

$\frac{1}{32}$ in. rubber sheet cost	$ 2.40/lb.
$\frac{1}{32}$ in. PTFE sheet cost	$40.00/lb.
$\frac{1}{2}$ in. diameter PMMA cost	$ 4.70/lb.
$\frac{1}{2}$ in. diameter nylon cost	$11.40/lb.
Glass bottles cost	$ 0.06–0.09/lb.
Window glass cost	$ 0.35/lb.
Textile glass fiber cost	$ 0.55/lb.

In Table 6.4, the modulus of elasticity, yield point, tensile strength, and elongation are the values normally determined from a tensile test run at a constant rate of crosshead motion, chosen to give a strain rate of about 0.20 to 0.25 in/min before yield, and up to 10 in/min afterward (ASTM D638-61T). Linear viscoelastic behavior, which may persist to higher strains, is not considered in determining the yield point. The only common way of describing creep resistance is at elevated temperature through the ASTM heat distortion temperature. This temperature is determined by loading a specimen $\frac{1}{2}$ in. wide, $\frac{1}{8}$ to $\frac{1}{2}$ in. thick, and 4 in. between supports as a simple beam to a stress of either 66 or 264 psi, heating the specimen in an oil bath whose temperature is rising at 2°C/min, and observing the temperature when the increase in deflection reaches 0.010 in. (ASTM D648-56). The corresponding strains are within the usual linear viscoelastic range (Prob. 9.38), but the strain rate and temperature are both well above those appropriate for structural service. The temperature given for continuous resistance to "heat" (i.e., temperature) is one above which chemical decomposition begins to be important, although the times and method of determination are not specified in the description of tables from which these data were taken.

6.9 SUMMARY

Monomers may be joined by the thousands into chain or network polymers by a variety of chemical reactions. Different spatial arrangements of the same repeat units affect the tendency to crystallize and thus the mechanical behavior. At low temperatures, polymers can be glassy, with a typical Young's modulus of one to ten million psi and a Poisson's ratio as low as 0.2. The glass transition temperature, at which the coefficient of thermal expansion increases abruptly, marks the beginning of increased viscoelastic deformation leading to creep, relaxation, and the elastic aftereffect. In very long-chain or in cross-linked polymers, the increasing viscous flow above the transition temperature is limited by restriction of the motion at points along the molecular chains, even though short sections can slide over each other as easily as in a liquid. This results in rubber elasticity. In non-cross-linked polymers, further increases in temperature finally allow flow as a viscous liquid. At stresses giving strains of greater than 1 or 2%, plastic flow can occur even at low temperatures (if fracture has not intervened). In a crystalline material, the transition to rubbery behavior is retarded, although viscous effects become much more pronounced above the glass transition temperature, which is of the order of one-half to two-thirds of the melting point. If a polymer decomposes before it becomes fluid, it is called thermosetting; if it can be softened without decomposition so that it recovers its original properties on cooling, it is thermoplastic.

A simple molecular picture of the viscous and viscoelastic behavior shows that the coefficient of viscosity varies with temperature according to an equation of the form

$$\mu = \mu_0 e^{u_a/kT}, \tag{6.6}$$

where the activation energy u_a is of the order of Eb^3, and b in turn is of the order of the molecular radius.

Since there is ideally no change in internal energy during isothermal deformation of rubber, its resistance to deformation depends on the decrease in entropy due to aligning molecular chains.

◀The stress-strain-temperature, or constitutive, equation for rubber is given in terms of the principal stretch ratios, the density, the universal gas constant R, and the molecular weight between chain pinning points M_c by

$$\sigma_{11} = \frac{\rho RT}{M_c}[\lambda_1^2 + c(\lambda_1, \lambda_2, \lambda_3)], \tag{6.15}$$

where c is a parameter chosen such that equations of the above form match the stress conditions in all three directions.▶

For strains less than 1 or 2%, the viscoelastic stress-strain relation is linear in the sense that doubling the stress history doubles the strain at a given time, even though the stress-strain curve from tensile tests at constant strain rate may not be linear. This linearity allows us to express the constitutive equations more concisely in the form of a compliance or a relaxation modulus:

$$J(t) = \epsilon(t)/\sigma, \quad (6.21)$$

or

$$E_r(t) = \sigma(t)/\epsilon. \quad (6.22)$$

From such data, superposition enables one to obtain the behavior under a complex stress or strain history. The viscoelastic behavior of a polymer can be represented by a number of different spring and dashpot models, which are equivalent for one temperature but not at different temperatures. A Maxwell representation consists of a spring of modulus E and dashpot of tensile viscosity η in series, and has the relaxation modulus

$$E_r(t) = Ee^{-Et/\eta} = Ee^{-t/\tau}. \quad \text{(Table 6.3)}$$

Specific equations are given for the behavior of models consisting of two Maxwell elements, and for the equivalence of these to models consisting of a series of Voigt elements, each of which consists of a spring and dashpot in parallel.

Polymers with regular structures can crystallize, often with a folded chain structure, forming sheets which grow radially outward from a nucleus to form a spherulite. When these polymers are deformed to large strains, the chains can be partially unfolded and pulled into a more or less parallel configuration, where they can again crystallize. The resulting stress-strain curve may have a dip in it, leading to localized deformation, or Lüders bands. The drawn fibers left behind by the moving band are much higher in strength and stiffness. Noncrystalline polymers may also exhibit this drawing behavior, although to a lesser degree. Under combined loading, the deformation may be confined to bands and slip lines similar to those observed in plastically deformed crystals.

In the glassy region, polymers typically have a modulus of 10^5 to 10^6 psi if organic and up to 10^7 psi if inorganic. Poisson's ratios are 0.25 to 0.4. Their range of linear behavior does not exceed 1 or 2% strain and fracture may follow immediately. In the rubbery region, the modulus drops to 10^2 to 10^3 psi, while the bulk modulus drops by only a factor of two, so Poisson's ratio rises to the order of 1/2. Where cold-drawing occurs, the strains may reach several hundred percent.

REFERENCES

The best single book covering both the chemistry and mechanical behavior of polymers is Billmeyer (1962). For a description of the chemical structure of polymers and the methods of polymerization, see Noller (1957) or Flory (1953). The molecular structure in polymers is well treated by Geil (1963). An early general book on the structure, mechanical behavior, and production and molding of plastics is Schmidt and Marlies (1948). Viscoelastic behavior is treated in a series of books by Alfrey (1948), Ferry (1961), and Tobolsky (1960), as well as in the more comprehensive treatments edited by Eirich (1958) and Stuart (1956). For current technical data on specific plastics, see Simonds (1959), Nielsen (1962), the references to Table 6.4, or current technical literature of the

American Society for Testing and Materials, the American Society of Mechanical Engineers, the Society of Plastics Engineers, or the Society of the Plastics Industry (Randolph, 1962). Important theoretical and scientific developments appear in the *Journal of Polymer Science* (approximately six volumes per year), the *Journal of Applied Polymer Science*, and the *Journal of Applied Physics*.

ALFREY, T., JR.	1948	*Mechanical Behavior of High Polymers*, Interscience, New York.
ASTM	1961(a)	"Method of Test for Deflection Temperature of Plastics Under Load," *ASTM Standards* **9**, 521, No. D648–56.
ASTM	1961(b)	"Method of Test for Tensile Properties of Plastics," *ASTM Standards* **9**, 448–459, No. D638-61T.
BECKER, G. W. RADEMACHER, H. J.	1962	"Mechanical Behavior of High Polymers under Deformations of Different Time Function, Type, and Magnitude," *J. Polymer Sci.* **58**, 621–631.
BERNAUER, F.	1930	*Gedrillte Kristalle*, Gebrüder Bornträger, Berlin.
BILLMEYER, F. W., JR.	1962	*Textbook of Polymer Science*, Interscience, New York.
BORN, J. W.	1961	"Effects of Nuclear Radiation on Rubber," *Mat. Res. Stand.* **1**, 280–286.
BORN, M. GREEN, H. S.	1949	*A General Kinetic Theory of Liquids*, Cambridge University Press, Cambridge, England.
CLAUSER, H. R. (ed.)	1962	*Materials in Design Engineering, Materials Selector Issue*, Reinhold, New York.
CORNING GLASS WORKS	1961	"Properties of Selected Commercial Glasses," *Bulletin B-83*.
EIRICH, F. R. (ed.)	1958	*Rheology: Theory and Applications*, Vol. II, Academic Press, New York.
FERRY, J. D.	1961	*Viscoelastic Properties of Polymers*, Wiley, New York.
FINDLEY, W. N.	1954	"Effect of Crystallinity and Crazing, Aging and Residual Stress on Creep of Monochloro Trifluoro Ethylene, Canvas Laminate, and Polyvinyl Chloride, Respectively," *Proc. ASTM* **54**, 1307–1312.
FLORY, P. J.	1953	*Principles of Polymer Chemistry*, Cornell University Press, Ithaca, New York.
FUJIWARA, Y.	1960	"The Superstructure of Melt-Crystallized Polyethylene, I: Screwlike Orientation of Unit Cell in Polyethylene Spherulites With Periodic Extension Rates," *J. Appl. Polymer Sci.* **4**, 10–15.

GEIL, P. H.	1963	*Polymer Single Crystals*, Interscience, New York.
GLASSTONE, S. LAIDLER, K. J. EYRING, H.	1941	*Theory of Rate Processes*, McGraw-Hill, New York.
HANDBOOK CHEM. PHYS.	1958	*Handbook of Chemistry and Physics*, 39th ed., C. D. Hodgman, ed., Chemical Rubber Co., Cleveland.
INTERNATIONAL CRITICAL TABLES	1930	*International Critical Tables*, E. W. Washburn, ed. McGraw-Hill, New York.
KINGERY, W. D.	1960	*Introduction to Ceramics*, Wiley, New York.
KOHLRAUSCH, F.	1876	"Experimental-Untersuchungen über die elastische Nachwirkung bei der Torsion, Ausdehnung, und Biegung," *Ann. Phys. Chemie*, F. Poggendorf, ed., Series 6, **8**, 337–375.
KUHN, W.	1936	"Beziehungen zwischen Molekülegrösse, statischer Molekülgestalt und elastischen Eigenschaften hochpolymerer Stoffe," *Kolloid Z.* **76**, 258–271.
LEHMANN, O.	1888	*Molekularphysik*, Vol. 1, Englemann, Leipzig, p. 385.
MCCANN, H. (ed.)	1961	*Modern Plastics, Encyclopedia Issue*, Breskin, Bristol, Conn.
MCLOUGHLIN, J. R. TOBOLSKY, A. V.	1952	"The Viscoelastic Behavior of Polymethylmethacrylate," *J. Colloid Sci.* **7**, 555–568.
MANN, J. ROLDAN-GONZALEZ, L.	1962	"Orientation in Nylon Spherulites: A Study by X-Ray Diffraction," *J. Polymer Sci.* **60**, 1–20.
MANUFACTURING CHEMISTS' ASSOC.	1957	*Technical Data on Plastics*, Manufacturing Chemists' Assoc., Washington, D. C.
MEYER, K. H. SUSICH, G. V. VALKO, E.	1932	"The Elastic Properties of Organic High Polymers and Their Significance," *Kolloid Z.* **59**, 208–216.
MOREY, G. W.	1954	*Properties of Glass*, Reinhold, New York.
NEBEL, R. W.	1959	"Advancing Frontiers of Nylon and Dacron Polyester Fibers," *Text. Res. J.* **29**, 777–786. See also U. S. Patents 2,931,068 and 2,937,380.
NIEGISCH, W. D.	1959	"The Nucleation of Polyethylene Spherulites by Single Crystals," *J. Polymer Sci.* **40**, 263–266.
NIELSEN, L. E.	1962	*Mechanical Properties of Polymers*, Reinhold, New York.
NOLLER, C. R.	1957	*Chemistry of Organic Compounds*, 2nd ed., Saunders, Philadelphia.

O'Connor, D. G. Findley, W. N.	1962	"Influence of Normal Stress on Creep in Tension and Compression of Polyethylene and Rigid Polyvinyl Chloride Copolymer," *Trans. ASME, J. Eng. Ind.* **B84,** 237–247.
Orowan, E.	1951	"Creep in Metallic and Nonmetallic Materials," *Proc. 1st U. S. Nat. Con. Appl. Mech.* 453–472.
Payne, A. R. Scott, J. R.	1960	*Engineering Design with Rubber*, Maclaren, London.
Randolph, A. F. (ed.)	1962	*SPI Plastics and Engineering Handbook*, Society of the Plastics Industry, Reinhold, New York.
Raumann, G. Saunders, D. W.	1960	"The Anisotropy of Young's Modulus in Drawn Polyethylene," *Proc. Phys. Soc.* **77,** 1028–1037.
Reding, F. P. Walter, E. R.	1959	"An Electron Microscope Study of the Growth and Structure of Spherulites in Polyethylene," *J. Poly. Sci.* **38,** 141–155.
Richard, K. Graube, E.	1956	"Die Kaltverstreckung bei Niederdruckpolyäthylen," *Kunststoffe* **46,** 262–269.
Schmidt, A. X. Marlies, C. A.	1948	*Principles of High-Polymer Theory and Practice*, McGraw-Hill, New York.
Schwarzl, F. Staverman, A. J.	1956	"Nonlinear Deformation Behavior of High Polymers," *Die Physik der Hoch polymeren*, Vol. 4, H. A. Stuart, ed., Springer, Berlin. pp. 126–164.
Simonds, H. R.	1959	*Source Book of the New Plastics*, Vol. 1, Reinhold, New York. See also Vol. 2, 1961.
Stanworth, J. E.	1953	*Physical Properties of Glass*, Clarendon Press, Oxford, England.
Stookey, S. D.	1960	"Glass-Ceramics," *Mech. Eng.* **82,** 65–68.
Stuart, H. A. (ed.)	1956	*Die Physik der Hochpolymeren*, Vol. 4, Springer, Berlin.
Tobolsky, A. V.	1960	*Properties and Structure of Polymers*, Wiley, New York.
Tobolsky, A. V.	1963	Personal communication.
Treloar, R. L. G.	1958	*The Physics of Rubber Elasticity*, 2nd ed., Oxford University Press, London.
Warren, B. E. Biscoe, J.	1938	"Fourier Analysis of X-Ray Patterns of Sodasilica Glass," *J. Am. Ceramic Soc.* **21,** 250–265.
Weissenberg, K.	1949	"Abnormal Substances and Abnormal Phenomena of Flow," *Proc. 1st Int. Con. Rheology*, North Holland, Amsterdam, pp. [29]-46.
Whitby, G. S. (ed.)	1954	*Synthetic Rubber*, Wiley, New York.

WHITNEY, W. 1963 "Observations of Deformation Bands in Amorphous Polymers," *J. Appl. Phys.* **34**, 3633–3634.

ZACHARIASEN, W. H. 1932 "The Atomic Arrangement in Glass," *J. Am. Chem. Soc.* **54**, 3841–3851.

ZAUKELIES, D. A. 1962 "Observation of Slip in Nylon 66 and 610 and Its Interpretation in Terms of a New Model," *J. Appl. Phys.* **33**, 2797–2803.

PROBLEMS

◀ 6.1. (a) Estimate the frequency of vibration of a segment of chain polymer relative to its surroundings, using typical values of the Young's modulus reported in Table 6.4.

(b) From the bond energy and atomic radius, estimate the frequency of vibration of a hydrogen atom relative to the polymer chain to which it is attached.

(c) Show that the thermal activation at room temperature can supply a quantum of energy for oscillation of the chain, but not of the hydrogen atom.

◀ 6.2. Derive Eq. 6.3, giving the change in strain energy as a pair of molecular segments undergoes unit strain, switching from an unfavorable to a favorable position under an applied stress.

◀ 6.3. Derive the viscoelastic shear strain, Eq. 6.5.

◀ 6.4. Derive the coefficient of viscosity, Eq. 6.6.

◀ 6.5. Derive the coefficient of viscosity directly, assuming the number of switching sites remains constant due to the generation of new sites by thermal activation.

◀ 6.6. Using Eq. 6.6 and the ideas in the paragraph following it, estimate the viscosity of some liquid. Compare your estimate with handbook data.

E 6.7. Note the change in temperature on adiabatically stretching a rubber band by an amount just short of failure. This can be done by holding the rubber band to your lip or forehead just before and after extension. Estimate, from the specific heat, the loss in internal energy required to return the rubber band to its original temperature. Estimate the work put into deforming the rubber band from estimates of the force and displacement involved, and show that the net change in internal energy on isothermally stretching a rubber band is small compared with the work done.

◀ 6.8. Derive the following expression for the standard deviation of the coordinate x_1 between ends of a molecular chain consisting of N_m links each of length l:

$$\sigma = \sqrt{N_m}\, l/\sqrt{3}.$$

◀ 6.9. Derive the equation for the free energy change, Eq. 6.14a, from Eq. 6.13 and other necessary equations.

◀ 6.10. Derive the constitutive equations for a rubbery material, Eqs. 6.15, from the expression for free energy and work per unit volume.

◀ 6.11. (a) Derive the equation for the behavior of rubber under uniaxial tension, Eq. 6.16, from the general constitutive relation.

(b) Derive the equation for rubber under biaxial tension, Eq. 6.17, from the general constitutive relation.

◀6.12. Derive the equation for simple shear ($u_1 = \gamma_{12} x_2$, $u_2 = u_3 = 0$), Eq. 6.18, by finding the principal stretch ratios from the construction of Fig. 6.16, and applying them to the general constitutive equations for a rubbery material. Use x_1, u_1 as variables.

◀6.13. Show qualitatively that for simple shear, as discussed in Prob. 6.12, a compressive stress will develop. In a round bar in torsion with no axial load, one would then expect the bar to become thinner.

E◀6.14. Twist a round bar of silicone putty to failure. Note the decrease in diameter and increase of length. Compare it to a similar effect in torsion of a rubber band at angles of twist below that for which buckling occurs. Repeat the experiment, twisting the silicone putty specimen at an angular rate of only about 1 rpm, frequently turning it end-to-end to prevent changes in shape due to gravity. Note that after the number of revolutions which produces fracture at high rates of twisting, there is negligible lengthening and pinching-off of the specimen. Why should this be so?

6.15. Show the equivalence of the Maxwell elements of Fig. 6.20 and the Voigt elements of Fig. 6.21 from a qualitative argument if the viscosity of the first Voigt dashpot and the stiffness of the last Voigt spring are zero. Consider the limiting cases of behavior for very long and very short times, and the case where the viscosities differ enough that an intermediate rubbery region can be found for each model.

◀6.16. (a) Derive Eq. 6.20 from the model of Fig. 6.20, and evaluate the coefficients in terms of the moduli and coefficients of viscosity of the elements.

(b) Derive Eq. 6.20 from the Voigt model of Fig. 6.21, taking the viscosity of the first dashpot and the stiffness of the last spring to be zero. Determine the coefficients of the differential equation in terms of the modulus of elasticity and the coefficients of viscosity of the various elements.

6.17. Show that for a constitutive equation such as Eq. 6.20, if the strain is found as a function of time for one stress history, then for another stress history with double the stress at each particular time, the strain at that time will also be doubled. Thus Eq. 6.20 is a *linear* constitutive equation.

6.18. Prove that a material having a constitutive equation of the form of Eq. 6.20 satisfies the Boltzmann superposition principle.

6.19. Sketch the shape of the stress-strain curves for constant strain-rate tests of Maxwell and Voigt elements.

6.20. For tests at constant strain rate, one might suppose that a stress-strain plot was equivalent to a stress-time plot, to which the principle of superposition might be applied to obtain the strain-time behavior under service loads. Show how this idea can, or cannot, be applied to (a) a Maxwell element, (b) a Voigt element, and (c) a general viscoelastic material.

◀6.21. Derive the creep compliance of a Maxwell and of a Voigt representation of a viscoelastic material, in each case considering a representation consisting of two elastic and two viscous elements. By comparison, express the characteristics of the Voigt elements in terms of those of the Maxwell elements, thus deriving Eqs. 6.23.

◀6.22. Derive the relaxation modulus for a Maxwell and for a Voigt representation, in each case consisting of two elastic and two viscous elements. Thus determine the expressions for the constants of the Maxwell elements in terms of those of the Voigt elements.

6.23. Derive the apparent activation energy spectrum of Fig. 6.26. Show that a decrease in activation energy resulting in a rapid change in viscosity leads to a high apparent activation energy.

6.24. Propose a simple series of tests which can be carried out on one specimen to determine the maximum strain at which the constitutive relations are still linear.

6.25. Why does the viscosity of liquid metals vary relatively slowly with temperature? [*Hint:* Consider the activation energy for liquid metals as compared with organic liquids.]

◀6.26. A liquid exhibiting appreciable strain-induced anisotropy is contained in the apparatus shown in Fig. 6.31, in which the floating platform is free to rise or fall but not rotate. What will happen as the cup is rotated?

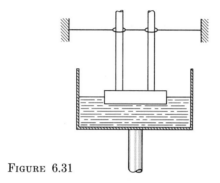

FIGURE 6.31

6.27. Estimate roughly the relaxation modulus for polyvinyl chloride from the creep compliance of Fig. 6.11.

6.28. Estimate the specific heats of some organic liquids not listed in Table 6.2, and check your estimates against data found in handbooks.

6.29. How can the heat distortion temperature be used to give an estimate of the long-time behavior of a polymer under constant loads?

◀6.30. Estimate the modulus of the cross-linked polystyrene of Fig. 6.8 from the mole fraction of cross-linking agent, assuming that each mole joins polystyrene chain segments of equal length.

◀6.31. In reporting on the effects of nuclear radiation on rubber, Born (1961) gives values of the initial dynamic modulus and of the 100% modulus, "that is, rubber modulus at 100% elongation":

Rubber	Modulus, psi	
	100%	Dynamic
Natural rubber	470	1280
Styrene-butadiene rubber	320	1320
Chloroprene rubber	510	2100

(a) What do you understand by the term 100% modulus? Illustrate your answer with a sketch. Are the results *qualitatively* in agreement with your definition?

(b) Are the results given in the table *quantitatively* in agreement with your definition? See Payne and Scott (1960, p. 125) for a definition, if you don't want to try your own.

E 6.32. Using a rubber band or a strip of rubber cut from a balloon, determine Poisson's ratio for rubber under the assumption that rubber is isotropic.

E◀6.33. Fasten a spring and a slightly stretched rubber band in series between two rigid points. Estimate the change in spring length due to the largest temperature change you can conveniently apply. Check your results experimentally.

E◀6.34. Determine the uniaxial stress-strain curve for a rubber band and compare it with the theoretical relation.

E◀6.35. Calculate the biaxial stress-stretch relations for rubber and use these to predict the pressure-volume curve for a balloon. Check the pressure-volume curve by inflating the balloon, placing a small weight with a large flat surface on the balloon, and measuring the contact area. The balloon may be slowly deflated, and simultaneous measurements made of the diameter of the balloon and the contact area.

E 6.36. What spring and dashpot model of a viscoelastic material is reasonable for silicone putty? Determine the steady-state viscosity by making a dumbell-shaped specimen and hanging it from one end. Measure the diameter as a function of time. From these data and the density, calculate the coefficient of viscosity. How can you prove that the strain rate varies linearly with the stress? Since it does turn out that the silicone putty has a steady-state viscosity, remember that it will, in sufficient time, creep through very small holes. Should these be the fabric of your clothing, you will be very disappointed, for silicone putty is quite insoluble. Moral: keep the silicone putty in an impervious container.

E 6.37. (a) Cut a yard-long strip of polyethylene from a dry cleaner's bag. Hang weights on it, as indicated in Fig. 6.32, and note the time-dependent deformation. For how large strains is the creep rate linear?

(b) Stretch a strip between your fingers for half a minute. Release the load suddenly until the polyethylene strip just becomes slack, and brace your fingers to hold that position. Note the elastic after-effect.

E 6.38. Cut a yard-long strip from a polyethylene bag, and by subjecting it to various histories of load or deformation, determine an appropriate viscoelastic model consisting of one or two springs and dashpots.

(a) What are the magnitudes of the elastic constants and coefficients of viscosity for your model?

(b) Are models derived from tests involving creep, relaxation, or elastic aftereffect consistent?

E 6.39. Stretch a piece of polyethylene about 1-in. square (or larger, if you can get a larger piece from a bread wrapper or dry cleaning envelope). Note that in some directions, a yield point and inhomogeneous deformation are observed. Whether or not this is the case, before the specimen fractures, try pulling it transversely. Note the markedly lower stress required for transverse flow. How is this related to the Bauschinger effect?

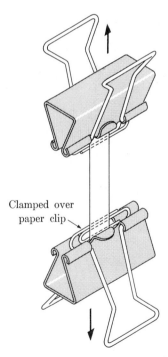

Clamped over paper clip

FIGURE 6.32

6.40. (a) Using a Maxwell model of a polymer with a single activation energy, propose a method of estimating the creep deflection at an arbitrary stress and temperature from the results of the heat distortion test.

(b) Would you expect your estimate to be high or low?

(c) From such a model with a reasonable estimate of the heat of activation, estimate the creep of polyvinyl chloride in 500 hours at room temperature from the heat distortion temperature. Compare your results with Fig. 6.11.

6.41. In order to predict the long-time creep behavior of a polymer, it is proposed to run short-time tests at the service temperature and at a slightly higher temperature. How can the service behavior be estimated from these data, using the idea of activation energy? Is this procedure likely to overestimate or underestimate the creep in service?

6.42. The stress-strain curve for rubber in Fig. 1.2 shows a slight residual strain on release of load. Judging from the mechanisms of deformation in rubber, is this likely to be a real effect or should one look for errors in the instrumentation?

6.43. Why does most of the hysteresis in rubber shown by Fig. 1.2 occur at large strains?

6.44. Discuss the possibilities of thermal nucleation of dislocations in polymers.

6.45. A particularly interesting example of the effect of stress history on viscoelastic materials is the Kohlrausch (1876) effect in which loading is applied for a long time in one direction, and then a short time in the opposite direction. What kinds of deformation could be observed on final release of load?

Part II

MECHANICS OF MATERIALS

CHAPTER 7

FUNDAMENTAL EQUATIONS OF CONTINUUM MECHANICS

7.1 SYNOPSIS

The preceding discussion of the physical mechanisms of deformation should make it clear that no one simple equation will give an exact description of the stress-strain-time-temperature, or constitutive, relations for a real material. In spite of this complexity, the engineer must combine the constitutive relations with the equilibrium and strain-displacement equations and use them to determine the load-deformation characteristics of structures, and the stress and strain at points within them. In order to obtain solutions or even good approximations for these problems, he must obtain relatively simple constitutive relations by idealizing the actual behavior of the material. Besides the elastic idealization already considered, we shall treat here the plastic, creep, and viscoelastic approximations. The appropriateness of each of these models depends on the problem at hand.

After a brief review of the equations of elasticity, we shall consider plastic behavior, in which the deformation is assumed to be entirely independent of time. Criteria for the initiation of the plastic flow under combined stress will be discussed, as well as the derivation of the stress-strain relations from these criteria. Ways of estimating strain-hardening will be considered, along with the limitations imposed by the Bauschinger effect.

At high temperatures and stresses, materials creep with time, and we shall consider appropriate modifications of the constitutive relations.

At low stress levels in polymers, the time effects may correspond to linear viscoelasticity. For such materials, we shall show the interrelation between test results with constant or oscillating stress or strain, and show how these results lead to three-dimensional constitutive relations. The stress distribution for an ideally viscoelastic material will be shown to be identical with that for an elastic material in many practically important cases.

At high rates of loading, inertia effects become important, so we must consider the effects of stress waves traveling through the material and their effects on the local strain for given applied loads and deformations.

7.2 IDEALIZATION OF MECHANICAL BEHAVIOR

The solution of a problem in the mechanics of continua requires not only that the components of stress be in equilibrium,

$$\frac{\partial \sigma_{11}}{\partial x_1} + \frac{\partial \sigma_{21}}{\partial x_2} + \frac{\partial \sigma_{31}}{\partial x_3} = 0,$$

$$\frac{\partial \sigma_{12}}{\partial x_1} + \frac{\partial \sigma_{22}}{\partial x_2} + \frac{\partial \sigma_{32}}{\partial x_3} = 0, \quad (2.11)$$

$$\frac{\partial \sigma_{13}}{\partial x_1} + \frac{\partial \sigma_{23}}{\partial x_2} + \frac{\partial \sigma_{33}}{\partial x_3} = 0,$$

and that the components of strain be derivable from displacements,

$$\epsilon_{11} = \frac{\partial u_1}{\partial x_1}, \quad \gamma_{23} = \frac{\partial u_2}{\partial x_3} + \frac{\partial u_3}{\partial x_2},$$

$$\epsilon_{22} = \frac{\partial u_2}{\partial x_2}, \quad \gamma_{31} = \frac{\partial u_3}{\partial x_1} + \frac{\partial u_1}{\partial x_3}, \quad (2.25)$$

$$\epsilon_{33} = \frac{\partial u_3}{\partial x_3}, \quad \gamma_{12} = \frac{\partial u_1}{\partial x_2} + \frac{\partial u_2}{\partial x_1},$$

but it also requires that some form of stress-strain relation be satisfied. For an isotropic elastic material, when the components of strain are linear functions of the components of stress, these relations are

$$\epsilon^e_{11} = \frac{\sigma_{11}}{E} - \frac{\nu \sigma_{22}}{E} - \frac{\nu \sigma_{33}}{E}, \quad \gamma_{23} = \frac{\sigma_{23}}{G},$$

$$\epsilon^e_{22} = -\frac{\nu \sigma_{11}}{E} + \frac{\sigma_{22}}{E} - \frac{\nu \sigma_{33}}{E}, \quad \gamma_{31} = \frac{\sigma_{31}}{G}, \quad (3.19)$$

$$\epsilon^e_{33} = -\frac{\nu \sigma_{11}}{E} - \frac{\nu \sigma_{22}}{E} + \frac{\sigma_{33}}{E}, \quad \gamma_{12} = \frac{\sigma_{12}}{G},$$

where

$$G = \frac{E}{2(1+\nu)}. \quad (3.18)$$

The elastic part of the strain in crystals is rather uniformly distributed through the lattice. Inelastic strain is less uniformly distributed, since it occurs by the motion of dislocations. This further strain must be added to the elastic part. It is the total strain which enters into the strain-displacement equations. For example, with plasticity and creep we write

$$d\epsilon_{11} = d\epsilon^e_{11} + d\epsilon^p_{11} + d\epsilon^c_{11}. \quad (7.1)$$

For polymers, this division into parts may not be applicable, because the elastic strain may not be homogeneous on the molecular level. In any event, as we have seen in Chapters 4, 5, and 6, the variables of time and temperature may become important when inelastic deformation begins. Also, stress effects may not be linear. The resulting constitutive equations are exceedingly complex. To obtain workable relations, it is necessary to idealize the behavior into a few limiting cases, between which the behavior of real materials is likely to lie. Stress-strain-time diagrams for some idealized kinds of behavior are given in Figs. 1.1 to 1.7.

For instance, *plastic* deformation is taken to be that occurring immediately upon application of the load and remaining on release of load. *Creep* is that part of the deformation which requires a finite time to develop. Most creep remains on release of load, but there may be also an elastic aftereffect. The limiting case in which the creep rate is directly proportional to the stress is Newtonian viscosity. This may also be combined with elasticity in various ways to give *linear viscoelasticity*, in which creep, relaxation, and aftereffects may be present, and in which, for example, doubling the applied loading history doubles the resulting strain at any given time. Linear viscoelasticity is usually encountered in polymers at strains of less than 1 or 2%, and in metals, only at impractically high temperatures and low stresses, if at all.

The appropriate choice between these idealizations depends not only on the material, but also on the kind of problem at hand. For instance, the *elastic* stress-strain relations are appropriate where a given deformation must be accommodated repeatedly and reproducibly, as in the case of springs, or where vibrations are of importance. These relations are also important in designing to prevent brittle fracture and fatigue, where the stress must be confined to the elastic regime throughout the entire part. Considering a body to be completely *rigid* is often satisfactory when one needs to know the overall motions or forces in machine parts. It is often satisfactory in describing wear and friction. A *rigid-plastic* approximation may be useful where strains are large, as in forging, machining, and the study of the stress and strain around a crack in a ductile material. It is also useful in designing safety devices to accommodate desired deformation and to limit a force or an acceleration, as in automobile bumpers, crash panels, shear pins, and blow-out diaphragms. The rigid-plastic approximation is also useful in the design of structures, for we shall see in Section 10.3 that for loads just less than those required for large-scale plastic deformation, the deformation is limited to the order of magnitude of the elastic deformation. To demonstrate this idea quantitatively requires an *elastic-plastic* analysis, which also may be important in studying crack propagation from a sharp notch when the average stress is below the yield strength. Elastic-plastic behavior must also be considered in studying residual stress arising from forming processes or service loads, although if the plastic strains are large, one may consider the material as rigid-plastic on loading and elastic on unloading. An elastic-plastic approximation is also required for studies of the springback encountered in sheet-metal forming. A *viscoelastic* approximation is useful for polymers at strains less than 1 or 2% where the relations are linear, but time effects play an important role.

7.3 PLASTIC YIELDING UNDER COMBINED STRESSES

The tendency of a metal to yield under combined stresses is usually given in terms of a function of the stress components, such that when this function reaches a stress level which would cause yielding in a tensile test, yielding will occur. This function is called the *equivalent stress*. At present it is not possible to derive the function for the equivalent stress theoretically from the behavior of single crystals. Instead, some conditions which the function must satisfy will be described, and then empirical functions meeting these conditions will be compared with experimental data.

Since yielding occurs by slip or twinning, and since these are caused by shear components of stress, it is to be expected that the equivalent stress causing yielding will depend on the shear components of stress, or on the differences of the normal components of stress. It is to be expected that yielding will depend on all three differences of the normal components of stress. For example, consider the bicrystal shown in Fig. 7.1. Even though the transverse normal stress component σ_{22} is intermediate in magnitude between the other two, it will still play a part in making the right-hand crystal deform, and thus make plastic deformation of the assembly easier.

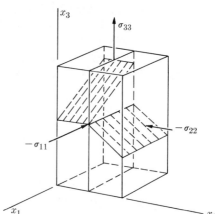

FIG. 7.1. Effect of intermediate stress σ_{22} on yielding of a bicrystal.

A polycrystal in which all grain orientations are equally likely will be isotropic. In this case, it may be convenient to choose the coordinate axes to be those giving principal components of stress, denoted by σ_1, σ_2, and σ_3, so that the conditions for yielding can be described in terms of only the pairs of differences of the normal components, without having to consider shear components.

The most commonly used equivalent stress function satisfying the physical requirements discussed above is

$$\bar{\sigma} = \sqrt{\tfrac{1}{2}[(\sigma_2 - \sigma_3)^2 + (\sigma_3 - \sigma_1)^2 + (\sigma_1 - \sigma_2)^2]}. \tag{7.2a}$$

This criterion is known by the names of a number of men who independently conceived it: von Mises, Huber, and Hencky (as well as Maxwell, who mentioned it in correspondence). It is also known as the distortion energy criterion or octahedral shear stress criterion. It has been shown to be proportional to the root mean square over all planes through a point of the maximum shear stress on each plane (Novoshilov, 1952), but we shall take its principal justification to be empirical. We shall refer to it as the *Mises yield criterion*.

It is often more convenient to work with this yield function in terms of arbitrary coordinate axes referred to which there may be shear components of stress. It can be shown that when only one component of shear stress, say σ_{23}, is present, the equivalent stress is (Prob. 7.1)

$$\bar{\sigma} = \sqrt{3\sigma_{23}^2}.$$

This suggests that in the presence of all components of stress, an appropriate equivalent stress is

$$\bar{\sigma} = \sqrt{\tfrac{1}{2}[(\sigma_{22} - \sigma_{33})^2 + (\sigma_{33} - \sigma_{11})^2 + (\sigma_{11} - \sigma_{22})^2] + 3\sigma_{23}^2 + 3\sigma_{31}^2 + 3\sigma_{12}^2}, \tag{7.2b}$$

◀ or in terms of stress deviators, Eq. 3.22, and the summation convention, Section 2.8 (Prob. 7.2),

$$\bar{\sigma} = \sqrt{\tfrac{3}{2} s_{ij} s_{ij}}. \tag{7.2b} ▶$$

It can be shown (see for example Prager and Hodge, 1951, pp. 16, 22, 33) that this equivalent stress function is indeed identical with that based on the principal components of stress, Eq. 7.1. The Mises yield criterion then is that yielding occurs when the equivalent stress reaches the yield strength in tension, Y:

$$\text{Yielding occurs when } \bar{\sigma} = Y. \tag{7.2c}$$

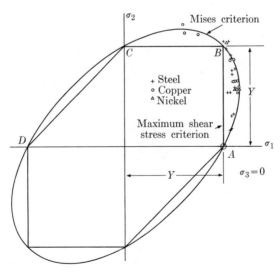

FIG. 7.2. Yielding of thin-walled tubes under combined stresses. (Lode, 1926; from Crandall and Dahl, 1959. Courtesy of McGraw-Hill.)

A second empirical equivalent stress is that based on the maximum difference of the principal stress components. Half the difference between the maximum and minimum principal stress components is the maximum shear stress referred to any coordinate axes. Thus this yield criterion is often called the *maximum shear stress criterion*. It also is known by the names of two men who proposed it, Tresca and Guest. According to this criterion, yielding occurs when the maximum shear stress equals the maximum shear stress in the tensile test,

$$\bar{\tau} = (\sigma_{max} - \sigma_{min})/2, \tag{7.3a}$$

$$\text{Yielding occurs when } \bar{\tau} = k. \tag{7.3b}$$

Experimental data to test these criteria can be obtained by testing thin-walled tubes under combined internal pressure and axial load or under combined torsion and tension (or by pulling strips grooved at an oblique angle, Prob. 7.14). Some of the early and still the best data are presented in Fig. 7.2. They fall between the Mises and the maximum shear stress yield criteria, although they are closer to the Mises value. The locus of points at which yielding occurs is called the yield locus. It is shown in Fig. 7.3 for the case in which all three principal stresses differ from zero.

◄In the case of anisotropic metals, the Mises and the maximum shear stress yield criteria no longer apply. A generalization of the Mises yield criterion has been proposed by Hill (1950) for an orthotropic metal:

$$\bar{\sigma} = \left\{ \frac{1}{H+G} \left[(\sigma_{22} - \sigma_{33})^2 + G(\sigma_{33} - \sigma_{11})^2 + H(\sigma_{11} - \sigma_{22})^2 \right] \right.$$
$$\left. + L\sigma_{23}^2 + M\sigma_{31}^2 + N\sigma_{12}^2 \right\}^{1/2}. \tag{7.4}$$

Information on the coefficients G and H relating the normal components of stress has been obtained, but there are few if any cases in which all five coefficients of Eq. 7.4 have been evaluated (see Prob. 7.3). ▶

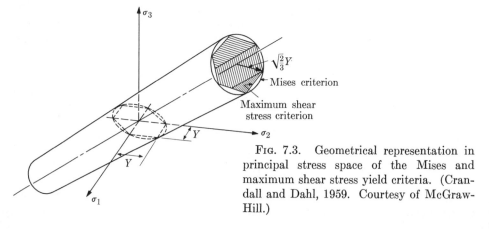

FIG. 7.3. Geometrical representation in principal stress space of the Mises and maximum shear stress yield criteria. (Crandall and Dahl, 1959. Courtesy of McGraw-Hill.)

The concept of a yield function or yield locus can be extended to a larger scale than the stress at a point. For example, in the analysis of bars and beams, it is often easier to work with the axial loads and the bending and twisting moments, rather than with the detailed distribution of the stress within the member itself. This simply is an extension of the idea that it is easier to work with average stress in a polycrystalline material rather than with the local stresses within grains or around dislocations. In a long bar, for example, one can plot the combinations of axial load and twisting moment required to produce yielding. The theoretical derivation of such plots may require a knowledge of the stress-strain behavior and will be discussed later.

7.4 EFFECT OF PLASTIC DEFORMATION ON THE EQUIVALENT FLOW STRESS

As crystals are plastically deformed, the stress for plastic flow, called the flow stress,* usually rises. The rise is due to the increased difficulty of dislocation motion due to other dislocations which are on the same or intersecting planes, or which are themselves blocked by solutes, precipitates, dislocation tangles, or by twin, phase, or grain boundaries. The dislocation structure depends on the previous history of plastic straining. The effect of strain on equivalent flow stress should therefore be considered in two stages (Kröner, 1963): the effect of plastic strain on dislocation structure and the dependence of the equivalent stress on that dislocation structure. Because our understanding of these stages is as yet incomplete, and because we can already see that it will be complex even when it is complete, we shall here combine the two stages and look for empirical relations giving the equivalent flow stress as a direct function of the plastic strain history. In doing so, we shall use our understanding of dislocations as a guide to the form of such a function, although historically the functions preceded the concept of a dislocation.

If it is assumed that the material remains reasonably isotropic, then the function can be expressed in terms of the principal components of plastic strain. Since the volume of the metal remains constant except for second-order effects in the core of dislocations, the strain-hardening should depend on some function of the differences between the components of plastic strain rather than their absolute values. Finally, we observe that strain-hardening occurs even if a metal is deformed in one direction and then deformed back to its original shape. Therefore, the strain-hardening does not depend directly on the total strains, but rather on a monotonically increasing integral of the plastic strain increments. These conditions are all met by assuming that the equivalent stress depends on the equivalent plastic strain, defined in terms of the components as is the equivalent stress, but with a numerical factor to make its product with the equivalent stress be the

---

* "Flow strength" would be a clearer term, since this is a characteristic of the state of the material, not to be confused with whatever force per unit area happens to be applied at the moment. Perhaps to avoid confusion with "fracture strength," however, most workers use the term "flow stress."

plastic work in uniaxial tension and pure shear (Prob. 7.4):*

$$\int d\bar{\epsilon}^p = \int \{\tfrac{2}{9}[(d\epsilon_{22}^p - d\epsilon_{33}^p)^2 + (\ )^2 + (\ )^2]$$
$$+ (d\gamma_{23}^p)^2/3 + (d\gamma_{31}^p)^2/3 + (d\gamma_{12}^p)^2/3\}^{1/2}. \quad (7.5a)$$

The factor 2/9 also makes the equivalent strain in the uniaxial tensile test equal to the axial normal component of strain (Prob. 7.5). The integral in Eq. 7.5a is taken over the entire deformation since the last annealing treatment.
◀ In terms of tensor components of plastic strain (Prob. 7.6),

$$\int d\bar{\epsilon}^p = \int \sqrt{\tfrac{2}{3}\, d\epsilon_{ij}^p\, d\epsilon_{ij}^p}. \quad (7.5b)$$

◀ For a material which yields according to the maximum shear stress yield criterion, it will be seen in Section 7.5 that it is again appropriate to define an equivalent strain such that its product with the equivalent stress will be the plastic work. We first define a principal slip component as a shear strain referred to a plane at 45° to a pair of the principal directions, as shown in Fig. 7.4. Any pair of principal slip components can give any combination of principal components of strain. For instance, if $|d\epsilon_1^p|$ is taken to be the maximum absolute principal component of strain, then by appropriate choice of the two principal slip components shown in Fig. 7.4, one can match arbitrary principal normal components of strain:

$$|d\gamma_3^p| = 2|d\epsilon_2^p|, \quad |d\gamma_2^p| = 2|d\epsilon_3^p|, \quad |d\gamma_2^p| + |d\gamma_3^p| = 2|d\epsilon_1^p|. \quad (7.6)$$

We take the equivalent shear strain to be the sum of the absolute values of these principal slip components:

$$\int d\bar{\gamma}^p = \int |d\gamma_2^p| + |d\gamma_3^p| = \int 2|d\epsilon_{\max}^p|. \quad (7.7)$$

The first of these equalities is more descriptive, the second more convenient. Equation 7.7 can be considered to be the definition of a *work equivalent shear strain*, since if the maximum shear stress $\bar{\tau}$ is required for flow on each of the two deforming principal slip planes, which will be shown to be the case in Section 7.5, the increment of plastic work is (Prob. 7.7)

$$dW^p = \bar{\tau}\, d\bar{\gamma}^p. \quad (7.8)$$

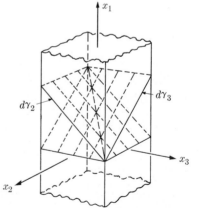

FIG. 7.4. Principal slip components.

---

* General proof requires the stress-strain relations of Eq. 7.16 (see Hill, 1950, p. 37).

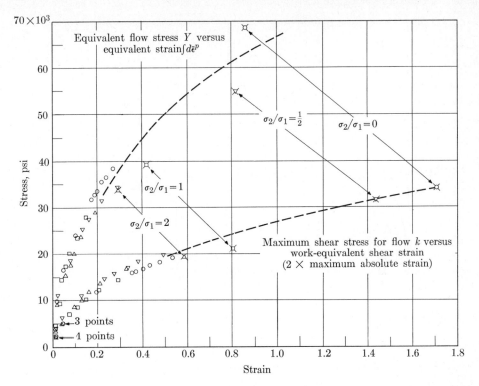

FIG. 7.5. Combined stress tests on annealed copper tubes, as correlated by equivalent strain $\int d\bar{\varepsilon}^p$ and work equivalent shear strain $\int d\bar{\gamma}^p$. (Data from Davis, 1943.)

On the other hand, if the shear stress were zero on the second principal slip plane, the work would be

$$dW^p = \bar{\tau}\, d\gamma_{\max}. \qquad (7.9) \blacktriangleright$$

We now postulate that the equivalent flow stress depends on the corresponding equivalent strain. That is, no yielding occurs unless the equivalent stress reaches the equivalent flow stress, which in turn depends on the equivalent strain.

For the Mises criterion, using either of two equivalent symbols for the equivalent flow stress, we postulate

$$\text{Yielding occurs when } \bar{\sigma} = Y\left(\int d\bar{\varepsilon}^p\right) = \bar{\sigma}_Y\left(\int d\bar{\varepsilon}^p\right). \qquad (7.10)$$

For the maximum shear stress criterion, we postulate

$$\text{Yielding occurs when } \bar{\tau} = k\left(\int d\bar{\gamma}^p\right) = \bar{\tau}_Y\left(\int d\bar{\gamma}^p\right). \qquad (7.11)$$

Data to check these equations are presented in Fig. 7.5, for annealed copper tubes under combined internal pressure and axial load. The correlation is within ±5% with the Mises yield criterion, giving slightly better results at the smaller strains,

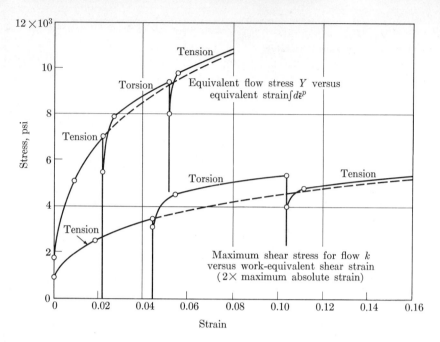

FIG. 7.6. Alternating tension and torsion as correlated by equivalent strain $\int d\bar{\varepsilon}^p$ and work equivalent shear strain $\int d\bar{\gamma}^p$. (Data from Sautter, Kochendörfer, and Dehlinger, 1953.)

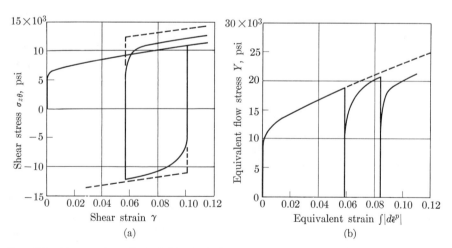

FIG. 7.7. Reversed torsion of thin-walled copper tubes, stress-relieved at 600°F before testing. 30 mm grain diameter. (Data from Meyer, 1957.)

but deviating more at the higher strains, perhaps as a result of anisotropy. As a matter of fact, a plot of maximum shear stress versus maximum shear strain, corresponding to Eq. 7.9, gives an even better fit at high strains. Further tests would be required to determine whether this is generally true or a fortuitous effect of anisotropy in this case.

A more crucial test of the concept of equivalent stress and strain is obtained when the ratios between the components change during the test, as is the case when first tension and then torsion is applied. Data obtained from this kind of test are presented in Fig. 7.6. The Mises criterion is better and provides a satisfactory correlation, except just where the stress is changed. An even more severe test of the correlation is obtained when the stress is completely reversed in sign. It is known from the observations of Bauschinger (Section 5.8) that the stress for reversed plastic flow is reduced. This is illustrated by the experiments on reversed torsion plotted in Fig. 7.7. The reduction in flow stress on reversal persists for the first few percent of reversed plastic strain, after which it has risen nearly to the value for the same equivalent strain without reversal. Thus the Bauschinger effect persists over strains large compared with the elastic strains, even though small compared with unity.

7.5 RELATIONS BETWEEN STRESS AND PLASTIC STRAIN INCREMENT

So far, the stress-strain relation has been limited to the relation between the equivalent strain and the equivalent flow stress. In other words, we have considered only a scalar magnitude of strain. We now turn to the question of how the magnitude is proportioned among the various components of strain increment. This proportionality depends on the components of stress, rather than on the equivalent stress alone. The relations between the components of the strain increments and the stress can be derived from the yield locus through an understanding of the micromechanisms of plastic flow.

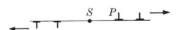

FIG. 7.8. Illustration of the dependence of the strain near P on the stress at the dislocation source S.

First, since only the current direction of dislocation motion is determined by the current stress, the current stress determines only the current strain increments and not their total values, as it does in elasticity. Second, one must examine the idea that the plastic flow or slip at a point is governed by the stress at that point, independent of the stress at nearby points. In a very small region, this postulate is no longer valid. Consider the situation shown in Fig. 7.8, where dislocations generated at S move along the plane for relatively long distances before they are blocked. If the stress required to activate the source is the limiting factor, the plastic flow at a point P, along the slip line, is governed not only by the stress at P, but also more importantly by the stress at the dislocation source S. In addition, the idea of plastic strain is based on average displacements over regions large enough to contain many slip lines. Therefore, the concepts of a critical shear stress for slip and of a stress-strain relation can be considered valid only when applied to a region which is large compared with both the mean spacing of dislocation sources and the mean distance of travel of a dislocation, or the length of a slip band produced by a cascade of dislocations.

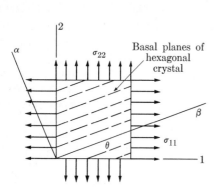

FIG. 7.9. Crystal under biaxial stress.

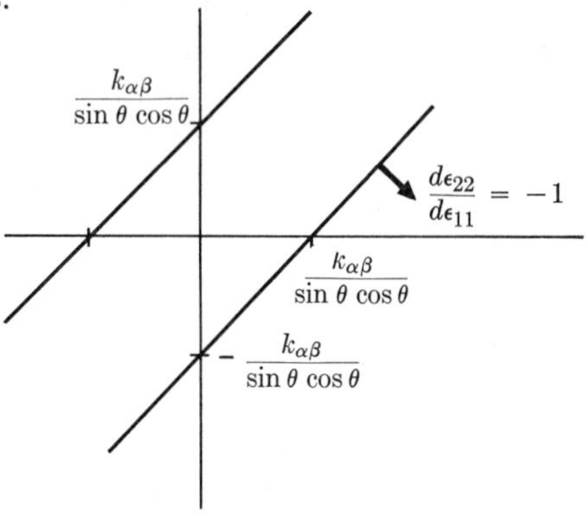

FIG. 7.10. Yield locus for stressed crystal of Fig. 7.9, with associated strain increment.

◀Assuming a region large enough so that the strain increment is dependent only on the local stress, we now turn to the relation between the yield locus and the stress-strain relations for a single crystal. Consider, for example, a crystal having a set of slip planes normal to the α axis and a slip direction parallel to the β axis, as indicated in Fig. 7.9. Consider the special case in which it is desired to predict the effect of two normal stress components in a coordinate system making an angle θ with the α, β coordinate system on the yielding of the crystal. If the shear stress for yielding on the α, β slip system is denoted by $k_{\alpha\beta}$, it can be shown that the yield locus for plastic deformation in terms of the normal components of strain is that shown in Fig. 7.10 (Prob. 7.8). The plastic deformation which results can also be represented (Fig. 7.10) by drawing from the point representing the state of stress a short vector whose components are proportional to the corresponding increments of strain, $d\epsilon_{11}$ and $d\epsilon_{22}$. These increments are determined from the fact that, as discussed in Section 4.10, when plastic deformation occurs at the resolved shear stress, say $\tau_{\alpha\beta}$, the strain increment produced is the corresponding one, $d\gamma_{\alpha\beta}$. When this is done, it turns out that the vector representing the plastic strain increment is normal to the yield locus (Prob. 7.9).

◀In general, slip or twinning on other crystallographic planes will provide further limitations on the stress that can be withstood without plastic deformation. These will appear in Fig. 7.10 as other pairs of lines, or in the three-dimensional case as other pairs of planes. The result is a polygonal yield locus. When the state of stress is represented by a point on one of the sides of the polygon, only one slip system will be active. If the state of stress is at one of the corners of the polygon, simultaneous slip can take place on the two crystallographic planes corresponding

to the adjacent sides of the polygon. A similar state of affairs holds for the six-dimensional polyhedron which arises when all components of stress are considered.
◀The fact that the strain increment components plotted as vectors in the six-dimensional stress space are always normal to the yield locus is equivalent to the *principle of maximum plastic resistance** which states: *For any plastic strain increment, the state of stress actually occurring gives an increment of work which equals or exceeds the work which would be done by that strain increment $d\epsilon_{ij}^p$ with any other state of stress within or on the yield locus.* In equation form, if σ_{ij}^* represents any state of stress within or on the yield locus (Prob. 7.10),

$$\sum_{i=1}^{3}\sum_{j=1}^{3}(\sigma_{ij}-\sigma_{ij}^*)\,d\epsilon_{ij}^p \geq 0. \tag{7.12}$$

◀It will now be shown, following Bishop and Hill (1951), that this principle of maximum plastic resistance holds not only when the strain is due to slip on a single crystallographic system, but also for single crystals when strain occurs simultaneously on a number of crystallographic systems and for whole aggregates of single crystals. The principle of maximum plastic resistance also holds if the work done is described either in terms of stress and strain or such generalized "stresses" and "strains" as forces, bending moments, extensions, and rotations.
◀The total work done when slip is occurring on a number of crystallographic planes, giving a total strain $d\epsilon_{ij}^P$, is

$$dW^p = \sum_{i}^{3}\sum_{j}^{3}\sigma_{ij}\,d\epsilon_{ij}^p. \tag{2.41}$$

The total strain can be expressed as the sum of the strains referred to the i,j coordinate axes due to slip on different α, β systems† each giving a shear strain component $d\epsilon_{\alpha\beta}$,

$$d\epsilon_{ij}^p = \sum_{\alpha,\beta}d\epsilon_{\alpha\beta}^p l_{i\alpha}l_{j\beta},$$

where the summation is taken over all slip systems rather than letting α, β range from 1 to 3. The work done against a stress σ_{ij} is then given by

$$dW^p = \sum_{\alpha,\beta}\sum_{i}^{3}\sum_{j}^{3}\sigma_{ij}l_{i\alpha}l_{j\beta}\,d\epsilon_{\alpha\beta}^p.$$

Introducing the law of transformation of the stress components to define the state of stress on each of the different crystallographic slip systems gives

$$dW^p = \sum_{\alpha,\beta}\sigma_{\alpha\beta}\,d\epsilon_{\alpha\beta}^p.$$

---

* This term has also been used with a different meaning, as mentioned in Section 10.5.
† For an arbitrary plastic strain, there are five independent components of strain (the sixth being determined by incompressibility). These therefore require at least five *independent* crystallographic slip systems. Rocksalt does not always have five independent systems and can be brittle in polycrystalline form. See for example Groves and Kelly (1963).

Carrying the process through for the work which would be done if the plastic deformation could occur under a state of stress $\sigma^*_{\alpha\beta}$ within the yield locus, we find

$$dW^{p*} = \sum_{\alpha,\beta} \sigma^*_{\alpha\beta}\, d\epsilon^p_{\alpha\beta}.$$

Thus, the difference between the work done by the actual state of stress and any other within or on the yield locus is given by

$$dW^p - dW^{p*} = \sum_{\alpha,\beta} (\sigma_{\alpha\beta} - \sigma^*_{\alpha\beta})\, d\epsilon^p_{\alpha\beta}. \tag{7.13}$$

Since all terms on the right-hand side of Eq. 7.13 are greater than zero, the principle of maximum plastic resistance has been proven for the case of simultaneous slip on several systems of a crystal. Now, if this principle holds for any element of a single crystal, it must hold for the sum of a large number of elements, provided that each is in equilibrium so it acquires no kinetic energy. We can further say that either for a polycrystalline aggregate or for an entire structure, the work done in a given plastic deformation will be greater than would be calculated for any fictitious state of stress which would be in equilibrium with the external forces and would be either within or on the yield locus at every point within the body. For example, consider a polycrystalline element large enough that the usual definition of stress and strain can be applied. The principle of maximum plastic resistance then is

$$\sum_i^3 \sum_j^3 (\sigma_{ij} - \sigma^*_{ij})\, d\epsilon^p_{ij} \geq 0. \tag{7.14}$$

◀Equation 7.14 is identical in form with Eq. 7.12, but different in meaning, for the stress and strain in Eq. 7.12 apply to a small region within a grain, whereas those here refer to averages over many grains.

◀The principle of maximum plastic resistance has also been derived from the postulate that for any increment of plastic deformation requiring a slight increase in the stress, the work done by the agent applying the stress *increase* acting through the displacements must always be positive (see Drucker, 1956). The derivation given above is preferred by the writer in view of its more direct connection to the dislocation mechanisms involved in plastic deformation.▶

However arrived at, the principle of maximum plastic resistance leads us to two important results. First, the yield locus can never be concave. Second, the yield locus determines the stress-strain relations. To obtain these results more explicitly, we must introduce the idea of representing stress and strain as vectors in a multidimensional space. These are vectors in the sense that they are comprised of sets of numbers called components, but they are not vectors in the sense that they transform in a certain way with change in coordinate axes, which in this context is not relevant. The plastic work done as a body undergoes a plastic strain can be thought of as the dot product of the stress and strain increment vectors according to Eqs. 2.37 or 2.41. In this sense, the principle of maximum plastic resistance is that the dot product of the stress and the strain increment vectors must always be positive. As shown in Fig. 7.11, this leads to the fact that the yield locus cannot be concave. Similarly, the strain increment vector associated with a given point on the yield locus must always be normal to the locus, except possibly where the yield locus has a sharp corner, in which case the strain increment vector is confined to a narrow region between the normals to the adjacent sides. This is illustrated in Fig. 7.12 for the Mises and maximum shear stress yield criteria. Similar arguments hold in the six-dimensional space in which all stress and strain components are represented.

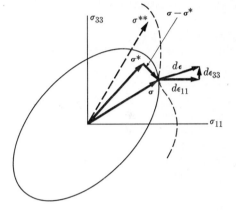

Fig. 7.11. Since $(\sigma - \sigma^*)\, d\epsilon^p > 0$, the strain increment must be normal to the yield locus, and the concave locus shown dashed, allowing the stress σ^{**}, is impossible.

The condition of normality has two important results. First, in experimental work, if one notes not only the fact of yielding, but also the resulting plastic strain increments, then one has not only a point on the yield locus, but also a line (that normal to the increment) behind which the yield locus must lie. Second, a stress-strain relation is implied because of the proportionality between the increments

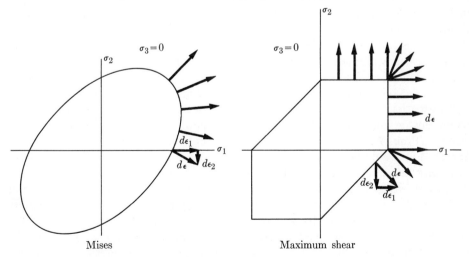

Fig. 7.12. Illustration of normality of strain increment vector to yield locus.

of strain and the components of the outward normal to the yield locus, which can be expressed in terms of its partial derivatives (Prob. 7.11),

$$\frac{d\epsilon_{11}^p}{\partial \bar{\sigma}/\partial \sigma_{11}} = \frac{d\epsilon_{22}^p}{\partial \bar{\sigma}/\partial \sigma_{22}} = \cdots = \frac{\partial \gamma_{23}^p}{\partial \bar{\sigma}/\partial \sigma_{23}} = \cdots = d\lambda, \qquad (7.15)$$

where $d\lambda$ is a scalar of proportionality. Equation 7.15 is called the *associated flow rule*. For the Mises yield criterion, taking the scalar $d\lambda$ such that the increment of equivalent stress corresponds to that found from the uniaxial strain-hardening curve, the associated flow rule reduces to (Prob. 7.12)

$$\begin{aligned} d\epsilon_{11}^p &= [\sigma_{11} - \tfrac{1}{2}(\sigma_{22} + \sigma_{33})]\, d\bar{\epsilon}^p/Y, \\ d\epsilon_{22}^p &= \\ d\epsilon_{33}^p &= \\ d\gamma_{23}^p &= 3\sigma_{23}\, d\bar{\epsilon}^p/Y, \\ d\gamma_{31}^p &= \\ d\gamma_{12}^p &= \end{aligned} \qquad (7.16a)$$

or in terms of stress deviators (Prob. 7.13),

$$d\epsilon_{ij}^p = 3s_{ij}\, d\bar{\epsilon}^p/2Y. \qquad (7.16b)$$

It is instructive to compare Eqs. 7.16a with the corresponding elastic stress-strain relations, Eqs. 3.19. First, the plastic equations give strain increments rather than strains. This comes from the physical fact that the present state of stress tells only what the next increment of dislocation motion will be; it does not tell about the past motion which led to the present state of strain. Second, the Poisson ratio of approximately 0.3 for elasticity becomes 0.5 for plastic deformation, because the motion of dislocations produces negligible volume change. Finally, the factor $d\bar{\epsilon}^p/Y$ is analogous to the reciprocal of the modulus of elasticity.

Equations 7.16 have been checked experimentally by tests on thin-walled tubes under various combinations of tension, torsion, and internal pressure (see Davis, 1943; Phillips and Gray, 1961, for example). An especially ingenious test is that in which flat-bottomed grooves are cut obliquely across a sheet, limiting the mode of deformation. Within these grooves, the ratios of two components of stress and strain can be determined with much simpler equipment, and anisotropy can be studied (Prob. 7.14). It must be borne in mind that the stress-strain relations discussed above do not take into account the Bauschinger effect. This is not only important on complete reversal of load, but may also be significant in changing the state of stress from, say, tension to shear, as discussed in Section 7.4. A good deal of work has been carried out to show how the shape of the yield locus changes under such conditions (see for example, Drucker, 1956; Phillips and Gray, 1961; Szczepinski, 1963; and their references). Work on the related effects on the stress-strain relations by Lensky (1960) has not been interpreted in the light of the associated flow rule. For strains of more than 1% since the last change in stress, however, Eqs. 7.16 remain a good approximation until the anisotropy produced by the cold-working becomes important. At that stage, provided that an anisotropic yield locus such as Eq. 7.4 can be determined, it can be differentiated to obtain the anisotropic stress-strain relations (Hill, 1950, p. 320).

◂It is awkward to give a corresponding expression for the stress-strain relations based on the maximum shear stress yield criterion and the associated flow rule. The simplest way to describe the strain is to say that it consists of pure shear referred to the same coordinates as the maximum shear stress. This is ambiguous at the corners of the maximum shear stress criterion, as shown in Fig. 7.12. For instance, in the tensile test, the same maximum shear stress occurs between two different pairs of principal components of stress. Thus any combination of the principal slip components could be activated. To allow for this, the concept of the work equivalent shear strain, Eq. 7.7, was developed.

◂In certain cases it may be mathematically more convenient to work with equations in terms of the components of strain themselves, rather than the strain increments. With this change, the Mises, or flow, equations (7.16) become the *Hencky*, or *deformation*, equations. In many cases, the change in the ratios of the components of the strain during deformation is small enough that solutions using the Hencky equations give useful approximations to engineering problems (Budiansky, 1959).▸

One of the powerful results of the associated flow rule is that it applies not only to the stress-strain relations, but also to the behavior of complete structures. As discussed in Section 7.3, the yield criterion for structural elements can be expressed in terms of generalized loads, which may be loads, bending moments, or twisting moments. In order to apply the principle of maximum plastic resistance, it is necessary to use corresponding components of deformation such that the product of the generalized load and the corresponding generalized deformation gives the work done on the structure. This choice of generalized loads and deformations is analogous to the choice of generalized coordinates in the use of Lagrange's equations in rigid-body mechanics, as discussed, for example, by Byerly (1916). We may thus state the following result.

Consider generalized loads such as forces, bending moments, or twisting moments, Q_i, and generalized plastic deformation increments, such as extensions, curvatures, or rotations, denoted by dq_i^p, such that the plastic work is given by

$$dW^p = \int \sum_i Q_i \, dq_i^p \, dV. \tag{7.17}$$

If the yield locus is in the form

$$\overline{Q}(Q_1, Q_2, \ldots) = 0, \tag{7.18}$$

the ratios between the increments of deformation are given by

$$\frac{dq_1^p}{\partial \overline{Q}/\partial Q_1} = \frac{dq_2^p}{\partial \overline{Q}/\partial Q_2} = \cdots = d\lambda. \tag{7.19}$$

An example is suggested in Problem 7.15.

7.6 CONSTITUTIVE RELATIONS FOR CREEP

Since creep is essentially time-dependent plastic flow, it is to be expected that the stress-strain relations for a creeping material will follow those of plastic deformation, except that creep rates rather than strain increments are described. First, the creep rate is approximately determined as a function of stress level and temperature from a tensile creep test. Under combined stress, this creep rate becomes an equivalent creep rate, and an equivalent stress is also used. The various components of creep rate can then be found by equations which are similar to the Mises stress-strain relations, except that since strain *rates* result, and since creep can occur at any stress, the equivalent stress $\bar{\sigma}$ is substituted for the equivalent flow stress Y:

$$\frac{d\epsilon_{11}^c}{dt} = \left[\sigma_{11} - \tfrac{1}{2}(\sigma_{22} + \sigma_{33})\right] \frac{d\bar{\epsilon}^c}{dt} \Big/ \bar{\sigma},$$
$$\vdots \qquad\qquad\qquad\qquad\qquad\qquad\qquad (7.20)$$
$$\frac{d\gamma_{23}^c}{dt} = 3\sigma_{23}\left(\frac{d\bar{\epsilon}^c}{dt} \Big/ \bar{\sigma}\right), \text{ etc.}$$

As in the case of plastic strain increments, these equations are valid only when Bauschinger-type effects can be neglected. Actually, for small strains, the dislocation structure depends on the past history, and so the shape of the yield locus depends on that previous history. In addition, in creep, if the metallurgical structure is not in equilibrium at the given temperature, there will be metallurgical changes occurring with time which will introduce another effect of history. Since, in many applications, creep strains of a few percent or even of the order of magnitude of elastic strains are important, the utility of these stress-strain relations for creep is much more restricted than those for plastic deformation. For further discussion, see Section 19.5, or reviews by Finnie and Heller (1959) and Odqvist and Hult (1962).

7.7 VISCOELASTIC CONSTITUTIVE EQUATIONS

As the temperature is increased, the energy due to thermal motion becomes a larger fraction of the bond strength. The effects of prior deformation are shaken out of the structure more rapidly. In the limiting case, the microstructure has a steady-state configuration, and the stress produces only a first-order bias in the thermal motion. The result is linear viscosity (Section 6.4). It may be obtained as a limiting case of the creep equations (7.20) in which the factor $(d\bar{\epsilon}^c/dt)/\bar{\sigma}$ becomes the normal coefficient of viscosity, η, proportional to the shear coefficient, $\mu = \eta/3$:

$$\frac{d\epsilon_{11}^v}{dt} = \frac{\sigma_{11} - \tfrac{1}{2}(\sigma_{22} + \sigma_{33})}{\eta}, \qquad \frac{d\gamma_{23}^v}{dt} = \frac{3\sigma_{23}}{\eta}$$
$$\frac{d\epsilon_{22}^v}{dt} = \qquad\qquad\qquad\qquad\qquad (7.21)$$
$$\vdots$$

7.7 VISCOELASTIC CONSTITUTIVE EQUATIONS

The only differences between these and the elastic equations are the substitution of strain rate for strain and a Poisson's ratio of 1/2 for the lower value usually encountered in elastic behavior. So far as the other basic equations of deformable media are concerned, the strain-displacement equations of elasticity become relations of identical form between strain rate and velocity, and the equilibrium equations remain unaffected. Thus, problems of determining stress distributions in a viscous material are identical with those for an elastic material, with the substitution of strain rates for strains and the use of 1/2 for Poisson's ratio. If the problem is one of determining a stress distribution and if Poisson's ratio does not come into the elastic solution of the problem, then the stress distribution in a viscous material will be identical with that in an elastic material. Cases in which this simple result holds are those in which loads rather than deformations are specified, as in the plane stress ($\sigma_{33} = 0$) of pulled, bent, or twisted bars and for those stress components with forces which lie in the plane of plane strain ($\epsilon_{33} = 0$), such as the longitudinal stress in plates bent to a single curvature (Section 9.5). Poisson's ratio would make a difference in the case of contact stress between spheres, press fits, or a spherical cavity under internal pressure.

As we have seen in Section 6.6, many polymers exhibit a mixture of viscous and elastic behavior which can be represented by a network of springs and dashpots. We shall now show that the similarity between viscoelastic and elastic stress analysis holds even for such general viscoelastic materials; in other words, provided that stress boundary conditions are applied, the viscoelastic stress distribution is constant in time and equal to the elastic one, when that does not depend on Poisson's ratio. In order to prove this statement, we must extend and organize our notion of a viscoelastic material.

◀ To simplify the description of viscoelastic stress-strain relations for three-dimensional states of stress and strain, it is desirable to express the components in terms of average normal stress σ, dilatational strain ϵ, and the deviators from them, as mentioned in Section 3.3:

$$s_{11} = \sigma_{11} - \sigma, \text{ etc.},$$
$$s_{23} = \sigma_{23}, \text{ etc.}, \qquad (3.22)$$
$$e_{11} = \epsilon_{11} - \epsilon/3, \text{ etc.},$$
$$e_{23} = \epsilon_{23}, \text{ etc.} \qquad (3.23)$$

In the isotropic elastic case,

$$e_{ij} = s_{ij}/2G, \qquad \epsilon = \sigma/B. \qquad (3.24)$$

Note that only one component of stress and one of strain appears in each equation. This simplification is especially convenient when the stress-strain relations involve the additional variable of time. As in the simple one-dimensional case of Eq. 6.20, the stress-strain-time relation for a viscoelastic material takes on the form of a linear differential equation with constant coefficients. Allowing for any degree of complexity in the network representing the material requires an arbitrary number of terms:

$$\sum_{m=0}^{a} p_m \frac{d^m s_{ij}}{dt^m} = \sum_{n=0}^{b} q_n \frac{d^n e_{ij}}{dt^n}. \tag{7.22}$$

In most viscoelastic materials, the compressibility is elastic, as given by Eq. 3.24. The viscoelastic behavior of the stress deviator combines with the elastic bulk behavior. For example, in a material having a Maxwell relation between stress and strain deviators,

$$\frac{de_{ij}}{dt} = \dot{e}_{ij} = \frac{\dot{s}_{ij}}{2G} + \frac{s_{ij}}{2\mu}, \tag{7.23}$$

the stress-strain behavior under uniaxial stress is (Prob. 7.16)

$$\dot{\epsilon}_{11} = \frac{\sigma_{11}}{3\mu} + \dot{\sigma}_{11}\left(\frac{1}{3G} + \frac{1}{9B}\right). \tag{7.24}$$

◀ The solution of problems in mechanics with time-dependent stress-strain relations might be expected to be quite difficult. Lee (1960a) has shown, however, that many problems can be simplified by using the Laplace transform, which for a time-dependent function $f(t)$ is defined as

$$\int_0^\infty e^{-st} f(t)\, dt = f^*(s). \tag{7.25}$$

Then for zero initial values of the function and its first $n-1$ derivatives, as given for example by Churchill (1958),

$$\int_0^\infty e^{-st} D^n f(t)\, dt = s^n f^*(s). \tag{7.26}$$

Thus the stress-strain relations for a general linear viscoelastic material, Eqs. 3.24 and 7.22, become (see also Alfrey, 1948, p. 545–547)

$$\bar{\sigma}^* = 3B\bar{\epsilon}^*, \tag{7.27}$$

$$\left(\sum_{m=0}^{a} p_m s^m\right) s_{ij}^* = \left(\sum_{n=0}^{b} q_n s^n\right) e_{ij}^*. \tag{7.28}$$

Since the Laplace transform of a partial derivative in space is the partial derivative of the Laplace transform (Churchill, 1958, p. 112), the equilibrium equations and the strain displacement relations transform simply to corresponding relations in the transformed strains and displacements:

$$\sum_{i=1}^{3} \frac{\partial \sigma_{ij}^*}{\partial x_i} = 0, \tag{7.29}$$

$$\epsilon_{ij}^* = \frac{1}{2}\left(\frac{\partial u_i^*}{\partial x_j} + \frac{\partial u_j^*}{\partial x_i}\right). \tag{7.30}$$

Likewise the boundary conditions can be transformed if they specify the forces acting at the surface:

$$\text{Given: } F_i^*.\tag{7.31}$$

Equations 7.27 through 7.31 are formally identical with the equations of elasticity, with a shear modulus which is a function of the transform parameter s according to Eq. 7.28:

$$2G = \sum_{n=0}^{b} q_n s^n \bigg/ \sum_{m=0}^{a} p_m s^m. \tag{7.32}$$

The solution to these equations can then be found from the corresponding elastic problem, and then the inverse Laplace transform can be applied to obtain the time-dependent stress and strain.

◀When the distribution of stress does not depend on the elastic constants, the Laplace transform parameter will not enter into the solution, and the inverse transform will show that for constant applied loads, the viscoelastic stress distribution will be independent of time and equal to the elastic stress distribution. Thus the similarity discussed above between viscous and elastic materials has been extended to viscoelastic ones.

TABLE 7.1

SOME VISCOELASTIC FUNCTIONS

(Stated in terms of uniaxial tests; similar relations hold for deviatoric components)

	Constant		Sinusoidal	
	Strain	Stress	Strain	Stress
Independent variable	$\epsilon = \text{const}$	$\sigma = \text{const}$	$\epsilon = \|\epsilon\|e^{i\omega t}$	$\sigma = \|\sigma\|e^{i\omega t}$
Dependent variable	$\sigma(t)$	$\epsilon(t)$	$\sigma(\omega)$	$\epsilon(\omega)$
Viscoelastic function	$E(t) = \sigma(t)/\epsilon$	$J(t) = \epsilon(t)/\sigma$	$\sigma(\omega) = [E'(\omega) + iE''(\omega)] \times \|\epsilon\|e^{i\omega t}$	$\epsilon(\omega) = [J'(\omega) - iJ''(\omega)] \times \|\sigma\|e^{i\omega t}$

◀Even in cases where the stress distribution is independent of time, it is still necessary to use data to determine the time-dependent deflection, which in many engineering problems must be limited to a reasonable value. As indicated in Table 7.1, the data are usually obtained from one or more of four different tests carried out with either a constant or a sinusoidally varying strain or stress. The steady-state experiments are most suitable for long times, whereas the sinusoidal experiments are easiest for studying high-frequency response.

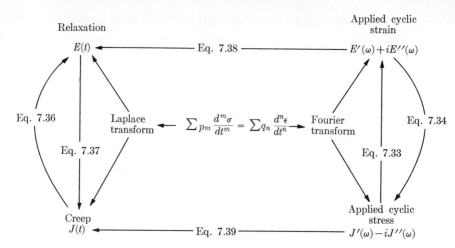

FIG. 7.13. Some conversions between viscoelastic functions.

◀ It is often desirable to convert from one to the other of the viscoelastic functions defined in Table 7.1. Some of the possibilities are shown in Fig. 7.13. In the case of sinusoidally varying conditions, the conversion is made through the use of complex variables (Prob. 7.17):

$$J' = E'/(E'^2 + E''^2), \qquad J'' = E''/(E'^2 + E''^2), \tag{7.33}$$

$$E' = J'/(J'^2 + J''^2), \qquad E'' = J''/(J'^2 + J''^2). \tag{7.34}$$

In addition, it is frequently desirable to define the phase angle between the input and output (Prob. 7.18):

$$\tan \delta = J''/J' = E''/E'. \tag{7.35}$$

Converting a creep compliance to a relaxation modulus, or vice versa, is more difficult. Theoretically this can be done with the superposition principle (Section 6.6), carrying out integrals over an infinite time range. In practice, approximate equations are often satisfactory and much more convenient to use. For instance, if the slope of a plot of the logarithm of creep compliance versus the logarithm of time has a nearly constant value m near the time t, the relaxation modulus can be found from the creep compliance by (Prob. 7.19)

$$E(t) = \frac{\sin m\pi}{m\pi J(t)}. \tag{7.36}$$

Similarly, if m is changed to denote the slope of the plot of the logarithm of relaxation modulus versus logarithm of time at time t, the creep compliance at that time is

$$J(t) = \frac{\sin m\pi}{m\pi E(t)}. \tag{7.37}$$

◀The process of going from frequency data to creep or relaxation behavior can be approximated by finding the complex modulus or compliance at several frequencies. Ferry (1961, Chap. 4) gives

$$E(t = 1/\omega) = E'(\omega) - 0.40\, E''(0.40\,\omega) + 0.014\, E''(10\,\omega), \qquad (7.38)$$

$$J(t = 1/\omega) = J'(\omega) + 0.40\, J''(0.40\,\omega) - 0.014\, J''(10\,\omega). \qquad (7.39)$$

The inverse process, of determining the complex modulus from compliance or relaxation data, is more involved. The reader is referred to Ferry (1961, Chap. 4) or Alfrey (1948, pp. 537–556) for further information. Ferry has plotted the viscoelastic functions for a variety of polymers, showing their interrelations in a graphical and instructive way. ▶

7.8 DYNAMIC STRESS ANALYSIS

◀In many cases, structures are subjected to rapid enough loadings, so that the inertia terms in the momentum equilibrium equation, Eq. 2.13, cannot be neglected. In this case, stress waves travel through the material. First consider a wave, resulting from an impact, traveling down a long rod of elastic-plastic material. For the element shown in Fig. 7.14, the forces acting on either end are described in terms of "engineering" stress, defined as force per unit of original area and the original coordinate x_1^o. The momentum equations reduce to

$$\frac{\partial \sigma_{11}^o}{\partial x_1^o} = \rho \frac{\partial^2 u_1}{\partial t^2}. \qquad (7.40)$$

This equation may be rewritten in terms of the rate of change of stress with "engineering" strain, defined as length change per unit original length, as

$$\frac{d\sigma_{11}^o}{d\epsilon_{11}^o} \frac{\partial \epsilon_{11}^o}{\partial x_1^o} = \frac{d\sigma_{11}^o}{d\epsilon_{11}^o} \frac{\partial^2 u_1}{\partial x_1^{o2}} = \rho \frac{\partial^2 u_1}{\partial t^2}. \qquad (7.41)$$

Equation 7.41 is similar to the equation for the propagation of elastic waves, except that here the rate of change of stress with strain, $d\sigma^o/d\epsilon^o$, is not a constant but a function of strain. It turns out that each increment of strain propagates with a

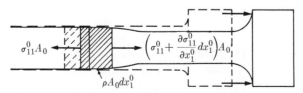

FIG. 7.14. Forces acting on an element of a bar under impact.

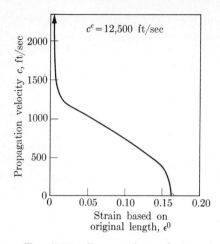

FIG. 7.15. Propagation velocity of plastic strain in annealed copper. (Duwez and Clark, 1947.)

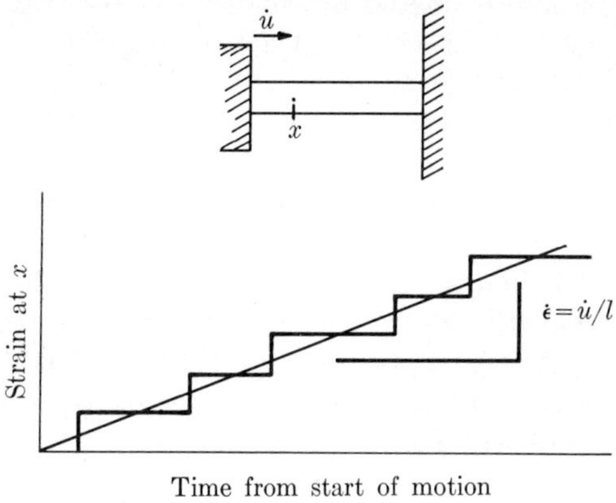

FIG. 7.16. Strain-time history for low-velocity impact.

velocity corresponding to the current slope of the stress-strain curve (Kolsky, 1953):

$$c(\epsilon^o) = \left(\frac{d[\sigma^o(\epsilon^o)]/d\epsilon^o}{\rho}\right)^{1/2}. \qquad (7.42)$$

◂The velocities corresponding to the stress-strain curve for an annealed copper wire are shown in Fig. 7.15. Note that at the uniform strain of 16%, where the slope of the stress-strain curve goes to zero, the velocity of wave propagation also goes to zero. This means that under an impact great enough to produce that strain, the plastic strain wave will not propagate, but necking will occur near the point of impact.

◂If a bar is struck with a velocity V, which is maintained constant, the magnitude of the strain must be such that the straining of material, as the elastic and plastic waves sweep over it, produces the applied velocity. A wave producing a strain increment $d\epsilon^o$, and traveling through a bar at velocity c, produces in the material which it sweeps over a velocity increment of

$$dV = c(\epsilon^o)\, d\epsilon^o. \qquad (7.43)$$

Thus the engineering strain ϵ_1^o required for an impact of velocity V is given by

$$V = \int_0^{\epsilon_1^o} c(\epsilon^o)\, d\epsilon^o. \qquad (7.44)$$

◂Equations 7.42 and 7.44 provide a means of estimating whether or not a process is truly dynamic. If the velocity of "impact" is low enough, the loading will build up through a series of elastic waves being reflected back and forth through the specimen. The strain at a point will increase with time as shown in Fig. 7.16. If the strain increments due to these waves are small compared with the final strain, then the usual static stress analysis can be applied, and the strain rate is approximately the "impact" velocity divided by the specimen length. Since the elastic velocity of sound in steel is about 16,000 ft/sec (Prob. 7.20), for steel the ordinary "impact test," in which the impacting head drops about four feet and breaks a tensile specimen, turns out to be a static test because it may undergo several waves before yielding (Prob. 7.21). Dynamic effects would be more important in lead because of its low yield strength (Prob. 7.22).

◀ In a three-dimensional body, different components of the stress waves travel with different velocities. In his review of stress waves in solids, Kolsky (1953) shows that in the interior of a solid, hydrostatic stress waves travel at the *dilatational or uniaxial strain wave velocity*

$$c_1 = \left(\frac{B + 4G/3}{\rho}\right)^{1/2}, \tag{7.45}$$

and shear stress waves travel at the slower *shear wave velocity*

$$c_2 = (G/\rho)^{1/2}. \tag{7.46}$$

A wave traveling along the surface of a semi-infinite body, on the other hand, travels with the *Rayleigh wave velocity*, equal to 0.92 to 0.95 of the shear wave velocity, for values of ν from 0.25 to 0.5.

◀ When a material exhibits time-dependent behavior, the above analysis must be modified. It works reasonably well if the effects of strain rate are small and if the deformations are large, as shown in Fig. 7.17. In the case of materials exhibiting a time or rate dependence, the theory is not so satisfactory, although it does predict

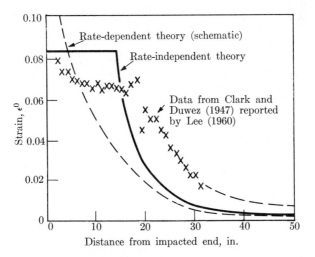

Fig. 7.17. Comparison of observed strain distribution with rate-independent theory and qualitative estimate of rate-dependent theory. Annealed copper wire 80 in. long impacted at 109 ft/sec for 1.5 millisec. (Data from Lee, 1960b.)

the gross effects satisfactorily. Sternglass and Stewart (1953) showed that strain-rate effects had to be taken into account in the effect of small added impacts on an already prestrained bar, apparently since the stress rise due to increased strain rate propagates elastically. An analysis for material with nonlinear strain-rate dependence has been given by Plass and Wang (1959), but as indicated in Fig. 7.17, this is not a much better fit in the case of a single impact. Another theory, based on the idea of an initial elastic wave followed by relaxation, has been proposed by Riparbelli (1953) and Lensky and Fomina (1959). Discrepancies between these theories and experimental work have not been fully settled, judging from reviews by Cristescu (1960), Kolsky (1960), Lee (1960b), and Craggs (1961). Of these, Lee's is suggested as the first to read. A more complete review of dynamic phenomena in impact is given by Goldsmith (1960), and more recent papers appear in Kolsky and Prager (1964). ▶

7.9 SUMMARY

In any material, the state of stress must satisfy the equilibrium or momentum equations

$$\frac{\partial \sigma_{11}}{\partial x_1} + \frac{\partial \sigma_{21}}{\partial x_2} + \frac{\partial \sigma_{31}}{\partial x_3} = \rho \frac{d^2 u_1}{dt^2}, \text{ etc.} \tag{2.13}$$

or

◀ $$\sigma_{ij,i} = \rho \frac{d^2 u_j}{dt^2}. \tag{2.13}$$ ▶

The state of strain or strain rate must be such that the components can be derived from three components of displacement or velocity:

$$\epsilon_{11} = \frac{\partial u_1}{\partial x_1}, \quad \gamma_{23} = \frac{\partial u_2}{\partial x_3} + \frac{\partial u_3}{\partial x_2}, \text{ etc.} \tag{2.25}$$

or

◀ $$\epsilon_{ij} = \frac{1}{2}\left(\frac{\partial u_i}{\partial x_j} + \frac{\partial u_j}{\partial x_i}\right). \tag{2.23}$$ ▶

The stress-strain-time-temperature, or constitutive, equations represent a third principle which must be brought to bear in finding the stress and strain distribution. For an isotropic elastic material, the equations are

$$\epsilon^e_{11} = (\sigma_{11} - \nu\sigma_{22} - \nu\sigma_{33})/E, \quad \gamma_{23} = \sigma_{23}/G, \text{ etc.,} \tag{3.19}$$

where

$$G = E/2(1+\nu). \tag{3.18}$$

◀ For a general anisotropic material with thermal expansion,

$$\epsilon^e_{ij} = S_{ijkl}\sigma_{kl} + \alpha_{ij}\Delta T. \tag{3.9}$$ ▶

7.9 SUMMARY

When metals deform plastically, one must add to the elastic strain, which is uniformly distributed through the crystal lattice, the plastic or creep contributions which are due to dislocations, twinning, etc.:

$$d\epsilon_{11} = d\epsilon^e_{11} + d\epsilon^p_{11} + d\epsilon^c_{11}. \tag{7.1}$$

General plastic deformation in a polycrystalline metal depends on the shear components of stress and the differences of the normal components, but not on the average of the normal stress components, since hydrostatic pressure and tension do not tend to produce dislocation motion. The Mises criterion, which satisfies this condition and is appropriate for an isotropic body, is that plastic flow occurs whenever an equivalent stress, defined by

$$\bar{\sigma} = \sqrt{\tfrac{1}{2}[(\sigma_{22} - \sigma_{33})^2 + (\sigma_{33} - \sigma_{11})^2 + (\sigma_{11} - \sigma_{22})^2] + 3\sigma^2_{23} + 3\sigma^2_{31} + 3\sigma^2_{12}}, \tag{7.2b}$$

◀ or in terms of stress deviators,

$$s_{ij} = \sigma_{ij} - \delta_{ij}\sigma_{kk}/3, \quad \text{or} \quad \bar{\sigma} = \sqrt{\tfrac{3}{2}s_{ij}s_{ij}}, \tag{7.2b} ▶$$

is equal to the uniaxial stress causing plastic flow in the tensile test, i.e.,

$$\text{Yielding occurs when } \bar{\sigma} = Y(\textstyle\int d\bar{\epsilon}^p) = \bar{\sigma}_Y(\textstyle\int d\bar{\epsilon}^p). \tag{7.10}$$

The equivalent plastic strain is defined as

$$\int d\bar{\epsilon}^p = \int \sqrt{\tfrac{2}{9}[(d\epsilon^p_{22} - d\epsilon^p_{33})^2 + (d\epsilon^p_{33} - d\epsilon^p_{11})^2 + (d\epsilon^p_{11} - d\epsilon^p_{22})^2]} \\ + (d\gamma^p_{23})^2/3 + (d\gamma^p_{31})^2/3 + (d\gamma^p_{12})^2/3. \tag{7.5a}$$

$$◀ \quad d\bar{\epsilon}^p = \sqrt{\tfrac{2}{3} d\epsilon^p_{ij} d\epsilon^p_{ij}}. \tag{7.5b} ▶$$

◀ An alternative yield criterion is based on the maximum shear stress: i.e.,

$$\text{Yielding occurs when } \bar{\tau} = (\sigma_{\max} - \sigma_{\min})/2 = k, \tag{7.3}$$

where it is assumed that the flow stress in shear rises with plastic strain as a function of the sum of the principal shear strains which could comprise the deformation:

$$k = k(\textstyle\int d\bar{\gamma}^p) = k(2\textstyle\int |d\epsilon^p_{\max}|). \tag{7.7, 7.11}$$

For yielding of anisotropic materials, one may use Eq. 7.4. ▶

The current state of stress determines only the current directions of dislocation motion and hence only the current strain *increments*. The total strain has to be found by integrating over the past history. The components of the strain increment for given stress depend on the shape of the yield locus, as a consequence of the principle of maximum plastic resistance. Stated in terms of generalized stress

Q_i and generalized components of strain increment dq_i such that the plastic work per unit volume is given by

$$dW^p = \int \sum_i Q_i \, dq_i^p \, dV, \qquad (7.17)$$

it can be shown from the mechanisms of deformation within a crystal that the strain increments must take on the form

$$dq_i^p = \frac{\partial \overline{Q}(Q_i)}{\partial Q_i} \, d\lambda. \qquad (7.19)$$

This relation is known as the associated flow rule. For the Mises yield criterion, it gives the following relations between stress and strain increment, which may also be developed by analogy from the elastic stress-strain relations:

$$\begin{aligned} d\epsilon_{11}^p &= [\sigma_{11} - \tfrac{1}{2}(\sigma_{22} + \sigma_{33})] \, d\bar{\epsilon}^p/Y, & d\gamma_{23}^p &= \sigma_{23}(3 \, d\bar{\epsilon}^p/Y) \\ d\epsilon_{22}^p &= \cdots, & d\gamma_{31}^p &= \cdots, \\ d\epsilon_{33}^p &= \cdots, & d\gamma_{12}^p &= \cdots, \end{aligned} \qquad (7.16a)$$

or

$$d\epsilon_{ij}^p = 3s_{ij} \, d\bar{\epsilon}^p/2Y. \qquad (7.16b)$$

This formulation of the plastic stress-strain increment relations implies that strain increments are independent of past history. Actually the dislocation structure becomes anisotropic after plastic flow, leading to changes in shape of the yield locus and the reduction of flow stress on reversal of load. This is known as the Bauschinger effect. Strain increments of 1 or 2% wipe out most of these previous effects and make the isotropic equations applicable. Plastic strains of the order of 20% or more produce preferred orientations, so that anisotropy must be taken into account.

A stress-strain relation for creeping materials can be obtained simply by changing the plastic strain increments to strain rates:

$$\begin{aligned} d\epsilon_{11}^c/dt &= [\sigma_{11} - \tfrac{1}{2}(\sigma_{22} + \sigma_{33})](d\bar{\epsilon}^c/dt)/\bar{\sigma} \\ &\vdots \\ d\gamma_{23}^c/dt &= [\sigma_{23}](3d\bar{\epsilon}^c/dt)/\bar{\sigma}. \end{aligned} \qquad (7.20)$$

This formulation is subject to the same limitation on past history as the plastic formulation, with the added effect of changes in temperature on structure, which may continue throughout a test, even for relatively large strains. To a first approximation, the factor $(d\bar{\epsilon}^c/dt)/\bar{\sigma}$ is a function of the temperature and equivalent stress, provided that a steady-state structure has been obtained.

When the rate parameter $(d\bar{\epsilon}^c/dt)/\bar{\sigma}$ of Eq. 7.20 is independent of stress, the equations become those for the viscous flow of an incompressible material, since for a given element, the strain rate is the same as a velocity gradient.

In a polycrystalline material, the creep in some grains may be more pronounced than that in others. In this case, as in polymers, creeping elements are in parallel with elastic elements. This type of structure leads to a delayed recovery or elastic aftereffect on release of load, as well as to an initial transient creep. In this situation, it is not possible to define the total strain as made up of different parts, elastic, plastic, and creep, each depending on the stress. Instead, the stress is made up of the sum of those parts carried by the creeping and the elastic material. The magnitude of the elastic aftereffect does not exceed a fraction of the elastic strain for most metals, but is much higher in polymers.

Since, in creep of metals, the structure may be dependent on temperature as well as on the strain history, there will be even more limitations on these equations for creep than on those for plastic flow. Furthermore, in most practical applications, the magnitude of the creep strains is small, and so the limitations are of more concern. This is especially true in relaxation, in which significant reductions of stress at constant total strain are produced by creep strains which are less than the initial elastic strain.

◀ The general three-dimensional viscoelastic equations are most easily expressed in terms of stress and strain deviators:

$$s_{ij} = \sigma_{ij} - \sigma\delta_{ij}, \tag{3.22}$$

$$e_{ij} = \epsilon_{ij} - \epsilon\delta_{ij}/3 \tag{3.23}$$

giving

$$\epsilon = \sigma/B, \tag{3.24}$$

$$\sum_{n=0}^{r} p_n \frac{d^n s_{ij}}{dt^n} = \sum_{m=0}^{t} q_m \frac{d^m e_{ij}}{dt^m}. \tag{7.22}$$

Since the viscoelastic equations are linear, solutions may be superposed according to the Boltzmann superposition principle, which is useful in converting among creep, relaxation, and oscillating load tests. Application of the Laplace transformation converts problems in viscoelasticity to ones in elasticity. Provided that the elastic constants do not enter into the stress distribution of the transformed problem, the distribution of stress in a viscoelastic material will be identical with that in an elastic material.

◀ When materials are subjected to high rates of strain, inertia effects become important. Stress waves *within* an elastic medium may travel with either of two velocities, that corresponding to a dilatational or uniaxial strain wave,

$$c_1 = [(B + 4G/3)/\rho]^{1/2}, \tag{7.45}$$

or that corresponding to a shear wave,

$$c_2 = (G/\rho)^{1/2}. \tag{7.46}$$

When waves travel near a surface, the velocities are somewhat different. For example, in a long slender bar the velocity is

$$c = \sqrt{E/\rho}.$$

For plastic deformation, the velocity of the plastic wavefront at which a strain ϵ is attained is similarly given by

$$c = \sqrt{(d\sigma^o/d\epsilon^o)\rho}, \tag{7.42}$$

with $d\sigma^o/d\epsilon^o$ the local slope of the engineering stress-strain plot. When a rod impacts a rigid body, the strain is that value such that the integral of the wave velocities over the corresponding strains is equal to the impact velocity:

$$V = \int_o^{\epsilon_{\max}} c(\epsilon^o)\, d\epsilon^o. \tag{7.44}$$

Thus for an elastic impact, the maximum stress is

$$\sigma = EV/c.$$

Although Eqs. 7.42 and 7.44 have satisfactorily correlated many data, more recent analyses and experiments have shown that it is sometimes necessary to take into account the strain-rate sensitivity of the material. ▶

REFERENCES

ALFREY, T., JR.	1948	*Mechanical Behavior of High Polymers*, Interscience, New York.
BISHOP, J. F. W. HILL, R.	1951	"A Theory of the Plastic Distortion of a Polycrystalline Aggregate Under Combined Stress," *Phil. Mag.* **42,** 414–427; see also 1298–1307.
BUDIANSKY, B.	1959	"A Reassessment of the Deformation Theories of Plasticity," *J. Appl. Mech.* **26,** 259–264.
BYERLY, W. E.	1916	*An Introduction to the Use of Generalized Coordinates in Mechanics and Physics*, Ginn, Boston.
CHURCHILL, R. V.	1958	*Operational Mathematics*, McGraw-Hill, New York.
CRAGGS, J. W.	1961	"Plastic Waves," *Progress in Solid Mechanics*, Vol. 2, I. N. Sneddon and R. Hill, eds., North Holland Publ. Co. Amsterdam, pp. 143–200.
CRANDALL, S. H. DAHL, N. C. (eds.)	1959	*An Introduction to the Mechanics of Solids*, McGraw-Hill, New York.
CRISTESCU, N.	1960	"European Contributions to Dynamic Loading and Plastic Waves," *Proc. 2nd Symp. Naval Struct. Mech.*, E. H. Lee and P. S. Symonds, eds., Pergamon Press, London.
DAVIS, E. A.	1943	"Increase of Stress with Permanent Strain and Stress-Strain Relations in the Plastic State for Copper under Combined Stresses," *Trans. ASME, J. Appl. Mech.* **65,** A187–A196.

DIENES, G. J. 1956 "Mechanical Properties and Imperfections in Crystals," *Rheology: Theory and Applications*, Vol. 1, F. R. Eirich, ed., Academic Press, New York, pp. 121–139.

DRUCKER, D. C. 1956 "Stress-Strain Relations in the Plastic Range of Metals—Experiments and Basic Concepts," *Rheology: Theory and Applications*, Vol. 1, F. R. Eirich, ed., Academic Press, New York, pp. 97–119.

DUWEZ, P.
CLARK, D. S. 1947 "An Experimental Study of the Propagation of Plastic Deformation Under Conditions of Longitudinal Impact," *Proc. ASTM* **47**, 502–532.

ELLINGTON, J. P. 1958 "An Investigation of Plastic Stress-Strain Relationships Using Grooved Tensile Specimens," *J. Mech. Phys. Solids* **6**, 276–281.

FERRY, J. D. 1961 *Viscoelastic Properties of Polymers*, Wiley, New York.

FINNIE, I.
HELLER, W. R. 1959 *Creep of Engineering Materials*, McGraw-Hill, New York.

GOLDSMITH, W. 1960 *Impact*, Arnold, London.

GROVES, G. W.
KELLY, A. 1963 "Independent Slip Systems in Crystals," *Phil. Mag.* **8**, 877–888.

HILL, R. 1950 *The Mathematical Theory of Plasticity*, Clarendon Press, Oxford, England.

KOLSKY, H. 1953 *Stress Waves in Solids*, Clarendon Press, Oxford, England.

KOLSKY, H. 1960 "Experimental Wave-Propagation in Solids," *Proc. 1st Symp. Naval Struct. Mech.*, J. N. Goodier and N. J. Hoff eds., Pergamon Press, London, pp. 233–262.

KOLSKY, H.
PRAGER, W. (eds.) 1964 *Stress Waves in Anelastic Solids*, Symposium at Brown University, Springer, Berlin.

KRÖNER, E. 1963 "Dislocation: A New Concept in the Continuum Theory of Plasticity," *J. Math. Phys*, **42**, 23–37.

LEE, E. H. 1960(a) "Visco-Elastic Stress Analysis," *Proc. 1st Symp. Naval Struct. Mech.*, J. N. Goodier and N. J. Hoff eds., Pergamon Press, London, pp. 456–482.

LEE, E. H. 1960(b) "The Theory of Wave Propagation in Anelastic Materials," *International Symposium on Stress Wave Propagation*, N. Davids, ed., Interscience, New York, pp. 199–228.

LEE, E. H.
TUPPER, S. J. 1954 "Analysis of Plastic Deformation in The Steel Cylinder Striking a Rigid Target," *J. Appl. Mech.* **21**, 63–70.

LENSKY, V. S. 1960 "Analysis of Plastic Behavior of Metals under Complex Loading," *Proc. 2nd Symp. Naval Struct. Mech.*, E. H. Lee and P. S. Symonds, eds., Pergamon Press, London, pp. 259–278.

Lensky, V. S. Fomina, L. H.	1959	"Propagation of One-Dimensional Waves in a Material with Delayed Yielding," *Izvest. Akad. Nauk SSSR, Otd. Tekh. Nauk, Mekh. i Masch.*, No. 3, 133–136.
Lode, W.	1926	"Versuche über den Einfluss der mittleren Hauptspannung auf das Fliessen der Metalle Eisen, Kupfer, und Nickel," *Z. Physik*. **36**, 913–939.
Meyer, J. A.	1957	Unpublished research, Dept. of Mechanical Engineering, M.I.T., Cambridge, Mass.
Novozhilov, V. V.	1952	"The Physical Meaning of the Stress Invariants of the Theory of Plasticity," *Prikl. Math. i. Mekh.* **16**, 617–619. SLA Translation R-4398.
Odqvist, F. K. G. Hult, J. A. H.	1962	*Kriechfestigkeit Metallischer Werkstoffe*, Springer, Berlin.
Phillips, A. Gray, G. A.	1961	"Experimental Investigation of Corners in the Yield Surface," *Trans. ASME, J. Basic Eng.* **83D**, 275–288.
Plass, H. J., Jr. Wang, N. M.	1959	"Longitudinal Plastic Waves in Long Rods of Strain-Rate Dependent Material," *Proc. 4th Midwest Conf. Solid Mech.* University of Texas, pp. 331–348.
Prager, W. Hodge, P. G., Jr.	1951	*Theory of Perfectly Plastic Solids*, Wiley, New York.
Riparbelli, C.	1953	"On the Time Lag of Plastic Deformation," *Proc. 1st. Midwest. Conf. Solid Mech.* Purdue University, pp. 148–157.
Sautter, W. Kochendörfer, A. Dehlinger, U.	1953	"Über die Gesetzmässigkeiten der plastischen Verformung von Metallen unter einem mehrachsigen Spannungzustand," *Z. Metallkunde* **44**, 553–565.
Sternglass, E. J. Stewart, D. A.	1953	"Experimental Study of the Propagation of Transient Longitudinal Deformation in Elastic-Plastic Media," *J. Appl. Mech.* **20**, 427–434.
Szczepinski, W.	1963	"On the Effect of Plastic Deformation on Yield Condition," *Arch. Mech. Stos.* **15**, 275–294.

PROBLEMS

7.1. Starting from the expression for equivalent stress in terms of normal components, derive the equation preceding Eq. 7.2b for the equivalent stress in terms of a single shear component.

◀ **7.2.** Show that the stress deviator form of Eq. 7.2b is identical with the first form, and also to Eq. 7.2a in terms of principal components of stress.

◀ **7.3.** Music wire is a high-carbon steel wire which can be drawn to a strength in tension of 450,000 psi. The 0.03% yield strength in shear, found by twisting the wire, is 212,000 psi. Give an expression for the equivalent stress for such a metal. Assuming that the yield strengths are at corresponding offsets, tell what you can about the coefficients of anisotropy.

7.4. Show that the numerical factor in (a) the first form and ◀(b) the deviator form of Eq. 7.5▶ makes the product of the equivalent stress and the equivalent strain increment equal to the increment of plastic work, in uniaxial tension and pure shear.

7.5. Show that the factor $\frac{2}{9}$ in Eq. 7.5a, giving the equivalent strain increment, makes the equivalent strain in the uniaxial tensile test equal to the axial normal component of strain.

◀ **7.6.** Show that the two forms of the equation for the equivalent strain, Eqs. 7.5a and 7.5b, are identical.

7.7. Derive the equation for the plastic work increment in terms of the equivalent shear stress and strain increment, Eq. 7.8.

◀ **7.8.** Show that the yield locus of Fig. 7.10 holds for the crystal of Fig. 7.9.

◀ **7.9.** Show that for the crystal of Fig. 7.9, the vector representing the plastic strain increment is normal to the yield locus.

◀**7.10.** Derive Eq. 7.12 from the statement of the principle of maximum plastic resistance.

7.11. Show that Eq. 7.15 follows from the normality of the strain increment vector to the yield locus.

7.12. Derive the plastic stress-strain relations associated with the Mises yield criterion, Eq. 7.16a.

◀**7.13.** Derive the stress deviator form of the plastic stress-strain relations, Eqs. 7.16b.

7.14. For a grooved plate, as shown in Fig. 7.18, calculate the ratios of the shear and normal stress and strain components referred to the axes of the groove, assuming the material is isotropic (Ellington, 1958).

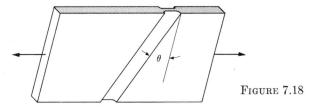

FIGURE 7.18

7.15. Demonstrate the validity of Eq. 7.19 in the special case of a thin-walled tube under combined torsion and tension. Plot the yield locus in terms of axial load and torque. Calculate the ratios of extension to twist and show that these define vectors normal to the yield locus.

◀7.16. Derive the uniaxial behavior of the viscoelastic material whose viscoelastic behavior in terms of stress deviators is given by Eq. 7.23.

◀7.17. (a) From Table 7.1, show that the complex compliance and modulus are related by

$$\mathbf{EJ} = [E'(\omega) + iE''(\omega)][J'(\omega) - iJ''(\omega)] = 1.$$

(b) Using the result of part (a), derive the equation for the components of the complex compliance in terms of those of the complex modulus, Eq. 7.33.

(c) As above, determine the components of the complex modulus in terms of those of the complex compliance.

◀7.18. Derive the equation for the phase angle in terms of the components of the complex modulus or compliance, Eq. 7.35.

◀7.19. Derive the equation for the relaxation modulus in terms of the creep compliance, Eq. 7.36.

◀7.20. Show that the velocity of sound in steel is about 16,000 ft/sec.

◀7.21. (a) Calculate the elastic strain required to accomodate the impact velocity which would result from a hammer falling through a height of 4 ft. What characteristics must the steel specimen have so it will be under a reasonably uniform state of stress at the time it begins to yield?

(b) By calculating the plastic wave velocities for steel after yielding, and the resulting strains under the impact from the given height, find the number of wavefronts which will pass a given point before the steel is strained to a critical value.

◀7.22. Estimate the impact velocity which lead could absorb without becoming plastic as the first wave travels into it.

E 7.23. Straighten out a paper clip. Load it lightly as a cantilever beam, as shown in Fig. 7.19, applying a horizontal load and adding a vertical one, or vice versa. For light loads, note that the final deflection is the same whichever load is applied first. Anneal the paper clip in a gas or kitchen-match flame and load it again, this time deflecting it with a load great enough for a permanent bending on the vertical loading. Note that now the deflection does depend on whether the vertical or horizontal component of load was applied first. This effect of history is a characteristic of plastic deformation.

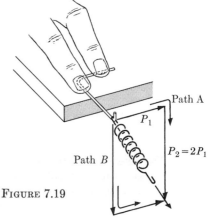

FIGURE 7.19

E 7.24. Form a bar with enlarged ends of silicone putty or Plasticine. Punch a transverse cylindrical hole in the specimen, and subject it to repeated torsion. Note the change in shape of the hole. Do the usual viscous or plastic equations predict this? Judging from what you have seen, estimate what would happen under repeated tension and compression. Are your predictions fulfilled?

7.25. In terms of dislocation motion, why should changing the mode of straining from tension to shear, for example, cause a dip in the plot of equivalent flow stress versus equivalent strain? How large a dip does your explanation predict?

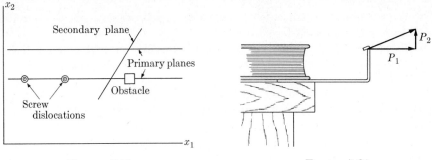

FIGURE 7.20   FIGURE 7.21

◀7.26. A possible exception to the associated flow rule may be the following example. As shown in Fig. 7.20, consider a primary slip plane, with an obstacle on it blocking dislocation motion. Assume that cross glide can occur on a secondary plane, and that after a little bit of cross glide, dislocation motion will continue on a primary plane. Note that the shear component σ_{23} will affect dislocation motion on both planes, but that the component σ_{13} will affect dislocation motion only on the cross glide plane.

(a) In a plot of σ_{23} versus σ_{13}, show the yield locus for small amounts of shear stress σ_{13}.

(b) Sketch the strain increment vector under the assumption that slip elsewhere in the crystal occurs on the plane normal to the x_2 axis. Note the apparent contradiction with the associated flow rule.

(c) Discuss experiments which might be run to test this contradiction, indicating the appropriate materials and methods of loading.

(d) Comment on the validity and significance of this counterexample.

◀7.27. Sometimes the Bauschinger effect is so strong that reverse yielding occurs on release of load. Is this result consistent with the principle of maximum plastic resistance? Discuss.

E 7.28. Determine the stress-strain curve of Plasticine, first in tension and then in compression. Subject the Plasticine to repeated small tension and compression cycles, and again determine a stress-strain curve. Does the equivalent strain correlate the increase in flow stress?

E◀7.29. Cut a strip of polyethylene from a bag. With ink, draw a circle or a cross on the strip. Stretch it, noting the yield zone. What are the magnitudes of the three principal components of strain during yielding? Study the anisotropy of the polyethylene sheet by cutting strips out at various angles and noting their deformation. How much can you determine about the shape of the yield locus from this information?

E 7.30. For a bent cantilever beam loaded as in Fig. 7.21, determine the shape of the yield locus in terms of the two components of force. Note the direction of motion for fully plastic flow. Are the incremental deformation vectors normal to the yield locus? That is, is the associated flow rule satisfied? For this experiment, it is easier to work with a low-strength material. Therefore, soften the paper clip in the flame of a large wooden match or in a gas flame before carrying out the test.

E◀7.31. From a compressive stress-strain curve for modelling clay, estimate the plastic wave velocity. Show from this that dynamic effects should be observed on dropping a bar endwise onto a table from a height of one foot. Note experimentally the nonuniform strain which is observed under these conditions. Calculate the maximum strain expected, and check your results experimentally. For more details, see the rigid-plastic solution suggested by Lee and Tupper (1954).

◀7.32. Using techniques developed or presented in this chapter, revise your estimate of the relaxation modulus of polyvinyl chloride from the creep compliance of Fig. 6.11 (see Prob. 6.27).

◀7.33. Dienes (1956) describes a "standard linear solid" as "characterized by a single time of relaxation. The behavior of the standard linear solid is governed by the differential equation

$$\frac{d\epsilon^n}{dt} = \frac{1}{\tau_r}(\Delta_M \epsilon^e - \epsilon^n),$$

where ϵ^n = nonelastic strain, ϵ^e = elastic strain, Δ_M = relaxation strength, and τ_r = relaxation time."

In this context, what is meant by the relaxation strength and what is a simple spring-dashpot representation of a "standard linear solid"?

7.34. In a problem in which the ratios between the components of stress are changing, suppose that a solution has been obtained both in terms of the equations for the strain increment, Eq. 7.16, and the Hencky equations for the strains. Explain whether the actual behavior should lie between these two solutions, or to one side or the other.

CHAPTER 8

TENSILE AND COMPRESSIVE DEFORMATION

8.1 SYNOPSIS

As a first example of the application of the fundamental equations of mechanics, we shall consider the deformation of an anisotropic bar in tension and contrast it with the more familiar isotropic case. We then turn to the beginning of plastic deformation, defining more carefully than before the terms by which it is usually described, and considering its possible localization into bands.

As plastic deformation proceeds, changes in shape of the specimen become important, playing an important role in the maximum load of a plastic material or the time to rupture of a viscous material. The resulting necking of plastic materials changes the state of stress and strain in a way we must consider in trying to understand the final fracture of the specimen. The terms by which the plastic deformation and final fracture are described are given along with a plea for more complete reporting of such data.

In compression, since the limitation is not so much one of fracture as of buckling, we shall consider the limitations imposed by this phenomenon on the load-carrying capacity in the elastic and plastic regimes. Finally, we shall give a few examples of stress-strain curves as a guide to the ranges of behavior that can be expected.

Fig. 8.1. Deformation of an anistropic bar in tension.

8.2 ELASTIC TENSILE DEFORMATION OF ANISOTROPIC MATERIALS

◀ Elastic deformation of a bar of anisotropic material subjected to a single normal component of stress σ_{33} gives the displacements shown in Fig. 8.1 (Prob. 8.1) (Voigt, 1928):

$$u_1 = \sigma_{33}(S_{1133}x_1 + S_{1233}x_2),$$
$$u_2 = \sigma_{33}(S_{1233}x_1 + S_{2233}x_2), \qquad (8.1)$$
$$u_3 = \sigma_{33}(2S_{1333}x_1 + 2S_{2333}x_2 + S_{3333}x_3).$$

Cross sections receive a tensor shear strain of $S_{1233}\sigma_{33}$ and are tilted relative to the x_3 axis by the angles $2S_{1333}\sigma_{33}$ and $2S_{2333}\sigma_{33}$. If the bar is loaded by clamping in a grip or by pulling on an enlarged section, the restraint of this tilting will require a bending moment and shearing force. These reactions can be avoided only by using long specimens and studying the behavior of the material at some distance from the grips, or by using specially designed grips which pivot near the beginning of the test section. ▶

8.3 THE BEGINNING OF PLASTIC FLOW

Plastic flow begins almost imperceptibly. For instance, in annealed mild steel, in regions which at first seem elastic, there are plastic strains of the order of 10^{-5} (Owen, et al., 1958). The gradual development of plastic strain has led to a number of different definitions of the stress at which plastic flow begins, depending on the use to which the data are to be put. Except when noted, the following definitions are those given by the American Society for Testing Materials (ASTM, 1961). Ideally the *elastic limit* is defined as the greatest stress which a material is capable of sustaining without any permanent strain remaining on complete release of the stress. The testing required to determine the elastic limit is illustrated in Fig. 8.2. The repeated unloading and reloading called for in determining the elastic limit is tedious and may affect the specimen, so it is desirable to have a definition that can be applied to a test in which the extension is increased monotonically. When the elastic strains are small enough so that the elastic stress-strain relation is linear, an equivalent definition is the *proportional limit*, defined as the greatest stress which a material is capable of sustaining without any deviation from proportionality of stress to strain. The proportional limit is illustrated in Fig. 8.3.

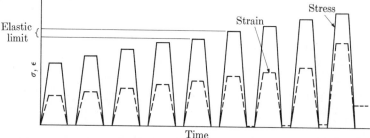

Fig. 8.2. Determination of elastic limit.

Both the definitions of elastic limit and proportional limit are unrealistic in that they are based on the absence of *any* plastic strain or deviation from proportionality, and therefore require perfect instrumentation. The beginning of plastic flow is more realistically defined in terms of the stress required for a measurable or practically significant amount of plastic flow. Specifically, the *yield strength* is the stress at which the material exhibits either a specified limiting deviation from the proportionality of stress to strain or a specified total strain. When the deviation

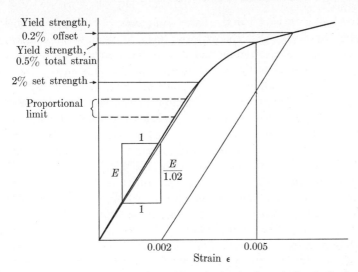

FIG. 8.3. Terms describing the beginning of plastic flow.

is expressed in terms of an increase in strain above the proportional value, as illustrated in Fig. 8.3, the result is termed an *offset* yield strength. The offset is usually specified to be 0.2% (a plastic strain of 0.002) for steel and aluminum alloys. For copper and its alloys, the yield strength is customarily based on a total strain, usually 0.5% (0.005). With this definition of the yield strength in terms of total strain, it is unnecessary to determine the elastic slope of the stress-strain curve and plot the offset. On the other hand, the yield strength based on total strain is unsatisfactory for high-strength alloys such as beryllium copper, in which the elastic strain itself can reach 0.5%. While the above values for offset or total strain may be suitable for many applications, Rosenfield and Averbach (1960) used a yield strength based on plastic strain of 2×10^{-6} as determined by the residual strain on unloading, since they were interested in studying the dislocation mechanisms associated with the very beginning of plastic flow. Even under these conditions, the flow that was measured was equivalent to the motion of some hundreds of dislocations across the specimen.

In springs, plastic deformation of the order of a small fraction of the elastic deformation may be of practical significance. Therefore, it is common in that industry, although not specified by the ASTM, to define a yield strength as the stress at which the plastic deformation or set on release of load is a small fraction, say 2%, of the total deflection under load. This is called the 2% set strength or the *yield strength at 2% set* (not "offset").

8.4 LOCALIZATION OF PLASTIC FLOW

Some materials, such as the annealed mild steel of Fig. 5.36 and the nylon of Fig. 6.28, exhibit a drop in load at small plastic strains. The maximum stress before the drop is called the *upper yield point*, and the minimum before the stress starts to rise again is called the *lower yield point*. The drop in flow stress may be described by saying that the material has a lower flow stress after plastic flow than before; this strain-softening may arise, for example, from the breakaway of dislocations from a cloud of interstitials (Section 5.9) or from the unfolding of chains in an initially crystalline polymer such as nylon (Section 6.7). Strain-softening leads to a localization of plastic flow in bands or necks, because the decrease in load in the first part to pass its peak load reduces the load on the remainder. Deformation is thus concentrated in the first region until it either hardens or breaks. Such Lüders band behavior is well illustrated by Nadai (1950), Green and Hundy (1956), and Hendrickson and Wood (1958).

◀ To analyze the localization of flow in more detail, consider the flow-stress curve to be continuous, with a maximum, a point of inflection, and a minimum. Take the stress to be uniaxial and neglect strain-rate effects. Such a stress-strain curve is shown in Fig. 8.4(a). Where changes in cross-sectional area are important, this curve may be based on load per unit original area (see Section 8.5). As a specimen of such material is extended, the stress rises until the weakest section reaches its upper yield point $\sigma_3 = \sigma_u$. Further extension causes a decrease in load in the first element to reach a maximum stress, thus partially unloading the remaining elements. If the first element to yield is infinitesimal in size, this drop in load will

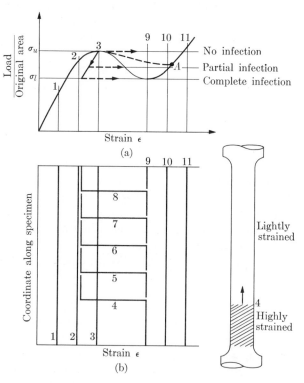

FIG. 8.4. Band formation in a material with upper and lower yield points. No strain-rate effect. (a) Stress-strain curve. (b) Coordinate-strain diagram.

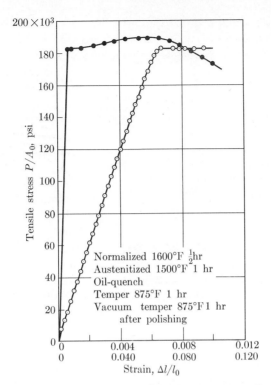

FIG. 8.5. Tensile stress-strain curves for AISI 4340 steel. Note: Use upper strain scale for open symbols which show the lower portion of stress-strain curves. (Suhre and Brock, 1959. Courtesy of University of Illinois.)

cause enough elastic contraction in the rest of the specimen to continue the stretching of the most-strained element. Equilibrium will finally be reestablished at some point A. The stress σ_A will be closer to σ_u, the smaller the yielded element is in comparison with the rest of the specimen.

◀Two possibilities now arise, even in the absence of strain-rate effects. If the flow stress at each point is uniquely determined by the strain at that point, as assumed in a stress-strain relation, then further extension will require increasing loads in both the lightly and heavily strained regions. When the stress in the lightly strained region has again reached σ_u, a second element will yield, and the process will be repeated. In the limit, as yielding occurs in very small elements, the drop in stress will be negligible and the yielding will occur at essentially constant stress, as shown in Fig. 8.5 for a 4340 steel. Since all lightly strained elements will be at the same stress, further yielding can occur randomly throughout the specimen.

◀On the other hand, it may be that yielding in one element is not caused solely by the stress in that element, but that the finite length of slip bands or the mean free path of dislocations will mean that plastic flow in one element will tend to spill over into the next, making plastic flow easier in that element. In this case,

314   TENSILE AND COMPRESSIVE DEFORMATION   8.4

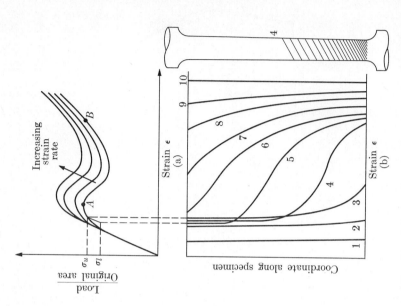

Fig. 8.7. Band formation in a strain-rate dependent material. (a) Stress-strain curve. (b) Coordinate-strain diagram.

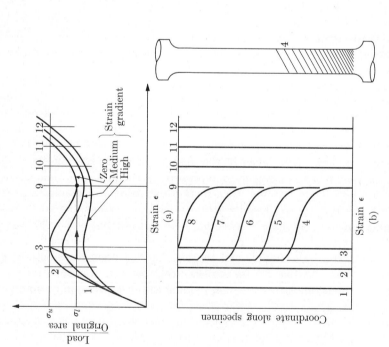

Fig. 8.6. Band formation in a material with infection due to effect of strain gradient on stress-strain curve. No strain-rate effect. (a) Stress-strain curve. (b) Coordinate-strain diagram.

the state will be unstable, for further yielding can occur in the unyielded element which is closest to the yielded zone, dropping the stress further. This process is repeated until in the limit the stress drops to the minimum on the stress-strain curve, σ_l. At this stage some definite fraction of the specimen will have yielded, depending on the shape of the stress-strain curve and the value of the elastic modulus. Further extension will occur by the growth of the yielded region at the expense of the unyielded region. Because of infection, this growth will occur at an interface, as sketched, for example, in Fig. 8.4(b). The stress will stabilize at σ_l until there is no more unyielded material, after which the stress will begin to rise along the hardening curve through points 10 and 11. In short, plastic deformation will occur by the growth of a band at the lower yield point σ_l. Other materials might exhibit a partial infection, so that the lower stress necessary to cause plastic deformation in an element adjacent to a yielded element is intermediate between σ_u and σ_l.

◀ The stress-strain curve shown in Fig. 8.4 leads to an infinitely sharp front between the two zones or an infinitely high strain gradient. Our knowledge of the mechanisms of plastic deformation tells us that the unyielded material will yield more easily because of the slip bands extending from the yielded material. The strain gradient may serve as a first measure of the effect of infection. Thus to improve the description, we might assume that yielding depended not only on the stress, but also on the strain gradient, so that in the presence of high strain gradients, the flow stress would be lower, as shown in Fig. 8.6(a). The resulting deformation is now more spread out, as shown in Fig. 8.6(b) (Prob. 8.2).

◀ Another possible improvement in the description of the yield phenomena can be seen by recognizing that for any finite rate of extension, an infinitely high strain gradient would require an infinitely high strain rate. Since this would certainly affect the flow stress, we consider the stress-strain curves to be rate-dependent, as shown in Fig. 8.7(a). For definiteness we assume a slight taper in the specimen, so one end initially has a slightly higher strain and strain rate than the other. When the wider part of the specimen has reached σ_u, the narrower point may already be at point A.

◀ One might at first think that the strain would jump discontinuously to the point B corresponding to the low strain rate. Such a path would be impossible, however, for it would require an infinitely high strain rate to deform the material instantaneously. Instead, the strain rate increases rapidly at the narrow end. As shown in Fig. 8.7(b), if enough strain occurs in the narrow end, there may have to be a decrease in strain in the rest of the specimen, and hence a decrease in load, in order to maintain the constant rate of overall elongation. The rapid strain in the narrow region soon brings that material over to the neighborhood of the point B, and for the deformation to continue, adjacent material must begin to extend rapidly. In the meantime, the strain rate gradually falls off to zero in the material which has been deformed. The deformation may attain a steady-state configuration, advancing at constant velocity. Finally, when all the material has been deformed, the strain rate must begin to rise in the deformed material. If this process is sudden enough, the required increase in stress will cause a sudden jump at the end of the lower yield region.

◀ If, on the other hand, the material is more strain-rate sensitive, or the specimen is shorter, there may be no flat lower yield region at all, but rather first a gradual decrease and then a gradual rise in the load, as shown for a particular sample of undrawn nylon in Fig. 6.28. The reason for this modified behavior is that the front spreads throughout the specimen instead of progressing in a steady state from one end to the other. Under such conditions, it is very hard to generalize from the experimental load-extension curve to obtain the constitutive relation plotted in Fig. 8.7. Such generalization is made even more difficult in fibers, where it is hard to prevent slippage or drawing in the grips. If the fronts are very steep, the process of infection may be important. If they are more gradual, and if the specimen is not uniformly tapered, but initially has an irregular diameter, then multiple necks can form in a material which creeps, apparently because the time dependence means that continued deformation occurs throughout the specimen. Eventually some other element thins down enough to reach the stress at which rapid straining occurs. The two resulting bands require a strain rate in each just half of what it was for one neck, and so a lower stress level is required. The above discussion has been limited to constant rate of extension, and would be different for constant load. The quantitive development of these ideas is as yet incomplete, but has been started for the sharp yield front by Hart (1955). Data on the upper yield point in mild steel are given by Hendrickson and Wood (1958). Data for nylon have been presented by Miklowitz (1947). ▶

8.5 FINITE STRAIN

Strain has been defined as the relative displacement between two points per unit of distance between them. When the strains are small, it makes little difference whether the distance between the two points is the current distance or the original distance. When the strains are no longer very small compared with unity, however, these two distances will differ. Furthermore, the strains defined in terms of partial derivatives of the displacements with respect to either the initial or final coordinate position will no longer vanish for rigid-body rotation of the part, as pointed out in Section 2.6. It is therefore necessary to extend the definition of strain. The appropriate definition depends on the kind of material involved. In rubber, as discussed in Section 6.5, the straightening out of the molecular chains is elastic, so the structure, and hence further deformation, depends on the original as well as on the current shape. In plastic deformation, however, the current motion of the dislocations is determined primarily by the current stress. The history of stress produces only the relatively small Bauschinger effect arising from the anisotropic dislocation structure. It is this dependence of dislocation motion on stress which led to the statement of the plastic stress-strain relations in terms of differential strain increments, which are by definition always small compared with unity. The relative unimportance of the Bauschinger effect led to the use of

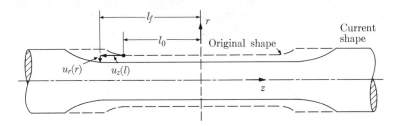

FIG. 8.8. Uniform deformation in a tensile specimen.

a scalar, the equivalent plastic strain, for defining the amount of strain-hardening. In other words, in plastic deformation, the material does not "remember" its original shape, so the definition of plastic strain is based on the current shape or dimensions of the part and not on the original ones. The correlation of the stress-strain behavior of specimens which have been subjected to large plastic deformation is made by calculating the equivalent strain increment at each stage of the plastic deformation and then summing the increments over the entire course of the deformation to obtain the equivalent strain. For common types of tests, it is convenient and possible to do this once and for all, and to express the resulting equivalent strain in terms of the initial and final dimensions of the part. In tensile and compressive tests, the equivalent strain can be related to the initial and final lengths, l_o and l_f, as follows. Let the z direction be axial, with the corresponding displacements $u_z(z)$, as shown in Fig. 8.8. Since the strain is constant along the length of the specimen, the axial normal component of the strain increment, $d\epsilon_{zz}$, can be written in terms of the displacement at the end of the specimen, $u_z(l)$.

$$d\epsilon_{zz} = d\left[\frac{\partial u_z(z)}{\partial z}\right] = d\left[\frac{u_z(l)}{l}\right]. \tag{8.2}$$

But the z displacement, $du_z(l)$, is just the change of length of the specimen, dl. With this substitution, the differential of the axial normal component of strain becomes

$$d\epsilon_{zz} = dl/l.$$

When the expressions for the other two normal components of strain are found and substituted into the expression for equivalent strain, there results, if elastic strains are negligible (Prob. 8.3),

$$d\bar{\epsilon}^p = dl/l. \tag{8.3}$$

Integration of Eq. 8.3 gives an expression for the equivalent strain in terms of the initial and final length:

$$\bar{\epsilon}^p = \ln l_f/l_o. \tag{8.4}$$

When elastic strains are taken into account, the result is (Prob. 8.4)

$$\bar{\epsilon}^p = \ln l_f/l_o - \sigma/E. \tag{8.5}$$

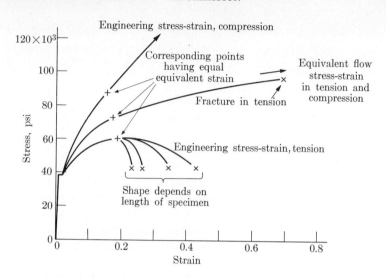

Fig. 8.9. Equivalent and engineering flow stress-strain curves in tension and compression for AISI 1020 steel, hot-rolled. (Crandall and Dahl, 1959. Courtesy of McGraw-Hill.)

The logarithmic form of Eq. 8.4 has led to the use of the term "logarithmic strain" instead of "equivalent plastic strain." In other kinds of loading, however, the idea of equivalent plastic strain does not lead to the logarithmic form, and hence the term "logarithmic" is not appropriate in general. For example, in torsion, as shown in Chapter 9, the equivalent strain is expressed in terms of the tangent of the angle through which an initially axial element is rotated. The equivalent plastic strain is also called, for brevity, the true strain, although it would not be a true (or rather convenient) basis for describing rubber. Likewise, the equivalent stress is sometimes called the true stress, although it may not be if there are local variations in stress across a section.

In describing the behavior of a plastically deforming material, as in Fig. 8.9, where the results of tension and compression tests are compared, equivalent stress and equivalent plastic strain give the simplest picture. For calculations of structures, however, it may be convenient to work with definitions of stress and strain based on the original dimensions of the part rather than on the current dimensions. Such values of the stress and strain have been called "engineering stress" and "engineering strain" because of their utility in structural calculations. In other cases of engineering interest, such as metal-forming and the stability of pressure vessels, the equivalent stress and equivalent plastic strain are more convenient, so the term "engineering" stress or strain is not always appropriate for stresses and strains based on the original area. It will be used here because of tradition, however, and because other expressions such as "approximate" or "nominal" can also be misleading. Engineering stress-strain curves are also shown in Fig. 8.9. The use of the engineering stress has the advantage that the load-carrying capacity of a part can be calculated directly from such a stress-strain curve and the dimensions to which the part was made. It has the disadvantage that the stress-strain curve is no longer a unique characteristic of the material, but depends on the mode of loading as well. One interesting feature of the engineering stress-strain curve in tension is that it shows that a maximum load is reached, even though the equivalent stress-strain curve of the material continues to rise. This phenomenon will be further discussed in the following section.

8.6 MAXIMUM LOAD IN TENSION

In a number of cases involving plastic deformation in tension, the load-carrying capacity of the part at first rises, but later reaches a maximum and begins to decrease. This phenomenon occurs in bars and strips subject to tension, in tubes under combined internal pressure and axial load, and in spheres under internal pressure. The load maximum arises from the fact that while the flow stress of the material is continuing to increase, the cross-sectional area is decreasing. A point can be reached where the decrease of cross-sectional area is more rapid than the increase of flow stress due to strain-hardening.

To investigate this phenomenon quantitatively, consider the load-carrying capacity of a round bar in tension, in terms of the equivalent flow stress* $\bar{\sigma}$, and the cross-sectional area A. The condition of maximum load is found by equating the differential of the load to zero:

$$dP = d\,A\sigma = A\,d\sigma + \sigma\,dA = 0. \tag{8.6}$$

The differential change in area can be expressed in terms of the transverse components of strain, and these in turn can be expressed in terms of the axial strains through the stress-strain relations. The process is simplified if the assumption is made, as is very nearly the case, that elastic strain is negligible and the only significant contribution is the plastic strains. By this method, the condition for maximum load can be found to be that point on the curve of equivalent plastic strain versus equivalent stress for which (Prob. 8.5)

$$\bar{\sigma} = d\bar{\sigma}/d\bar{\epsilon}^p. \tag{8.7}$$

If the equivalent stress-strain curve can be fitted by an equation of the form

$$\bar{\sigma} = \bar{\sigma}_1(\bar{\epsilon}^p)^n, \tag{8.8}$$

the maximum load in tension occurs at the stress $\bar{\sigma}_1 n^n$ and the strain n (Prob. 8.6).

---

* Since in tensile testing the strain is always being increased, the applied equivalent stress $\bar{\sigma}$ is always equal to the equivalent flow stress (flow strength) $\bar{\sigma}_Y(\int d\bar{\epsilon}^p)$ [alternatively $Y(\int d\bar{\epsilon}^p)$]. The distinction between $\bar{\sigma}$ and $\bar{\sigma}_Y$ is therefore often dropped, as we shall do from here on.

The value of the maximum load per unit area which is obtained in the tensile test is called the ultimate tensile strength or simply the *tensile strength* (T. S.). It is one of the most commonly quoted mechanical properties of materials. The term tensile "strength" is something of a misnomer, for the actual stress is higher than the tensile strength, because the actual cross-sectional area is less than the original area. Furthermore, the stress continues to rise as the test proceeds, for the specimen has not yet fractured. For the stress-strain law of Eq. 8.8 the tensile strength is $\bar{\sigma}_1 (n/e)^n$ (Prob. 8.7), where e is 2.718 ....

◀ For a thin-walled sphere of radius r and thickness t, the condition of maximum internal pressure is (Prob. 8.8)

$$d\bar{\sigma}/d\bar{\epsilon}^p = 3\bar{\sigma}/2. \tag{8.9}$$

For a thin-walled cylinder with closed ends, the maximum internal pressure occurs when (Prob. 8.9)

$$d\bar{\sigma}/d\bar{\epsilon}^p = \sqrt{3}\,\bar{\sigma}, \tag{8.10}$$

while for one with the end loads carried externally, the condition is the same as that for a sphere (Prob. 8.10). ▶

8.7 NECKING

Under uniaxial loading, a condition of maximum load is followed by nonuniform deformation, discussed in Section 8.3. If the load does not again rise, this localization leads to failure, as shown in Fig. 1.8. Because of the nonuniform strain which develops after the point of maximum load in a tensile test, the strain at maximum load is sometimes called the *uniform strain*. Although this strain is not commonly used in specifications in this country, it is used for structural metals in Denmark and Switzerland. The chief reason for using the uniform strain as a measure of ductility of the material is that in very long rods subject to tension, most of the rod undergoes only this uniform strain. The added strain before fracture occurs only in a small region and contributes almost nothing to the overall ductility of the structure. Another reason for considering the uniform strain is that it is the strain at maximum load in the tensile test. Further extensions of the part will result in a decrease of load-carrying capacity.

The uniform strain can be determined after fracture, if the specimen was long enough that part of it was unaffected by the subsequent neck. This requires an initial length-to-diameter ratio of about 8, although a fair approximation can be obtained with the usual one of 5. For a material which strain-hardens according to the power law (Eq. 8.8), the uniform strain is equal to the exponent n (Prob. 8.11)
◀ The necking of a strip takes place in a different fashion from that in a bar. One might at first expect a neck to form straight across the strip as indicated in Fig. 8.10. Detailed consideration of the resulting deformation shows that such is not possible, for if the strip were to be formed in the manner indicated, the unyielding material on either side of the neck would prevent transverse contraction except

at the edges and require a condition of plane strain. Under these conditions the components of strain would be given by

$$d\epsilon_{11} = -d\epsilon_{33}, \quad d\epsilon_{22} = 0. \quad (8.11)$$

From the plastic stress-strain relations and the Mises yield criterion, it can be shown that under these conditions, the axial component of stress would be greater than the flow stress (Prob. 8.12):

$$\sigma_{33} = \bar{\sigma}(2/\sqrt{3}). \quad (8.12)$$

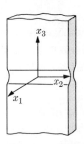

FIG. 8.10. Impossible necking configuration.

From Eq. 8.12 it can be seen that a higher axial stress is required to cause the local deformation in the neck than is required to cause general deformation throughout the strip. Therefore, localization will not occur unless the strip is artificially machined down to a thickness of less than $\sqrt{3}/2$ times that of the adjacent material. This technique is sometimes used to study the yield criterion and stress-strain relations, especially in anisotropic materials (Ellington, 1958).

◀ To find how necking in strips can occur, we must look for a mode of deformation in which plane strain is possible. It is desired to find a direction such that plane strain can be afforded; that is, a condition for which the ratios of the stresses are given by

$$\sigma_{3'3'}/\sigma_{2'2'} = 2. \quad (8.13)$$

It can be seen from the Mohr's circle construction in Fig. 8.11 that such a choice of coordinates is possible. Necking should occur along a line making an angle of 54.7° with the axial direction as shown in Fig. 8.12 (Prob. 8.13). While this form of necking does not require an increase of stress, it turns out that the area does not decrease as fast as with homogeneous uniaxial strain, and until the stress-strain relation reaches the condition $d\bar{\sigma}/d\bar{\epsilon} = \bar{\sigma}/2$, the homogeneous strain will persist (Hill, 1952). When the strip is rolled up into a tube and subjected to pure tension, the mode of necking discussed above is no longer possible and a much more irregular pattern is seen.

◀ In a sphere subjected to internal pressure, necking cannot occur in any strip without local plane strain conditions or deformation elsewhere in the spherical

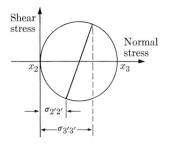

FIG. 8.11. Mohr's circle for tension.

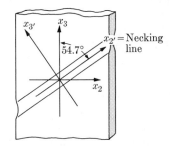

FIG. 8.12. Necking in a flat strip.

shell (Prob. 8.14). This does not, of course, prove that some form of localized asymmetrical deformation is not possible, but as a matter of fact Marin et al. (1948) have extended spheres well beyond the point of maximum pressure without finding localized plastic flow.

◀ The mechanics of necking and the formation of Lüders bands at an upper yield point are identical, and the possibility of localization into bands depends on the existence of directions in the material across which discontinuities in strain or strain gradient can occur. In mathematical terms, the problem is one of finding the so-called "characteristics" of the partial differential equations governing flow (Hill, 1950, pp. 323, 345).

◀ The tendency to necking is reduced in the presence of creep. This effect may be considered by turning to the extreme case of a material behaving in a linearly viscous fashion. First we find the rate of change of area for a given load P. If the changes in cross-sectional area are gradual enough along the specimen, the strain and stress distributions may still be considered uniaxial, and the strain rate is given by

$$d\epsilon_{33}/dt = P/\eta A. \tag{8.14}$$

The axial strain rate can be expressed in terms of the rate of decrease of the cross-sectional area, taking into account the incompressibility of the material:

$$dV = 0 = d(Al) = l\, dA + A\, dl, \qquad d\epsilon_{33} = -dA/A. \tag{8.15}$$

Combining these equations gives the rate of decrease of cross-sectional area:

$$dA/dt = -P/\eta. \tag{8.16}$$

Thus the rate of decrease of area is independent of current area. Any initial difference in area will be maintained but not increased as the test proceeds, so no pronounced necking will develop. This lack of necking in a viscous material can be observed in the stringing out of a drop of very viscous syrup, or the drawing of a glass fiber. The time to rupture is finite (Prob. 8.15):

$$t_r = \eta A_o/P = 1/\dot{\epsilon}_o. \tag{8.17}$$

For a material which creeps according to a power law, an intermediate amount of necking is observed (Prob. 8.16). ▶

8.8 DISTRIBUTION OF STRAIN AND STRESS IN THE NECK OF A CIRCULAR SPECIMEN

After a plastic specimen has begun to neck, it is difficult to determine the equivalent strain in terms of the gauge length according to Eq. 8.4, because an infinitesimal gauge length would be required, and it would be necessary in advance to mount a strain gauge just where the specimen was going to neck. Instead it is more convenient to use the incompressibility of plastic flow to express the axial strain in terms of the change in cross-sectional area or diameter. Since Bridgman (1952) has shown experimentally that the strain is constant across the neck of a tensile specimen, diameter measurements give not only the average, but even the true local plastic strain. Neglecting elastic components of strain, it can be shown that the relation between the equivalent strain and the initial area A_o and the final area A_f is given by (Prob. 8.17)

$$\bar{\epsilon}^p = \ln \frac{l_f}{l_o} = 2 \ln \frac{d_o}{d_f}. \tag{8.18}$$

In the case of some high-strength materials, the elastic strain can be of the order of 1% and perhaps should not be neglected. In this case, the expression for the equivalent strain in terms of the diameters is

$$\bar{\epsilon}^p = 2 \left[\ln \frac{d_o}{d_f} - \nu \frac{\sigma_{33}}{E} \right]. \tag{8.19}$$

This correction is only approximate, because as we shall see below, the stress varies across the neck; but when it does the plastic strain is large and the correction unimportant.

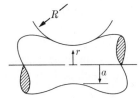

Fig. 8.13. Coordinates for necked tensile specimen.

On either side of the neck, the forces have a radial component, leading to transverse components of stress inside the neck. In order to obtain the local stress and the equivalent flow stress in a necked specimen, Bridgman (1952) worked out a semiempirical analysis in terms of the radius of curvature at the neck and the diameter of the neck. This analysis is based on the previously mentioned experimental observation that the axial strain component, ϵ_{zz}, is constant across the minimum section. Compatibility considerations then lead to the conclusion that $\epsilon_{rr} = \epsilon_{\theta\theta}$. This strain distribution, with the stress-strain relations, the yield criterion, and the equilibrium equations, along with reasonable assumptions as to the nature of the variation of the directions of principal stresses in the neighborhood of the neck, leads to the following result for the stress components in terms of the coordinates of Fig. 8.13:

$$\frac{\sigma_{zz}}{\bar{\sigma}} = 1 + \ln\left(\frac{a}{2R} + 1 - \frac{r^2}{2aR}\right), \quad \sigma_{\theta\theta} = \sigma_{rr} = \sigma_{zz} - \bar{\sigma}. \tag{8.20}$$

The state of stress corresponds to a constant uniaxial tension across the neck with a superimposed hydrostatic tension building up toward the center. Thus the load carried by the hydrostatic stress must be subtracted from the total load carried by the cross section in order to find the equivalent flow stress. This correction is

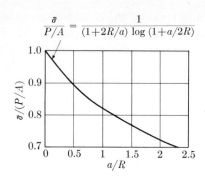

Fig. 8.14. Ratio of equivalent to nominal stress in a necked tensile specimen.

presented as the ratio of equivalent flow stress to average axial stress, P/A, in Fig. 8.14.

Determination of the longitudinal radius R is inconvenient, although it can be done by sliding a cone along the neck until the curvature matches that of the specimen and then measuring the radius of curvature of the cone at that point (Marshall and Shaw, 1952). Bridgman (1952) has found that this can often be avoided by using an empirical relation based on the data shown in Fig. 8.15, although Marshall's and Shaw's work indicate that the linear growth of a/R often has an intercept greater than 0.1.

◀ That Bridgman's analysis indeed gives the correct stress distribution in the neck of the tensile specimen is not yet entirely settled. As an independent check on the validity of this equation, Marshall and Shaw (1952) ran tensile tests on specimens which were machined to arbitrary values of longitudinal curvature at

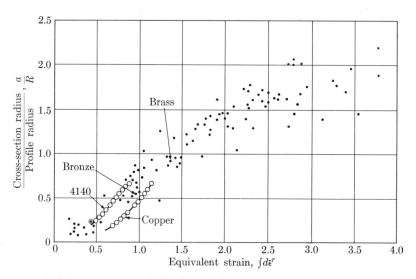

Fig. 8.15. Necking radius as a function of strain. Dots are from Bridgman (1952) and are for steel except as noted. Circles are for 4140 hot-rolled, 256 Brinell hardness; and electrolytic copper, annealed one hour at 500°C, 64 Brinell hardness. (From Marshall and Shaw, 1951.)

different stages. They found that the results they obtained could be correlated into a smooth curve by applying the Bridgman correction. On the other hand, Parker et al. (1946) determined the stress distribution in the neck of a tensile specimen by unloading a necked specimen, calculating the stress change on unloading from elasticity theory, and then boring out the specimen to determine the residual stress. From the residual stress and the stress change due to unloading it was possible to determine what the state of stress under plastic deformation had been. They disagreed with Bridgman's results, although their precision is not high. The authors feel confident enough about the Bridgman correction to have used it wherever possible in reporting tensile stress-strain curves in this book.

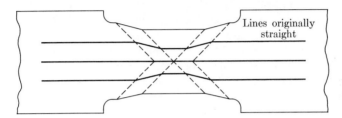

FIG. 8.16. Hypothetical mode of necking in plane strain.

◀ In the case of an artificially necked specimen subject to plane strain, it is possible to carry the analysis out in more detail, but even here the research is not yet complete. Onat and Prager (1954) proposed the mode of deformation shown in Fig. 8.16. Here the state of stress is constant throughout the specimen and deformation is assumed to occur by shear at 45° at the bases of the two triangular wedges which are deformed as they draw into the neck. This solution satisfies the equilibrium, compatibility, and stress-strain relations for a non-strain-hardening material. It turns out, however, that it is not the only solution which does so. Using Hill's work (1957), Cowper and Onat (1962) showed how strain-hardening provides the correct field, but even they could not obtain an exact solution where the length of the specimen was twice the thickness. This is an example of an unsolved problem in plasticity. ▶

8.9 DUCTILITY

As necking continues, fracture finally occurs. In a necked tensile specimen, fracture starts at the center, where the stress is highest, and spreads outward, as discussed further in Section 16.5. The strain to fracture of a material is called its *ductility*. One measure of the ductility is the *reduction of area* (R. A.), defined as the ratio of the change in area to the initial area. One can show that for round bars, the fractional reduction of area is related to the equivalent strain to fracture, $\bar{\epsilon}^p_f$, by the equation (Prob. 8.18)

$$\text{R.A.} = 1 - e^{-\bar{\epsilon}^p_f}. \tag{8.21}$$

Since the state of stress in a necked specimen at fracture depends on the shape, e.g., on whether the specimen is a strip or a bar; since in some kinds of tests such as those giving equal biaxial stress no necking occurs at all; and since fracture in all probability depends not only on the state of stress and strain, but also on the history of their development, the strain at fracture will not be the same under all conditions of testing. Therefore, the reduction of area cannot be considered to be a property of the material, but only a characterization of its behavior in a certain mode of testing.

A more frequently quoted measure of the ductility of a material is the *elongation*, defined as the change in gauge length to fracture divided by the original gauge length (that is, the approximate strain at fracture). As a measure of ductility of a material, the elongation has a disadvantage that it is an engineering strain rather than an equivalent strain, and furthermore, it consists of some sort of weighted average of the uniform strain in the unnecked portion of the specimen and the higher strain in the necked region. As such, it depends both on the length and on the cross-sectional dimensions of the specimen (Prob. 8.19).

8.10 DATA

With the above background, we turn to the stress-strain curves of a number of materials shown in Fig. 8.17 to 8.21. In these diagrams, the Bridgman correction for necking is taken into account, either exactly or approximately, depending on the available data. While there is a wide variation in the yield strength and

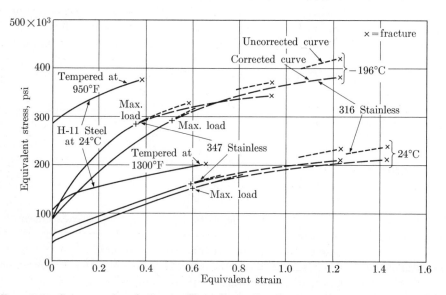

FIG. 8.17. Stress-strain relation for H–11 die steel and at two temperatures for stainless steels, with Bridgman correction for stress. (Data from Nunes and Larson, 1961, and DeSisto, 1962.)

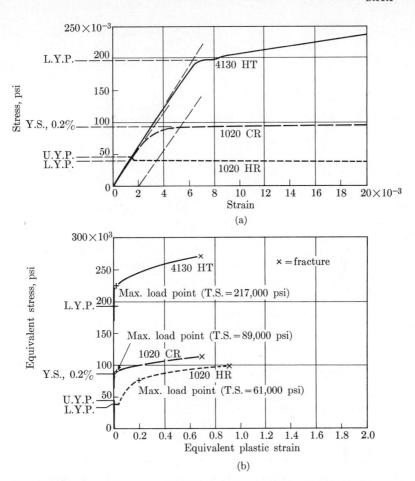

FIG. 8.18. (a) Small strains. (b) Large strains. Stress-strain curves for three steels. Dotted curve: mild steel, hot-rolled (1020 HR). Dashed curve: mild steel, cold-rolled (1020 CR). Solid curve: 0.3%C, 0.5% Mn, 0.25% Si, 0.9% Cr, balance Fe (4130 HT), oil-quenched from 1600°F, tempered at 600°F. (Crandall and Dahl, 1959. Courtesy of McGraw-Hill.)

the strain to fracture, the rate of strain-hardening after the first 20% of strain becomes relatively constant at $E/150$ to $E/700$, as mentioned in Section 5.4.

It is tedious to prepare equivalent stress-strain curves and expensive to present them. Since load and length can be recorded automatically, it is possible to obtain engineering stress-strain curves without hand labor. An equivalent stress-strain curve, on the other hand, especially after the point of necking, requires hunting for the minimum section. This is facilitated by the recording caliper gauge reported by Powell et al. (1955), but its use has not become common. A method for obtaining points on the curve before the point of maximum load is to use a slightly tapered specimen, so that the maximum load will give different applied

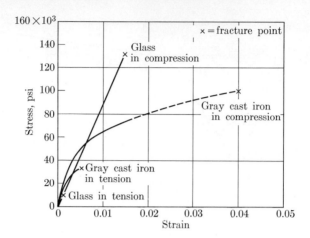

Fig. 8.19. Stress-strain curves for cast iron and glass. (Data for cast iron from Coffin, 1950, and Ingberg and Sale, 1926. Data for glass from Morey, 1954.)

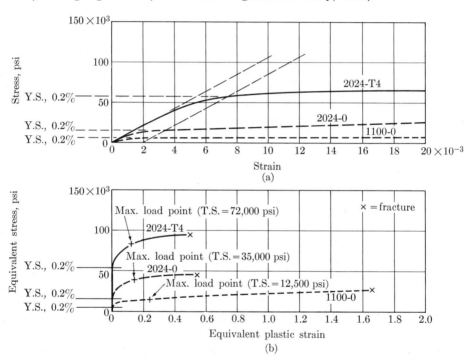

Fig. 8.20. (a) Small strains. (b) Large strains. Stress-strain curves for commercially pure aluminum and an aluminum alloy with two treatments. Dotted curve: commercially pure aluminum annealed (1100–O). Dashed curve: 4.6% Cu, 1.5% Mg, 0.7% Mn, balance Al, annealed. Solid curve: 4.6% Cu, 1.5% Mg, 0.7% Mn, balance Al (2024–T4), water-quenched from 915°F, aged 24 hours at 250°F. (Crandall and Dahl, 1959. Courtesy of McGraw-Hill.)

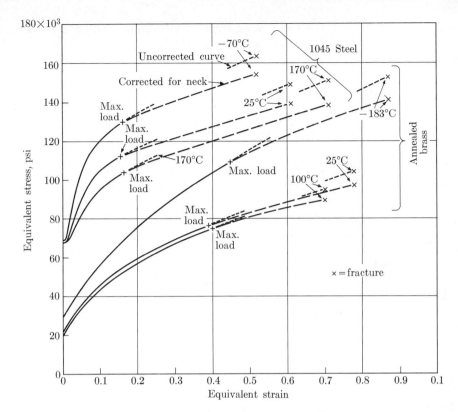

FIG. 8.21. Effect of temperature on stress-strain curves, with estimated Bridgman correction. 1045 HR: annealed 1 hour at 1415°F, furnace-cooled. Brass: CR, annealed 1 hour at 800°F, furnace-cooled. (Data from MacGregor and Fisher, 1945.)

stress at different points along the bar. The diameters can be measured at leisure after the test is over, and the equivalent stress and strain calculated.

In many cases, it would be better not to present curves at all, but rather only tabular data from which the curves could be approximately reconstructed. As a minimum, one would like to have a yield strength, the tensile strength, the uniform strain, the fracture stress, and the reduction of area. From these it is possible to get three points which, with the modulus of elasticity, define the stress-strain curve fairly closely (Prob. 8.20).

Note that the added data of uniform strain and reduction of area do not even require additional instrumentation during the tensile test, since the uniform elongation can be measured from the diameter change of a relatively straight portion of the specimen, and the reduction of area can be measured after fracture, while the fracture stress requires only the added observation of the load at fracture. Unfortunately, the uniform strain and fracture stress are almost never given, and the reduction of area is frequently omitted. This is especially serious because of the value of the strain at fracture on monotonic loading in estimating the life of a part under low-cycle fatigue, as will be discussed in Section 18.6. Note that even from the complete engineering stress-strain curve, it is not possible to reconstruct an equivalent stress-strain curve without making some assumptions about its shape (Prob. 8.21).

8.11 INSTABILITY IN TENSION AND COMPRESSION

If a tensile specimen is subjected to a *load* which is controlled as a function of time, it is in a condition of unstable equilibrium after the maximum load has been reached. That is, any slight further extension will cause a decrease of load-carrying capacity, so that some of the applied load would be available for relative acceleration of the ends of the specimen. The extension would then rapidly increase. If the *extension* is controlled as a function of time, however, the specimen is perfectly stable. Intermediate cases are encountered in practice. In the usual testing machine, for example, even if there is no time for flow of oil in a hydraulic machine or for a rotation of the gears which move the crosshead of a mechanical machine, the elasticity of the framework prevents the applied load from dropping to zero as the specimen extends an infinitesimal amount. The elasticity of the machine can be described by a spring constant K, which gives the rate of decrease of force exerted by the machine per unit extension of the specimen for no change in the controlled position of the crosshead. Then as shown in Fig. 8.22, instability depends on the rate of decrease of load with extension for the specimen and occurs whenever

$$-dP/du_z \geq K. \tag{8.22}$$

Typical testing machines have stiffnesses of the order of 300,000 lb/in.

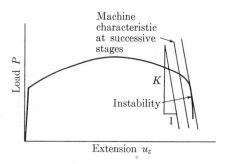

FIG. 8.22. Specimen-machine instability.

◀Instabilities can occur in compression testing as well; the reader is familiar with the fact that edgewise compression of a sheet of paper leads to excessive deformation, not by any permanent change in the material itself, but rather by a lateral deflection of the paper, called *buckling*. In general, if a column is subject to any slight eccentricity of axial loads, to transverse loads, or to applied bending

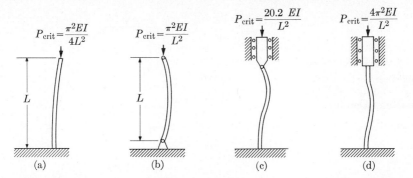

Fig. 8.23. Loads for elastic buckling of uniform columns. (Crandall and Dahl, 1959. Courtesy of McGraw-Hill.)

moments, then under increased axial load, the lateral deflection will increase, gradually at first but later very rapidly as a critical load is reached. This critical load is called the *buckling load*. For a uniform column fixed at one end, free at the other, made of material with modulus E, and having a cross section with an area moment of inertia of I, the buckling load is given by (see, for example, Crandall and Dahl, 1959)

$$P = \pi^2 EI/4L^2. \tag{8.23}$$

◀ It is seen from this equation that if a column is long enough, buckling is determined only by the stiffness of the material and is independent of its yield strength. Other configurations are shown in Fig. 8.23.

◀ The approach of buckling can be detected by measurement of increasing lateral deflection, decreasing lateral stiffness, or decreased natural frequency of lateral vibration. Of these, the easiest way to get a rough indication of the approach to the critical load is by striking the member in question to compare its natural frequency with the natural frequency of an unloaded member. The basis for this test is the relation given by Den Hartog (1952, p. 297) between unloaded natural frequency ω_c, loaded natural frequency ω, and the ratio of current to critical load

$$P/P_c + (\omega/\omega_c)^2 = 1. \tag{8.24}$$

This equation is exact for a pin-ended column, but only approximate for more complicated configurations. Lack of firm pinning at the ends of the column may invalidate this method. A refinement is to measure the lateral stiffness of the column under the application of a known small transverse load. Southwell (1932) has developed a method by which the buckling load P_c can be estimated from a series of measurements of strain in a pin-ended column. The strain at the center, ϵ, turns out to be related to the current load P, the buckling load P_{crit}, the amplitude of the initial half-sinusoidal runout a, the moment of inertia I, and the thickness of the cross section h by

$$\frac{\epsilon}{P} = \frac{\epsilon}{P_{\text{crit}}} + \frac{ah}{2EI}. \tag{8.25}$$

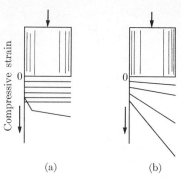

Fig. 8.24. Development of strain in a column. (a) Concentric loading followed by bending. (b) Simultaneous compression and bending.

For a further treatment of buckling, see Den Hartog (1952) and Timoshenko and Gere (1961).

◀ The explanation of the buckling load for a column loaded into the plastic regime has been given only relatively recently (Shanley, 1946). Initially, attempts had been made to predict the buckling load from the idea that as the lateral deflections increased, the compressive strain would decrease on one side, resulting in an elastic unloading, as shown in Fig. 8.24(a). This led to loads about double those observed. Shanley pointed out, however, that large deflections could occur without strain reversal if they developed gradually as the column is loaded, as shown in Fig. 8.24(b). This idea justified a previously proposed equation for buckling in the plastic regime similar to that for the elastic regime, except that the modulus of elasticity is replaced by the local slope of the stress-strain curve. For a column fixed at one end (Fig. 8.23a):

$$P_c = \frac{\pi^2 (d\sigma/d\epsilon) I}{4L^2}.$$ (8.26)

If the column is loaded without allowing any lateral deflection, it will be stable when the constraint is removed, even though the above load is exceeded, provided that the load is still less than about double that value, which corresponds to the analysis based on strain reversal.

◀ A column of material which creeps will exceed any given deflection if it is subject to an initial curvature or a transverse load. The time required may be finite. For example, if the material creeps at a steady rate corresponding to

$$d\epsilon/dt = \dot{\epsilon} = c\sigma^3,$$

and if $\dot{\epsilon}_i$ is the initial strain rate calculated from this equation at $\sigma = P/A$, the time for infinite deflection is given in terms of the following variables (Hoff, 1958):

ϵ^e = elastic strain at elastic buckling, equal to $\pi^2 I/AL^2$ for a column pinned at both ends,
r = radius of gyration of section = $\sqrt{I/A}$,
a_i = amplitude of the initial half-sinusoidal runout

$$t_\infty = \tfrac{1}{3}(\epsilon^e/\dot{\epsilon}_i) \ln (r/a_i).$$ (8.27)

Note that for a relatively large runout, the critical time is roughly that for the creep strain to reach the elastic buckling strain at the initial strain rate. Hoff's article (1958), from which this example was taken, provides a good review of the subject, as well as an introduction to other phenomena which would not be anticipated from the elastic buckling of columns. ▶

8.12 SUMMARY

◀ In anisotropic materials under a uniaxial tensile stress, there develop not only the usual axial extension and lateral contraction, but also shear strains causing a tilting of the ends of the specimen, making it difficult to apply the desired uniform axial stress. ▶

The beginning of plastic deformation is described by the elastic limit, proportional limit, and yield strength or upper or lower yield point. When the load-extension curve drops, as in the case of annealed low-carbon steel or undrawn nylon, Lüders bands appear, indicating inhomogeneous deformation. The strain in these bands is of the order of 3% in steel at room temperature, but several hundred percent in nylon. The details of the growth of these bands depend on the proportions of the specimen, the strain-rate sensitivity of the materials, and the tendency of the deformed region to "infect" neighboring undeformed regions. In isotropic materials under uniaxial stress, the bands tend to occur in oblique zones at 55° to the tensile direction.

Under large plastic strains, so long as the deformation remains uniform over the final gauge length l_f, the equivalent plastic strain which determines the equivalent flow stress is given by

$$\int d\bar{\epsilon}^p = \ln l_f/l_o. \tag{8.4}$$

In tension, the increase in stress due to strain-hardening is offset more and more by a decrease in cross-sectional area until a maximum load is reached. The engineering stress at this load is called the tensile strength. For uniaxial testing, it occurs when

$$d\bar{\sigma}/d\bar{\epsilon}^p = \bar{\sigma}. \tag{8.7}$$

As in the case of inhomogeneous yielding, the maximum load gives a localized deformation called necking. The stress in the neck of the specimen is no longer uniaxial, and corrections must be applied to calculate the flow stress from the equivalent load and the current cross-sectional area. In specimens that creep, there is less necking.

◀ In the limiting case of a viscous material, all sections thin down at the same rate of area decrease per unit time. The time required for infinite strain is given in terms of the viscosity, initial cross-sectional area, and load by

$$t_r = \eta A_o/P. \tag{8.17}$$ ▶

Typical metals have yield strengths varying from $E/100$ to $E/2000$ and rates of strain-hardening of $E/200$ to $E/600$, aside from possible lower yield points, or except for regions of higher hardening rate at strains up to 0.10 to 0.20.

◀ Instabilities can occur in both tension and compression, depending on the stiffness of the machine. Both elastic and plastic buckling can occur, at loads for a cantilever column of

$$P_c = \frac{\pi^2 (d\sigma/d\epsilon) I}{4L^2}. \qquad (8.26) ▶$$

REFERENCES

ASTM	1961	"Definition of Terms Relating to Methods of Mechanical Testing," *ASTM Standards* **3**, 215–226, No. E6-61.
BRIDGMAN, P. W.	1952	*Studies in Large Plastic Flow and Fracture*, McGraw-Hill, New York.
COFFIN, L. F., JR.	1950	"The Flow and Fracture of a Brittle Material," *J. Appl. Mech.* **17**, 233–248.
COWPER, G. R. ONAT, E. T.	1962	"The Initiation of Necking and Buckling in Plane Plastic Flow," *Proc. 4th U. S. Nat. Con. Appl. Mech., ASME*, pp. 1023–1030.
CRANDALL, S. H. DAHL, N. C. (eds.)	1959	*An Introduction to the Mechanics of Solids*, McGraw-Hill, New York.
DEN HARTOG, J. P.	1952	*Advanced Strength of Materials*, McGraw-Hill, New York.
DESISTO, T. S.	1962	"Low Temperature Charpy, True Stress-Strain, and Notched Tensile Properties of Base and Weld Deposits of AISI Types 301, 310, 316, and 347 Stainless Steels," *Proc. ASTM*, **62**, 756–764.
DEWHIRST, D. L. SIDEBOTTOM, O. M.	1962	"Inelastic Design of Load Carrying Members, V, Theoretical and Experimental Creep Analyses of Beam-Columns," *WADD Tech. Rept. 60-580*, Aeronautical Systems Division, U. S. Air Force, Wright-Patterson Air Force Base, Ohio.
ELLINGTON, J. P.	1958	"An Investigation of Plastic Stress-Strain Relationships Using Grooved Tensile Specimens," *J. Mech. Phys. Solids* **6**, 276–281.
GREEN, A. P. HUNDY, B. B.	1956	"The Initial Plastic Yielding in Notched Bend Test," *J. Mech. Phys. Solids* **4**, 128–144.
HART, E. W.	1955	"A Uniaxial Strain Model for a Lüders Band," *Acta Met.* **3**, 146–149.
HENDRICKSON, J. A. WOOD, D. S.	1958	"The Effect of Rate of Stress Application and Temperature on the Upper Yield Stress of Annealed Mild Steel," *Trans. ASM* **50**, 498–516.

HILL, R.	1950	*The Mathematical Theory of Plasticity*, Clarendon Press, Oxford, England.
HILL, R.	1952	"On Discontinuous Plastic States with Special Reference to Localized Necking in Thin Sheets," *J. Mech. Phys. Solids* **1**, 19–30. See also S. P. KEELER and W. A. BACKOFEN, "Plastic Instability and Fracture in Sheets Stretched over Rigid Punches," *Trans. Quart. ASM* **56**, 25–48.
HILL, R.	1957	"On the Problem of Uniqueness in the Theory of a Rigid Plastic Solid IV," *J. Mech. Phys. Solids* **5**, 302–307.
HOFF, N. J.	1958	"A Survey of the Theories of Creep Buckling," *Proc. 3rd U. S. Nat. Con. Appl. Mech.*, pp. 29–49.
INGBERG, S. H. SALE, P. D.	1926	"Compressive Strength and Deformation of Structural Steel and Cast Iron Shapes at Temperatures up to 950°C," *Proc. ASTM* **26**, Part 2, 33–51.
LESSELLS, J. M.	1954	*Strength and Resistance of Metals*, Wiley, New York.
MACGREGOR, C. W. FISHER, J. C.	1945	"Tension Tests at Constant Strain Rates," *J. Appl. Mech.* **12**, 217–227.
MARIN, J. DUTTON, V. L. FAUPEL, J. H.	1948	"Tests of Spherical Shells in the Plastic Range," *Welding Res. Suppl.* **27**, 593s–607s.
MARSHALL, E. R. SHAW, M. C.	1952	"The Determination of Flow Stress from a Tensile Specimen," *Trans. ASM* **44**, 705–725.
MIKLOWITZ, J.	1947	"The Initiation and Propagation of the Plastic Zone Along a Tension Specimen of Nylon," *J. Colloid Sci.* **2**, 193–215.
MOREY, G. W.	1954	*Properties of Glass*, Reinhold, New York.
NADAI, A.	1950	*Theory of Flow and Fracture of Solids*, 2nd ed., McGraw-Hill, New York.
NUNES, J. LARSON, F. R.	1961	"A Method for Determining the Plastic Flow Properties of Sheet and Round Tensile Test Specimens," *Proc. ASTM* **61**, 1349–1361.
ONAT, E. T. PRAGER, W.	1954	"The Necking of a Tension Specimen in Plane Plastic Flow," *J. Appl. Phys.* **25**, 491–493.
OWEN, W. S. COHEN, M. AVERBACH, B. L.	1958	"Some Aspects of Pre-Yield Phenomena in Mild Steel at Low Temperatures," *Trans. ASM* **50**, 517–540.
PARKER, E. R. DAVIS, H. E. FLANIGAN, A. E.	1946	"A Study of the Tension Test," *Proc. ASTM* **46**, 1159–1174.
POWELL, G. W. MARSHALL, E. R. BACKOFEN, W. A.	1955	"Diameter Gage and Dynamometer for True Stress-Strain Tension Tests at Constant True Strain Rate," *Proc. ASTM* **55**, 797–809.
ROSENFIELD, A. R. AVERBACH, B. L.	1960	"Initial Stages of Plastic Deformation in Copper and Aluminum," *Acta Met.* **8**, 624–629.

Shanley, F. R.	1946	"The Column Paradox," *J. Aero. Sci.* **13,** 678. See also "Inelastic Column Theory," *J. Aero. Sci.* **14,** 261–268.
Southwell, R. V.	1932	"On the Analysis of Experimental Observations in Problems of Elastic Stability," *Proc. Roy. Soc. (London)* **A135,** 601–616.
Suhre, J. R. Brock, G. W.	1959	"A Comparison of the Fatigue Behavior of Leaded and Non-Leaded AISI 4340 Steel at High Hardness Levels," *Theo. and Appl. Mech. Dept. Rep. 570,* University of Illinois, Urbana, Illinois.
Timoshenko, S. Gere, J. M.	1961	*Theory of Elastic Stability,* 2nd ed., McGraw-Hill, New York.
Voigt, W.	1928	*Lehrbuch der Kristallphysik,* 2nd ed., Teubner, Leipzig.

PROBLEMS

◄ 8.1. Show that the displacements of Eq. 8.1 satisfy the necessary fundamental equations.

◄ 8.2. Show how the stress-strain curves of Fig. 8.6(a) lead to the points of inflection for the strain distribution in Fig. 8.6(b).

8.3. Neglecting elastic strain, derive the equation for the equivalent strain in a tensile test, Eq. 8.3.

8.4. Considering elastic strain, derive the equation for equivalent strain in a tensile test in terms of length, Eq. 8.5.

8.5. Derive the equation for the equivalent strain at maximum load in a tensile test, Eq. 8.7.

8.6. For a power-law type of equivalent stress-strain curve, such as Eq. 8.8, show that the maximum load in tension occurs at the strain n and the equivalent stress $\bar{\sigma}_1 n^n$.

8.7. Show that for the power-law stress-strain relation, Eq. 8.8, the tensile strength is $\bar{\sigma}_1(n/e)^n$.

8.8. Derive the equation for the maximum pressure in a thin-walled sphere, Eq. 8.9.

8.9. Derive the equation for the maximum internal pressure in a thin-walled cylinder with closed ends, Eq. 8.10.

8.10. Show that the maximum pressure for a cylinder with the end loads carried externally occurs at the same point of the stress-strain curve as the maximum load in a thin sphere.

8.11. Show that for a material which strain-hardens according to the power law, Eq. 8.8, the uniform strain is equal to the exponent n.

◄8.12. Derive the equation for the axial stress required for a hypothetical transverse neck, Eq. 8.12.

◄8.13. Show that necking or Lüders bands should occur along lines at an angle of 54.7° with the axial direction, as shown in Fig. 8.11.

◄8.14. Show that necking cannot occur in any strip in a spherical shell without either local plane strain conditions or deformation elsewhere as well.

◀8.15. Show that the time required for a viscous specimen to rupture is given by Eq. 8.17.

◀8.16. Derive an expression for the shape of an initially tapered specimen which creeps according to a power law of the form

$$\sigma = \sigma_1(\dot{\epsilon}/\dot{\epsilon}_1)^n.$$

8.17. Derive the equation for the equivalent strain in terms of the diameter ratio, Eq. 8.18.

8.18. Derive the equation for the reduction of area in terms of the equivalent strain at fracture, Eq. 8.21.

8.19. For highly necked specimens, it is reasonable to assume that the specimen diameter varies linearly with the distance from the central plane of the neck. Under these conditions, derive an expression for the elongation in terms of the reduction of area, relevant specimen dimensions, and the uniform elongation (Lessells, 1954).

8.20. Derive expressions for the parameters of a power-law stress-strain curve, Eq. 8.8, in terms of observable items of data, such as (a) tensile strength and uniform elongation, or (b) tensile strength, fracture strength, and reduction of area.

8.21. Estimate the reduction of area of a specimen from information that can be obtained from the engineering stress-strain curve, along with a plausible assumption about the shape of the equivalent stress-strain curve.

◀8.22. Compare the angle of the Lüders band in Fig. 1.35 with that expected theoretically. Consider various possible reasons for any discrepancy, indicating whether they are quantitatively reasonable.

◀8.23. Estimate the tensile strength and uniform strain to be expected after a 90% reduction of area by drawing of 1100 aluminum wire.

8.24. If the slope of the strain-hardening curve of a material is of the order of $E/200$ to $E/600$, what is the maximum strength the material can have without losing all its uniform elongation?

8.25. Single crystals may have a stress-strain curve of the type shown in Fig. 8.25. For what stress or strain would the specimen begin to neck?

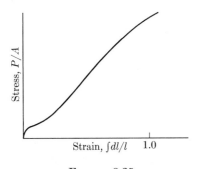

FIGURE 8.25

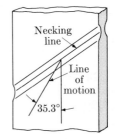

FIGURE 8.26

◀8.26. Show that the vector expressing the relative motion across the oblique strip which is necking in a thin sheet in tension (Fig. 8.26) makes an angle of 35.3° with the axial direction.

◀8.27. Derive a relation between the reduction of area and the equivalent strain at fracture for necking in thin sheets.

8.28. Show that for testing machines with very low stiffness, Eq. 8.22 predicts that instability will occur at the tensile strength of the specimen.

8.29. The elasticity of a tensile specimen itself, in the region away from the neck, may provide enough extension to contribute to the instability. Derive an expression for the maximum slope of the stress-strain curve which can be followed without instability, in terms of the relevant dimensions of the specimen.

◀8.30. Show that the buckling loads of Figs. 8.23(b) and 8.23(d) can be obtained by assuming the columns to consist of an assembly in series of columns of the type of Fig. 8.23(a). Why does this procedure not work for a column such as shown in Fig. 8.23(c)?

E◀8.31. Pull tensile specimens of Plasticine apart at the lowest and the highest rates you can. Note the difference in the shape of the neck. From what you can determine from this necking, what is the nature of the rate effects in the constitutive equations for Plasticine?

E◀8.32. Estimate and measure the maximum compressive load which can be obtained in a given piece of spring wire.

E◀8.33. Determine a stress-strain curve for modeling clay. From the curve, predict the maximum length a column may have without undergoing plastic buckling. Try compressing columns of about that length. How good was your prediction?

◀8.34. Dewhirst and Sidebottom (1962) ran compression tests on ½-in. square bars, 12 in. long, of 17-7PH stainless steel at 972°F. At this temperature the stress-strain curve was approximately

$$\epsilon = \frac{(\sigma, \text{psi})}{15 \times 10^6} + (0.001)(t, \text{hours})\left(\frac{\sigma, \text{psi}}{40{,}000}\right)^5.$$

Estimate the collapse time of such bars loaded concentrically (but with an initial deflection of 0.02 in. as soon as the load was applied). A compressive stress of 35,000 psi was maintained. Set up an expression for your result, including all intuitive correction factors, before evaluating it numerically to see how close you come to the observed time of 25 min. Dewhirst and Sidebottom predicted the load to 10% (corresponding to about a factor of 2 in time) but needed a computer to do it.

CHAPTER 9

BENDING AND TORSION

9.1 SYNOPSIS

In this chapter we consider problems in the mechanics of solids in which it is possible to obtain relatively simple, although sometimes approximate, solutions. The torsion of round cylindrical bars and the bending of beams and plates fall in this category. In these cases, symmetry and the fact that conditions must be the same along the length of the member lead to a strain distribution which varies linearly across a section. The corresponding stress can be found from the stress-strain relation and integrated to give the twisting or bending moment, not only for elasticity, but also for plasticity, viscoelasticity, or creep. In wide plates, the transverse curvature present in beams is prevented, and instead transverse stress develops. We shall investigate the width of a flat plate required for this transition. For anisotropic materials, more components of strain are present, although they still vary linearly across the section. These added components mean that a bending arises from a twisting moment, and vice versa. We shall investigate the possibility of deriving stress-strain information from the torque-twist data available from experiments.

We shall also study two sources of stress which may be present in the absence of external loads. Residual stresses develop when plastic strain produces a non-linear stress distribution, since the linearly distributed elastic change in stress on unloading does not reduce the stress everywhere to zero. Likewise, temperature gradients or a variation of coefficients of thermal expansion across a section give rise to a curvature of a flat strip. As in the isothermal case, plane sections remain plane, and the extension and curvature of the strip must be chosen to maintain zero axial force and be in equilibrium with the bending moment, which in this case is zero.

9.2 TORSION OF ROUND CYLINDRICAL BARS

For a long, round, cylindrical bar subjected to a uniform twisting moment, uniformity of conditions along the length of the bar and symmetry arguments lead to a strain distribution which varies linearly outward from the center of the shaft (Crandall and Dahl, 1959). Furthermore, symmetry and isotropy show that if twist in one direction were to lengthen the bar, so would twist in the other direction. This would require a quadratic rather than a linear type of stress-strain relation. But for small elastic strain or for strain increments in plastic flow and creep, the dependence on the components of stress is linear (Eqs. 3.19, 7.16, and 7.20). Therefore, as long as elastic strains are small or plastic flow does not produce important anisotropy, there will be no change in the length of the bar. Symmetry and linearity also show that there will not be any other normal component of strain (Prob. 9.1). Thus in terms of the coordinates of Fig. 9.1, the displacements depend on the angle of twist per unit length, $d\phi/dz$, as follows:

$$u_r = u_z = 0, \quad u_\theta = rz(d\phi/dz). \quad (9.1)$$

From the definition of strain, Eq. 2.28, the strain distribution is

$$\epsilon_{zz} = \frac{\partial u_z}{\partial z} = 0, \quad \epsilon_{rr} = \frac{\partial u_r}{\partial r} = 0,$$

$$\epsilon_{\theta\theta} = \frac{1}{r}\frac{\partial u_\theta}{\partial \theta} + \frac{u_r}{r} = 0,$$

$$\gamma_{zr} = \frac{\partial u_r}{\partial z} + \frac{\partial u_z}{\partial r} = 0, \quad (9.2)$$

$$\gamma_{r\theta} = \frac{\partial u_\theta}{\partial r} + \frac{1}{r}\frac{\partial u_r}{\partial \theta} - \frac{u_\theta}{r} = 0,$$

$$\gamma_{z\theta} = \frac{1}{r}\frac{\partial u_z}{\partial \theta} + \frac{\partial u_\theta}{\partial z} = r\frac{d\phi}{dz}.$$

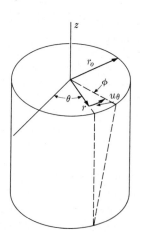

FIG. 9.1. Deformation of a round bar in torsion.

For isotropic materials, the presence of only one shear component of strain leads to the presence of only the corresponding component of stress, which is found from the appropriate elastic, plastic, or creep stress-strain relation. The one component of stress satisfies the equilibrium equations, Eq. 2.14 (Prob. 9.2). Having satisfied compatibility, stress-strain relations, and equilibrium at each point, we now have *a* solution to the problem. From uniqueness theorems, we know it to be *the* solution except for residual stresses or the effects of gross changes in geometry such as buckling. The shear stress is then integrated across the section to find the twisting moment:

$$M_{zz} = \int_0^{r_o} 2\pi r^2 \sigma_{\theta z}(\gamma_{\theta z})\, dr, \quad (9.3)$$

where the strain is found as a function of radius from Eq. 9.2. Equation 9.3 can be integrated explicitly if the stress-strain relation for the material can be stated in a simple enough form.

◀Several examples are given below. (a) For an elastic material with shear modulus G, the twisting moment is (Prob. 9.3)

$$M_{zz} = \frac{G\pi r_o^4}{2}\frac{d\phi}{dz} = \frac{\pi r_o^3}{2}\sigma_{z\theta\max}. \quad (9.4)$$

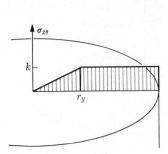
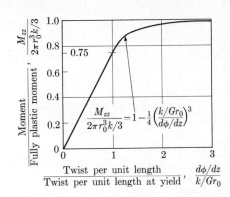

Fig. 9.2. Stress distribution for a non-strainhardening material in torsion.

Fig. 9.3. Torque-twist curve for a non-strainhardening material.

◂(b) For an elastic-plastic, non-strainhardening material with shear modulus G and yield strength in shear k, the shear stress varies with radius, as shown in Fig. 9.2, and gives the torque-twist relation plotted in Fig. 9.3 (Prob. 9.4). For twists large compared with those required for initial yielding, the twisting moment approaches the limit moment (Prob. 9.5):

$$M_{zz} = 2\pi r^3 k/3. \tag{9.5}$$

Note that although plastic flow begins at 0.75 of the limit moment, the contribution of the plastic flow to the twist does not equal the elastic contribution until 95% of the limit moment is reached. If elastic deflections are unimportant, then a design can be based on a moment nearly equal to the limit moment. This illustrates the value of finding a limit load, often by approximate methods as discussed in Sections 10.5 and 10.6.

◂(c) For a strain-hardening material, the stress-strain relation can sometimes be expressed by a power relation of the form $\sigma_{z\theta} = k_1(\gamma_{z\theta})^n$. The twisting moment can then be found in terms of the angle of twist by (Prob. 9.6)

$$M_{zz} = \frac{2\pi k}{3+n} r_o^{3+n} \left(\frac{d\phi}{dz}\right)^n = \frac{2\pi r_o^3}{3+n} \sigma_{z\theta\max}. \tag{9.6}$$

In many cases the integration must be performed numerically.

◂(d) For a viscous material with a coefficient of viscosity μ, the torque is related to the rate of twisting according to (Prob. 9.7)

$$M_{zz} = \frac{\mu \pi r_o^4}{2} \frac{\partial}{\partial t}\left(\frac{\partial \phi}{\partial z}\right) = \frac{\pi r_o^3}{2} \sigma_{z\theta\max}. \tag{9.7} ▸$$

The similarity between the elastic and linearly viscous cases, Eqs. 9.4 and 9.7, is a special case of the general result for viscoelastic materials discussed in Section

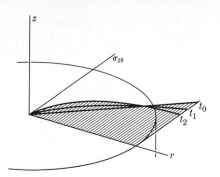 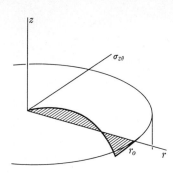

Fig. 9.4. Change in stress distribution during creep.

Fig. 9.5. Residual stress due to elastic unloading at time t_2 from the stress distribution of Fig. 9.4.

7.7. Wherever the elastic problem deals with displacements or strains, the corresponding viscous problem deals with velocities or strain rates. The equilibrium equations are identical. Thus, any solution to a problem for a linearly viscous material can be found by analogy from the corresponding problem for a linearly elastic material, if the Poisson's ratios, which are different, do not enter.

For a material which creeps with a nonlinear stress dependence, the torque can still be found by integrating the stresses as determined from the strain rates if the rate of twisting is given. In practice, a more common problem is to find the rate of twisting under constant torque. When the torque is first applied, the stress distribution is elastic or plastic. Let us consider low stresses, so the stress distribution is nearly elastic, and thus varies linearly across the section. In creep under steady-state conditions, the stress distribution will no longer be linear but will vary across the section as shown, for example, in Fig. 9.4 (Prob. 9.8). Thus when the torque is held constant, the stress at the outside will decrease with time, while that at the center increases. To predict this effect, it is necessary to know the rate of creep under varying stress. Analytical expressions or even data are seldom available for these conditions. If a steady state of creep is attained later, initial transients can be neglected or considered in an approximate way, as done for the data shown in Fig. 9.6 (Prob. 9.9). Much of the data could have been fitted with a viscoelastic approximation (Prob. 9.10).

On release of load, the change in stress is nearly elastic and thus linear. When this linear change in stress is superimposed on an originally nonlinear stress distribution resulting from plastic flow or creep, a stress remains, even when the torque is reduced to zero (Fig. 9.5). Such *residual stress* will be discussed in Chapter 12.

◀For bars of noncircular cross section, symmetry no longer guarantees that plane sections remain plane. In fact, warping will occur, and the strain varies around the periphery. At external edges, it can be shown from the vanishing of stress on the surface, from the fact that $\sigma_{12} = \sigma_{21}$, and from the transformation

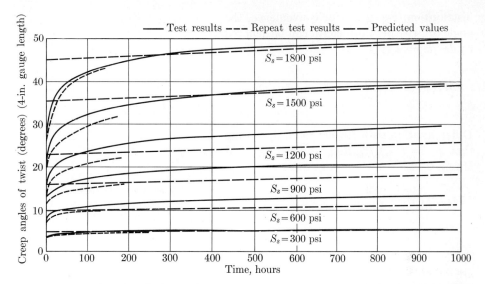

FIG. 9.6. Predicted and observed creep of nylon in torsion at 77°F and 50% relative humidity. (Marin, Webber, and Weissmann, 1954. Courtesy of American Society for Testing Materials.)

law for shear components acting on a plane, that the stress and hence strain must vanish (Prob. 9.11). For a further discussion of the elastic case see, for example, Timoshenko and Goodier (1951). For a discussion of the plastic case, see Nadai (1950) or Hill (1950) for the stress distribution, and Prager and Hodge (1951) for the strain and the warping of the cross section. Anisotropic effects are most easily considered after studying bending. ▶

9.3 DETERMINING THE STRESS-STRAIN RELATION FROM THE TORQUE-TWIST CURVE

◀Because stress-strain relations are the same at all points, in round rods it is possible to obtain the stress-strain relation of the material from the torque-twist curve. Even though the relation is the same, the stress and strain themselves vary across the section of a bar under torsion. Determining the stress-strain relation from the torque-twist curve therefore requires a more careful analysis than does the corresponding problem in the tensile test. As frequently happens, there is a key point in the analysis. In this case, the key point is that the stress across the section can be regarded as a function of either radius or strain, and the dependence of stress on strain is the same whether different angles of twist are considered at one radius as the test proceeds, or different radii are considered at the same stage in the test. We may thus change the variable of integration of the moment equation, Eq. 9.3, from radius to strain, with the aid of the equation for the strain distribution, Eq. 9.2. For convenience, the angle of twist per unit length can be denoted by $d\phi/dz = \phi_{,z}$:

$$M_{zz} = 2\pi\phi_{,z}^{-3} \int_0^{(\gamma_{\theta z})_o} \gamma_{\theta z}^2 \sigma_{\theta z}(\gamma_{\theta z}) \, d\gamma_{\theta z}. \tag{9.8}$$

Differentiation with respect to $(\gamma_{\theta z})_o$, noting that $\phi_{,z} = (\gamma_{\theta z})_o/r_o$, and rearranging yield (Prob. 9.12)

$$\sigma_{\theta z}((\gamma_{\theta z})_o) = \frac{3}{2\pi r_o^3}\left(M_{zz} + \frac{\phi_{,z}}{3}\frac{dM_{zz}}{d\phi_{,z}}\right). \tag{9.9}$$

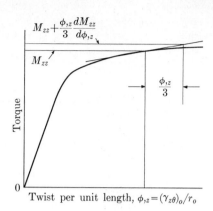

FIG. 9.7. Determination of factor for determining shear stress from torque-twist curve.

To evaluate Eq. 9.9, M_{zz} and $dM_{zz}/d\phi_{,z}$ are taken at the point on the torque-twist curve for which $\phi_{,z} = (\gamma_{\theta z})_o/r_o$, as shown in Fig. 9.7.

◀When the material exhibits significant rate dependence, the above analysis is only approximate, for since the strain rate varies with radius, the stress-strain curves at all radii will not be identical. Fields and Backofen (1957) have shown how the rate dependence can be determined from tests at a variety of twist rates.▶

9.4 BENDING OF BEAMS OF SYMMETRICAL CROSS SECTION

The analysis of the distribution of stress and strain in a beam under uniform bending is simplified by considering the limitations on possible modes of deformation which are imposed by symmetry, isotropy, and constancy of conditions along the length. From these it follows that plane sections remain plane and normal to the axis of the beam. This result can be demonstrated by showing that any other mode of deformation contradicts the assumed symmetry, isotropy, and axial uniformity (Prob. 9.13).

Since plane sections remain plane and normal to the axis, the axial normal component of strain varies linearly across a section in proportion to the distance from some line across the section, called the *neutral axis*. With a beam bent about an axis of symmetry, as shown in Fig. 9.8, it is convenient to take the x_1 direction as parallel to the neutral axis. Then, in terms of the undetermined coordinate to the neutral axis, a,

$$\epsilon_{33} = (x_2 - a)/R_1. \tag{9.10}$$

Contrary to the case of torsion, there are some displacements in the plane of the cross section, and thus more than one component of strain. These other components are found most easily after considering the stress, and depend on whether elastic or plastic stress-strain relations are appropriate.

In view of the thinness of the beam and the fact that there is no stress on its lateral surfaces, it seems reasonable to assume that only the axial normal component of stress is present. The equilibrium equations in terms of rectangular coordinates (Eqs. 2.11) are satisfied. If the curvature is appreciable, the boundary conditions of zero stress on the top and bottom surfaces are more easily expressed in forms of cylindrical coordinates, Eqs. 2.14. From these, it can be shown that a transverse pressure $-\sigma_{22}$ should develop, which depends on the depth of the beam, h. Roughly speaking, it is of the order of $h\sigma_{11}/R_1$ (Prob. 9.14). This term is usually negligible, but in thin tubes or the flanges of I-beams, there are surfaces on which it cannot be present at all, and appreciable distortion of the section and loss of stiffness may result.

Verification of the assumption of only one component of stress depends on being able to satisfy the stress-strain and strain-displacement equations as well as equilibrium conditions. For elastic deformation, the stress-strain relations lead to the presence of three components of strain:

$$\epsilon_{33} = (x_2 - a)/R_1,$$
$$\epsilon_{22} = \epsilon_{11} = -\nu\epsilon_{33} = -\nu(x_2 - a)/R_1. \quad (9.11)$$

The displacements corresponding to these components of strain are illustrated in Fig. 9.9, and are given analytically by (Prob. 9.15)

$$u_3 = (x_2 - a)x_3/R_1,$$
$$u_2 = -x_3^2/2R_1 - \nu(x_2 - a)^2/2R_1 + \nu x_1^2/2R_1, \quad (9.12)$$
$$u_1 = -\nu x_1(x_2 - a)/R_1.$$

It is well to relate the various terms in Eq. 9.12 to the features of the deformation shown in Figs. 9.8 and 9.9 (Prob. 9.16).

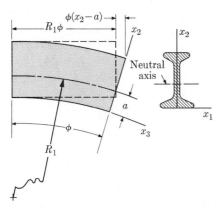

Fig. 9.8. Longitudinal strain in pure bending, $\epsilon_{33} = \phi(x_2 - a)/\phi R_1$.

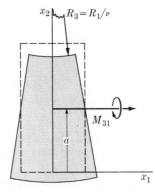

Fig. 9.9. Deformation of a cross section of a rectangular beam in pure bending.

We have now a solution satisfying the strain-displacement equations, the stress-strain relations, and at least approximately the equilibrium equations, along with the appropriate boundary conditions, provided that the displacements in a cross section are small compared with the dimensions of the cross section. Just what is meant by "small" displacements will be discussed further in Section 9.5. Once the stress distribution is found, the bending moments can be found by integration, using the appropriate stress-strain relation and Eq. 9.10 for the strain distribution. The unknown coordinate of the neutral axis, a, is found by equating the axial force to zero in the absence of applied loads:

$$F_3 = \iint \sigma_{33}(\epsilon_{33})\, dx_1\, dx_2 = 0,$$

and

$$M_{31} = \iint \sigma_{33}(\epsilon_{33}) x_2\, dx_1\, dx_2. \tag{9.13}$$

For the symmetry considered here,

$$M_{32} = \iint \sigma_{33}(\epsilon_{33}) x_1\, dx_1\, dx_2 = 0.$$

For the elastic stress distribution, the integrals may be expressed in terms of geometrical properties of the section, such as the centroid, the area moment of inertia, and the product of inertia,

$$\bar{x}_2 = \iint x_2\, dx_1\, dx_2 \Big/ \iint dx_1\, dx_2,$$

$$I_{22} = \iint (x_2 - \bar{x}_2)^2\, dx_1\, dx_2, \tag{9.14}$$

$$I_{12} = \iint (x_1 - \bar{x}_1)(x_2 - \bar{x}_2)\, dx_1\, dx_2,$$

giving (Prob. 9.17)

$$\begin{aligned} F_3 &= 0 = E\bar{x}_2/R_1 - Ea/R_1 \quad \text{or} \quad a = \bar{x}_2, \\ M_{31} &= EI_{22}/R_1, \\ M_{32} &= EI_{12}/R_1. \end{aligned} \tag{9.15}$$

The bending moments are often expressed in terms of the maximum stress, $\sigma_{max} = E|x_2 - a|_{max}/R_1$ (Prob. 9.18)

$$\begin{aligned} M_{31} &= \frac{|\sigma_{max}| I_{22}}{|x_2 - a|_{max}} \\ M_{32} &= \frac{|\sigma_{max}| I_{12}}{|x_2 - a|_{max}}. \end{aligned} \tag{9.16}$$

For bending of beams of unsymmetrical cross section, see Crandall and Dahl (1959, Section 7.7).

For anisotropic material, the analysis of beams is easiest if it is assumed that the stress distribution is the same as for isotropic material. This assumption is later shown to be correct. The stress-strain relations then may produce every component of strain. These strains lead to warping and twisting, as well as to bending (Prob. 9.19). The converse effect occurs in torsion (Prob. 9.20).

In fully plastic deformation, a set of displacements similar to those for the elastic displacement can be found, except that the transverse curvature is different due to the effective Poisson's ratio of one-half for plastic deformation. This differing curvature leads to an incompatibility when the elastic-plastic case is being treated, and transverse components of stress must develop. The exact solution is not known.

It is possible to determine the stress-strain relation from a moment-curvature plot, neglecting elastic strains (Nadai, 1950, pp. 357–359), but the required experiments are more difficult to execute than those in torsion. Residual stresses again arise on unloading from plastic deformation, as shown in Fig. 9.10. Composite beams can be analyzed as well in either the elastic or fully plastic regimes. The only difference from the homogeneous case is that the variation in stiffness or strength from one constituent to another will cause a variation in stress which must be taken into account in the integration to determine the moment.

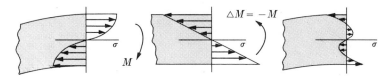

FIG. 9.10. Nonlinear applied stress plus a linear stress change on unloading leaves a residual stress.

The plastic bending of a rectangular beam about axes other than those of symmetry affords a relatively simple opportunity to check the associated flow rule, relating the deformation to the yield criterion for generalized stress and strain. In this case, the generalized "stresses" are taken to be the components of bending moment about the two axes of symmetry, as shown in Fig. 9.11. For an isotropic, non-strainhardening material bent far enough into the plastic region so that the elastic strains may be neglected, the yield locus can be found by calculating the components of the bending moment for bending about an axis making the angle θ with the axis of symmetry (Prob. 9.21)

$$M_{31} = \frac{Ybh^2}{4}\left(1 - \frac{b^2}{3h^2}\tan^2\theta\right) \quad \text{for} \quad -\frac{h}{b} < \tan\theta < \frac{h}{b},$$

$$M_{32} = \frac{Ybh^2}{4}\left(\frac{2}{3}\frac{b^2}{h^2}\tan\theta\right) \quad \text{for} \quad -\frac{h}{b} < \tan\theta < \frac{h}{b}. \tag{9.17}$$

The generalized components of strain must be chosen so that when these are multiplied by the respective bending moments and summed, they will give the plastic work. In this case, the generalized strains are the changes in bending angle:

$$d\omega_2/d\omega_1 = \tan\theta. \tag{9.18}$$

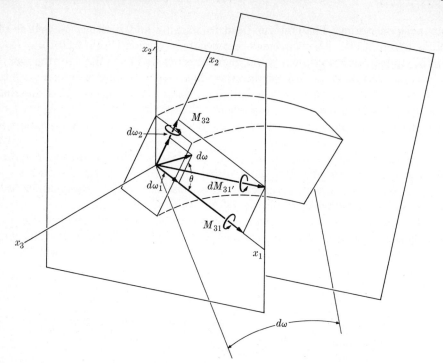

Fig. 9.11. Rectangular beam bent about an asymmetrical axis.

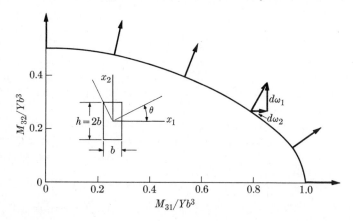

Fig. 9.12. Generalized yield locus showing normality of generalized strains, which are components of angle of bending.

When these components of bending are represented as a vector drawn from the yield locus at the corresponding value of θ, they are normal to it, as shown in Fig. 9.12 (Prob. 9.22).

As with torsion, the analysis can be extended to cases involving viscoelasticity or primary and secondary creep.

9.5 BENDING OF PLATES

◀As wider and wider beams are considered, the warping of the cross section can become comparable to the thickness of the beam. When this transverse curvature is combined with the longitudinal curvature which the beam develops, it turns out that equilibrium is no longer satisfied under the assumed distribution of stress. The situation is illustrated in Fig. 9.13. The u_2 deflection of the cross section means that the edges of the member tend to be in tension, while the central part tends to be in compression. This results in a moment on the x_3 section with a component in the x_2 direction, M_{32}. Due to the curvature of the beam, moments on nearby cross sections do not cancel, but require a counteracting moment. The only place such a moment can arise is from a moment acting on the x_1 surface with a component in the $-x_3$ direction, M_{1-3}. This moment tends to prevent the transverse curvature. If the plate member is wide enough, the transverse curvature will be confined to a region near the edge and may be neglected.

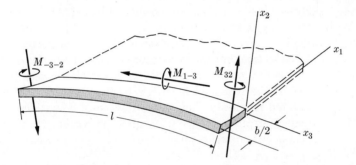

FIG. 9.13. Development of a transverse moment from a transverse curvature.

◀In this case, the displacement can be expressed in terms of a constant c to be determined:

$$u_3 = x_3 x_2 / R_1,$$
$$u_2 = -x_3^2/2R_1 - cx_2^2/2R_1, \qquad (9.19)$$
$$u_1 = 0.$$

Differentiation of these displacements gives the strain distribution

$$\epsilon_{33} = x_2/R_1, \qquad \epsilon_{22} = -cx_2/R_1, \qquad \epsilon_{11} = 0. \qquad (9.20)$$

The stress-strain relations then provide the stress (Prob. 9.23) and the condition that $\sigma_{22} = 0$ affords determination of the constant c (Prob. 9.24):

$$\sigma_{11} = \nu \sigma_{33},$$
$$\sigma_{33} = Ex_2/R_1(1-\nu^2), \qquad (9.21)$$
$$c = \nu/(1-\nu).$$

The stress may be integrated to give the bending moments (Prob. 9.25):

$$M_{31} = Ebh^3/12R_1(1-\nu^2), \qquad M_{13}/l = -\nu Eh^3/12R_1(1-\nu^2). \qquad (9.22)$$

Note that as a result of neglecting edge effects, the boundary condition of no transverse moment on the edges of the plate is no longer satisfied.

◀A rough estimate of the total width b required to develop the transverse moment may be obtained in the following way. Starting from the point shown as the origin in Fig. 9.13, where the neutral axis is at the center of the section, the transverse curvature increases toward a maximum value of ν/R_1, where R_1 is the longitudinal radius of curvature. Assume that this variation is linear. It then turns out that the width b required to develop full transverse moment per unit length, M_{13}/l, is given in terms of the maximum normal strain component $\epsilon_{33\text{max}}$ by (Prob. 9.26)

$$b = \sqrt{5}h/\sqrt{\epsilon_{33\text{max}}}. \qquad (9.23)$$

A more exact analysis gives a numerical coefficient of about 3 and has been checked experimentally by Ashwell (1950).

◀When the bending is so great that elastic strains may be neglected, strains still vary linearly across the section, but the transverse stress is one-half the longitudinal stress instead of ν times the longitudinal stress as in the elastic case (Prob. 9.27). Note that according to the Mises yield criterion, the transverse stress will increase the stress required for plastic flow above that for a beam, due both to the fact that the equivalent strain is higher for a given curvature (Prob. 9.28), and the axial normal component of stress is higher for a given equivalent stress, giving $\sigma_{33} = 2Y/\sqrt{3}$, $\sigma_{11} = Y/\sqrt{3}$ (Prob. 9.29).

◀The elastic-plastic analysis is complicated by the fact that when yielding first occurs, the transverse stress increases from ν to one-half times the longitudinal stress as plastic flow moves in toward the center. This changes the equivalent stress for a given longitudinal stress. The effect is small, and the reader is referred to Hill (1950, p. 79) for details.

◀In a viscoelastic material, the lateral constraint alters the stress distribution. For example, consider a constant bending moment applied to a plate of material, with elastic bulk behavior and Maxwell shear behavior given in terms of mean stress, dilatation, and the stress and strain deviators by Eqs. 3.24 and 7.23:

$$\epsilon = \sigma/B, \qquad \dot{e}_{ij} = \dot{s}_{ij}/2G + s_{ij}/2\mu.$$

For short times the viscous effects may be neglected, and the stress distribution is elastic, with a transverse stress given by

$$\sigma_{11} = \nu\sigma_{33}. \qquad (9.24)$$

For long times (i.e., $t \gg \mu/G$), a steady state of stress will be reached, and the deformation will be essentially viscous, incompressible flow, for which (Prob. 9.30)

$$\sigma_{11} = \sigma_{33}/2. \qquad (9.25)$$

This increase in transverse stress will reduce the strain rate, giving an initial transient similar to the four-element model of a viscoelastic material. Likewise on unloading, the transverse residual stress will cause an elastic aftereffect. Thus the behavior of a structure can differ from that of the material of which it is composed. The initial transient in the transverse stress can be worked out quantitatively with the aid of the Laplace transform for the stress-strain relations and the elastic solution for the bending of a plate (Prob. 9.31). ▶

9.6 THERMAL EFFECTS IN ISOTROPIC MATERIALS

◀An analysis similar to that for beams and plates holds for bimetallic strips subject to temperature changes, except that here two different stress-strain relations must be written for the two different materials constituting the strip. The axial strain must be chosen so that the axial force is zero. This will lead to a distribution of stress such as that shown in Fig. 9.14. For example, in a composite beam of two materials of thickness h_1 and h_2, Young's modulus E_1 and E_2, and coefficient of thermal expansion α_1 and α_2, respectively, the curvature due to thermal expansion is given by (Prob. 9.32)

$$\frac{1}{R} = \frac{\frac{3(\alpha_2 - \alpha_1)\,\Delta T}{2h_1} \frac{E_2}{E_1} \frac{h_2}{h_1}\left(1 + \frac{h_2}{h_1}\right)}{\left(1 + \frac{h_2}{h_1}\frac{E_2}{E_1}\right)\left[1 + \left(\frac{h_2}{h_1}\right)^3 \frac{E_2}{E_1}\right] - \frac{3}{4}\left[1 - \left(\frac{h_2}{h_1}\right)^2 \frac{E_2}{E_1}\right]^2}. \qquad (9.26)$$

◀The deformation caused by thermal shock can also be developed theoretically. The derivation is easiest when the material is so thick that the underlying material prevents any strain other than the component normal to the surface. A rise in temperature then causes a biaxial compression in the plane of the surface. If the temperature rise is sufficient, plastic deformation will occur. The only further rise in the stress will be due to strain-hardening, if any. On uniform cooling, the thermal contraction will be more than enough to provide the elastic strain needed to drop the stress to zero, and hence a reversed state of tension will develop on the surface. For the ideal case of a non-strainhardening material with constant Young's modulus E, coefficient of thermal expansion α, and a yield strength Y, the temperature rise

$$(T - T_0)_Y = Y(1 - \nu)/E\alpha \qquad (9.27)$$

can be obtained before yielding (Prob. 9.33). The residual stress left after a greater

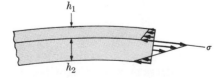

FIG. 9.14. Stress distribution in a uniformly heated bimetallic strip.

increase in temperature is (Prob. 9.34)

$$\sigma_{11'} = \sigma_{22'} = \frac{E\alpha(T - T_0)}{(1 - \nu)} - Y \quad \text{for} \quad \frac{Y(1 - \nu)}{E\alpha} < T - T_0 < \frac{2(1 - \nu)Y}{E\alpha},$$

$$\sigma_{11r} = \sigma_{22r} = Y \quad \text{for} \quad T - T_0 > \frac{2(1 - \nu)Y}{E\alpha}. \tag{9.28}$$

If the temperature is increased further, reversed plastic flow will occur on cooling to the initial temperature. In any case, the residual tensile stress may produce cracking. If a thin enough plate is used, the thermal stress on the surface will cause a general curvature in the plate, which can be estimated from the known history of thermal shocking. For further information, see Manson (1954).
◀Boley (1959) has reviewed thermal stress in general, and Weiner and Landau's (1960) review article gives further developments in elastic-plastic thermal stress. Boley and Weiner (1960) is a general reference book on thermal stress.▶

9.7 SUMMARY

For torsion of round cylindrical bars of isotropic materials, symmetry and the fact that conditions must be constant along the length of the bar give the shear strain in terms of the angle of twist per unit length:

$$\gamma_{z\theta} = r\, d\phi/dz. \tag{9.2}$$

Likewise, in pure bending of an isotropic bar to a longitudinal radius of curvature R_1, the axial normal component of strain is given in terms of the distance from the neutral plane by

$$\epsilon_{33} = x_2/R_1. \tag{9.11}$$

There turns out to be only one corresponding component of stress, and the equilibrium equations are identically satisfied.

Integration of the stress gives the twisting moment in torsion:

$$M_{zz} = \int_0^{r_0} 2\pi r^2 \sigma_{\theta z}(\gamma_{\theta z})\, dr.$$

In bending, the axial load and bending moments are given by

$$F_3 = \iint \sigma_{33}(\epsilon_{33})\, dx_1\, dx_2 = 0,$$

$$M_{31} = \iint \sigma_{33}(\epsilon_{33}) x_2\, dx_1\, dx_2, \tag{9.13}$$

$$M_{32} = \iint \sigma_{33}(\epsilon_{33}) x_1\, dx_1\, dx_2.$$

When plastic strain or creep produces a nonlinear stress distribution, the linearly distributed elastic change of stress on unloading will not reduce the stress to zero, but will leave residual stress.

◂For anisotropic materials, a single component of stress suffices; but more components of strain arise so that the application of the twisting moment may produce bending, and vice versa. For plates with width b, wide enough so that

$$b/h > \sqrt{5/\epsilon_{max}}, \qquad (9.23)$$

a curling up of the edges of the plate produces a transverse bending moment which holds the plate flat, so the deformation is one of plane strain.
◂In both torsion and bending, one can calculate the plastic stress-strain curve from the torque-twist or the bending moment-curvature plot. With rate-dependent materials, however, variation of strain rate through the bar makes it necessary to run tests at a variety of twist rates to get the stress-strain relation for even one strain rate.
◂Temperature gradients across a section, or layers of materials of different coefficients of expansion, give rise to curvature of a flat strip. As in the isothermal case, plane sections remain plane and the extension and curvature of the strip are chosen to maintain the axial force and bending moment equal to zero.▸

In linearly viscoelastic materials, the stress distribution is linear across the section, as in the elastic case, and there is no residual stress except for the transverse stress in a wide plate.

REFERENCES

ASHWELL, D. G.	1950	"The Anticlastic Curvature of Rectangular Beams and Plates," *J. Roy. Aero. Soc.* **54,** 708–15.
BOLEY, B. A.	1960	"Thermal Stresses," *Structural Mechanics (Proc. 1st Symp. Naval Struct. Mech.)*, J. N. Goodier and N. J. Hoff, eds., Pergamon Press, London, 378–406.
BOLEY, B. A. WEINER, J. H.	1960	*Theory of Thermal Stresses*, Wiley, New York.
CRANDALL, S. H. DAHL, N. C. (eds.)	1959	*An Introduction to the Mechanics of Solids*, McGraw-Hill, New York.
FIELDS, D. S., JR. BACKOFEN, W. A.	1957	"Determination of Strain-Hardening Characteristics by Torsion Testing," *Proc. ASTM* **57,** 1259–1272.
HILL, R.	1950	*The Mathematical Theory of Plasticity*, Clarendon Press, Oxford, England.
MANSON, S. S.	1954	"Behavior of Materials Under Conditions of Thermal Stress," *NACA Technical Rept. 1170*.
MARIN, J. WEBBER, A. C. WEISSMAN, G. F.	1954	"Creep-Time Relations for Nylon in Tension, Compression, Bending and Torsion," *Proc. ASTM* **54,** 1313–1343.
NADAI, A.	1950	*Theory of Flow and Fracture of Solids*, 2nd ed., McGraw Hill, New York.

Prager, W. Hodge, P. G., Jr.	1951	*Theory of Perfectly Elastic Solids*, Wiley, New York.	
Timoshenko, S. Goodier, J. N.	1951	*Theory of Elasticity*, McGraw-Hill, New York.	
Voigt, W.	1928	*Lehrbuch der Kristallphysik*, 2nd ed., Teubner, Leipzig.	
Weiner, J. H. Landau, H. G.	1960	"Thermal Stresses in Elastic-Plastic Bodies," *Plasticity* (*Proc. 2nd Symp. Naval Struct. Mech.*), E. H. Lee and P. S. Symonds, eds., Pergamon Press, London, pp. 369–384.	

PROBLEMS

9.1. Show from symmetry and linearity arguments that there will be no transverse components of strain for a round bar in torsion.

9.2. Show that the one component of shear stress in the case of torsion satisfies the equilibrium equations.

9.3. Derive the equation for the twisting moment in an elastic round bar in torsion, Eq. 9.4.

◀ 9.4. Derive the torque-twist relation of Fig. 9.3 for an elastic-plastic bar in torsion.

9.5. Derive the fully plastic twisting moment, Eq. 9.5.

9.6. Derive the equation (9.6) for the twisting moment of a material which strain-hardens according to the rule $\sigma_{z\theta} = k_1(\gamma_{z\theta})^n$.

9.7. By direct derivation, obtain the equation for the twisting moment for a bar of viscous material, Eq. 9.7.

9.8. Show qualitatively how the stress distribution in a round bar during creep should change, as indicated in Fig. 9.4.

◀ 9.9. Did the authors of the article on which Fig. 9.6 was based consider the transient creep to be recoverable or not?

9.10. How much of the data presented in Fig. 9.6 could have been fitted with a viscoelastic approximation? What further information is required to determine the numerical value of the strain at which the linear viscoelastic approximation breaks down?

9.11. Show that the shear components of strain must vanish along the edges of a twisted bar of polygonal cross section. (Along grooves the strain becomes infinite.)

◀9.12. Carry out the derivation of Eq. 9.9, giving the stress at a given point on a torque-twist curve (Nadai, 1950, p. 348).

9.13. Choose some nonplanar mode of deformation for a beam in bending and show that it contradicts the assumed symmetry, isotropy, and constancy of conditions along the length of the beam. (For an answer see, for example, Crandall and Dahl, 1959, p. 283.)

9.14. By considering equilibrium of an appropriate element of a long rectangular beam of depth h bent into the shape of a semicircle, or from the equations of equilibrium in cylindrical coordinates, show that a transverse pressure will develop across the neutral axis of the beam and be of the order of magnitude of $h\sigma_{11}/R_1$.

9.15. Show that the displacements of Eq. 9.12 give the strains of Eq. 9.11.

9.16. Explain which of the terms in Eq. 9.12 is associated with which features of the deformation shown in Figs. 9.8 and 9.9.

9.17. Derive the location of the neutral axis and the moment-curvature relations for a beam in bending, Eqs. 9.15.

9.18. Derive the relation between bending moment and maximum stress, Eq. 9.16.

◀9.19. Show that the following displacements are correct for bending of a bar of anisotropic material. Identify the physical meaning associated with the different terms (Voigt, 1928):

$$\frac{u_1 I_{22}}{M_{31}} = S_{1133}x_2x_1 + S_{1233}x_2^2 + S_{3133}x_2x_3,$$

$$\frac{u_2 I_{22}}{M_{31}} = \frac{S_{2233}x_2^2}{2} - \frac{S_{3333}x_3^2}{2} - S_{3133}x_1x_3 - \frac{S_{1133}x_1^2}{2},$$

$$\frac{u_3 I_{22}}{M_{31}} = S_{3333}x_2x_3 + S_{3133}x_1x_2 + S_{2333}x_2^2.$$

◀9.20. Show that the following set of displacements is correct for a round bar of anisotropic material under torsion. Identify the physical meaning of the different terms:

$$\frac{u_1 I}{M_{33}} = S_{1123}x_1^2 - 2S_{1131}x_1x_2 - (S_{2223} + 2S_{1231})x_2^2$$
$$- 2(S_{2323} + S_{3131})x_2x_3 - S_{3323}x_3^2,$$

$$\frac{u_2 I}{M_{33}} = 2S_{2223}x_1x_2 - S_{2231}x_2^2 + S_{3331}x_3^2 + 2(S_{2323} + S_{3131})x_3x_1$$
$$+ (S_{1131} + 2S_{1223})x_1^2,$$

$$\frac{u_3 I}{M_{33}} = 2S_{3323}x_1x_3 - 2S_{3331}x_2x_3 + 2S_{3123}x_1^2 + 2(S_{2323} - S_{3131})x_1x_2 - 2S_{2331}x_2^2.$$

9.21. Derive the equation for the bending moment in a plastic beam bent around an arbitrary axis, Eq. 9.17.

9.22. Show that the generalized strain increment vector for the beam of Prob. 9.21 is normal to the yield locus, as shown in Fig. 9.12.

◀9.23. Obtain the stress distribution for an elastic plate in bending from Eq. 9.20.

◀9.24. Evaluate the constant c in Eq. 9.20 from the condition of zero stress normal to the plate.

◀9.25. Derive the equation for the bending moments on a plate being bent, Eq. 9.22.

◀9.26. Derive the approximate width required to develop the transverse bending moment, Eq. 9.23.

◀9.27. Derive, from fundamental equations, the transverse stress for a plate being bent plastically.

◀9.28. Show that in the bending of a plate, the equivalent strain is higher for a given curvature than in the bending of a bar.

◀9.29. Show that in bending a flat plate, the axial normal component of stress is higher for a given equivalent stress than it would be in bending a bar, so that $\sigma_{33} = 2Y/\sqrt{3}$, $\sigma_{11} = Y/\sqrt{3}$.

◀9.30. Derive the equation for the final transverse state of stress in a beam of Maxwell material being bent as a flat plate, Eq. 9.25.

◀9.31. Derive the initial transient in the transverse stress in a plate of viscoelastic material whose stress deviatiors follow the Maxwell relation.

◀9.32. Derive the equation for the curvature in a bimetallic strip in the special case where the thickness and the moduli of elasticity of the two materials are equal (Eq. 9.26). As an alternative to a direct derivation, one can assume a temperature change without allowing any curvature and calculate the required restraining moment. The curvature on release of this moment can then be obtained with the usual equations for the isothermal case.

◀9.33. Derive the equation for the maximum temperature change that can be imposed without causing yielding in an infinitely thick slab, Eq. 9.27.

◀9.34. Derive the equation (9.28) for the residual stress left in a slab after a higher thermal shock than that given by Eq. 9.27.

◀9.35. In the tensile testing of single crystals, anisotropy may cause a tilting of the ends of the specimen relative to the test section. Is it possible at all, or under any special conditions, to eliminate this tilting by a combination of bending, torsion, and axial load?

9.36. Show why residual stress will or will not affect the analysis given for the determination of a shear stress-strain curve from the torque-twist curve of a round bar in torsion.

9.37. Under what conditions is a residual stress expected from the creep tests on nylon in torsion shown in Fig. 9.6?

9.38. Find the strain at the final temperature in the heat-distortion test for polymers described in Section 6.8.

E 9.39. Make a spring dynamometer for use in desk-top experiments. Wrap 0.026 in. diameter music wire tightly around a pencil for about ten turns. Be very careful to avoid letting the end slip through your fingers, since it can cut. For future reference, note the diameter of the pencil and the diameter of the coil after spring back. To break off the wire, clamp it between two pennies as a means of gripping it while you are bending it back and forth. Form a loop at each end by first bending the loop to the desired shape while it is still in the plane of the end coil, and then grasping the loop between two pennies and bending it into a plane containing the axis, as shown in Fig. 9.15. (a) Estimate the spring constant. (If necessary see Crandall and Dahl (1959, p. 278).) Compare this estimate with the spring constant for your dynamometer. (b) What is the maximum allowable load? What do you find experimentally?

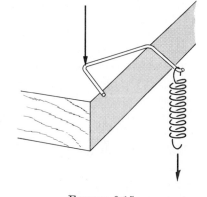

FIGURE 9.15

E 9.40. Opne up a paper clip into the form of a U, as shown in Fig. 9.15. What will the stiffness be on loading one leg of the U as indicated? Check this with your spring dynamometer.

E 9.41. Determine the direction of motion and the spring constant of the L-shaped cantilever beam in Fig. 9.16 under loads such that the elastic limit is not exceeded.

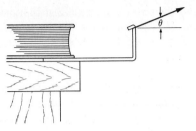
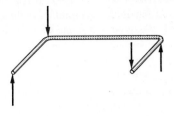

FIGURE 9.16 FIGURE 9.17

E 9.42. (a) Anneal a paper clip in a wooden match or gas flame. Bend the paper clip and feel the upper yield point. After reannealing it, measure the load with a spring to see whether the load actually drops. What must the ratio of upper to lower yield point be in order for the bending moment to drop? Continue bending the paper clip with a constant bending moment until you have nearly formed a complete loop. Determine the yield strain at the lower yield point.

(b) Will the upper yield point be more likely to be demonstrated in bending or torsion? After bending a paper clip into the shape of a U, anneal it and twist it as shown in Fig. 9.17. Note that the extent of the yield zone can be measured by observing the region in which the scale has cracked off. Determine the yield strain in torsion. How does it compare with the tensile value?

E 9.43. Straighten out a paper clip and load it as a cantilever beam. Determine the load for initial plastic deformation and the load for fully plastic deformation. How do these depend on the history of bending the beam at the point of maximum moment? Predict these stresses in view of the expected residual stress at that point.

E 9.44. Suppose that a paper clip is opened up into a rectangular shape and loaded as indicated in Fig. 9.15. Note that at the corner nearest the applied load, the bending moments and twisting moments are equal. Will yielding occur first in bending or torsion? Which has the greater fully plastic moment? Consider the effects of history as in Prob. 9.43.

E 9.45. Determine the bending moment required for fully plastic flow of 0.026-in. diameter music wire. How much spring back do you expect when this wire is wrapped around a pencil 0.3 in. in diameter? How much difference does it make whether the pencil is hexagonal or circular in cross section? What does this result tell you about the design of dies for forming sheet metal?

E 9.46. Estimate the elastic limit in tension of a spring which is made by coiling 0.026 in. diameter music wire around an 0.3-in. diameter pencil. What extension of the spring will cause initial yielding? If the spring is "set out" by straining it well into the plastic region, by how much is the elastic range increased?

E 9.47. A torsion spring is one whose ends are rotated about the center line. Note that when the spring is in torsion, the wire is in bending. Is it preferable to use the torsion spring in the direction tending to open it or to close it? How much will the residual stresses reduce the allowable rotation of the ends of the spring when it is used in torsion? Check your results experimentally.

E 9.48. From the hardness, the tensile strength, or from compressive tests on modeling clay, predict the length of a cantilever beam that can support its own weight. Check your result experimentally.

9.49. Determine the generalized yield locus and the stress and strain distributions for a round bar under combined bending and torsion. [*Hints:* Start from the displacements, as nearly as you can judge them. The integrals do not come out in closed form. Do you recognize the special functions?]

9.50. A specimen of 2024-T4 aluminum alloy is tested in bending in the form of a sheet 1.25 in. wide and 1.50 in. between supports with a knife edge applied in the center. The thickness of the sheet is 0.125 in. Estimate the maximum load and compare it with the observed value of 1080 lbs. Give possible reasons for the discrepancy, including quantitative estimates of the magnitudes of the discrepancies expected from various sources.

CHAPTER 10

APPROXIMATE STRESS ANALYSIS

10.1 SYNOPSIS

The difficulty of obtaining exact solutions for many problems of interest in the mechanics of the forming, testing, and use of materials leads us to look for approximate solutions. In this chapter we shall consider the loads required for given deflections, both in the elastic and fully plastic conditions. The estimation of local states of stress and strain will be deferred to the next chapter. Roughly speaking, lower bounds on the exact loads are found by ignoring the compatibility or geometrical requirements but satisfying the equilibrium equations, and, in plasticity, the yield criterion. Upper bounds on the exact load are found by assuming modes of deformation and choosing loads high enough to supply either the total elastic strain energy or the total plastic dissipation of energy, ignoring the fact that the resulting stress distribution may not satisy equilibrium. As the deformation is increased in a ductile, non-strainhardening material, the load reaches a constant value called the limit load, if changes in shape are not important and if no fracture occurs.

We shall estimate the plastic limit load for notched specimens, finding that in a doubly grooved plate, the limit load per unit minimum area can be roughly three times the tensile strength, whereas in a singly grooved plate, the limit load per unit minimum area nearly equals the tensile strength. Thus, aspects of the geometry which would make little difference in the elastic case become crucial under plastic deformation. For plastic problems which can be approximated by plane strain conditions, equilibrium and compatibility equations are reformulated in terms of curvilinear coordinates which are locally parallel to directions of maximum shear stress.

Using the tensile test as an example, we show that the strain distribution is not necessarily determined by the externally applied loads, although the stress distribution is unique within any region which may be deforming. A number of other useful corollaries of the limit load theorems are given.

In some cases of practical interest, it is awkward to obtain upper and lower bounds exactly; compatibility and equilibrium conditions may then be satisfied over large elements rather than at each point.

10.2 ENERGY METHODS IN ELASTICITY

◀Energy methods are occasionally used in elasticity to determine the true or compatible stress distribution from among a variety of stress distributions satisfying equilibrium and the external loads. This is possible because an energy function of stress distribution turns out to be a minimum for the exact solution. Our interest in such theorems will not be so much in this use, however, as in their use in estimating the overall stiffness of a structure and in their similarity to corresponding theorems which are helpful for estimating the limit load of a plastic material.

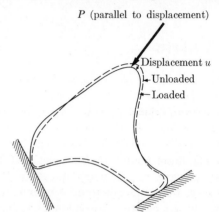

Fig. 10.1. Displacement of a body under a load P.

◀**A. Theorem of Minimum Potential Energy.** Consider the problem of determining the stiffness of a body against a displacement at some point, as indicated in Fig. 10.1. Assume a possible field of displacements throughout the body compatible with its support or other constraints. From this field of displacements, the strains can be determined at each point. From the components of strain and the elastic constants, one can determine the strain energy density at each point:

$$U = \sum_i \sum_j \sum_k \sum_l \frac{C_{ijkl}\epsilon_{kl}\epsilon_{ij}}{2}. \qquad (2.37),\ (3.4)$$

The theorem of minimum energy states that:

The strain energy obtained from displacements compatible with any boundary conditions, integrated over the entire volume, will be a minimum for the exact displacement distribution.

For proof see Sokolnikoff (1956, p. 382), who includes in a potential energy theorem not only the strain energy but also the potential energy of any external agency that may also be applying loads. We can express the strain energy exactly in terms of the actual stiffness, $K = P/u$. Then the principle of minimum strain energy becomes, using the summation convention (Section 2.8),

$$\frac{Ku^2}{2} \leq \int \frac{C_{ijkl}\epsilon_{ij}\epsilon_{kl}}{2}\, dV. \qquad (10.1)$$

An upper bound to the stiffness is thus given by

$$K \leq K_{\text{u.b.}} = \int C_{ijkl} \epsilon_{ij} \epsilon_{kl} \, dV/u^2. \tag{10.2}$$

◀ As an example of the application of Eq. 10.2, consider the extension of a long cylindrical bar of cross-sectional area A. In our ignorance, let us assume the presence of a quadratic as well as a linear term in the expression for the displacements:

$$u_3 = ax_3 + bx_3^2.$$

(We assume here we do know that there will be no strain energy associated with the transverse displacements and hence we neglect them.) The coefficients in the above equation must be chosen so that the displacement at the lower end is u:

$$u_3(l) = u = al + bl^2 \quad \text{or} \quad a = (u - bl^2)/l.$$

Corresponding upper bounds to the stiffness are then given by (Prob. 10.1)

$$K_{\text{u.b.}} = \frac{AE}{l}\left(1 + \frac{b^2 l^2}{3u^2}\right).$$

It should be no surprise that the lowest upper bound is found for $b = 0$, giving uniform strain. We therefore conclude that

$$K \leq AE/l.$$

◀ While this application of the principle may seem very trivial, it does find important use in far more difficult problems, such as determining the elastic constants of polycrystalline materials from the elastic constants of the single crystals. In this case, we assume the strain to be uniform throughout the polycrystals, even though we know it not to be the correct solution. The anisotropic elastic constants, referred to the specimen axes, vary from point to point according to a random orientation of the grains. For uniaxial loading, the stiffness of a unit cube becomes the modulus E. The resulting estimate of E is then, if anything, somewhat above the correct value. We would really like to know how far above the correct answer this estimate lies. While an exact solution would be required to answer this problem completely, we can obtain a partial answer by finding a lower bound to the stiffness. This is done with the aid of the principle of minimum complementary energy. ▶

◀ **B. Theorem of Minimum Co-Energy.** The co-energy or complementary energy in a material is defined as the integral of the strain over the stress:

$$U^{\text{co}} = \sum_i \sum_j \int \epsilon_{ij} \, d\sigma_{ij}.$$

Actually, for the linear elastic systems to which the concept of a spring constant applies, the complementary energy is equal to the strain energy (Prob. 10.2).

Again if we are interested only in the stiffness under loading at a single point, and where we now consider a given load at that point, the theorem of minimum complementary energy is that:

Among those stress distributions satisfying the equilibrium equations at each point, and in equilibrium with the external load, the strain energy found from such stress distributions is a minimum for the exact stress distribution.

◀ To apply this theorem in order to find the stiffness of the body, we note that if a load P is applied at a point where the stiffness is K, the work done, which for an elastic body will be the increase in its potential energy, is equal to $P^2/2K$. Since this energy is less than or equal to the complementary energy obtained from a possibly incorrect stress distribution, we have obtained a lower bound to the stiffness, that is, a number which must be smaller than the actual stiffness:

$$K \geq K_{\text{l.b.}} = \frac{P^2}{\int S_{ijkl}\sigma_{ij}\sigma_{kl}\,dV}. \tag{10.3}$$

◀ As an example, consider again the problem of the stiffness of a uniform bar in tension. We assume, in ignorance of the correct solution, that the stress distribution is uniform along the length, but varies from the surface to the center of the rod according to an equation of the form

$$\sigma_{33} = \left[\frac{P}{A}a - b\left(\frac{r}{r_o}\right)^2\right].$$

From this family of stress distributions a series of lower bounds to the stiffness can be obtained, depending on the parameter b (Prob. 10.3). The greatest of these gives the best estimate of the stiffness, and turns out to be the one for which $b = 0$, corresponding to uniform stress across the section. In this case, we have therefore found the stiffness exactly:

$$EA/l = K_{\text{l.b.}} \leq K \leq K_{\text{u.b.}} = EA/l.$$

In most cases of interest, we would not be so fortunate as to include the exact solution in a family of assumed displacements or stress distributions, and so we would not find the stiffness exactly. If the bounds were close enough together, however, we might well have an answer close enough for practical purposes.
◀ This theory can be used to obtain a lower bound to the elastic constants of polycrystalline material if one simply assumes that the stress distribution is uniform throughout all grains, regardless of the fact that due to anisotropy, such uniformly stressed grains would no longer fit together. The upper and lower bounds for the polycrystalline elastic constants obtained by these two theorems are close enough together for the most practical purposes, and an even better estimate can be taken from the root mean square of the two. This last estimate has the disadvantage that one does not know whether the result will be high or low. ▶

10.3 THE PLASTIC LIMIT LOAD

The concept of stiffness is no longer applicable in plastic flow, because the ratio of load to deformation changes as the deformation increases. To see what concept takes its place, consider the two tensile tests in Fig. 10.2, one on a smooth cylindrical specimen, and the other on a similar specimen containing a transverse hole. For this aluminum alloy, and in fact, for structural metals in general, the strain-hardening coefficient is much less than the elastic modulus, so that soon after plastic deformation begins, the load-deformation curve becomes nearly horizontal, compared with the elastic curve. The corresponding load is called the *limit load*. If a number of loads are simultaneously applied, in constant proportion to each other, any one of them can be referred to as the limit load, in terms of which the others are given. In parts whose deflection must be no more than two or three times the elastic deflection, the load must be limited to just a little less than the limit load. In the case of a member containing a hole, the strain concentration around the hole will result in the beginning of plastic yielding at loads which are only a fraction of those causing yielding in the smooth specimen. The plastic flow cannot contribute much to the overall elongation, however, because it is localized and in parallel with material that is still elastic. Only when deformation requires no further elastic strains (that is, plastic flow extends at least across the entire section), does the deformation increase markedly. Again, if it is desired to design parts whose deflections are of the order of the elastic deflection, it is only necessary to restrict the load to slightly less than the limit load. We now turn to the problem of estimating the limit load, after some necessary mathematical preliminaries.

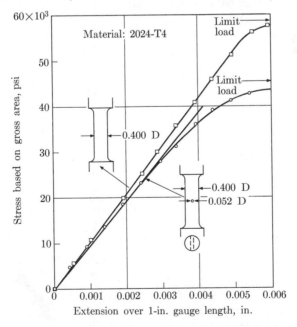

Fig. 10.2. Effect of strain concentration on deflection.

10.4 THE PRINCIPLE OF VIRTUAL WORK

◀In proving the limit analysis theorems of plasticity, as well as in working with other problems involving solids in equilibrium, it is convenient to relate the work done by external forces to the work per unit volume. This requires the transformation of surface to volume integrals with the Green-Gauss theorem (Eq. 2.39) in a way similar to that used in deriving the work per unit volume in terms of the strain. The result is called the *principle of virtual work*, and may be stated as follows:

Assume a possibly fictitious set of stress components σ_{ij}^a, in equilibrium throughout a body and in equilibrium with the external forces, which can either be represented by stresses on the surfaces, or by concentrated external loads or moments P_k^a. Also assume a virtual or arbitrary displacement field u_i^b in the body, which may have nothing to do with the loads through any stress-strain relation, but such that the strains $d\epsilon_{ij}^b$ in the body are compatible with all applied deformations, dp_k^b. These applied deformations may be displacements on the surface under the loads, or the rotations under the moments. The virtual work done by the external forces moving through the virtual displacements is equal to the volume integral of the stress times the corresponding strain increments (Prob. 10.4):

$$\int \sum_{i=1}^{3} \sum_{j=1}^{3} \sigma_{ij}^a \, d\epsilon_{ij}^b \, dV = \sum_k P_k^a \, dp_k^b. \qquad (10.4)$$

◀Sometimes in working with beams or plates it is more convenient to use overall forces and bending moments on a section instead of stresses. In terms of such generalized stresses Q_i^a acting on a section, and the corresponding generalized strains q_i^b, defined so the product of the two gives the work per unit length or area, the principle of virtual work is (Prob. 10.5)

$$\int \sum_{i=1}^{3} \sum_{j=1}^{3} \sigma_{ij}^a \, d\epsilon_{ij}^b \, dV = \int \sum_i Q_i^a q_i^b \, dl = \sum_k P_k^a \, dp_k^b. \qquad (10.5) \blacktriangleright$$

10.5 THE THEOREMS OF LIMIT ANALYSIS

Exact determination of the limit load, even for a non-strainhardening material, is usually a very difficult problem, requiring that the stress-strain relations, the strain-displacement or compatibility relations, the equilibrium equations, and the yield criterion be satisfied simultaneously. Not too surprisingly, exact solutions are known only for a few very simple cases. It is therefore useful for the engineer and even for the research worker in the mechanical behavior of materials to be able to get approximate solutions. Furthermore, it is important to know whether these estimates are too high or too low. In some problems involving the structural use of metals, it is important to be able to make estimates which are, if anything, less than the limit load. Such estimates are *lower bounds* to the limit load. If a lower bound is obtained, then any loads less than this lower bound will not result in excessive deformation of the structure. In metal-forming processes, on the other hand, it is important that enough force be available to accomplish the deformation. In this case, it is desirable to have an *upper bound* to the limit load. Since there is no guarantee as to how far one of these bounds is from the actual limit load, it is often desirable to have both upper and lower bounds. If the upper and lower bounds to the limit load do not differ by a practically significant amount, then one can regard the limit load as being determined for practical purposes.

In spite of the desirability of obtaining estimates which bound the limit load, it was only in 1951 that rigorous and general theorems became available (Drucker, et al., 1951; Hill, 1951). For this reason these theorems are not usually given in undergraduate texts on mechanics, and yet as we shall see, the applications of these theorems to many problems can be very simple. In addition, the theorems provide a freer rein to one's imagination and inventiveness, for they allow one to obtain useful estimates from admittedly wrong assumptions.

In developing the theorems of limit analysis, we shall assume a rigid-plastic material. The theorems are also valid for an elastic-plastic material, providing they are applied to situations in which the plastic deformation has proceeded far enough so that further deformation does not produce any change in the stress distribution. From that point on, the elastic material does not deform and can be considered to be rigid. When the stress distribution is relatively uniform, this point occurs soon after yielding; but where there are large strain concentrations, the local deformation may be much greater than at yield. In applying the theorems to a case of this type, it is important to ensure that this steady state of stress can indeed be reached, and that a lower maximum load has not already been reached due to large-scale changes in geometry arising from buckling, extensive local plastic flow around a notch, or cracking.

Roughly speaking, lower bounds to the limit load are obtained from stress distributions which satisfy equilibrium and the yield conditions, and upper bounds are obtained from the work required for displacement distributions which satisfy compatibility. Specifically, the lower bound theorem is:

> *In a rigid-plastic continuum there can be no plastic deformation under loads for which a stress distribution can be found which*
> (a) *everywhere satisfies the equilibrium equations,*
> (b) *is in equilibrium with external loads, and*
> (c) *is everywhere within the yield locus.*

That is, under given loads, if a stress distribution σ_{ij} can be found in equilibrium internally and with the given loads, and everywhere within the yield locus, then plastic deformation cannot occur. The power of the theorem lies in the fact that the stress distribution chosen need not be the correct one.

It is this theorem of limit analysis that is the reason that the techniques of elementary stress analysis give satisfactory results for members containing irregularities, such as fillets and oil holes. If an analysis is based on a straight member of the same dimensions as the net cross section of the member, and if maximum stress nowhere exceeds the yield strength, then it satisfies the requirements of the theorem for the lower bound. This is of course not the correct stress distribution, for it is known that localized plastic flow will occur in the vicinity of the stress concentrations. Nevertheless, the lower bound limit theorem proves that *general* plastic flow cannot occur at this load.*

◀Prager (1959) proved the lower bound theorem by contradiction. Assume that plastic deformation *can* occur under loads for which a stress distribution can be found which is in equilibrium and within the yield locus. Call the nonyielding stress distribution σ_{ij}^n. Call the distribution of hypothetical strain increments $d\epsilon_{ij}^h$, and the corresponding displacements under the loads dp_k^h. From the principle of virtual work, we have

$$\int \sum_{i=1}^{3} \sum_{j=1}^{3} \sigma_{ij}^n \, d\epsilon_{ij}^h \, dV = \sum_k P_k \, dp_k^h. \tag{10.6}$$

◀If plastic flow actually did occur, there would exist a stress distribution σ_{ij}^h corresponding to the hypothetical strains. This hypothetical stress distribution would have to satisfy equilibrium. Applying the principle of virtual work to the hypothetical distribution of stress and strain, we obtain

$$\int \sum_{i=1}^{3} \sum_{j=1}^{3} \sigma_{ij}^h \, d\epsilon_{ij}^h \, dV = \sum_k P_k \, dp_k^h. \tag{10.7}$$

From Eqs. 10.6 and 10.7, the work done per unit volume integrated over the volume for the nonyielding stress distribution is equal to that for the hypothetical stress distribution. Since plastic flow is assumed to occur under the hypothetical stress distribution, the stress must be on the yield locus at some points, at least. From the principle of maximum plastic resistance, Eq. 7.14, for the hypothetical strain increment to occur it is necessary that the work done by the hypothetical stress distribution be greater than that done by the nonyielding state of stress, σ_{ij}^n, which is within the yield locus at each point in the body:

$$\sum_{i=1}^{3} \sum_{j=1}^{3} \sigma_{ij}^h \, d\epsilon_{ij}^h > \sum_{i=1}^{3} \sum_{j=1}^{3} \sigma_{ij}^n \, d\epsilon_{ij}^h. \tag{10.8}$$

Integration of this inequality over the entire volume leads to a contradiction with

---

* The lower bound theorem was at one time called the principle of maximum plastic resistance, but that term now applies to the more recent principle by that name which led in Section 7.5 to the associated flow rule. There, for a given yield locus and strain increment, the stress maximized the work. Here, the name arose from the fact that in a structure, among all stress distributions satisfying equilibrium and the yield criterion, the exact distribution was that which maximized one of the generalized loads with the others held constant, i.e., which maximized the lower bound (Sadowsky, 1943).

the equality of the total work obtained from Eqs. 10.6 and 10.7. Therefore, the hypothetical exception to the theorem must be false and in turn the theorem must be true. ▶

How high the limit load may be is still not known from lower bounds. Therefore, it is desirable to have a means of estimating a load which is *certain* to cause plastic deformation. This estimate can be provided by the upper bound theorem:

In a rigid-plastic continuum, deformation must occur under any system of loads P_k, for which a distribution of displacements can be found such that
(a) the displacement boundary conditions, if any, are satisfied,
(b) the displacements can be differentiated to give a strain, with no change in volume anywhere, and
(c) the resulting plastic work done throughout the material, found from the resulting equivalent strain, is less than the work done by the external loads acting through the assumed displacements:

$$\sum_k P_k \, dp_k > \int_V \bar{\sigma}_Y \, d\bar{\epsilon}^p \, dV \qquad (10.9)$$

where $\bar{\sigma}_Y$ is the equivalent flow stress.

The power of this theorem lies in the fact that the assumed displacements need not be the correct ones.

◀ This theorem also is proved by contradiction, first expressing the plastic work in terms of the components rather than equivalent stress and strain (Eq. 2.41). Assume that under a system of loads satisfying the inequality of Eq. 10.9, an equilibrium stress distribution can be found such that plastic flow does not occur, that is, such that the stress distribution σ_{ij}^n always lies within or on the yield locus. From the principle of virtual work applied to this stress distribution and the strains $d\epsilon_{ij}^y$ accompanying the yielding postulated in the inequality, Eq. 10.9, there results

$$\int_V \sum_{i=1}^{3} \sum_{i=1}^{3} \sigma_{ij}^n \, d\epsilon_{ij}^y \, dV = \sum_k P_k \, dp_k^y. \qquad (10.10)$$

Since the virtual work done by the external forces is the same in Eqs. 10.9 and 10.10, there is an inequality between the two volume integrals:

$$\int_V \sum_{i=1}^{3} \sum_{j=1}^{3} \sigma_{ij}^y \, d\epsilon_{ij}^y \, dV < \int_V \sum_{i=1}^{3} \sum_{j=1}^{3} \sigma_{ij}^n \, d\epsilon_{ij}^y \, dV. \qquad (10.11)$$

But according to the principle of maximum plastic resistance, at each point in the body

$$\sum_{i=1}^{3} \sum_{j=1}^{3} \sigma_{ij}^y \, d\epsilon_{ij}^y > \sum_{i=1}^{3} \sum_{j=1}^{3} \sigma_{ij}^n \, d\epsilon_{ij}^y. \qquad (10.12)$$

When this inequality is integrated throughout the volume, it contradicts the inequality obtained in Eq. 10.11. Therefore, the assumed nonyielding stress distribution cannot exist. That is, it is impossible to have a stress distribution which will not allow yielding in the presence of loads large enough so the surface work done by them exceeds the plastic work per unit volume under some assumed strain distribution. ▶

These limit theorems of plasticity are similar to the ones giving estimates of upper and lower bounds to the stiffness of an elastic body as discussed in Section 10.2. Similar theorems hold for the resistance of a body to viscous deformation. For further discussion, including creep, see Drucker (1960, p. 425) and Calladine and Drucker (1962).

10.6 APPLICATIONS OF THE THEOREMS OF LIMIT ANALYSIS

As an example of the application of the theorems of limit analysis, consider the loads required to pull apart a flat plate with a groove on one side of it, as shown in Fig. 10.3. Under elastic loading there will be very high concentrations of stress and strain at the root of the notch. As the load is increased, yielding will occur first at the root of the notch and then spread outward. Until the specimen becomes plastic all the way through, the deformation will not be very large. The problem of calculating just what the distributions of stress and strain are during this elastic-plastic process is so difficult that it has not yet been carried out. But it is possible to calculate the limit load, that is, the load at which plastic flow is unrestrained by elastic material. This is the same as the maximum load in nonhardening materials, provided that fracture does not occur.

We first look for a lower bound. We must find a stress distribution which satisfies the equilibrium equations and is everywhere less than the yield strength, and then calculate the load corresponding to this distribution. The equilibrium equations are

$$\frac{\partial \sigma_{11}}{\partial x_1} + \frac{\partial \sigma_{21}}{\partial x_2} + \frac{\partial \sigma_{31}}{\partial x_3} = 0,$$

$$\frac{\partial \sigma_{12}}{\partial x_1} + \frac{\partial \sigma_{22}}{\partial x_2} + \cdots = 0, \qquad (2.13)$$

$$\frac{\partial \sigma_{13}}{\partial x_1} + \cdots + \cdots = 0.$$

First, choose the simplest possible stress distribution. Such a stress distribution is shown in Fig. 10.4. The σ_{33} component of stress is equal to the yield strength, and all other components of stress are zero within the layer of material to the left of the root of the notch. In the material to the right of the notch, that is, in the shoulders, all stress components are assumed zero. Now does this stress distribution satisfy the equilibrium equations? The only stress gradient present is the variation of σ_{33} in the x_1 direction across the interface between the supposedly stressed and the supposedly unstressed material. But this derivative, $\partial \sigma_{33}/\partial x_1$,

10.6 APPLICATIONS OF THEOREMS OF LIMIT ANALYSIS

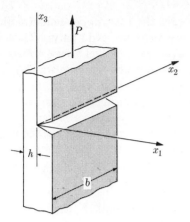

Fig. 10.3. Singly grooved plate in tension.

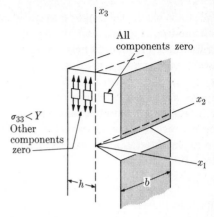

Fig. 10.4. Assumed stress distribution for lower bound to the limit load.

does not appear in the equilibrium equations. All other gradients are zero, and so the equilibrium equations are satisfied identically. The load corresponding to this stress distribution is simply σ_{33} times the minimum area of the specimen. Thus, if σ_{33} is less than the yield strength Y, the stress will be everywhere within the yield locus. We have therefore satisfied the three requirements of the lower bound theorem and may conclude that a lower bound can be within an infinitesimal amount of Ybh. Since the limit load must be greater than or equal to any lower bound,

$$P_{\text{limit}} \geq P_{\text{l.b.}} = Ybh. \tag{10.13}$$

Now we turn to the problem of estimating upper bounds, that is, loads that are certain to produce plastic deformation in the rigid-plastic material. It is necessary to assume some mode of deformation in the plate. A simple mode is that shown in

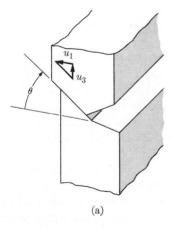
(a)

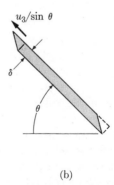
(b)

Fig. 10.5. Assumed strain distribution for upper bound to the limit load. (a) Overall view. (b) Detail of sheared layer.

Fig. 10.5, where the angle θ may be chosen to get the least upper bound. Shear deformation is concentrated in a thin zone between the two rigid parts. Thus, the strains are zero everywhere except in a thin layer, say of thickness δ, where the shear strain is of the magnitude $u_3/\delta \sin \theta$. This is the only region in which plastic deformation is occurring. We shall need to calculate the work done in the sheared layer. We could calculate the work from the stress-strain relations and the yield criterion, but it is much easier to make use of the fact proven in Problem 7.4 that the plastic work is the product of the equivalent stress multiplied by the equivalent plastic strain increment. The equivalent plastic strain increment is, from Eq. 7.5,

$$d\bar{\varepsilon}^p = (d\gamma_{1'3'}/\sqrt{3}) = du_3/\sqrt{3}\delta \sin \theta.$$

Since for plastic flow the equivalent stress is the yield strength Y, the total plastic work is

$$dW^p = \int \bar{\sigma}\, d\bar{\varepsilon}^p\, dV = Y(du_3/\sqrt{3}\delta \sin \theta)(\delta bh/\cos \theta)$$
$$= (Y/\sqrt{3})(bh/\cos \theta)(du_3/\sin \theta).$$

Note that the plastic work is independent of the thickness of the sheared layer and is simply the yield strength in shear times the sheared area and the shear displacement. An upper bound to the limit load is given by the load such that the work done by it equals this plastic work. For a displacement du_3 of the lower end relative to the upper, the total work done by the external loads will be $P\, du_3$. We have now satisfied the three conditions for an upper bound, and therefore

$$P_{\text{u.b.}} = (Y/\sqrt{3})bh/\sin \theta \cos \theta.$$

To get a close estimate of the limit load we want the lowest possible upper bound. This is attained when $\theta = 45°$. The limit load must be less than this upper bound. Thus

$$P_{\text{limit}} \leq P_{\text{u.b.}} = \frac{2}{\sqrt{3}}\, Ybh. \tag{10.14}$$

This result is about 15% higher than the lower bound we previously obtained, Eq. 10.13. Perhaps we can find a lower bound which is somewhat higher. Note that we previously assumed uniaxial stress. In the tensile test, where the stress is uniaxial, there is a transverse contraction. This would also occur if we were testing not a grooved plate, but a stack of thin sheets as shown in Fig. 10.6. To prevent this, a transverse stress σ_{22} would be required. Looking at the Mises yield criterion, shown in Fig. 7.2, or analyzing the yield criterion, Eq. 7.2, we see that the maximum value of σ_{33} is obtained when the transverse stress σ_{22} is equal to $\sigma_{33}/2$.

In finding a lower bound, we do not need to inquire just how the transverse stress arises, or indeed about whether it really develops at all. But for those whose curiosity demands an answer at this point, we can suggest that it could arise from the elastic stress distribution around a notch under plane strain, found from Eq. 11.23, or it could develop out of the limited plastic flow which occurs as the limit

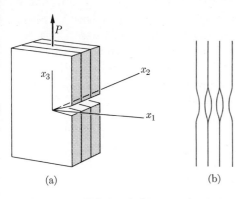 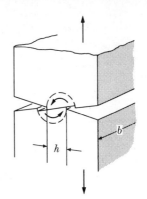

FIG. 10.6. Effect of absence of a transverse stress. (a) Notched sheets. (b) Section showing thinning near notch.

FIG. 10.7. Simple flow field for a doubly grooved plate.

load is approached. Likewise, a stress σ_{11} might develop due to the constraint of the shoulder. This could not extend to the back of the specimen, however, and will be neglected. Thus as a tentative refinement, assuming $\sigma_{11} = 0$, $\sigma_{22} = \frac{1}{2}\sigma_{33}$, we find that the value of σ_{33} before yielding can be as high as

$$\sigma_{33} = 2Y/\sqrt{3}.$$

Whether the stress actually can rise this high under plane strain conditions is not relevant, since we are concerned only with a lower bound. This assumed presence of σ_{22} results in the assumption of some forces on the sides of the plate. They cannot of course be present at the very outside edges, but the transverse stress will quickly build up a little way in from the sides. The resulting higher value of the axial stress now leads us to a lower bound which gives

$$P_{\text{limit}} \geq P_{\text{l.b.}} = 2bhY/\sqrt{3}. \tag{10.15}$$

Combining this with Eq. 10.14, we see that the equality must hold, so the limit load is uniquely defined. Note that the strain distribution is not unique; we could equally well have assumed shearing on the other 45° plane from the notch, or a combination of the two. (See Section 10.8 for further discussion of such lack of uniqueness.)

When a plate is grooved in opposite faces, as shown in Fig. 10.7, it becomes easier for the material to flow around the corner from the sides of the groove than for slip to break through to the face of the specimen as in the singly grooved case. A simple upper bound is obtained by assuming the deformation to consist solely of rigid-body rotation of the two half cylinders indicated in Fig. 10.7, causing shear on the dashed lines, but no strains elsewhere. The resulting upper bound for the limit load is given in terms of the yield strength Y of the material as (Prob. 10.6)

$$P_{\text{limit}} \leq P_{\text{u.b.}} = \frac{2\pi}{\sqrt{3}} Ybh. \tag{10.16}$$

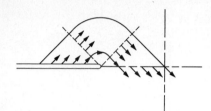

Fig. 10.8. Detail of flow field for a doubly grooved plate. All displacements are equal in magnitude. (Hill, 1950.)

This bound is over three times as high as the load that would be carried by a uniformly stressed plate having the same minimum cross section. In order to find a lower bound to the limit load it is necessary to find a stress distribution satisfying the equilibrium equations throughout the body and nowhere exceeding the yield strength. A simple one is found as before by assuming a uniform stress in a central strip:

$$P_{\text{limit}} \geq P_{\text{l.b.}} = \frac{2}{\sqrt{3}}\, Ybh.$$

This differs so much from the upper bound that neither is of much use. Refinement requires a study of the general form of the equilibrium and yield condition for plane strain, considered in Section 10.7, which shows that the flow field of Fig. 10.8 gives an upper bound of

$$P_{\text{limit}} \leq P_{\text{u.b.}} = \frac{2}{\sqrt{3}}\left(1 + \frac{\pi}{2}\right) Ybh, \qquad (10.17)$$

and does satisfy the equilibrium conditions and yield criterion. To prove that it also provides a lower bound, it is necessary to show an equilibrium stress distribution not exceeding the yield *anywhere* in the body, even outside the deforming region. For example, if the total thickness is less than $(1 + \pi)$ times the minimum, slip can occur as shown in Fig. 10.9 (Prob. 10.7). The exact solution is: for a sharp notch the total thickness to face must be 10.0 times the thickness at the root of the notch if flow is not to break through to a face (Bishop, 1953).

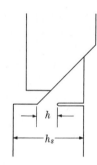

Fig. 10.9. Hypothetical displacements for a shallow notch.

The limit load has now been exactly determined, but the question remains as to whether the stress distribution is the only one possible. It is indeed the only one, as shown by the uniqueness theorem of Section 10.8.

Equation 10.17 states that the load carried by a doubly grooved specimen in plane strain can be nearly three times that carried by a smooth specimen. A similar result has been obtained for circumferentially grooved specimens by Levin (1955) and Shield (1955). These results are also useful in interpreting the hardness test, for if the direction of load is reversed and only the lower half of the specimen is considered, the problem is similar to that of a hardness indentation,

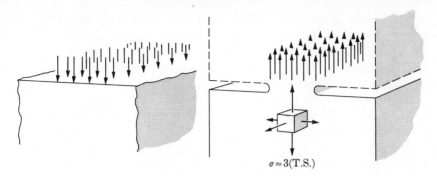

FIG. 10.10. Similarity of stresses under hardness indentation and in plastic tension of a doubly grooved plate.

as indicated in Fig. 10.10. In this case, too, the average stress is three times the flow stress of the material. When moderate amounts of strain-hardening are taken into account, one may approximate this relation by the rule of thumb given in Section 1.2: the hardness expressed in force per unit area is three times the tensile strength. Further use of this relation will be made in the chapter on hardness (Chapter 13).

The high value of the stress under a hardness indenter is sometimes called an example of *plastic constraint*, but this phrase is somewhat misleading, for it is the rigid material that does the constraining, and, as we shall see in Section 11.6, there is a similarly high state of transverse and lateral stress in front of a sharp notch in an elastic material.

FIG. 10.11. Hardened cap screw with minimum area $A_u + A_d$, decarburized over the area A_d.

As an example of the application of the theorems of limit analysis to a problem in design, consider the question of strength to be expected from the decarburized* machine screw shown in Fig. 10.11. As a lower bound, one may simply assume a uniform stress Y_u in the minimum unaffected zone A_u and a stress Y_d in the decarburized zone A_d, both areas being taken at the roots of threads. The stress in the threads is assumed zero. This gives a lower bound load

$$P_{\text{limit}} \geq P_{\text{l.b.}} = Y_u A_u + Y_d A_d.$$

---

* As we shall see in Prob. 22.12, the object of the decarburization is to retard fracture in the thread roots.

Actually, this is not a true lower bound, for we have not satisfied equilibrium *everywhere*, namely, in the head and in the threads, where the load is transferred to the nut. This is a lower bound, however, for the load at which deformation will occur in the shank. As an upper bound for deformation in the shank we may assume uniform strain not only in the shank, but also in the threads, that is, throughout the entire volume V. We know this assumption to be incorrect, for it would require stresses acting on the surfaces of the threads, but even though the estimate is incorrect, it still provides useful information. The limit load is thus found to be

$$Y_u A_u + Y_d A_d \leq P_{\text{shank limit}} \leq V_u Y_u / L + V_d Y_d / L. \qquad (10.18\text{a})$$

We now have upper and lower bounds between which the limit load (the load required for extended plastic deformation of the shank) must lie. For data on the actual result, see Prob. 22.12. We can also obtain an upper bound to the load required to shear the threads by simply assuming a mode of deformation in which the threads are sheared off:

$$P_{\text{thread limit}} \leq A_s Y_d / \sqrt{3}. \qquad (10.18\text{b})$$

One is tempted to compare Eqs. 10.18a and 10.18b to determine the ratio of thread to shank area which must be engaged to ensure failure of the shank before failure of the thread. Actually, this is not possible at this stage, since we would need to show that the limit load for the threads was higher than the limit load for the shank. This would require a lower bound for the threads to be higher than an upper bound for the shank. Determining such a lower bound for the threads is a problem as yet unsolved. It illustrates one of the basic difficulties encountered with limit analysis: the fact that lower bounds, which are often the ones desired, are so difficult to obtain. Fortunately it usually turns out that the upper bounds are closer to the correct solution, so that in practice this limitation is not so serious as it might first seem.

Another example of the theorems of limit analysis is found in trying to estimate the polycrystalline stress-strain curve from data on single slip in single crystals. Taylor (1938) based his calculations on the collection of slip systems $d\gamma^{(i)}$, which minimized the amount of dissipated plastic work on the slip systems of all the grains, under the assumption that the total strains resulting from the various slip systems were equal in all grains. It is left to the reader to decide whether this gives a lower or an upper bound or neither (Prob. 10.8).

◀There are a number of corollaries of the limit analysis theorems which are helpful in applying the theorems of limit analysis. Some of these corollaries have been used implicitly in the preceding examples. Except for the first, these corollaries were summarized by Drucker (1960). Their proof is left to the reader (Prob. 10.9).

1. Elastic solutions based on the net sections, neglecting stress-concentrating discontinuities such as holes and grooves, constitute lower bounds if the equivalent stress is everywhere less than the equivalent flow stress.

2. The limit analysis theorems apply to a strain-hardening material if the local flow stress is used (although it may be difficult to find).

10.7 BOUNDS FOR PROBLEMS IN PLANE STRAIN

3. Initial stresses or deformations have no effect on the limit load, provided the geometry is essentially unaltered.

4. Addition of (weightless) material, without any change in the position of the applied load, cannot reduce the limit load, and conversely removal cannot increase it.

5. An upper bound computed from a convex yield locus which circumscribes the actual yield locus will be an upper bound on the actual limit load. Likewise a lower bound computed from an inscribed yield locus will be a lower bound on the limit load. This corollary allows us to simplify the yield locus in some cases and still get useful results.

6. Adding (weightless) material to a nondeforming boundary cannot increase the limit load. ▶

10.7 BOUNDS FOR PROBLEMS IN PLANE STRAIN

◀ To provide a lower bound solution for a general three-dimensional problem is a difficult task. In fact, one of the largest obstacles to a wider application of limit analysis is lack of a means by which one can obtain lower bound solutions and progressively modify them until a reasonably high value is attained. In rigid-plastic plane strain, however, it is possible to restate the yield criterion and the equilibrium equations so that bounds can more readily be determined. In the first place, since the transverse strain is zero, the transverse stress must be the average of the other two (Prob. 10.10). Both the maximum shear stress and the Mises yield criteria then reduce to the result that the difference of the principal stresses must be a constant (Prob. 10.11). In other words, one can express the yield criterion by stating that the radius of Mohr's circle must be a constant equal to the yield strength k in shear. The state of stress can then be defined in terms of the mean normal stress σ and the orientation of the direction of principal stress relative to the coordinate axes. It is convenient to use curvilinear coordinates, where the axes are taken to be those of maximum shear stress. Calling these the α and β axes, as illustrated in Fig. 10.12, the angular coordinate of the α axis is denoted by ϕ.

◀ In terms of k and σ, the stress components relative to the x_1 and x_2 axes are

$$\sigma_{11} = \sigma - k \sin 2\phi,$$
$$\sigma_{22} = \sigma + k \sin 2\phi, \quad (10.19)$$
$$\sigma_{12} = k \cos 2\phi.$$

Substitution of these expressions into the equilibrium equations (2.11), differentiating and *then* considering the x_1 and x_2 axes to be locally parallel to the curvilinear coordinates so that $\phi = 0$, gives the result (Prob. 10.12)

$$\frac{\partial \sigma}{\partial x_1} - 2k \frac{\partial \phi}{\partial x_1} = 0, \quad \frac{\partial \sigma}{\partial x_2} + 2k \frac{\partial \phi}{\partial x_2} = 0.$$

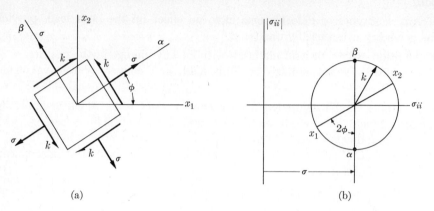

Fig. 10.12. Definition of curvilinear coordinates of maximum shear. (a) Physical plane. (b) Mohr's circle or stress plane.

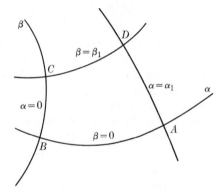

Fig. 10.13. Diagram for proof of Hencky's first theorem.

The partial differential equation of equilibrium has now been separated into ordinary differential equations along one or the other of two particular directions. This is an application of the method of characteristics, a mathematical technique which is also of value in problems of compressible flow and nonsteady fluid flow. For a concise introduction to the method, see Hill (1950, pp. 345–348). These ordinary differential equations can now be integrated along the characteristic directions to give

$$\text{along an } \alpha \text{ line: } \sigma - 2k\phi = \text{constant,}$$
$$\text{along a } \beta \text{ line: } \sigma + 2k\phi = \text{constant.} \tag{10.20}$$

◂An important feature of the curvilinear coordinates describing the stress field is found by considering the four adjacent points shown in Fig. 10.13. Since the change in mean stress between points A and C must be the same by either the path ABC or the path ADC, Eq. 10.20 indicates that (Prob. 10.13)

$$\phi_B - \phi_A = \phi_C - \phi_D. \tag{10.21}$$

◂ Equation 10.21, known as Hencky's first theorem, amounts to the statement that if two curvilinear lines of one family intersect a pair of lines from the other family, then the angular change between points of intersection is the same for each of the two intersected lines. Any set of orthogonal curvilinear coordinates having this property constitutes a stress distribution which satisfies the equilibrium equations, provided that the radius of Mohr's circle is constant along it. The reader is referred to Hill (1950, pp. 132–137, 144–149) for further treatment of the theory of these fields. The flow field for the doubly grooved plane-strain specimen shown in Fig. 10.8, satisfies these requirements (Prob. 10.14). Equations 10.20 lead to the lower bound for the load in a doubly grooved specimen in plane strain given by Eq. 10.17 (Prob. 10.15) provided that the shoulders are wide enough that an equilibrium field can be extended into them. This analysis was first carried out by Bishop (1953), and shows that the total thickness must be at least 10.0 times the ligament thickness.

◂ The equations for the displacements can be also simplified if they are expressed relative to the α, β curvilinear coordinates. The reason for this simplification is that in these coordinates, the normal components of strain vanish, leaving only shear components. In terms of displacement increments δu and δv in the directions of the α and β lines, respectively, the displacements referred to the rectangular coordinates are

$$\delta u_1 = \delta u \cos \phi - \delta v \sin \phi,$$
$$\delta u_2 = \delta u \sin \phi + \delta v \cos \phi. \quad (10.22)$$

The absence of normal strains and incompressibility mean that since there is plane strain,

$$\frac{\partial u_1}{\partial x_1} + \frac{\partial u_2}{\partial x_2} = 0. \quad (10.23)$$

Substitution of Eq. 10.22, differentiation, and then taking the cartesian coordinates locally parallel to the curvilinear coordinates so that $\phi = 0$ gives (Prob. 10.16)

$$\text{along an } \alpha \text{ line: } d(\delta u) - \delta v \, d\phi = 0,$$
$$\text{along a } \beta \text{ line: } d(\delta v) + \delta u \, d\phi = 0. \quad (10.24)$$

These are known as the Geiringer equations. It is one of the indications of the relative youth of the field of plasticity that these fundamental equations, which must be satisfied in order to assure a solution, were not derived until 1930, and they did not become generally used until about 1950. These equations can always be integrated for a rigid-plastic material, provided that a path along a characteristic α or β line can be found from each point under the prescribed deformation to a free surface. For an elastic-plastic material, these characteristics could end on the elastic-plastic boundary, but the resulting interaction of the elastic and plastic displacements has prevented solution of the problem except in cases where at least one end of each characteristic direction lies on the free surface. Where both ends lie on an elastic boundary, the interaction between the plastic deformation and the elastic deformation has led to a problem which, up to the present, seems to have been ignored rather than solved (see Hill, 1950, p. 243). The displacement fields of the doubly grooved, plane-strain specimen (Fig. 10.8) do satisfy these equations (Prob. 10.17). For further discussion of the integration of Eqs. 10.24, see Hill (1950, pp. 149–151). ▶

10.8 UNIQUENESS OF STRAIN AND STRESS DISTRIBUTIONS

The strain distribution in plastic flow may not be unique; that is, it may be possible to find two different strain distributions which satisfy the applied loads, the stress-strain relations of the material, and the equilibrium and compatibility conditions. In other words, the strain distribution is not unique. This rather surprising fact can be seen by considering the tensile test specimen at the condition of maximum load, when necking can begin anywhere along the length. With sufficient strain-hardening, the point of maximum load is postponed (along with necking) and the strain remains uniquely determined.

Another example that shows how strain-hardening provides uniqueness is the plane-strain, singly grooved tensile specimen discussed in Section 10.6. If shear initially occurred along only one 45° plane from the root of the groove, any strain-hardening would cause the next shear to occur along the other plane. One is thus led to expect the symmetrical strain pattern shown in Fig. 10.14, in which an average shear strain of unity occurs in the deforming bands (Prob. 10.18). One might expect that strain-hardening would cause other changes in the deformation. Even in annealed aluminum, however, this general mode of deformation occurs, although the angle is different from 45°, as shown in Fig. 10.15.

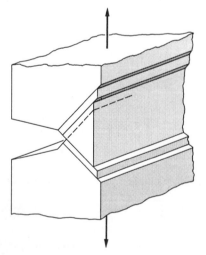

Fig. 10.14. Alternating shear in a singly grooved specimen.

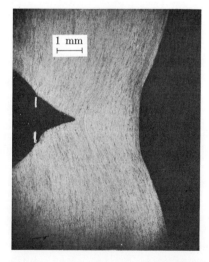

Fig. 10.15. Cross section showing deformation in a singly grooved plate of annealed aluminum after vertical tension. Marks indicate original notch depth.

FIG. 10.16. Prandtl's (1920) flow field for a doubly grooved plate.

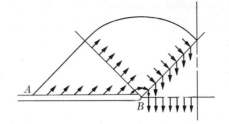

Still another example is the doubly grooved, plane-strain tensile specimen. Both the flow field of Fig. 10.16 and that discussed earlier and presented in Fig. 10.8 satisfy all the relevant equations. In this case, Neimark (1959) has proved that an infinitesimal amount of strain-hardening provides a unique solution, and he has found that the displacements vary approximately linearly along the line AB, giving a solution intermediate between those of Figs. 10.8 and 10.16. The strain in the central region is of the order of the specimen extension divided by the ligament half thickness, although much higher strains are reached in the fans just above and below the tip of the crack.

The regions which remain rigid do have a certain uniqueness, however. Suppose one can find a complete solution, giving a strain distribution satisfying the yield criterion, compatibility, and the stress-strain relations in the deforming region, and satisfying equilibrium and nowhere exceeding the yield in the rigid region. The stress distribution in the rigid region may not be the actual one in an elastic-plastic solution. Even if it is not, however, the region found to be rigid in that solution must indeed be rigid in any solution, including the exact one. In short, the rigid region is uniquely determined in the sense that it must be rigid in all solutions (Bishop et al., 1956). For a general treatment of uniqueness of displacements, see Hill (1958).

Even though the strain distributions are not aways unique, the stress distributions in the deforming region are unique for a rigid-plastic nonhardening material, as indicated by the following uniqueness theorem.

If the upper and lower bounds coincide, and if two or more solutions are found which satisfy all equations in the deforming region, the stress distributions corresponding to the solutions must be identical, except in a region that is rigid in all solutions.

◀ Hill (1951) first proved this theorem as follows. If the theorem were not so, there could be two different distributions of stress in the specimen, σ_{ij}^a and σ_{ij}^b, corresponding to a single load P. The principle of virtual work can be applied to the difference between the two solutions:

$$\int \sum_{i=1}^{3} \sum_{j=1}^{3} (\sigma_{ij}^a - \sigma_{ij}^b)(d\epsilon_{ij}^a - d\epsilon_{ij}^b)\, dV = \sum_{k} (P_k - P_k)(dp_k^a - dp_k^b). \quad (10.25)$$

Since $(P_k - P_k)$ is zero, two possibilities exist: either the integrand changes sign from one region to another, or else it is always zero. As a result of the principle of maximum plastic resistance, Eq. 7.14, the integrand cannot be negative, since

$$\sum_{i=1}^{3}\sum_{j=1}^{3}(\sigma_{ij}^a - \sigma_{ij}^b)\,d\epsilon_{ij}^a > 0$$

and

$$\sum_{i=1}^{3}\sum_{j=1}^{3}(\sigma_{ij}^b - \sigma_{ij}^a)\,d\epsilon_{ij}^b = \sum_{i=1}^{3}\sum_{j=1}^{3}(\sigma_{ij}^a - \sigma_{ij}^b)(-d\epsilon_{ij}^b) > 0.$$

Thus, $(\sigma_{ij}^a - \sigma_{ij}^b)(d\epsilon_{ij}^a - d\epsilon_{ij}^b) = 0$ everywhere. If the region is deforming in one solution, it can deform by an arbitrary amount, so $d\epsilon_{ij}^a - d\epsilon_{ij}^b$ cannot always be zero. Therefore

$$\sigma_{ij}^a - \sigma_{ij}^b = 0 \quad \text{or} \quad \sigma_{ij}^a = \sigma_{ij}^b.$$

That is, the stress is unique (neglecting differences in hydrostatic component), with the possible exception of differences along a straight element of a yield locus, where $(d\epsilon_{ij}^a - d\epsilon_{ij}^b) = 0$.

◀ The possible ambiguity about differences in hydrostatic stress or differences in states of stress along the same line element of a straight-sided yield locus, such as the maximum shear stress criterion, can be removed by considering boundary conditions on the stress where the plastic flow breaks through to a free surface. Of course, the stress distribution may be changed by the superposition of a hydrostatic pressure on the entire problem, but this change does not affect the strains or the load required. ▶

10.9 LESS EXACT APPROXIMATIONS TO THE LIMIT LOAD

Even bounds to the limit load may be rather difficult to obtain. Sometimes, therefore, estimates of the loads required for plastic deformation are made by satisfying the equilibrium or compatibility equations only in a gross way. The results are not rigorous lower and upper bounds, but may be very helpful in estimating the order of magnitude of forces that are to be expected.

As a first example, consider the forging of the bar shown in Fig. 10.17. Assume that it is long enough so that further length changes do not occur, but rather the material is squeezed out sideways. Thus the deformation is one of plane-strain. One might at first be inclined to assume homogeneous deformation in order to obtain an upper bound, requiring the following load to supply the work:

$$P = \frac{2}{\sqrt{3}}\,Ybl. \tag{10.26}$$

Equation 10.26 does not give a true upper bound, for some work goes into friction, as well as into plastic deformation. Handling cases involving friction is outside the scope of limit analysis, for to estimate the work, as desired for an upper bound,

10.9 LESS EXACT APPROXIMATIONS TO THE LIMIT LOAD

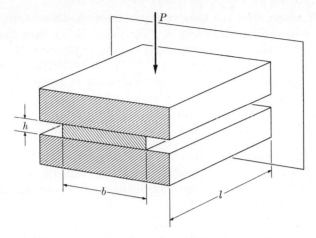

FIG. 10.17. Pressing a bar under plane strain.

requires a knowledge of the normal stress in order to multiply it by the coefficient of friction and get the shear force against which work is done during the frictional sliding. Unfortunately, however, a knowledge of the normal stress is not available in an upper bound solution (see Drucker and Prager, 1952). For a lower bound for this problem, the presence of friction would require a rather complicated stress distribution throughout the part. Instead of satisfying equilibrium in detail, however, one may consider only overall equilibrium for elements consisting of an entire cross section, as shown in Fig. 10.18, where the origin of coordinates is taken at the center of a free face. For this element, the partial differential equations can be changed to ordinary differential equations:

$$\frac{\partial \sigma_{11}}{\partial x_1} + \frac{\partial \sigma_{31}}{\partial x_3} \rightarrow \frac{d\sigma_{11}^{(m)}}{dx_1} + \frac{2\tau}{h} = 0 \quad \text{or} \quad \frac{d(-\sigma_{11}^{(m)})}{dx_1} = \frac{2\tau}{h}. \tag{10.27}$$

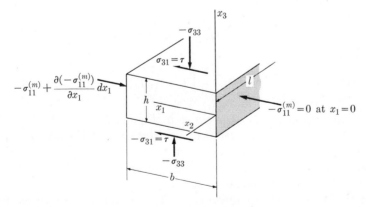

FIG. 10.18. Element of a forged bar to illustrate application of equilibrium in terms of mean stress $\sigma_{11}^{(m)}$.

Equation 10.27 shows that the transverse compressive stress is built up by the friction on the top and bottom surfaces. The stress in the long direction of the bar is found from the plane-strain condition, in terms of the other two components, and turns out to be their average (Prob. 10.10). If other shear components of stress can be neglected, the yield criterion expressed in terms of average stress across the thickness is

$$\sqrt{(\sigma_{11}^{(m)} - \sigma_{33})^2 + 2\tau^2} = 2Y/\sqrt{3}. \tag{10.28}$$

For a value of the coefficient of friction f low enough so that the friction force is small compared with the yield strength of the material, its effect on the yield criterion, Eq. 10.28, can be neglected. With the overall equilibrium equation in the form of Eq. 10.27, then (Prob. 10.19),

$$-\sigma_{33} = \frac{2Y}{\sqrt{3}} e^{2fx_1/h}. \tag{10.29}$$

Integration of this equation gives (Prob. 10.20)

$$P = \frac{2Y}{\sqrt{3}} (e^{fb/h} - 1) \frac{lh}{f}, \tag{10.30}$$

which is valid for

$$(f\sigma_{33})^2 \ll Y^2$$

or (Prob. 10.21)

$$x_1/h \ll -(\ln f)/f. \tag{10.31}$$

The rapid rate of increase of compressive stress given by Eq. 10.29 will soon lead to a shear stress at the interface between the dies and work which is nearly equal to the shear strength. In that case, $\sigma_{33} \to \sigma_{11}$. The equilibrium equation now indicates a linear variation of compressive stress with distance:

$$\frac{d(-\sigma_{11}^{(m)})}{dx_1} = \frac{2Y}{\sqrt{3}} \frac{1}{h}. \tag{10.32}$$

Integration gives the load

$$P = \frac{2Y}{\sqrt{3}} \left(1 + \frac{b}{4h}\right) bl. \tag{10.33}$$

These approximate results are compared in Fig. 10.19 with the exact solution reported by Hill (1950, p. 228) for the case where no sliding occurs between the die and the work piece. Since in actual practice the work will slide relative to the die to some extent out near the end, the approximate solution, which satisfies equilibrium only in a gross sense, is as likely to be satisfactory as is the exact solution of the problem with inexact boundary conditions. The reliability of the solution depends largely on the skill and experience of the analyst in choosing an appropriate approximation to the problem.

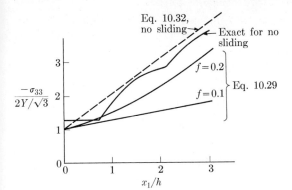

FIG. 10.19. Normal stress on a slab being pressed.

As a second example of approximate plastic analysis satisfying overall, but not local, conditions consider the drawing process shown in Fig. 10.20. From the indicated flow, one could obtain an upper bound. Because the central core has a shorter distance to travel, there will be a shearing as it pulls out ahead of the outer layer. The corresponding strain distribution is tedious to calculate. Therefore one often simply postulates that the reduction in area occurs as it would in the tensile test. From this the energy required per unit of material drawn can be estimated and then the drawing force can be calculated to be large enough to provide this work during the given displacement (Prob. 10.22):

$$P = A_2 Y \ln (A_1/A_2). \tag{10.34}$$

In actual fact, of course, the strain will not be simply one of pure extension and so the load calculated by this procedure will be low, rather than being an upper bound. Furthermore, the added forces associated with friction must be included.

A similar approximate analysis could be applied to extrusion processes, such as shown in Fig. 10.21, but here the results would be even more approximate because of the greater shear strain at the die opening, and because much more friction would be present. It is interesting to note that for the two-dimensional case of drawing (or extrusion) without friction, a die design has just recently been found which

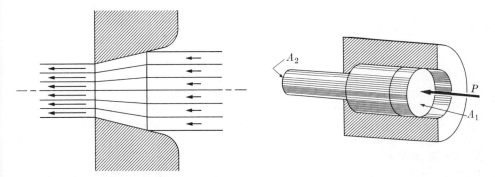

FIG. 10.20. Hypothetical flow field for drawing.

FIG. 10.21. Extrusion.

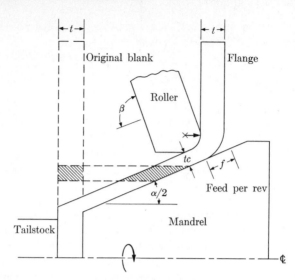

Fig. 10.22. Shear spinning. (Kalpakcioğlu, 1961. Courtesy of American Society of Mechanical Engineers.)

is 100% efficient, so that Eq. 10.34 provides the exact solution (to the theoretical case with no friction) (Richmond and Devenpeck, 1962).

As a final example of approximate plastic analysis using overall equations, consider the shear spinning of a cone, shown in Fig. 10.22. The deformation is primarily shear of the small element passing under the roller. If we assume the deformation to be pure shear (which cannot be exact under these stress boundary conditions, but turns out to be closely approximated in practice), we find the following expression for the force in the tangential direction on the roller, which controls the power requirements (Prob. 10.23):

$$F_T = tf \frac{\bar{\sigma}}{\sqrt{3}} \gamma \sin \frac{\alpha}{2}. \tag{10.35}$$

This equation gives results which are about 20% too low, according to experiments by Kalpakcioğlu (1961).

10.10 SUMMARY

Because exact solutions to problems in the mechanics of deformable solids are so difficult to obtain, it is often useful to seek approximate solutions. Estimates giving upper or lower bounds to the stiffness of an elastic body, or to the limit load of a plastic one, are obtained by finding solutions which satisfy some, but not all, of the basic equations. In the plastic case, if the prior elastic and plastic deformation does not lead to important changes in geometry, such as fracture, buckling, or extensive plastic flow around a notch, the limit load for elastic-plastic materials is the same as that for rigid-plastic materials. Two theorems give lower and upper bounds to the limit load:

1. Lower bounds to the limit load: In a rigid-plastic continuum, there can be no plastic deformation under loads for which a stress distribution can be found which

   (a) everywhere satisfies the equilibrium equations,
   (b) is in equilibrium with the external loads, and
   (c) is everywhere within the yield locus.

2. Upper bounds to the limit load: In a rigid plastic continuum, deformation *must* occur under any system of loads for which a distribution of displacements can be found such that

   (a) displacement boundary conditions, if any, are satisfied
   (b) the displacements can be differentiated to give a strain with no change in volume anywhere, and
   (c) the plastic work required by the strain, and integrated throughout the material is less than the work done by the external loads acting through the assumed displacements:

$$\sum_k P_k \, dp_k > \int_V \bar{\sigma}_Y \, d\bar{\epsilon}^p \, dv. \qquad (10.9)$$

From the first theorem, any solution using elastic stress-strain relations gives a lower bound to the limit load, provided the maximum stress is set equal to the yield strength of the material. This lower bound may be needlessly low, if stress concentrations are high. Upper bounds to the limit load are provided by the second theorem. These theorems are also applicable to strain-hardening materials, provided that appropriate lower or upper bounds to the flow stress at each point can be estimated.

With the aid of these two theorems, it is often possible to obtain practical estimates of the load-carrying capacity of structures and components that would be very difficult to analyze exactly. One of the most important results is that in a plate with deep grooves, or under a hardness indenter, the nominal stress across the minimum cross-sectional area or contact area can be about three times the tensile strength of the material. This high value is due to the fact that adjacent undeforming material provides transverse stresses which tend to prevent deformation under the primary stress. With many other shapes, including a singly grooved plate, the limit load is nearly the same as that of a smooth specimen with the same minimum section.

If one is fortunate enough to choose stress and displacement fields so that the upper and lower bounds given by the two theorems coincide, the limit load is known exactly. The stress distribution is then uniquely determined in any region which is assumed to deform in the upper bound solution. The strain distribution is not uniquely determined, however, the most striking example being that necking can occur anywhere along the length of a long tensile specimen.

◀ In plastic flow under conditions of plane strain, it is convenient to refer the equilibrium and compatibility equations to the curvilinear coordinates along lines of maximum shear stress, giving, in terms of ϕ, the counterclockwise rotation of the α line from the x axis,

$$\begin{aligned}\text{along an } \alpha \text{ line, } \sigma - 2k\phi &= \text{constant,} \\ \text{along a } \beta \text{ line, } \sigma + 2k\phi &= \text{constant,}\end{aligned} \quad (10.20)$$

$$\begin{aligned}\text{along an } \alpha \text{ line, } d(\delta u) - \delta v \, d\phi &= 0, \\ \text{along a } \beta \text{ line, } d(\delta v) + \delta u \, d\phi &= 0.\end{aligned} \quad (10.24) \blacktriangleright$$

Even more approximate methods for estimating the loads for plastic deformation are often useful. For example, in drawing and extrusion, the applied forces can be estimated from the work per unit volume required to produce the same reduction of area as in the tensile test, giving

$$P = A_2 Y \ln (A_1/A_2). \tag{10.34}$$

This is not an upper bound to the force because the compatibility equations have not been satisfied; in fact, it underestimates the loads because the actual history of straining is not as direct as in the tensile test. Alternatively, in a case such as the forging of thin disks, estimates of the forces are sometimes made on the basis of satisfying the equilibrium conditions for relatively large elements. These do not constitute a true lower bound, for the equilibrium conditions have not necessarily been satisfied at each point. When the friction is high enough to cause shear in the material, the load for forging a plate of width b and thickness h, with no change in length, is

◀
$$P = \frac{2Y}{\sqrt{3}} \left(1 + \frac{b}{4h}\right) bl. \tag{10.33} \blacktriangleright$$

REFERENCES

For further information on the limit analysis theorems, including applications to design, see the small book by Prager (1959). The analysis of slip line fields under plane-strain conditions is also presented there, although the treatment by Hill (1950, pp. 128–138) may be preferred. A concise but complete review of plasticity was given by Drucker (1960).

Bishop, J. F. W.	1953	"On the Complete Solution to Problems of Deformation of a Plastic-Rigid Material," *J. Mech. Phys. Solids* **2**, 43–53.
Bishop, J. F. W. Green, A. P. Hill, R.	1956	"A Note on the Deformable Region in a Rigid Plastic Body," *J. Mech. Phys. Solids* **4**, 256–258.

CALLADINE, C. R. DRUCKER, D. C.	1962	"A Bound Method for Creep Analysis of Structures: Direct Use of Solutions in Elasticity and Plasticity," *J. Mech. Eng. Sci.* **4**, 1–11.
DRUCKER, D. C.	1960	"Plasticity," *Structural Mechanics*, E. H. Lee and P. S. Symonds, eds., Pergamon Press, London, pp. 407–456.
DRUCKER, D. C. GREENBERG, H. J. PRAGER, W.	1951	"The Safety Factor of an Elastic-Plastic Body in Plane Strain," *J. Appl. Mech.* **18**, 371–378.
DRUCKER, D. C. PRAGER, W.	1952	"Soil Mechanics and Plastic Analysis of Limit Design," *Quart. Appl. Math.* **10**, 157–165.
ELY, R. E.	1963	"Ultimate Strength of Grooved Flat Bars for Simulated Biaxial Loading," *J. Exp. Mech.* **3**, 152.
HILL, R.	1950	*The Mathematical Theory of Plasticity*, Clarendon Press, Oxford, England.
HILL, R.	1951	"On the State of Stress in a Plastic-Rigid Body at the Yield Point," *Phil. Mag.*, Series 7, **42**, 868–875.
HILL, R.	1958	"A General Theory of Uniqueness and Stability in Elastic-Plastic Solids," *J. Mech. Phys. Solids* **6**, 236–249.
HOFFMAN, O. SACHS, G.	1953	*Introduction to the Theory of Plasticity for Engineers*, McGraw-Hill, New York.
KALPAKCIOĞLU, S.	1961	"The Dynamics of the Shear Spinning Process," *Trans. ASME, J. Eng. Ind.* **B83**, 125–130.
LEVIN, F.	1955	"Indentation Pressure of a Smooth Circular Punch," *Quart. Appl. Math.* **13**, 133–137.
NEIMARK, J. E.	1959	"The Initiation of Ductile Fracture in Tension," Sc.D. Thesis, M.I.T., Cambridge, Mass.
PRAGER, W.	1959	*An Introduction to Plasticity*, Addison-Wesley, Reading, Mass.
PRANDTL, L.	1920	"Über die Härte Plastischer Körper," *Nachrichten von der Königlichen Gesellschaft der Wissenschaften zu Göttingen, Mathematisch-physikalische Klasse*, pp. 74–85.
RICHMOND, O. DEVENPECK, L. M.	1962	"A Die Profile for Maximum Efficiency in Strip Drawing," *Proc. 4th U.S. Nat. Con. Appl. Mech.*, pp. 1053–1057.
SADOWSKY, M. A.	1943	"A Principle of Maximum Plastic Resistance," *J. Appl. Mech.*, **10**, *Trans. A.S.M.E.*, **65**, A65–A68.
SEELY, F. B. SMITH, J. O.	1952	*Advanced Mechanics of Materials*, Wiley, New York, pp. 576–8. See also U.S. Patent Application No. 779252, 1958.
SHIELD, R. T.	1955	"On the Plastic Flow of Metals under Conditions of Axial Symmetry," *Proc. Roy. Soc. (London)* **A233**, 276–286.
SOKOLNIKOFF, I. S.	1956	*Mathematical Theory of Elasticity*, 2nd ed., McGraw-Hill, New York.
TAYLOR, G. I.	1938	"Plastic Strain in Metals," *J. Inst. Met.* **62**, 307–324.

PROBLEMS

◀ 10.1. In the example following Eq. 10.2, derive the expression for the upper bound to the stiffness of a long bar which is being extended.

◀ 10.2. For linear elastic systems, show that the complementary energy is equal to the strain energy.

◀ 10.3. In the example following Eq. 10.3, derive lower bounds for the stiffness as a function of the stress distribution parameter b.

◀ 10.4. Prove the principle of virtual work, using the Green-Gauss theorem applied to a product of the streses and displacement increments, equilibrium, and the strain-displacement equations.

◀ 10.5. Derive the first equality of Eq. 10.5 for a beam in pure bending, with the bending moment as a generalized stress, and an appropriate generalized strain.

10.6. Derive the upper bound given in Eq. 10.16 for the doubly grooved plate.

10.7. Derive the lower bound width of $(1 + \pi)$ times the minimum thickness to prevent flow in the shoulders of a doubly grooved plate, as indicated in Fig. 10.9.

◀ 10.8. Is the analysis of Taylor (1938), referred to in Section 10.6 for the flow stress of a polycrystal in terms of the single crystallographic slip systems, based on an upper bound, a lower bound, or some other estimate? Why did he choose the combination of slip systems which minimized the work for a given strain?

◀ 10.9. Prove one of the corollaries of the limit analysis theorems given at the end of Section 10.6.

10.10. Prove that under plane-strain conditions, the normal stress component in the direction of zero strain must be the average of the other two, providing that the fully plastic condition has been attained.

10.11. Prove that under plane strain, the maximum shear stress and the Mises yield criteria are identical.

◀10.12. Following the ideas just after Eqs. 10.19, derive the equilibrium equations in terms of coordinates locally parallel to the coordinates of maximum shear stress.

◀10.13. Derive Hencky's first theorem, Eq. 10.21.

10.14. Prove that the flow field for the doubly grooved, plane-strain specimen shown in Fig. 10.8 satisfies the equilibrium equations in the form of Eq. 10.20.

◀10.15. Show that Eq. 10.17 provides a lower bound for the doubly grooved specimen, provided that there is no yielding in the shoulders.

◀10.16. Derive the Geiringer equations (10.24).

◀10.17. Show that the displacement field of the doubly grooved, plane-strain specimen shown in Fig. 10.8 satisfies the Geiringer equations.

10.18. In the singly grooved, plane-strain tensile specimen shown in Fig. 10.14, prove that there must be an average shear strain of unity in the deformation bands, referred to the direction of those bands.

10.19. Derive the equation for the normal stress on a bar being forged, Eq. 10.29.

10.20. Derive the equation for the load corresponding to the above problem.

10.21. Derive the limitation on the specimen dimensions for the results of the Problems 10.19 and 10.20 to be valid.

10.22. Derive the equation for the minimum load required in drawing, Eq. 10.34.

10.23. Derive the equation for the force on the roller in shear spinning, Eq. 10.35.

10.24. A $\frac{1}{4}$-in. thick plate of 6061-T6 aluminum alloy has notches cut into it as shown in Fig. 10.23. How would the maximum load depend on the angle θ of the ligament? (See Ely, 1963.)

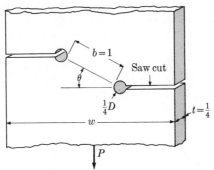

FIGURE 10.23

◀**10.25.** In the lower bound estimate of the stiffness of a bar under an axial load, show that if the axial stress did vary from the inner to the outer layer, compatibility would not be satisfied.

10.26. A round bar has four saw cuts in one cross section, leaving a square section at the root of the cuts. If the material is 2024-T3 aluminum alloy with a Brinell hardness of 140, what can you say about the torque required to twist it off? What assumptions are you making? It may be helpful to know that for a non-strainhardening material, the torque required for fully plastic torsion of a straight rod of square cross section is given in terms of the length h of a side of the square, and the yield strength k in shear, by $kh^3/3$.

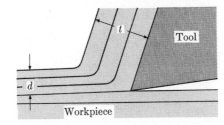

FIGURE 10.24

10.27. Estimate the order of magnitude of the force required in machining to produce a chip of given dimensions, as shown in Fig. 10.24. Note that the shape is not known a priori, and that limit analysis does not apply if the dimensions are not known.

10.28. A new fastening device consists of spiral coils of wire woven into the edge of two pieces of cloth. The coils are slipped into each other and a long pin or wire is pushed through the intersection of the coils, holding them together, as shown in Fig. 10.25. If the coils are $\frac{5}{16}$ in. in diameter, made with 0.028-in. diameter wire, of cold-rolled type 302

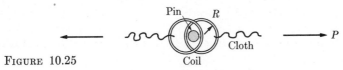

FIGURE 10.25

stainless, what is the maximum load they can carry before excessive deformation occurs? A value of the yield strength of 250,000 psi is given for this material. Note that elastically the maximum moment is 0.318 PR, as given by Seely and Smith (1952). As a partial check on your answer, it is known experimentally that loads of 14 lb per coil can be carried without visible deformation.

E 10.29. Estimate the limit load of a bar of modeling clay containing a transverse hole. What diameter must a smooth bar have to give it approximately the same limit load as the bar with the transverse hole? Check it experimentally by pulling a bar having these two sections in series.

E 10.30. Repeat Prob. 10.29 with a specimen having a single notch cut deeply into one side, and also with two notches cut in from opposite sides. How deep must the pair of notches be in order to prevent plastic deformation in the shoulder? Is your model approximately plane stress, plane strain, or a mixture? Do you get the same result with different kinds of modeling clay? If not, why not?

E 10.31. A bar of Plasticine, $\frac{1}{2}$ in. in diameter and 2 in. long is to be reduced by forging between plates in a single loading to a thickness of 0.2 in. What will the width be? What is the required load, judging from hardness tests or other means of determining the flow stress of Plasticine? Experimental results: a pencil weighing 0.2 ounces produced an 0.02-in. diameter indentation. On a specimen twice as large in every dimension as the forging described here, the observed load required was 95 lb. Do these data check your prediction?

10.32. Show that the limit load of a circumferentially notched round bar under torsion is identical with that of a round bar with the same minimum diameter and no notch.

10.33. Estimate the loss in tensile limit load of a round bar of diameter D containing a transverse oil hole of diameter d. Compare your result with Fig. 10.2.

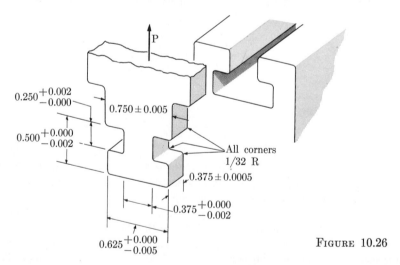

FIGURE 10.26

10.34. Estimate the limit load of a root fastening of a turbine blade of the type shown in Fig. 10.26. As a check on the expected load, tests were run on a prototype of hot-rolled steel having a Rockwell B-scale hardness of 81, which corresponds to a tensile strength of approximately 75,000 psi. The maximum load was found to be 8650 lb. How close to this does your estimate come?

10.35 A 1018 steel sheet with a Brinell hardness of 254 kg/mm² and a thickness of $\frac{1}{8}$ in. is to be punched with an 0.500-in. diameter punch through an 0.506-in. diameter die. Estimate the maximum load. Does it have a lower or upper bound? Compare your result with the experimentally observed value of 11,300 lb. Is this consistent with your analysis? If not, can you explain the reason? How important are bending or friction, in addition to the shearing action which one would ideally expect with a die?

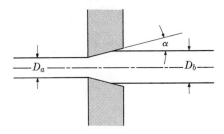

FIGURE 10.27

10.36. Sachs (See Hoffman and Sachs, 1953, p. 180.) gives the following formula for the stress required in wire drawing (see Fig. 10.27):

$$\frac{\sigma_{x_a}}{Y} = \frac{1+B}{B}\left[1 - \left(\frac{D_a}{D_b}\right)^{2B}\right]$$

where $B = f/\tan \alpha$. Show that in the limit as B becomes much less than unity (low friction), this expression reduces to the result that the stress in the drawn wire is equal to the yield strength times the strain corresponding to the reduction in area.

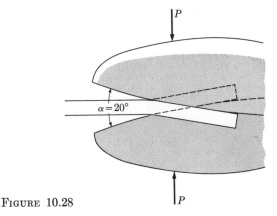

FIGURE 10.28

10.37. Estimate the force required to shear aluminum alloy sheet, as shown in Fig. 10.28. The sheet is 0.045 in. thick and has a Rockwell hardness of R_{30T}, which corresponds to a tensile strength of about 45,000 psi. Neglect the effect of bending. Aside from bending effects, explain why you have an upper or a lower bound, or some other approximation. (The actual force was 160 lb.)

10.38. The Charpy bending impact specimen shown in Fig. 10.29 is made of aluminum. A Rockwell superficial hardness test was performed, giving a hardness number 15T-45 (15-kg major load, $\frac{1}{16}$-in. diameter indenter, penetrating 0.045 mm). The diameter of the impression was also measured and found to be 0.070 cm.

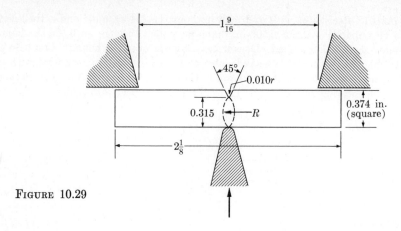

FIGURE 10.29

Estimate how much energy is absorbed by the specimen in forcing it through the supports. You may assume the mode of deformation shown by the dotted lines, where all the slip is concentrated on cylindrical surfaces of radius of curvature R (R may be varied to extremize the results). Other assumptions for modes of deformation are possible. Would your estimate be likely too high or too low? The actual energy absorbed was 64 ft-lb. Explain quantitatively the reasons for differences.

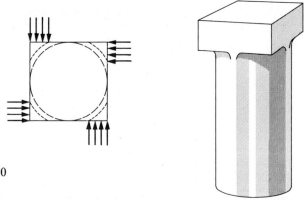

FIGURE 10.30

10.39. A square-headed bolt (Fig. 10.30) has a dimension across the flats equal to the diameter d of the bolt. Due to the design of the wrench, the load is applied only across the outer quarter of the flats, as indicated in the figure. Consider the possibility of the corners shearing off along the indicated dashed circular arcs. What does this tell about the thickness t which the head should have to prevent rounding of the corners before twisting off the shank? Explain your answer on the basis of limit analysis.

10.40. Under what conditions will or may the addition of material raise the limit load?

CHAPTER 11

STRESS AND STRAIN CONCENTRATIONS

11.1 SYNOPSIS

An exact solution for the distributions of stress and strain around a cylindrical hole in an elastic body under biaxial tension leads to several results which are characteristic of strain concentrations in general. The first is that the concentrations of stress and strain die off at distances of the order of a few hole diameters, and thus the effects of geometric discontinuities are localized. This is a special case of St. Venant's principle that if a set of self-equilibrating forces is applied over a given region, the stress due to these forces dies off at distances of the order of the size of the region. This means that the local stresses in structures can be found by determining the nominal stress distribution, neglecting local discontinuities such as fillets and oil holes, and then multiplying the nominal stress by a stress concentration factor. Another result is that the stress distributions in plane stress and plane strain are identical. Examination of the corresponding problem for a plastic material shows that the stress concentration is limited by plastic flow, but the strain concentration may be greater than in the elastic case. The strain concentration is then the more important quantity. Plastic strain concentrations are not only different from elastic strain concentrations, but also the plastic strain concentration in plane stress is quite different from that in plane strain. The strain concentration may die out more slowly with distance, so that St. Venant's principle does not hold for plastic deformation.

Results are given for other forms of stress concentrations, arising from spherical holes, elliptical holes, cracks under various loadings, and the contact of spheres. The spherical hole is helpful in determining density changes in solid-solution alloys and the stress field that a dislocation encounters in the neighborhood of a solute. The case of contact of spheres is useful in interpreting hardness tests on metals and rubber, as well as in understanding the life of ball bearings. A simple formula is given by which one can estimate roughly the magnitude of other kinds of elastic strain concentrations. We conclude by pointing out that strain concentrations have little effect on the initial deflection of a part and sometimes its limit load, but that they do facilitate fracture. Several ways of reducing strain concentrations are also pointed out.

11.2 INTRODUCTION

Many parts of structures and machines are roughly similar to beams, bars, tubes, or plates. Where the similarity is close enough, the distributions of stress and strain and the deformation of these members can be found from the solutions available in the theory of elasticity and plasticity. In most cases, however, there are important differences between the actual conditions in the member and those in the simplified model. For example, the loads may be applied at concentrated points instead of distributed over a section, and there may be fillets at abrupt changes in section where members are fastened together, or where shoulders are required for positioning. The exact analysis of the distribution of stress and strain for the more complicated cases is extremely difficult. Therefore, calculations are usually based on the simplified shapes or loadings, giving *nominal* stresses or strains. The effects of the concentrated loads, holes, or fillets are considered separately. As we shall see, this procedure is possible in elastic stress analysis because discontinuities in loading or cross section affect only a local region. The ratio of the maximum stress near one of these discontinuities to the nominal stress is called the *stress concentration factor*. Once the stress concentration factor is known, the actual stress at a point is found by multiplying the nominal stress by the stress concentration factor. Concentrations of strain can be handled in a similar fashion.

The question arises as to whether the stress or the strain concentration is of more utility, and in fact, which components of either. The answer depends on whether one is concerned with local yielding, brittle fracture, short- or long-life fatigue, or ductile fracture. For these reasons it is well to examine all components of both stress and strain, in both the elastic and plastic cases.

11.3 STRESS AND STRAIN DISTRIBUTIONS AROUND CYLINDRICAL HOLES UNDER BIAXIAL STRESS

The concentration of stress around a small hole in a thick plate under biaxial tension, shown in Fig. 11.1, not only is of practical importance and relatively easy to solve, but also provides an insight into the general character of stress and strain concentration. First we consider the elastic case. Assume that the plate is so thick compared with the diameter of the hole that the strain in the axial direction is determined by the strain in the plate at a distance large compared with the hole diameter. This assumption determines the axial displacement u_z in the plate (Prob. 11.1):

$$u_z = \epsilon_{zz} z = [\sigma_{zz\infty} - 2\nu\sigma_{rr\infty}]z/E. \quad (11.1)$$

Symmetry conditions limit the other displacements to a radial one, depending only on the radius:

$$u_r = u_r(r). \quad (11.2)$$

These displacements may be differentiated

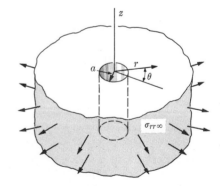

Fig. 11.1. Hole in a thick plate.

to obtain the strain from Eqs. 2.28 and combined with the expressions for stress as a function of strain, Eqs. 3.25, to obtain the components of stress. These components of stress must, in turn, satisfy the equilibrium equations, Eqs. 2.14. Combination of these equations eventually leads to a single differential equation for the radial displacements (Prob. 11.2):

$$\frac{d^2 u_r}{dr^2} + \frac{1}{r}\frac{du_r}{dr} - \frac{u_r}{r^2} = 0. \tag{11.3}$$

The displacement and hence the radial components of stress can be obtained in terms of two constants of integration (Prob. 11.3):

$$\frac{u}{r} = A + \frac{B}{r^2}, \tag{11.4}$$

$$\sigma_{rr} = \frac{E}{1 - \nu - 2\nu^2}\left[A + \nu\epsilon_{zz}\right] - \frac{B}{1+\nu}\frac{1}{r^2},$$
$$\sigma_{\theta\theta} = \frac{E}{1 - \nu - 2\nu^2}\left[A + \nu\epsilon_{zz}\right] + \frac{B}{1+\nu}\frac{1}{r^2}. \tag{11.5}$$

These equations are useful for a variety of problems, including stress distributions around shrink fits, the behavior of thick-walled cylinders under internal pressure, and fracture by the increase of porosity, depending on how the constants of integration are evaluated. In our case, they are found by satisfying the boundary conditions on the radial components of stress: zero at the surface of the hole, and the applied stress $\sigma_{rr\infty}$ at $r = \infty$, giving

$$\sigma_{rr} = \sigma_{rr\infty}(1 - a^2/r^2),$$
$$\sigma_{\theta\theta} = \sigma_{rr\infty}(1 + a^2/r^2), \tag{11.6}$$
$$\sigma_{zz} = E\epsilon_{zz\infty} + 2\nu\sigma_{rr\infty}.$$

These equations show that the radial stress increases monotonically with radius while the tangential stress decreases, both reaching the value of the applied biaxial stress at infinity. The axial stress turns out to be a constant. The axial strain does not affect the radial or tangential components of stress. Thus, the stress components in the plane are the same for plane strain ($\epsilon_{zz} = 0$) and plane stress ($\sigma_{zz} = 0$). This is a general result in elasticity theory, as discussed by Timoshenko and Goodier (1951, p. 25).

Various resulting stress and strain concentrations are given in Table 11.1 (Prob. 11.4). Before discussing these further, we shall turn to the plastic case.

◀In considering plastic states of stress and strain around holes, the strain distribution in a rigid-plastic material under plane strain is a good starting point. Here the axial component of strain vanishes; for incompressibility the radial and tangential normal components of strain must then be equal and opposite. From the definition of strain in terms of displacement, noting that the tangential dis-

TABLE 11.1

Stress and Strain Concentration Factors for Round Holes under Biaxial Stress

(Poisson's ratio $\nu = 0.3$ for numerical examples)

	Plane Stress	Plane Strain
Elastic		
Maximum stress, $\sigma_{\theta\theta max}/\sigma_{rr\infty}$	2	2
Equivalent stress, $\bar{\sigma}_{max}/\bar{\sigma}_\infty$	2	$\dfrac{2\sqrt{1 - \nu + \nu^2}}{1 - 2\nu} = 4.4$
Maximum strain, $\epsilon_{\theta\theta}/\epsilon_{rr\infty}$	$\dfrac{2}{1 - \nu} = 2.9$	$\dfrac{2(1 - \nu)}{(1 - 2\nu)} = 3.5$
Equivalent strain,* $\bar{\epsilon}_{max}/\bar{\epsilon}_\infty$	2	$\dfrac{2\sqrt{1 - \nu + \nu^2}}{1 - 2\nu} = 4.4$
Rigid plastic, max. shear stress yield criterion and associated flow rule		
Maximum stress, $(\sigma_{\theta\theta})_{r=a}/(\sigma_{rr})_{r=r_2}$	1	$1/\ln(r_2/a)$
Equivalent stress	1	1
Maximum strain, $\epsilon_{\theta\theta max}/(\epsilon_{rr})_{r=r_2}$	r_2/a	$(r_2/a)^2$
Equivalent strain	r_2/a	$(r_2/a)^2$

* Based on form of Eq. 7.5 but with factor $1/2(1 + \nu)^2$ replacing $\tfrac{2}{9}$ to give a value corresponding to uniaxial strain in the tensile test.

placements must vanish from symmetry conditions, we find that

$$\partial u/\partial r + u/r = 0. \tag{11.7}$$

On integration and substitution of the boundary condition $u = u_2$ at a large radius $r = r_2$, the strain distribution can be found (Prob. 11.5):

$$\epsilon_{\theta\theta} = -\epsilon_{rr} = u_2(r_2/r)^2. \tag{11.8}$$

Note that in this case the strain distribution could be determined without solving for the stress distribution.
◀The stress distribution is found from the equilibrium equations and the yield criterion. The simplest solution is obtained by assuming the maximum shear stress yield criterion (Fig. 7.2). The associated flow rule and the strain distribution confine the stress to the segment of the yield locus for which

$$\sigma_{\theta\theta} - \sigma_{rr} = 2k.$$

Integration of the equilibrium equations with this yield criterion, subject to the boundary condition of zero stress at the periphery of the hole, $r = a$, gives the stress distribution (Prob. 11.6)

$$\sigma_{rr} = 2k \ln r/a,$$
$$\sigma_{\theta\theta} = 2k(1 + \ln r/a). \tag{11.9}$$

The strain and stress concentration factors corresponding to Eqs. 11.8 and 11.9 are presented in Table 11.1, and will be discussed after deriving the plane stress case.

◀ Under plane stress conditions, with the maximum shear stress yield criterion and the associated flow rule, one might guess that the axial and tangential stress components will control the yielding, so the tangential stress will have to be constant at Y. Integration of the equilibrium equations shows that the radial stress is indeed intermediate between the other two, justifying the guess and giving the result (Prob. 11.7)

$$\sigma_{rr} = Y(1 - a/r),$$
$$\sigma_{\theta\theta} = Y, \tag{11.10}$$
$$\sigma_{zz} = 0.$$

◀ Since the radial stress is intermediate, the flow rule associated with the maximum shear stress yield criterion shows that the deformation will consist of axial contraction and circumferential elongation, with no radial strain. Noting from symmetry that the tangential components of the displacement must vanish, and introducing the strain displacement equation, gives (Prob. 11.8)

$$\epsilon_{rr} = 0$$
$$\epsilon_{\theta\theta} = -\epsilon_{zz} = (\epsilon_{\theta\theta})_{r=r_2}(r_2/r). \tag{11.11}$$

For elastic-plastic solutions, the reader is referred to Hill (1950) or Prager and Hodge (1951). ▶

The stress and strain concentrations discussed above and presented in Table 11.1 illustrate a number of generalizations which also apply to more complicated situations. First, provided that the part is large compared with the radius of the hole, the elastic stress and strain concentration factors are constants. Reference to Eq. 11.6 shows that in this case "large enough" is of the order of a distance equal to three times the radius of the hole. This is a special case of the general result that if a set of forces acting on a given boundary is redistributed without changing the resulting force or moment, the effect of the change will die out at a distance equal to a few times the dimension of the boundary. This principle, known as *St. Venant's principle*, was first postulated intuitively and is still being discussed both qualitatively (Donnell, 1962) and quantitatively (Horvay, 1957). In this case, one may apply St. Venant's principle by thinking of a plate without a hole to which counterbalancing forces are applied where the hole would be, as shown in Fig. 11.2.

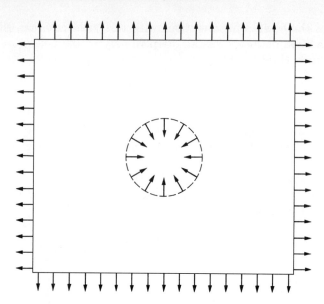

Fig. 11.2. Effect of a hole simulated by a self-equilibrating set of forces.

In an incompressible ($\nu = \frac{1}{2}$) material under plane strain, the strain concentration is infinite (Prob. 11.9).

In the plastic condition, the stress and strain concentration factors depend on the geometry of the piece, partly because of the incompressibility mentioned above. The equivalent stress is an exception, of course, for it must be constant throughout a plastically deforming region. The concept of a strain concentration factor is of relatively little value because it depends on the macroscopic geometry of the part, rather than depending solely on the local geometry of the stress raiser.

Another difference between the elastic and plastic cases is that for the elastic solutions based on stress components in the plane, there is no difference between the plane stress and the plane strain factors. In the plastic case, there is a marked difference between the two. To gain further insight into stress concentrations, we now consider holes under uniaxial stress.

11.4 STRESS AND STRAIN DISTRIBUTIONS AROUND CYLINDRICAL HOLES UNDER UNIAXIAL STRESS

The stress around a hole with *uniaxial* rather than biaxial tension applied at infinity is of more interest, because from it, by the principle of superposition (Section 3.4), the elastic stress distributions can be obtained for any applied state of stress (Prob. 11.10). The solution to the uniaxial elastic case is more conveniently carried out with advanced techniques, and the interested reader is referred to Timoshenko and Goodier (1951, p. 80). The results are given in terms of the angle θ measured from the axis of tension:

$$\frac{\sigma_{rr}}{\sigma_\infty} = \frac{1}{2}\left(1 - \frac{a^2}{r^2}\right) + \frac{1}{2}\left(1 - 4\frac{a^2}{r^2} + 3\frac{a^4}{r^4}\right)\cos 2\theta,$$

$$\frac{\sigma_{\theta\theta}}{\sigma_\infty} = \frac{1}{2}\left(1 + \frac{a^2}{r^2}\right) - \frac{1}{2}\left(1 + 3\frac{a^4}{r^4}\right)\cos 2\theta,$$

$$\frac{\sigma_{r\theta}}{\sigma_\infty} = -\frac{1}{2}\left(1 + 2\frac{a^2}{r^2} - 3\frac{a^4}{r^4}\right)\sin 2\theta,$$

$$\frac{\sigma_{zz}}{\sigma_\infty} = \nu + \frac{E\epsilon_{zz\infty}}{\sigma_\infty}.$$

(11.12)

Here the maximum stress concentration factor is 3. Along the edge of the hole at $\theta = 0$ and $180°$ there exists a compressive stress; this stress can produce buckling in thin sheets.

For a hole in a rigid-plastic, non-strainhardening material under uniaxial stress at infinity, for the conditions to be those corresponding to plane stress it is necessary that the diameter of the hole be large compared with the thickness of the sheet. Under these conditions, there will be no stress concentration, and the resulting deformation will be made up of the modes which cause necking in a thin sheet, as discussed in Section 8.7 (Prob. 11.11).

In a plate thick compared with the diameter of the hole, one may assume plane strain conditions. The mode of deformation becomes that associated with the plane strain deformation of a singly grooved plate, discussed in Section 10.8. The stress in front of the notch is constant at the plane strain yield stress, $2Y/\sqrt{3}$, while the shear strain in the deforming region referred to the 45° coordinates is unity, as discussed in Section 10.8. It is interesting to note that in this case the strain does not increase with extension; instead the region of deformation grows. Stress and strain concentration factors for round holes under uniaxial stress are summarized in Table 11.2.

The example of the round hole under uniaxial stress further demonstrates the applicability of St. Venant's principle in the elastic regime and its breakdown in the plastic regime. In contrast to biaxial stress, the uniaxial stress concentration factors are really the same for all elastic cases. It is for this reason that one seldom sees them distinguished in the literature, since in most cases there is at least one free surface at infinity. The lack of *stress* concentration in non-strainhardening, rigid-plastic material may be noted again. The *strain* concentrations all become infinite, because for a specimen which is infinitely long the average strain goes to zero with a finite deformation in the narrow deforming band. It is further worth noting that the character of the strain concentration is entirely different in the plane stress and plane strain regimes, much more so that in the elastic case. In the fully plastic condition, therefore, a concept of strain concentration is not really applicable, and it is better to define the strain directly.

TABLE 11.2

Stress and Strain Concentration Factors for Round Holes under Uniaxial Stress
(Poisson's ratio $\nu = 0.3$ for numerical examples)

	Plane stress	Plane strain
Elastic		
Maximum stress	3	3
Equivalent stress	3	3
Maximum strain	3	3
Equivalent strain	3	3
Rigid-Plastic, Non-strain-hardening		
Maximum stress	1	1
Equivalent stress	1	1
Maximum strain (absolute)	$\approx u_1/t$	$\gamma_{45°} = 1$
Equivalent strain, $\bar{\gamma}$ (absolute)	$\approx 2u_1/t$	1

One is tempted to generalize, and conclude that the plastic strain concentration factor is always greater than the elastic. This is not necessarily so, for consider a tensile specimen with fillets at the shoulders. In the elastic regime, the fillets cause a strain concentration, which reaches a maximum on the curved portion. When fully plastic flow develops, as indicated in Fig. 8.16, the presence of the shoulder prevents plastic flow there. Thus for this specimen, the plastic strain concentration factor ($= 0$) and the plastic stress concentration factor (< 1) are both less than the elastic ones.

The plastic strain distributions discussed above are valid only for non-strain-hardening materials, since, as might be imagined, the presence of a small amount of strain-hardening will spread the plastic deformation out into adjacent unhardened material. Since actual materials do strain-harden somewhat, their behavior will be intermediate between the elastic behavior and that for a nonhardening material. Solutions for some hardening materials have been summarized by Hill (1950) and further work has been reported by Budiansky and Mangasarian (1960) and Davis (1963). As pointed out by McClintock and Rhee (1962), and as will be seen in Section 13.4 in connection with indentation tests, where the plastic strain distribution is already spread out, one can interpolate between the elastic and the nonhardening solutions fairly satisfactorily. If, on the other hand, the plastic deformation is theoretically confined to a small band, then a relatively small amount of strain-hardening may cause large deviations from the nonhardening results. An extreme example is a circumferentially notched bar in torsion, in which for a nonhardening material all the twist is concentrated across the single plane containing the root of the notch.

11.5 STRESS AND STRAIN CONCENTRATIONS AROUND SPHERICAL HOLES

◀The distribution of stress around a spherical hole is found by working from the equilibrium equations and the strain displacement equations in terms of spherical coordinates, Eqs. 2.15 and 2.29. The differential equation for the displacement is

$$\frac{d^2 u_r}{dr^2} + \frac{2}{r}\frac{du_r}{dr} - \frac{2u_r}{r^2} = 0. \tag{11.13}$$

On integration, one finds that the displacements must be of the form (Prob. 11.12)

$$u_r = \frac{C_1}{r^2} + C_2 r. \tag{11.14}$$

With appropriate boundary conditions this gives the displacement around a misfitting spherically symmetrical solute atom, Eq. 5.1. The stress distribution is (Prob. 11.13)

$$\sigma_{rr} = -\frac{2EC_1}{(1+\nu)r^3} + \frac{EC_2}{1-2\nu},$$

$$\sigma_{\theta\theta} = \frac{EC_1}{(1+\nu)r^3} + \frac{EC_2}{1-2\nu}. \tag{11.15}$$

◀For a hole in material subject to a uniform triaxial tensile stress at infinity, the maximum stress concentration factor is $\frac{3}{2}$ (Prob. 11.14). For uniaxial tension, with $\nu = 0.3$, it turns out to be $\frac{45}{22}$ (Timoshenko and Goodier, 1951, pp. 359–361). These factors are lower than the corresponding ones for the cylindrical hole, which is not surprising, since the force which must be diverted around the hole can now "flow" around it on *all* sides rather than only on two.

◀In studying the effects of combined stress, one needs to know not only the maximum stress but also the other component at the equator and the stress at the pole. In terms of the coordinates of Fig. 11.3 and Poisson's ratio ν,

$$\sigma_{\theta\theta}^A = \sigma_{\phi\phi}^A = -\frac{3+15\nu}{2(7-5\nu)}\sigma_\infty,$$

$$\sigma_{\theta\theta}^B = \frac{27-15\nu}{2(7-5\nu)}\sigma_\infty, \tag{11.16}$$

$$\sigma_{\phi\phi}^B = \frac{15\nu-3}{2(7-5\nu)}\sigma_\infty.$$

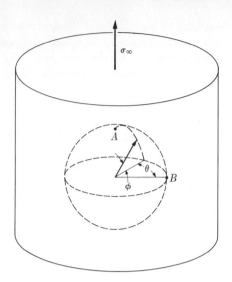

Fig. 11.3. Coordinates for stress around a spherical hole.

◂Changes in lattice spacing in solid solutions can be estimated with the aid of Eqs. 11.14 and 11.15 (Friedel, 1955; Eshelby, 1954). Suppose that solute atoms with radius a_2, modulus E_2, and Poisson ratio ν_2 are substituted into holes of radius a_1 in a solvent having modulus E_1 and Poisson ratio ν_1. Assume that the materials are isotropic, that the misfit is small enough so that the small strain theory of elasticity may be used, and that the electronic interactions between the unlike atoms cause no further changes in size. Then, for the solvent and solute to fit, their radial displacements at their common boundary $r = a$ must add up to the original misfit, and the radial components of stress at the common boundary must be equal. For a solute atom in an infinite matrix with no stress at infinity, this leads to (Prob. 11.15)

$$u_r = \frac{\dfrac{(a_2 - a_1)}{a} \dfrac{a^3}{r^2}}{1 + \dfrac{2E_1}{(1 + \nu_1)} \dfrac{(1 - 2\nu_2)}{E_2}} \qquad (r > a). \tag{11.17}$$

The volume change of the material initially within a radius R is then $4\pi R^2 u_r(R)$, leading to a volumetric strain of the corresponding sphere of $3u_r/R$ or an average linear strain of the lattice points within the sphere of u_r/R. This is the strain due to one solute atom; one might estimate that the strain due to an atomic concentration c would be greater in proportion to the number of solute atoms

$$n = \frac{\frac{4}{3}\pi R^3 c}{\frac{4}{3}\pi a^3}.$$

The average lattice strain is then finally

$$\epsilon = \frac{c(a_2 - a_1)/a}{1 + \dfrac{2E_1}{(1 + \nu_1)} \dfrac{(1 - 2\nu_2)}{E_2}}. \tag{11.18}$$

Actually, the strain is somewhat greater than this because in a finite sphere the stress drops to zero at the free boundary at a finite radius. When this is taken into account for a sphere of radius R, the strain is greater by the factor (Prob. 11.16; Eshelby, 1954)

$$\frac{3(1-\nu_1)}{1+\nu_1} = 1.5 \qquad (\nu_1 = \tfrac{1}{3}). \tag{11.19}$$

Eshelby also shows that the result is not affected if the solute atoms are randomly distributed, rather than concentrated at the center as assumed here. These equations may now be used to predict deviations from the empirical equation that the lattice spacing l varies linearly with concentration (Vegard's law; Barrett, 1952, p. 221):

$$l - l_1 = c(a_2 - a_1). \tag{11.20}$$

Such a comparison has been made by Friedel (1955), who used this theory to predict the deviations from Vegard's law for different solutes in Al, Cu, Ag, and Au. The fit was within a factor of two in 16 of all 32 cases, and also in 8 of the 9 cases in which the solvent and solute came from the same column of the periodic table.

◀ The stress analysis for a spherical hole also provides information on the elastic energy associated with the formation of a precipitate particle from a solid solution. It is interesting to note that for either a circularly cylindrical or a spherical precipitate or inclusion, the stress and strain are constant within the inclusion, even though they are changing without. A similar result has been observed by Eshelby (1957) for elliptical inclusions. ▶

11.6 STRESS AND STRAIN DISTRIBUTIONS AROUND ELLIPTICAL HOLES AND SHARP CRACKS

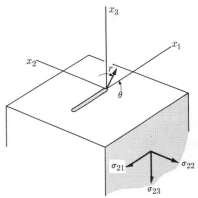

FIG. 11.4. A crack in an infinite solid.

A. Elastic Material. As one might expect from the effect of a circular hole, an elliptical hole or a crack will produce a high stress concentration. The stress distribution for an arbitrary mode of loading and shape of body and crack would be quite difficult to determine. Fortunately it has recently been recognized that near the tip of the crack, essentially only three things can occur: the faces can be pulled apart, or sheared perpendicular or parallel to the leading edge of the crack. These correspond to the three stress components indicated for the crack in Fig. 11.4. By studying these three modes and superimposing them to determine the state for more complex loading, we can obtain valuable insight into fracture processes. The stress distri-

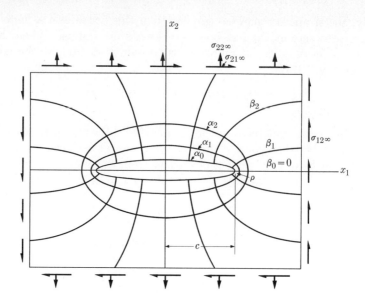

Fig. 11.5. Elliptical coordinates for a crack.

butions will be given in detail below; suffice it to say here that the stress concentrations under tension and under shear parallel to the leading edge produce maxima at the very tip, while shear normal to the leading edge gives a tension peak on one side and a compression peak on the other side of the plane of the crack. An approximate method of estimating the stress concentration will be given in Section 11.8.

◀The first solution was that of Inglis (1913) for an elliptical hole, which in the limit becomes a sharp crack. The analysis is more complex than that for a circular hole, but, again, it is carried out in terms of curvilinear coordinates which simplify the boundary conditions. The elliptical coordinates shown in Fig. 11.5 are used. While the general expression for this stress distribution is complicated, the primary interest is in the stress at the surface of the crack. Here the only component will be the normal component parallel to the surface, $\sigma_{\beta\beta}$. The elliptical coordinates are defined in terms of a crack half-length c, the radius of curvature a at the tip, and the slope dx_2/dx_1 of the surface, by

$$\alpha_0 = (a/c)^{1/2}, \quad \beta = -\tan^{-1}[(\tanh \alpha_0)/(dx_2/dx_1)]. \tag{11.21}$$

For a very sharp crack, the radius of curvature of the tip is small compared with the crack half-length, and hence the coordinate α_0 is small compared with unity. The maximum stress is then found to occur close to the tip of the crack where the values of β are also small compared with unity. Significant stress concentrations arise from the normal component $\sigma_{22\infty}$ and the shear component $\sigma_{21\infty}$. In this case the stress component $\sigma_{\beta\beta_0}$ at the surface is approximately given by

$$\sigma_{\beta\beta_0} = \frac{2\alpha_0\sigma_{22\infty} - 2\beta\sigma_{21\infty}}{\alpha_0^2 + \beta^2}. \tag{11.22}$$

The stress distribution in and around elliptical inhomogeneities has been studied by Eshelby (1957). Oddly enough, it turns out that while there is a stress concentration in the matrix, which for sharp inclusions is about the ratio of Young's moduli, in the inclusion itself there is no concentration of stress.

◀ The stress distribution beneath the surface of the crack is more easily seen by considering the limiting case of a sharp crack, shown in Fig. 11.4. The three kinds of loading give rise to three components of stress concentration: a component symmetrical about the plane of the crack, an antisymmetrical component arising from shear transverse to the edge of the crack, and a longitudinal shear component along the leading edge of the crack.* The magnitude of each component is given by a stress intensity factor k.† For the symmetrical case (Williams, 1957), we have

$$\sigma_{rr} = \frac{k_1}{\sqrt{2r}} \left[+\frac{5}{4} \cos \frac{\theta}{2} - \frac{1}{4} \cos \frac{3\theta}{2} \right],$$

$$\sigma_{\theta\theta} = \frac{k_1}{\sqrt{2r}} \left[+\frac{3}{4} \cos \frac{\theta}{2} + \frac{1}{4} \cos \frac{3\theta}{2} \right], \tag{11.23}$$

$$\sigma_{r\theta} = \frac{k_1}{\sqrt{2r}} \left[+\frac{1}{4} \sin \frac{\theta}{2} + \frac{1}{4} \sin \frac{3\theta}{2} \right].$$

In the antisymmetrical case (Williams, 1957), we get

$$\sigma_{rr} = \frac{k_2}{\sqrt{2r}} \left[-\frac{5}{4} \sin \frac{\theta}{2} + \frac{3}{4} \sin \frac{3\theta}{2} \right],$$

$$\sigma_{\theta\theta} = \frac{k_2}{\sqrt{2r}} \left[-\frac{3}{4} \sin \frac{\theta}{2} - \frac{3}{4} \sin \frac{3\theta}{2} \right], \tag{11.24}$$

$$\sigma_{r\theta} = \frac{k_2}{\sqrt{2r}} \left[\frac{1}{4} \cos \frac{\theta}{2} + \frac{3}{4} \cos \frac{3\theta}{2} \right].$$

These two cases correspond to the normal and shear components applied to Inglis' elliptical crack. In the antiplane strain case, where the tip of the crack is under longitudinal shear, the stress distribution can be found from a Schwarz-Christoffel

---

* The antisymmetrical and longitudinal shear components are sometimes called "skew-symmetric" and "antiplane" strain, respectively. In working with long cracks in thin plates, Erdoğan and Sih (1963) have used opposite names, calling them longitudinal and transverse shear (relative to the crack, not the leading edge as here).

† The symbols $\mathcal{K}$, $K/\sqrt{\pi}$, and $a\sqrt{2}$ are also used for the stress intensity factor k.

TABLE 11.3

Stress Intensity Factors, k_i*

Configuration	k_1	k_3	Ref.
(through crack in infinite plate, width $2c$, tension σ_∞)	$\sigma_\infty c^{1/2}$	$\sigma_\infty c^{1/2}$	Irwin (1957)
(edge crack of depth c, semi-infinite plate, tension σ_∞)	$1.1 \sigma_\infty c^{1/2}$	$\sigma_\infty c^{1/2}$	Irwin (1962a) Wigglesworth (1957)
(central crack $2c$ in finite plate width w, tension σ_∞)	$\approx \sigma_\infty c^{1/2} \left(\dfrac{w}{\pi c} \tan \dfrac{\pi c}{w} \right)^{1/2}$	$= \sigma_\infty c^{1/2} \left(\dfrac{w}{\pi c} \tan \dfrac{\pi c}{w} \right)^{1/2}$	Irwin (1957)
(two edge cracks of depth c each in plate width w, tension σ_∞)	$\approx \sigma c^{1/2} \sqrt{\dfrac{w}{\pi c} \tan \dfrac{\pi c}{w}} \times \sqrt{1 + 0.2 \cos \dfrac{\pi c}{w}}$	$= \sigma_\infty c^{1/2} \left(\dfrac{w}{\pi c} \tan \dfrac{\pi c}{w} \right)$	ASTM (1960)
(edge crack with remote circular boundary, tension σ_∞)	$\approx 1.1 \sigma c^{1/2}$		Irwin (1962b)
(edge crack with bending stress distribution, $\sigma_{\infty\max}$)	plot in Ref.		Winne and Wundt (1958)

*See also Paris and Sih in Fracture Toughness Testing, ASTM, 1965.

TABLE 11.3 cont.

Configuration	k_1	k_3	Ref.
	see Ref. p. 649		Irwin et al. (1958)
	$\dfrac{P}{2ct}c^{1/2}\left[\dfrac{2}{\pi}\sqrt{\dfrac{c+b}{c-b}}\right]$		Irwin (1957)

transformation (see, for example, Hult and McClintock, 1957):

$$\sigma_{\theta z} = \frac{k_3 \cos(\theta/2)}{\sqrt{2r}}, \qquad \sigma_{rz} = \frac{k_3 \sin(\theta/2)}{\sqrt{2r}}. \tag{11.25}$$

◀ The stress intensity factors k_i in Eqs. 11.23 to 11.25 are determined by the loads, the general shape of the body, and the path of the crack. For a crack of length $2c$ in an infinite solid, the coefficients are

$$k_1 = \sigma_{22\infty}\sqrt{c}, \qquad k_2 = \sigma_{21\infty}\sqrt{c}, \qquad k_3 = \sigma_{23\infty}\sqrt{c}. \tag{11.26}$$

Other cases are given in Table 11.3. Further examples are given in recent work by Erdoğan et al. (1962), Ang et al. (1963), and references given in those papers. Another method is to work from information on stress concentration factors obtained for notches similar in shape to the cracks being studied, but with finite root radii. For instance, by comparison with the solution for an elliptical hole in a plate under shear or tension, giving a maximum stress $\sigma_{\beta\beta\text{max}}$ near the tip of a notch with radius a (Prob. 11.17),

$$k_1 = \frac{\sigma_{\beta\beta\text{max}}\sqrt{a}}{2}, \qquad k_2 = \sigma_{\beta\beta\text{max}}\sqrt{a}. \tag{11.27}$$

Likewise, for grooves under longitudinal shear with a maximum stress $\sigma_{23\text{max}}$ at a root of radius a (Prob. 11.18),

$$k_3 = \sigma_{23\text{max}}\sqrt{a}. \tag{11.28}$$

Equations 11.27 and 11.28 hold not only for the cracks in an infinite medium, for which they were derived, but also for other cases, since they deal with the local stress distribution at the tip of cracks or notches. They may therefore be used to estimate crack tip stress intensities from stress concentration data, and conversely.

◀We now return to a detailed examination of the character of the stress in front of a crack, considering the symmetrical case corresponding to tension. Williams noted that the maximum value of the shear stress at a given radius was obtained at 90°, and the maximum equivalent stress was obtained at 70° to the direction of the crack; thus, plastic flow will tend to occur on either side of the crack rather than directly in front of it. The reason that the maximum equivalent stress occurs off to the side is that the radial component of stress directly ahead of the crack is equal to the tangential component. At first sight this may seem to be in contradiction with the fact that at the very tip of the crack the radial component of stress must be zero, but a detailed examination of the stress distribution for the elliptical case shows that the radial component of stress rises very rapidly from the surface and soon reaches a value equal to the tangential component. Under plane strain conditions, this biaxial tension in front of the crack will lead to a high transverse stress as well. Thus the high triaxial tension in front of a notch, which has been observed in the plastic case (Section 10.6), is not unique to the plastic condition. The tendency for brittle fracture might be represented by the maximum value of the tensile stress, by the maximum normal component across a radial line, or by the maximum value of the dilatation (volumetric strain) at a certain radius from the tip of the crack. The first of these reaches a maximum at an angle of 60° from the plane of the crack, whereas the other two are maximized along the plane of the crack.▶

B. Plastic Material. The fully plastic strain distribution for a non-strainhardening material containing a sharp crack depends fundamentally on whether the crack is an internal or an external one (Fig. 11.6). As discussed in Section 10.8, with the internal crack the strain is constant throughout a band that widens as the test proceeds. For the external crack, the strain is distributed throughout the ligament and increases with specimen extension. A plane stress condition does not exist for the sharp crack, because any thickness of material is large compared with a zero radius of curvature.

◀An explicit elastic-plastic solution has only been obtained for the case of longitudinal shear, parallel to the leading edge of the crack. As indicated in Fig. 11.7, at low stress levels the plastic zone is circular in shape, just touching the tip of the

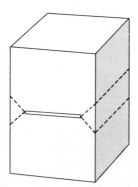
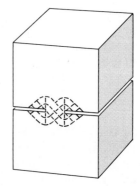

Fig. 11.6. Contrast in plane strain plastic flow field for internal and external cracks.

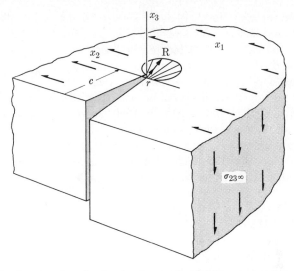

FIG. 11.7. Plastic zone for elastic-plastic, nonhardening material under longitudinal shear.

crack. The maximum radius from the tip of the notch to the end of the plastic zone, which is directly along the line of a notch, is given approximately in terms of the crack depth c, the applied stress at infinity $\sigma_{23\infty}$, and the yield strength in shear k of the material, by (Prob. 11.19)

$$R_{\max} = c(\sigma_{23\infty}/k)^2 = k_3^2/k^2. \qquad (11.29)$$

Within this plastic zone, the distribution of strain is given, in terms of the yield strain in shear k/G, by (Prob. 11.20)

$$\gamma_{z\theta} = \left(\frac{k}{G}\right)\left(\frac{R}{r}\right). \qquad (11.30)$$

At higher stress levels, the plastic zone becomes elongated, and its extent must be determined by numerical integration or in some cases an elliptic integral (Koskinen, 1963; Bilby, et al., 1963).

◀ The strain concentration for a groove of finite radius under longitudinal shear can be found from Eqs. 11.29 and 11.30 by noting that the directions of the shear stress components define a series of stress-free surfaces with various root radii. The corresponding stress and strain concentrations turn out to be (Prob. 11.21)

$$\frac{\gamma_{23\max}}{(\sigma_{23\infty}/G)} = \frac{c}{a}\frac{\sigma_{23\infty}}{k}, \qquad \frac{\sigma_{23\max}}{\sigma_{23\infty}} = \frac{k}{\sigma_{23\infty}}. \qquad (11.31)$$

Note that as in the case of the hole under biaxial tension, the stress and strain concentrations are no longer constants, but the strain concentration increases while the stress concentration decreases with increasing applied stress. Such observations led Neuber (1961) to the conclusion that the elastic stress or strain concentration factor was the root mean square of the elastic-plastic stress and strain concentration factors. This result holds here (Prob. 11.22), although as we have seen from the case of a filleted tensile specimen, discussed in Section 11.4, the result is not of general validity as the fully plastic case is approached.

◀ In elastic-plastic tension, only a few numerical solutions have been obtained and these have not given strain distributions or detailed plastic zones at the very tip of a sharp crack (Allen and Southwell, 1950; Jacobs, 1950; Hendrickson et al., 1958; Stimson and Eaton, 1961; Christensen and Denke, 1961; and Harper and Ang, 1963). A number of solutions have been based on the assumption that the plastic strain distribution is identical with the elastic, but this result is not likely to be valid, judging from the results in shear. ▶

11.7 CONTACT STRESS

◀ The problem of determining the deformation and distribution of stress in two contacting bodies was first solved by Hertz in 1881, but a convenient reference is given by Timoshenko and Goodier (1951, pp. 372–382). In this analysis the two bodies are considered to have surfaces which locally can be represented as spheres of radii R_1 and R_2, as shown in Fig. 11.8. The radius of the area of contact is

$$a = \sqrt[3]{\frac{3}{4} \frac{PR_1R_2}{R_1 + R_2}\left(\frac{1 - \nu_1^2}{E_1} + \frac{1 - \nu_2^2}{E_2}\right)}, \qquad (11.32)$$

and the deflection between the two bodies is

$$u_2 - u_1 = \sqrt[3]{\frac{9}{16} \frac{P^2(R_1 + R_2)}{R_1 R_2}\left(\frac{1 - \nu_1^2}{E_1} + \frac{1 - \nu_2^2}{E_2}\right)^2}. \qquad (11.33)$$

The maximum stress is at the center of contact and equal to $\frac{3}{2}$ the average stress:

$$\sigma_{33\mathrm{max}} = 3P/2\pi a^2. \qquad (11.34)$$

In a ductile material, yielding depends on the equivalent stress. The maximum value of the equivalent stress occurs on the axis through the center of contact and about one-half contact radius below the surface. The equivalent stress for $\nu = 0.3$ is about 0.62 times the maximum stress at the center of the contacting area. In a brittle material, fracture depends on the maximum tensile component of stress, which is the

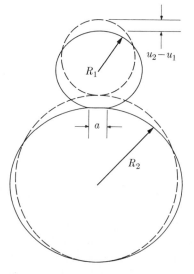

FIG. 11.8. Spherical bodies in contact.

radial stress component at the edge of the contact circle. This produces ring cracking under a spherical indenter. The value of the maximum tensile stress is $(1 - 2\nu)/3$ times the maximum stress at the center.

◀ The nature of the stress analysis for this case is essentially different from the usual type of analysis encountered in the theory of elasticity in that the geometry of loading, in particular the radius of the contacting circle, depends on the magnitude of the load. Thus the problem is nonlinear, and the stress distribution depends on the elastic constants as well as on the loads.

◀ Timoshenko and Goodier also present equations for contact of ellipsoidal surfaces. The case of a spherical ball in a spherical seat can be treated by changing the sign of the radius of curvature R_2 in the above equations. ▶

11.8 ESTIMATING ELASTIC STRESS CONCENTRATIONS

Relatively few of the many possible kinds of strain concentrations have been studied exactly. Instead, what has been done is to obtain a few exact solutions, check these against photoelastic stress analysis, and from them make estimates of what the strain concentration should be in a variety of other practical cases. Summaries of such work are given by Roark (1954), Peterson (1953), Neuber (1959), and Savin (1961), to which the reader is referred for specific cases. For a rough approximation, however, it is possible to characterize the results of most investigations by an empirical equation.

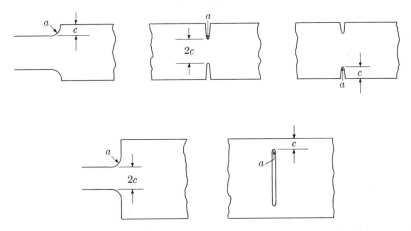

Fig. 11.9. Definition of terms for approximate equation for stress concentration.

Consider the variety of stress concentration shapes shown in Fig. 11.9. In each of these, there is a radius of curvature a at the root of the notch and a relevant dimension which may be taken to be either the half-thickness of the remaining material, the half-length of a two-sided crack, the length of a single-ended crack, or the height of a shoulder. Let the least of such dimensions be called c. Then the strain or stress concentration factor is given approximately by

$$\frac{\epsilon_{max}}{\epsilon_{nom}} \approx \frac{\sigma_{max}}{\sigma_{nom}} \approx 1 + (0.5 \text{ to } 2)\sqrt{\frac{c}{a}}. \tag{11.35}$$

From Eqs. 11.21 and 11.22, the factor 2 is exact for stress concentrations for plane stress tension applied to an elliptical hole; otherwise, the numerical coefficient should be chosen at the low end of the range for fillets with generous radii and for bending and torsion. It should be chosen at the high end for tension, particularly of circumferentially notched tubes. The lower values should also be chosen for larger angles between the two flanks of the notch. In spite of this vagueness, Eq. 11.35 does provide a handy working approximation to elastic stress and strain concentrations (Prob. 11.23).

11.9 EFFECTS OF STRAIN CONCENTRATIONS

Strain concentrations tend to increase the deflection of a part, to affect its load-carrying capacity, and to facilitate fracture.

The effect of strain concentrations on deflections has been studied very little, probably because its effect is small. For example, consider the load-extension curve of a specimen containing a transverse hole, shown in Fig. 10.2. The increased elastic deflection is little more than would be estimated from considering the stress and strain to be uniformly distributed across each section (Prob. 11.24). As a second example, the added deflection of a built-in cantilever beam, due to distortion of its base, gives a rotation equivalent to an added length of the beam of only $(16/15\pi)(1 - \nu^2)d$, where d is the diameter of the beam (Brown and Hall, 1962). The increased deflection may be important when it is due to the growth of a crack. It could conceivably be dangerous if it tends to cause buckling or change the natural frequency of the member so that it is vibrating near its critical mode. This is not likely, for if the small added deflection due to a hole would cause the natural frequency to become critical or would cause buckling, the structure is probably too close to these conditions for ordinary design tolerances.

Although the elastic limit is reduced by the hole, as expected, the increase in deflection due to plastic flow is small until the limit load is reached, as shown again by the example of Fig. 10.2. The effect of a strain concentration on the limit load capacity of a part is never more than the effect due to the reduction of area alone, and it is sometimes less where plastic constraint by the shoulders produces a state of triaxial stress in the minimum section and retards plastic deformation.

Fracture, on the other hand, is promoted by strain concentrations even when plastic flow occurs. This result has been obscured by the fact that for materials that can deform plastically as compared with those that cannot, the plastic flow retards fracture. For a given material, however, a hole or notch will reduce the amount of deformation to fracture, as will be discussed later in Chapter 16 on ductile fracture.

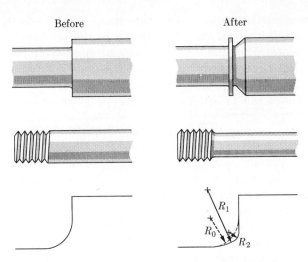

FIG. 11.10. Means of reducing strain concentrations.

11.10 AMELIORATION OF STRAIN CONCENTRATION

Strain concentrations can be reduced, of course, by insuring that no reentrant corners have small radii of curvature. In some aircraft companies this is considered so important that machinists are forbidden to have sharp-nosed tools in their possession at all. In ductile materials, the more strain-hardening that is present the less will be the strain concentration factor. Where it is necessary to have a shoulder for positioning of a part, the effect can be reduced by cutting away material behind the shoulder; similarly, material can be removed at the end of the screw threads, as shown in Fig. 11.10. A fillet can also be undercut or made with an elliptical shape (Berkey, 1944) so that the largest radius of curvature is found near the end where the stress concentration would otherwise be greatest.

11.11 SUMMARY

Calculations of the load-carrying capacity of parts are usually based on nominal stresses, obtained from analysis of simplified shapes, so that concentrations of strain caused by fillets, grooves, holes, or other abrupt changes of section are neglected. In analogy to St. Venant's principle, in elasticity such local changes produce only local effects which can be described in terms of stress concentration factors. The magnitude of elastic stress or strain concentrations can be given roughly in terms of the radius of curvature a, at the tip of a notch, and a relevant dimension c, such as notch depth, crack half-length, or ligament half-thickness, whichever is least:

$$\frac{\epsilon_{max}}{\epsilon_{nom}} \approx \frac{\sigma_{max}}{\sigma_{nom}} \approx 1 + (0.5 \text{ to } 2)\sqrt{\frac{c}{a}}. \qquad (11.35)$$

Although strain concentrations reduce the elastic limit of a part, the resulting plastic flow is at first localized enough so that it hardly produces any appreciable decreases in the force-deformation curve. The *stress* concentration is limited by plastic flow, but the *strain* concentration is often higher than for elasticity.

Strain concentrations have little effect on deflection and on limit loads. Elastic strain concentrations can be reduced by adjacent holes and grooves which effectively soften the shoulders.

◀Stress and strain concentration factors for circular holes under uniaxial and biaxial tension, presented in Tables 11.1 and 11.2, show the influence of such factors as stress or strain concentration, plane stress or plane strain conditions, elastic or plastic flow, and the particular component of stress or strain. The stress and strain distribution around a spherical hole, discussed in Section 11.5, provides insight into the lattice distortion in solid solutions.

◀For an elliptical hole, the stress distribution at the surface is best given in terms of the elliptical coordinates, defined in terms of the crack half-length c, the tip radius of curvature a, and the slope of the surface dx_2/dx_1 by

$$\alpha_0 = (a/c)^{1/2}, \qquad \beta = \tan^{-1}[(\tanh \alpha_0)/(dx_2/dx_1)]. \tag{11.21}$$

Under a normal stress $\sigma_{22\infty}$ and a shear stress $\sigma_{21\infty}$ the normal stress at the surface of the crack near its tip is given for sharp cracks ($\alpha_0, \beta \ll 1$) by

$$\sigma_{\beta\beta 0} = \frac{2\alpha_0 \sigma_{22\infty} - 2\beta \sigma_{21\infty}}{\alpha_0^2 + \beta^2}. \tag{11.22}$$

In front of a sharp crack, the stress can be described in terms of symmetric, antisymmetric, and longitudinal shear components of the stress singularity. For instance, for a crack under tension, the normal components in the plane of the crack are (from 11.23)

$$\sigma_{rr} = \sigma_{\theta\theta} = k_1/\sqrt{2r},$$

where the stress intensity factor k_1 is given by the applied stress at infinity and the crack half-length c by

$$k_1 = \sigma_{22\infty}\sqrt{c}. \tag{11.26}$$

Complete equations for all components in the neighborhood of the crack tip are given, along with values of the stress intensity factors for other loading conditions (Table 11.3). The relation between the stress intensity factor and the stress concentration factor for a similar notch with a finite tip radius is given in Eqs. 11.27 and 11.28.

◀For sharp grooves under longitudinal shear at stress levels low compared with the shear strength k, the plastic zone at the root of the notch is circular with maximum extent

$$R_{\max} = c(\sigma_{23\infty}/k)^2 = k_3^2/k^2. \tag{11.29}$$

Within this plastic zone the strain is

$$\gamma = (k/G)R/r. \qquad (11.30)$$

In this case, where the applied stress is low, plastic flow is constrained by an elastic region; the elastic stress concentration factor is the root mean square of the plastic stress and strain concentration factors.

◀The equations for the contact of two spherical bodies are given as Eqs. 11.32, 11.33, and 11.34 for their later value in connection with hardness and rolling contact fatigue.▶

REFERENCES

ALLEN, D. N. DE G. SOUTHWELL, R. V.	1950	"Relaxation Methods Applied to Engineering Problems—XIV Plastic Straining in Two-Dimensional Stress Systems," *Phil. Trans. Roy. Soc. (London)* **A242**, 379–414.
ANG, D. D. FOLIAS, E. S. WILLIAMS, M. L.	1963	"The Bending Stress in a Cracked Plate on an Elastic Foundation," *J. Appl. Mech.* **30**, 245–251.
ASTM	1960	"Fracture Testing of High-Strength Sheet Materials," ASTM Committee on Fracture Testing of High-Strength Testing Materials, *Bull. ASTM*, Jan. 1960, 29–40.
BARRETT, C. S.	1952	*Structure of Metals*, 2nd ed., McGraw-Hill, New York.
BERKEY, D. C.	1944	"Reducing Stress Concentrations with Elliptical Fillets," *Proc. Soc. Exp. Stress Anal.* **1**, 2, 56–60.
BILBY, B. A. COTTRELL, A. H. SWINDEN, K. H.	1963	"The Spread of Plastic Yield from a Notch," *Proc. Roy. Soc. (London)* **A272**, 304–314.
BROWN, J. M. HALL, A. S.	1962	"Bending Deflection in a Circular Shaft Terminating in a Semi-Infinite Body," *J. Appl. Mech.* **29**, 86–90.
BUDIANSKY, B. MANGASARIAN, O. L.	1960	"Plastic Stress Concentrations at a Circular Hole in an Infinite Sheet Subjected to Equal Biaxial Tension," *J. Appl. Mech.* **27**, 59–64.
CHRISTENSEN, R. H. DENKE, P. H.	1961	"Crack Strength and Crack Propagation Characteristics of High-Strength Metals," *ASD Tech. Rept. 61–207*, Aeronautical Systems Division, U.S. Air Force, Wright-Patterson Air Force Base, Ohio.
DAVIS, E. A.	1963	"Extension of Iteration Method for Determining Strain Distributions to the Uniformly Stressed Plate with a Hole," *J. Appl. Mech.* **30**, 210–214.
DONNELL, L. H.	1962	"About St. Venant's Principle," *J. Appl. Mech.* **29**, 752–753.

Erdoğan, F. Sih, G. C.	1963	"On the Crack Extension in Plates under Plane Loading and Transverse Shear," *Trans. ASME, J. Basic Eng.* **D85**, 519–527.
Erdoğan, F. Tuncel, O. Paris, P. C.	1962	"An Experimental Investigation of the Crack Tip Stress Intensity Factors in Plates under Cylindrical Bending," *Trans. ASME, J. Basic Eng.* **D84**, 542–546.
Eshelby, J. D.	1954	"Distortion of a Crystal by Point Imperfections," *J. Appl. Phys.* **25**, 255–261.
Eshelby, J. D.	1957	"The Determination of the Elastic Field of an Ellipsoidal Inclusion and Related Problems," *Proc. Roy. Soc.* (London) **A241**, 376–396.
Friedel, J.	1955	"Deviations from Vegard's Law," *Phil. Mag.*, Series 7, **46**, 514–516.
Harper, G. N. Ang, A.	1963	"A Numerical Procedure for the Analysis of Contained Plastic Flow Problems," *Civil Engineering Structural Research Series No. 266*, University of Illinois, Urbana.
Hendrickson, J. A. Wood, D. S. Clark, D. S.	1958	"The Initiation of Brittle Fracture in Mild Steel," *Trans. ASM*, **50**, 656–681.
Hill, R.	1950	*The Mathematical Theory of Plasticity*, Clarendon Press, Oxford, England.
Horvay, G.	1957	"Some Aspects of St. Venant's Principle," *J. Mech. Phys. Solids* **5**, 77–94.
Hult, J. A. H. McClintock, F. A.	1957	"Elastic-Plastic Stress and Strain Distributions Around Sharp Notches Under Repeated Shear," *IXe Congrés International de Mécanique Appliquée, Actes* **8**, 51–58.
Inglis, C. E.	1913	"Stresses in a Plate Due to the Presence of Cracks and Sharp Corners," *Trans. Naval Arch.* **60**, 219–230.
Irwin, G. R.	1957	"Analysis of Stresses and Strains near End of a Crack," *J. Appl. Mech.* **24**, 361–364.
Irwin, G. R.	1962(a)	"Analytical Aspects of Crack Stress Field Problems," *Theo. and Appl. Mech. Dept., Rept. 13*, University of Illinois, Urbana.
Irwin, G. R.	1962(b)	"Crack Extension Force for a Part-Through Crack in a Plate," *J. Appl. Mech.* **29**, 651–654.
Irwin, G. R. Kies, J. A. Smith, H. L.	1958	"Fracture Strengths Relative to Onset and Arrest of Crack Propagation," *Proc. ASTM* **58**, 640–660.
Jacobs, J. A.	1950	"Relaxation Methods Applied to Problems of Plastic Flow," *Phil. Mag.* **41**, 349–361.
Koskinen, M. F.	1963	"Elastic-Plastic Deformation of a Single Grooved Flat Plate Under Longitudinal Shear," *Trans. ASME, J. Basic Eng.* **D85**, 585–594.

McClintock, F. A. Rhee, S. S.	1962	"On Effects of Strain-Hardening on Strain Concentrations," *Proc. 4th U.S. Nat. Con. Appl. Mech.*, 1007–1013.
Neuber, H.	1959	*Kerbspannungslehre*, 2nd ed., Springer, Berlin. See also translation of 1st ed., 1937, "Theory of Notch Stresses," Edwards, Ann Arbor (1946).
Neuber, H.	1961	"Theory of Stress Concentration for Shear-Strained Prismatical Bodies with Arbitrary Non-Linear Stress-Strain Law," *J. Appl. Mech.* **28**, 544–551.
Peterson, R. E.	1953	*Stress Concentration Design Factors*, Wiley, New York.
Prager, W. Hodge, P. G., Jr.	1951	*Theory of Perfectly Plastic Solids*, Wiley, New York.
Roark, R. J.	1954	*Formulas for Stress and Strain*, McGraw-Hill, New York.
Savin, G. N.	1961	*Stress Concentration Around Holes*, Pergamon Press, London (Translated from the Russian).
Stimpson, L. D. Eaton, D.	1961	"The Extent of Elastic-Plastic Yielding at the Crack Point of an Externally Notched Plane Stress Tensile Specimen," *ARL Tech. Rept. 24*, Aeronautical Research Laboratory, Wright-Patterson Air Force Base, Ohio. See also M. L. Williams, "Some Observations Regarding the Stress Field near the Point of a Crack," *Proc. Crack Propagation Symp.*, The College of Aeronautics, Cranfield Vol. 1, 130–165.
Timoshenko, S. Goodier, J. N.	1951	*Theory of Elasticity*, McGraw-Hill, New York.
Wigglesworth, L. A.	1957	"Stress Distribution in a Notched Plate," *Mathematika* **4**, 76–96.
Williams, M. L.	1957	"On the Stress Distribution at the Base of a Stationary Crack," *J. Appl. Mech.* **24**, 109–114.
Winne, D. H. Wundt, B. M.	1958	"Application of the Griffith-Irwin Theory of Crack Propagation to Bursting Behavior of Discs Including Analytical and Experimental Studies," *Trans. ASME* **80**, 1643–1658.

PROBLEMS

11.1. Derive the equation for the axial strain at infinity due to a biaxial stress, Eq. 11.1.

11.2. Derive the elastic displacement distribution, Eq. 11.3.

11.3. (a) Integrate Eq. 11.3 to obtain the radial displacement distribution of Eq. 11.4.
(b). Derive Eq. 11.5 for the stress distribution in terms of two constants of integration.

◀ 11.4. Derive the elastic plane strain concentration factors of Table 11.1.

◀ 11.5. Derive the strain distribution around a hole in a rigid-plastic material under biaxial tension and plane strain, Eq. 11.5.

◀ 11.6. Derive the stress distribution around a cylindrical hole in a material under biaxial tension and plane strain, yielding according to the maximum shear stress yield criterion, Eq. 11.9.

◀ 11.7. Derive the stress distribution for plane stress conditions with the maximum shear stress yield criterion and associated flow role, Eq. 11.10.

◀ 11.8. Derive the corresponding strain distribution, Eq. 11.11.

◀ 11.9. Explain in physical terms why the strain concentration should become infinite as Poisson's ratio approaches one-half, for the plane strain condition. An example of such a state of strain is found in a thick walled rubber vacuum hose of fixed length.

◀11.10. Show how the state of stress around a hole with an arbitrary applied stress at infinity can be obtained by superposition of equations of the form of Eq. 11.12.

◀11.11. Sketch the mode of deformation for plane stress uniaxial tension of a sheet containing a hole. Prove your answer correct by demonstrating a lower bound solution throughout the specimen including the nondeforming regions. If no other mode of deformation is compatible with a complete lower bound solution, you must have the unique strain distribution.

◀11.12. Derive the displacement distribution around spherical holes under biaxial tension, Eqs. 11.13 and 11.14.

◀11.13. Derive the stress distribution in terms of the constants of integration of the displacement, Eq. 11.15.

◀11.14. Show that for a spherical hole in an elastic material under a uniform triaxial tensile stress at infinity, the maximum stress concentration factor is $\frac{3}{2}$.

◀11.15. Derive the equation for the radial displacement around a solute atom, Eq. 11.17.

◀11.16. Derive the correction due to a sphere of finite radius, Eq. 11.19.

◀11.17. Derive the relation between the stress intensity factors for a sharp crack and the maximum stress near the tip of an elliptical hole, Eq. 11.27 (Irwin, 1958).

◀11.18. Derive the relation between stress intensity factor and stress concentration factor near the tip of a rounded groove under longitudinal shear, Eq. 11.28.

◀11.19. Derive Eq. 11.29 by carrying out the following steps:

(a) Show that in the plastic region of Fig. 11.7, the assumption of constant shear stress normal to radial lines satisfies the equilibrium equations and the yield criterion.

(b) Show that at the boundary of the plastic zone the stress distribution is exactly that expected in an elastic material whose crack tip is at the center of the plastic zone, provided the stress intensity factor is appropriately chosen (Koskinen, 1963).

◀11.20. Show that within the plastic zone of Fig. 11.6, the strain distribution is given by Eq. 11.30. [*Hint:* Starting from the elastic-plastic boundary, determine the relative displacements along nearby radial lines.]

◀11.21. Derive the equation for the stress and strain concentrations in a crack under longitudinal shear, Eq. 11.31.

◀11.22. Show that the root mean square of the plastic stress and strain concentration factors given in Eq. 11.31 is the elastic strain concentration factor for this shape.

11.23. Estimate the strain concentration in a cantilever beam 1.80 in. deep by 8 in. long with a $\frac{1}{4}$-in. radius fillet at the end where it is attached to a 4-in. deep support (photoelastic observations give a factor of 1.65).

11.24. Calculate the increased elastic deflection of the bar shown in Fig. 10.2 from the decrease in section over the corresponding length. Should further experiments be carried out?

11.25. In Section 10.6 it was stated that the addition of material could not reduce the limit load of a part. Show that the addition of material can reduce the strength of a brittle part which fails when the maximum stress reaches a critical value.

◀11.26. Using the principle of superposition, show how the stress intensity factor for the internal crack in an infinite plate can be obtained from the expression given in Table 11.3 for the stress intensity factor due to a pair of loads P applied to the face of the crack a distance b from its center.

E 11.27. Prepare specimens with a notch or a hole in them from silicone putty. Press screening into them to impose a grid. Estimate the strain concentrations for your configuration from elastic values. Compare them with your observations.

E 11.28. Flatten out a piece of modeling clay with a hole in it. Impose a grid on the surface by impressing it with a piece of screening. Stretch or compress the plate and compare the plastic strain concentration with that predicted for elastic and viscous materials. Would you expect it to be higher or lower?

11.29. Show that for a hole under uniaxial tension, the maximum equivalent stress does occur at the surface of the hole. (Use plausibility arguments if desired.)

CHAPTER 12

RESIDUAL STRESS

12.1 SYNOPSIS

Residual stresses are defined as those remaining in a material in the absence of loads or of changes in temperature. They may be associated with the microstructure of the material. On a macroscopic scale, they can arise from metallurgical changes. Examples are shown of their production from nonuniform plastic flow or creep, due to forming operations or temperature changes.

It is pointed out that the effects of residual stress are the same as those of any other stress except that they can be relieved by small amounts of relaxation or plastic strain. Various methods of measuring residual stress are given, followed by a discussion of methods of stress relief.

12.2 CLASSIFICATION OF RESIDUAL STRESSES

A stress system satisfying internal equilibrium, with no external loads or temperature gradients, will be called a *residual stress*. Residual stresses may be classified into two broad and somewhat overlapping groups.

The *macroscopic* residual stresses are of long range, extending over a macroscopic dimension of a part. These stresses, which arise from large-scale deformation of a part considered as a continuum, will be discussed in later sections.

The *microscopic* residual stresses are of short range and are usually confined to parts of a material where a homogeneous continuum view of strain is no longer possible. Therefore, proper understanding of microresidual stresses requires a familiarity with structural details. On the smallest scale, microresidual stresses arise from misfitting solute atoms and individual dislocations, as discussed in Sections 11.5 and 4.4. The hardening that arises from the interaction of these microresidual stresses was discussed in Section 5.2.

On a somewhat larger scale, residual stresses arise from dislocation pile-ups, kink boundaries, deformation-induced tilt boundaries, deformation twins, and other diffusionless shear transformations. Such configurations and their residual stresses play fundamental roles in plastic deformation, in crack initiation in brittle fracture, in fatigue crack growth, and in the Bauschinger effect.

In materials containing inhomogeneities, the homogeneous deformation of the matrix is prevented by the hard particles or precipitates. The deforming matrix produces a drag effect on the inhomogeneity during deformation. When deformation ceases, residual stresses are set up. These residual stresses arising from inhomogeneities are a borderline case between microscopic and macroscopic residual stresses.

12.3 MECHANICAL AND THERMAL SOURCES OF RESIDUAL STRESS

Residual stress may be introduced either intentionally or accidentally during assembly of structures. For example, in assembling the members of a frame building, as shown in Fig. 12.1, mild misalignment of the rivet holes may be corrected by driving a taper pin through a pair of holes in mating members to bring them into line and then riveting the adjacent holes. Another example is the manufacture of prestressed concrete beams, in which the reinforcing wire is put in tension before the concrete is poured. After the concrete has hardened, release of the load on the wires leaves the stress distribution shown in Fig. 12.2.

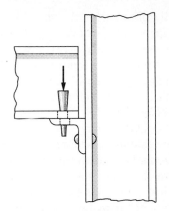

FIG. 12.1. Residual stress due to misalignment in assembly.

The concrete can then carry a greater bending moment both because it is pre-bent in the opposite direction and because it now operates in the compressive regime in which the elastic range of stress is greater. A third example of residual stress due to assembly is that of the shrink or pressed fit, for example, when a disk is placed on a shaft whose diameter is somewhat greater than the hole in the disk. This is accomplished either by pressing it on directly or by heating the disk and cooling the shaft until there is no interference, sliding the two together, and letting them reach the same temperature. The radial compressive stress at the interface may be enough to hold the two parts together in service without further fastening (see for example, Timoshenko, 1941, p. 241).

Loads sufficient to produce plastic deformation can result in residual stress if the elastic change in stress during unloading is not identical with the plastic distribution under load. Examples of this are the bending and torsion discussed in Chapter 9.

Residual stress can also be introduced by creep or time-dependent deformation caused by loads. Again, the condition for creep to cause residual stress is that the stress distribution under the conditions of creep be different from the elastic stress

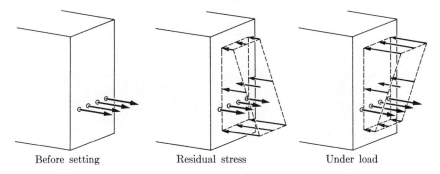

FIG. 12.2. Residual stress in prestressed concrete beam.

distribution for the same load. In this case, release of load will superimpose an elastic change of stress which will not exactly cancel out the original stress but will leave a residual stress. Such residual stress is particularly important in materials used for high-temperature applications. These materials often have sufficient ductility at elevated temperatures but very limited ductility at low temperatures. They are therefore subject to fracture in the presence of residual stress when cooled.

Metal-forming operations are a frequent source of residual stress. The residual stress due to forming sheet metal by bending with a single radius of curvature can be analyzed with the aid of the theory of bending of plates presented in Section 9.5. For instance, consider a plate of nonhardening material bent until almost the entire section has flowed plastically. In terms of the coordinate axes shown in Fig. 9.13, if the metal is nonhardening, with flow stress Y, the stress under load is, from Prob. 9.29,

$$\sigma_{33} = 2Y/\sqrt{3} \quad \text{and} \quad \sigma_{11} = Y/\sqrt{3},$$

for which the applied moment is

$$M_{31} = (bh^2/4)(2Y/\sqrt{3}).$$

Unloading may be thought of as adding a reversed bending moment. If the process is assumed elastic, the added stress is that of ordinary elastic bending, given by Eqs. 9.21 and 9.22. The resulting residual stress on the convex surface is (Prob. 12.1)

$$\sigma_{33} = -Y/\sqrt{3}, \quad \sigma_{11} = -(Y/\sqrt{3})(3\nu - 1). \tag{12.1}$$

Since this state of stress is within the yield locus, the assumption of elasticity on unloading is justified. Conditions under which reverse plastic flow will *not* occur, or will terminate on repeated loading, have been discussed by Hodge (1954).

Other metal-forming operations, such as drawing or spinning, have not yet been successfully analyzed, although they have been studied experimentally.

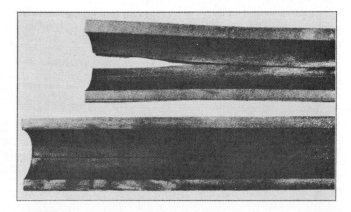

Fig. 12.3. Residual stress causing longitudinal cracks originating from inner wall of tube. (Horger, 1950, p. 506. Courtesy of Wiley.)

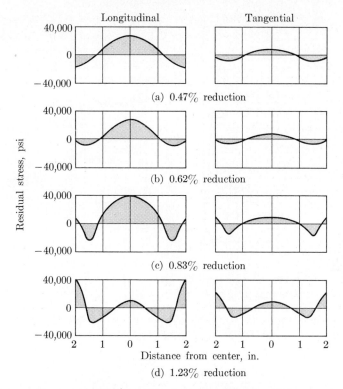

Fig. 12.4. Change in sign of residual stress with increased reduction. (Bühler and Buchholtz, 1934. From Baldwin, 1949. Courtesy of ASTM.)

One of these, the drawing of cartridge cases, provided much of the early impetus to the study of residual stress, since cartridge cases often failed due to stress corrosion cracking under the influence of residual stress (ASTM and AIME, 1945). In tubing with heavier sections, such as shown in Fig. 12.3, the residual stress may be high enough to cause fracture spontaneously.

In solid bar and strip produced by rolling, the character of the residual stress depends strongly on the depth of the reduction. As shown in Fig. 12.4, in light rolling the plastic flow does not penetrate all the way across the section, and hence the surface is left in compression with the core in tension. For heavier reductions, the entire cross section becomes plastic. While the analysis cannot be carried out exactly, it turns out that in a variety of metal-forming processes, including rolling, drawing, and extruding, the residual stress is tensile at the surface. Baldwin (1949) has also pointed out that the degree of penetration in a plastic zone depends on the ratio of the roll diameter to the diameter of the work piece, the residual stress tending to be compressive for small-diameter rolls and tensile for large-diameter rolls. Perhaps with the larger rolls the plastic zone penetrates more deeply because there is less chance for the material to be pushed to the back or front.

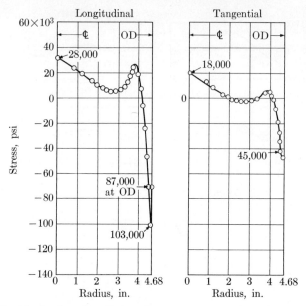

Fig. 12.5. Residual stress in rolled well-fit portion of 9.5-in. diameter crank pins with 12,000-lb roller pressure. (Horger, 1950. Courtesy of Wiley.)

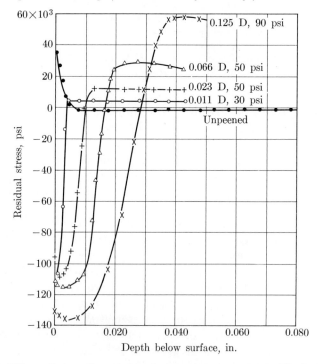

Fig. 12.6. Effect of shot diameter and air pressure on stress distribution for steel of hardness $R_C 42$. (Lessells and Broderick, 1956. Courtesy of Institution of Mechanical Engineers.)

The compressive residual stress at the surface which results from a light rolling or drawing operation is favorable in many cases in which fracture may occur. Compressive residual stress may be produced in other ways, for example, by surface rolling, as shown in Fig. 12.5, and by shot-peening, a process in which cast iron or steel shot is impelled at a surface with an air blast. Shot-peening is one of the most important commercial methods of producing parts with favorable residual stress, and has been extensively studied. Examples of the influence of some of the variables on the intensity and depth of residual stress are given in Fig. 12.6. For a bibliography of the subject and a description of the technology associated with carrying it out in practice, see publications by the Society of Automotive Engineers (SAE, 1952; Huang, 1954).

Residual stress may be caused by machining. According to the ASM Committee on Residual Stress (ASM, 1955), there are compressive stresses a few thousandths of an inch deep, increasing with depth of cut and dullness of tool. Detailed measurements by Field and Zlatin, reported by Henriksen (1957), indicate, however, that the stress at the very surface is tensile, but becomes compressive a little deeper. Dyachenko and Podosenova (1954) reported that the residual stress changed sign at intermediate cutting speeds, partly as a result of the effects of thermal stress on the residual stresses.

Grinding is a form of machining in which thermal effects are more pronounced. Frisch and Thomsen (1951) reported tensile residual stress due to grinding in SAE 1020 steel. Tarasov et al. (1957), on the other hand, reported compressive stress at the very surface with more moderate tensile stress (especially in view of their harder steel), at greater depths (see Fig. 12.7). They ascribe the change in sign of stress to a combination of thermal and mechanical effects.

Changes in temperature can produce residual stress if they are great enough and localized enough to produce plastic strain. For example, the surface layer of steel may be suddenly heated by electromagnetic induction and then quenched by

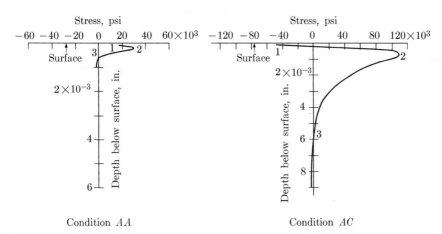

Fig. 12.7. Effect of grinding severity on residual stress in AISI type 52100 steel, $R_C 59$. Condition AA: soft wheel, 0.1×10^{-3} in. feed. Condition AC: hard wheel, 2×10^{-3} in. feed. (Tarasov, Hyler, and Letner, 1957. Courtesy of ASM.)

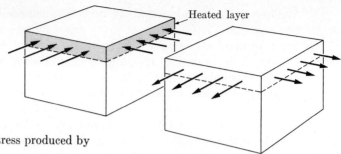

Fig. 12.8. Residual stress produced by thermal shock in a slab.

rapid heat conduction to the cool interior of the solid, as shown in Fig. 12.8. If the surface layer is thin enough, displacements parallel to the surface are prevented by the bulk of material below. For a non-strainhardening material with a linear coefficient of expansion α and yield strength Y, in uniaxial tension, the temperature change required to produce any yielding is given in Section 9.6 as

$$\Delta T = (1 - \nu)Y/E\alpha. \qquad (9.27)$$

The residual stress due to a temperature rise ΔT is

$$\sigma^r_{11} = -Y + \frac{E\alpha\,\Delta T}{1 - \nu} \quad \text{for} \quad \frac{(1-\nu)Y}{E\alpha} < \Delta T < \frac{2(1-\nu)Y}{E\alpha},$$

$$\sigma^r_{11} = Y \quad \text{for} \quad \Delta T > \frac{2(1-\nu)Y}{E\alpha}. \qquad (9.28)$$

Part of the effect of cutting speed on residual stress can be predicted by this equation (Prob. 12.2).

TABLE 12.1*

EFFECT OF QUENCHING ON RESIDUAL STRESS AND MECHANICAL PROPERTIES

Steel	Heat treatment	0.2% Yield strength, psi	Tensile strength, psi	Reduction of area, %	Endurance limit, psi	Residual surface stress, psi	
						Long.	Tang.
C34 (0.34%C)	Furnace-cool from 1110°F.	49,600	91,500	54	39,800		
	Oil (175°F) quench from 1110°F	54,900	94,700	54	42,700	−29,000	−29,200
	Ice-water quench from 1110°F	51,800	96,700	54	46,900	−45,500	−45,500
	Salt-water (−5°F) quench from 1110°F	50,200	93,400	55	48,400	−48,400	−45,500

* Horger (1950); from Bühler and Buchholtz (1933).

As a further example of the effect of thermal shocking on residual stress, consider the data given in Table 12.1. The smaller residual stress due to oil-quenching arises not only from the higher temperature to which the steel is quenched, but also from the lower heat transfer coefficient on the surface, so that the temperature gradient in the steel is not as steep as in the case of a salt-water quench.

The production of tempered glass is a case where residual stress arising from inhomogeneous thermal strain is utilized to make glass stronger. The process consists of cooling a glass plate in an air blast. The surface hardens and deforms the still hot interior. As the interior cools, the contraction leaves compressive residual stress (about 20,000 psi) in the surface and a low tensile residual stress in the interior. Since nearly all the strength-impairing defects are on the surface (see Chapter 15), their propagation is prevented until the externally applied loads produce a surface tensile stress in excess of the residual compressive stress.

12.4 METALLURGICAL AND CHEMICAL SOURCES OF RESIDUAL STRESS

As mentioned in Section 12.2, residual stress may be present in solids as a result of the differences in shape and density of various phases compared with the matrix from which they came. When a new phase precipitates out of a solid solution, the resulting residual stress field will affect the energy required to activate the reaction (Cohen, 1958). Furthermore the new lattice may bear some definite crystallographic relation to the old one, so that both shear and normal stress components may occur.

Residual stress may arise from inhomogeneous transformations in a part. For example; the austenite to martensite transformation in steel results in an appreciable increase in volume. Since the inner part of a specimen cools and transforms last, the resultant expansion often puts the surface in tension severe enough to cause cracks.

The surface of steel may be decarburized in the presence of an oxidizing atmosphere, resulting in a decrease in volume and a tendency to tensile residual

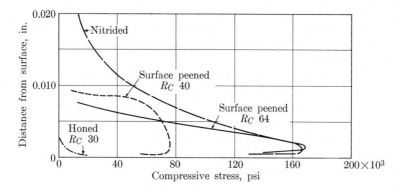

Fig. 12.9. Magnitude and depth of stress imposed by various surface treatments. (Almen, 1943. Courtesy of ASM.)

stress in the thin layer where decarburization occurred. If the temperature is high enough, creep will relax this residual stress, but furnace-cooling from 850°C may not allow enough relaxation, perhaps due to phase transformations during cooling (Dickie, 1931). Another method for the introduction of residual stress is by nitriding, in which heating in an ammonia atmosphere allows nitride formation, thus hardening and slightly swelling the steel. As shown in Fig. 12.9, a residual stress of the order of 150,000 psi may arise.

Essentially all electroplating processes produce residual stresses (sometimes tensile, sometimes compressive). Impurities and plating conditions (temperature, current density, and Ph, for example) have a marked influence on the residual stresses. An example of the effect of impurities is shown in Fig. 12.10. In chromium plating, the tensile stress may be high enough to produce cracking.

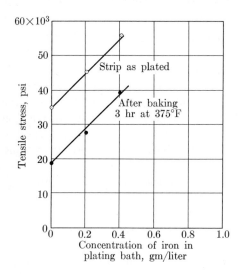

FIG. 12.10. Increase of residual stress in nickel plate as a result of iron contamination in chloride-sulfate bath. (Noble, 1955. From *ASM*, 1955. Courtesy of ASM.)

A related and more familiar form of residual stress is that arising from the effects of changes in moisture content in wood, where uneven drying can develop sufficient residual stresses to split or check the wood, as well as to warp it. Likewise, a loss of plasticizer from polymers can introduce a significant residual stress which is important in brittle polymers.

An interesting application of residual stress arising from chemical effects is the chemomechanical strengthening of glassware, known as the "Chemcor"* process. In the surface layer of the glass network some ions are replaced by another type of ion. A subsequent heat treatment produces a new phase with a volume increase. This sets up a compressive residual stress in a thin surface layer. This stress may reach 100,000 psi, and impart to the glass an extraordinary resistance to mechanical damage, as will be discussed further in Section 15.5.

---

* This is a trade name of the Corning Glass Company, where the process was invented.

12.5 COMPOSITE SOURCES OF RESIDUAL STRESS

In most cases, residual stresses arise from the simultaneous action of a number of sources. For example, in casting, welding, and heat treatment in general, residual stresses are due to both phase transformations and temperature gradients. In forging, both plastic deformation and temperature gradients contribute. Numerous other examples are given by Horger (1950). It is not uncommon to have residual stresses of the order of the yield strength of the material, as for example, in the quenching reported in Table 12.1.

12.6 MEASUREMENT OF RESIDUAL STRESS

Residual stresses are usually measured from strains observed during removal of material, or from the elastic strain in the crystal lattice, determined from x-rays.

The various mechanical methods of measuring residual stress are illustrated in Fig. 12.11. Basically, all methods consist of measuring the changes in shape or the strains in a part due to the removal of metal. From a knowledge of the mechanics of deformation, it is possible to calculate what the stresses must have been in the

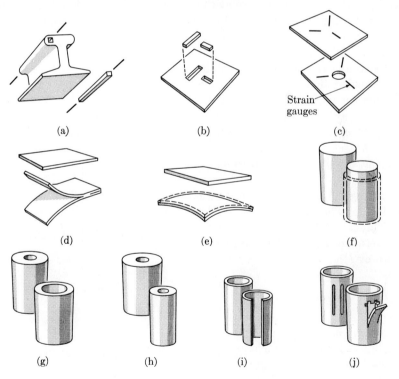

FIG. 12.11. Schematic representation of methods for measuring residual stresses. (*ASM*, 1955. Courtesy of ASM.)

metal that has been removed. When a series of layers is removed, it must be borne in mind that the removal of one layer changes the stress in the remainder of the part. This change must be taken into account in calculating the original stress.

Where uniaxial stresses predominate, as in the rail shown in Fig. 12.11(a), a direct measurement of the change in length on slitting will suffice. In sheet, where biaxial stress may be present, it may be determined by measuring the strain in coupons cut from the sheet, provided that the stress is constant across the thickness. A more elegant method is to measure the strains in the vicinity of a hole drilled in the plate, as shown in Fig. 12.11(c). Where the residual stress varies across the thickness of the plate, slitting will give a rough idea of the residual bending stress, as shown in Fig. 12.11(d). A more careful measurement requires the removal of successive layers, as indicated in Fig. 12.11(e).

In bar stock, when only longitudinal stress is present, the change in length on turning down will suffice (Fig. 12.11f). A more precise result is obtained by first boring a central hole and measuring the strains on progressive boring out or turning down of the part, with strains measured on the surfaces that are not being machined (Fig. 12.11g or h). When the shell is thin enough, the techniques illustrated in Fig. 12.11(i) and (j) can be used.

◀ As an illustration of one of these methods of determining residual stress, consider the section of beam shown in Fig. 12.12. Due to the residual stress, there is a force dF_3 in the layer that is to be removed. A counterbalancing force and moment exist on the remainder of the section as indicated. Provided that no further plastic deformation occurs, one can think of the process of removal of the layer in two stages: first, the layer is removed without change of shape, and second, the force and moment which are required to maintain the shape of the remaining metal are relaxed, giving the deformed shape, shown in Fig. 12.12. The residual stress in the removed layer ($dh < 0$ for removal) is given in terms of the change in radius of curvature by (Prob. 12.3)

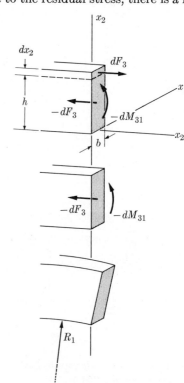

FIG. 12.12. Deformation due to removal of a layer containing residual stress.

$$d\left(\frac{1}{R_1}\right) = -\frac{12\, dM_{31}}{Ebh^3} = -\frac{6\sigma_{33}\, dh}{Eh^2}.$$
(12.2)

The increase in stress at the point x_2 due to relaxation of bending moment and load on removal of the layer dh is given by

$$d\sigma_{33}(h, x_2) = E\, d\epsilon_{33}(h, x_2)$$

$$= E(x_2 - h/2)\, d\left(\frac{1}{R_1}\right) - \frac{\sigma_{33}(h, h)\, dh}{h}.$$
(12.3)

The original stress in the layer just removed is found by integration over the successive layers, dh, which have been removed since beginning the removal (Osgood, 1954, p. 276) (Prob. 12.4):

$$\sigma_{33}(h_0, x_2) = -\frac{Ex_2^2}{6}\left(\frac{d(1/R_1)}{dh}\right)_{h=x_2} - \frac{Ex_2}{R_{1,h=x_2}} + \frac{E}{3}\int_{h_0}^{x_2} h\,\frac{d(1/R_1)}{dh}\,dh. \quad (12.4)$$

Note that the first term is positive, since $dh < 0$. The analysis for other techniques is summarized by ASM (1955) and given more completely by Heindlhofer (1948). ▶

Another method of measuring residual stress, which is sometimes more convenient although less precise, is to measure the normal pressure required to initiate a hardness indentation in a surface (Setty et al., 1957). This method suffers the disadvantage that a variety of stress states in the surface will produce the same tendency to yield under normal pressure (Prob. 12.5). Not all these states of stress are equally likely to cause trouble, as, for example, by yielding under the application of an additional uniaxial stress (Prob. 12.6).

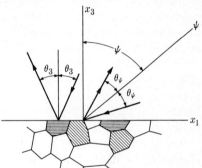

FIG. 12.13. Residual stress measurement with x-rays.

The only entirely nondestructive method of measuring residual stress in metals is through the use of x-rays, which give a direct measure of the elastic strain in the lattice. For example, the stress in the x_1 direction of the surface can be found by taking diffraction patterns of two sets of planes, one parallel to the surface and one making a steep angle with the surface, as shown in Fig. 12.13. The interplanar spacings are determined from the Bragg equation:

$$d_3 = n\lambda/2\cos\theta_3, \qquad d_\psi = n\lambda/2\cos\theta_\psi. \quad (12.5)$$

The differences between these spacings and the stress-free value d_0 give the strains ϵ_{33} and $\epsilon_{\psi\psi}$. From the transformation rule for components of strain, a relation can be obtained between the two measured strains and the component of normal strain ϵ_{11} lying in the plane. Combination with the elastic stress-strain relation, assuming the material to be isotropic, then yields (Prob. 12.7)

$$\sigma_{11} = \frac{E(\epsilon_{\psi\psi} - \epsilon_{33})}{(1+\nu)\sin^2\psi} = \frac{E(d_\psi - d_3)}{(1+\nu)d_0\sin^2\psi}. \quad (12.6)$$

After repeating the process twice more to obtain three normal components lying in the plane of the plate, the entire state of stress can be determined.

◀ In deriving the above equation for the stress distribution, it was assumed that the isotropic elastic constants could be used. This is not strictly correct because the stiffness of the grains varies with orientation. The effect can be partly compensated for by experimentally determining the coefficient $E/(1 + \nu)$ in Eq. 12.6. The sensitivity of this method is of the order of 1000 psi, and while it is most applicable to annealed materials, improved techniques have made it possible to measure residual stresses even with hardened steels. For further information on the techniques of the measurement of stress by x-rays, see the report by the Society of Automotive Engineers (SAE, 1960). ▶

12.7 EFFECTS OF RESIDUAL STRESS

The effects of residual stress are the same as of any other stress, except that they can be eliminated with relatively small amounts of plastic flow. Residual stress may cause deformation and fracture, it accelerates certain phase transformations, it may accelerate corrosion, and it may increase internal friction.

In the elastic region, residual stress may increase the tendency to buckling. For example, consider the plate shown in Fig. 12.14 which contains a high residual tensile stress in the center and compressive stress along each edge. This plate will twist more easily than one without residual stress, since for large deflections of the plate, the edges will have an opportunity to elongate and the center to contract, thus relieving the residual stress. The corresponding decrease in strain energy will tend to offset the increase in strain energy due to the shear stress of twisting. Buckling can thus occur more readily.

Residual stress allows yielding to occur at lower values of the applied load, as illustrated in Fig. 12.15 (Prob. 12.8). This in turn leads to earlier buckling of columns in the elastic-plastic range (Osgood, 1952). Likewise, residual stress

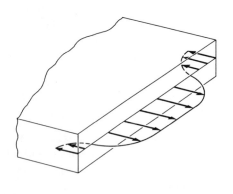

Fig. 12.14. Bar containing residual stresses tending to torsional buckling.

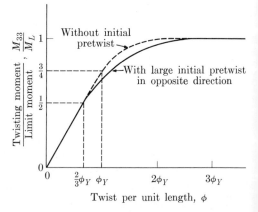

Fig. 12.15. Effect of residual stress on torque-twist curve. (Crandall and Dahl, 1959. Courtesy of McGraw-Hill.)

promotes creep of materials, but its effects die out when the inelastic strain approaches the elastic strain. One of the more significant effects of residual stress is on the distortion of parts that have been machined. The removal of layers containing residual stress leaves the part no longer in equilibrium in its original shape. This is the basis of the techniques of measurement of residual stress discussed in Section 12.6.

Residual stress also promotes fracture, particularly those modes of fracture in which relatively small plastic deformation is involved, or in which, due to the presence of notches, large plastic strains are obtained with small overall deflections. Such modes of fracture include fatigue, stress corrosion cracking, and brittle fracture.

The effect of residual stress on damping is due to raising of the total stress level to the point where dislocation motion can more readily occur under slight additional oscillatory loads. The resulting dislocation motion will tend to relieve the residual stress.

12.8 RELIEF OF RESIDUAL STRESS

Residual stress can be relieved by plastic strain or creep large enough to offset the elastic strain which gives rise to the residual stress. Examples are given by the stress relief illustrated in Fig. 12.16. If the cold-drawn bar were unloaded after stretching to a strain of 0.006, it would be found that the residual stress had almost disappeared. In general, tensile strain of the order of $2Y/E$ will relieve residual stress (Prob. 12.9). The absence of a lowered yield point in the stress-relieved material of Fig. 12.16 indicates that enough creep occurred during stress relief to relieve the residual stress (Prob. 12.14). In the presence of transformations such as tempering, the relaxation of residual stress can be even more abrupt (Maloof et al., 1953).

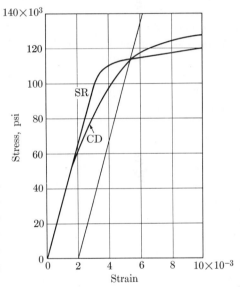

Fig. 12.16. Effect of stress relief at 900°F on yield strength of a cold-drawn, 0.45% carbon steel. CD, as drawn; SR, as strain-relieved. (Nachtman, 1955. Courtesy of ASME.)

Residual stresses can also be relieved by repeated stress, as indicated at the end of Section 12.7, and shown by Bühler and Buchholtz (1933). As pointed out by Ross and Morrow (1960), most of the relaxation occurs in the first few cycles. Both groups, as well as Fuchs and Mattson (1946), find relaxation of the order of 30 to 50% with stress amplitudes corresponding to the endurance limit. In softer materials, further reductions in residual stress are likely.

12.9 SUMMARY

Residual stresses are those remaining in a material in the absence of loads or of changes in temperature. They occur on a microscopic scale due to the anisotropic plastic behavior of individual grains, due to inclusions, and due to the presence of dislocations.

Residual stresses on a macroscopic scale can arise from forced alignment of parts during assembly, from loads causing nonuniform plastic flow or creep, from metal-forming, from nonuniform temperature changes large enough to cause plastic deformation or creep, and from volume changes due to metallurgical and chemical processes. Several of these sources are combined in causing the residual stresses due to heat treatment, welding, and casting. Macroscopic residual stress can be as high as the yield strength.

The effects of residual stress are the same as those of any other stress except that they can be relieved by small amounts of relaxation or plastic strain. Residual stresses can cause or augment deformation by buckling, plastic flow, and creep. In addition, they cause changes in dimension when parts are machined. Residual stress may be an aid to metallurgical transformations. As will be discussed later, residual stress interacts seriously with corrosion and can produce fracture.

The most common method of measuring residual stress is to observe the change in strain in the neighborhood of an area where metal has been removed by etching, milling, boring, or trepanning. Residual stress can be measured directly by measuring the distortion of the crystal lattice from its unstressed state. Another method is to measure the change in the hardness as measured by the pressure required to initiate an indentation.

Residual stresses are relieved by small amounts of permanent strain, of the order of the yield strain. These strains may be produced by uniform tension, by repeated stress (which in some cases produces creep), or by heating so that thermal creep occurs.

REFERENCES

Baldwin (1949) and ASM (1955) provide a further introduction to residual stress. The most comprehensive collection of data is that reported by Horger (1950), containing almost two hundred references. For an introduction to the determination of residual stress, see Heindlhofer (1948). The book edited by Osgood (1954) is a good mixture of theory and practice, especially in connection with structures. For other topics, consult the specific references given and the corresponding current literature, or the 1500 indexed abstracts of Huang (1954).

ALMEN, J. O.	1943	"Peened Surfaces Improve Endurance of Machined Parts," *Metal Prog.* **43**, 209–215, 270.
ASM	1955	"Residual Stresses," *Metal Prog.*, Second Supplement to 1948 *Metals Handbook* **68**, 2A, 89–96.

ASTM, AIME	1945	*Symposium on Stress Corrosion Cracking of Metals*, Philadelphia.
BALDWIN, W. M., JR.	1949	"Residual Stresses in Metals," *Proc. ASTM* **49**, 539–583.
BÜHLER, H. BUCHHOLTZ, H.	1933	"The Effect of Residual Stress on the Dynamic Bending Strength," *Mitteilungen Forschungs-Institut, Dortmund* **3**, No. 8, 235–248.
BÜHLER, H. BUCHHOLTZ, H.	1934	"Effect of Reduction and Cross Section by Cold Drawing on Residual Stresses in Rods," *Archiv für das Eisenhüttenwesen* **7**, 427–430.
COHEN, M.	1958	"Nucleation of Solid State Transformations," *Trans. Metall. Soc., AIME* **212**, 171–183.
CRANDALL, S. H. DAHL, N. C. (eds.)	1959	*An Introduction to the Mechanics of Solids*, McGraw Hill, New York.
DICKIE, H. A.	1931	Discussion of "Effect of Surface Conditions Produced by Heat Treatment on the Fatigue Resistance of Spring Steels," by G. A. Hankins and M. L. Becker, *J. Iron and Steel Inst.* **124**, 387–460.
DYACHENKO, P. E. PODOSENOVA, N. A.	1954	"Work-Hardening and Residual Stresses in Boring of Structural Steels," *Vestnik Machinostroyenya* **34**, 45–47, after Nachtman (1955).
FRISCH, J. THOMSEN, E. G.	1951	"Residual Grinding Stresses in Mild Steel," *Trans. ASME* **73**, 337–346.
FUCHS, H. O. MATTSON, R. L.	1946	"Measurement of Residual Stresses in Torsion Bar Springs," *Proc. Soc. Exp. Stress Anal.* **4**, 1, 64–73.
HEINDLHOFER, K.	1948	*Evaluation of Residual Stress*, McGraw-Hill, New York.
HENRIKSEN, E. K.	1957	"Hidden Troublemakers—Residual Machining Stresses," *Tool Engineer* **38**, 92–96.
HODGE, P. G., JR.,	1954	"Shakedown of Elastic-Plastic Structures," *Residual Stress in Metals and Metal Construction*, W. R. Osgood, ed.; prepared for Ship Structure Committee of National Academy of Sciences—National Research Council, Reinhold, New York, pp. 1–21.
HORGER, O. J.	1950	"Residual Stresses," *Handbook of Experimental Stress Analysis*, M. Hetenyi, ed., Wiley, New York, pp. 459–578.
HUANG, T. C.	1954	"Bibliography on Residual Stress," *Special Publication 125*, SAE.
LESSELLS, J. M. BRODRICK, R. F.	1956	"Shot-Peening as Protection of Surface Damaged Propeller Blade Materials," *Int. Conf. on Fatigue of Metals*, Inst. Mech. Engrs, London, 617–627.
MALOOF, S. R. ERARD, H. R. STEELE, R. K.	1953	Discussion to "The Effect of Quenching and Tempering on Residual Stresses in Manganese Oil Hardening Tool Steel," H. J. Snyder, *Trans. ASM* **45**, 605–616.

NACHTMAN, E. S.	1955	"Residual Stresses in Cold Finished Steel Bars," *Mech. Eng.* **77**, 886–889.
NOBLE, H. J.	1955	Quoted by ASM (1955).
OSGOOD, W. R.	1952	"The Effect of Residual Stresses on Column Strength," *Proc. of the 1st U.S. Nat. Con. Appl. Mech., ASME*, New York, 415–418.
OSGOOD, W. R. (ed.)	1954	*Residual Stresses in Metals and Metal Construction*, prepared for Ship Structure Committee of National Academy of Sciences—National Research Council, Reinhold Publishing Corp., New York.
ROSS, A. S. MORROW, J.	1960	"Cycle-Dependent Stress Relaxation of A286 Alloy," *Trans. ASME, J. Basic Eng.* **D82**, 654–660.
SAE	1952	"Manual on Shot-Peening," SAE *Special Pub. 84*.
SAE	1960	"The Measurement of Stress by X-Rays," Subcommittee on X-ray Procedures Division IV, Residual Stresses, of the Iron and Steel Technical Committee, SAE.
SAE	1964	*SAE Handbook*, Soc. Automotive Engrs., New York.
SETTY, S. K. LAPSLEY, J. T. THOMSEN, E. G.	1957	"Stresses Alter Hardness," *Mech. Eng.* **79**, 1127–1129.
TARASOV, L. P. HYLER, W. S. LETNER, H. R.	1957	"Effect of Grinding Conditions and Resultant Residual Stresses on the Fatigue Strength of Hardened Steel," *Proc. ASTM* **57**, 601–622.
TIMOSHENKO, S.	1941	*Strength of Materials*, Part II, 2nd ed., Van Nostrand, New York.

PROBLEMS

◀ 12.1. Derive the equation for the residual stress on the surface of a plate, Eq. 12.1.

12.2. What does the residual stress given by Eq. 9.28 predict for the effect of cutting speed on residual stress? See Dyachenko and Podosenova (1954).

◀ 12.3. Derive the equation for the surface stress in terms of the change in radius of curvature on removal of a surface layer, Eq. 12.2.

◀ 12.4. Derive the integral equation (12.4) for a point underneath the surface, taking into account the changes in stress at that point as previous layers were removed.

12.5. Show different states of residual stress in a surface layer which would lead to the same normal pressure required to initiate a hardness indentation.

12.6. Show two different residual stress patterns having an equal tendency to yield under normal pressure, but an unequal tendency to yield under the addition of a uniaxial stress.

◀ 12.7. Derive the equation for the residual stress in terms of change of interplanar spacing, Eq. 12.6.

12.8. (a) Justify the effect of residual stress on the points of initial deviation from linearity of the torque-twist curves of Fig. 12.15.

◀ (b) Derive the curve shown in Fig. 12.15 for the reversed torque-twist curve for a round bar after a large initial pretwist.

12.9. Show that tensile strains of the order of $2Y/E$ will relieve residual stresses in bars.

12.10. Discuss the effect of hardness on the residual stress shown in Fig. 12.17.

12.11. Compare the magnitudes of the residual stress reported in Table 12.1 with those which would be calculated from the idealized analysis summarized in Eq. 9.28.

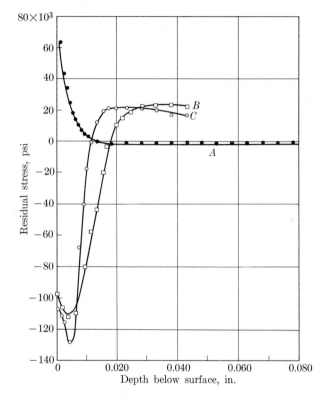

Symbol	Hardness	Coverage	Shot diameter, in.	Air pressure, psi
•	R_C 31, R_C 52	Not peened	0	0
□	R_C 31	High	0.039	50
○	R_C 52	High	0.039	50

FIG. 12.17. Effect of hardness on stress distribution. (Lessells and Broderick, 1956. Courtesy of Institution of Mechanical Engineers.)

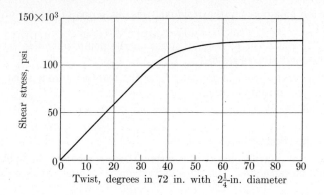

Fig. 12.18. Shear stress versus twist curve for steel torsion bar. (Fuchs and Mattson, 1946. Courtesy of Society for Experimental Stress Analysis.)

◀12.12. Fuchs and Mattson (1946) have given an example of a residual stress produced by twisting the torsion-bar springs used in a military vehicle. The shear stress-strain curve for their material, which was deduced from torque-twist data in the manner discussed in Section 9.3, is shown in Fig. 12.18. The bar was 6 ft long, $2\frac{1}{4}$ in. in diameter, made of SAE 9262 steel, with a Jominy end-quench hardenability of J-50 at 16/16. The bars were machined all over, hardened without decarburization, drawn to R_C49, and shot-peened with 0.055 in. diameter cast iron shot to an arc height of 0.014 in. on a C2 Almen strip, leaving biaxial residual compressive stress in a thin layer near the surface. (If desired, an explanation of the descriptions of hardenability and shot-peening may be found in the SAE Handbook, 1964.) Residual stress was introduced by twisting the bar 90° and unloading it. How much residual stress would be expected for this amount of pretwist? In carrying out the analysis, neglect the effect of surface residual stresses produced by the shot-peening. The spring-back observed by the authors was about 52°, and the actual residual stress is shown in Fig. 12.19.

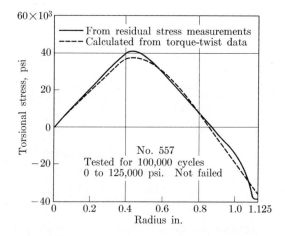

Fig. 12.19. Residual stresses in torsion spring. (Fuchs and Mattson, 1946. Courtesy of Society for Experimental Stress Analysis.)

12.13. Explain the effect of shot-peening on the residual shear stress of the torsion spring discussed by Fuchs and Mattson (1946) and referred to in Prob. 12.12. Does your expectation agree with the results reported in Fig. 12.19?

12.14. Estimate the order of magnitude of residual stresses before and after the stress relief illustrated in Fig. 12.16. If the part was at the stress-relief temperatures for about 30 minutes, estimate the order of magnitude of the creep rate and compare this with creep data (see Chapter 19).

12.15. Occasionally I beams split apart along the web (Osgood, 1954) as shown in Fig. 12.20.

(a) What longitudinal stress distribution must be present?

(b) Taking into account the boundary conditions at the end of the beam and St. Venant's principle, sketch the distribution of shear stress on the web where it joins the flange. From the equilibrium equations, what would you then expect for the transverse normal stress in the web near the end?

(c) How could the rolling procedure be modified to eliminate this residual stress?

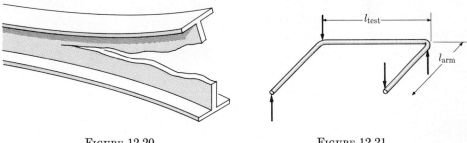

FIGURE 12.20   FIGURE 12.21

E 12.16. Coil a paper clip and a piece of spring wire around a pencil. Will they tend to coil up or uncoil on stress relief in a kitchen match flame? Try it and see. What will happen in torsion? To observe the effect, what proportion of test section to twisting arm will give the greatest deflection? What do you find? Can you develop a quantitative theory for it?

PART III
APPLICATIONS

CHAPTER 13

HARDNESS

13.1 SYNOPSIS

Hardness is a term that has different meanings to different people. For example, hardness is associated with:

1. Resistance to plastic deformation or resistance to wear to a design engineer.
2. Resistance to indentation or tensile strength to a tester of materials.
3. Resistance to cutting to a machinist.
4. Scratch resistance to a mineralogist.
5. Resistance to elastic deformation to the user of rubber.

While these meanings may appear to differ greatly from each other, the equivalent plastic flow stress Y of the material is a common factor in the first four. Much of the discussion to follow is concerned with relating the hardness values obtained from different testing machines to the flow stress of the material as measured in a tension or compression test. This not only gives insight into hardness-testing procedures, but makes it possible to derive conversion equations which relate one hardness scale to another.

In indentation hardness tests, a pyramid, cone, or ball is pressed into a flat surface. The restraint of the surrounding undeformed material means that most of the applied load merely develops hydrostatic compressive stress which is incapable of causing a metal to flow plastically. The approximate plastic stress analysis discussed in Chapter 10 shows that the hardness, measured as mean normal stress under an indenter, is about three times the tensile strength of a ductile material. More precisely, for strain-hardening materials, we shall see that the hardness is about three times the flow stress at 8% strain.

In the case of rubber, we shall see how the elastic equations for contact stress lead to a relation between the resistance to penetration or hardness of rubber and its modulus of elasticity.

13.2 INDENTATION TESTS

Indentation tests differ not only with regard to the shape of the indenter* (ball, cone, or pyramid) but in addition may differ as follows:

1. By use of a fixed load P, and measurement of the resulting diameter d, or width of the impression at the surface (Brinell, Vickers).

---

* "Penetrator," "indent*or*," "indent*er*:" "Penetrator," although used in standards, implies passage *through* the specimen rather than only *into* it. The suffix "-er" is common on newer and less legal words, even with Latin roots, e.g., "computer" and "propeller."

2. By use of the contact area A_c in computing the mean unit load on the indenter, P/A_c, or by use of the projected area of the impression at the surface A.

3. By use of a fixed load P, and measurement of the resulting depth of impression t (Rockwell).

4. By use of a variable load P to produce a given depth of impression t (Monotron).

5. By use of a large (3000 kg), medium, or small (1 to 300 gm) load on the indenter to produce impressions that range in size from macroscopic to microscopic. Large impressions may constitute a destructive test for some applications, while *microhardness* tests allow investigation of the individual constituents of an alloy.

The standard Brinell test, introduced by the Swedish engineer Brinell (1900), uses a 10-mm diameter steel or tungsten carbide sphere that is pressed into the flat surface of a test specimen under a load of 3000 kg. The load is maintained for 30 seconds and the diameter, d, of the remaining impression at the surface is measured with a microscope (10 ×) after the ball has been removed. The Brinell hardness number B (or BHN) is defined in kg/mm² as the ratio of load to contact area A_c, which is in turn given in terms of the ball diameter D and the impression diameter d or the depth of the impression t:

$$B = \text{BHN} = \frac{P}{A_c} = \frac{2P}{\pi D(D - \sqrt{D^2 - d^2})} = \frac{P}{\pi D t}. \tag{13.1}$$

Tables of the Brinell hardness for different values of the impression diameter are available (ASTM, 1961; ASM, 1961).

Standard values of load and ball diameter for use with steel are 3000 kg and 10 mm, respectively. For softer materials, or where a large-diameter impression may be objectionable, the load and ball diameter should be adjusted to keep the ratio d/D within the range 0.3 to 0.5. A tungsten carbide ball should be used in place of a steel ball for Brinell hardness greater than 500 kg/mm² in order to avoid false (low) readings due to deformation of the ball (Prob. 13.1).

The nearest edge of the specimen should be no closer than $2\frac{1}{2}$ impression diameters, and the thickness more than one diameter (Prob. 13.2). Since d/D is normally less than 0.5, this means that for a 10-mm ball, the uninterrupted width and depth of specimen may have to be as great as 1 and $\frac{1}{2}$ in., respectively, to avoid spurious side and bottom effects.

Like many tests and concepts adopted early in the development of a technology, the Brinell test has certain disadvantages.

1. It is cumbersome in using the contact area A_c in place of the projected area A.

2. It employs an indenter that does not provide geometrical similitude for impressions of different size d. The Brinell hardness of a given material is not constant for all values of d, as for a conical or pyramidal indenter, but will be a function of d/D due to the varying inclination at the top of a spherical sector.

3. It usually employs large balls and large loads and gives large impressions that are sometimes objectionable.

To overcome the first of these disadvantages, Meyer (1908) proposed that the projected area A be used in place of the contact area A_c. The hardness value obtained by use of the projected area, with a ball indenter, is often called the Meyer ball hardness:

$$\mathrm{M_B} = \frac{4P}{\pi d^2} \text{ kg/mm}^2. \tag{13.2}$$

This method not only allows the hardness value to be more easily computed, but provides a number that represents a better approximation to the average pressure on the area of contact. This is due to the fact that the lateral forces on the sloping surface tend to counteract each other. If there is no friction and the pressure between ball and material is uniform, the pressure is exactly given by Eq. 13.2 (Prob. 13.3).

◀To provide geometrical similitude under different values of load, Smith and Sandland (1922) proposed that a pyramid be substituted for a ball. The angle between opposite faces of their indenter was made 136°, as shown in Fig. 13.1. The indenter was made of diamond and relatively low loads (< 120 kg) were employed to provide impressions of a smaller size. The Vickers hardness value V is obtained by dividing the load P on such an indenter in kilograms by the contact area A_c of the impression in mm², which can in turn be found from the mean diagonal d_1 (Prob. 13.4):

$$V = \frac{P}{A_c} = \frac{1.854P}{d_1^2}. \tag{13.3}$$

The corresponding Meyer-Vickers hardness $\mathrm{M_V}$ is

$$\mathrm{M_V} = \frac{2P}{d_1^2} = \frac{V}{0.927} \text{ kg/mm}^2. \tag{13.4}$$

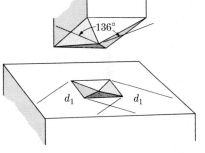

FIG. 13.1. The Vickers pyramidal indenter.

◀The lack of geometrical similarity found using a spherical indenter at constant load can also be overcome by measuring the load required to produce an indentation of constant depth, as is done with the Monotron machines. A diamond point with a spherical tip of 0.375-mm radius is impressed to a standard depth of 0.046 mm, and the required load in kilograms is reported. Equating force per unit projected area, using Eq. 13.1, one would expect the Brinell hardness to be about 9.2 times the Monotron hardness, although the manufacturer's curves give a factor of between 7 and 8. An interesting application of the Monotron machine is in the simultaneous measurement of force and penetration on a small, trepanned cylinder, with its lower end fixed in the specimen. One can thus obtain a stress-

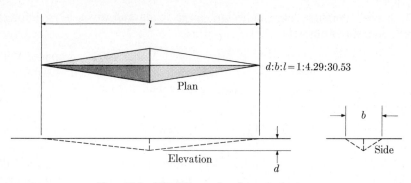

FIG. 13.2. The Knoop hardness indenter.

strain curve in compression from a very small specimen. Orowan (1958) has suggested the use of this machine for convenient studies of the plastic anisotropy of materials.

◀When the hardness of a thin specimen or of a brittle material has to be determined, it is desirable that the impression be shallow and of small volume. It is also necessary to have a small impression when determining the hardness of microstructural constituents of an alloy, or when a large impression would be objectionable in a part to be put into service. Mott (1956) has described and discussed in detail a large number of microhardness tests. The Vickers indenter may be used for microhardness tests. Alternatively, to give a long dimension to measure for a minimum indentation, Knoop et al. (1939) proposed the geometry shown in Fig. 13.2. This provides an impression diagonal in one direction that is 7.11 times as long as that in the other direction. The long diagonal l is the one that is used in computing the hardness number. Unlike the Vickers and Brinell tests, the Knoop hardness number is based on the projected area (ASM, 1948, p. 96) and is given by

$$K = P/A = 14.2\, P/l^2. \tag{13.5}$$

◀The advantages of the Knoop indenter over the Vickers in microhardness testing lie in the fact that a longer diagonal is obtained for a given depth of impression or a given volume of material deformed. It can be shown that for the same measured diagonal length (d_1 for Vickers impression = l for Knoop), the depth and area of the Knoop indentation are only 15% of those for the Vickers. The Knoop indenter is thus advantageous when shallow specimens or thin, hard layers must be tested. The Knoop indenter is also desirable for brittle substances where the tendency for fracture to occur is related to the area of stressed material.

◀The indenting apparatus for microhardness testing is usually built into a microscope used to locate the point to be indented and to measure the indentation. When working at the lightest loads it is important that the load be applied without vibration, and that the same point be returned to the center of the field of the microscope without difficulty. Careful polishing in advance is required for even, sharp impressions.▶

To eliminate the time required for microscopic observation of the indentation, Rockwell (1922) introduced a test in which a spherical or conical indenter is forced into a surface and the depth of the penetration is measured by a dial gauge. The dial gauge is set to a reference mark while a minor load is applied, in order to provide a firm contact with the surface. The major (total) load is then applied. To eliminate elastic effects, the net change in penetration is measured after returning the load to the minor value. To provide a scale of values which increases with hardness, the Rockwell hardness number is defined by an equation of the following form, and can be read directly from a suitably inscribed dial gauge:

$$R = C_1 - C_2 \, \Delta t_{mm}. \tag{13.6}$$

Various scales use different indenters, loads, and coefficients in Eq. 13.6 and are given in Table 13.1. Of these the most common are the Rockwell C (cone) and B (ball) scales, used for hard and soft metals, respectively. (In case of doubt, the C scale is used first to avoid damaging the ball.) The superficial scales are used with thin specimens or to minimize the impression. For instance, the Rockwell B scale may require a thickness of at least 0.04 in., whereas superficial scales are generally suitable for specimens half as thick (ASTM, 1961; see also Prob. 13.2).

Although the Rockwell hardness numbers increase with Brinell hardness, the two are not proportional and the dimensions of Rockwell hardness are not force per unit area. In fact the Rockwell hardness number cannot be assigned any dimensions, since it is defined in terms of an arbitrary equation (13.6). In order to use Rockwell hardness numbers in engineering computations, they must first be converted into Brinell or Meyer hardness numbers, which do have dimensions and physical significance.

◀Rockwell (1922) suggested such a conversion, based on an analysis in order to determine the *form* of the equation, and then more precise determination of the constants in this equation by the use of empirical data. Referring to Eq. 13.1, we find that

$$t = \frac{A_c}{\pi D}. \tag{13.7}$$

From Eqs. 13.1 and 13.6

$$R = C_3 - \frac{C_2 \, \Delta P}{\pi B D}. \tag{13.8}$$

For the Rockwell B scale, $C_3 = 130$, $\Delta P = 90$ kg, and $D = 1.59$ mm ($\frac{1}{16}$ in.), so

$$R_B = 130 - \frac{9000}{B}. \tag{13.9}$$

While this should be close to the correct *form* of the relation between Rockwell B and Brinell hardness, we should not be surprised to find somewhat different numerical coefficients, since several approximations have been made in deriving Eq. 13.9 (Prob. 13.5). When values of R_B are plotted against the corresponding values of $1/B$ (Fig. 13.3) it would appear that Eq. 13.9 is of the correct form and

TABLE 13.1

Scales for Rockwell Hardness Numbers

Symbol Indenter	Minor (Pre-) load, kg	Major (Total) load, kg	Coefficients in $R = C_1 - C_2 \Delta t$		C_3	Approximate coefficients in $B = C_4/(C_3 - R)$ or $B = [C_5/(C_3 - R)]^2$	
			C_1[1]	C_2 mm^{-1}		C_4 kg mm^{-2}	C_5 kg$^{1/2}$ mm^{-1}
Normal scales							
R_B $\tfrac{1}{16}$ ball[2]	10	100	130	500	134	6700	—[5]
R_C cone[3]	10	150	100	500	115	—	1500[5]
R_A cone	10	60	100	500	100	—	750
R_D cone	10	100	100	500	100	—	1110
R_E $\tfrac{1}{8}$ ball	10	100	130	500	130	6000	—
R_F $\tfrac{1}{16}$ ball	10	60	130	500	130	12,000	—
R_G $\tfrac{1}{16}$ ball	10	150	130	500	130	14,000	—
Superficial scales							
R_{15N} cone[4]	3	15	100	1000	100	—	690
R_{30N} cone	3	30	100	1000	100	—	1240
R_{45N} cone	3	45	100	1000	100	—	1630
R_{15T} $\tfrac{1}{16}$ ball	3	15	100	1000	100	2400	—
R_{30T} $\tfrac{1}{16}$ ball	3	30	100	1000	100	5400	—
R_{45T} $\tfrac{1}{16}$ ball	3	45	100	1000	100	8400	—

[1] Red scale for $C_1 = 130$; black scale for $C_1 = 100$.
[2] Ball is steel of the diameter shown in inches.
[3] Normal cone is diamond with 120° included angle and a spherical apex of 0.2-mm radius.
[4] Superficial cone is similar to normal cone but not interchangeable.
[5] These equations are empirical and good to about 20%; other coefficients are based on load per unit area and other approximations, and should be checked against experience such as ASM (1948, pp. 98–101).

that more accurate values for the coefficients are

$$R_B = 134 - \frac{6700}{B} \pm 7. \tag{13.10}$$

The range of ±7 includes 95% of all the test results for values of Brinell hardness ranging from 80 to over 500 kg/mm².

◀For conical indenters, even cruder approximations show the Rockwell hardness to be given in terms of the included half angle θ by (Prob. 13.6):

$$R_C = C_3 - C_2 \sqrt{\frac{P_{\text{major}}}{\pi B \tan^2 \theta}} \left(1 - \sqrt{\frac{P_{\text{minor}}}{P_{\text{major}}}}\right) = C_3 - \frac{C_5}{\sqrt{B}}. \tag{13.11}$$

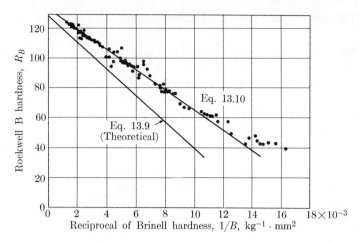

FIG. 13.3. Variation of Rockwell B hardness with the reciprocal of Brinell hardness (Data from de la Macorra, 1923.)

For the C scale, the coefficients in Eq. 13.11 are 100 and 1480, respectively, whereas coefficients determined empirically by Cowdrey and Adams (1944) are 122 and 1590, and by Tabor (1951) are 124 and 1510.

◀ The effect of rounding the apex can be neglected for combinations of metals and hardness scales such that the Brinell hardness is less than 30 times the minor load in kilograms (Prob. 13.7). When the Brinell hardness is greater than 30 times the major load, the behavior becomes that of a ball indenter (Prob. 13.8). More accurate empirical tables of conversions between hardness numbers are given in many places, for example by the ASM (1948) and the ASTM (1961), and are plotted in Fig. 13.4. ▶

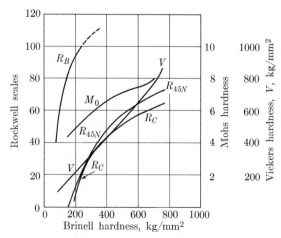

FIG. 13.4. Hardness numbers as functions of Brinell hardness. Tungsten carbide ball used above 500 kg/mm². Note that roughly $V = B = 10R_C = 10R_{45N}$. (Data from ASM, 1961 and Tabor, 1951.)

13.3 OTHER MEASURES OF PLASTIC HARDNESS

A. File Hardness. The *file hardness* test is a useful one that is partly of the cutting and partly of the scratching variety. A fully hardened file ($R_C 66$ or $B = 700$ kg/mm^2) is drawn by hand across a piece of steel. The file ceases to cut and merely rubs the hardened surface when the hardness is within 20% of the hardness of the file (i.e., $B = 580$ kg/mm^2 or harder). In quenched but not tempered steel, a skilled operator can immediately detect a soft decarburized layer and estimate its depth by means of a micrometer.

◀ **B. Scratch Hardness.** Mohs (1822) devised a hardness scale for use by mineralogists. He carefully selected minerals of increasing hardness and numbered them from 1 to 10. If an unknown mineral would scratch one of the standard materials but not the next harder one, its hardness was identified. Such tests define *scratch hardness*.

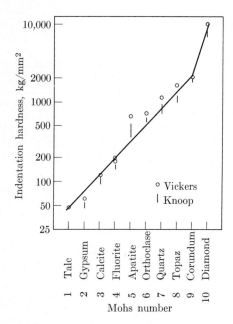

FIG. 13.5. Relation between Mohs hardness and indentation hardness. (Tabor, 1954. Courtesy of Imperial Chemical Industries. Data from Taylor, 1949 and Winchell, 1945.)

◀Tabor (1956) studied the meaning of scratch-hardness tests by preparing long steel surfaces having uniformly varying hardness from one end to the other. When a pointed member was dragged across the specimen it was found to cease plowing when its hardness was just 20% higher than the hardness of the plate. This was found to be the case regardless of the geometry of the scratching point and its hardness level. These experiments suggest that it should be possible to devise a *uniform* hardness scale of the Mohs type and that the minimum hardness ratio between adjacent members of the scale should be 1.2. Tabor then set about to study the characteristics of the actual Mohs scale. He plotted the log of the indentation hardness (Vickers or Knoop) against the Mohs hardness number M_0 (Fig. 13.5) and found that except for diamond, the hardness intervals were equal.

Each member was 60% harder than the preceeding one, and for the first nine standards ($M_0 = 1$ to 9),

$$V = V_T(1.6)^{M_0-1}, \qquad (13.12)$$

where V is the Vickers hardness (kg/mm^2) and V_T is the Vickers hardness for talc (kg/mm^2). It would have been physically possible for Mohs to have introduced two additional standards between each of those he adopted.

◀ These experiments of Tabor clearly show the close relationship between scratch hardness of the Mohs type and indentation hardness. Both hardness tests measure primarily the same thing—the relative resistance to *plastic flow*. It is important to note that this is so even though most of the Mohs standards are normally brittle minerals. As we shall see in Chapter 15 on brittle fracture, due to the very high hydrostatic pressure at the tip of the scratcher and the small volume deformed in making a scratch (size effect), materials that are normally brittle may behave in a plastic manner.

◀ Of the several microscratch tests for hardness that have been devised, the Bierbaum (1930) microcharacter is probably the most widely known. This instrument draws a diamond across a polished metallographic surface under a load of 3 gm. The width λ of the resulting groove is measured with a microscope in microns. The microhardness number is then

$$B_i = 10^4/\lambda^2. \qquad (13.13)$$

This method is used to study the hardness of metallographic constituents in alloys. The relation between Bierbaum and Vickers hardness is roughly (Prob. 13.9)

$$V = B_i. \qquad (13.14)$$

◀ **C. Cutting Hardness.** When a cutting tool (Fig. 13.6) is caused to produce a chip of fixed dimensions at low speed, it is found that the force in the direction of motion F_p is related to the indentation hardness of the material. Because the contact area is difficult to measure, it is customary in such tests to record the force per unit undeformed chip area (i.e., F_p/bt). This quantity is really the work done per unit volume, u:

$$u = \frac{\text{Work per unit time}}{\text{Volume removed per unit time}} = \frac{F_p V}{Vbt} = \frac{F_p}{bt}. \qquad (13.15)$$

It should be noted that in this type of hardness test, friction is not negligible. Experiments show that about one-fourth of the energy per unit volume is associated with friction and three-fourths with plastic flow.

◀ Values of specific work u in kg · mm/mm^3 are presented in Table 13.2 for values of b and t of 2.5 and 0.25 mm, respectively. These will be compared with other hardness numbers in Section 13.4, but as they stand, they can be used to estimate cutting forces and powers. There is a restriction, however, to speeds below 1 fpm or above about 500 fpm. At intermediate speeds a built-up nose tends to form. This nose seriously alters the geometry and sometimes the forces.

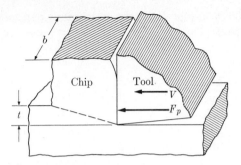

Fig. 13.6. Metal-cutting process.

◀ **D. Rebound Tests.** An object of standard mass and dimensions is dropped through a given distance onto the test surface and the height of rebound is determined. The height of rebound increases with the hardness of the test surface, since less energy is consumed in plastic deformation. The Shore scleroscope is an instrument of this type. For a discussion of the mechanics of these tests see Goldsmith and Yew (1962).

◀ **E. Damping Tests.** A sphere resting on the test surface is used to support a rocking arm of fixed mass and geometry. The rocking pendulum arm is set into periodic motion and the rate of decrease in amplitude due to damping on the spherical surface is noted. The degree of damping decreases with increased hardness. The Herbert (1923) pendulum hardness test, which uses a 1-mm diamond sphere, is a test of this type.

TABLE 13.2*

Specific Work u Associated with Metal Cutting
(Cut 2.5 mm wide by 0.25 mm deep)

Metal	Specific work u, kg/mm^2
Lead	18
Magnesium	42
Aluminum	56
Copper	280
Silver	120
Nickel	245
Cobalt	175
Titanium	210
Cast Iron ($B = 200$)	280
Steel ($B = 200$)	245
Steel ($B = 300$)	280
Steel ($B = 400$)	400

* From Shaw (1960). Courtesy of M.I.T. Press.

◀ **F. Abrasion Tests.** A small high-speed abrasive wheel is pressed against the test surface under a fixed load for a standard time. The size of the wear scar is the measure of hardness in this test.

◀ **G. Erosion Tests.** Sand or abrasive grain is caused to impinge upon a polished test surface under standard conditions. The loss of reflectivity of the surface is a measure of hardness in this case. ▶

13.4 INDENTATION TESTS AND STRESS-STRAIN BEHAVIOR

The hardness test is simple to execute, but to determine the flow stress from it requires in principle a solution for the deformation field around the indentation. Such solutions depend on the stress-strain relations of the material. If rate effects are negligible, the stress-strain curve can sometimes be roughly approximated by an equation of the form

$$\bar{\sigma}_Y = Y_1 \epsilon^n. \tag{13.16}$$

Analytical solutions have been found only in two extreme cases: the non-strain-hardening material ($n = 0$), and the elastic material ($n = 1$) which approximates a linearly strain-hardening material.

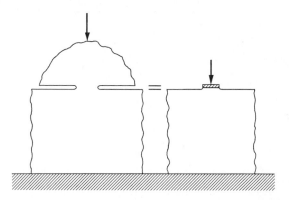

Fig. 13.7. Equivalence of notched compression and hardness tests.

A. Conical and Pyramidal Indenters. The simplest way of estimating the hardness of a material whose flow stress is independent of strain is to regard a hardness test as the inverse of the notched tensile test discussed in Section 10.6. As indicated in Fig. 13.7, this can be done by first thinking of the notched compression test, and then replacing the top half of the specimen by the hardness indenter. We thus conclude that the hardness, measured as force per unit area, should be approximately

$$H = P/A \approx 3Y. \tag{13.17}$$

◀ More precisely, by assuming plane strain conditions, a flat punch, and neglecting the piling-up of the metal which changes the shape around the indentation, we can apply the Prandtl solution (Fig. 10.16) (Prob. 13.10):

$$H = (2 + \pi)Y/\sqrt{3} = 2.97\ Y. \tag{13.18}$$

◀The flow fields can also be worked out for a wedge of finite angle with friction. The results are shown in Fig. 13.8 for a coefficient of friction which is just at the transition from a sliding to a sticking mode along the face of the wedge. Lee (1952) has pointed out that in this sticking mode the plastic zone must extend ahead of the tip of the indenter because the material there could not remain rigid in the presence of a plastic stress field around it. Thus a built-up edge forms in front of the indenter in the area marked OAC in Fig. 13.8. The mean pressure is slightly higher with sticking, $2.81Y$ as compared to $2.78Y$, because of a very small change in angle of the free surface (Prob. 13.11).

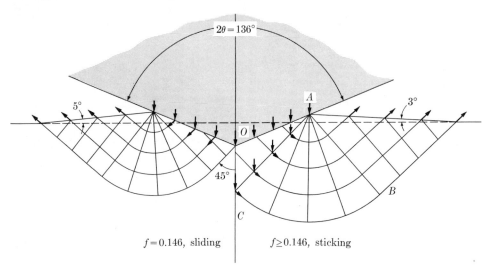

FIG. 13.8. Slip-line fields for wedge indentation at the transition between sliding and sticking.

◀As for other shapes of indenters, a solution has been obtained for a circular flat punch, using the maximum shear stress yield criterion, which gives $H = 2.9Y$ (Shield, 1955; Levin, 1955). Where the solutions indicate a rigid nose ahead of the indenter, they are also applicable to other shapes of indenter which would lie within the nose, provided that the friction is high enough to maintain rigidity.

◀These various results confirm Eq. 13.17 for a non-strainhardening material. To see the effects of strain-hardening, we go to the limiting case of linear hardening, $n = 1$ in Eq. 13.16, and neglect the effects of changes in direction of the stress components acting on any element, so that the Hencky equations of plasticity (Section 7.5), or the elastic equations, become applicable. From Sneddon (1951), one can obtain the Meyer hardness for the plane strain wedge and for the cone

TABLE 13.3*
Ratio of Hardness to Yield Strength for Cold-Worked Metals

Metal	Y, kg/mm^2	M_V, kg/mm^2	M_V/Y
Tellurium-lead	2.1	6.7	3.2
Aluminum	12.3	39.5	3.2
Copper	27	88	3.3
Mild steel	70	227	3.2

* From Tabor (1956).

indentation in terms of the wedge or cone half angle θ (Prob. 13.12):

$$M_{\text{wedge}} = [1 + \ln 2][E \cot \theta/2]/2(1 - \nu^2),$$
$$M_{\text{cone}} = [E \cot \theta/2]/2(1 - \nu^2). \tag{13.19}$$

Taking $\nu = \frac{1}{2}$ for incompressibility, these equations indicate that (Prob. 13.13)

For the wedge, $M/3 =$ uniaxial flow stress at $\epsilon = 0.152$.

For the cone, $M/3 =$ uniaxial flow stress at $\epsilon = 0.090$. (13.20) ▶

Experimental results for non-strainhardening materials obtained using a Vickers diamond indenter on fully cold-worked materials are presented in Table 13.3. The relation is nearly $H = 3.2Y$ for a variety of materials of widely different hardness.

The equation for strain-hardening materials, Eq. 13.20, was actually first found experimentally by Tabor (1951) by comparing the Meyer hardness before straining with the tensile flow stress $\sigma(\epsilon)$ for the same material after it had been subjected to a tensile plastic strain of ϵ. Regardless of the initial state of strain-hardening, the ratio $M_V/\sigma(\epsilon)$ had a value close to the value $M_V/Y = 3.2$ for a fully strain-hardened material, when ϵ was 0.08. Tabor's results for $\epsilon = 0.08$ are given in Table 13.4 and indicate good agreement with the value of 0.090 of Eq. 13.20.

◀ It is also of interest to analyze the action of a hardness indenter from the point of view of energy, since it provides an estimate of the volume of metal plastically deformed by an indenter, and since the energy u per unit volume absorbed by the plastically deformed metal is a convenient alternative to strain in describing the intensity of plastic deformation in other hardness tests, in particular the cutting test.

◀ For a pyramidal or conical indenter, similarity indicates that the load will vary in proportion to the area. The average work per unit volume, $\bar{u}$, will, for non-hardening materials, be the yield strength in tension multiplied by the average equivalent shear strain $\bar{\epsilon}$. Equating the total work done by the indenter to the deformed volume V_d times the average plastic work per unit volume, and noting the ratio between yield strength and hardness, it can be shown that the ratio of

TABLE 13.4*

Ratio of Hardness to Flow Stress at 8% Strain

Initial strain ϵ_0	Final strain $\epsilon = \epsilon_0 + 0.08$	$\sigma(\epsilon)$, kg/mm^2	3.13 $\sigma(\epsilon)$	$(M_V)_0$, kg/mm^2
		For Mild Steel		
0 (annealed)	0.08	55	172	168
0.06	0.14	62	194	191
0.10	0.18	66	206	202
0.13	0.21	67	210	208
0.25	0.33	73	228	226
		For Annealed Copper		
0	0.08	15	48.5	42.0
0.06	0.14	20	64.7	62.5
0.125	0.205	23.3	75.5	74.5
0.175	0.255	25	81.0	82.0
0.25	0.33	26.6	86.0	87.5

* From Tabor (1951).

deformed volume to the volume of the indentation is approximately (Prob. 13.14)

$$V_d/V_i = M_V/Y\bar{\epsilon} = 3.2/\bar{\epsilon}. \qquad (13.21)$$

Thus for a Vickers indenter for which the average strain is about 0.08, the *linear* dimensions of the deformed volume will be about 3.4 times those of the indentation. While this may seem excessive in view of the two-dimensional, fully plastic fields of Fig. 13.8, it has been pointed out by Hill (1950, p. 256) that outside the fully plastic region there is a much larger region of material that has undergone plastic strain of the elastic order of magnitude, and which must be present for the slipline field to hold (Prob. 13.2).

◀ The energy per unit volume absorbed in the Vickers test is further seen to be but $2\frac{1}{2}\%$ of the Meyer hardness M_V. Thus, if M_V is 200 kg/mm^2, $\bar{u}$ is but 5 kg/mm^2. In metal cutting, Table 13.2 showed the energy $\bar{u}$ per unit volume to be approximately equal to the Meyer hardness M_V, or 40 times larger than the energy per unit volume in the Vickers test. The reason for this does not lie in the presence of friction in the cutting hardness tests, but is due primarily to the difference in plastic strain in the two cases. Cutting involves plastic strain of the order of 3 while the Vickers test involves a strain of but 0.08. It is evident that the cutting test may be more useful than ordinary indentation tests when values of hardness at very high strain are needed. ▶

B. Spherical Indenters. For non-strainhardening materials, the presence of a built-up nose makes the conical solutions valid for spherical indenters if the friction is high enough or the penetration small enough. A ratio $M_B/Y = 3.2$ is again expected.

◀ For a strain-hardening material, we turn to the limiting case where $n = 1$, so that elastic equations are approximately valid. The Hertz equation (11.32) then gives the load P in terms of the indenter diameter D and indentation diameter d (Prob. 13.15)

$$P = \frac{4}{9}\frac{Ed^3}{D}. \tag{13.22}$$

Note that now the load varies as the cube, rather than as the square, of the indentation diameter, as a combined result of strain-hardening and progressive changes in shape. The relation between hardness and flow stress becomes (Prob. 13.16)

$$M_B/3 = \text{uniaxial flow stress at } \epsilon = (16/27\pi)(d/D)$$
$$\text{i.e., at } \epsilon = 0.189\,(d/D), \tag{13.23}$$

indicating a dependence on the degree of deformation.

◀ Experimentally, Meyer (1908) had found that when the applied load P was plotted against impression diameter d for a ball of diameter D, a straight line was obtained on log-log coordinates, indicating a relation of the form

$$P = C\left(\frac{d}{D}\right)^m, \tag{13.24}$$

where C is a constant and m is a quantity now referred to as the Meyer index. As the theoretical analysis indicates, this might be expected to vary from 2 for a nonhardening material to 3 for a linearly hardening material. Table 13.5 indicates that linear interpolation between these extremes is a good approximation to the actual data.

TABLE 13.5*

COMPARISON OF MEYER EXPONENT m
AND STRAIN-HARDENING EXPONENT n

Material	$m - 2$	n
Mild steel A	0.25	0.259
Yellow brass	0.44	0.404
Yellow brass, cold-drawn	0.10	0.194
Copper L	0.45	0.414
Steel 1A	0.25	0.24
Steel 6A	0.28	0.18
Nickel, annealed	0.50	0.43
Nickel, cold-rolled	0.14	0.07
Aluminum, annealed	0.20	0.15

* From O'Neill (1944).

When experiments similar to those leading to Table 13.4 are performed to find the equivalent strain under a spherical indenter, it is found that the strain varies linearly with d/D and that

$$B/3.2 = \text{uniaxial flow stress at } \epsilon = 0.2(d/D), \tag{13.25}$$

which is very close to that predicted by Eq. 13.23, and becomes equal to that for a Vickers indentation (0.08) when $d/D = 0.4$.

◀While the lack of geometric similarity introduced by a spherical indenter represents a complication in hardness interpretation, the Meyer index provides an approximate, although insensitive, measure of the tendency for a material to strain-harden. The difficulty is that we confine our study to very small strains (usually 0.1 or less). The tensile test (corrected to uniaxial stress) provides a much better measure of strain-hardening, since it is normally carried to strains of 0.5 or more before the test is terminated by fracture.

◀Besides the effects on the Meyer index, solutions for nonhardening and linearly hardening materials differ in that the nonhardening materials pile up around the indenter, whereas in the linearly hardening ones the indenter sinks in. Experiments show this characteristic also varies linearly between the two extremes (McClintock and Rhee, 1962). Although this method of interpolation between linearly hardening and nonhardening solutions is subject to some limitations, as discussed in Section 11.4 in connection with strain concentrations, its success in this case gives one more confidence in estimating complicated situations by interpolation between idealized extreme cases. As an added philosophical point, it is worth noting that the empirical equations preceded the theoretical ones, but that the presence of the theory now gives us more confidence in using the equations in new or more extreme situations. ▶

In conclusion, the ratio of hardness to flow stress is close to 3.2, provided that the flow stress is evaluated at a strain of 0.08 for conical indenters, and $0.2 \, d/D$ for spherical indenters. For materials whose tensile strength occurs at small strains, and for pyramidal indenters or for spherical indenters with d/D in the neighborhood of 0.4, the tensile strength in psi is about 500 times the hardness in kg/mm^2, so that a metal having a Brinell hardness of 200 kg/mm^2 will have a tensile strength of about 100,000 psi (Prob. 13.17).

13.5 DATA AND EFFECTS OF VARIABLES

Hardness data for a variety of materials are given in Tables 13.6 and 13.7. As to the effects of variables, a hardness test is a mode of plastic deformation, so variables that affect plastic deformation have similar effects on hardness.

A size effect arises from the small volume affected, particularly in the microhardness test. So long as the size of the impression is large compared with the mean slip-band length, normal hardness values will be obtained. However, as the load is decreased to the point where the impression size approaches the slip-band length at the equivalent strain, the hardness values begin to increase. Another source of the effect will be observed when the strain gradients around the impression require dislocation densities larger than those normally encountered in cold-worked metals at comparable levels of strain (Prob. 13.18). Values of Knoop hardness versus load are shown in Fig. 13.9 for several hard materials. When hardness is plotted against depth of impression in microns (1 micron = 10^{-6} m) all curves are found to be of similar shape and to turn upward for depths of impression less than 1 or 2 microns.

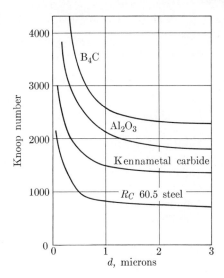

Fig. 13.9. Variation of Knoop hardness with depth of impression for several hard materials. (Shaw, 1952. Courtesy of Franklin Institute. Data from Thibault and Nyquist, 1947.)

In brittle materials such as glass an even more striking size effect is observed. As will be further discussed in Chapter 15, brittle materials such as glass contain many surface cracks. If the indenter engages one or more of these, the specimen will be broken before a fully developed impression can be made. If an impression can be made smaller than the mean crack spacing, it is possible to produce a more fully developed indentation in a brittle material. Thus brittleness is a function of specimen size. Marble, which is so brittle that it will break into many pieces if a Vickers test is attempted, may be indented successfully if the load is sufficiently small. The three impressions shown in Fig. 13.10 were made at loads of 15, 10, and 5 gm, respectively. While impressions (a) and (b) show cracks, that at (c) is free of cracks. There is some evidence that the cracks in specimens such as (b) are not present while loaded, but appear on unloading.

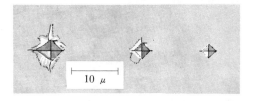

FIG. 13.10. Hardness indentations in marble under 15, 10, and 5-gm loads, respectively. (Shaw, 1954. Courtesy of National Academy of Sciences.)

In many practical applications the variation of hardness with temperature is important. This is the case with metal-cutting tools which must retain their hardness in the temperature range from 1000 to 1500°F. The manner in which

TABLE 13.6*

Typical Brinell Hardness Numbers

Material†	B, kg/mm²	$\dfrac{T.S., \text{psi}}{B, \text{kg/mm}^2}$
Soft rubber	0.02	—
Sodium	0.07	—
Hard rubber	1	—
Indium	1	—
Lead	4	670
Tin	5	625
Cellulose acetate	8	—
Vinyl resin	15	—
Polystyrene	20	—
Bakelite	55	—
Pure aluminum	15	490
cold-rolled	40	505
Cadmium	23	535
Silver	25	805
94 Al-6 Mg	28	640
Zinc	30	—
Sodium chloride	30	—
Gold	30	670
Magnesium	30	500
Copper	35	705
cold-rolled	110	510
Platinum	40	500
Duralumin (4.4 Cu, 1.5 Mg, 0.6 Mn, Bal Al)	40	505
aged	100	560
cold-rolled	120	560
Iron	70	640
Nickel	75	600
90 Cu-10 Sn Bronze	80	530
60 Cu-40 Zn Brass	90	620
cold-drawn	140	480
70 Cu-30 Zn Brass	90	495
Steels		
0.08C	105	480
0.25C	120	560
0.25C, cold-drawn	250	450
0.5C	140	600
0.5C, cold-drawn	350	450
18 Cr-8 Ni SS	250	—
Gray cast iron	190	—
Titanium	200	390

* Primarily from ASM (1948).
† Annealed structure unless otherwise noted.

TABLE 13.7*

Typical Vickers Hardness Numbers

Material	V, kg/mm^2
Structural Constituents of Steel	
Austenite	400
Cementite	1100
Pure iron	70
Ferrite	80
Graphite	10
Martensite	800
Tempered martensite	250–800
Pearlite (eutectoid)	250
Spheroidite	175
Sorbite	275
Bainite	485
Troostite	550
Fully hardened tool steel	650–700
Carbides and Oxides	
B_4C	2800
Cr_3C	1200
Iron-vanadium carbide	2700
Chromium-tungsten carbide	2000
Molybdenum-tungsten carbide	2100
SiC	2400
Mo_2C	2000
TaC	1800
TiC	2400
WC	1600
WC + 6 Co	1400
WC + 13 Co	1300
VC	2800
Al_2O_3	2100
Fe_2O_3	1100
Fe_3O_4	650
Diamond	8000
Quartz	1100
Glass	400–600
Hard chromium plate	1000
Nickel plate	340

* Shaw (1960). Courtesy of M.I.T. Press.

462   HARDNESS

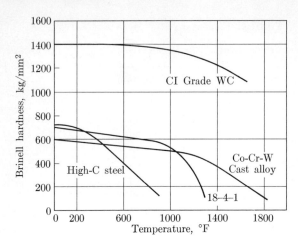

FIG. 13.11. Variation of hardness with temperature for a plain-carbon steel tool, an 18-4-1 high-speed steel tool, a cast alloy tool, and a cast iron grade cemented tungsten carbide tool. (Shaw, 1960. Courtesy of Massachusetts Institute of Technology Press.)

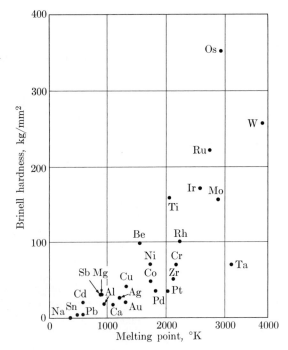

FIG. 13.12. Room temperature hardness of pure metals as a function of melting point. (Westbrook, 1953. Courtesy of ASM.)

hardness decreases with temperature is shown in Fig. 13.11 for several common cutting-tool materials.

The variation of hardness with temperature has been studied extensively by Westbrook (1953). Figure 13.12 shows the variation of Brinell hardness at room temperature with the absolute melting temperature for a number of pure metals. Such an approximate correlation suggested to Ludwik (1916) that the variation in hardness with temperature may depend on a temperature he called the *homol-*

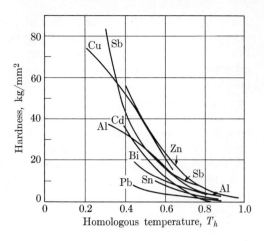

Fig. 13.13. Hardness of pure metals as a function of the homologous temperature $T_h = T/T_{mp}$. (Westbrook, 1953. Courtesy of ASM.)

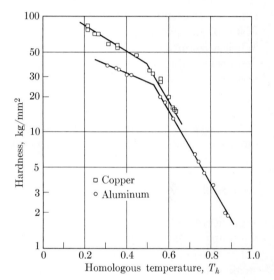

Fig. 13.14. Variation of hardness with homologous temperature for copper and aluminum. (Data from Ludwik, 1916.)

ogous temperature T_h, which is the ratio of the temperature to the melting temperature T_m:

$$T_h = T/T_m. \qquad (13.26)$$

Figure 13.13 shows that Ludwik's idea is only partly successful.

Logarithmic plots, such as Fig. 13.14, show two regions: a low-temperature cold-working region where metals deform plastically and a high-temperature hot-working region where *creep* is more important. The transition between the two hardness regimes was found to occur close to $T_h = 0.55 T_m$ for all metals. It may be noted that this homologous temperature is close to the recrystallization temperature. Like the value for recrystallization, the hardness transition temperature T_h is not an absolute quantity but depends on grain size, extent of strain,

strain rate, impurities present, etc. Mulhearn and Tabor (1961) have made an extensive study of the indentation hardness of indium and lead from the temperature of liquid air to 50°C and over a rate of loading of from 10^{-4} to 10^3 sec. Below a temperature of 0.6 T_m, creep was found to be negligible. Near the temperature at which liquid air boils, the mode of deformation of indium was found to shift from slip to twinning.

The influence of hydrostatic pressure on Brinell hardness was studied by Bridgman (1952) who ran tests on specimens subjected to pressures as high as 400,000 psi. He used 0.25-in. spheres and a load of 1500 kg. His results showed that depending on the metal tested, the Brinell hardness increased by from 2.4 to 6.2% for each 100,000 psi of hydrostatic pressure applied. This effect is small compared with the influence of hydrostatic pressure on ductility.

Setty et al. (1957) have studied the effect of a tensile stress on hardness by loading a specimen in tension and then making the hardness determination. A uniform tensile stress was found to decrease the apparent hardness of mild steel, the maximum reduction occurring at the yield point. When a tensile stress of 35,000 was applied, the hardness was found to decrease from $R_{30T}60$ to $R_{30T}50$. When a compressive stress was similarly applied before making the hardness indentation, mild steel appeared to be harder. The change in hardness was less for a specimen in bending than it was for a specimen under uniform stress, and the compressive effect was greater than the tensile effect. Presence of a tensile or compressive stress had no influence on the scleroscope (dynamic) hardness. These results suggest that microhardness values obtained on specimens containing high values of surface residual stress must be used with caution.

13.6 HARDNESS OF RUBBER

Since rubber is not plastic, its hardness must be related to its elastic stiffness, or Young's modulus, rather than to a flow stress. Young's modulus for rubber is not a constant, but varies with strain as discussed in Section 6.5. For strains small compared with unity it can be considered constant, however, since the deviations from linearity are small. Because elastic stiffness is being measured, the depth of penetration of an indenter must always be measured while the load is still applied. The harder the rubber, the smaller the penetration for a given load.

There are several instruments for measuring the hardness of rubber (ASTM, 1961, Vol. 11). These use indenters of different sizes and shapes and minor and major loads of different magnitudes. Most instruments also employ a pressure foot which surrounds the indenter and provides a reference surface to make the test less sensitive to the extent of the specimen. A surrounding pressure foot is not needed in metals, since there one measures the plastic deformation after return to original load, rather than the elastic deformation under load.

The only rubber hardness test to be considered here is that proposed by the International Standards Organization (ASTM, 1961). One version uses a spherical indenter 2.50 mm in diameter and a surrounding pressure foot 20 mm in diameter, with a 5-mm hole through which the indenter protrudes. The load on the pressure foot corresponds to 250 gm/cm^2, while the minor and major loads on the sphere are 30 and 580 gm.

The relation between the ISO penetration Δt, Young's modulus E, the tip radius R, and the load P, all in consistent units, is taken as

$$E = P/(1.9 R^{0.65} \Delta t^{1.35}). \tag{13.27}$$

An arbitrary table then gives the hardness in "degrees." Zero degrees corresponds to zero Young's modulus and 100 degrees to an infinite Young's modulus.

A theoretical relation can be obtained from the Hertz equations (11.32, 11.33). The resulting relation between ISO penetration and modulus of rubber, neglecting the minor load, is (Prob. 13.19)

$$E = 9P/(16 R^{0.5} t^{1.5}) = P/(1.78 R^{0.5} t^{1.5}), \tag{13.28}$$

which is reasonably close to the empirical equation (13.27).

13.7 SUMMARY

The hardness of plastically deforming materials, taken as force per unit area, is approximately 3.2 times the flow stress. As a first approximation, the flow stress may be taken to be the tensile strength, but more precisely, it should be the flow stress at 8% strain for the Vickers hardness test and at a strain of $\epsilon = 0.2\, d/D$ for the Brinell test. The Rockwell test is arbitrary, but R_C and R_{45N} numbers are roughly 10% of Brinell or Vickers numbers. Other common tests include the scratch and microhardness tests.

Since hardness is a measure of resistance to plastic deformation, it is affected by such variables as size, temperature, and rate in the same way that plastic flow is, and it provides a quick way of giving a rough characterization of that behavior. Microhardness tests clearly demonstrate that brittleness depends on the volume of the zone that must be plastically deformed, since normally brittle materials can be quite ductile under microhardness test conditions. A ductile size effect is also observed when the depth of penetration is less than 1 micron.

Scratch-hardness tests are closely related to indentation tests despite their apparently widely different test conditions, as indicated by Eq. 13.14. Cutting hardness is a convenient means for measuring the resistance materials offer to plastic deformation in the regime of very large plastic strain (3 to 5). The energy per unit volume involved in such tests is about 40 times that in an indentation test, due to the large difference in the plastic strain levels in these two types of tests.

The hardness of rubber comes from its resistance to elastic deformation, and the relation between hardness and elastic modulus is given by Eq. 13.28.

REFERENCES

Tabor (1951) gave a summary of the scientific studies at the time, and later in a short clear article (1956) summarized the physical meaning of hardness. For an introduction to the techniques and machines used in hardness testing, see the ASM Handbook (1948), which also contains tables of conversion between hardness measurements. Authoritative standards for the various tests are given by the American Society for Testing Materials (ASTM, 1961). The equipment and detailed test conditions for most of the hardness tests in use today may be found fully described in the following references: O'Neill (1934), Williams (1942), Von Weingraber (1952), Mott (1956), Small (1960), and ASTM (1961).

ASM	1948	"Hardness Tests," *Metals Handbook*, T. Lyman, ed., ASM, Novelty, Ohio, pp. 93–105.
ASM	1961	*Metals Handbook*, 8th ed., T. Lyman, ed., ASM, Novelty, Ohio, pp. 1234–1236.
ASTM	1961	*ASTM Standards*, Philadelphia. See in particular "Brinell Hardness of Metallic Materials," *ASTM Standard E 10-61*, **3**, 30–38; "Rockwell and Rockwell Superficial Hardness of Metallic Materials; *ASTM Standard E 18-61*, **3**, 39–52; "International Standard Hardness of Vulcanized Natural and Synthetic Rubbers," *ASTM Standard D 1415-56T*, **11**, 615–618.
Bierbaum, C. H.	1930	"The Microcharacter," *Trans. Am. Soc. Steel Treat.* **18**, 1009–1025.
Bridgman, P. W.	1952	*Studies in Large Plastic Flow and Fracture*, McGraw-Hill, New York.
Brinell, J. A.	1900	*Congrès International des Méthodes d'Essai*, Paris.
Cowdrey, I. H. Adams, R. G.	1944	*Materials Testing*, 2nd. ed., Wiley, New York.
de la Macorra, F.	1923	"Relation Between the Rockwell and Brinell Numeral for Testing Hardness of Metal," B. S. Thesis, M.I.T., Cambridge, Mass.
Goldsmith, W. Yew, C. H.	1962	"Penetration of Conical Indenters into Plane Metal Surfaces," *Proc. 4th U.S. Nat. Con. Appl. Mech., ASME*, Vol. 1., 177–188.
Herbert, E. G.	1923	"The 'Pendulum' Hardness Tester," *The Engineer* **135**, 390–391, 444–446, 686.
Knoop, F. Peters, G. Emerson, W. B.	1939	"A Sensitive Pyramidal-Diamond Tool for Indentation Measurements," *J. Res. Nat. Bureau of Stand.* **23**, 39–61.
Lee, E. H.	1952	"The Theoretical Analysis of Metal Forming Problems in Plane Strain," *J. Appl. Mech.* **19**, 97–103.
Levin, E.	1955	"Indentation Pressure of a Smooth Circular Notch," *Quart. Appl. Mech.* **13**, 133–137.

Ludwik, P.	1916	"Über die Änderung der Inneren Reibung der Metalle mit der Temperatur," *Z. Physik. Chemie* **91**, 232–247.
McClintock, F. A. Rhee, S. S.	1962	"On the Effects of Strain-Hardening on Strain Concentrations," *Proc. 4th U.S. Nat. Con. Appl. Mech., ASME*, pp. 1007–1013.
Meyer, E.	1908	"Untersuchungen über Prüfung und Härte," *Verein Deutscher Ingenieure Z.* **52**, 645–654.
Mohs, F.	1822	*Grundriss der Mineralogie*, Dresden. English trans. by W. Haidinger, Treatise on Mineralogy, Constable and Co., Edinburgh, 1925.
Mott, B. W.	1956	*Micro-Indentation Hardness Testing*, Butterworths, London, p. 272.
Mulhearn, T. O. Tabor, D.	1961	"Creep and Hardness of Metals: A Physical Study," *J. Met.* **89**, 7–12.
O'Neill, H.	1934	*Hardness of Metals and Its Measurement*, Chapman and Hall, London, p. 292.
O'Neill, H.	1944	"Significance of Tensile and Other Mechanical Tests of the Properties of Metals," *Proc. Inst. Mech. Engrs.* **151**, 116–130.
Orowan, E.	1958	Unpublished research.
Rockwell, S. P.	1922	"The Testing of Metals for Hardness," *Trans. Am. Soc. Steel Treat.* **2**, 1013–1033.
Setty, S. K. Lapsley, J. T. Thomsen, E. G.	1957	"Stresses Alter Hardness," *Mech. Eng.* **79**, 1127–1129.
Shaw, M. C.	1952	"A Yield Criterion for Ductile Metals Based upon Atomic Structure," *J. Franklin Inst.* **254**, 109–126.
Shaw, M. C.	1954	"Plastic Flow in the Cutting and Grinding of Materials," *Proc. Nat. Acad. Sci.* **40**, 394–401.
Shaw, M. C.	1960	*Metal Cutting Principles*, 3rd. ed., M.I.T. Press, Cambridge, Mass.
Shield, R. T.	1955	"On the Plastic Flow of Metals under Conditions of Axial Symmetry," *Proc. Roy. Soc. (London)* **A233**, 267–286.
Small, L.	1960	*Hardness Theory and Practice, Part I*, Service Diamond Tool Co., Ferndale, Michigan.
Smith, R. Sandland, G.	1922	"An Accurate Method of Determining the Hardness of Metals with Particular Reference to Those of a High Degree of Hardness," *Proc. Inst. Mech. Eng.* London, **1**, 623–641.
Sneddon, I. N.	1951	*Fourier Transforms*, McGraw-Hill, New York.
Tabor, D.	1951	*The Hardness of Metals*, Oxford University Press, London, p. 175.

Tabor, D.	1956	"The Physical Meaning of Indentation and Scratch Hardness," *British J. Appl. Phys.* **7,** 159–166.
Taylor, E. W.	1949	"Correlation of the Mohs Scale of Hardness with the Vickers Hardness Numbers," *Mineral. Mag.* **28,** 718–721.
Thibault, N. W. Nyquist, H. L.	1947	"The Measured Knoop Hardness of Hard Substances and Factors Affecting Its Determination," *Trans. ASM* **38,** 271–330.
Von Weingraber, H.	1952	*Technische Hartemessung*, Carl Hauser Verlag, Munich, p. 384.
Westbrook, J. H.	1953	"Temperature Dependence of the Hardness of Pure Metals," *Trans. ASM* **45,** 221–248.
Williams, S. R.	1942	*Hardness and Hardness Measurements*, ASM, Cleveland, Ohio.
Winchell, H.	1945	"The Knoop Microhardness Tester as a Mineralogical Tool," *Am. Mineral.* **30,** 583–595.

PROBLEMS

13.1. Explain how the deformation of the ball in a Brinell test will lead to low readings.

13.2. (a) Using the ideas of limit analysis, explain why the nearest edge of a specimen under a hardness indentation should be no closer than 2 to 3 impression diameters away from the indenter, and why the thickness about one diameter. (b) Compare and contrast the above limitation with the empirical one that for a material with hardness $R_B 28$, the thickness should be at least 0.036 in., whereas for a superficial test giving a hardness of about $R_{15T} 69$ the thickness need be only half as great (ASTM Standard E 18–61, 1961, Table 2).

13.3. Show that Eq. 13.2 gives the mean normal pressure on the surface of the material under an indenting sphere.

13.4. Derive the equation for the Vickers hardness number, Eq. 13.3.

13.5. Discuss the approximations made in deriving Eq. 13.9.

13.6. Derive the equation for the relation between Rockwell C scales and Brinell scales, Eq. 13.11. List the assumptions and limitations of your analysis.

13.7. Show that the effect of rounding the apex of a conical indenter can be neglected when the Brinell hardness is less than 30 times the minor load in kilograms.

13.8. Show that when the Brinell hardness is greater than 30 times the major load, the behavior of the Rockwell cone indenter becomes that of a ball indenter.

◄ 13.9. Derive the relation between the Bierbaum and Vickers hardness, Eq. 13.14.

◄13.10. Explain why the Prandtl flow field of Fig. 10.16 should be more appropriate than that of Fig. 10.8 for hardness indentations.

◄13.11. Show that for the flow field of Fig. 13.8, the Meyer hardnesses are $2.91Y$ and $2.88Y$ with and without sticking.

◀13.12. Derive the equation for the Meyer hardness of a material hardening at a linear rate E, Eq. 13.19. (Note omission of term in last equation in Sneddon (1951, p. 46), compared to the last equation on the previous page, and the definition of "total applied pressure").

◀13.13. Derive a theoretical equation (13.20) between the hardness and flow stress from Eq. 13.19.

◀13.14. Derive the ratio of deformed volume to volume of indentation, Eq. 13.21.

◀13.15. Derive the equation for the load on indenting a linearly hardening material with a sphere, Eq. 13.22.

◀13.16. Derive the equation for the relation between hardness and flow stress for a spherical indenter, Eq. 13.23.

13.17. Derive the relation that the tensile strength in psi is about 500 times the hardness in kg/mm^2.

13.18. Compare the typical dislocation density in a metal strained to 10% strain with that required for the strain gradients encountered when the hardness indentation is one micron in size.

◀13.19. Derive the relation between ISO penetration and modulus of rubber, Eq. 13.28.

13.20. Compare the work done per unit volume in a hardness test where the strain is approximately 0.08 with (a) that in rolling where the reduction of area may be as high as 10 to 1; and (b) that in an elastic body in tension where the strain at the elastic limit is about 0.003. The material under consideration in this case is steel having a Brinell hardness of 200 kg/mm^2.

13.21. There is one value of Brinell hardness B that should correspond numerically with the Vickers hardness V. (a) Determine this value of Brinell hardness. (b) For values of Brinell hardness greater than this value, should the Vickers or Brinell hardness value be the larger?

◀13.22. Explain briefly how there can be two entirely different slip-line field solutions to the flat-punch problem, each satisfying all of the conditions of the slip-line field theory for ideally plastic materials.

E 13.23. Design the simplest experiment you can think of to investigate the existence of a built-up nose on indenters having an included angle greater than 90°, as postulated by Prandtl and Lee.

13.24. Demonstrate that the Meyer index is two for a fully strain-hardened material.

13.25. If the Rockwell hardness of a material measured on the B scale is 85, estimate the ultimate tensile strength of this material in psi.

13.26. On the data sheet of a manufacturer of silicone bouncing putty ("silly putty"), the hardness was specified as being equal to "a penetration of 1.5 to 4.0 mm in 5 seconds of a $\frac{1}{4}$-inch diameter flat foot." Discuss the adequacy of the test specification given by this statement and suggest a more suitable specification.

◀13.27. Explain why the square of the scratch width λ is used to determine the Bierbaum microhardness number.

13.28. Estimate the force F_p required to make a cut of width $b = 0.05$ in. and depth $t = 0.005$ in., if the "cutting hardness" u of the material is 200 kg/mm^2.

13.29. Explain briefly why the strain in a cutting operation is so much higher than that in an indentation hardness test.

E 13.30. Using modeling clay, find experimentally the effects on apparent hardness of tests near edges or on thin specimens.

E 13.31. Using an indenter of chalk, compare the hardness values obtained for modeling clay:

(a) with a spherical indenter with $d/D = 0.3$,
(b) with a spherical indenter with $d/D = 0.5$,
(c) with a conical indenter of 136° included angle, and
(d) with a conical indenter of 20° included angle (corresponding to that of a pencil).

CHAPTER 14

DAMPING

14.1 SYNOPSIS

The terminology of damping is not so familiar as that for most testing, so we shall first discuss it in the terms of a simple spring and dashpot model. Any given mechanism of damping remains stiff above a certain frequency and is completely relaxed below a somewhat lower frequency. At intermediate frequencies, a damping peak occurs whose magnitude is related to the change in modulus between the two frequencies. Likewise, at a given frequency, an increase in temperature may allow some mechanism to relax, resulting in a damping peak at some particular temperature, with a slightly higher modulus of elasticity below the temperature and a slightly lower modulus above it.

We shall next discuss some of the mechanisms leading to damping at low stress levels, such as thermal currents, stress diffusion of solute atoms, grain boundary sliding, and simple dislocation interactions. Damping due to large-scale dislocation motion is not quantitatively understood as yet, although this damping, which may reach a fractional energy loss per cycle of 0.1 at a stress level near the fatigue limit, is often the most important in practical applications.

14.2 MODELS OF MATERIALS WITH DAMPING

Even within the range normally thought of as elastic, most materials exhibit a hysteresis or dissipation of free energy on cycling. This effect is called *damping* or *internal friction*, and may arise from either time-dependent phenomena or the irreversibility of plastic flow. A simple spring and dashpot model of a linearly rate-dependent material is shown in Fig. 14.1. The stress-strain relation has the form (Prob. 14.1)

$$\sigma + \dot{\sigma}\left(\frac{\eta_c}{E_c + E_d}\right) = \frac{1}{(1/E_c + 1/E_d)}\left(\epsilon + \frac{\eta_c}{E_c}\dot{\epsilon}\right). \quad (14.1)$$

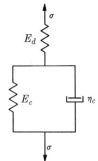

Alternatively, since η/E has the dimensions of time, the relation can be expressed in terms of two time constants, τ_1 and τ_2, and a modulus E_2:

$$\sigma + \tau_1\dot{\sigma} = E_2(\epsilon + \tau_2\dot{\epsilon}). \quad (14.2)$$

A material with such a stress-strain relation is called a *standard linear solid*.

Fig. 14.1. Standard linear solid.

The two time constants τ_1 and τ_2 are those for relaxation (constant strain) and transient creep (constant stress), respectively (Prob. 14.2). For our purposes here, the character of the stress-strain relation is best seen by considering cyclic tests. At very low rates of cycling, the second term on each side may be neglected, and the behavior is simply elastic:

$$\sigma = E_2 \epsilon = \frac{1}{(1/E_c + 1/E_d)} \epsilon. \tag{14.3}$$

The quantity E_2 is called the *relaxed* modulus, since for low frequencies all the force in the dashpot of Fig. 14.1 is relaxed. At very high rates of cycling, the first term on each side may be neglected in comparison with the second, and the behavior is again elastic:

$$\dot{\sigma} = \frac{\tau_2 E_2}{\tau_1} \dot{\epsilon} = E_1 \dot{\epsilon}, \qquad \sigma = E_1 \epsilon = E_d \epsilon. \tag{14.4}$$

The quantity E_1 is called the *unrelaxed* modulus, since for high frequencies there is no time for displacements across the dashpot of Fig. 14.1 to occur. It can be shown from the equations, or seen more directly from the figure, that

$$E_1 > E_2,$$

and so from Eq. 14.4,

$$\tau_2/\tau_1 = E_1/E_2 > 1.$$

At these extreme frequencies there is no hysteresis. To see what happens at intermediate frequencies, it is convenient to describe the sinusoidally varying stress and strain as the real parts of complex numbers $\underline{\sigma} \exp(i\omega t)$ and $\underline{\epsilon} \exp(i\omega t)$:

$$\sigma = \operatorname{Re}\underline{\sigma} e^{i\omega t} = \operatorname{Re}(\sigma' + i\sigma'')(\cos \omega t + i \sin \omega t) = \sigma' \cos \omega t - \sigma'' \sin \omega t,$$

$$\epsilon = \operatorname{Re}\underline{\epsilon} e^{i\omega t} = \epsilon' \cos \omega t - \epsilon'' \sin \omega t.$$

The stress-strain relation is found by substitution into the differential equation (14.2):

$$(1 + i\omega\tau_1)\underline{\sigma} = E_2(1 + i\omega\tau_2)\underline{\epsilon},$$

giving the complex modulus

$$E = E' + iE'' = \frac{\underline{\sigma}}{\underline{\epsilon}} = \frac{E_2(1 + i\omega\tau_2)}{1 + i\omega\tau_1} = \frac{E_2[1 + \omega^2\tau_2\tau_1 + i\omega(\tau_2 - \tau_1)]}{1 + \omega^2\tau_1^2}. \tag{14.5}$$

There is a phase difference δ between stress and strain given by

$$\tan \delta = \frac{\omega(\tau_2 - \tau_1)}{1 + \omega^2\tau_1\tau_2}. \tag{14.6}$$

Note that for very high and low frequencies the modulus approaches the unrelaxed and relaxed values, respectively, and the phase difference goes to zero, as shown

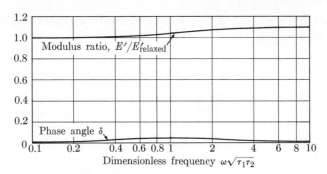

Fig. 14.2. Modulus change and phase angle for a standard linear solid with $\delta_{max} = 0.05$.

in Fig. 14.2. The maximum value of the phase difference is given by (Prob. 14.3)

$$\tan \delta_{max} = (\tau_2 - \tau_1)/2\sqrt{\tau_1 \tau_2},$$

when

$$\omega = 1/\sqrt{\tau_2 \tau_1}.$$

Since in most cases the stress and strain are nearly in phase (small hysteresis loops), $\tau_2 - \tau_1 \ll \sqrt{\tau_2 \tau_1}$, and we have approximately

$$\delta_{max} = \frac{\tau_2 - \tau_1}{2\sqrt{\tau_2 \tau_1}} \approx \frac{(E_1 - E_2)}{2E_2}. \tag{14.7}$$

Various measures are used to describe the amount of damping. The *phase angle* δ described by Eq. 14.6 is one of these. Electrical engineers often use a *quality factor*, Q, defined as the reciprocal of the phase angle.

The energy dissipated per cycle is given in terms of stress and strain amplitude σ_a and ϵ_a by (Prob. 14.4)

$$\Delta U = \int_V \int_0^{2\pi/\omega} \sigma \dot{\epsilon} \, dt \, dV = \int_V \pi \sigma_a \epsilon_a \sin \delta \, dV,$$

so the relative energy dissipation or *relative damping*, referred to the peak strain energy during the cycle, is

$$\Delta U/U = 2\pi \sin \delta \simeq 2\pi \delta. \tag{14.8}$$

Note that for the standard linear solid, the relative damping is independent of amplitude.

In structural work, damping is described in terms of the *specific damping energy*, which is the energy loss per cycle per unit volume, $D = d(\Delta U)/dV$.

The decay in amplitude of freely vibrating systems is another measure. If a_1 and a_2 are the amplitudes of successive oscillations, the *logarithmic decrement* is (Prob. 14.5)

$$\Lambda = \ln a_1/a_2 \approx \Delta a/a = \Delta U/2U. \tag{14.9}$$

We therefore have the following relations between measures of the damping of a standard linear solid:

$$\Delta U/U = \left(\int D\, dV\right)/U = 2\Lambda = 2\pi\delta = 2\pi Q^{-1}. \tag{14.10}$$

The quantities in Eq. 14.10 depend on frequency, but are independent of amplitude. Their maximum values are related to the relative change in modulus by Eq. 14.7.

In real materials there are usually many sources of damping, each with a characteristic frequency at which the damping reaches a peak. The representation of such a material requires a network of spring and dashpot elements, providing a series of relaxation times. The series of damping peaks can be presented in the form of an internal-friction spectrum, shown schematically in Fig. 14.3. To each peak in the internal-friction spectrum, there corresponds a fractional decrease in modulus equal to twice the maximum phase angle, according to Eq. 14.7. At any one frequency, however, the relations between the measures of damping given by Eq. 14.10 are valid for any linear viscoelastic material, and not only for a standard linear solid.

◀At higher stress levels, viscoelastic models may no longer be good approximations. For ideally plastic (rate-independent) flow, an analysis can be carried out by describing the stress-strain curve after any reversal of stress by an equation of the form

$$\epsilon = \sigma/E + (\sigma/\sigma_1)^{1/n}, \tag{14.11}$$

where $0 < n < 1$. The concept of a phase angle is no longer applicable, but can be replaced by a lag, defined as shown in Fig. 14.4, to which it is equivalent in the

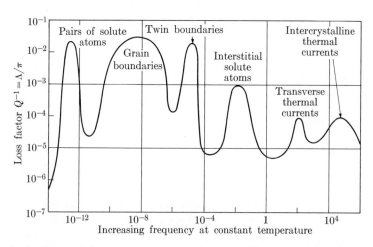

Fig. 14.3. Typical damping spectrum of a crystalline material at low stress. (After B. J. Lazan, 1959.)

viscoelastic case (Prob. 14.6). For small damping with uniform stress, the relations between the measures of damping become (Prob. 14.7)

$$\frac{\Delta U}{U} = \frac{\int D\, dV}{U} = 2\Lambda$$

$$= 8\,\frac{(1-n)/(1+n)}{1-(\tfrac{1}{2})^{(1-n)/n}} \quad \text{(lag)}.$$

(14.12) ▶

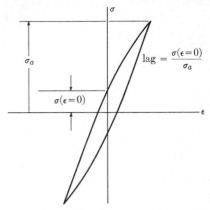

FIG. 14.4. Definition of lag.

14.3 SOURCES OF INTERNAL FRICTION

There are a large number of mechanisms which may cause damping; some of these are now well understood, others less so. Roughly speaking, internal friction can be divided into two main categories: those sources that are viscous or rate-dependent and those that are plastic or rate-independent. For linear viscoelasticity, as discussed above the damping is independent of amplitude, whereas plastic damping usually increases with stress amplitude. These two categories are further discussed in Table 14.1, which summarizes some of the characteristics of each class. We shall now discuss a few illustrative examples, some of which will show that linear viscoelastic damping can occur in crystalline materials.

A. Thermal Currents. The elastic compression of a material is commonly associated with heating effects. If the compression takes place sufficiently rapidly, there is no opportunity for heat to be conducted away, and the observed compressibility is the isentropic compressibility $1/B_s$ (corresponding to our previous unrelaxed modulus). For slow compression, temperature fluctuations are eliminated by thermal conduction, and we have the isothermal compressibility $1/B_T$ (or the relaxed modulus).

The difference in compressibilities can be derived from thermodynamics as was Eq. 3.11, which gave the difference between the isothermal and the isentropic moduli of elasticity. In terms of the absolute temperature T, the linear coefficient of expansion α, the density ρ, and the specific heat at constant pressure, C_p, the difference in compressibilities is

$$1/B_T - 1/B_s = T(3\alpha)^2/\rho C_p. \qquad (14.13)$$

Indeed, Eq. 14.13 gives the physical basis for Eq. 3.11, for linearity and symmetry arguments show that any thermal effects must arise from the dilatational but not the distortional parts of the strain.

Now consider an oscillating beam. We must use the adiabatic or isothermal modulus according to whether we are working at very high or very low frequencies;

TABLE 14.1*

Classification of Types of Damping of Materials

Name used here	Types of material damping		
	Rate-dependent	Rate-independent	
Other names	Viscoelastic, rheological, and "dynamic" hysteresis	Plastic, plastic flow, plastic strain and "static" hysteresis	
Nature of stress-strain laws	Essentially linear. Differential equation involving stress, strain, and their time derivatives	Essentially nonlinear, but excludes time derivatives of stress or strain	
Simplest representative mechanical model	Voigt unit / Maxwell unit (Anelasticity)	(mechanical model diagrams)	
Frequency dependence	Critically at relaxation peaks	No, unless other mechanisms present	
Primary mechanisms	Solute atoms, grain boundaries. Micro- and macrothermal and eddy currents. Molecular curling and uncurling in polymers	Magnetoelasticity	Plastic strain
Value of m in $D = C\sigma^m$	2	3 up to coercive force	2–3 up to σ_L* 2 to >30 above σ_L
Variation of δ with stress	No change, since $m - 2 = 0$	Proportional to σ since $m - 2 = 1$	Small incr. up to σ_L† Large incr. above σ_L
Typical values for δ	Anelasticity:‡ <.001 to .01 Viscoelasticity: <0.1 to >1.5	0.01 to 0.08	0.001 to 0.05 up to σ_L 0.001 to >0.1 above σ_L
Stress range of eng. importance	Anelasticity - low stress Viscoelasticity - all stresses	Low and medium Sometimes high	Medium and high stress
Effect of fatigue cycles	No effect	No effect	No effect up to σ_L Large changes above σ_L
Effect of temperature	Critical effects near relaxation peaks	Damping disappears at Curie temperature	Mixed. Depends on type of comparison
Effect of static preload		Large reduction for small coercive force	Either little effect or increase

* From Lazan (1959). Courtesy ASME.

† σ_L is a limiting stress somewhat below the fatigue limit, above which the damping becomes markedly dependent on the history.

‡ Defined by Zener (1948) as the deviation from a unique relation between stress and strain before large-scale plastic flow.

in neither of these extreme cases is there any damping. A complete solution of the problem at intermediate frequencies would require a consideration of the heating effects and the resulting thermal currents. To avoid this we note that if the damping is viscoelastic so that Eq. 14.1 applies, there are three constants. We have now two (the moduli), and need find only one more, which we take to be the frequency $1/\tau = 1/\sqrt{\tau_1 \tau_2}$ for which the damping is a maximum. Now we expect τ to be the time taken for heat to diffuse from one side of the beam to the other, i.e., over a distance h, the thickness of the beam. From dimensional analysis, the time must be related to the thermal diffusivity $D = k/\rho C_p$ by

$$\tau = \text{constant } h^2/D.$$

A detailed calculation gives the value of the constant so that (Zener, 1948, p. 79)

$$\tau = h^2/\pi^2 D. \tag{14.14}$$

Then with τ known, for the phase angle at frequency ω in terms of E, we have

$$\delta = \frac{1}{Q} = \frac{E_T - E_s}{E_T} \frac{\omega\tau}{1 + \omega^2 \tau^2}. \tag{14.15}$$

This relation, which contains no adjustable parameter, agrees well with the experimental results, Fig. 14.5.

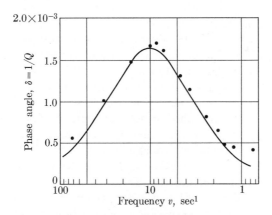

FIG. 14.5. Comparison of experiment and theory for damping of transverse vibrations in a beam of German silver by thermal currents. (Bennewitz and Rötger, 1938.)

A similar treatment applies to the deformation of a polycrystalline material. Owing to the different orientations of the grains, they will be deformed by different amounts and so will tend to different temperatures. A thermal current will therefore flow from one grain to another, which will give a damping effect. The maximum damping will again occur at a frequency $1/\tau$ from Eq. 14.14, but h now must be taken as the grain size. Since this is smaller than the specimen size, we

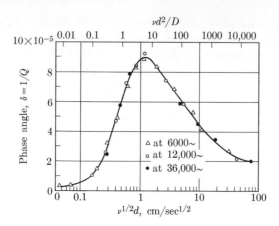

FIG. 14.6. Interrelation between grain size and frequency in damping of polycrystalline brass. (Randall, Rose, and Zener, 1939. Courtesy of American Institute of Physics.)

obtain a damping peak at a higher frequency, Fig. 14.6. At still higher frequencies heat flows only near the grain interfaces (Zener, 1948), giving shorter distances for diffusion and less damping as the frequency is raised further.

B. Snoek Damping in Body-Centered Cubic Metals. In α-iron containing carbon or nitrogen in solution, the solute atoms occupy interstitial positions at the centers of the cube edges (or, what is crystallographically equivalent, at the centers of the cube faces). The presence of the interstitials tends to distort the lattice slightly, extending the edge on which the impurity lies and contracting the transverse edges. In the absence of an applied stress all the possible sites for the interstitial are equivalent, but when a stress is applied it will be energetically favorable for the interstitials to occupy the edge most nearly parallel to the direction of maximum elongation. That is, an applied stress will tend to make the interstitials rearrange themselves. At very low frequencies this rearrangement can take place almost completely; at high frequencies virtually none will be able to occur. In both these cases the strain will be in phase with the applied stress, and there will be no damping associated with the process. The damping will be a maximum when the frequency of the stress is equal to the frequency at which the interstitials jump from one site to another. Observation of the position of the internal friction peak provides a convenient way of measuring this jump frequency, and the magnitude of the peak provides an indication of the number of interstitial atoms present.

Now the movement of an interstitial is a thermally activated process, so the average time until an interstitial jumps, at temperature T, will be

$$\tau = \tau_0 \exp(u/KT), \qquad (14.16)$$

where τ_0 is a constant and u is the activation energy; we may expect τ to be approximately equal to either of the relaxation times of Eq. 14.2.

Hitherto we have regarded τ as fixed and supposed that the frequency varied, but we can equally well suppose that τ is varied and the frequency ω is kept fixed. This is often convenient experimentally, since the specimen may then form part of a resonating system. From Eq. 14.16 we see that τ may be varied simply by changing the temperature. Damping measurements are therefore frequently reported as functions of temperature for a fixed frequency. Because of the exponential dependence on temperature, a change of frequency makes only a relatively small change in the temperature at which the maximum damping occurs. For the Snoek damping by interstitials, this maximum occurs at about 40°C for frequencies in the acoustic range (Snoek, 1941). This effect reportedly can be observed by heating a piece of annealed mild steel; near the temperature of maximum damping the specimen no longer rings clearly when struck. Ringing is found after further heating. Snoek damping is one of the most important mechanisms, reaching values as high as $\delta = 0.1$.

◀ **C. Point-Defect Flipping.** The Snoek effect in iron is a prominent example of a general mechanism of damping by short-range flipping of point defects or loose sites in crystalline and amorphous networks. The oscillation of the axis of symmetry of di-vacancies, the flipping of solute atom pairs, and the like, can all produce damping in a crystal lattice.

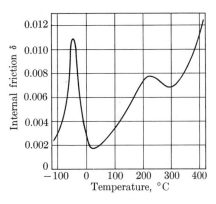

FIG. 14.7. Changes in internal friction with temperature in a chilled soda-silica glass. Two obvious maxima occur; the one at the left (−43°) is caused by sodium ion diffusion. Percent glass composition: Na_2O 24, SiO_2 76. (Fitzgerald, Laing, and Bachman, 1952. Courtesy of Society of Glass Technology.)

◀ In glassy networks the structural configurations that are responsible for the delayed elasticity and viscous flow (See Section 6.4) also lead to damping. Figure 14.7 shows the variation of damping in a chilled soda-silica glass with temperature at a frequency of 0.16 cycles per second. The activation energies for the two peaks, determined from the shift of the peaks with a change in frequency, have been found to be 16,000 and 34,600 calories per mole (Fitzgerald et al., 1952). These values are close to the energies for diffusion of sodium (20 kcal/gm-mole) and oxygen (59 kcal/gm-mole) in glass. It is therefore possible to associate these peaks with the short-range flipping of sodium and oxygen ions in the glass network. The general rise of the internal damping with increasing temperature in Fig. 14.7 is probably due to a gradual loosening of the glass network where some of the more loosely bound SiO_4 tetrahedra begin to flip positions. That this might be so would follow from the absence of the two low-temperature peaks in fused quartz, where only a continuous rise in damping with temperature is observed (Argon, 1956). The study of internal damping, especially in glassy materials, by identifying the various damping peaks (their height and breadth) is often a very powerful tool for a better understanding of the mechanisms of inelastic deformation. ◀ For a more lengthy discussion of internal damping arising from point defects, the reader is referred to the proceedings of a conference on internal friction edited by Leurgans (1962). ▶

D. Grain-Boundary Slip. By comparing the internal friction curves of single crystal and polycrystalline aluminum, Kê (1947) has shown that there is a relaxation process in the polycrystal not present in the single crystal. The activation energy associated with this process is found to be rather less than the activation energy of self-diffusion. It is suggested that the process is one of grain-boundary slip by the migration of atoms in the boundary; in this case the activation energy should be that for grain-boundary diffusion, approximately half the normal self-diffusion energy. Since this occurs at temperatures too high for use of aluminum as a structural material, we shall consider it further under the topic of creep in Chapter 19.

E. Internal Friction Associated with Plastic Deformation. At the stress levels commonly used in structures, there is a component of the internal friction which is clearly associated with the plastic deformation process, since its magnitude depends on the amount of previous strain, the annealing treatment, etc. We first consider some of the mechanisms which may contribute to this effect.

A crystal which has undergone plastic deformation will exhibit a number of internal damping peaks. Some of these damping peaks are due to the point defects which always accompany plastic deformation where intersecting slip systems are activated. Other damping peaks arise from the various modes of oscillation of dislocations. A very common dislocation configuration which arises both in single slip and in intersecting slip is the edge-dislocation dipole and its degenerate form, the dipole trail, where the edge dislocations are on neighboring planes (see

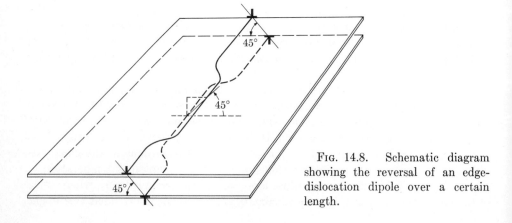

FIG. 14.8. Schematic diagram showing the reversal of an edge-dislocation dipole over a certain length.

Sections 4.8 and 5.3). As shown in Fig. 4.15, a dislocation dipole has two positions of equilibrium separated by an energy barrier, which, per unit length, is the area under the curve between the point $x/y = 1$ and the origin (Fig. 4.15). When a shear stress is applied, there will be a certain probability for some unfavorably oriented dipole segments to flip into favorable configurations, as shown in Fig. 14.8. When the external frequency coincides with the natural frequency of such dipole oscillations, an internal damping peak should arise. Gilman (1962) has compared the calculated activation energy for dipole flipping in Cu, Al, LiF, Ta, Mo, and W, and found them to be close to the activation energies of the so-called Bordoni internal friction peaks. If some of the internal friction is due to the oscillation of the dipole trails in body-centered cubic metals, the presence of Bordoni damping at very low stress amplitudes does not imply a very low lattice friction stress (see Section 5.4), since the lattice friction stress may not enter for dislocation dipole oscillations on adjacent planes.

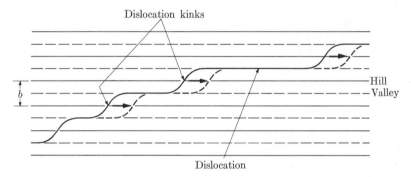

FIG. 14.9. Motion of kinks in a dislocation line.

Damping in deformed metals could also be due to an anisotropy of the lattice friction stress which a dislocation encounters in its slip plane. In the slip plane of a crystal of high symmetry, the periodic nature of the lattice would give a series of parallel potential hills and valleys. The energy of a dislocation would be minimum in a valley and maximum on a hill. Motion of the dislocation to a neighboring valley would require the overcoming of the lattice friction stress. Since dislocations are curved, the minimum energy configuration would be a series of straight segments lying in potential valleys with short bridging *kinks* draped over the potential hills as shown in Fig. 14.9. Although the motion of the straight segments requires overcoming the lattice friction stress, the small kinks can move along the straight segments with relative ease. Thus, Bordoni damping in deformed b.c.c. metals may result from the back-and-forth oscillation of such dislocation kinks over shallow secondary potential contours, as illustrated in Fig. 14.9*

At higher stress levels, damping may be due to the oscillation of dislocations and their breaking free from pinning points (see Niblett and Wilks, 1960).

---

* This is a modification of a damping mechanism discussed by Seeger (1956).

The mechanisms involving plastic deformation at high stress levels are likely to give relative damping coefficients which depend on the amplitude of the oscillation, and so are not given by a simple linear equation such as Eq. 14.2. The relative damping now depends more on the strain amplitude than on the frequency. (Although this statement is frequently made in the literature, experimental evidence for it is hard to find.)

Values of the specific damping energy for a variety of structural materials are shown in Fig. 14.10. There is a large increase in the plastic damping when a stress level of 50 to 100% of the fatigue limit is reached. Below this value, because of the simplification resulting from assuming relations of the form of Eq. 14.2, it is often worthwhile to represent the data by an equation with an exponent of 2, corresponding to a linear model. The line shown in Fig. 14.10 corresponds to a phase angle of 0.001 to 0.03, depending on the material (Prob. 14.8). The fact that the exponent in Fig. 14.10 is greater than two means that the magnitude of the damping depends on the amplitude of the oscillation. Furthermore, the damping then depends on the distribution of stress within the structure so that in specifying the damping it is necessary to take into account the character of the distribution of stress within the member. Of two parts with the same peak stress, the one with the more uniform stress distribution will exhibit higher damping (Prob. 14.9). This analysis has been worked out for a number of cases by Lazan (1954) and Podnieks and Lazan (1957).

It is worth emphasizing that the increase in damping occurs below the endurance limit. This effect is strong enough so that Föppl (1936) was able to run a steel specimen with a relative energy dissipation of 0.5 for a billion cycles without

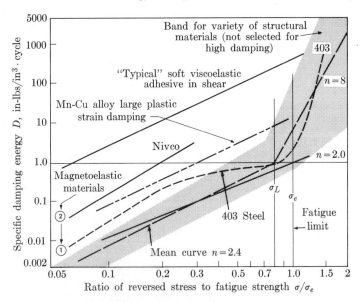

Fig. 14.10. Range of specific damping energy for a variety of structural materials. (Lazan, 1959. Courtesy of ASME. See also Lazan, 1954.)

failure. This damping caused a temperature rise of over 100°C above ambient throughout the three-year period.

Torvik et al. (1963) have presented data from combined torsion and tension tests on the effects of combined stress which show that for mild steel and copper the equivalent stress provides a correlation, whereas for alloys whose damping arises primarily from magnetic effects, the maximum stress tends to provide a better correlation. Such differences are not surprising in view of the differences in the mechanisms producing the damping.

Most experiments (Lazan, 1954; Mason, 1957) have been carried out at only one temperature and frequency, in spite of the fact that in early work Lazan found a decrease in specific damping of 5 in-lb/in^3 cycle, on going from 20 to 1750 cpm. Also Mason's results on 17S-T4 aluminum alloy at 17,200 cps indicated a phase angle of only 0.0015 at 70% of the endurance limit, which corresponds to a specific damping of 0.1 in-lb/in^3 cycle, well below that reported by Lazan (1954) for similar materials at 0.3 cps. Effects of frequency and stress amplitude on damping in cantilever beams at their natural frequencies have been studied by Crandall (1962), whose analysis is valuable, although subsequent work indicates air damping may have been important.

The present lack of knowledge concerning damping and the effects of temperature and frequency on it is an example of the gap that often develops between scientific and engineering work. The scientists, in order to simplify the problem for easier understanding, have in general restricted themselves to low stress levels, which are primarily of interest in acoustical work. The engineers, in taking data in the range they needed for structural work, have largely overlooked some of the important ideas and variables—in this case, the temperature and frequency dependence. Here is an area where more work needs to be done, and it seems likely that it will not only be of practical value, but can, if properly planned, help to shed light on the dislocation mechanisms of plastic flow.

14.4 STRUCTURAL DAMPING

Structural damping is the term applied to damping which arises from the form of the structure, as opposed to the structural material itself. One form of structural damping is slip damping in which a moderate amount of clamping pressure at a joint allows for some slip, but the pressure is still high enough so that such slip will result in appreciable dissipation. This method, while frequently very important, cannot always be relied upon in design, since the slipping at the joint may lead to fretting corrosion (Lazan, 1961). On the other hand, in experimental work, slip damping frequently obscures the results sought.

A more effective and commonly used form of structural damping consists of coating the part with a high-damping material. The problems that arise here are, first, to get a material that exhibits damping over a wide range of frequencies and temperatures, and to get this material in a region where the strain is high enough so that it will absorb a significant amount of energy. The strain concentration

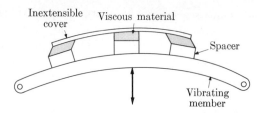

FIG. 14.11. Structural damping with viscous material under intensified shear.

can be accomplished by using a thick layer of the damping material or by using the material as a bonding agent between the structural member and a relatively inextensible cover layer, shown in Fig. 14.11.

A problem of obtaining a broad-band damping spectrum depends on finding a material with a large number of internal damping mechanisms having a narrow range of the same activation energies. This problem has been discussed by Bohn et al. (1961). They report values of phase angle running from 0.3 for a temperature bandwidth of 20°C to 0.06 for a temperature bandwidth of 100°C for some polymeric materials.

14.5 SUMMARY

At low stress levels, damping is due to a variety of mechanisms which cause damping peaks at different temperatures or frequencies. The damping is described in terms of a number of different parameters:

(a) The phase angle δ between stress and strain ($=Q^{-1}$ for $\delta \ll 1$).
(b) The relative damping, or fractional energy loss per cycle, $\Delta U/U$.
(c) The logarithm of the amplitudes in successive cycles, or logarithmic decrement, $\Lambda = \ln a_2/a_1$.
(d) The specific damping, or energy loss per unit volume, D, which is important when the other measures of damping depend on the applied stress, and the stress varies through the part.
(e) The loss in modulus of elasticity going from a lower to a higher temperature or from a higher to a lower frequency so that the mechanism causing the damping can relax.

For viscoelastic damping, these measures of damping are related by the following equations:

$$\Delta U/U = \left(\int_V D\, dV\right)/U = 2\Lambda = 2\pi\delta = 2\pi Q^{-1}. \qquad (14.10)$$

In the low-stress region, the fractional energy loss per cycle is typically of the order of 10^{-3} to 10^{-2}. This value gradually increases with stress up to a stress of 50 to 100% of the fatigue limit, above which it rises sharply in most metals and becomes history dependent.

At low stresses, damping may be due to thermal currents arising from stress variations across a specimen or from grain to grain (Eqs. 14.13 to 14.15), grain-boundary sliding, and switching of solute atoms from one preferred site to another. The damping due to plastic flow, which is important at high stress, is not completely understood as yet. Mechanisms of damping in plastically deformed metals include switching of dislocation pairs and oscillation of dislocation kinks. In glass there is a low-temperature peak of damping, apparently corresponding to the diffusion of sodium or oxygen ions, but at room temperature the phase angle is very low, of the order of 2×10^{-3}.

REFERENCES

A good review article is Nowick (1953), and the techniques required for good measurements are described by Entwistle (1960).

ARGON, A. S.	1956	Unpublished observations on anelasticity of quartz filaments in torsion.
BENNEWITZ, K. RÖTGER, H.	1938	"On the Internal Friction of Solids," *Z. Tech. Physik* **19**, 521.
BOHN, L. LINHARDT, F. OBERST, H.	1961	"Progress in the Development of Vibration Damping Materials," *Conf. on Acoustical Fatigue*, WADC TR 59-676, W. J. Trapp and D. M. Forney, Jr., eds., U. S. Air Force, Office of Technical Services, U. S. Department of Commerce, pp. 185-206.
CRANDALL, S. H.	1962	"Scaling Laws for Material Damping," *NASA Tech. Note D*-1467, Washington, D. C.
ENTWISTLE, K. M.	1960	"The Damping Capacity of Metals," *The Physical Examination of Metals*, 2nd ed., B. Chalmers and A. G. Quarrell, eds., Arnold, London, pp. 487-558.
FITZGERALD, J. V. LAING, K. M. BACHMAN, G. S.	1952	"Temperature Variation of the Elastic Moduli of Glass," *Trans. Soc. Glass Tech.*, **36**, 90-104.
FÖPPL, O.	1936	"The Practical Importance of Damping Capacity in Metals, Especially Steels," *J. Iron and Steel Inst.* **134**, 393-455.
GILMAN, J. J.	1962	"Debris Dipoles and Low Temperature Internal Friction in Crystals," *Phil. Mag.* **7**, 1779-1782.
KÊ, T. S.	1947	"Experimental Evidence of the Viscous Behavior of Grain Boundaries in Metals," *Phys. Rev.* **71**, 533-546.
LAZAN, B. J.	1954	"Fatigue Failure under Resonant Vibration Conditions," *Fatigue*, ASM, Novelty, Ohio, pp. 36-76.
LAZAN, B. J.	1959	"Energy Dissipation Mechanisms in Structures with Particular Reference to Material Damping," *Structural Damping*, J. E. Ruzicka, ed., ASME, New York, pp. 1-34.

LAZAN, B. J.	1961	"Review of Structural Damping Mechanisms," *Conf. on Acoustical Fatigue*, WADC TR 59-676, W. J. Trapp and D. M. Forney, Jr., eds., U. S. Air Force, Office of Technical Services, U. S. Department of Commerce, pp. 168–184.
LEURGANS, P. J. (ed.)	1962	"Conf. on Internal Frictions Due to Crystal Lattice Imperfections," *Acta Met.* **10**, 267–500.
MASON, W. P.	1957	"Internal Friction and Fatigue in Metals at a Large Strain Amplitude," *Proc. 9th Int. Cong. Appl. Mech.* Vol. 5, Brussells, 1956, pp. 379–390.
NIBLETT, D. H. WILKS, J.	1960	"Dislocation Damping in Metals," *Advances in Physics* **9**, 1–88.
NOWICK, A. S.	1953	"Internal Friction in Metals," *Prog. Metal Phys.* **4**, 1–70.
PODNIEKS, E. R. LAZAN, B. J.	1957	"Analytical Methods for Determining Specific Damping Energy Considering Stress Distribution," WADC TR 56–44, U. S. Air Force, Office of Technical Services, U. S. Department of Commerce.
RANDALL, R. H. ROSE, F. C. ZENER, C.	1939	"Intercrystalline Thermal Currents as a Source of Internal Friction," *Phys. Rev.* **56**, 343–348.
SEEGER, A.	1956	"On the Theory of the Low Temperature Internal Friction Peak Observed in Metals," *Phil. Mag.* **1**, 651–662.
SNOEK, J.	1941	"Effects of Small Quantities of Carbon and Nitrogen on the Elastic and Plastic Properties of Iron," *Physica* **8**, 711–733.
TORVIK, P. J. CHI, S. H. LAZAN, B. J.	1963	*Damping of Materials Under Biaxial Stress*, ASD Tech. Doc. Rept. *TDR*-62-1030, Aeronautical Systems Division, U. S. Air Force, Wright-Patterson Air Force Base, Ohio.
ZENER, C.	1948	*Elasticity and Anelasticity of Metals*, University of Chicago Press, Chicago.

PROBLEMS

14.1. (a) Derive the stress-strain relation of a standard linear solid, Eq. 14.1,

(b) What other spring and dashpot models have the same stress-strain relation?

14.2. Show that for a standard linear solid, the time constants τ_1 and τ_2 of Eq. 14.2 are those for $1/e$ of the transient to occur in relaxation (constant strain) or creep (constant stress), respectively.

14.3. Show that the maximum value of phase difference for a standard linear solid is given by

$$\tan \delta_{max} = (\tau_2 - \tau_1)/2\sqrt{\tau_2 \tau_1} \quad \text{when} \quad \omega = 1/\sqrt{\tau_2 \tau_1}.$$

14.4. Show that the energy dissipated per cycle for a standard linear solid is

$$\Delta U = \pi \sigma_a \epsilon_a \sin \delta,$$

so that the relative energy dissipation is as given by Eq. 14.8.

14.5. Derive the relation between logarithmic decrement and relative energy dissipation, Eq. 14.9.

14.6. Show that the lag of Fig. 14.4 is equivalent to the phase angle in a viscoelastic material.

14.7. Derive the relation between the measures of damping of a plastic material, Eq. 14.12. [*Hint:* Consider the integral of stress with respect to strain, and note that for small damping, the lag in stress is equal to the lag in strain.]

14.8. Show that the curve drawn on Fig. 14.10 for a viscoelastic material ($n = 2$) might correspond to a damping phase angle of 0.001 to 0.03, depending on the other characteristics of the alloy it represents.

14.9. Show that for amplitude-dependent damping of two parts having the same peak stress, the one with the more uniform stress distribution will exhibit higher damping.

14.10. In correlating the results of tests by others on the damping of cantilever beams, Crandall (1962) reported that for cold-rolled C1018 steel the damping expected from transverse thermal currents reached a peak of $\delta = 0.0014$ at a frequency of 12 cps. From this, what can you tell about the dimensions of the beam? Is this a reasonable value? At a frequency of 50 cps, the phase angle calculated from test results varied from 0.003 to 0.010 as the stress was increased from 20,000 to 40,000 psi.

CHAPTER 15

BRITTLE FRACTURE

15.1 SYNOPSIS

When subjected to increasing loads, solids may fracture. Here we shall discuss the simplest case, namely, the brittle fracture of glassy solids where there is almost no accompanying inelastic deformation. According to Griffith, brittle fracture is due to minute cracklike defects. Although the ideal strength is very high for a glassy solid whose atoms are held together with primary bonds, the severe stress concentration at sharp cracks can reduce this strength to ordinarily observed low values. We first consider the uniaxial case and then, following Griffith, we shall discuss brittle fracture under biaxial stresses and derive a brittle-fracture locus. From these results we shall establish the conditions of brittle fracture under triaxial stresses.

We shall present experimental confirmation of the theory and evidence for the presence of defects that are responsible for the great reduction in strength. Furthermore, it will be possible to furnish some plausible explanation for the origin of such cracks.

The problem of static fatigue, which is responsible for a strength reduction under prolonged loading, will be taken up both for glassy materials and briefly for steel, where the most common example is hydrogen embrittlement.

We shall consider the phenomena associated with rapidly moving cracks, such as acceleration, terminal velocity of traveling cracks, and distortion of stress fields due to the motion of cracks.

Further, the fracture surface markings on glass will be discussed and their relation to crack speed elucidated. This will enable us to identify the origin, direction of propagation, and to some extent the speed of fracture.

Finally we shall consider the size effect, which is a consequence of the flaws which lead to brittle fracture, giving an example based on the statistics of extreme values.

15.2 BRITTLE FRACTURE UNDER UNIAXIAL STRESS

When a solid is subjected to increasing loads, the resulting stresses will, at a certain stage, become high enough to cause the solid to break apart. If such breakage comes about before the piece has thinned down to zero thickness, it is called *fracture*; and if the amount of permanent deformation preceding fracture is negligible, it is called *brittle fracture*. Experiments show that brittle fracture takes place without any detectable warning and advances across planes of maximum tensile stress at speeds of the order of that of sound in the medium undergoing fracture. Inorganic glasses, polymers, and some body-centered cubic metals are all susceptible to brittle fracture at temperatures low enough where plastic or viscous modes of deformation cannot easily take place.

Before a theory for brittle fracture is presented, it is of interest to see just how strong a solid could be. Although an exact calculation of the ideal cohesive strength is possible for simple cases such as ionic crystals (Zwicky, 1923), approximate estimates are easier and far more direct. Considering the interatomic stress curve of a crystal, such as the one shown in Fig. 1.10, replaced by a half sine wave

$$\sigma = \sigma_c \sin \frac{2\pi x}{a} = \sigma_c \sin 2\pi\epsilon, \qquad (15.1)$$

we can estimate the ideal cohesive strength by matching its slope at $\epsilon = 0$ with Young's modulus. This gives

$$\sigma_c = E/2\pi. \qquad (15.2)$$

More exact estimates of the ideal cohesive strength, using more realistic interatomic force laws, give a value which ranges between 0.05 and $0.1E$ (Prob. 15.2). For most solids held together with primary bonds, Young's modulus E is of the order of 10^{12} dyne/cm^2. This compares well with a strength of 10^{11} dyne/cm^2 calculated from the heats of vaporization of transition metals. It is evident, however, that the generally observed strength values of 10^9 to 10^{10} dyne/cm^2 of most commercially available brittle materials fall far short of this ideal strength.

This discrepancy was resolved by Griffith (1920), who indicated that the ideal cohesive strength need not be reached over the entire solid but only at the root of a sharp and narrow crack. He showed that such a crack in a plate can become unstable at relatively low nominal stresses and extend lengthwise, thus progressively fracturing the plate as it advances.

Griffith based his analysis of the brittle-fracture criterion on an energy method to avoid making assumptions about the root radius of the crack which he proposed was responsible for the relatively low strength. A more satisfactory and simpler derivation than that of Griffith is possible by stating that the gap between the observed macroscopic strength and the ideal tensile cohesion is bridged by the stress-concentrating action of a sharp and long crack. Thus, if an atomically sharp crack (radius of curvature at crack tip $\rho = a$) of length $2c$ is situated with its plane normal to the direction of a uniaxial tensile stress σ_0 in a plate having Young's modulus E, the local tensile stress σ at the crack root acting across the extension of the plane of the crack is given by Inglis' (1913) formula (see Eq. 11.35)

$$\sigma = \sigma_0(1 + 2\sqrt{c/a}). \qquad (15.3)$$

The brittle-fracture criterion is then that the local stress reach the ideal cohesive strength, say

$$0.1E = \sigma_0(1 + 2\sqrt{c/a}).$$

When this is solved for the macroscopic strength σ_0 for very sharp and long cracks, one obtains

$$\sigma_0 = \frac{E}{20}\sqrt{\frac{a}{c}}. \qquad (15.4)$$

This derivation is due to Orowan (1934). The result differs little from the original one of Griffith. Griffith derived the condition for brittle fracture from an energy argument, which states that fracture occurs when for an infinitesimal extension of the crack there would be more elastic energy released than is required for the specific surface energy α of the new surfaces. That is, at this point the rate of draining of elastic energy from the stressed plate by the advancing crack is just enough to balance the rate of increase of surface energy, and the crack propagation reaches a self-sustaining point.

The fracture stress turns out to be

$$\sigma_0 = \sqrt{2\alpha E/\pi c}. \qquad (15.5)$$

Extension of the crack beyond a length $2c$ provides an ever-increasing driving force for further unstable extension of the crack.

The surface energy α can be estimated from the force law, Eq. 15.1. For brittle solids it is of the order of 10^3 ergs/cm^2. Thus the equation derived by Griffith, although different in form, gives an answer of the same magnitude as Eq. 15.4. A discussion of the stress method and energy method in deriving the equation of brittle fracture will be deferred to the end of the next section.

The analysis of Griffith and Orowan is for plane stress. Cracks in a real material would most likely not be very long in comparison with the major axis $2c$ of their cross section, but rather of uniform dimensions, like the shape of a penny. The stress-concentrating effect of a penny-shaped crack in an infinite elastic solid was solved by Sack (1946) and Sneddon (1946). Their results show that in this case the Griffith criterion is altered only by a small factor depending on Poisson's ratio ν:

$$\sigma_0 = \sqrt{\frac{\pi \alpha E}{2(1-\nu^2)c}}. \qquad (15.6)$$

15.3 BRITTLE FRACTURE UNDER BIAXIAL AND TRIAXIAL STRESS

In most instances a part is subjected to a state of stress more complex than uniaxial. Therefore, it is of interest to investigate the conditions for brittle fracture under biaxial and triaxial stress. Griffith (1924) gave the solution for the two-dimensional case where cracks of identical size are assumed to have their planes parallel to the unstressed direction but are otherwise randomly oriented. He considered one of these cracks in a biaxial stress field σ_1 and σ_2, at an arbitrary orientation θ with respect to the axis of σ_2 as seen in Fig. 15.1, and calculated the maximum value of the peripheral stress on the free surface of the crack as a function of the orientation angle θ. He then postulated that fracture would result whenever this maximum peripheral stress would reach locally the ideal value σ_c in

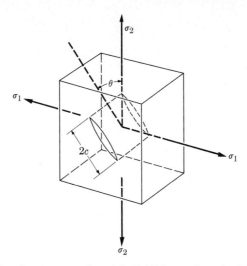

FIG. 15.1. An elementary cube with Griffith crack under biaxial stress.

uniaxial tension given by Eq. 15.2. Thus he found that while the smaller stress σ_2 is in the range $-3\sigma_0 \leq \sigma_2 \leq \sigma_1$, fracture takes place when the algebraically larger stress σ_1 reaches the value

$$\sigma_1 = \sigma_0 = \frac{E}{20}\sqrt{\frac{a}{c}}, \tag{15.7}$$

as if the stress σ_2 were not present at all. When σ_2 is in the range $\sigma_2 < -3\sigma_0$, however, fracture takes place when (Prob. 15.16)

$$(\sigma_1 - \sigma_2)^2 + 8\sigma_0(\sigma_1 + \sigma_2) = 0. \tag{15.8}$$

The relation between σ_1 and σ_2 is illustrated graphically in Fig. 15.2. If the condition that all cracks have their planes parallel to the unstressed axis is removed, it can be expected that the locus is limited by the broken lines. This would result from the fact that when the third stress σ_3, is zero, fracture in the region outside the broken line would always result from cracks that have the normals to their planes in the 2, 3 or 1, 3 planes. The important result of this analysis is that the uniaxial compressive strength of a brittle material is about 8 times its tensile strength. Other defects give different results (Probs. 15.3, 15.4).

For fracture under triaxial stress, it is again desired to find the crack with the most critical orientation. It seems reasonable to assume that such a crack will be parallel to the axis of the intermediate principal stress. Since normal stress parallel to the plane of the crack produces almost no stress concentration, the fracture will be determined by the maximum and minimum principal stresses. Using the biaxial theory, one can thus construct the fracture locus of Fig. 15.3, which has threefold axial symmetry. So long as both σ_2 and σ_3, the algebraically smaller stresses, are greater than $-3\sigma_0$, then fracture will occur when σ_1 reaches a critical value σ_0.

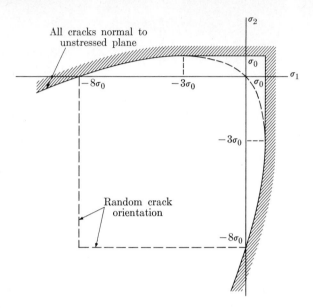

Fig. 15.2. Brittle fracture under biaxial stress.

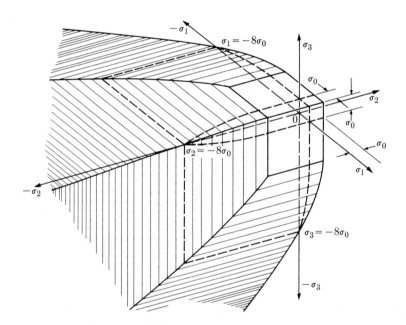

Fig. 15.3. Brittle fracture solid representing the states of triaxial stress for brittle fracture. A stress state falling outside solid will result in fracture.

TABLE 15.1

Bridgman's Compressive Strength Experiments on Pyrex Under Triaxial Stress

Experiment	σ_1, psi	$\sigma_2 = \sigma_3$, psi	σ_0, psi	Remarks
1	—	—	—	Separation between glass and pull rod
2	−271,000	−355,000	1410	Surface untreated, fractured irregularly
3	−200,000	−360,000	5720	Surface untreated, fracture normal to σ_1
4	−220,000	−396,000	6300	Surface ground and polished, fracture normal to σ_1
5	−35,000	−380,000	35,800	Surface fire-polished, fracture normal to σ_1

The experiments of Bridgman are in qualitative agreement with the three-dimensional fracture condition. Bridgman (1947) carried out experiments on small Pyrex glass cylinders under hydrostatic compression where the axial compression was gradually relieved by backing off the compression plates. Fracture always occurred when the net axial stress was still predominantly compressive. When his results, consisting of five experiments, are evaluated according to the fracture condition (15.8) (where it has to be kept in mind that for this experiment $\sigma_2 = \sigma_3$), the values of σ_0 are found to be in reasonable agreement with the bulk tensile strength of Pyrex, which ranges from about 2000 to 20,000 psi, depending on the surface condition (Argon et al., 1960); the calculated values of σ_0 are given in Table 15.1.

◀ The brittle-fracture condition for triaxial stresses must be used with caution, however, since in some cases it may not apply. For instance, sometimes all surface cracks on glass can be made to have their planes perpendicular to the surface, making the pressure normal to the surface not a significant parameter (Prob. 15.5).

◀ Furthermore, in the above discussion of fracture under multiaxial stress, it was tacitly assumed that the cracks will not close up under hydrostatic pressure and will, therefore, not transmit normal or tangential surface tractions. In reality, it can be expected that cracks will close up under compression, transmitting some normal and tangential stress and thereby altering the fracture conditions. McClintock and Walsh (1962) modified the Griffith (1924) solution by considering the effect of normal and frictional forces on the crack surfaces. Assuming that the cracks close under negligible compressive stress and that a coefficient of friction of unity is valid, these investigators found that the biaxial fracture locus is as shown in Fig. 15.4. This modified fracture locus agrees better with published results on the fracture of rocks.

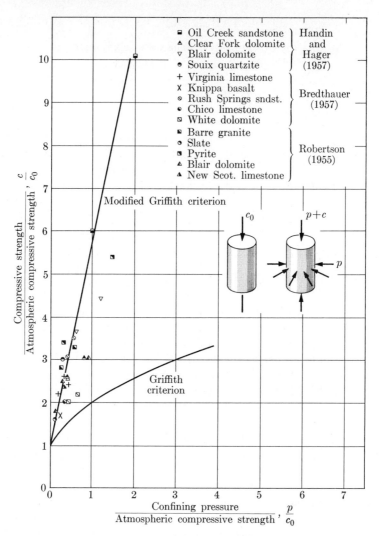

Fig. 15.4. Effect of a coefficient of friction of unity across the surface of a Griffith crack on the compressive strength of a brittle solid. Crack assumed to close as soon as pressure is applied across it. (After McClintock and Walsh, 1962.)

◀It is now possible to give a more critical discussion of the stress and energy methods of deriving a condition for brittle fracture.

◀The derivation of the brittle-fracture condition given by Griffith and based on energy balances is only a necessary, but not a sufficient condition. It states that the required surface energy must be present in the vicinity of the crack, but this alone does not guarantee propagation of a crack. Thus, an extension of a shear crack in its plane could release enough elastic energy to become unstable in the manner of Griffith's (1920) derivation; yet a shear crack never extends in its plane, but rather propagates into an S-shape across planes of maximum tensile stress (Erdoğan and Sih, 1963). The stress condition, on the other hand, states that fracture *will* commence from a defect when the local tensile stress reaches the ideal cohesive strength. It is therefore not only a necessary but also a sufficient condition. The stress condition is also able to predict fracture under biaxial stress as well as the point of fracture initiation and the course of the fracture if the stress field around a crack of complicated shape can be solved. ▶

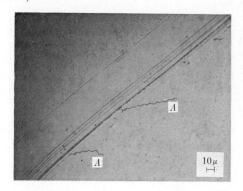

FIG. 15.5. Loose machining chips (A) resulting from dragging a very sharp tungsten carbide tool over a microscope slide.

As is shown in Fig. 15.3, brittle fracture should not occur in hydrostatic (equal triaxial) compression. A condition of this type is achieved at the tip of a very sharp tungsten carbide tool as it is dragged over a glass surface under its own weight. Under the superimposed hydrostatic pressure (which can easily reach magnitudes equal to the ideal tensile strength, Prob. 15.6), the shear stresses resulting from the dragging operation can gouge the glass surface and produce long chips by a process of flow without intervening fracture, as can be seen in Fig. 15.5. [See also Fig. 20.8(a), Ryschkewitch, 1942, and Joos, 1957.]

15.4 EXPERIMENTAL VERIFICATION OF THE GRIFFITH THEORY

◀ As a verification of his theory, Griffith (1920) introduced cracks of known lengths into large, thin-walled, glass bulbs and subsequently broke them under internal pressure. The breaking strengths were found to be inversely proportional to the square root of the crack length, while the proportionality factor was very close to that given by Eq. 15.7. To show that the ultimate strength of a brittle material in the absence of flaws is indeed of the same magnitude as the theoretically estimated ideal strength, Griffith tested freshly drawn glass rods in bending and found their strength to be over 10^6 psi. Such high strengths were also measured by Orowan (1933) on mica sheets with stress-free edges.

◀ Although the mechanism of the formation of the cracks in initially flawless brittle solids is not fully understood at the present, it is clear that all such materials must have a number of the relatively largest cracks in them to give the uniformly low values encountered in everyday practice. It is easy to demonstrate, however,

TABLE 15.2

Variation of Strength of Glass Fibers as a Function of Their Length as Observed by Anderegg

Length of fiber, mm	Diameter of fiber, mm	Tensile strength kg/mm^2
5	13	150
10	13.5	122
20	12.5	121
45	13	115
90	12.7	76
183	12.7	87
1560	13	72

that even these weak solids are very strong if very small parts can be tested, since there would be less likelihood of finding cracks. Anderegg (1939) has tested soda lime glass filaments of various lengths from 5 to 1560 mm but all of the same diameter. His results are shown in Table 15.2, and are consistent with the idea of a distribution of cracks of various sizes.

◄The existence of strength-impairing cracks on the surfaces of glass plates was demonstrated more directly by exposing them at elevated temperatures to sodium vapor (Andrade and Tsien, 1937, Argon, 1959a), or to molten alkali salts (Ernsberger, 1960). In both of these cases a thin layer on the surface of approximately 0.001-in. thickness is transformed into a glass of much higher coefficient of expansion than the parent glass. Upon subsequent cooling of the plate, a network of cracks, bearing a strong resemblance to those on drying mud, develops on the surface as a result of differential contraction. The origins of such networks (Fig. 15.6) appear to be the Griffith cracks, since networks do not form on glass surfaces

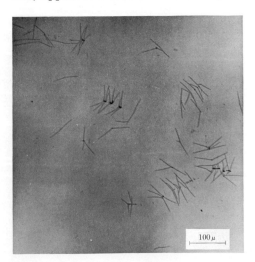

Fig. 15.6. Origins of a network of superficial cracks developed by melting a eutectic salt of KNO$_3$ and LiNO$_3$ on the surface. (Ernsberger, 1960. Courtesy of Royal Society of London.)

which have been previously acid polished with HF to remove a thin surface layer, and with it the Griffith cracks.

◄ Attempts to observe cracks under the light microscope have not been successful, probably because the opening at the surface of a crack with elliptical cross section and of sufficient severity to cause a strength reduction to one percent of the ideal strength would be too small to be resolved (Prob. 15.7). Electron-microscopical techniques have been unsuccessful, apparently because of the difficulty in making replicas which would penetrate existing surface cracks. ►

15.5 ORIGIN OF STRENGTH-IMPAIRING CRACKS

The cause of strength-impairing cracks is still not completely understood. Griffith (1920) observed that freshly drawn fibers of soda glass lose their high strength when aged. The very high stresses that result from the contact of solid objects are also known to reduce the strength of glass fibers and plates drastically (Griffith, 1920; Ernsberger, 1960). Inclusions resulting from the adhesion of dust particles to the surfaces of solidifying glass impair mechanical properties (Holloway, 1959). It has been observed (Murgatroyd, 1944) that annealing of super-strong thin glass fibers which were previously drawn from the melt reduces their strength even in the absence of mechanical damage. On the other hand, it is known that such annealing results in a slight increase in density (Bateson, 1958). Furthermore, indentation experiments on freshly cleaved smooth fracture surfaces of massive plate glass show that there is a definite upper limit for fracture stress of 3×10^5 psi (considerably below the ideal strength) and that aging of these surfaces does not change this upper limit (Argon et al., 1960). From these details, the following possible explanation emerges: thin fibers drawn from a hot melt are cooled very rapidly, preserving a supercooled high-temperature structure of lower atomic order and coordination in which the distortion of atomic bonds is spread throughout the structure, producing no concentration of severed bonds. An isotropically high strength results. There is no evidence that the high strength in these fibers results from the alignment of "chains" by drawing, as it was once assumed, nor from temper stresses (Anderson, 1958). If the glass is cooled slowly or if chilled glass is stabilized, compaction with an increase in coordination of atoms appears to start at many places at once. As these compacted regions grow together, the atomic bonds in the "interface" accommodate a greater degree of distortion and extension than in the high-temperature structure, with the result that more bonds may be highly extended in those interfacial regions. The electron-microscopical observations of Warshaw (1960) on fracture surfaces produced in vacuum reveal a cellular fine structure with dimensions ranging from about 30 to 300 A. A typical example is shown in Fig. 15.7. It is likely that these are the cells inferred from the indentation experiments. It may, furthermore, not be unreasonable to expect that the more mobile sodium ions in the glass migrate to and accumulate at these more highly strained interfacial regions. This would explain the upper limit in the indentation strength of bulk glass. Moreover, it is known that the sodium ion

FIG. 15.7. Electron micrograph of a fracture surface of Pyrex glass showing a cellular fine structure. (Warshaw, 1960. Courtesy of American Ceramic Society.)

plays a critical role in the corrosion of glass surfaces by water vapor (Charles, 1958). Therefore, it is likely that the weak connecting layer of sodium ions between compacted regions is preferentially attacked by water vapor where it breaks into the surface. In this fashion more severe and possibly longer cracks can arise on free surfaces. The higher resistance to aging of fused quartz may be due to the absence of this.

In general, however, experiments show that the severe cracks which are responsible for most of the failures in glass result from mechanical damage that is introduced either in processing or in handling. This is clear from the fact that the strength of nearly all glass articles can be materially increased (tenfold increases are common) when a thin surface layer is etched off with hydrofluoric acid. Such higher strengths, once established, are preserved in the absence of new mechanical damage and chemical corrosion.

Plate glasses are often very weak, the cause being an insufficient polish. When such glass surfaces are lightly etched in dilute hydrofluoric acid, the large fissures of rough polishing can be made visible, as shown in Fig. 15.8. This very simple procedure is used by the glass companies as a quick test for the quality of the polish.

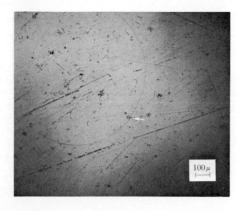

FIG. 15.8. Grinding cracks, developed by a light hydrofluoric acid etch, on insufficiently polished Pyrex plate glass, causing the indentation fracture.

Many glass articles that have a nominal strength of 1000 to 10,000 psi can be materially strengthened by the introduction of compressive residual surface stresses by quenching. It is easily possible to set up residual compressive stresses of 20,000 psi by this method. The magnitude of the compressive stress is often limited by the strength of the glass while it is still partly hot and when the surface is under tension. A new method avoids this problem by building up a state of compressive stress in a thin surface layer by ion-exchange methods and suitable heat treatments without ever developing a tensile stress on the surface. Such glassware, identified by the trade name Chemcor, with compressive residual surface stresses of the order of 100,000 psi, is extremely resistant to mechanical damage.

15.6 STATIC FATIGUE

When glass is subjected to a constant stress somewhat below the average expected fracture stress value obtained in rapid loading (10^{-5} sec^{-1}), fracture will nevertheless occur after a certain period of delay which depends on the stress margin. Thus, for instance, Holland and Turner (1940) have observed a reduction in strength of 70% in bent glass plates when the loading time was increased from 34 seconds to 34 hours. Prolonged loading at stresses lower than 30% of the short-term strength values did not lead to any fracture, giving a kind of endurance limit in the absence of any mechanical damage. In the other direction, Baker and Preston (1946) have found that the decrease of the duration of the load from 10 seconds to 10^{-2} second increased the fracture strength of soda lime glass nearly twofold. This dependence of fracture stress on the period of loading has been called *static fatigue*. Experiments have shown that this effect disappears in the absence of water vapor, as in a vacuum.

Experiments of Charles (1958) have shown that water corrodes glass surfaces by attacking the loosely bound sodium ion in the glassy network, which in turn plays a catalyzing role in the attack on the silicon-oxygen bond by the hydroxyl ion. This attack is accelerated under tensile stress, and its rate is governed by the rate of diffusion of the sodium ion to the centers of attack. Thus the highly stressed roots of the Griffith cracks become preferential sites where the corrosion is most severe. When the crack has grown in length to a size to satisfy the Griffith condition at that stress level, rapid fracture occurs.

Since this process is based on the diffusion of the sodium ion through the glassy network, its rate, which can be estimated from measurements of the delay time between the application of the stress and fracture, should depend strongly on the absolute temperature, through a Boltzmann factor. Good confirmation of this was obtained by comparing the activation energy $U_a = 18.8$ kcal/mole of the delay time with the value 20 kcal/mole for the activation energy of diffusion of the sodium ion in a soda glass network. The absence of delayed fracture at liquid nitrogen temperature (Kropschot and Mikesell, 1957) and in vacuo after being fully baked out (Baker and Preston, 1946) are all in support of the above theory.

In practice, static fatigue often causes puzzling breakages. Poor quality glassware on occasions has been observed to break while it is under no apparent stress. In such cases the cause may have been a gradual extension of a particularly long crack under the action of residual or thermal stresses. To prevent such breakage all glassware of complicated shape is commonly fully annealed.

A similar static-fatigue process also occurs in steel, where it results from a supersaturation of hydrogen and is known as hydrogen embrittlement. Although the understanding of hydrogen embrittlement is far from being complete, it appears that excess hydrogen gas entrapped in steel tends to segregate and nucleate cracks even in the absence of any applied stress, and thereby embrittle it. (For a recent review see Cotterill, 1961, and also more recent work by Tetelman, 1963.) Parts which have been electroplated are especially susceptible to hydrogen embrittlement.

15.7 PHENOMENA OF FRACTURE PROPAGATION

◀ When the Griffith equation is satisfied and a crack begins to propagate through a truly brittle material, the stress concentration due to the elongating crack steadily increases, and the released elastic energy exceeds the surface energy by increasing amounts, so that the fracture process accelerates and reaches very high speeds in a short time. A simplified analysis due to Mott (1948) indicates that the maximum velocity of a brittle crack becomes limited by the speed with which an elastic field can rearrange itself.

◀ The first studies of the problem (Yoffe, 1951; Craggs, 1960; and McClintock and Sukhatme, 1960) dealt with some form of traveling cracks or stress fields, and showed that as the velocity increased, the maximum stress shifted to either side of the plane of the crack. This would tend to cause forking in fast cracks, slowing them down. Craggs found the limiting velocity to be 0.66 times the shear wave velocity. Figure 15.9 shows the bifurcation resulting from a series of increasingly energetic blows at the edge of a plate. Very strong fibers show similar effects, sometimes completely shattering from the sudden release of elastic waves.

◀ The analysis of Craggs indicates that fast-moving brittle cracks may be to some extent self-propagating, since it indicates that the driving stress for faster-moving cracks is less. This could explain the puzzling observations of Schardin (1959) and co-workers that fast-moving brittle cracks rarely if ever slow down, but either keep on moving or stop suddenly. When such a crack, being driven by a very low stress, is disturbed by an obstacle or hard region, its speed momentarily decreases, requiring a larger stress than is available to propagate it at this velocity, causing the crack to stop suddenly.

◀ More recently, solutions have been found for growing cracks (Akita and Ikeda, 1959; Cotterell, 1964) indicating that the limiting velocity is the Rayleigh wave velocity ($v = 0.92/G$, see Section 7.8). Experiments by Schardin (1959) show that the ratio of the velocity of brittle cracks to the shear wave velocity varies with the composition in various glasses. The highest ratio was 0.614 for a pure

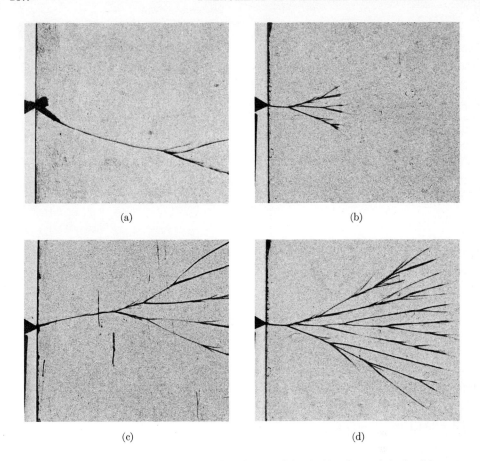

Fig. 15.9. Fracture in glass plate produced by striking knife edge at left; load increases from (a) to (d). (Schardin, 1959. Courtesy of M.I.T. Press.)

fused quartz glass. When a network modifier, especially PbO, was added, the ratio rapidly fell close to the value around 0.38. Since fused quartz is one of the most perfectly elastic solids, the elastic velocity ratio would appear to be close to 0.614. The decrease with addition of network modifiers suggests a certain amount of dissipative deformation at the crack tip. This is qualitatively in agreement with observations of increased amounts of irreversible flow on glass surfaces treated with alkali vapors (Ernsberger, 1960).

◀Surprisingly, the more recent theories fail to indicate forking. In addition, the stress singularity required for propagation varies with velocity in a different way. Unfortunately, no one author has compared all the analyses, nor the data, so questions remain unresolved. The reader may find those references known to the authors of this book from the references mentioned in the above papers, and will have to form his own conclusions.▶

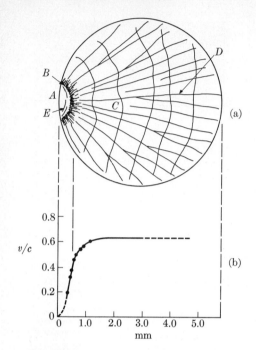
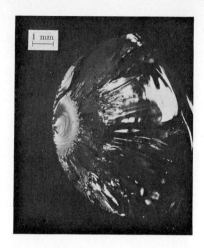

Fig. 15.11. Origin of fracture in a rod. Note Wallner lines in the fracture mirror.

Fig. 15.10. (a) Schematic drawing of the appearance of the fracture surface of a rod (6-mm diameter) broken in bending. (b) Velocity of the crack at various points of the fracture. (After Smekal, 1952.)

15.8 FRACTURE SURFACES

The surfaces produced by brittle fracture bear certain characteristic markings that are of great interest in reconstructing the fracture process and determining the condition under which it occurred. Figure 15.10(a) shows a sketch of Fig. 15.11, which is a typical fracture surface of a rod broken in simple tension. Three different regions are clearly visible. Such a fracture surface can be correlated with typical curves of crack velocity as a function of distance, such as Fig. 15.10(b). The region marked A, presenting a mirror smooth appearance, is where the crack became unstable and started to accelerate, still traveling with a slow speed. As the speed of the crack increases, the surface becomes rougher and has a somewhat matte appearance to the unaided eye as seen in region B. The maximum velocity is reached rapidly and the region B blends into region C, where the crack now moves with its steady-state velocity. As the crack reaches high velocities, the above-mentioned distortion of the stress field causes portions of the crack front to deviate randomly above or below the extension of the crack plane. Such departures from planar fracture are then joined by relatively steep ridges that form the radiating straight lines commonly called *hackle marks* (D in Fig. 15.10a). Hackle marks have on the other hand also been observed in slow fracture of materials such as plastics and "Jello," where they can often be explained by the change in

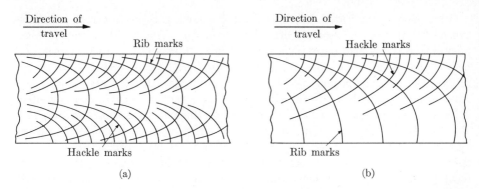

FIG. 15.12. Rib and hackle marks (a) on a fracture surface of plate glass with surface compression, and (b) on fracture surface of plate broken in bending about axis lying in plate.

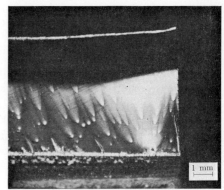

FIG. 15.13. Hyperbolas on the fracture surface of Plexiglass. Main fracture originated at the lower right-hand edge.

the direction of the maximum tensile stress during fracture. Thus the rate effect is not the only cause of fracture patterns. Further distortions of the stress field at higher crack velocities cause the type of large-scale forking that was mentioned earlier. The direction of the hackle marks serves as a convenient tool in tracing complicated fractures to their origins. Small disturbances in normal stress often leave recognizable traces of the fracture front, delineating its position at various stages, see E in Fig. 15.10(a).* In plates that are not fully annealed, but have a small amount of residual compressive stress on the surface, the fracture front advances more rapidly in the center and drops behind at the surface, leaving convex traces that have been called *rib markings* (Fig. 15.12a). When the plate is fractured by bending about an axis in the plane, these rib markings are strongly antisymmetrical as in Fig. 15.12(b). Clearly, by their nature, hackle marks and rib marks form orthogonal intersections.

---

* These should not be confused with Wallner lines that arise from the interference of the crack front with elastic waves released by the fracture itself. Such lines cut the fracture front at a variety of angles depending on their origin, and have been used by Smekal (1952) in the determination of fracture velocities. See Fig. 15.11.

Other more complicated fractures show the same general markings. Fractures caused by thermal stresses are often very smooth. This results from the highly nonuniform stress field permitting the slow growth of a fracture with perhaps the help of water-vapor diffusion from the atmosphere. Such fractures, which may progress with interruptions, will show only rib marks. Brittle fracture surfaces in plastics show, in addition to the above markings, certain other interesting effects. A fracture surface of polymethyl methacrylate (plexiglass) is shown in Fig. 15.13. The hyperbolas on the surface are evidences of cracking ahead of the main fracture front. They are caused by the interaction of the main fracture front and the spreading fracture front of localized cracking that has evidently started at the focal point of the hyperbola. Such hyperbolas show very strong preferential specular reflections of red and green colors, which are very sharply delineated under low magnification.

15.9 THE SIZE EFFECT AND STATISTICAL ASPECTS OF BRITTLE FRACTURE

It has been known for a long time that the strength of a fine glass fiber or metal wire is usually higher than that of a larger rod. In materials that exhibit brittle fracture, this *size effect* is particularly marked. This is evidently due to the fact that the strength of a brittle material is governed by the stress at the root of its most dangerous crack, and that it becomes more and more likely to find dangerous cracks in specimens as their size increases. The effect of length has been shown in Table 15.2. Reinkober (1931) showed the effect by breaking silica glass fibers, then the two fragments, then the fragments of the fragments, each time obtaining higher strength values. It must be remarked here that the great rise in strength observed by Griffith (1920) and Anderegg (1939) in filaments with decreasing diameters and attributed for a long time to the limitation imposed on crack size by the diminishing fiber diameter has been shown by Otto (1955) to be due to different molecular structure, independent of accidentally introduced cracks. Otto demonstrated that finer filaments, which must necessarily be drawn from a hotter melt, have a much stronger structure than thicker filaments, drawn from cooler melts. He also showed that filaments of different diameters drawn from the same high-temperature melt were all equally strong.

◀A quantitative theory for the size effect in brittle materials depends on the distribution of flaws throughout the material. In some brittle materials, such as inorganic glasses, cracks are formed only at the free surfaces; in others, such as cast iron, they are uniformly distributed in the volume; in others still, such as polymers, there are different types of flaws in the surface and in the volume. In a long wire of ductile metal, length may be the important variable, since necking can occur anywhere along the length, but requires the cooperation of all elements across a section. For convenience in this discussion, we shall consider the case in which the flaws are distributed throughout the area, and leave the dependence on length or volume to be obtained by an appropriate change of variable.

◂Assume, therefore, that in a unit area there are $g(S)\,dS$ flaws giving strengths in the range S to $S + dS$.* The probability of fracture taking place in an infinitesimal area δA, below a stress S_1, is

$$\delta\phi = \delta A \int_0^{S_1} g(S)\,dS. \tag{15.9}$$

The probability for no fracture below S_1 is then

$$(1 - \delta\phi) = 1 - \delta A \int_0^{S_1} g(S)\,dS. \tag{15.10}$$

If there are n infinitesimal areas in the total specimen ($n = A/\delta A$), the probability Φ that none of them will fail below a stress S_1 is the product of their individual probabilities, since these are independent events:

$$1 - \Phi = (1 - \delta\phi)^n = \left[1 - \delta A \int_0^{S_1} g(S)\,dS\right]^{A/\delta A}. \tag{15.11}$$

If the volume elements are chosen smaller and smaller, increasing the number n in the limit to infinity, the probability of fracture below S_1 will become (Prob. 15.8)

$$\Phi(S_1) = 1 - \exp\left[-A \int_0^{S_1} g(S)\,dS\right]. \tag{15.12}$$

Note that as $S_1 \to \infty$, the exponent will become more negative, so that $\Phi(S_1) \to 1$. This is necessary, since any area A must fail at sufficiently high values of the applied stress. Furthermore, for a given stress the probability of failure increases toward unity as the area increases. This is a statistical statement of the size effect in brittle fracture.

◂The specific strength distribution $g(S)$ has been obtained for surface cracks in various glasses by Argon (1959b) using ball indenters. Results for window glass shown in Fig. 15.14 are typical in showing that there are few large cracks, but many small ones, although the peak is not always found. A similar conclusion for a variety of glasses was reached by Sucov (1962).

◂In large specimens, only the largest cracks contribute to the failures. Thus only the lowest part of the strength distribution of Fig. 15.14 is of interest. Below $S_0 = 8 \times 10^{10}$ dynes/cm^2, the distribution may be approximated by

$$g(S) = a[(S - S_l)/(S_0 - S_l)]^n, \tag{15.13}$$

where $S_l \approx 4.8 \times 10^{10}$ dynes/cm^2, and $n \approx 2.0$. The resulting probability distribution of specimen strengths can be found by the substitution of Eq. 15.13 into Eq. 15.12 and integration from S_l to S_1 to obtain the probability of no failure below S_1,

$$\Phi(S_1) = 1 - \exp\left\{-\frac{aA}{n+1}\left[\frac{(S_1 - S_l)^{n+1}}{(S_0 - S_l)^n}\right]\right\}. \tag{15.14}$$

---

* Here S is used to denote stress so that σ may be used for standard deviation.

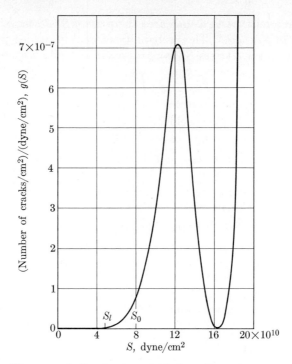

Fig. 15.14. Specific strength distribution of window glass. (After Argon, 1959(b).)

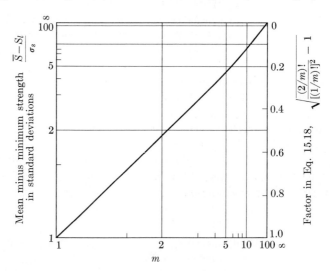

Fig. 15.15. Parameter relating the least value to the exponent m for extreme-value distributions.

◀ Equation 15.14 can be differentiated with respect to S_1 to obtain the frequency function of specimen strengths, and then the mean, standard deviation, and skewness can be found as described below. In evaluating the required integrals, the factorial or gamma function appears:

$$x! = \Gamma(x+1) = \int_0^\infty p^x e^{-p}\, dp. \tag{15.15}$$

At the same time, it is convenient to redefine the specific strength distribution in terms of $m = n + 1$ and an area A_0:

$$g(S) = a\left(\frac{S - S_l}{S_0 - S_l}\right)^n = \frac{[(1/m)!](S - S_l)^{m-1}}{mA_0(S_0 - S_l)^m}. \tag{15.16}$$

In terms of this, the mean strength of specimens with an area A at constant stress is

$$\bar{S} = \int_{S_l}^\infty Sf(S)\, dS = S_l + (S_0 - S_l)(A_0/A)^{1/m}. \tag{15.17}$$

The reason for introducing A_0 can now be seen, since, if S_0 represents roughly the largest strength for which Eq. 15.13 is a reasonable fit, A_0 is the area of the specimen having the mean strength S_0 and is the smallest area of specimen to which this extreme-value theory can be applied.

◀ The standard deviation of strength of specimens of area A is

$$\sigma_S = \sqrt{\int_{S_l}^\infty (S - \bar{S})^2 f(S)\, dS} = (S_0 - S_l)\left(\frac{A_0}{A}\right)^{1/m} \sqrt{\frac{(2/m)!}{[(1/m)!]^2} - 1}. \tag{15.18}$$

The amount by which the mean strength exceeds the minimum is shown in Fig. 15.15. Finally, the skewness of the distribution of strength of the specimens is

$$\alpha_3 = \frac{1}{\sigma_S^3}\int_{S_l}^\infty (S - \bar{S})^3 f(S)\, dS = \frac{(3/m)! - 3(2/m)!(1/m)! + 2[(1/m)!]^2}{\{(2/m)! - [(1/m)!]^2\}^{3/2}}, \tag{15.19}$$

as shown in Fig. 15.16.

◀ Equations 15.17 and 15.18 provide quantitative information on the size effect. For instance, suppose we are interested in predicting the performance of a part with an area A_2 under uniform stress, from laboratory data on geometrically similar parts with the area A_1 under uniform stress. The difference in mean strengths is

$$\frac{\bar{S}_2 - \bar{S}_1}{\sigma_1} = -\frac{1 - (A_1/A_2)^{1/m}}{\sqrt{\frac{(2/m)!}{[(1/m)!]^2} - 1}}. \tag{15.20}$$

The ratio of the standard deviations of the two sizes is

$$\sigma_2/\sigma_1 = (A_1/A_2)^{1/m}. \tag{15.21}$$

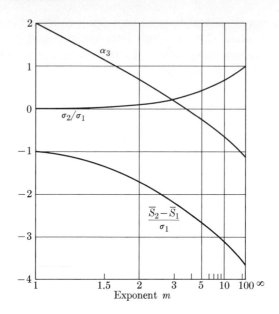

FIG. 15.16. Effect of size on standard deviation and mean strength for area ratio $A_2/A_1 = 100$.

This size effect is plotted in Fig. 15.16 for an area ratio $A_2/A_1 = 100$. Note that for typical values of m of 2 to 6, the mean strength is reduced by a few standard deviations on going from the smaller to the larger specimens.

◀ If large enough, the size effect can be used to estimate the parameters of the extreme-value distribution, by first finding the effect of size on either the mean or standard deviation and determining the corresponding value of the exponent m from Fig. 15.14. Once this is done, the least strength S_l can be found from the mean $\bar{S}$ and the standard deviation σ of the strength from Eq. 15.17. One could in principle estimate m from the skewness, but in practice sufficient data are rarely available for a reasonably accurate estimate. The size effect for specimens with nonuniform stress will be discussed in Section 18.6 on notch sensitivity in fatigue. ▶

15.10 SUMMARY

Brittle fracture is the sudden separation of a stressed body into two or more parts without any measurable inelastic deformation. This kind of fracture is thus accompanied by minimal energy dissipation and manifests itself by fracture velocities that are close to those of sound in the same medium.

Brittle fracture results from submicroscopic cracks with atomically sharp roots, where the stress can become concentrated beyond the capacity of the body to resist it.

Griffith (1920) derived the brittle-fracture relation known by his name by recognizing that an infinitely sharp elliptical crack of semi-major axis c, normal to the direction of uniaxial tensile stress σ_0, in a plate with Young's modulus E, can reach a point of instability. At this point, the increase in surface energy necessary for the growth of the crack can be supplied by the elastic strain energy from parts of the stressed plate through which the fracture traverses. The nominal stress at this point becomes

$$\sigma_0 = \sqrt{2\alpha E/\pi c}, \qquad (15.5)$$

where α is the specific surface energy. The nominal stress σ_0 for fracture can be derived more readily from the stress concentration around an elliptical crack of length $2c$ with an atomically sharp radius a by requiring that this stress concentration raise the nominal stress σ_0 to the ideal cohesive strength $\sigma_c = 2\alpha\pi/a = E/2\pi$ at the root of the crack, i.e.,

$$\sigma_0 \approx \frac{\sigma_c}{2}\sqrt{\frac{a}{c}}. \qquad (15.3)$$

Cracks normally present in commercial glasses limit their strengths to the order of 10,000 psi. Higher values can be obtained by superposing residual compressive stresses which prevent the growth of surface cracks.

When a thin plate with a randomly oriented crack is biaxially loaded by principal stresses σ_1 and σ_2, brittle fracture will result when the peripheral stress somewhere near the tip of the crack reaches the ideal strength. The locus of the various combinations of the principal stresses σ_1 and σ_2 giving brittle fracture is illustrated in Fig. 15.2.

One important conclusion to be drawn from this is that the compressive strength c_0 of an ideally brittle material is eight times the tensile strength.

◀Consideration of closing of elliptical cracks under compressive forces, with resulting transmission of normal and shear stresses across crack faces, has given better agreement between theory and experiment for the fracture of rocks under pressure.

◀Although direct observations of the bulk of the so-called Griffith cracks have still not been possible, very strong evidence for their presence has accumulated, especially by recent experiments on superficially stressing glass surfaces by alkali ion diffusion. Recent experiments on strength of fresh surfaces have also given a better understanding of the genesis of cracks, bridging the gap between the ideal strength and strength of mechanically damaged surfaces. It has been established with sufficient certainty that the most dangerous cracks in brittle glassy substances result from mechanical damage.▶

Under prolonged loading below its short-term strength value, a glass may still break if the cracks on the surface can progressively extend by a process of stress corrosion under the effect of water vapor. This effect is known as static fatigue and manifests itself by a delay time between the application of load and fracture. A similar process operates in steel where hydrogen gas in supersaturation can produce cracks by segregation. This is known as hydrogen embrittlement.

◀When brittle cracks become unstable and accelerate toward eventual fracture, they gather velocities that reach an appreciable fraction of the velocity of sound in the medium. Under certain conditions, especially at high stress, the crack front can split into a number of diverging fronts. ▶

The surfaces of brittle fracture show a number of characteristic markings which make it possible to trace the fracture to its origin and tell something about its velocity at different stages, as well as the stress state that has caused it.

Brittle materials can be expected to have a whole family of cracks in them or on them. Although fracture would result from the most severe crack when the solid is uniformly stressed, nonhomogeneous stresses as well as different sizes of specimens could show different strengths depending on which crack would be favored under each condition. Hence, a dependence of the strength on the size of the stressed portion will result. Statistical approaches to brittle fracture relating flaw densities and distributions to specimen strength are possible.

REFERENCES

Orowan (1949) gives a general critical review of fracture in solids. For glasses, in particular static fatigue, see Chapters 3 and 4 of Stanworth (1950). A more recent review article discussing the Griffith theory of brittle fracture of glass and its development over the past 40 years is given by Anderson (1959). In regard to the statistics of brittle fracture, see Hoel (1954) for a general introduction to statistics, Weibull (1939 a, b) and Fisher and Hollomon (1947) for early development and applications, and Gumbel (1954) for the theoretical background of the extreme value distribution.

AKITA, Y. IKEDA, K.	1959	"Stress Field for a Propagating Slit at Constant Speed," *Transp. Tech. Res. Inst., Rept.* 37, Tokyo.
ANDEREGG, F. O.	1939	"Strength of Glass Fibers," *Ind. Eng. Chem.* **31**, 290–298.
ANDERSON, O. L.	1958	"Cooling Time of Glass Fibers," *J. Appl. Phys.* **29**, 9–12.
ANDERSON, O. L.	1959	"The Griffith Criterion for Glass Fracture," *Fracture*, B. L. Averbach et al., eds., M.I.T. Press, Cambridge, Mass., and Wiley, New York, pp. 331–353.
ANDRADE, E. N. DA C. TSIEN, L. C.	1937	"On Surface Cracks in Glasses," *Proc. Roy. Soc. (London)* **A159**, 346–354.
ARGON, A. S.	1959(a)	"Surface Cracks on Glass," *Proc. Roy. Soc. (London)* **A250**, 472–482.
ARGON, A. S.	1959(b)	"Distribution of Cracks on Glass Surfaces," *Proc. Roy. Soc. (London)* **A250**, 483–492.
ARGON, A. S. HORI, Y. OROWAN, E.	1960	"Indentation Strength of Glass," *J. Am. Ceramic Soc.* **43**, 86–96.

Baker, T. C. Preston, F. W.	1946	"The Effect of Water on the Strength of Glass," *J. Appl. Phys.* **17,** 179–188.
Bateson, S.	1958	"Critical Study of the Optical and Mechanical Properties of Glass Fibers," *J. Appl. Phys.* **29,** 13–21.
Bredthauer, R. D.	1957	"Strength Characteristics of Rock Samples under Hydrostatic Pressure," *Trans. ASME* **79,** 695–708.
Bridgman, P. W.	1947	"The Effect of Hydrostatic Pressure on the Fracture of Brittle Substances," *J. Appl. Phys.* **18,** 246–258.
Charles, R. J.	1958	"Dynamic Fatigue of Glass," *J. Appl. Phys.* **29,** 1657–1662.
Cotterell, B.	1964	"On the Nature of Moving Cracks," *J. Appl. Mech.* **31,** 12–16.
Cotterill, P.	1961	"The Hydrogen Embrittlement of Metals," *Progress in Materials Science*, Vol. 9, Chalmers and King, eds., Pergamon Press, London.
Craggs, J. W.	1960	"On the Propagation of a Crack in an Elastic Brittle Material," *J. Mech. Phys. Solids* **8,** 66–75.
Erdoğan, F. Sih, G. C.	1963	"On the Crack Extension in Plates Under Plane Loading and Transverse Shear," *Trans. ASME, J. Basic Eng.* **D85,** 519–527.
Ernsberger, F. M.	1960	"Detection of Strength Impairing Surface Flaws in Glass," *Proc. Roy. Soc. (London)* **A257,** 213–223.
Fisher, J. C. Hollomon, J. H.	1947	"A Statistical Theory of Fracture," *Trans. AIME*, **171,** 546–561.
Griffith, A. A.	1920	"The Phenomena of Rupture and Flow in Solids," *Phil. Trans. Roy. Soc. (London)* **A221,** 163–198.
Griffith, A. A.	1924	"The Theory of Rupture," *Proc. 1st Int. Con. Appl. Mech.* (Delft), 55–63.
Gumbel, E. J.	1954	"Statistical Theory of Extreme Values and Some Practical Applications," *National Bureau of Standards Appl. Math. Series*, No. 33.
Handin, J. Hager, R. V.	1957	"Experimental Deformation of Sedimentary Rocks under Confining Pressure: Tests at Room Temperature on Dry Samples," *Bull. Am. Assn. Petroleum Geologists* **41,** 1–50.
Hoel, P. G.	1954	*Introduction to Mathematical Statistics*, 2nd. ed., Wiley, New York.
Holland, A. J. Turner, W. E. S.	1940	"The Effect of Sustained Loading on the Breaking Strength of Sheet Glass," *J. Soc. Glass Tech.* **24,** 46–57.
Holloway, D. G.	1959	"The Strength of Glass Fibers," *Phil. Mag.* **4,** 1101–1106.
Inglis, C. E.	1913	"Stresses in a Plate Due to the Presence of Cracks and Sharp Corners," *Trans. Inst. Naval Arch.* **55,** 219–230.

Joos, P. 1957 "Über die Mikrohärte von Glasoberflächen" *Z. Angew. Phys.* **9,** 556–561.

Kropschot, R. H. 1957 "Strength and Fatigue of Glass at Very Low Temperatures," *J. Appl. Phys.* **28,** 610–614.
Mikesell, R. P.

McClintock, F. A. 1960 "Travelling Cracks in Elastic Materials under Longitudinal Shear," *J. Mech. Phys. Solids* **8,** 187–193.
Sukhatme, S. P.

McClintock, F. A. 1962 "Friction on Griffith Cracks in Rocks Under Pressure," *Proc. 4th U. S. Nat. Cong. Appl. Mech.* Vol. 2, pp. 1015–1022.
Walsh, J. B.

Mott, N. F. 1948 "Brittle Fracture in Mild Steel Plates," *Engineering* **165,** 16–18.

Murgatroyd, J. B. 1944 "The Strength of Glass Fibers, Part II The Effect of Heat Treatment on Strength," *J. Soc. Glass Tech.* **28,** 388–405.

Orowan, E. 1933 "The Tensile Strength of Mica and the Problem of the Technical Strength," *Z. Physik* **82,** 235–266.

Orowan, E. 1934 "The Mechanical Strength Properties and the Real Structure of Crystals," *Z. Kristallographie* **89,** 327–343.

Orowan, E. 1949 "Fracture and Strength of Solids," *Rep. Prog. Phys.* **12,** 185–232.

Otto, W. H. 1955 "Relationship of Tensile Strength of Glass Fibers to Diameter," *J. Am. Ceramic Soc.* **38,** 122–124.

Reinkober, O. 1931 "The Tear Strength of Thin Quartz Filaments," *Physik. Z.* **32,** 243–250.

Robertson, E. C. 1955 "Experimental Study of the Strength of Rocks," *Bull. Geol. Soc. Am.* **66,** 1275–1314.

Ryschkewitch, E. 1942 "The Plasticity of Brittle Materials," *Indust. Diamond Rev.* **6** (1946), 339–344. Translated and abstracted from *Glastech. Ber.* **20,** 166–174.

Sack, R. A. 1946 "Extension of Griffith's Theory of Rupture to Three Dimensions," *Proc. Phys. Soc. (London)* **58,** 729–736.

Schardin, H. 1959 "Velocity Effects in Fracture," *Fracture*, B. L. Averbach et al., eds., M.I.T. Press, Cambridge, Mass., and Wiley, New York, pp. 297–330.

Smekal, A. G. 1952 "Dynamics of the Brittle Tensile Fracture of Cylindrical Glass Bars," *Sitzungsberichte, Österreichische Akademie der Wissenschaften* **161,** 361–373.

Sneddon, I. N. 1946 "The Distribution of Stress in the Neighborhood of a Crack in an Elastic Solid," *Proc. Roy. Soc. (London)* **A187,** 229–260.

Stanworth, J. E. 1950 *Physical Properties of Glass*, Oxford University Press, London, pp. 65–115.

Sucov, E. W.	1962	"New Statistical Treatment of Ball Indentation Data to Determine Distribution of Flaws in Glass," *J. Am. Ceramic Soc.* **45,** 214–218.
Tetelman, A. S.	1963	"The Hydrogen Embrittlement of Ferrous Alloys," *Fracture in Solids*, AIME Conference Series, Vol. 20, Interscience, New York, pp., 671–708.
Warshaw, I.	1960	"Structural Implications of the Electron Microscopy of Glass Surfaces," *J. Am. Ceramic Soc.* **43,** 4–9.
Weibull, W.	1939(a)	"A Statistical Theory of the Strength of Materials," *Ingeniörs Vetenskaps Akademien, Handlingar*, No. 151.
Weibull, W.	1939(b)	"The Phenomenon of Rupture in Solids," *Ingeniörs Vetenskaps Akademien, Handlingar*, No. 153.
Yoffe, E. H.	1951	"The Moving Griffith Crack," *Phil. Mag.*, 7th Ser., **42,** 739–750.
Zwicky, F.	1923	"Cleavage Strength of Rock Salt," *Physik. Z.* **24,** 131–137.

PROBLEMS

15.1. When impurities are eliminated from simple liquids to make nucleation of the vapor phase difficult, it has been observed that the liquids can support large hydrostatic tension (negative pressure) without fracturing.

Considering the fracture of a liquid as an instability in the growth of a bubble, derive an expression for the ideal strength of an ideally pure liquid (with zero vapor pressure) by analyzing the balance of surface energy of the bubble and the work done by the hydrostatic tensile forces. Evaluate your results by assuming reasonable quantities for the relevant parameters. How does your estimate compare with the ideal strength of solids?

15.2. Using the approximate relation for the stress between atomic layers

$$\sigma = A[(r_0/r)^2 - (r_0/r)^9],$$

where r_0 represents the equilibrium distance between two atomic layers and r the distance under stress, derive an expression for the ideal cohesive strength as a function of Young's modulus, of which the constant A is a function.

15.3. Chalk frequently has pores of approximately spherical shape. In an infinite solid under uniaxial tension σ, there are stress concentrations at points A and B of a spherical hole, as given in Fig. 15.17. Assuming that fracture in tension results when a tensile stress component equals a value K, calculate and plot a fracture locus for biaxial stress.

15.4. Cast iron, a brittle material, has a fracture envelope for biaxial stress as shown in Fig. 15.18. What is the largest torque that can be sustained by a hollow cast-iron cylinder 2 in. in diameter with a wall thickness of 0.1 in.?

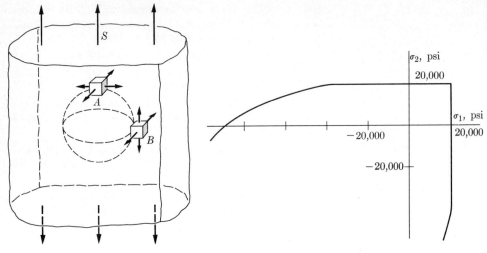

FIGURE 15.17

FIGURE 15.18

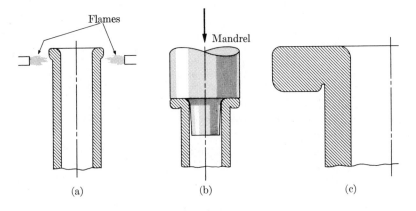

(a)   (b)   (c)

FIGURE 15.19

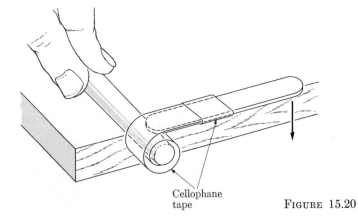

FIGURE 15.20

15.5. In brittle fracture under triaxial stresses σ_1, σ_2, σ_3, the intermediate stress σ_2 usually plays no role when cracks are randomly oriented. In glass, however, cracks usually are perpendicular to the surface, making any stress normal to the surface not a significant parameter. Propose a method of producing a triaxial compressive stress in which the most negative stress, σ_3, acts normal to the surface, thus making σ_2 a significant stress.

15.6. The tungsten carbide tool bit which was dragged over the surface of the glass plate in Fig. 15.5 weighed 25 gm. From this information and from the width of the major grooves of Fig. 15.5, estimate the hydrostatic pressure at the tip of the tool bit. Explain why cracks are formed in profusion when the pressure on the tool is increased further.

15.7. Show that an atomically sharp elliptical crack of a stress concentration factor of 100 cannot be resolved in the optical microscope. Note that the smallest resolvable distance in the field of view of a microscope is $\delta = \lambda/\text{N.A.}$, where λ is the wavelength of light used and $\text{N.A.} = n \sin \alpha$ is the numerical aperature of the objective, admitting a cone of light with apex angle 2α through a medium of refractive index n between specimen and objective. For the best high magnification objectives N.A. is around unity. [*Hint:* Use Eq. 15.3].

15.8. Show that in Eq. 15.11 when $A/\delta A \to \infty$ the expression in the limit goes to that shown in Eq. 15.12.

15.9. Glass hypodermic syringes are produced by one manufacturer on an automatic machine in the following fashion. (a) The ends of the cut tubes are smoothed and slightly flared out by rapid rotation (Fig. 15.19a). (b) A formed mandrel finishes the flaring operation by forcing out the edges (Fig. 15.19b). Some of the resulting tubes had strongly undercut, reentrant corners, as shown in Fig. 15.19(c). These were found to be so fragile that they would lose their flared edges before reaching the customary annealing treatment. Once annealed, their strength would be considerably improved, though they would still be susceptible to early fracture in service. Discuss the possible causes of this failure and recommend improvements that could increase the tube strength.

E 15.10. Determine the breaking strength of chalk in bending. From this, estimate the strength of chalk in torsion, and verify your results experimentally. To load the chalk in torsion, use the arrangement shown in Fig. 15.20. Explain the shape of the fracture surface. Compare your results with Problem 15.3 as well as with the Griffith theory.

E 15.11. Notch specimens of chalk in different ways. Estimate the stress concentration factors and from them the strength in bending or torsion (use the arrangement shown in Fig. 15.20). Note the rather pronounced size effect.

E 15.12. Produce a brittle fracture in silicone putty by sudden loading. Can you tell the origin of fracture? If in doubt, intentionally introduce a small nick or apply bending to the specimen. Note the markings on the fracture surface, and their relation to the origin of fracture.

E 15.13. Determine the fracture strength of silicone putty by molding a dumbell-shaped specimen and using it to attempt to lift weights of progressively larger sizes.

E 15.14. Make tensile specimens out of thin strips of paper, and fasten them with cellophane tape to tongue depressors for gripping. Measure the tensile strength as a function of length of specimen. Introduce notches of various acuities into the paper specimens, keeping the minimum width the same. Is the loss in strength what you would expect from the stress concentration factors? Repeat the experiment with cellophane tape. Examine the fractures in each case. Why is the cellophane tape so much more notch-sensitive? What would be the behavior of geometrically similar specimens of larger size?

15.15. A rod of brittle material, subject to a longitudinal stress σ, contains a penny-shaped crack of radius c (Fig. 15.21). The plane of the crack is normal to the axis of the rod. If the radius of the crack grows by an amount dc, assume that the elastic strain energy of a spherical shell of radius c and thickness dc is converted to surface free energy, and that no external work or heat transfer occurs. Assume the initial elastic energy per unit volume of the shell to be the same as if the crack were not present. Find the necessary value of σ where the crack is just able to grow spontaneously from the initial radius c.

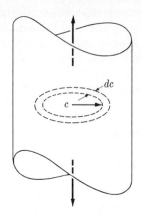

FIGURE 15.21

◄**15.16.** Derive the Griffith condition for fracture under combined stress, Eqs. 15.7, and 15.8. (a) Choose coordinate axes as shown in Fig. 15.22, and for convenience let $(\sigma_1 + \sigma_2)/2 = \sigma$ and $(\sigma_1 - \sigma_2)/2 = \tau$. Express the stress near the tip of the crack in terms of these variables and the angle $\xi = 2\theta$, using Eq. 11.22 for the stress concentration around an elliptical crack:

$$\sigma_{\beta\beta} = \frac{2\alpha_0[\sigma + \tau \cos \xi] + 2\beta\tau \sin \xi}{\alpha_0^2 + \beta^2}.$$

(b) It is easiest to first find the angle giving the highest stress for a given β. Show that this is $\tan \xi = \beta/\alpha_0$.

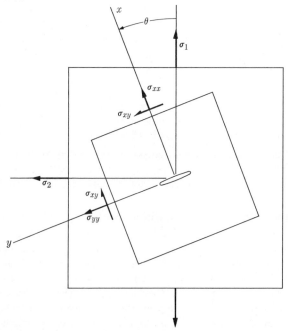

FIGURE 15.22

(c) Substitute the result of part (b) back to eliminate ξ, and then find the point on the crack where the stress is maximum (do not discard any valid solutions).

(d) Substitute the result of part (c) into the expression for $\sigma_{\beta\beta}$ to eliminate β. Then equate $\sigma_{\beta\beta}$ to the ideal cohesive strength.

(e) Sketch the resulting fracture locus in terms of the principal stress coordinates, showing it agrees with that given in Eq. 15.8 and Fig. 15.2.

CHAPTER 16

DUCTILE FRACTURE

16.1 SYNOPSIS

In contrast to brittle fracture, a large amount of permanent strain may occur before fracture. The limiting case of separation by viscous or plastic flow until the cross section vanishes is called *rupture*. If appreciable plastic flow occurs, but there is a fracture surface and the deformation is less than for rupture, the process is termed *ductile fracture*. We shall first examine loads and extensions expected for rupture from plasticity theory for a variety of smooth and notched specimens. We shall next examine some mechanisms by which holes and cracks can develop in a ductile material before rupture. These include the separation around dislocations at high triaxial stress, the growth of voids from inclusions or other inhomogeneities, and the yet unexplained growth of fine cracks. Another mode of fracture, which may occur even after considerable deformation, is the cleavage of the crystal lattice along crystallographic planes, to be discussed in Chapter 17.

Once fracture is nucleated under certain combinations of stress and strain, an analysis of simple cases based on cylindrical holes leads us to a fracture criterion which depends much more strongly on the entire history of stress and strain than do those for yielding. It also indicates a strong dependence on the anisotropy of the inclusions from which the holes grew, and on the size of the part relative to the inclusion spacing.

With the very large strain concentrations that result from cracks, ductile fracture can occur at nominal stresses below the yield strength. Then even though the material is ductile, the structure is brittle in that it experiences little overall plastic deformation before fracture. The mechanics of this situation will be developed for the simple case of cracks under longitudinal shear. The implication of these results for the more practical case involving tension will be examined.

16.2 RUPTURE

The continued extension of a part may gradually reduce the cross-sectional area to the vanishing point. This process is called *rupture*. If the separation of a part into two pieces occurs with less elongation than required for rupture, the process is called *fracture*. When fracture intervenes, the amount by which the extension falls short of the rupture elongation is a measure of the sensitivity to fracture.

In single crystals which slip on only one slip plane, rupture can in principle occur, after the maximum load is reached, by the localization of strain on a single plane so that the two parts of a specimen simply slide off each other, as shown for a zinc crystal in Fig. 4.31. The elongation to rupture is then determined by the orientation of the slip system.

In polycrystals, it should also be possible to calculate the elongation to rupture. In practice, there are relatively few cases in which the calculation has been made, even for non-strainhardening materials. One of the cases in which the elongation to rupture can be calculated is the singly grooved, plane strain, tensile specimen, discussed in Sections 10.6 and 10.8. Alternating shear on the two active shear planes gives an extension at rupture equal to the minimum thickness of the specimen (Prob. 16.1). The best available results for various other notched specimens of non-strainhardening material are summarized in Table 16.1. Note that in general the extension to rupture of notched or grooved specimens is typically of the order of one-half to twice the minimum dimension. If the material strain-hardens, greater extensions will occur as a result of the larger region of plastic deformation. As an example, recall from Prob. 8.6 that the uniform tensile elongation is equal to the exponent n in a material which strain-hardens according to $\sigma = \sigma_1 \epsilon^n$.

Rupture cannot occur in homogeneous compression, bending, or torsion because the cross section never vanishes.

TABLE 16.1

MAXIMUM LOAD P_{max} AND RUPTURE EXTENSION D FOR DUCTILE MATERIALS WITH YIELD STRENGTH Y AND TENSILE STRENGTH T (McCLINTOCK, 1961)

Specimen	Data	References
	Limitations: $t/a_n \gg 1$ for $\omega = 0$, $\rho = 0$, $a_s/a_n > (8.67)(2/\sqrt{3})(T/Y)F_L$ for other cases see McClintock (1961) $P_{max} = \left(\dfrac{2T}{\sqrt{3}}\right)(2a_n t) \times \left[\left(1 + \dfrac{\pi}{2} - \dfrac{\omega}{2}\right) - \dfrac{\rho}{a_n}\left(e^{(\pi-\omega)/2} - 1 - \dfrac{\pi}{2} + \dfrac{\omega}{2}\right)\right]$ for $\rho/a_n \leq e^{(\pi-\omega)/2} - 1$ $P_{max} = \left(\dfrac{2T}{\sqrt{3}}\right)(2a_n t)\left[\left(1 + \dfrac{\rho}{a_n}\right)\ln\left(1 + \dfrac{\rho}{a_n}\right)\right]$ for $\rho/a_n \geq e^{(\pi-\omega)/2} - 1$ $D = a_n$ for $\omega \leq \pi/2$, $\rho/a_n < 0.2$	Bishop (1953) Hill (1950) McClintock (1961)
	Limitations: $t/a_n \gg 1$ $a_s/a_n > (2/\sqrt{3})(T/Y)F_L$ $P_{max} = \left(\dfrac{2T}{\sqrt{3}}\right)(a_n t)$ $D = a_n$	McClintock (1961) Prob. 16.1

(cont.)

TABLE 16.1 (cont.)

Specimen	Data	References
	Limitations: $t/a_n \ll 1$, Limitation on a_s/a_n not known but see Hill (1952) $P_{\max} = \left(\dfrac{2T}{\sqrt{3}}\right)(2a_n t)$ for $\rho/a_n < \tfrac{1}{2}$, $\omega < 141°$ $D = t$	Hill (1952) McClintock (1961)
	Limitations: $r_s/r_n > 3.20\sqrt{(T/Y)F_L}$ $P_{\max} = 2.9(T\pi r_n^2)$ for $\rho = 0$, $\omega = 0$ (Tresca yield criterion) D not known	Shield (1955) Shield (1955) and Levin (1955)
	Limitations: $t/a_n \gg 1$ a_s/a_n limitation not known $M_{\max} = \left(\dfrac{2T}{\sqrt{3}}\dfrac{a_n^2 t}{4}\right)(1.38)$ for $\omega < 88°$, $\rho/a_n = 0$ $\left(\dfrac{2T}{\sqrt{3}}\dfrac{a_n^2 t}{4}\right)(1.31)$ for $\rho/a_n = 0.17$, $\omega < 62°$ Does not separate	Green (1953)
	Limitations: $t/a_n \gg 1$ $a_s/a_n > 1.39(T/Y)F_L$ for $\rho/a_n = 0$, $\omega = 60°$ $M_{\max} = \left(\dfrac{2T}{\sqrt{3}}\dfrac{a_n^2 t}{4}\right)(1.26)$ for $\omega < 60°$, $\rho/a_n = 0$, which falls off about linearly with ρ/a_n through $M_{\max} = \left(\dfrac{2T}{\sqrt{3}}\dfrac{a_n^2 t}{4}\right)(1.16)$ for $\omega < 98°$, $\rho/a_n = 0.25$ Does not separate	Green and Hundy (1956)
	Limitations: $t/a_n \ll 1$, a_s/a_n limitation unknown $M_{\max} = 1.07\left(T\dfrac{a_n^2 t}{4}\right)$ for $\rho/a_n < 0.1$, $\omega < 141°$ Does not separate. Doubly notched case not given but could be found from above reference.	Ford and Lianis (1957)
	Limitations: none $M_{\max} = \dfrac{T}{\sqrt{3}}\left(\dfrac{2\pi r_n^3}{3}\right)$ Does not separate	Walsh and Mackenzie (1959)

16.3 MECHANISMS OF DUCTILE FRACTURE

Brittle fracture and rupture are extreme forms of fracture. Between them lies *ductile fracture*, defined as separation in the presence of some plastic flow, but with a distinct surface of fracture and less extension than rupture. Sometimes, as in steel at low temperature, the only plastic flow is that required at grain boundaries to join together the brittle cleavage cracks in the individual grains. Such fracture will be described in more detail in Chapter 17. The cases discussed here will be ones in which macroscopic plastic deformation plays a more fundamental part.

A. Initiation of Ductile Fracture. In a tensile test the force-deformation curve sometimes drops off suddenly just before fracture. If the test can be stopped before fracture, sectioning will then reveal an internal crack, such as shown in Fig. 16.1. Closer examination often shows voids opening up, in the case of Fig. 16.2 from oxide inclusions. In steel the sources may be cracks in the relatively brittle pearlite (Tipper, 1949). Grain boundaries containing impurities or inclusions may be another source (see Section 17.4). Some of these sources may be present initially, in the as-processed metal. Others develop only after considerable plastic deformation has occurred. Here the initiation of fracture depends on the stress on the brittle elements.

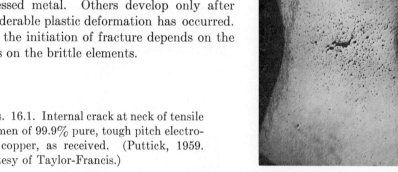

Fig. 16.1. Internal crack at neck of tensile specimen of 99.9% pure, tough pitch electrolytic copper, as received. (Puttick, 1959. Courtesy of Taylor-Francis.)

Sometimes fracture is preceded by such an extreme localization of strain that the mechanism of final fracture is irrelevant. An example is shown in Fig. 16.3, where intense shear occurred along a diagonal plane. Fortunately for observation, the fracture deviated from this plane, leaving behind evidence of the intensity of

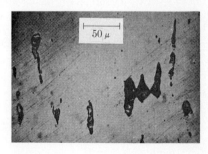

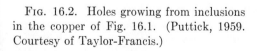

Fig. 16.2. Holes growing from inclusions in the copper of Fig. 16.1. (Puttick, 1959. Courtesy of Taylor-Francis.)

Fig. 16.3. Shearing at cup and cone fracture in copper of Fig. 16.1. (a) As polished. (b) Etched. (Puttick, 1960. Courtesy of Taylor-Francis.)

Fig. 16.4. Deformation bands and a crack near a notch in 7075-T6 aluminum-zinc alloy. The deformation was by shear with displacements normal to the plane of the paper.

the shear. Another example is given by the precipitation-hardened alloy of Fig. 16.4. In this case torsion produced displacements normal to the plane of the section. Discontinuities in scratch marks on the surface showed that the shear strain in the deformation bands was greater than 4. It is very likely that strain-softening of the alloy is occurring inside the deformation bands, perhaps due to overaging. Under such high strains, fracture may occur in the bands by a number of mechanisms. If so, once the rate of strain-hardening is low enough so that such high strain concentrations can occur, little additional deformation of the specimen will be required for fracture. The critical point in such a process is then not the fracture itself, but the strain required to bring the rate of strain-hardening down to the critical value. This mode of crack initation should depend primarily on strain rather than on stress.

◀ The question may arise as to whether fracture could begin in a pure metal with no defects other than dislocations. Various such possibilities will be discussed in Chapter 17, but we shall consider a limiting case here: the opening up of a void from a single dislocation under biaxial strain, as shown in the bubble-raft model of a crystal in Fig. 16.5 (see McClintock and O'Day, 1965). A strain of several percent was needed, but it was lower than that required from considerations of the ideal strength of a perfect lattice. Once the crack was formed, fracture followed quickly by other dislocations running off from it, opening up a void. In any event, here the controlling variable is a high triaxial stress. ▶

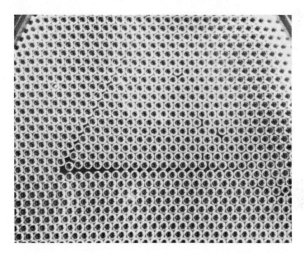

FIG. 16.5. Cracking from a dislocation in a bubble raft under biaxial strain. Bubble diameter 1.0 mm.

B. Development of Ductile Fracture. If the initiation of ductile fracture is due to cracks in brittle phases or inclusions, the development of fracture is more critical than its initiation, for the cracks are soon blunted, leaving a number of small holes. This problem will be treated more quantitatively in the next section, but suffice it to say here that the growth and coalescence of the holes will depend on the amount of further plastic strain and on the transverse stress which tends to open up the holes instead of letting them stretch out into long harmless strings. If the directions of principal stress rotate, the rotation of the holes will also be important. A fracture criterion will have to depend on the history of stress and strain, rather than primarily on their current values, as does yielding.

When fracture initiation is due to extremely localized strain concentrations, as a result of strain-softening, its development is simply a continuation of its initia-

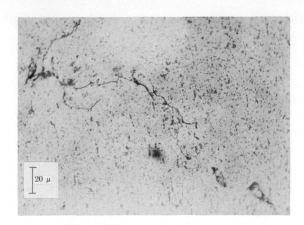

Fig. 16.6. Fine cracks in the center of a doubly grooved specimen of 1100 aluminum. (Neimark, 1959.)

tion. The cases observed above arose in shear, but conceivably such a process also leads to the growth of the very fine cracks in tension, shown in Fig. 16.6, observed by Neimark (1959) in the doubly grooved specimens of aluminum. Fine cracks have also been seen in iron by Puttick (1959). Their fineness indicates a very rapid growth per unit extension. The mechanism of growth of such sharp cracks is not understood. At the atomic level, any dislocation motion associated with the expected stress at the tip of the crack would lead to blunting, which, if repeated without other mechanisms, would lead to a macroscopically blunt crack (Prob. 16.2). Parker (1957) did find the fracture surface to contain the slip direction.

On a continuum scale, the strain and deformation around the tip of a crack in uniaxial tension has not yet been found but again one would expect blunting, unless, perhaps, strain-softening leads to highly localized plastic flow. If so, this mode of fracture will depend on the strain to bring the material to its critical state of hardening, or softening.

The above discussion should indicate that there will probably never be any one criterion for ductile fracture, but rather several competing ones, depending on various combinations of stress, strain, and perhaps their history. We shall now examine one of these, namely fracture by the growth of holes, after which we shall turn to observations and descriptions of fracture on a larger scale.

16.4 DUCTILE FRACTURE BY GROWTH OF HOLES

◀ As a starting point for an analysis of the growth of holes, consider a metal containing a large number of cylindrical holes at a spacing l. We shall assume that the holes are initially far enough apart so that they do not interact appreciably. Their growth can be determined by studying the growth of a single hole in a cylinder of diameter equal to the average spacing between the holes, as indicated in Fig. 16.7, while the length and diameter of the cylinder change according to the macroscopic strain in the material.

◀ When the ratio of hole radius to spacing increases enough, say by a factor F, it will be assumed that the holes are close enough together so that they begin to interact. Fracture by local rupture will then quickly follow. The factor F should be of the order of the ratio of initial hole spacing to radius, which for a reasonably sound metal will be of the order of hundreds. The degree of damage is best described in terms of the relative increase in radius of the hole a relative to the diameter of an entire element, so that fracture occurs at a value of damage, η, equal to unity:

$$d\eta = -d \ln(a/l)/\ln F. \qquad (16.1)$$

◀ We wish to describe the rate of damage in terms of the components of stress and strain increment applied to the specimen as a whole. If the transverse stress components

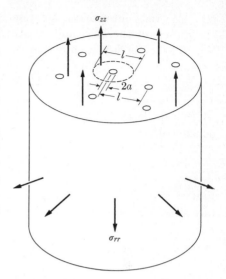

FIG. 16.7. Schematic diagram of cylindrical holes.

are equal, the holes will remain circular in section and their growth can be studied by a relatively simple analysis. The change in hole spacing can be expressed in terms of either of the applied strain components at infinity if one neglects the changes in density associated with the hole growth:

$$dl/l = d\epsilon_{rr}^{\infty} = -d\epsilon_{zz}^{\infty}/2 = -d\epsilon_{zz}/2. \qquad (16.2)$$

The change in radius of the hole requires an analysis of the strain distribution around a hole under biaxial tension, similar to the plastic analysis carried out in Section 11.3, but with an arbitrary axial strain, corresponding to extension of the specimen as a whole. The Mises yield criterion and associated flow rule is more convenient than the maximum shear-stress criterion. It turns out to be easiest to seek the nominal radial stress required for a given circumferential strain. From the equilibrium equations and the stress-strain relations,

$$\frac{\partial \sigma_{rr}}{\partial r} = -\frac{\sigma_{rr} - \sigma_{\theta\theta}}{r} = -\frac{2\bar{\sigma}}{3r}\left(\frac{d\epsilon_{rr}}{d\bar{\epsilon}} - \frac{d\epsilon_{\theta\theta}}{d\bar{\epsilon}}\right). \qquad (16.3)$$

The ratios between the strain increments are found by integrating the strain-displacement equations, along with the condition of incompressibility. In terms of the tangential strain at the hole ($d\epsilon_{\theta\theta a} = da/a$), the equations are (Prob. 16.3)

$$d\epsilon_{rr} = -[(d\epsilon_{\theta\theta})_a + d\epsilon_{zz}/2]a^2/r^2 - d\epsilon_{zz}/2,$$
$$d\epsilon_{\theta\theta} = [(d\epsilon_{\theta\theta})_a + d\epsilon_{zz}/2]a^2/r^2 - d\epsilon_{zz}/2. \qquad (16.4)$$

Substitution of these equations and the definitions of equivalent stress and strain into Eq. 16.3, integration, solution for the circumferential strain increment at the surface of the hole, and substitution of the result along with Eq. 16.2 into Eq. 16.1 for the definition of damage gives the damage rate (Prob. 16.4):

$$\frac{d\eta}{d\epsilon_{zz}} = \frac{\sqrt{3}}{2} \frac{\sinh[\sqrt{3}\sigma_{rr}^\infty/(\sigma_{zz}^\infty - \sigma_{rr}^\infty)]}{\ln F}, \quad (16.5a)$$

which can be integrated directly to give the fracture strain if the stress components are constant:

$$\epsilon_{zz}^f = \frac{2}{\sqrt{3}} \frac{\ln F}{\sinh[\sqrt{3}\sigma_{rr}^\infty/(\sigma_{zz}^\infty - \sigma_{rr}^\infty)]}. \quad (16.5b)$$

Equation 16.5b, plotted in Fig. 16.8, indicates the very large dependence of fracture strain on transverse stress.

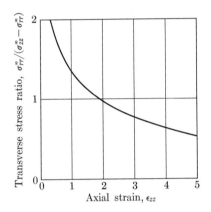

FIG. 16.8. Axial strain required for fracture according to Eq. 16.5b, for a relative hole growth factor $F = 100$.

◀ When the three principal components of strain are all different, the holes become more or less elliptical. An analysis for viscous materials has been made by Berg (1962). By comparison of the viscous and plastic behavior for circular holes, an estimate has been made for plastic holes with unequal transverse stress components σ_{aa} and σ_{bb}. The effects of strain-hardening were approximated in terms of the exponent n in $\sigma = \sigma_1 \epsilon^n$. The resulting equations for damage and fracture, analogous to Eqs. 16.5, are approximately (McClintock, 1965)

$$\frac{d\eta_{zb}}{d\bar{\epsilon}} = \frac{\sinh[(1-n)(\sigma_{aa}^\infty + \sigma_{bb}^\infty)/(2\bar{\sigma}/\sqrt{3})]}{(1-n)\ln F_{zb}}, \quad (16.6a)$$

and for a constant stress ratio

$$\bar{\epsilon}^f = \frac{(1-n)\ln F_{zb}}{\sinh[(1-n)(\sigma_{aa}^\infty + \sigma_{bb}^\infty)/(2\bar{\sigma}/\sqrt{3})]}. \quad (16.6b)$$

Under high triaxial tension there is a strong effect of strain-hardening due to the

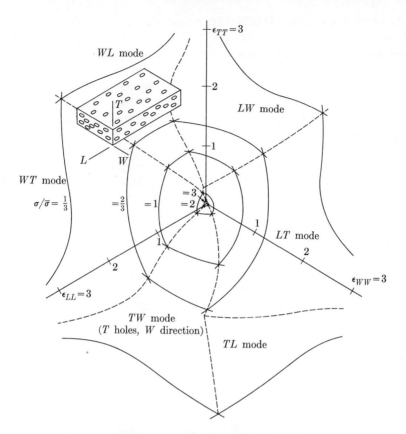

FIG. 16.9. Dependence of fracture strain on mean normal stress in a hypothetical rolled slab. Modes of fracture are denoted by the axis of the holes and the growth direction which would be critical, assuming the following arbitrary growth factors for each mode: F_{WT}, 40; F_{TW}, 80; F_{TL}, 120; F_{LT}, 20; F_{LW}, 30; F_{WL}, 60.

hyperbolic sine. An example of a fracture locus for a particular nonhardening material is shown in Fig. 16.9. The three principal strain components are plotted in one plane as triangular coordinates, which is possible because of the approximate incompressibility of the material. The ratio of mean normal stress to equivalent stress is plotted as the third coordinate. It should be emphasized that this locus is constructed for a history with the stress ratio constant. In most cases, as in the necking of a tensile specimen, the stress ratio varies during the test, and hence the equation for the damage rate must be integrated with the stress history taken into account. Furthermore, if the directions of the principal stress components are rotating, the axes of the elliptical holes also rotate, and this will affect the coalescence of the holes (McClintock et al., 1965).

◀ In spite of its limitations, a criterion for ductile fracture by the growth of holes is valuable in that it and the Griffith criterion for brittle fracture provide limiting cases with which other mechanisms of ductile fracture may be compared. ▶

16.5 FRACTURE OF UNNOTCHED SPECIMENS

A. Modes of Fracture. Examples of different modes of fracture found in unnotched tensile testing are shown in Fig. 16.10. The cup and cone fracture, Fig. 16.10(b), will be discussed first, since it is the one most frequently encountered in ductile metals. As mentioned in Section 8.7, the strain is relatively uniform across the neck of a tensile specimen, but a transverse stress and an accompanying higher axial stress develops in the center, so that unless the fracture is particularly susceptible to initiation at the surface, it begins at the center, as shown in Fig. 16.1. At first the fracture spreads outward on a surface roughly normal to the direction of maximum tensile stress. As the crack approaches the outside edge, the configuration becomes similar to that of a singly grooved plate in tension, allowing a shear strain of unity to occur along both of two 45° planes from the root of the crack (see Sections 10.6 and 11.6). If the material cannot withstand such a strain along with a normal stress of half the flow stress, a crack will propagate along the plane, leading to the *cup and cone* appearance of Fig. 16.10(b). There has been speculation that this shearing part of the cup and cone fracture is due to adiabatic heating of the specimen, but early reports of its suppression with a stiff testing machine have not been verified by later work. For example, specimens of mild steel, pulled so slowly that adiabatic heating is negligible throughout the fracture process, have shown deformation similar to that of Fig. 16.3.

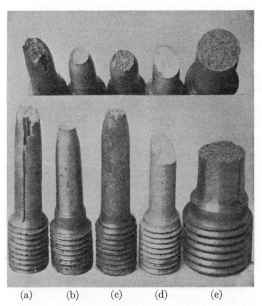

FIG. 16.10. Typical tensile fractures. (a) Wrought iron (ragged fracture showing slag). (b) Heat-treated alloy steel (perfect cup and cone). (c) Heat-treated alloy steel (rosette fracture). (d) Duralumin, heat-treated (shear fracture). (e) Gray cast iron (granular fracture). (Cowdrey and Adams, 1935. Courtesy of Wiley.)

If the material can withstand the shear and normal stress on the 45° plane, the deformation will proceed as indicated in Fig. 10.14 (with a slight circumferential strain superimposed). This gives the *double cup* fracture resulting from rupture of the remaining material, as observed by Rogers (1960).

Apparent shear fracture may occur in specimens that exhibit relatively little necking before fracture, as shown in Fig. 16.10(d). There is actually a normal stress component in addition to the shear acting across the plane of the crack. Such fractures are more likely in specimens cut from sheet or plate in which there is an anisotropy due to rolling, although the reasons are not known.

In relatively brittle materials, fracture may run *normal* to the direction of maximum tensile stress, not only in the center, but all the way to the edge of the specimen, as in the cast iron specimen in Fig. 16.10(e).

Long strings of inclusions in the axial direction can, by hole growth, give fracture surfaces in planes parallel to the axis of the specimen. Examples of such axial fractures are the *woody* and *rosette* fractures shown in Fig. 16.10(a) and (c). After the specimen is in effect shredded, the transverse stress disappears, the axial load-carrying ability drops off, and fracture follows immediately by the apparent shearing process. Woody fractures may also be encountered in hard-drawn wire, even without obvious inclusions.

In torsion the usual type of failure is a shearing-off normal to the axis of the specimen, as shown in Fig. 16.11(a). The appearance of the early part of the fracture is obscured by rubbing as the crack progresses inward, but the region of final fracture in the center has a matte appearance. In more brittle materials, a helical fracture may be observed, as shown in Fig. 16.11(b), indicating the greater importance of the normal stress. One advantage of the torsion specimen over the tensile specimen is that the strain and stress are both greatest at the surface, so it is possible to watch the initiation of the fracture with a microscope. Further studies of this nature would no doubt be rewarding.

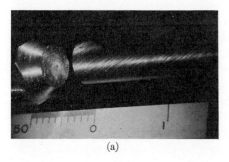
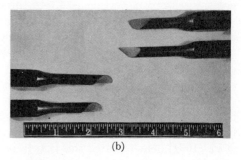

(a) (b)

Fig. 16.11. (a) Ductile fracture in torsion. 2024-T4 aluminum alloy. (b) Brittle fracture in torsion. (Ross, Sernka, and Jominy, 1956. Courtesy of ASM.)

B. Effects of Combined Stress. The effect of transverse stress on the initiation of the crack in a cup and cone fracture has been shown by Uzhik (1948), who found that eliminating the transverse stress in the center by drilling a small axial hole in the specimen changed the fracture from cup and cone to shear.

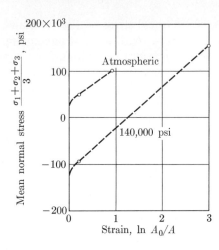

Fig. 16.12. Mean stress as a function of strain at the center of a specimen under hydrostatic pressure. Replotted from Bridgman (1952, p. 51).

A more thorough study of pressure effects was that by Bridgman (1952), who superimposed high hydrostatic pressure during tensile testing. As might be expected, he found that high pressures had relatively little effect on the flow stress of the material, but did increase the reduction of area markedly. In fact, the effect was so great that the specimen would neck down to the point where the curvature produced an average normal tensile stress at the center of the neck which was greater than that in a test conducted at atmospheric pressure (see Fig. 16.12). Apparently, the prior history of plastic deformation had healed the material in some way so that even once the stress at the center of the specimen became tensile, fracture did not occur until a further strain had occurred which was even greater than the strain required for fracture in the test at atmospheric pressure. The final fractures tended to be of the shear rather than the cup and cone type. Both of these results are consistent with fracture by the growth of holes, discussed in Section 16.4. In these tests not only is a high degree of cold-work attained, but also the resulting material would be expected to be markedly anisotropic because of the preferred orientation developed within the grains of the polycrystalline material. The marked improvement in ductility might lead one to suspect that if a material could be worked under high hydrostatic pressure, other properties, such as fatigue resistance, might also be improved.

An experimentally simple comparison of different stress states can be made between tension and torsion tests. The interpretation is not so simple, since the torsion produces a rotation of many voids which may be growing. Backofen et al. (1954) demonstrated this clearly in copper by showing that a helical, "wolf's ear" fracture develops on torsion followed by tension, but that a reversed torsion to bring the net rotation to zero restores the usual tensile fracture. Torsion and tension data are given in Table 16.2. As expected, the equivalent strain does not provide a good correlation.

Clearly there is no one simple explanation for these differing results, and it is likely that several fracture mechanisms are operative. Sometimes with strain-softening the twist becomes localized, leading to higher strains than calculated

TABLE 16.2*

Comparison of Fracture Strain in Tension and Torsion

Material	Tensile strength, psi	Equivalent plastic strain	
		Tension	Torsion
1100–0 aluminum, annealed 660°F	13,000	2.62	7.3
60–40 brass as-rolled	56,000	0.68	0.51
7075-T6 aluminum alloy	84,000	0.37	0.34
4340 steel, O.Q. from 1520°F, temp. at 400°F	271,000	0.52	0.16

* From Halford and Morrow (1962).

from overall measurements, but it is not known whether this occurred in these tests.

We next turn to biaxial tension. Davis (1943) studied thin-walled copper tubes under axial load and internal pressure, obtaining the results summarized in Table 16.3. The division of the results into two distinct groups is not surprising in view of the fact that local necking accompanied the circumferential fractures, and so the stress ratio at fracture was more nearly that corresponding to plane strain conditions than indicated by the nominal applied stress ratio (Prob. 16.5). There is also evidence of anisotropy. Terry and McClaren (1962) used special cross-shaped specimens to study fracture under biaxial stress in high-strength sheet materials, obtaining the results shown in Table 16.4. The presence of any transverse stress markedly reduces the equivalent plastic strain at fracture, especially with the relatively more ductile materials. Oddly enough, however, the higher

TABLE 16.3*

Effect of Combined Stress on Local Equivalent Plastic Strain at Fracture in Copper

$\dfrac{\text{Circumferential stress}}{\text{Axial stress}}$	Crack direction	$\epsilon_{\theta\theta}^p$	ϵ_{rr}^p	$\bar{\epsilon}^p$
0	circumferential	−0.33	−0.51	0.85
$\frac{1}{4}$	circumferential	−0.06	−0.73	0.88
$\frac{3}{8}$	circumferential	−0.02	−0.76	0.89
$\frac{1}{2}$	circumferential	0.02	−0.69	0.81
$\frac{3}{4}$	longitudinal	0.12	−0.38	0.53
1	longitudinal	0.18	−0.40	0.60
2	longitudinal	0.20	−0.29	0.49

* From Davis (1943).

TABLE 16.4*

Equivalent Plastic Strain at Fracture
in High-Strength Sheet Materials
(Maximum principal stress transverse to rolling direction)

Material	Tensile strength, psi	Stress ratio, axial: trans.	Equivalent plastic strain, %	Number of specimens
2014-T6 aluminum	71,000	1:0	11, 13	2
		1:$\frac{1}{2}$	2$\frac{1}{2}$–4$\frac{1}{2}$	3
		1:1	5–9	5
B-120VCA titanium	182,000–197,000	1:0	7–14	4
		1:$\frac{1}{2}$	2.1–2.3	3
		1:1	3–6	3
5 Cr,Mo,V steel, high strength	272,000–286,000	1:0	3–20	6
		1:$\frac{1}{2}$	3$\frac{1}{2}$–4$\frac{1}{2}$	3
		1:1	5–7	3
5 Cr,Mo,V steel, low strength	216,000–230,000	1:0	25, 28	2
		1:$\frac{1}{2}$	3, 4$\frac{1}{2}$	2
		1:1	5–6	3
Airsteel X-200	296,000	1:0	3–9	4
		1:$\frac{1}{2}$	1–4$\frac{1}{2}$	2
		1:1	2–5	4

* From Terry and McClaren (1962).

transverse stress gives a higher equivalent strain (although a lower maximum principal strain). The value of these various mechanical tests would have been enhanced by accompanying metallographic studies.

Using doubly grooved plane strain specimens of 7075-T6 aluminum zinc alloy, Neimark (1959) found the equivalent strain at fracture to fall off with mean normal stress as shown in Table 16.5.

Another way in which transverse tension can be produced is by brazing a thin layer of soft material between two hard steel blocks. When these blocks are pulled apart, a large amount of triaxiality is developed in the ductile material, which may produce separation by the coalescing of a series of holes (Moffatt and Wulff, 1963).

An extreme case of triaxiality is that carried out dynamically by O'Brien and Davis (1961). They detonated an explosive at one side of a thick plate, generating a compression wave with a sharp front and slow decay. This compression wave traveled through the plate, reflected from the free side as a tensile wave, and generated a plane tensile wave. If the peak tensile stress in a plane wave rises to σ_1 in a material in which the equivalent flow stress is Y, the transverse tensile stress required for plane motion is $\sigma_1 - Y$.

In aluminum alloys ranging in composition from 2024-T3 aluminum-copper alloy through commercially pure aluminum to high-purity single crystals, O'Brien

TABLE 16.5*
Equivalent Plastic Strain at Fracture for 7075-T6 Aluminum Alloy

Mode of testing	Mean normal stress, psi	Equivalent plastic strain, $\bar{\epsilon}^p$
Torsion	0	0.48
Unnotched	55,000	0.44
120° notch	118,000	0.22
90° notch	147,000	0.20

*From Neimark (1959).

and Davis found that the stress required for fracture by opening up a series of holes was about 200,000 psi in all cases, even though the initial yield strengths differed by a factor of 5000. There are at least two reasons that contribute to the remarkable constancy of fracture stress. First, the very high compressive wave going through the material causes plastic deformation and rather high strain-hardening. Second, the fact that the stress levels are so nearly uniform may be due to one mechanism, common to all forms of aluminum, being active. This could be, for example, the growth first of cracks and then holes from single dislocations (Fig. 16.5) or from the intersection of extended dislocations.

In summary, it seems that because of the variety of fracture mechanisms, no general rule can be given for the effect of pressure on the equivalent strain for ductile fracture. Higher mean or transverse pressures tend to suppress fracture, but exceptions may exist.

16.6 NOTCH SENSITIVITY

In a tensile test, the reduction of area is a useful measure of ductility, with 100% reduction of area corresponding to pure rupture. In the presence of a notch, large local strains may produce the kinds of fracture discussed in the preceding section, even when the extension and perhaps the maximum load are below the values given for pure rupture in Table 16.1. We therefore compare the performance of a section of material containing a notch with a standard based on the performance of a rigid-plastic, non-strainhardening material whose yield strength equals the tensile strength of the material being tested. The tensile strength is used as a standard since some strain-hardening is expected before the maximum load is reached. The load and deformation factors of notched sections are defined as follows (McClintock, 1961):

$$\text{Load factor } F_L = \frac{\text{Actual maximum load}}{\text{Standard load}},$$

$$\text{Deformation factor } F_D = \frac{\text{Actual extension across notch at fracture}}{\text{Standard extension of rupture}}.$$

(16.7)

The notch sensitivity has often been described in terms of the ratio of maximum load to that for an unnotched specimen of the same minimum area, but this can be misleading (Prob. 16.6).

For pure aluminum, the load and deformation factors for various notches and sizes were found to range between 0.7 and 1.2. For a 7075-T6 singly grooved specimen deformation factors were nearly zero. A load factor of 0.65 dropped to 0.44 for a specimen 10 times the size in linear dimensions. This size effect is important in fracture. Recall that in determining the resistance of structures to *deformation*, the absolute size of the structure need not be taken into account, and scale models can be used. In determining the resistance to *fracture*, however, there may be a definite size effect, so that with two geometrically similar configurations, the large one has the lower load or deformation factor This can be an important consideration in using laboratory tests to estimate the resistance to fracture of large structures. When ductile fracture occurs under nominally elastic conditions, the size effect can be predicted theoretically, as we shall see in Section 16.7. Under generally plastic conditions, not enough is known of stress and strain distributions to predict load and deformation factors theoretically. Thus in using them to estimate the performance of a section of some different material or in some different shape, one must modify any data in the light of the present qualitative understanding of the various factors affecting fracture.

16.7 THE MECHANICS OF ELASTIC-PLASTIC FRACTURE

With more extreme combinations of geometry and material there may be a rather small plastic zone which is completely surrounded and controlled by one of the standard types of elastic stress singularity discussed in Section 11.6. Thus the fracture, which depends on the stress and strain within the plastic zone, will depend on the stress-intensity parameter (and its history) which controls the plastic zone. Ideally there exists a series of regions around a crack, each providing the boundary condition for the next, as shown in Fig. 16.13. At large distances there is a nominal applied stress, and the material is in the elastic regime. Near the tip of the crack, there will be stress concentrations whose distributions can be described in terms of the three stress-intensity factors discussed in Section 11.6. Even closer to the crack there is a plastic region in which the stress and strain magnitudes will depend on the surrounding elastic singularities in some way as yet unknown. Finally there will be a region where the material will become inhomogeneous, first due to grain boundaries, then subgrains, and finally dislocations and atoms. A difficulty is that these regions may overlap. For example, the nominal stress may be so high that the plastic zone is too large to be considered as surrounded by a simple elastic singularity because that in turn is affected by the boundary of the specimen. On the other hand, the plastic zone may be so small that it is markedly affected by the structure of the material. In the first example one might need a better elastic-plastic solution; in the second case, one would try to go directly from the elastic stress distribution to whatever description could be given of the subgrain deformation.

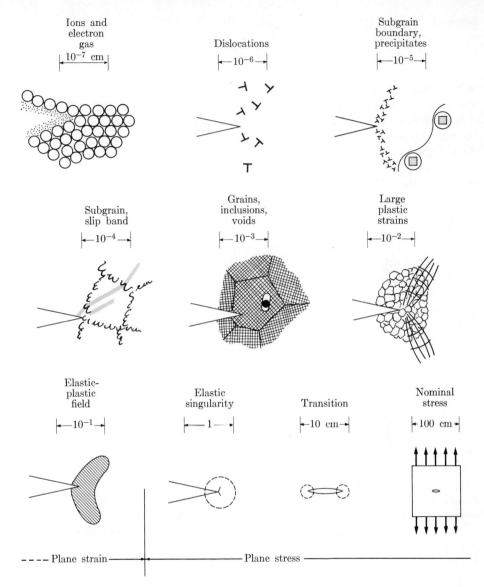

FIG. 16.13. Example of a crack tip viewed at progressively coarser scales.

◀Even where separate regions with simple stress distributions exist, there are few solutions. For a growing crack in the elastic-plastic regime, the only solution available is for a crack under longitudinal shear (parallel to the crack and its leading edge). This is the limiting case of a circumferential groove in a round bar under torsion, or of a crack growing from a spline in a shaft.

◀Before the crack starts to grow, the distributions of stress and strain are as given in Section 11.6. There is a plastic zone extending a distance R in front of the crack. Within it, the strain at the point r from the tip of the crack is given in terms of the yield strain in shear, $\gamma_y = k/G$ by Eq. 11.30:

$$\gamma = \frac{\gamma_y R}{r}. \tag{11.30}$$

◀ In a material which does not strain-harden, fracture cannot depend on the stress, which is constant throughout the plastic region. Furthermore, in pure shear triaxiality is not a variable. As a first approximation, one might assume that fracture depends only on the strain, and that fracture would occur whenever the strain reached a critical value. This fracture criterion would indicate that fracture would occur under arbitrarily low applied stress levels, since the strain is always infinite at the tip of a sharp notch where $r = 0$. This infinite strain appears in a vanishingly small region. Actually for fracture by the growth of voids we would require some critical strain over a region comparable to the spacing of void nuclei. Alternatively, if fracture results from excessive deformation in bands within grains, a critical strain would have to occur throughout a region consisting of most of a grain. The linear size of such regions may be called the structural size of the material, ρ_s. As a simple postulate consistent with these ideas, assume that fracture occurs only after the plastic strain at fracture γ_f^p has been attained everywhere along a line of length ρ_s directly in front of the crack. From Eqs. 11.29 and 11.30, giving the strain distribution for low applied stress levels, the stress for crack initiation from a groove of depth c is (Prob. 16.7)

$$\sigma_{23\infty} = k\sqrt{\rho_s[(\gamma_f^p/\gamma_y) + 1]/c}. \tag{16.8}$$

◀ Once the crack starts to grow from the notch, if Eq. 16.8 were still to apply, the crack would be immediately unstable, because the stress for propagation would drop. But in deriving Eq. 16.8 it was assumed that the load was applied while the crack or notch was at a fixed length. Actually, with plastic flow, there turns out to be less strain at a point in front of a crack if most of the load was applied while the crack was farther away (see Probs. 16.15 and 16.16). Therefore the crack stops until there is a further increase in load. The process of strain due to initial loading, strain due to crack growth, and further strain increment due to the load increment required to prepare the specimen for the next increment of crack growth is illustrated in Fig. 16.14. Eventually, a stage is reached in which the strain increment due to the growth of the crack is sufficient to bring the material one structural distance ρ_s ahead of the crack up to the fracture strain. The crack now becomes unstable, since it can grow without further increase in applied stress. The radius of the plastic zone for which this can happen must be calculated numerically. A very close approximation is found by calculating the crack length for instability when a crack is cut under constant load. In this case, an exact calculation is possible but leads to a rather complicated expression. In both a notch subject to increasing load and a notch cut under constant load, a useful approximation to the radius of the plastic zone at low stress levels is (McClintock, 1958)

$$R = \rho_s \exp[\sqrt{2\gamma_f^p/\gamma_y} - 1]. \tag{16.9}$$

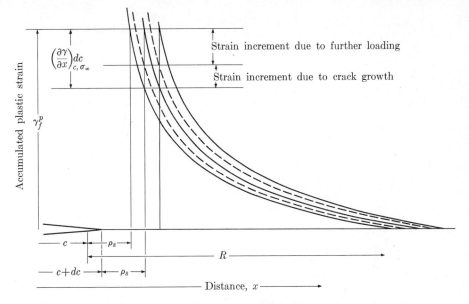

FIG. 16.14. Accumulation of strain during crack growth.

At low stress levels, the applied stress for instability is (Prob. 16.8)

$$\sigma_{23\infty} = G\gamma_y \sqrt{\frac{\rho_s \exp{(\sqrt{2\gamma_f^p/\gamma_y} - 1)}}{c}}. \quad (16.10)$$

As the plastic strain goes to zero, a more precise approximation than Eq. 16.10 shows that the stress for instability drops to the stress for initiation, and both approach the Griffith-Orowan equation (15.4) for brittle fracture (Prob. 16.9). The elastic stress-intensity term, found from

$$k_3 = \sigma_{23\infty}\sqrt{c}, \quad (11.26)$$

gives the stress distribution that, if present in the surrounding region, will cause the crack to become unstable. This term may be used with Table 11.3 to correlate the fracture stress for parts of different shapes if the stress level is low enough so that the radius of the plastic zone is small compared with the crack length or distance to the boundary of the part. In this case, R, the extent of the plastic zone, is found directly from an elastic-plastic analysis, as described in Section 11.6.

◀ Corresponding solutions of the mechanics of fracture have not been obtained for tension. A direct analogy, changing shear components of stress and strain to normal components, does seem to provide a good correlation for cracks in foil and thin sheet (McClintock, 1958, 1960). As shown in Fig. 16.15, the features of an initially stable crack growth followed by instability, and the near coincidence of instability on increasing load as compared with that on extending the crack by cutting it under constant load, indicate that the theory does explain the principal features of ductile fracture at a nominal applied stress that is below the yield.

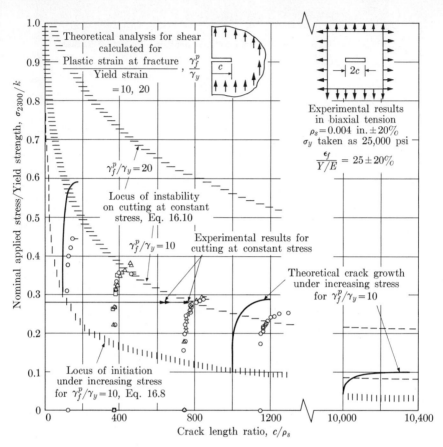

Fig. 16.15. Crack growth to instability. (McClintock, 1958. Courtesy of ASME.)

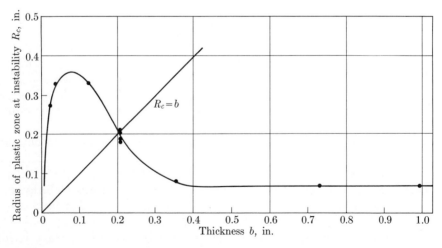

Fig. 16.16. Dependence of radius of plastic zone at instability on sheet thickness for 7075-T6 aluminum-zinc alloy. (ASTM, 1960. Courtesy of ASTM.)

TABLE 16.6

Stress Singularity Parameters Describing Fracture*

Name	Symbol	Definition in terms of k_1	Reference
Stress singularity intensity factor	$k_1, \mathcal{K}_1$ K_1 a_1	k_1 $k_1\sqrt{\pi}$ $-k_1/\sqrt{2}$	Irwin, et al. (1958) ASTM (1960) Williams (1957)
Nominal plastic zone radius	R	$(k_1/\text{T.S.})^2$	McClintock (1960)
Equivalent surface energy	α, γ	Plane stress $\pi k_1^2/2E$ Plane strain $\pi k_1^2(1-\nu^2)/2E$	Irwin (1948), Orowan (1949)
Crack extension force (strain energy release rate from elastic field)	$G, \mathcal{G}$	Plane stress $\pi k_1^2/E$ Plane strain $\pi k_1^2(1-\nu^2)/E$	Irwin, et al. (1958)

* In terms of $k_1 = \sigma_\infty \sqrt{c}$ for crack of length $2c$ in an infinite plate where $(\sigma_{\theta\theta})_{\theta=0} = k_1/\sqrt{2r} +$ terms of order r^0; subscript 1 denotes crack under tension, subscript c denotes critical value at crack instability.

TABLE 16.7*

Stress Singularity for Unstable Fracture

Material	Yield strength, psi	Thickness, in.	Temperature, °F	R_{1c},† in.	k_{1c}, lb·in$^{-3/2}$	Fracture type
Polymethyl-methacrylate						
cast	10,000	0.05	Room	0.0055	750	Normal
hot-stretched				0.035–0.055	1900–2400	
2024-T3	47,000	0.032	Room	0.51	31,000	Shear
		0.25	Room	0.44	31,000	Shear
		4	Room	0.49	33,000	Normal
7075-T6	68,000	0.032	Room	0.078	19,000	Shear
		0.750	Room	0.088	20,000	Normal
		2-in. bar	Room		20,000	Normal
Low-C steel (ABS ship steel C)	105,000	0.75	−112	0.080	30,000	Normal
	100,000	1.00	−40	0.067–86	26,000–29,000	Normal
	36,000	0.75	32	1.085	37,000	Shear
NiMoV steel for rotors	101,000	6	Room	<1.0	<100,000	Normal
	89,000	6	Room	1.8–2.4	120,000–140,000	Normal
4340 steel	230,000	0.25	Room	0.028	38,000	Normal
6150 steel	260,000	0.75	Room	0.32	150,000	50% shear

* Data from Irwin et al. (1958).
† R_{1c} calculated from $(k_{1c}/\text{Y.S.})^2$.

◀For thicker sheets and plates, once the plate thickness becomes large compared with the plastic zone, plane strain conditions occur. The transverse stress is likely to reduce the fracture strain markedly, in turn reducing the extent of the plastic zone required for fracture. An example is shown in Fig. 16.16 where the plastic zone was calculated from an approximate equation analogous to Eq. 11.29

$$R = c(\sigma/\text{T.S.})^2. \tag{16.11}$$

At the same time that the extent of the plastic zone becomes less than the plate thickness, the shape of fracture changes from shear to normal, with perhaps only a small shear lip.

◀Under plane strain conditions, the fracture criterion will depend on normal stress as well as the strain assumed in the theory leading to Eq. 16.9. Pending further theoretical developments, it is necessary to correlate results on the basis of a parameter such as the approximate extent of the plastic zone calculated from Eq. 16.11 or an elastic stress-intensity coefficient k_1, assuming the elastic-plastic zone is small enough relative to the boundaries of the part so that the corresponding elastic stress singularity can exist between them. Other parameters are given in Table 16.6. Values of the approximate extent R of the plastic zone, and the corresponding stress intensity factor k_1 are given in Table 16.7 for a number of materials.▶

16.8 SUMMARY

For a non-strainhardening material, the extension to rupture is of the order of one-half to two times the minimum dimension of the part. Strain-hardening, creep, or viscous flow may increase this figure. The maximum load for rupture of plastic materials is determined by the fully plastic stress distribution, and for various grooved specimens may be one to three times the load carried by a straight specimen of the same cross-sectional area.

Ductile fracture occurs by the opening up and coalescing of holes formed at inclusions or other inhomogeneities, by the growth of fine cracks, by strain-softening and the formation of intense bands of shear, or under very high triaxial tensile stress by hole growth from dislocations or their intersections. A theory for ductile fracture based on the growth of holes indicates that such fracture depends on the history of strain and the hydrostatic component of stress, on the ratio of part size to inclusion spacing, and on the anisotropy of inclusions.

These different mechanisms and the macroscopic stress and strain distributions lead to a variety of macroscopic forms of fracture: cup and cone, shear, normal, rosette, and woody.

The notch sensitivity of a fairly ductile part can be described by comparing it with an ideally plastic case. With a large strain concentration, as at the tip of a crack, ductile fracture may occur at nominal stresses below the yield strength.

◀In shear if fracture can be assumed to occur when the total strain γ_f is obtained over a region ρ_s in a material with shear strain at yielding γ_y, a longitudinal shear crack of length c will become unstable when the longitudinal shear stress $\sigma_{23\infty}$ reaches a critical value

$$\sigma_{23\infty} = G\gamma_y \sqrt{\rho_s[\exp(\sqrt{2\gamma_f^p/\gamma_y} - 1)]/c}. \qquad (16.10)$$

This tendency to crack instability can be described in terms of the approximate extent of the plastic zone or intensity of the surrounding elastic stress field, k_1. Typical values are given in Table 16.7. ▶

REFERENCES

Orowan's summary of fracture (1949) remains a good introduction. Cottrell (1959) has sharpened attention on the role of inclusions in ductile fracture. The mechanics by which tests on predominantly elastic structures are correlated has been reviewed by Irwin (1960), and the American Society for Testing Materials (ASTM, 1960). For other topics, such as ductile fracture by fine cracks, the mechanics of rupture, and the mechanics of plastic flow around cracks, see specific references in the text.

ASTM	1960	"Fracture Testing of High Strength Sheet Materials," *Bull. ASTM*, No. 243, 29–40.
BACKOFEN, W. A. SHALER, A. J. HUNDY, B. B.	1954	"Mechanical Anisotropy in Copper," *Trans. ASM* **46,** 655–680.
BERG, C. A.	1962	"The Motion of Cracks in Plane Viscous Deformation," *Proc. 4th U.S. Nat. Con. Appl. Mech.*, Vol. 2, pp. 885–892.
BISHOP, J. F. W.	1953	"On the Complete Solution to Problems of Deformation of a Plastic-Rigid Material," *J. Mech. Phys. Solids* **2,** 43–53.
BRIDGMAN, P. W.	1952	*Studies in Large Plastic Flow and Fracture*, McGraw-Hill, New York.
COTTRELL, A. H.	1959	"Theoretical Aspects of Fracture," *Fracture*, B. L. Averbach et al., eds., M.I.T. Press, Cambridge, Mass., and Wiley, New York, pp. 20–53.
COWDREY, I. H. ADAMS, R. G.	1935	*Materials Testing*, Wiley, New York.
DAVIS, E. A.	1943	"Increase of Stress with Permanent Strain and Stress-Strain Relations in the Plastic Flow of Copper under Combined Stresses," *Trans. ASME* **65,** A187–A196.
FORD, H. LIANIS, G.	1957	"Plastic Yielding of Notched Strip under Conditions of Plane Stress," *Z. Ang. Math. Phys.* **8,** 360–382.
GREEN, A. P.	1953	"The Plastic Yielding of Notched Bars Due to Bending," *Quart. J. Mech. Appl. Math.* **6,** 223–239.
GREEN, A. P. HUNDY, B. B.	1956	"Initial Plastic Yielding in Notch Bend Tests," *J. Mech. Phys. Solids* **4,** 128–144.

Halford, G. R. Morrow, J.	1962	"Low Cycle Fatigue in Torsion," *Proc. ASTM* **62**, 695–709.
Hill, R.	1950	*The Mathematical Theory of Plasticity*, Oxford University Press, London, p. 250.
Hill, R.	1952	"On Discontinuous Stress States with Special Reference to Necking in Thin Sheets," *J. Mech. Phys. Solids* **1**, 19–30.
Irwin, G. R.	1948	"Fracture Dynamics," *Fracturing of Metals*, ASM, Novelty, Ohio, pp. 147–166.
Irwin, G. R.	1960	"Fracture Mechanics," *Structural Mechanics*, J. N. Goodier and N. J. Hoff, eds., Pergamon Press, London, pp. 557–594.
Irwin, G. R. Kies, J. A. Smith, H. L.	1958	"Fracture Strength Relative to Onset and Arrest of Crack Propagation," *Proc. ASTM* **58**, 640–660.
Levin, E.	1955	"Indentation Pressure of a Smooth Circular Punch," *Quart. J. Appl. Math.* **13**, 133–137.
McClintock, F. A.	1958	"Ductile Fracture Instability in Shear," *J. Appl. Mech.* **25**, 581–588.
McClintock, F. A.	1960	"Discussion of Irwin (1960)," *Trans. ASME J. Basic Eng.* **82D**, 423–425. See also discussion in *Mat. Res. Stand.* **1**, 277–279 (1961).
McClintock, F. A.	1961	"On Notch Sensitivity," *Welding J. Res. Suppl.* **26**, 202–208.
McClintock, F. A.	1965	"Effects of Root Radius, Stress, Crack Growth and Rate on Fracture Instability," *Proc. Roy. Soc.* **A285**, 58–72.
McClintock, F. A. Kaplan, S. M. Berg, C. A.	1965	"Ductile Fracture by Hole Growth in Shear Bands," unpublished paper.
McClintock, F. A. O'Day, W. R. Jr.	1965	"Biaxial Tension, Distributed Dislocation Cores, and Fracture in Bubble Rafts," submitted to International Conference on Fracture, Sendai, Japan.
Moffatt, W. G. Wulff, J.	1963	"Tensile Deformation in Fracture of Brazed Joints," *Welding J. Res. Suppl.* **42**, 115s–125s.
Neimark, J. E.	1959	"The Initiation of Ductile Fracture in Tension," ScD Thesis, M.I.T., Cambridge, Mass.
O'Brien, J. L. Davis, R. S.	1961	"On the Fracture of Solids under Impulsive Loading Conditions," *Response of Metals to High Velocity Deformation*, Am. Inst. Mining, Met. and Pet. Eng., Metallurgical Conferences, Vol. 9, Interscience, New York, pp. 371–388.
Orowan, E.	1949	"Fracture and Strength of Solids," *Rept. Prog. Phys.* **12**, 186–232.
Parker, E. R.	1957	*Brittle Behavior of Engineering Structures*. Wiley, New York, p. 67.
Puttick, K. E.	1959	"Ductile Fracture in Metals," *Phil. Mag.*, **4**, 964–969.

PUTTICK, K. E.	1960	"Shear Component of Ductile Fracture," *Phil. Mag.* **5**, 759–762.
ROGERS, H. C.	1960	"Tensile Fracture of Ductile Metals," *Trans. Met. Soc. AIME* **218**, 498–506.
ROSS, S. T. SERNKA, R. F. JOMINY, W. E.	1956	"Some Relationships between Endurance Limit and Torsional Properties of Steel," *Trans. ASM* **48**, 119–148.
SHIELD, R. T.	1955	"On the Plastic Flow of Metals under Conditions of Axial Symmetry," *Proc. Roy. Soc. (London)* **A233**, 267–286.
TERRY, E. L. McCLAREN, S. W.	1962	"Biaxial Stress and Strain Data on High Strength Alloys for Design of Pressurized Components," *Tech. Doc. Rept. No. ASD-TDR-62-401*, Air Force Systems Command, Office of Technical Services, U.S. Depart. of Commerce, Washington 25, D.C.
TIPPER, C. F.	1949	"The Fracture of Metals," *Metallurgia*, **39**, 133–137.
UZHIK, G. V.	1948	*Resistance to Brittle Rupture and Strength of Metals*, Moscow.
WALSH, J. B. MACKENZIE, A. C.	1959	"Elastic-Plastic Torsion of a Circumferentially Notched Bar," *J. Mech. Phys. Solids* **1**, 247–257.
WILLIAMS, M. L.	1957	"On the Stress Distribution at the Base of a Stationary Crack," *J. Appl. Mech.* **24**, 109–114.

PROBLEMS

16.1. Show that for the singly grooved, plane strain tensile specimen of a nonhardening plastic material, the extension to rupture will be equal to the minimum thickness of the specimen.

16.2. Show that under tensile stresses across a crack tip, dislocation motion, either away from the tip or into it, will lead to a blunting of the crack.

◀16.3. Derive the equation for the strain increments around a cylindrical hole with axial strain $d\epsilon_{zz}$, Eq. 16.4

◀16.4. Derive the equation for the rate of damage per unit axial strain due to growth of a cylindrical hole in a plastic material, Eq. 16.5a.

16.5. Find the ratios of the principal components of stress expected during the formation of a circumferential neck in a thin-walled tube, as tested by Davis (see Table 16.3).

16.6. Show that for some shapes of specimens, it is possible for the ratio of maximum load to that for an unnotched specimen of the same minimum area to exceed unity, whereas the corresponding ratio for other shapes of specimens cannot exceed unity. A value of unity for this ratio is thus not sufficient evidence for the absence of notch sensitivity.

◀16.7. Derive the equation for the nominal applied stress required to initiate ductile fracture at the root of a groove under longitudinal shear, Eq. 16.8.

◀16.8. Derive the equation for the nominal applied stress required for instability of a crack growing under longitudinal shear in an elastic-plastic, non-strainhardening material, Eq. 16.10, from Eq. 16.9.

◀16.9. An accurate approximation to the load for instability is

$$\sigma_{23\infty} = G\gamma_y \sqrt{\rho_s[\exp(\sqrt{1 + 2\gamma_f^p/\gamma_y} - 1)]/c}.$$

Show that in the limiting case as the plastic strain goes to zero, the stresses for initiation and instability become equal, and correspond to those which would be expected for a brittle material. What further assumptions are required?

16.10. Judging from the data in Table 16.2 what can you conclude about the mechanisms leading to fracture in the four materials for which data are given?

16.11. Calculate the increase in the reduction of area due to a pressure of 140,000 psi in Bridgman's data given in Fig. 16.12.

◀16.12. Make a plot of load factor versus specimen size for one of the materials listed in Table 16.7. At what stress do you feel your curve is no longer valid?

E 16.13. See how many of the phenomena of ductile fracture you can simulate by tests on Plasticine in which you have punched holes, perhaps using a straightened-out paper clip.

16.14. (a) What is the Burgers vector of the dislocation shown in Fig. 16.5?

◀(b) Why should the width of the dislocation (i.e., the number of highly disturbed bubbles) increase under biaxial tension (McClintock and O'Day, 1965)?

◀(c) Discuss the effect of solute atoms on the ideal strength of a perfect lattice.

◀16.15. The effect of history of loading and geometry on the tendency to fracture can be illustrated by considering the extreme case of a rigid-plastic bar under torsion. Consider two loading paths, each starting with a bar of length l and diameter D and finishing with the ends of the bar twisted through an angle θ and the central portion of length, say $l/2$, turned down to a diameter $d/2$. In the first path, the bar is twisted through the angle θ and then machined to the final dimension. In the second, the bar is first machined and then twisted through the angle θ. Sketch the strain distribution in the minimum section for each history. Sketch the stress distribution just as further plastic flow begins in each case. Which makes fracture more likely, twisting followed by growth of the notch, or growth of the notch followed by twisting?

◀16.16. The tendency of the strain to remain constant ahead of a crack growing under longitudinal shear, as indicated by the analogy of Prob. 16.15, is modified by the tendency for strain to increase with the growth of the crack as a result of redistribution of stress within the plastic zone and slight displacements at its boundary as the zone spreads. It turns out that the resulting strain increment at the point x_1, directly ahead of a crack of length c, when the crack grows at constant stress (approximately constant R/c) is (from McClintock, 1965)

$$d\gamma = \frac{\gamma_y}{x_1 - c}\left(1 + \frac{R}{c} + \ln\frac{R}{x_1 - c}\right)dc. \qquad (a)$$

To determine the initial stability of the crack, apply the fracture criterion of Section 16.7 both before and after crack growth by dc at constant applied stress. Show that for the points ρ_s ahead of the respective crack tips the difference in strain is

$$d\gamma = \frac{\gamma_y}{\rho_s}\left[-\frac{R}{\rho_s} + 1 + \frac{R}{c} + \ln\frac{R}{\rho_s}\right]dc. \tag{b}$$

As an example, consider a material with at least a little ductility, say $\gamma_f/\gamma_y = 2$, so that at the initiation of cracking Eq. 11.30 gives $R/\rho_s = 2$. Show that Eq. (b) indicates a decrease of strain for cracks longer than about $6\frac{1}{2}\,\rho_s$, so such cracks are initially stable. Sketch the history of loading on a plot of R/c (or applied stress) vs. crack length, and contrast it with the path followed in deriving Eq. 16.8, to the same final R/c and crack length. Note that the fracture criterion indicates further cracking would occur spontaneously after one path but not the other.

CHAPTER **17**

TRANSITIONAL MODES OF FRACTURE

17.1 SYNOPSIS

Brittle fractures in engineering structures have at times caused large-scale damage. Merchant vessels have broken in two in calm weather while being in harbor, (Fig. 17.1), bridges have collapsed, pipe lines and gas storage tanks have ripped open. Although quantitative prediction of brittle fracture is not yet possible, many of its features are reasonably well understood.

FIG. 17.1. T-2 tanker that failed at pier. (Parker, 1957. Courtesy of Wiley.)

Some body-centered cubic metals, such as low-carbon steel and tungsten, and glassy materials, such as inorganic glasses and polymers, will deform plastically or viscously at elevated temperatures or at slow strain rates, but at low temperatures or under high strain rates, or when notched, they will undergo brittle fracture.

Considering the simpler case of inorganic glasses first, we shall show that their transition from ductile to brittle behavior is governed by the very rapid rise of viscosity with decreasing temperature.

We shall discuss various mechanisms for crack nucleation by plastic deformation in crystalline materials and investigate the conditions for the brittle propagation of a crack. This will be followed by a discussion of the factors such as strain rate, notches, triaxial stresses, etc., which affect the transition temperature. These effects will be summarized by a convenient diagramatic representation known as the Davidenkov diagram.

We shall consider briefly the markings on the surfaces of fracture and how these can be used to trace the fracture to its origin. As a final topic we shall discuss the common engineering tests which are used to determine the temperature where ductile to brittle transition occurs in the fracture behavior. This will be accompanied by some data for a variety of common ferrous materials.

17.2 TRANSITION BETWEEN FRACTURE MODES IN GLASSY SOLIDS

The phenomenon of transition of fracture mode in glassy materials is of the simplest type and will therefore be studied first.

It was shown in Section 6.6 that the mechanical response of many glassy materials can be adequately represented by an idealized model of springs and dashpots. One of the two simplest ways to idealize a material having both elasticity and viscosity is with a spring and dashpot in series, which is known as a *Maxwell material* (see Table 6.3). Here the spring represents the instantaneous Young's modulus E of the material while η stands for its steady-state tensile viscosity. If such a material is uniaxially deformed at an imposed strain rate $d\epsilon/dt$, the stress in it will satisfy the differential equation

$$\frac{1}{E}\frac{d\sigma}{dt} + \frac{\sigma}{\eta} = \frac{d\epsilon}{dt}. \qquad (17.1)$$

We shall further assume that the Maxwell material has penny-shaped ellipsoidal cracks either in its interior or on its surface. In such a material, brittle fracture may take place under high strain rates when there is insufficient time for viscous deformation to significantly alter the shape of the ellipsoidal cracks. If the viscosity is very low, however, cracks will lose their shape before the elastic strains can be concentrated at their tips. Thus, fracture will not take place, but the material will rupture instead by thinning down to a filament.

These effects can be discussed qualitatively with reference to Eq. 17.1. When this equation is rewritten for finite increments of strain, stress, and time, $\Delta\epsilon$, $\Delta\sigma$, Δt, respectively, and the material is considered to be initially stress free, then

$$E\frac{\Delta\epsilon}{\Delta t} = \frac{\Delta\sigma}{\eta/E} + \frac{\Delta\sigma}{\Delta t}. \qquad (17.2)$$

Now if the time of loading $\Delta t \ll \eta/E$, then the first term on the right-hand side of Eq. 17.2 can be neglected, giving a purely elastic loading. Thus, for loading times much shorter than η/E the material will behave as though it were perfectly elastic. In this case, the increment of local stress is governed by the stress concentrations of the cracks, and brittle fracture would then result if this local stress at the tip of a crack reached the value of the ideal tensile strength, as discussed in Section 15.2. Alternatively, if the quantity η/E is very much smaller than Δt, the second term on the right-hand side of Eq. 17.2 can be neglected in comparison with the first term, giving the equation of Newtonian viscous flow. This indicates that for very small strain rates, there is always enough time for the elastic strains to relax. In this case, viscous deformation will alter the shape and orientation of the cracks and render them ineffective, making brittle fracture impossible.

From the above discussion, we see that brittle fracture in glassy materials is governed by the relative magnitudes of the time constant of loading (a measure of which may conveniently be taken as the reciprocal of the strain rate) and the quantity η/E, the *relaxation time* of the material. The relaxation time, being proportional to the viscosity, is a strong function of temperature (see Eqs. 6.6 and 6.9):

$$t = \frac{\eta}{E} = \frac{\eta_0}{E} e^{u_a/kT}. \tag{17.3}$$

Here η_0 is a constant with the same dimensions as the viscosity and u_a is an activation energy of the molecular process responsible for the viscosity.

Since the relaxation time decreases rapidly with increasing temperature, at a certain range of temperature, for a given strain rate, the time constant of loading will become comparable to the relaxation time, and the fracture mode will go from brittle fracture to rupture. The temperature at which the fracture mode shows its most rapid change is called the *fracture transition temperature*. In practice, the transition temperature is often determined by plotting the energy absorbed in the deformation process prior to fracture as a function of temperature, and picking the temperature where this energy shows its most rapid change. The transition temperature is a function of the rate of increase of stress; it will decrease with slower rates and increase with higher rates.

The above discussion of fracture-mode transition is true only for uniaxial stresses. If the stress state is of a triaxial nature, the triaxiality component does not promote viscous deformation, while it will be instrumental in brittle fracture. The triaxial component of *stress* is therefore of importance. It can be expected that a triaxial tensile stress component will increase the transition temperature, while a triaxial compressive stress will suppress the transition temperature. In many practical cases, the fracture begins at a free surface. Then the most that can be done by a notch is to raise the intermediate principal stress. As we have seen, however, the intermediate principal stress has little effect on brittle fracture. Therefore where fracture starts at the surface the principal effect of a notch seems to be not so much one of producing triaxiality as one of simply increasing the local strain rate. If fracture can be nucleated beneath the surface, then the triaxiality produced by a notch is important (see Section 16.4).

17.3 TRANSITION BETWEEN FRACTURE MODES IN CRYSTALLINE MATERIALS

Certain crystalline materials, whose flow stress rises rapidly with decreasing temperature and with increasing strain rate, exhibit a fracture-mode transition similar to the one in glassy materials discussed above. Body-centered cubic metals such as iron and tungsten, and ionic salts such as NaCl, AgCl, and MgO, fall into this category. Another, and perhaps more unifying, feature of these materials is their cleavage along well-defined crystallographic planes. Although the fracture-mode transition in these materials is similar to that in glassy materials, the basic mechanisms of fracture initiation and fracture propagation are markedly different.

In structurally stable (or at least metastable) glassy materials, defects of the type that were assumed to exist in the treatment of the previous section can be formed only by mechanical damage or chemical corrosion but not primarily by the deformation process itself. *Perfect* glassy materials, therefore, would either fracture in a brittle manner when the nominal stress in them reached the ideal strength, or would viscously deform and rupture. In both cases, the energy absorbed prior to fracture would be large. A fracture-mode transition based on an energy criterion would not apply. In crystalline materials exhibiting a fracture-mode transition, there is the additional possibility of formation of defects such as microcracks in the process of slip or twinning. Hence, these materials are susceptible to brittle fracture at relatively low stresses, regardless of the initial perfection of the material. Once such defects are established in the process of plastic deformation, the occurrence of brittle or ductile fracture will be governed by whether or not the conditions of fracture propagation are satisfied, i.e., whether or not further plastic deformation will level out the stress concentrations at the defects before the local conditions will satisfy a criterion of a brittle type of fracture propagation. Thus, in these materials there are two distinct phases of the fracture process, namely the phase of crack initiation and the phase of crack propagation, which will be discussed separately in the following sections.

17.4 CRACK FORMATION BY PLASTIC DEFORMATION

A variety of mechanisms of crack formation in plastic deformation have been observed and analyzed. The ones we shall discuss here, which are not likely to exhaust the possibilities, will be grouped into two main categories: crack formation by dislocation pile-ups and by twin intersections.

A. Crack Formation by a Dislocation Pile-Up. *1. Pile-up at a Strong Obstacle.* The stress-concentration at a dislocation pile-up, which results when the progress of a train of edge dislocations is stopped by a strong obstacle, was discussed in Section 4.11. It was pointed out by Zener (1949) that as the number of dislocations in a pile-up increases, the dislocations at the head of the pile-up may become so close that they can fuse together to nucleate a microcrack, as shown in Fig. 17.2. Stroh (1954, 1955), who has calculated this possibility, showed that once the two dislocations at the head of a dislocation pile-up fuse together, the atomic-size microcrack can grow into the high tensile stress field steadily until all the remaining dislocations of the pile-up have run into the crack as shown in Fig. 17.2(b). ◀In a quantitative study of this model, Stroh (1954) has shown that a pile-up of n edge dislocations (Fig. 17.3) not only concentrates by a factor n the applied net shear stress at the tip of the pile-up (see section 4.11), but, in addition, the pile-up produces a concentration of normal stress above and below the slip plane, which also varies linearly with the number n (see Fig. 17.3). If the applied shear stress

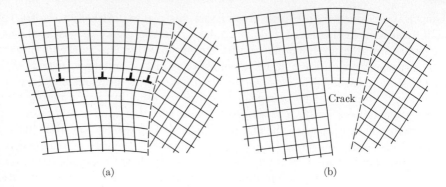

Fig. 17.2. Formation of a microcrack by fusing together of dislocations at the tip of a pile-up. (After Zener, 1949, and Stroh, 1954, 1955.)

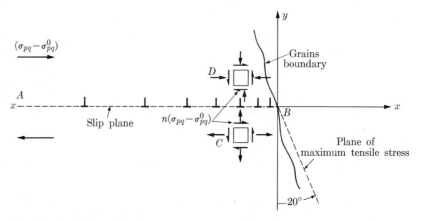

Fig. 17.3. A dislocation pile-up at a grain boundary B. The shear stresses as well as normal stresses are highly concentrated at the point B.

on the slip plane p in the slip direction q is σ_{pq}, and if there is a friction stress σ_{pq}^0 that has to be overcome in order to move a dislocation, then the maximum normal stress σ_n produced at the obstacle by a pile-up of n dislocations and acting across a plane making an angle of 20° with the negative y axis in Fig. 17.3 is, according to Stroh,

$$\sigma_n = \frac{2}{\sqrt{3}} n(\sigma_{pq} - \sigma_{pq}^0). \tag{17.4}$$

◀ If a crack is to be nucleated by such a normal stress the stress must not only equal the ideal strength but also it must not decay too rapidly, so that as the nucleated crack grows, enough strain energy can be drained into it to provide for the energy of the crack surfaces. Stroh has shown that this is indeed the case for the stress field around the tip of a dislocation pile-up. Thus, the condition for crack nucleation is merely (see Section 15.2)

$$\sigma_n = \sigma_c = 2\pi\alpha/b = E/2\pi. \tag{17.5}$$

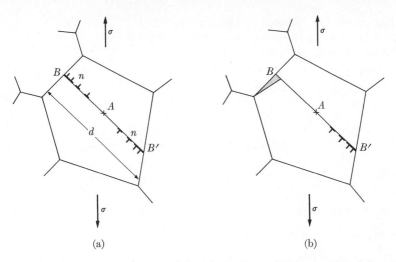

Fig. 17.4. Crack nucleation from a dislocation pile-up inside a grain of diameter d.

Neglecting for the moment the friction stress, we see that to nucleate a crack, the required number n of dislocations are about in the ratio of the ideal strength to the yield strength Y in tension, i.e.,

$$n \cong \sigma_c/Y. \qquad (17.6)$$

In most body-centered cubic metals this ratio is near 300 and does not exceed 1000.

◀Although the normal stress produced by a dislocation pile-up is highest across a plane making an angle of 70° with the plane of the pile-up, cracks may actually be nucleated along the nearest cleavage plane in crystals which can undergo cleavage fracture. If the material is polycrystalline, the stress necessary for crack nucleation will depend on the grain size. This is because the size of the grain will determine the length of the dislocation pile-up and hence the intensity of the stress concentration.

◀Consider a grain such as that shown in Fig. 17.4 with a dislocation mill situated in its center at point A. If the number of dislocations arrested by the grain boundaries at points B and B' are n, their stress field will tend to oppose the applied shear stress at the mill at point A. When the total shear stress due to the piled-up dislocations equals the applied stress, the mill will cease operation. From the analysis of Eshelby et al. (1951) it is known that the effect at large distances of the n dislocations in a pile-up is about the same as a giant dislocation of strength nb located near the tip of the pile-up at $\frac{1}{4}$ the distance between the first and the last dislocations. From these relations it is possible to calculate the net shear stress that can hold a number n of dislocations in equilibrium inside a grain of diameter d. This is (Prob. 17.1)

$$\sigma_{pq} - \sigma_{pq}^0 = \frac{8nbG}{3\pi(1-\nu)d}. \qquad (17.7)$$

Elimination of the number n between Eqs. 17.4 and 17.7 and utilization of Eq. 17.5 gives the condition for crack initiation in a polycrystal of grain diameter d:

$$\sigma_{pq} - \sigma_{pq}^0 = \sqrt{\frac{4G\alpha}{\sqrt{3}(1-\nu)d}}. \quad (17.8)$$

◀When there is a minimum normal stress σ_3, as shown in Fig. 17.5, the resolved shear stress difference appearing in Eq. 17.7 is given in terms of the principal stress components by

$$\sigma_{pq} - \sigma_{pq}^0 = \tfrac{1}{2}[\sigma_1 - \sigma_3 - \sigma_0]$$
$$= \tfrac{1}{2}(q\sigma_1 - \sigma_0), \quad (17.8\text{a})$$

where σ_0 is the apparent friction stress observed in a tension test, and the factor q is 0 for pure hydrostatic stress and 2 for a state of pure shear. The condition for cleavage-crack initiation by confined plastic deformation in a grain can now be obtained by combining Eqs. 17.7, 17.8, and 17.8a:

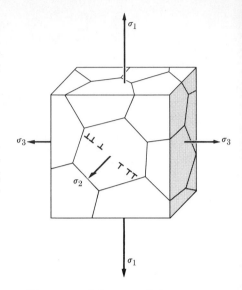

FIG. 17.5. Pile-up of dislocations in a grain subjected to a triaxial state of stress.

$$q\sigma_1 = \sigma_0 + \sqrt{\frac{16G\alpha}{\sqrt{3}(1-\nu)d}} = \sigma_0 + K\,d^{-1/2}, \quad (17.9)$$

where K has a magnitude of about 10^8 dyne $\cdot$ cm$^{-3/2}$, for σ_0, σ_1, and d given in cgs units.

◀The discussion of Eq. 17.9 and its implications will be deferred to Section 17.5 when its interrelation with the fracture propagation condition, to be discussed in Section 17.5, can be better understood.▶

The mode of crack nucleation by a dislocation pile-up presupposes that there is no possibility of accommodation of the stress concentration by slip or twinning in the vicinity of the tip of the dislocation pile-up and that crack nucleation is the only possible mode of accommodation. Johnston et al. (1962) have observed such cracks along the grain boundary in magnesium oxide bi-crystals compressed at room temperature as shown in Fig. 17.6. In this photograph the remaining dislocations in the pile-up have been revealed as pits by a suitable etchant, and the microcracks penetrating into the adjacent grains are clearly visible.

In coarse-grained "high-purity" iron McMahon (1963) has observed that in deformation at room temperature many carbide particles at grain boundaries fracture, apparently while the matrix undergoes the Lüders strain. Figure 17.7 shows a typical case. It is likely that these carbide particles fracture as a result of the stress concentration of dislocation pile-ups rather than by stresses arising from drag forces exerted by the deformation of the surrounding ferrite (Prob. 17.2).

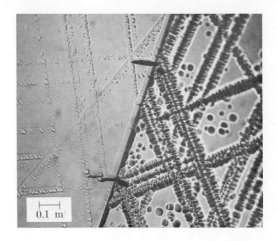

Fig. 17.6. Formation of microcrack at end of dislocation pile-up in polycrystalline MgO. The slanting rows of pits in the right-hand grain are dislocation pile-ups. The two horizontal streaks in the grain with light pits are cleavage microcracks. (Johnston et al., 1962. Courtesy of Taylor-Francis.)

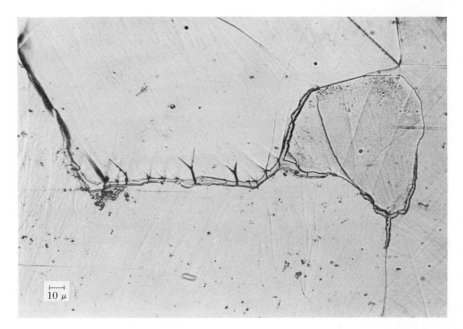

Fig. 17.7. A cracked carbide particle at a grain boundary of a deformed "high-purity" iron specimen. (Courtesy of McMahon, 1963.)

Thus it appears in this case that the undeformable carbide particles, when present at grain boundaries, can accommodate only by fracture. In uncontaminated grain boundaries, however, the stress concentration of dislocation pile-ups can always be relieved by slip in one grain or the other.

Particle precipitation in the grain boundaries can also lead to brittleness by grain-boundary cracking (Adcock and Bristow, 1936). Crack initiation is also associated with pearlite platelets in the ordinary tensile test of mild steel, but only after strains of the order of unity.

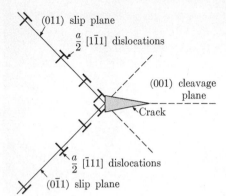

Fig. 17.8. Production of microcrack at the intersection of two slip planes in the b.c.c. lattice under a tensile stress in the [001] direction. (After Cottrell, 1958.)

2. Pile-up at Sessile Dislocations. Cottrell (1958) has proposed another dislocation pile-up mechanism of crack nucleation in the body-centered cubic lattice. As shown in Fig. 17.8, in this mechanism the two leading dislocations on two intersecting slip planes are required to combine to form an immobile (sessile) dislocation which then serves as an obstacle to the following dislocations. Electron microscopic examinations of thin iron foils have revealed that such sessile dislocations, although numerous, are rather short (Carrington et al., 1960). Furthermore, calculations by Stroh (1959) have indicated that such dislocations would dissociate under the force of the piling-up dislocations before the stress field can be sufficiently intensified to generate a crack. Thus it is unlikely that this mechanism plays an important role.

3. Pile-up in the Form of a Low-Angle Boundary. If a low-angle tilt boundary is sheared through as shown in the sequence of Fig. 17.9, or alternatively, if a finite low-angle tilt boundary is formed by the vertical alignment of edge dislocations as a result of slip on parallel planes, a high stress concentration may result at the termination of the boundary. If the local normal stress reaches the level of the ideal strength and if the stress concentration does not decay too rapidly with distance, a crack

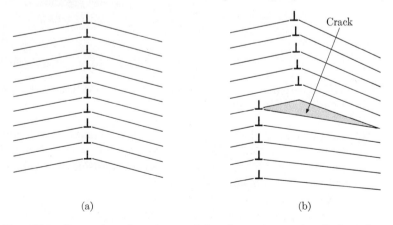

Fig. 17.9. Formation of a microcrack by shear of part of a tilt boundary.

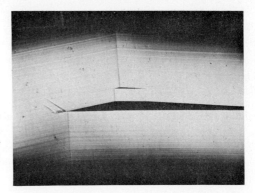

Fig. 17.10. Cleavage fractures in a single crystal of zinc, approximately ten times magnified. (Gilman, 1954. Courtesy of Am. Inst. Mining and Metallurgical Engineers.)

can be nucleated as shown in the zinc crystal of Fig. 17.10. This mode of crack nucleation is very common in inhomogeneously deformed hexagonal metals and also in single crystals of ionic compounds such as magnesium oxide. Considering this mechanism, Stroh (1958) has calculated an orientation dependence of tensile strength for zinc crystals which agrees well with the experimental results of Deruyttere and Greenough (1956).

B. Crack Formation by the Intersection of Deformation Twins. Under high enough stress, formation of deformation twins is possible in many structures of high symmetry. When such twin bands intersect, high stress concentrations can result. This is especially true in body-centered cubic crystals where the shear strain inside a twin band is as high as $1/\sqrt{2}$ (Prob. 17.3). Thus if the progress of an extending twin band is interrupted by another twin in its path, as sketched in Fig. 17.11, the large shear displacement across a twin band could result in crack formation. Such cracks have been observed by many investigators. Crack nucleation by this mechanism in a single crystal of an iron-silicon solid solution was

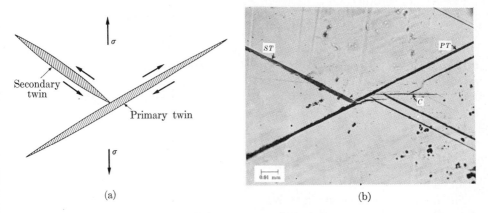

Fig. 17.11. (a) Crack nucleation by the intersection of two deformation twins. (b) The effect as observed in silicon iron: PT, primary twin; ST, secondary twin; C, crack emanating from intersection. (Hull, 1960. Courtesy of Pergamon Press.)

observed by Hull (1960). It is surprising that in spite of the many twin intersections which are normally observed, only a few lead to crack nucleation. This indicates that in the majority of cases the stress concentrations are alleviated by various mechanisms of accommodation involving slip inside or outside the twins, before cracks can be initiated.

17.5 THE PROCESS OF CRACK PROPAGATION

The nucleation of a crack by plastic deformation does not automatically assure fracture of a large part, since the conditions of crack initiation and crack propagation can be different. In a single crystal, if a strong enough obstacle could be found to initiate a crack by a dislocation pile-up, or by intersection of twin bands, under an applied tensile stress, the nucleated crack will almost always be long enough to satisfy Griffith's equation of crack propagation, provided, of course, that the crack will encounter no obstacles which can stop and blunt it by plastic deformation. In polycrystalline materials, on the other hand, a cleavage crack, once nucleated, will very likely be stopped by a grain boundary where the cleavage planes will undergo a large and discontinuous change in orientation into the neighboring grain. Once a crack is stopped at a grain boundary, plastic deformation could occur to some extent in the vicinity of the crack root and blunt it. Figure 17.12 shows a split grain in low-carbon steel deformed at $-195°C$; the localized deformation around the crack tip at the grain boundary is clearly visible. For a crack to propagate from one grain to others, to eventual fracture, it is necessary that the elastic energy which is released by the propagating crack should provide not only the energy for the surfaces of the crack but also the energy for all additional fracture work. This is to say that the energy of the jagged cleavage surfaces and the work of plastic deformation at the points where the crack hesitates must both be provided from the elastic energy which the propagating crack releases. Under certain conditions where the amount of plastic deformation is limited to a thin layer of the fracture surface, it is possible to consider the localized fracture work per unit area as a kind of pseudo surface energy α' to be added to the surface energy α in the Griffith equation (Orowan, 1948). Since the specific fracture work, however, is normally 10^6 to 10^7 erg/cm^2, two to three orders of magnitude greater than the surface energy α (for some values see Tables 16.6 and 16.7), the latter can be neglected altogether, giving a relation for crack propagation between the applied uniaxial stress σ and the crack size c:

$$\sigma = \sqrt{E\alpha'/c} = K^* c^{-1/2}. \qquad (17.10)$$

In a polycrystal, where the initial crack is of the size of the average grain diameter, c can be replaced by d.

The study of low-energy fractures in steel plates by Orowan (1945) has disclosed that the plastically distorted layer on the fracture surfaces is of a thickness of a fraction of a millimeter (0.2 to 0.4 mm). Such plastic zones are so near the grain size that the continuum plasticity theory of Section 16.7 may not apply

Fig. 17.12. Cleavage crack in a grain of coarse-grained ferrite deformed 8% at −140°C. (Hahn et al., 1959. Courtesy of M.I.T. Press.)

(McClintock, 1960). Furthermore, Tipper (1945) observed that fracture propagation is in these cases not a continuous process, and that new microcracks may arise ahead of the main fracture front inside the plastic zone. This would then result in a process of bridging a number of microcracks by part cleavage and part plastic deformation as the main fracture front advances through them.

17.6 VARIABLES AFFECTING THE FRACTURE-MODE TRANSITION

Examination of experimental results, such as in Fig. 17.13, indicates that there are three regimes of fracture in uniform tension. At room temperature and above, fracture in a slow tension test is entirely ductile after a reduction of area of the specimen of the order of 50%. In region A of the figure, which is the upper end

558 TRANSITIONAL MODES OF FRACTURE 17.6

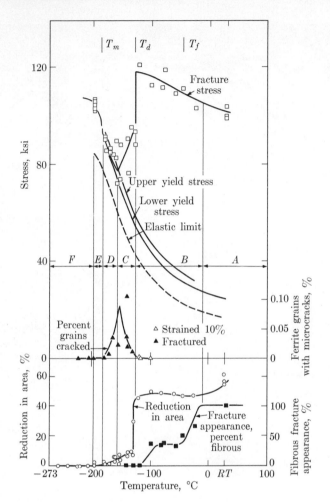

Fig. 17.13. Summary of tensile properties, fracture appearance, and microcrack data for coarse-grained ($d = 0.10$ mm) project steel E. (Hahn et al., 1959. Courtesy of M.I.T. Press.)

of regime I, the flow stress is low so that the stress concentrations around pile-ups can always be dissipated by slip in adjacent grains. In region B of the first regime, a large rise in the lower yield stress, and hence in σ_0, occurs. Fracture in this region is still preceded by a large reduction in area, and starts out as ductile fracture; it does, however, tend to become cleavage fracture as the fracture propagates. Toward the lower-temperature boundary of region B, dormant microcracks can be detected on the surface and the fracture appearance becomes almost completely of a cleavage type everywhere but at its origin. At the low-temperature boundary of this regime, it appears that Eq. 17.9 begins to be satisfied.

The resulting microcracks are confined within grains, however, and cannot spread until the stress level is raised by strain-hardening to satisfy the propagation

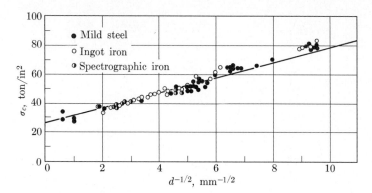

FIG. 17.14. Dependence of the brittle fracture stress on grain size at 77°K. (Petch, 1953. Courtesy of British Iron and Steel Institute.)

condition Eq. 17.10. This is the boundary of the first regime. In regime II, the yield strength continues its rapid rise with dropping temperature. Here, cracks are nucleated with increasing frequency, and the stress level needs to be raised only a small amount by strain-hardening for the cracks to propagate. A discontinuous drop in the reduction of area is observed. The temperature where this discontinuity in ductility occurs is called the *energy-transition temperature*. In this regime, which is made up of regions C, D, and E, the observed number of dormant microcracks first increases and then decreases as the apparent friction stress σ_0 rises. Toward the lower boundary of this regime, the conditions for crack nucleation and propagation begin to be satisfied simultaneously. This comes about from the rapid rise of the friction stress with dropping temperature. The applied stress σ_1 now becomes large enough so that the first microcrack can be propagated immediately. Equation 17.9 thus becomes not only a condition for fracture initiation but also for fracture propagation. Petch (1953) has performed experiments on mild steel, ingot iron, and spectroscopically pure iron at liquid nitrogen temperatures (77°K) and found that the brittle fracture strength obeys a relation of the type of Eq. 17.9, as shown in Fig. 17.14 (Prob. 17.4).

In regime III, which is shown in Fig. 17.13 by region F, microcracks can no longer be seen. Here fracture is very sudden and accompanied by only a minute amount of plastic deformation in the form of twinning. Cracks that lead to fracture appear to result from twin band intersections. In this regime, the macroscopic yield strength σ_0 is now above the stress necessary for twinning, and therefore slip and microcrack formation resulting from dislocation pile-ups become unobservable, i.e., Eq. 17.9 no longer governs.

As can be seen from Eqs. 17.9 and 17.10, in a coarse-grained specimen the boundary between regimes I and II (the energy-transition temperature) as well as that between II and III will be pushed up in the temperature scale, while they will be suppressed for fine-grained specimens.

In a study of the effect of grain size on the fracture stress, Low (1954) has observed (Fig. 17.15) that for grain sizes smaller than the critical grain size satisfy-

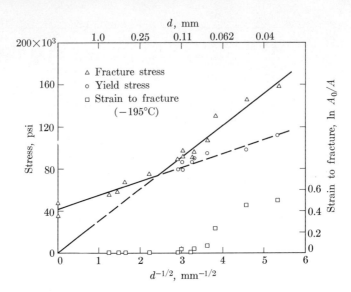

Fig. 17.15. Yield and fracture stresses at 77°K as a function of grain size for a low-carbon steel. Single crystal cleavage stresses are plotted at $d^{-1/2} = 0$. (Low, 1954. Courtesy of ASM.)

ing both Eqs. 17.9 and 17.10, fracture is always preceded by some plastic deformation. For grain sizes larger than the critical size, fracture takes place when Eq. 17.9 is satisfied, without any macroscopic plastic deformation, although presumably some microscopic plastic deformation takes place in isolated grains. It is to be noted that in this region, Eq. 17.10 is always satisfied before Eq. 17.9, and fracture is contingent on the formation of a microcrack according to Eq. 17.9.

An increase in the strain rate is known to raise the friction stress. This should therefore produce the same behavior as a lower temperature, that is, it should raise the transition temperature. The amount of rise is of the order of 15°C for a factor of 10 rise in the strain rate.

The presence of large triaxial stresses will lower q and make slip possible only at higher stresses σ_1, i.e., make simultaneous satisfaction of Eqs. 17.9 and 17.10 possible; hence raise the transition temperature. The last two factors explain the transition in fracture appearance in region B of Fig. 17.13. As the ductile fracture progresses slowly with accompanying formation of extensive plastic zones, it may produce a stress state with a high triaxiality, which raises the local transition temperature and makes cleavage fracture possible. Once such rapid cleavage fracture starts, the high strain rate of the limited plastic deformation at the traveling crack root is sufficient to maintain a cleavage-type fracture propagation. If such fracture propagation should for some reason stop, it cannot be made to start in a brittle fashion again until further extensive plastic deformation and some ductile fracture raises the triaxiality to a required high level, as was observed by Felbeck and Orowan (1955) and is illustrated in Fig. 17.16. The strong effect of strain rate on the transition temperature can be easily appreciated by comparing

Fig. 17.16. Fracture surfaces of ship plate showing stepwise propagation of a crack. (Felbeck and Orowan, 1955. Courtesy of AWS.)

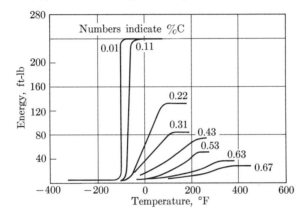

Fig. 17.17. Effect of carbon content on the shape of the fracture energy-transition temperature curve. (Rinebolt and Harris, 1951. Courtesy of ASM.)

the transition temperature in a slow tension test such as that of Fig. 17.13 with the transition temperature of a similar steel in a notched Charpy impact test, as that of Fig. 17.17.

The observations of Knott and Cottrell (1963) on notched Charpy specimens in slow bending are in substantial agreement with those of Hahn, Averbach, Owen, and Cohen. Knott and Cottrell showed that fracture is preceded by some plastic deformation at all temperatures at the root of the notch. While this plastic deformation is in the form of twinning and is almost immeasurably small at 100°K, it is 100% at room temperature and spreads across the whole narrow region of the specimen. Figure 17.18 shows how an initially ductile fracture transforms into a cleavage fracture in a notched bending specimen.

The effect of impurities, especially of interstitials such as carbon, on the transition temperature is striking. It has been demonstrated that zone-refined iron can undergo as much as 90% reduction in area at liquid helium temperatures when tested in slow tension (Smith and Rutherford, 1957). The very significant rise of transition temperature in the notched Charpy test with increasing carbon content

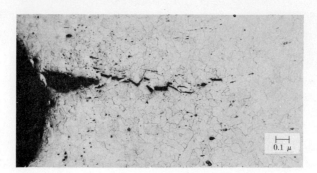

Fig. 17.18. Change from ductile to cleavage fracture at 290°K. (Knott and Cottrell, 1963. Courtesy of British Iron and Steel Institute.)

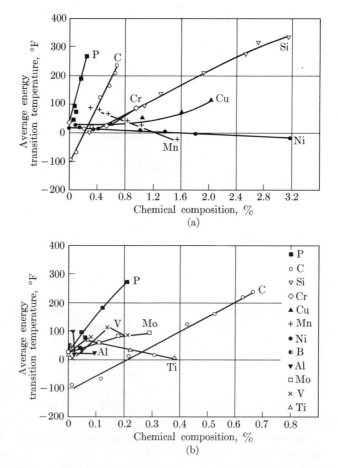

Fig. 17.19. Effect of chemical composition on average energy transition temperature. (Rinebolt and Harris, 1951. Courtesy of ASM.) (See note in Prob. 17.5.)

is shown in Fig. 17.19. It is fairly evident that the effect must be chiefly due to hardening resulting in the immobilization of possible slip sources which could potentially relieve the stress concentrations of isolated dislocation pile-ups. Other impurities that are not very effective in producing upper yield points, but nevertheless result in some degree of solution-hardening, can have similar but less striking effects than carbon, as is shown in Fig. 17.19. Only manganese and titanium have a beneficial effect.

Large doses of high-energy radiation such as neutrons can cause a rise in the friction stress σ_0 and therefore raise the transition temperature. Wilson and Billington (1955) have observed an 80°C rise in the transition temperature of pressure-vessel steel when exposed to a reactor neutron dose of $2.5 \times 10^{19}/\text{cm}^2$. Other less obvious effects are also present. The effective friction stress σ_0 as well as the specific fracture work α' are strong functions of plastic strain; i.e., if Eqs. 17.9 and 17.10 cannot be satisfied at the start of plastic deformation, they will become progressively more difficult to satisfy. This is because plastic deformation with its accompanying lattice bending and rotations upsets cleavage planes and sharply increases the necessary cleavage work. This so-called *strain-strengthening* effect was demonstrated by Gilman (1960), who measured the specific fracture work for a variety of undeformed as well as deformed crystals, and found very significant increases in the latter. This is also the reason why prior plastic deformation with enough superimposed hydrostatic compression lowers the transition temperature. In notched impact tests, the effect of prior plastic deformation, however, rather often results in a rise in transition temperature. The reason for this is evidently due to the rise in the flow stress through strain-hardening, compounded by the high triaxial stress state. Williams and Hughes (1961), however, found a decrease in transition temperature with cold work (cf. also Section 17.7).

These effects are tabulated for easy reference in Table 17.1. They can be further summarized by stating that any influence that raises the flow stress without too great an increase in α' will raise the transition temperature.

TABLE 17.1

Factors Influencing the Fracture-Energy Transition

High strain rate	Raises the flow stress	Raises T_R
High concentration of interstitial impurities	Raises the flow stress	Raises T_R
Large triaxial stress components	Requires the application of large normal stresses	Raises T_R
Strain aging	Raises yield point	Raises T_R
Prior plastic deformation	Causes strain strengthening (raised α')*	Lowers T_R
	Raises flow stress*	Raises T_R
Radiation damage	Raises flow stress	Raises T_R

* These two effects of prior plastic deformation are of comparable magnitude and are often responsible for erratic results.

In engineering practice it is naturally desirable to keep the transition temperature as low as possible. For this reason it is necessary to avoid any sharp corners or notches which can, in unavoidable periods of overstraining, increase the triaxiality and the strain rate, and hence could locally raise the transition temperature to dangerous levels. Mild steels with a lower transition temperature may be preferable to stronger higher-carbon steels with, however, higher transition temperatures. When even a low-carbon steel cannot properly fill the low-temperature application, a higher-manganese steel or even stainless steel may become necessary. The latter, having a face-centered cubic structure, does not exhibit a fracture-mode transition. Processes of plastic-working with an inherently high component of hydrostatic compressive stress, such as rolling and drawing, raise the specific fracture work α' and reduce the transition temperature, and are therefore beneficial; but the raised flow stress makes them more notch-sensitive.

Whenever possible, a small grain size is desirable, since it serves to limit the size of a microcrack that may be formed by plastic deformation.

Often, in welded structures, it has been found that although the virgin material used in manufacturing the structure had a comfortably low transition temperature, the structure itself failed under normal operating conditions without exhibiting any appreciable ductility. In such cases, the origin of the fracture has been traced to vicinities of welds. It has been demonstrated by Rockey et al. (1962) that the thermomechanical conditions around a weld, namely high temperature and thermal stresses, cause a structural embrittlement which is most undesirable. It is therefore necessary to perform the materials evaluation tests on samples that have undergone treatments characteristic of the manufacturing processes.

The materials which undergo a fracture transition are those in which the yield stress rises sharply with decreasing temperature and possess a relatively easy cleavage type fracture mode which, once started, can propagate with very low accompanying plastic work. Face-centered cubic metals generally have a low temperature sensitivity of the yield stress and do not undergo a fracture transition.

17.7 THE DAVIDENKOV DIAGRAM

The behavior shown in Fig. 17.13 has been idealized by Davidenkov (1936) in a very convenient diagram which bears his name, and is shown in Fig. 17.20. The variation of the initial yield strength with temperature is plotted against absolute temperature as the curve Y. The line B represents the brittle strength of the steel at low temperatures in the absence of general yielding. The preceding discussion of crack nucleation requires that a small amount of local plastic deformation occurs for the establishment of a brittle strength B even in the absence of general yielding. These two curves (Y and B) intersect at a temperature $T_R^{(1)}$. Below $T_R^{(1)}$ the specimen will reach the brittle strength before it reaches the stress for general yielding, and hence will undergo brittle fracture. Above $T_R^{(1)}$ it will yield first, and will therefore undergo more or less plastic deformation. As was mentioned earlier, an increased strain rate, a large triaxial tension component, impurities,

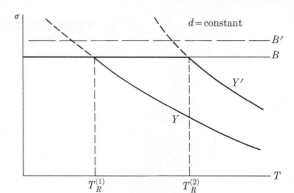

FIG. 17.20. The Davidenkov diagram showing a rise in transition temperature due to a rise in the flow stress Y.

and radiation damage all raise the yield stress. This effect can be represented by the raised curve Y' which may amount to as much as a threefold increase of Y due to maximum triaxiality in an infinitely sharp crack. The intersection between the curves B and Y' is at a higher temperature $T_R^{(2)}$; i.e., the fracture transition temperature has been raised from $T_R^{(1)}$ to $T_R^{(2)}$. Prior plastic deformation at moderate temperatures, on the other hand, raises the fracture work α' and therefore the brittle strength from B to B', thereby reducing the transition temperature for typical changes in Y. A decrease in grain size will raise the brittle strength.

17.8 FRACTURE MARKINGS

The surfaces of brittle fracture in crystalline materials like iron and ionic crystals show a number of characteristic markings in addition to those discussed in brittle fracture of glassy materials. There are still rib marks and hackle marks (which, on steel are better known as *chevron* markings); they identify the origin of fracture. These are more pronounced on fine-grained polycrystalline materials, where the effects of the crystalline nature of fracture are overshadowed. In Fig. 17.21, for instance, the origin of a brittle fracture in type 4340 steel is shown, while

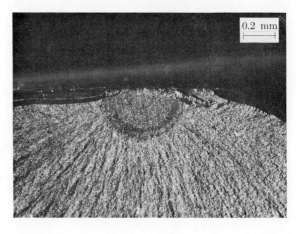

FIG. 17.21. Photomicrograph of a fracture origin in 4340 steel. (Courtesy of McMillan and Pelloux, Boeing Scientific Research Laboratory.)

Fig. 17.22. Chevron markings on the fracture surface of a steel plate, slightly reduced. (Shank, 1954. Courtesy of Welding Research Council.)

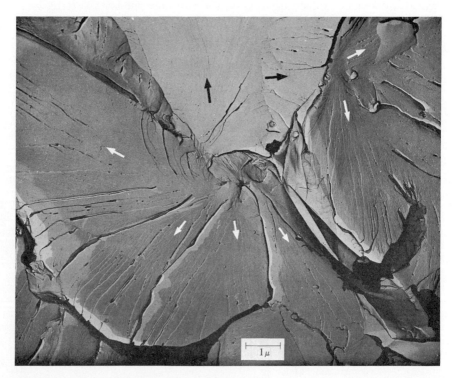

Fig. 17.23. River markings on the cleavage surface of polycrystalline iron fractured at 77°K. (Low, 1959. Courtesy of M.I.T. Press.)

FIG. 17.24. Cleavage steps on a cube plane of a KCl crystal at 400°C.

Fig. 17.22 shows the brittle fracture surface in a steel plate, with well-developed chevron marks pointing toward the origin. Examination of individual grains, or more conveniently, fracture surfaces of single crystals, reveals a set of hackle-like marks which are cleavage steps running normal to the crack front. These steps are formed when different parts of the crack front running on parallel cleavage planes are joined. When the crack front encounters low-angle twist boundaries or when a twisting moment is present in addition to a tension across the plane of the crack, a new set of cleavage steps is formed to accommodate the twist of the cleavage plane. Usually such cleavage steps coalesce, giving a single large step. At other times, smaller cleavage steps of opposite sign will annihilate each other. These markings are sometimes called *river marks*, since tributaries join up in the direction of the advancing crack front. These markings also aid in determining the origin of fracture. Figure 17.23 shows river markings on the fracture surface of polycrystalline iron, while Fig. 17.24 shows the cleavage steps on a cleaved surface of a KCl crystal.

17.9 COMMON TESTS FOR THE ENGINEERING EVALUATION OF BRITTLE-FRACTURE TENDENCIES IN METALS

From the engineering standpoint it often becomes necessary to evaluate the sensitivity to brittle fracture of various materials. Certainly the tension test is not directly suitable for such purposes. The introduction of a sharp circumferential

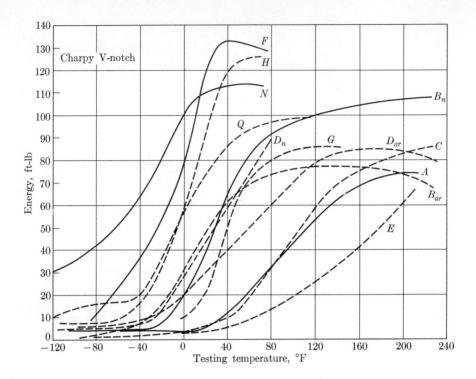

FIG. 17.25. Energy-temperature curves obtained by Charpy V-notch tests of twelve steels. (Broodberg et al., 1948. Courtesy of AWS.)

notch in a tension specimen, however, gives high local strain rates and introduces a triaxial stress component, both of which enhance brittle-fracture tendencies.

The most commonly used test for the investigation of brittle fracture of normally ductile materials is the impact test. Here a massive pendulum is raised to a standard height and is subsequently released to strike a standard prismatic specimen ($1 \times 1 \times 5$ cm) either as a cantilever beam (Izod), or under three-point bending (Charpy). These specimens may in addition have standard notches on their tension faces to increase the strain rate and triaxiality. In another test, the released pendulum may strike a circumferentially notched cylindrical specimen axially to load it in tension (tensile impact). In all these cases, the energy absorption during fracture is measured by the loss of potential energy of the pendulum. Such impact tests have been carried out at different temperatures for various materials. The resulting curve of energy absorption as a function of temperature shows clearly the temperature where transition in fracture mode occurs. The effects of various factors on the transition temperature can be found conveniently in this fashion. Figure 17.25 gives the impact energy-temperature curves for various low-carbon steels in a Charpy V-notch impact test. Table 17.2 gives the composition and tensile data. For some materials the energy absorption at the fracture-energy transition temperature may be very high. For certain applications, 15 ft-lb of energy absorption in the Charpy impact experiment has been considered sufficient. This

TABLE 17.2

Chemical Composition and Tensile Data for Steels of Fig. 17.25

Steel*	C	Mn	Si	P	S	Ni	Al	Cu	Cr	Mo	Sn	N
A	0.26	0.50	0.03	0.012	0.039	0.20	0.012	0.03	0.03	0.006	0.003	0.004
B	0.18	0.73	0.07	0.008	0.030	0.05	0.015	0.07	0.03	0.006	0.012	0.005
C	0.24	0.48	0.05	0.012	0.026	0.02	0.016	0.03	0.03	0.005	0.003	0.009
D	0.22	0.55	0.21	0.013	0.024	0.16	0.020	0.22	0.12	0.022	0.023	0.005
E	0.20	0.33	0.01	0.013	0.020	0.15	0.009	0.18	0.09	0.018	0.024	0.005
F	0.18	0.82	0.15	0.012	0.031	0.04	0.054	0.05	0.03	0.008	0.021	0.006
G	0.20	0.86	0.19	0.020	0.020	0.08	0.045	0.15	0.04	0.018	0.012	0.006
H	0.18	0.76	0.16	0.012	0.019	0.05	0.053	0.09	0.04	0.006	0.004	0.004
N	0.17	0.53	0.25	0.011	0.020	3.39	0.077	0.19	0.06	0.025	0.017	0.005
Q	0.22	1.13	0.05	0.011	0.030	0.05	0.008	0.13	0.03	0.006	0.018	0.006

Steel	Condition	Yield point, psi	Tensile strength, psi	Elongation in 2 in., %	Elongation in 8 in., %	Reduction of area, %	Rockwell B hardness
A	As-rolled	36,000	59,000	41	34	58	60
B_{ar}	As-rolled	33,000	57,500	44	34	64	61
B_n	Normalized	36,000	57,000	44	34	63	60
C	As-rolled	36,000	65,000	39	30	56	67
D_{ar}	As-rolled	37,500	65,000	..	30	54	..
D_n	Normalized	35,000	60,000	..	32	59	..
E_{ar}	As-rolled	30,000	57,000	..	32	56	..
E_n	Normalized	35,000	57,500	..	31	56	..
F	As-rolled	34,000	61,000	..	31	62	..
G	As-rolled	41,500	70,000	..	28	56	..
H	As-rolled	36,000	63,500	42	30	63	70
N	As-rolled	58,000	80,000	35	26	65	84
Q	Quenched and tempered	46,000	72,000	45	23	62	81

* Steels A, B, and C, semikilled; D, F, G, and H, fully deoxidized; E, rimmed.

would permit the use of a material somewhat below its energy transition temperature. As mentioned earlier, the fracture after some amount of plastic deformation can be ductile, fibrous, or of cleavage type. A cleavage fracture generally attains high velocity and dissipates less energy in the fracture process than the ductile or fibrous modes which, as mentioned in Section 16.3, can be rupture on a fine scale. Thus, although a large amount of plastic work may precede the initiation of a cleavage as well as of a ductile mode of crack propagation, in inhomogeneously stressed structures of incipient cleavage type the crack lengthens and extends rapidly into previously undeformed material. For this reason, the appearance of the fracture surface is often studied and the percentage of ductile fracture in the fractured Charpy specimen is plotted against the test temperature. This gives a fracture-appearance transition temperature which is normally about 100°C higher than the energy transition temperature in the notched and unnotched tension (cf. T_f versus T_d in Fig. 17.13).

Often a quick test of the material properties is desired. In such cases a prismatic bar can be notched with a saw, clamped in a vise, and hit with a hammer. The shape of the impacted specimen as well as the appearance of the fracture surface can then tell a great deal about the tendency of the material to brittle fracture.

To study the effect of strain rate alone on brittle fracture, it is necessary to start the fracture in a brittle manner without going through large amounts of preparatory plastic deformation. This can be achieved in steel specimens by forming a hard surface layer by nitriding. The fracture will then start in this layer and gather enough velocity to go through the soft matrix in a brittle manner (Newhouse and Wundt, 1960).

17.10 SUMMARY

Some materials that undergo ductile fracture may, under conditions of temperature and loading which raise the stress, show an abrupt fracture-mode transition to a relatively brittle fracture. Most glassy materials and some engineering metals fall into this category, which includes most body-centered cubic metals, but not face-centered cubic metals.

In glassy materials the fracture-mode transition is governed by the very rapid rise in viscosity with decreasing temperature, so material that could deform freely at elevated temperatures becomes susceptible to the brittle propagation of cracklike defects as it changes its character from an essentially viscous material to an elastic one.

In body-centered cubic metals, which show a large rise in yield strength with dropping temperature and increased strain rates, brittle propagation of cracks along {100} planes becomes possible. In such materials, defects may also be formed in the process of plastic deformation. A pile-up of edge dislocations at a strong obstacle such as a grain boundary can produce very high normal stresses at the obstacle. If such normal stresses cannot be relieved by local plastic deformation, a crack can arise when the normal stresses reach the ideal strength and enough strain energy can be released by the crack to provide for the energy of the crack surfaces. In a polycrystal where the grain diameter d limits the size of dislocation pile-ups, the condition for the applied tensile stress σ_1 for nucleating a crack by this mechanism is

$$q\sigma_1 = \sigma_0 + K\,d^{-1/2}, \qquad (17.9)$$

where q is a factor associated with triaxiality, σ_0 is a tensile friction stress, and K is a material constant of magnitude 10^8 dyne·cm$^{-3/2}$. In addition to the dislocation pile-up, which can manifest itself in various forms, stress associated with the intersection of deformation twins can also give rise to crack nuclei.

Once microcracks span a grain, they will propagate in a brittle manner when a modified Griffith equation

$$\sigma_1 = (E\alpha'/d)^{1/2} \qquad (17.10)$$

is satisfied. Here α' is the specific fracture work, which includes the work to join cleavage cracks in neighboring grains, and, as can be determined from Table 16.7, is several orders of magnitude larger than α, the actual surface energy.

At high temperatures the flow stress is low. The stress concentrations around pile-ups can always be relieved by local plastic deformation, and there is no brittle fracture. At low temperatures the yield stress for slip will be high so that deformation will start in the form of twinning, giving rise to crack production by twin intersections and subsequent brittle fracture. At an intermediate range of temperature where the processes of crack production and accommodation by local slip compete, a transition from brittle to ductile behavior will occur. The median temperature of this narrow range is called the transition temperature. An increased strain rate, solution-hardening, work-hardening (for notched specimens), or radiation-hardening will raise the flow stress and thereby raise the transition temperature. A small grain size limits the size of defects and therefore suppresses the transition temperature. Under favorable circumstances the lattice distortions due to plastic strain can be beneficial in deflecting and blunting cracks and making their propagation more difficult.

Markings on brittle fracture surfaces make it possible to identify the origin of the fracture. These markings are called *chevron markings* and *river markings;* of these the former bears a certain similarity to hackle marks on glass. In body-centered cubic crystals the fracture is cleavage on a {100} plane in each grain, leaving a shiny, faceted appearance.

The susceptibility to brittle fracture of engineering materials is generally tested by creating extreme conditions of strain rate and triaxiality in the form of a notched impact test. A variation of the test temperature then produces a transition temperature which is for nearly all purposes conservatively high enough to be used in design. A less conservative criterion for design is to require only 15 ft-lb of energy absorption in the Charpy impact experiment. The temperature where the energy absorption is 15 ft-lb is normally below the energy transition temperature. In some cases, such as welding, the metallurgical microstructure can be altered radically, changing the properties of the materials which may have behaved well in a notched impact test. It is therefore necessary to perform tests on samples given typical manufacturing treatments.

REFERENCES

For a general treatment of brittle behavior of engineering structures, Parker (1957) is recommended. For reviews of metallurgical and microstructural aspects of transitional fracture, Petch (1954) and Low (1963), and for a theory of fracture of metals, Stroh (1957), should be consulted. International conferences on fracture are held with a frequency of about one in three years; for these see Averbach et al. (1959) and Drucker and Gilman (1963).

Adcock, F. Bristow, C. A.	1936	"Iron of High Purity," *Proc. Roy. Soc. (London)* **A153,** 172–200.
Averbach, B. L. Felbeck, D. K. Hahn, G. T. Thomas, D. A., Eds.	1959	*Fracture, Proc. Int. Conf., Swampscott*, M.I.T. Press and J. Wiley and Sons, New York.
Broodberg, A. Davis, H. E. Parker, E. R. Troxell, G. E.	1948	"Causes of Cleavage Fracture in Ship Plate—Tests of Wide Notched Plates," *Welding J. Res. Suppl.* **27,** 186–199.
Carrington, W. Hale, F. K. McLean, D.	1960	"Arrangement of Dislocations in Iron," *Proc. Roy. Soc. (London)* **A259,** 203–227.
Cottrell, A. H.	1958	"Theory of Brittle Fracture in Steel and Similar Metals," *Trans. AIME* **212,** 192–203.
Davidenkov, N. N.	1936	*Dinamicheskaya Ispytania Metallov*, Moscow.
Deruyttere, A. Greenough, G. B.	1956	"The Criterion for the Cleavage Fracture of Zinc Single Crystals," *J. Inst. Met.* **84,** 337–345.
Drucker, D. C. Gilman, J. J., Eds.	1963	"Fracture of Solids," *Met. Soc. AIME Conf. 20*, Interscience, New York.
Eshelby, J. D. Frank, F. C. Nabarro, F. R. N.	1951	"The Equilibrium of Linear Arrays of Dislocations," *Phil. Mag.* **42,** 351–364.
Felbeck, D. K. Orowan, E.	1955	"Experiments on Brittle Fracture of Steel Plates," *Welding J. Res. Suppl.* **34,** 570s–575s.
Gilman, J. J.	1954	"Mechanism of Ortho Kink Band Formation in Compressed Zinc Monocrystals," *J. Metals* **6,** 621–629.
Gilman, J. J.	1960	"Direct Measurement of the Surface Energies of Crystals," *J. Appl. Phys.* **31,** 2208–2218.
Hahn, G. T. Averbach, B. L. Owen, W. S. Cohen, M.	1959	"Initiation of Cleavage Microcracks in Polycrystalline Iron and Steel," *Fracture*, Averbach et al. Eds. M.I.T. Press, Cambridge, Mass., and Wiley, New York, pp. 91–116.
Hull, D.	1960	"Twinning and Fracture of Single Crystals of 3% Silicon Iron," *Acta Met.* **8,** 11–18.

JOHNSTON, T. L. STOKES, R. J. LI, C. H.	1962	"Crack Nucleation in Magnesium Oxide Bi-Crystals Under Compression," *Phil. Mag.* **7,** 23–24.
KNOTT, J. F. COTTRELL, A. H.	1963	"Notch Brittleness in Mild Steel," *J. Iron and Steel Inst.* **201,** 249–260.
LOW, J. R.	1954	"The Relation of Microstructure to Brittle Fracture," *Relation of Properties to Microstructure*, ASM, Novelty, Ohio, pp. 163–179.
LOW, J. R.	1959	"Review of the Microstructural Aspects of Cleavage Fracture," *Fracture*, B. L. Averbach et al., eds., M.I.T. Press, Cambridge, Mass., and Wiley, New York, pp. 68–90.
LOW, J. R.	1963	"The Fracture of Metals," *Progress in Material Science* **12,** Chalmers and King, Eds., Pergamon Press, New York.
MCCLINTOCK, F. A.	1960	"Discussion of 'Physical Nature of Plastic Flow and Fracture,' by J. J. Gilman," *Plasticity: Proc. 2nd Symp. Naval Struct. Mech.* E. H. Lee and P. S. Symands, eds., Pergamon Press, London, pp. 43–99.
MCMAHON, C.	1963	*Micromechanisms of Cleavage Fracture in Polycrystalline Iron*, Sc.D. Thesis, M.I.T., Cambridge, Mass.
MCMILLAN, C. PELLOUX, R. M. N.	1962	"The Analysis of Fracture Surfaces by Electron Microscopy," *Boeing Sci. Res. Labs. Rpt.* D1-82-0169.
NEWHOUSE, D. L. WUNDT, B. M.	1960	"A New Fracture Test for Alloy Steels," *Metal Prog.* **77,** 81–83.
OROWAN, E.	1945	"Notch Brittleness and the Strength of Metals," *Trans. Inst. Engrs. Shipbuilders Scotland* **89,** 165–215.
OROWAN, E.	1948	"Fracture and Strength of Solids," *Rept. Prog. Phys.* **12,** 185–232.
PARKER, E. R.	1957	*Brittle Behavior of Engineering Structures*, J. Wiley & Sons, New York.
PETCH, N. J.	1953	"The Cleavage Strength of Polycrystals," *J. Iron and Steel Inst.* **174,** 25–28.
PETCH, N. J.	1954	"The Fracture of Metals," *Progress in Metal Physics* **5,** Chalmers and King, Eds., 1–52, Pergamon Press, New York.
RINEBOLT, J. A. HARRIS, W. J. JR.	1951	"Effect of Alloying Elements on Notch Toughness of Pearlitic Steels," *Trans. ASM* **43,** 1175–1214.
ROCKEY, K. C. LUDLEY, J. H. MYLONAS, C.	1962	"Exhaustion of Extensional Ductility Determined by Reversed Bending of Five Steels," *Proc. ASTM* **62,** 1120–1133.
SHANK, M. E.	1954	"A Critical Survey of Brittle Failure in Carbon Plate Steel Structures Other than Ships," *Welding Res. Council Bulletin Series*, No. 17, New York.
SMITH, R. L. RUTHERFORD, J. L.	1957	"Further Considerations on the Ductility of Iron at 4.2°K," *Acta Met.* **5,** 761–762.

Stroh, A. N.	1954	"The Formation of Cracks as a Result of Plastic Flow," *Proc. Roy. Soc. (London)* **A223**, 404–414.
Stroh, A. N.	1955	"The Formation of Cracks in Plastic Flow II," *Proc. Roy. Soc. (London)* **A232**, 548–560.
Stroh, A. N.	1957	"A Theory of the Fracture of Metals," *Advances in Physics* **6**, 418–465.
Stroh, A. N.	1958	"Cleavage of Metal Single Crystals," *Phil. Mag.* **3**, 597–606.
Stroh, A. N.	1959	"Crack Nucleation in Body Centered Cubic Metals," *Fracture*, B. L. Averbach et al., eds., M.I.T. Press, Cambridge, Mass., and Wiley, New York, pp. 117–122.
Tipper, C. F.	1945	*Report of Conference on Brittle Fracture in Steel Plates* **24**, British Iron and Steel Research Association, Cambridge, England.
Williams, T. R. G. Hughes, D. H.	1961	"The Effect of Plastic Deformation and Strain Aging on the Transition Temperature of Mild Steel," *Metallurgia* **63**, 233–237.
Wilson, J. C. Billington, D. S.	1955	"Effect of Nuclear Radiation on Structural Materials," *Problems in Nuclear Engineering*, Pergamon Press, London, pp. 97–106.
Zener, C.	1949	"Micro Mechanism of Fracture," *Fracturing of Metals*, ASM, Novelty, Ohio, pp. 3–31.

PROBLEMS

◀17.1. Following the arguments preceding Eq. 17.7, derive the relation between the net shear stress and the number of dislocations inside a grain of diameter d.

17.2. The cracking of a carbide particle as shown in Fig. 17.7 may be attributable to the tensile stresses generated by a shear drag of a plastically deforming matrix as shown in Fig. 17.26. If the matrix has a critical macroscopic shear strength k and the platelet is considered completely rigid, calculate the length to width ratio of the platelet to give a critical tensile stress σ_c at its center. Substituting reasonable values for the strengths k and σ_c, calculate the ratio l/w. Comparing this value with the photograph of Fig. 17.7 speculate about the likelihood of this mechanism.

17.3. Calculate the shear strain inside a twin band in a body-centered cubic lattice from Fig. 1.20.

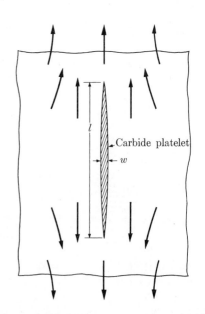

Figure 17.26

17.4. For the steel of Fig. 17.14 determine the friction stress, and by comparing the slope of the line with the factor K in Eq. 17.9, calculate the value α. Does this have a reasonable magnitude?

17.5. A gasoline storage vessel is to be used in a climate where the temperature may vary between $-30°F$ and $100°F$. The vessel is to be fabricated by welding. Two types of steel are being considered (a) An 1018 steel with the following composition: 0.18% C, 0.68% Mn, 0.02% P, 0.01% S, 0.08% Si; (b) A type M high-manganese steel of the following composition: 0.16% C, 1.30% Mn, 0.10% P, 0.024% S, 0.024% Si. With the information in Fig. 17.19, decide whether or not either of these steels is suitable. Discuss the risks involved and what inspection processes and additional tests should be conducted. [*Note:* The base composition of the steel used for Fig. 17.19 is 0.30% C, 0.30% Si, 1.00% Mn, balance Fe, and its initial transition temperature for V-notch Charpy was 25°F.]

CHAPTER 18

FATIGUE

18.1 SYNOPSIS

Fracture under repeated application of stress or strain is called fatigue. It first became serious in early railway equipment, and has been studied in the laboratory for over a hundred years. Since fatigue is essentially the formation and growth of cracks, we shall look at recent studies of the basic mechanisms by which these two stages can occur. With these in mind, we turn to the effects of stress or strain amplitude, mean stress, combined stress, various stress histories, and the speed of testing. The nature of the material affects the fatigue life, first of all through the general level of hardness, but also through the metallurgical structure, the level and distribution of impurities, and the surface condition. A third class of variables is associated with the environment: its temperature, humidity, or special composition.

As an example of the interaction of these various effects, we consider rolling contact fatigue like that encountered in ball bearings. From this we turn to a more general discussion of engineering considerations, indicating how fatigue failures can be prevented in design or controlled in service.

18.2 THE HISTORY AND METHODS OF STUDY OF FATIGUE

Man has known since ancient times that he could break wood or metal by repeated bending back and forth if the amplitude is large enough. It was a surprise, however, that fracture occurred even after the application of stresses within the apparent elastic range. This discovery came in the mid-nineteenth century when fatigue failures of railway axles in Europe became widespread, probably because only then were parts subjected to the tens or hundreds of thousands of cycles required for fatigue failure without general yielding. As is usually the case with an unexplained service failure, the first step was to reproduce it in the laboratory. Wöhler in Germany and Fairbairn in England made extensive studies as early as 1860. Typical data from Wöhler's early work, as well as more recent investigations, are shown in Figs. 18.1 and 18.2. For steels and titanium there exists a safe stress amplitude, called the *fatigue limit* (or endurance limit), below which no fracture will occur regardless of the number of stress reversals. For most metals, no such limit exists. The entire diagram is commonly called an S-N curve.

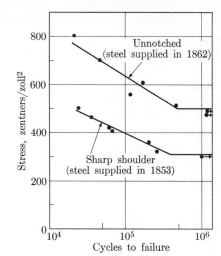

Fig. 18.1. Wöhler's S-N curves for Krupp axle steel. (From Peterson, 1963.) [Note: 1 zentner = 50 kg; 1 zoll = 1 in.; 1 zentner/zoll² = 110 psi.]

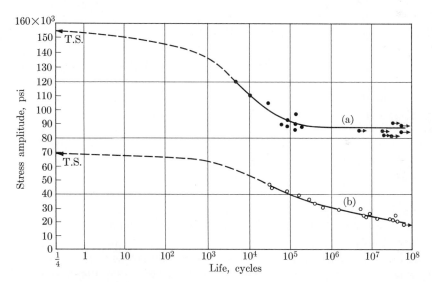

Fig. 18.2. Typical fatigue data in rotating bending. (a) Titanium alloy RC 130B, 155,000 psi T.S. (Sinclair et al., 1957.) (b) Aluminum alloy 2024-T4, 68,000 psi T.S. (MacGregor and Grossman, 1952.)

The first important microscopic observations on the process of fatigue date back to the investigation of Ewing and Humphrey (1903) on the surfaces of wrought-iron specimens, fatigued in rotating bending. These observations disclosed that at stress amplitudes below the yield point, slip bands appear in some grains within the first thousand cycles. Continued testing adds only a few more slip bands to the previous ones, while some bands become progressively more intense. A characteristic feature which distinguishes such slip activity from monotonic straining is that it is confined to a few slip bands within a few grains.

As the slip bands become more intense, many of them develop fine cracks. These cracks gradually spread, often joining other cracks. A large crack then propagates rather rapidly, resulting in a fatigue fracture.

With one or two exceptions, particularly that of Gough (1928, 1933), not much fundamental work was done on fatigue for the next fifty years. Although vast amounts of experimental data were collected on fatigue behavior, very few people took time to put a specimen under a microscope and look at the process in any detail.

In recent years, however, there has been a renewed interest in microscopic observation and also in the use of x-ray and electron-micrographic techniques, which we shall now consider in connection with crack initiation.

18.3 MECHANISMS OF CRACK INITIATION

A series of microscopic observations on copper are shown in Fig. 18.3. As shown in (a), even after less than a tenth of the final life, electropolishing failed to leave a completely smooth surface. Further cycling redeveloped the slip markings (b), and even more was left after a second repolishing (c). This process is repeated, and in (f) it is clear that an actual crack has formed, since the neighboring areas were undeformed by the previous cycling. Similar slip markings develop below the fatigue limit, as shown in Fig. 18.4, indicating that whatever causes the fatigue limit, it is not the absence of slip. Another important feature of the fatigue process is its local nature. For these two reasons it has proven very difficult to determine by such methods as the measurement of damping, magnetic hysteresis, or x-ray line broadening, which determine average behavior over many slip bands, whether or not a particular part is being run above its fatigue limit. In other ductile materials a variety of kinds of subgrain structure have been observed (Forsyth and Stubbington, 1955). Detailed observations of slip bands occasionally show extrusions from one part and intrusions (depressions) at another (Fig. 18.5). Such intrusions and extrusions sometimes occur in pairs adjacent to each other. The intrusion is especially dangerous because it may act as the start of a surface crack which later propagates to cause failure. In aluminum alloys, rows of holes have been observed, as shown in Fig. 18.6.

Damage below the surface can be observed by sectioning. Wood (1959) and Wood et al. (1963) have used taper sectioning to observe the development of cracks in the surface, and also to observe apparent, although controversial, damage below the crack tips. Because of the effective magnification of the taper sectioning, the cracks appear far sharper in the section than they are in reality, and fail to indicate directly any connections of holes to the surface. In 7075-T6 aluminum, sectioning has revealed deformation bands similar to those for monotonic loading shown in Fig. 16.4 (McClintock, 1956).

◀ Use of the electron microscope in the study of fatigue requires either replication of the surface or sectioning to form a thin foil ($\sim$1000 angstroms thick). A series of electron-transmission micrographs for aluminum are given in Fig. 18.8. The

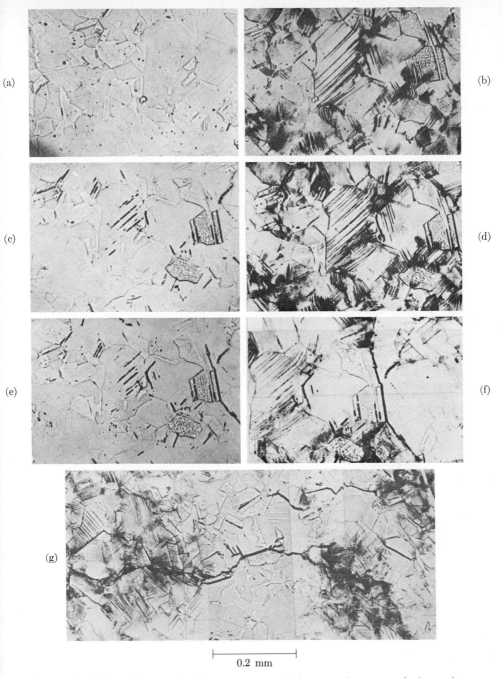

0.2 mm

FIG. 18.3. Photomicrographs of a pure copper fatigue specimen tested above the endurance limit. (a) Electropolished after 0.27×10^6 cycles. (b) Fatigued to 1.54×10^6 cycles. (c) As in (b) after electropolishing. (d) Fatigued to 2.8×10^6 cycles. (e) As in (d) after electropolishing. (f) Fatigued to fracture at 3.65×10^6 cycles. (g) Another field of the specimen after fracture. (Thompson, Wadsworth, and Louat; from Smith, 1957. Courtesy of the Royal Society.)

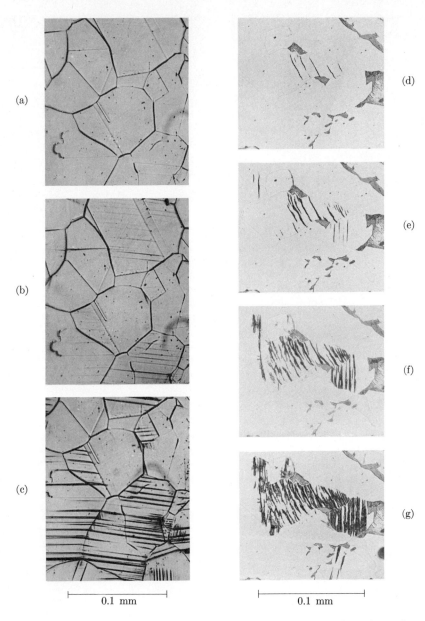

FIG. 18.4. Photomicrographs of fatigue specimens tested just below the endurance limit. (a) Pure nickel at 10,000 cycles. (b) Pure nickel at 50,000 cycles. (c) Pure nickel at 270,000 cycles. (d) Mild steel at 50,000 cycles (0.19% C). (e) Mild steel at 500,000 cycles. (f) Mild steel at 6.5×10^6 cycles. (g) Mild steel at 46×10^6 cycles. (Smith, 1957. Courtesy of the Royal Society; and Hempel, 1956. Courtesy of Instit. of Mech. Eng.)

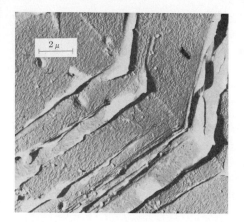 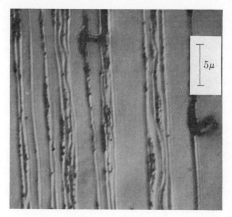

Fig. 18.5. Formvar-carbon replica, shadowed left to right so that intrusions appear as outcroppings which cast shadows. (Cottrell and Hull, 1957. Courtesy of the Royal Society.)

Fig. 18.6. Slip traces and holes revealed after electropolishing. Aluminum, 7.5% Zn, 2.5% Mg alloy. (Forsyth, 1959. Courtesy of the Royal Aircraft Establishment.)

results are both complex and varied. There is a tendency for subgrain formation as the deformation cycling proceeds, but as Feltner (1963) has pointed out, this also occurs to some extent on monotonic loading. There is little that can be associated with specific fatigue damage, indicating that the important phenomena are occurring near the surface and near the crack. It is difficult to obtain thin foil sections there because the chemical or electropolishing required for thinning eats away the edges of the specimens which contain the regions of interest. Although the subgrain formation of Fig. 18.8(g) was not reported by Segall (1963), who found loops (Fig. 18.8f) to persist, subgrains have been observed by x-ray diffraction techniques, where sizes of the order of 1 to $8\,\mu$ have been reported in aluminum and steel (Grosskreutz, 1962; Holden, 1961). In other metals, Segall reports patterns more like Fig. 18.8(e) or, alternatively, like an exaggerated form of the linear structure of Fig. 18.8(b). As in monotonic loading, the dislocation structures found in foil sections show relatively little connection with the slip bands seen on the surface. An exception is the work of Forsyth (1963) on thin films of aluminum alloys; he has found dislocation tangles, subgrains, and regions where the precipitate is depleted in bands such as those shown in Fig. 18.6.

◀The natural desire to provide a theoretical explanation for the development of fatigue cracks has led to several dislocation models (Probs. 18.1, 18.2, 18.3, 18.4), although these seem of necessity to be oversimplified. ▶

Whatever the specific dislocation mechanism, replica observations of steel by Hempel (1956) show cracking in regions of roughening within slip bands. Roughening also occurs on a macroscopic scale.

Before considering roughening, let us examine the stress-strain behavior in fatigue. The presence of persistent slip bands indicates that slip is continuing along a few bands instead of there being strain-hardening and more general slip as in monotonic loading. Wood and Davies (1953) have demonstrated this with the torque-twist curves of Fig. 18.7, in which the twisting moment is plotted versus the total accumulated plastic twist without regard for sign for a variety of amplitudes of reverse twisting. This showed that for small angles of reverse twisting, very little strain-hardening results. This is a direct consequence of the Bauschinger effect. If the data are extrapolated to the very small strain levels associated with a fatigue test, very little, if any, strain-hardening should be expected on the active slip planes. It was also noted that the total accumulated strain required for fracture increased greatly as the strain amplitude was decreased. From this it was inferred that for very small plastic strain amplitudes, it might be possible to accumulate plastic strain indefinitely without fracture. See also the results of Föppl, referred to in Section 14.3.

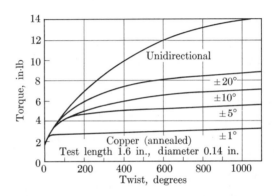

FIG. 18.7. Torque versus twist curves for different amplitudes of alternating torsion as indicated (Wood, 1956. Courtesy of Academic Press.)

◀ The question arises as to whether reversed plastic straining is stable. It appears that it is not, judging from the large strain experiments of Raymond and Coffin (1963), who observed progressive necking of aluminum specimens in alternating tension and compression. An analysis of this phenomenon for the simple one-dimensional case was carried out by McClintock (1963), who found that if a steady-state, stable, hysteresis loop developed for the stress-strain curve of the material, kinematically irreversible flow occurs, resulting in a localization of strain. This progressive roughening can be readily observed in low-cycle fatigue even in non-metals such as Plasticine and silicone putty (Prob. 18.5). Specifically, the increase in inequality of strain amplitude between two adjacent sections in series is a measure of the development of roughness. If the hysteresis loop can be described by two pairs of straight lines corresponding to a modulus of elasticity E and a strain-hardening per unit plastic strain h, and with a maximum stress σ, the logarithmic increase per cycle in the difference in strain amplitude between the two sections is given by

$$\frac{d(\ln \delta\epsilon)}{dn} = \frac{\sigma^2}{h^2 - \sigma^2}. \tag{18.1}$$

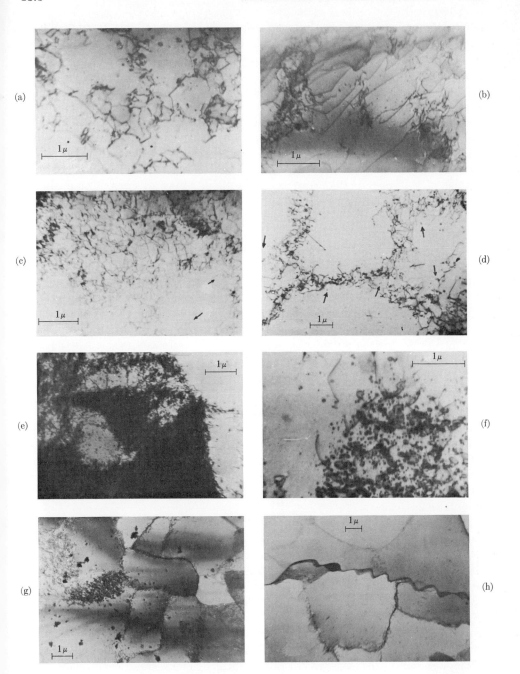

FIG. 18.8. Electron-transmission micrographs of pure aluminum foil after fatigue at a constant strain amplitude. (a, b) $\frac{1}{2}$ cycle; (c, d) 100 cycles; (e, f) 3×10^5 cycles; (g) 1.7×10^7 cycles; (h) 2×10^7 cycles. (Wilson and Forsyth, 1959. Courtesy of the Royal Aircraft Establishment.)

While this analysis deals with macroscopic phenomena, there is no reason to believe that similar effects will not occur on a microscopic scale, leading to progressive roughening in a slip band and ultimate crack formation, as actually observed by Hempel (1956) and others.

◀The presence of a steady-state hysteresis loop and the variation of fracture strain with strain amplitude necessitate modifications in a theory of the S-N curve and fatigue limit based on progressive hardening in plastic spots imbedded in an elastic material (Orowan, 1939), although it is still true, as this model indicates, that strain-hardening in the early stages of a test can raise the fatigue limit above the yield point, e.g., in annealed low-carbon steel or ingot iron (Prob. 18.6)▶

Most of the micromechanisms discussed above apply to crystals, but similar behavior is found in polymers, with a fatigue limit at perhaps 20 to 30% of the tensile strength. For a recent summary, see Nielsen (1962).

In summarizing the observations on the initiation of fatigue cracks, we note that they develop in local regions of continuing cyclic plastic flow. The Bauschinger effect which leads to this localization is present in all materials which fail by fatigue, but is accentuated by overaging and strain-softening. In crystals, the accompanying dislocation structure differs in degree but not in kind from that found in monotonic loading. Specific mechanisms of crack initiation, which may occur at the scale of dislocations, slip bands, or the continuum, are the formation of extrusions and intrusions, surface roughening, and hole growth. In any event, it is possible to impede crack initiation by forming alloys which are harder and more stable, especially at the surface.

18.4 MECHANISMS OF CRACK PROPAGATION

Microcracks form in slip bands at only a small fraction, say 1 to 30%, of the fatigue life, so the time required for propagation is of practical importance if cracking cannot be totally avoided. In principle, crack propagation is similar to initiation except that the processes are occurring at a folded surface rather than on a flat one.

The resulting strain concentration requires study of the same kinds of regions shown in Fig. 16.13—the stress distribution at long distances, a local elastic region in which the distribution can be characterized by the stress intensity factors, a plastic zone contained within these regions where the deformation becomes inhomogeneous because of grains or subgrains, and finally dislocations, atoms, and even electrons (in a corrosive environment). The relative importance of these regions may shift from case to case or as one crack grows, and perhaps a region will entirely disappear. Nevertheless, all will be important for a complete understanding of crack growth, particularly in a corrosive environment.

The recognition that the growth rate of fatigue cracks could be correlated with the elastic stress-intensity factors came from Paris, Gomez, and Anderson (1961).*

---

* Their work was published in a less well-known periodical after being rejected by others, but the word spread rapidly.

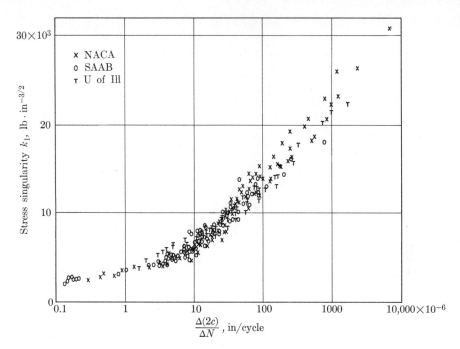

Fig. 18.9. Crack extension-rate data for 2024-T3 aluminum alloy. (Paris et al., 1961. Courtesy of the University of Washington.)

An example of the correlations is shown in Fig. 18.9, and an extensive set of data was given by Donaldson and Anderson (1961).

At the microscopic level, Laird and Smith (1962) recognized that crack propagation can occur by a kinematically irreversible opening and closing of the crack, with the surface folding into a new position at each cycle, much as in the case of crack initiation, but on a much more intensive scale, since the folding is located at the root of the crack. This process has not yet been analyzed, but evidences for it are the characteristic markings left by each cycle during the growth, as shown in Fig. 18.10. The curvature and remarkable regularity of these markings indicate that even at this scale, there is little influence of the subgrain or crystallographic structure. Continuum plasticity should be an aid to understanding in lieu of a solution entirely in terms of the many dislocations involved.

Another limiting form of crack propagation is the development of more intense subgrain structure, or, in the case of precipitation-hardened alloys, overaging and strain-softening. Damage, or even new cracks (Holden, 1961; Wood et al. 1963) occur ahead of the main crack. Here fracture depends on a given strain being acquired ahead of the crack in a region of size perhaps equal to that of a subgrain, in order for crack propagation to occur. For cracks growing under longitudinal shear, such a theory has been developed along the lines discussed in Section 16.7 for ductile fracture. When the plastic zone is large compared with the subgrain size but small compared with the crack length, the theory predicts the crack growth

Fig. 18.10. Early stage of fatigue fracture in Rene 41 plate. Nominal stress 45,000 psi. (Christensen and Denke, 1961. Courtesy U S Air Force.)

rate quantitatively in a way which varies as the square of the crack length and the fourth power of the applied stress (McClintock, 1963). If crack growth were entirely by kinematically irreversible shape change, dimensional analysis would show it to vary directly as the size of the plastic zone (directly as the crack length and as the square of the applied stress).

18.5 FATIGUE AS AN ENGINEERING PROBLEM

Fatigue fracture is the most unexpected service failure. Corrosion and wear probably are more frequent, but they are gradual and anticipated. Presently, as for strain-hardening and ductile fracture, there is no quantitative theory of fatigue based on fundamental principles. The theoretical models do, however, provide a guide to interpreting the great wealth of practical information available. Proper understanding and use of this information can prevent many of the failures which do now occur.

Even without a quantitative understanding of the process of fatigue, the engineer must design safe, reliable structures. He must therefore resort to testing, first of all the complex materials of engineering interest. This job has been done well enough so that most failures do not arise from a lack of knowledge of fatigue strength. Next he must test components, for the stress patterns in complicated parts are uncertain, and high local stresses may be present, either from the geometry of the part or from small cracks left from machining and forming operations.

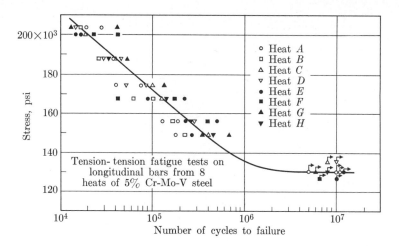

FIG. 18.11. Fatigue data on several heats of 5% Cr-Mo-V steel bars heat-treated to 240 to 265,000 psi tensile strength. (Simkovich and Loria, 1961. Courtesy of ASM.)

Finally he must obtain data from parts in service, for the loads in service are often not known. Once one knows the life from service tests, it is not too hard to redesign to avoid further failures.

The experimental engineering work that has been done can be grouped into two broad areas. The first involves collecting fatigue data on real engineering materials and also on full-size components in order to help the designer. Various types of loading and environments are included. Since the number of real materials is so large and the number of possible test conditions infinite, such data are bound to be incomplete; and the chances of finding data to fit a particular situation are slim. This forces the engineer to use his experience and intuition in making the best possible prediction from the available information.

The most common type of apparatus for collecting data is a rotating shaft loaded in bending, with a complete stress cycle each revolution. A variety of other machines are described by the American Society for Testing Materials (ASTM, 1949).

The collection of fatigue data can, in the long-life region, be a time-consuming problem. Because of this time factor, sufficient data at long lives are not often available. A machine running full time at 10,000 rpm will accumulate about 5×10^9 cycles per year. The typical test program of a material will consist of testing ten to fifteeen specimens at various stress levels. Any specimen that lasts 10^8 cycles is usually considered to be below the fatigue limit. The results of such a program are as in Fig. 18.11. Note that within several of the heats, the variation appears less than for the whole group. If one tests a really large number of specimens near the supposed fatigue limit, the results are as in Fig. 18.12. An exact value for the endurance limit cannot be given. In view of the fact that the specimens used in these tests were much more carefully machined and homogeneous than any practical part is likely to be, one should not use working stresses as

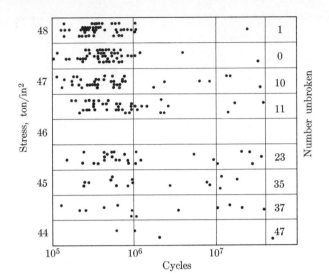

FIG. 18.12. Four hundred polished specimens tested near the endurance limit EN-24 steel. [*Note:* 1 British ton = 2240 lb.] (Clayton-Cave et al., 1955. Courtesy of Iron and Steel Inst.)

high as reported data might appear to permit. The scatter is not as bad as a first glance would suggest, however. At 48 long* tons/in^2 (tsi), only one specimen lasted longer than 10^6 cycles, while at 44 tsi only one specimen broke before 10^6 cycles. For the designer, then, a 10% drop in stress almost eliminates fatigue failures. McClintock (1955) has developed statistical approaches to determine when extraneous scatter has been eliminated so that only that due to the material remains. He shows the scatter in life is no greater than might be expected from the usual variation in the strength of material, say the yield strength. (*Metals Handbook*, ASM, 1961, pp. 938–949, reports standard deviations of 3 to 5% for the yield strength of aluminum alloys.)

As was seen in Fig. 18.2, curve (b), many materials do not show a fatigue limit. Since many tables may list the stress for 10^8 cycles as a fatigue limit, one should be careful in using such information for high-speed machinery, where more cycles may occur.

18.6 EFFECTS OF STRESS

A. Effect of Stress Amplitude. Since the stress amplitude is the primary variable in fatigue testing and since lives are so long, the results of fatigue tests are usually reported in graphs of stress or its logarithm versus the logarithm of the number of loading cycles, such as the *S-N* curves of Figs. 18.1 and 18.2.

---

* The British commonly report stress levels in tons as long tons (2240 lb).

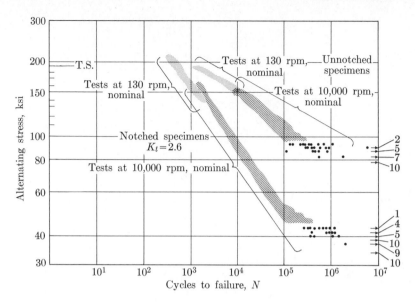

FIG. 18.13. Tests of SAE 4340 steel, T.S. 190 ksi. Notched specimens, C.C. Moore rotating bending tests. (Data from Cummings et al., 1957.)

Inspection of the representative fatigue behavior of metals illustrated in Fig. 18.13 shows that the fatigue strength is almost equal to the tensile strength up to 10^3 cycles of fatigue life, i.e., the S-N curve is flat in the high stress-amplitude region. In this range, known as low-cycle or short-life fatigue, the life is governed by the plastic strain amplitude and will be discussed below. The most characteristic region of a fatigue curve is between a life of about 10^2 to 10^3 and 10^6 to 10^7 cycles, where the fatigue life is a very strong function of the stress amplitude. As Fig. 18.13 suggests, the relationship between the stress amplitude σ and the number of cycles to fracture N in this range, where the stress drops from the tensile strength to about half that value, is given by

$$\sigma^a N = \text{constant}, \qquad (18.2)$$

where the exponent a can range from 8 to 15 (Prob. 18.7). Around 10^6 to 10^7 cycles, ferrous metals reach a fatigue limit, while the allowable stress amplitude for non-ferrous materials continues to drop. For nonferrous materials, the strength for a life of 10^8 cycles is normally reported, and is known as the *fatigue strength* at 10^8 cycles, as distinguished from the fatigue limit of ferrous materials.

As the strain amplitude is increased to the level of the tensile strength, general yielding begins and fatigue is better described in terms of strain. The strain amplitude is more closely related to damage than is the stress amplitude. In the limiting case of monotonic load, one might consider fracture to be caused by one quarter cycle of a strain whose amplitude is the equivalent plastic strain at frac-

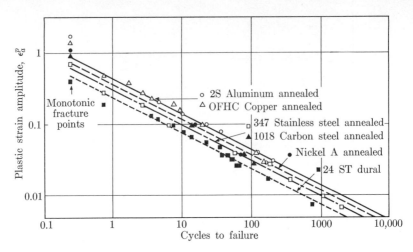

FIG. 18.14. Strain-cycle plot for low-cycle fatigue. Composite data for several materials. (Tavernelli and Coffin, 1959. Courtesy of ASM.)

ture. Starting from this extreme case, one may plot strain amplitude as a function of the number of cycles instead of the older plot of stress amplitude. It turns out that at least in the plastic regime, this plot on a logarithmic basis is remarkably constant, with a slope of minus one-half, as shown for a wide range of materials in Fig. 18.14. As suggested by Tavernelli and Coffin (1962), this relation can be extended nearer to the fatigue limit by including the elastic strain at the fatigue limit, ϵ_l. Writing it in terms of strain, we have,

$$\epsilon_a = \frac{\epsilon_f}{2(N)^{1/2}} + \epsilon_l. \tag{18.3}$$

As pointed out in a discussion by Manson of the work by Tavernelli and Coffin, this equation can be relied on to predict allowable strain amplitudes at a given life within a factor of two or better. This correlation suggests that perhaps one should plot ϵ-N curves rather than S-N curves to represent fatigue behavior, since the strain is more descriptive in the short-life regime, and equally easy to calculate in the elastic regime (Prob. 18.8).

B. Effect of Mean Stress. Many practical applications involve complex states of stress. In order to see how the resulting behavior might be predicted from simple bending fatigue tests, recall that basically, fatigue results from alternating shear on a slip plane. The strain amplitude on the slip plane will be larger the larger the shear stress amplitude on that plane, and the nearer the environment is to general yielding. Normal stress across the slip plane may be important in crack initiation and during the early states of crack growth, although later the maximum normal stress on any plane will determine the direction of the crack. With these ideas in mind, consider the effect of an alternating stress amplitude superimposed on a time-average, or mean stress, as shown in Fig. 18.15. Cross-plotting of the S-N

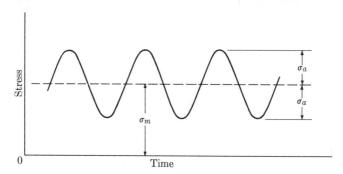

FIG. 18.15. Manner of defining time-mean stress σ_m and alternating stress σ_a.

curves gives the alternating stress amplitude versus the mean stress for different lives, as shown in Fig. 18.16. For long life, the effect of mean stress *per se* is not large. For short life, however, the incidence of general plastic deformation makes the effect of mean stress more important (Prob. 18.9).

If data such as given in Fig. 18.16 are not available, one may assume either of two limiting conditions: that the allowable stress amplitude is constant until the time-mean stress is great enough so that the maximum stress of a cycle reaches the tensile strength, or that there is a linear relation between stress amplitude and time-mean stress. Both of these possibilities are sketched in Fig. 18.17, along with the limitations associated with other modes of failure than fatigue. In equation form, the alternative criteria for safety are

$$\sigma_a/\sigma_l \leq 1 \quad \text{and} \quad (\sigma_a + \sigma_m)/(\text{T.S.}) \leq 1 \tag{18.4a}$$

and

$$\sigma_a/\sigma_l + \sigma_m/(\text{T.S.}) \leq 1. \tag{18.4b}$$

The slanting straight line, called a Goodman or Soderberg (1930) relation, is usually felt to give a conservative estimate, since most data fall above it (Prob. 18.10). Exceptions are shown by some brittle materials and some notched aluminum alloys (Lazan and Blatherwick, 1952).

Residual stress, discussed in Chapter 12, is a common source of mean stress, although it may gradually relax under cyclic stressing. Shot-peening is often used to obtain beneficial compressive residual stress. Appropriate grinding (Tarasov and Hyler, 1957) and rolling operations can also result in compressive residual stress. The opposite is possible, however, if the wrong conditions are used.

◀ **C. Fatigue under Combined Stress.** Even fewer data are available when several components of stress are present; for example, under mean torsion and alternating tension. Ideally four factors would be considered: the alternating stress on the most critical slip plane, the normal stress during that shear stress, the nearness of general yield, and the anisotropy of the material. Such a criterion would be difficult to evaluate and use, and data to check it have not been obtained. We shall simplify the problem by neglecting anisotropy and the effect of normal stress. The slip plane with the maximum shear stress amplitude $(\tau_a)_{\max}$ can then be determined

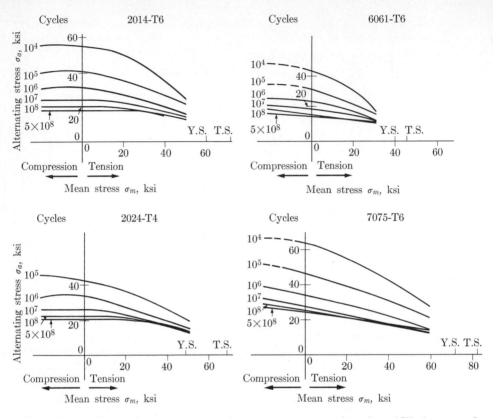

Fig. 18.16. Alternating stress versus time-mean stress as a function of life for several aluminum alloys. (Data from Howell and Miller, 1955.)

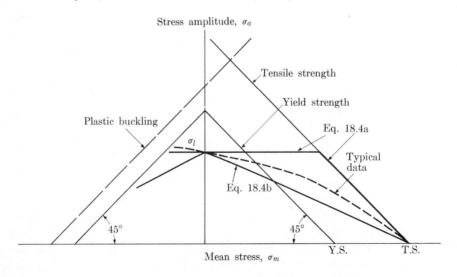

Fig. 18.17. Criteria for fatigue under mean stress.

readily. The nearness to general yield can be expressed by using the maximum shear stress on any plane at any part of the cycle, τ_{max}. Again taking the amplitude and maximum stress separately, or a linear function related to the mean stress, as alternative criteria which reduce to Eqs. 18.4a and 18.4b, respectively (Prob. 18.11):

$$(\tau_a)_{max}/\tau_l \leq 1 \quad \text{and} \quad \tau_{max}/\tfrac{1}{2}(\text{T.S.}) \leq 1 \tag{18.5a}$$

and

$$\frac{(\tau_a)_{max}}{\tau_l} + \frac{\tau_{max} - (\tau_a)_{max}}{\tfrac{1}{2}(\text{T.S.})} \leq 1. \tag{18.5b}$$

Depending on the mode of maximum loading, the fracture strength or buckling stress may be substituted for the tensile strength in Eqs. 18.5. These equations may be unduly conservative under high compressive stress, and in any event are hypothetical, since they have been compared with data only in special cases. For the fatigue limit of smooth specimens under combined bending and torsion, these ideas lead to an equation of the form suggested by Ransom and Mehl (1952) for no mean stress:

$$(\sigma_a/\sigma_l)^2 + (\tau_a/\tau_l)^2 \leq 1, \tag{18.5c}$$

where σ_a and τ_a are the bending and torsional stress amplitudes, respectively, and σ_l and τ_l, the fatigue limits in bending and torsion, are in the ratio $\sigma_l/\tau_l = 2$ (Prob. 18.12). Anisotropy may markedly affect this ratio, as shown in Fig. 18.18. The σ_l/τ_l ratios vary from about one to nine. However, if one eliminates the obviously anisotropic plastic laminates, the notched and splined specimens for which the stress concentration factors are not equal in tension and torsion, and finally the cast iron specimens which have many stress concentrations in the form of graphite flakes, the remaining values group near two, as expected. The scatter

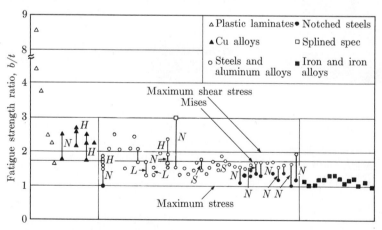

Fig. 18.18. Variation in the ratio of torsion (t) to bending (b) fatigue limits for various types of material (letters indicate different heat treatment H, size and shapes S, location L, and notches N). Ratio as predicted by various theories as indicated. (Findley, 1957. Courtesy of ASME.)

could well be due to anisotropy caused by the forming process. A rolled rod, for instance, has inclusions stretched out along its length. These inclusions will be at 45° to the maximum shear stress in a bending test but will be parallel to the maximum shear stress in a torsion-test.

◀ Equation 18.5b has sometimes been stated using the amplitude of the equivalent stress rather than the maximum shear stress amplitude on any plane. However, in reviewing the problem of combined stress, Findley et al. (1961) have shown that in the center of a roller that is being squeezed between two other rollers, the equivalent stress is constant even though the stress components rotate in the material. Thus the amplitude of the equivalent stress is zero, and fracture would not be predicted. It actually does occur, as expected from the Eq. 18.5b, where the maximum shear stress amplitude on any slip plane is used.

◀ When the components of alternating stress have different frequencies, a non-sinusoidal shear stress occurs on any plane. This situation will be discussed in Section 18.6E. ▶

D. Stress Concentrations and Size Effects. The most important consideration in fatigue is that it depends on the local stress or strain in some small area. In designing against deformation, strain concentrations make no appreciable contribution to the overall deformation, but under repeated strain they can initiate cracks which then spread throughout the entire specimen. In addition, due to the small

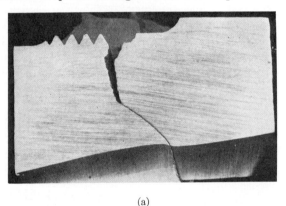

(a)

(b)

(c)

FIG. 18.19. Fatigue failures due to mistakes. (a) *Design error:* omission of an undercut just beyond the threads which would reduce the stress concentration at the base of the last thread. Hard steel BHN 285. Fracture by fatigue down to base of open V followed by sudden partly ductile fracture. This error caused a flight failure and the pilot died in the crash. (b) *Rough machining:* steel bolt of 125,000 psi T.S. threads are badly torn. (c) *Machining fault:* connecting rod ring designed with good fillet. Fillet was cut away in notch to clear a socket wrench. Crack started at base of notch. (Battelle, 1941. Courtesy of Wiley.)

TABLE 18.1*

Effect of Surface Finish on the Fatigue Limit in Rotating (Cantilever) Bending 0.33% C Steel

Finish	Fatigue limit, psi
High polish (longitudinal)	41,500
F F emery	40,500
No. 1 emery	40,000
Coarse emery	39,000
Smooth file	38,500
Turned	36,500
Bastard file	35,500
Coarse file	33–34,000

* From Thomas (1923).

slope of the S-N curve, areas at only slightly greater stress than average will fail in a much shorter time. This means that any sort of stress concentration will be much more important in fatigue than in ductile fracture. Compare for instance the data for unnotched and notched specimens of 4340 steel of Fig. 18.13. To increase fatigue life it is essential to avoid stress concentrations, or to minimize their effect as much as possible.

The use of fillets, rounded corners, avoidance of sharp changes in cross section, stress relief, etc., is always wise. (See Caswell, 1947, or Battelle, 1941, for good design practice.) Figure 18.19 shows some failures which might have been avoided.

On a fine scale, the surface finish is important in reducing local stress concentrations as indicated by the data in Table 18.1. Further aspects of the surface behavior will be discussed in Section 18.7.

In reporting tests on notched specimens, one may calculate a *fatigue strength reduction factor* K_f, defined as the ratio of unnotched to notched fatigue limit. This value is less than the theoretical elastic stress concentration factor K_t, but increases with size, as shown in Fig. 18.20. This size effect arises from several sources. First, there is a statistical variability which may be described in terms of the ratio of standard deviation in stress, σ_S, to the mean stress, $\bar{S}$, for specimens all having the same life. The highly stressed region at the tip of a notch is small, so there is a smaller probability of finding a weak spot. As discussed below, for typical notches, the resulting notch insensitivity is in the range

$$(K_t/K_f) - 1 = (3 \text{ to } 30)\sigma_S/\bar{S}. \tag{18.6}$$

A second source of notch insensitivity is that in regions small compared with the length of slip bands in which cracks form, the constraint of the surrounding material at low stress will reduce plastic deformation in the band, and hence increase

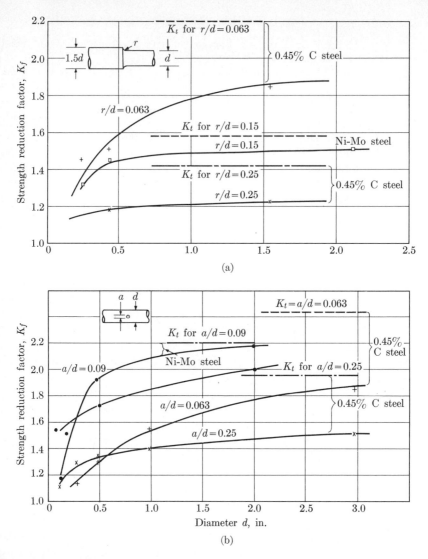

Fig. 18.20. (a) Relation between K_t and K_f for fillets, specimens geometrically similar. (b) Relation between K_t and K_f for holes, specimens geometrically similar. (Peterson and Wahl, 1936. Courtesy of ASME.)

the fatigue limit. The size effect is aggravated by difficulties in forging, casting, and heat-treating large parts. Other factors also affect the value of K_f (Probs. 18.13, 18.14). By much the same arguments, high stress gradients favor a longer fatigue life for a part with a given maximum stress.

One of the surprising phenomena associated with notched specimens is the fact that fatigue cracks may often start to grow from the root of notches but then stop

(Frost and Phillips, 1957). At first glance, one might expect that this would be due to the fact that the stress amplitude can be greater at the tip of a notch, where both tensile and compressive displacements can occur, than at the tip of a crack, where the closure of the crack faces prevent a complete strain reversal. That this explanation is not complete is shown by the existence of nonpropagating cracks even when the mean load greatly exceeds the alternating load so that no closure is expected (Frost and Phillips, 1957). Solution of this problem very likely awaits the determination of the stress and strain history around a crack growing under tension.

◀ For an explicit development of the statistical contribution to notch insensitivity, assume that the three-dimensional stress distribution in a specimen is given in terms of the maximum stress S_m, the stress S_l below which the probability of failure goes to zero (see Eq. 15.14), the coordinates x_1, x_2, and x_3 from the point of maximum stress, and corresponding scale factors l_1, l_2, and l_3. A form of stress distribution is chosen which is convenient for statistics if awkward for stress analysis:

$$S - S_l = (S_m - S_l)\left[1 - \left(\frac{x_1}{l_1}\right)^{p_1}\right]\left[1 - \left(\frac{x_2}{l_2}\right)^{p_2}\right]\left[1 - \left(\frac{x_3}{l_3}\right)^{p_3}\right], \quad (18.7)$$

where the exponents p_i are chosen as 0, 1, or 2 for stress distributions which are constant, fall off linearly, or fall off parabolically from the maximum. Substitution of Eq. 18.7 into Eq. 15.11 gives a result similar to Eq. 15.12, except that the area is replaced by a product of effective lengths l_i^e, which are $l_i^e = l_i$, l_i/m, or $l_i/\sqrt{m}$, for $p = 0$, 1, or 2. The product of these effective lengths gives an effective area (or volume, if that is more relevant) which is used in the size-effect equations (15.17 and 15.18).

◀ If two circular, notched fatigue specimens have the same circumference, a notch insensitivity term will arise from the different effective lengths of the smooth and notched specimens, l_s^e and l_n^e. The mean value of the stress at fracture in a number of notched specimens, in terms of that of smooth specimens, is then

$$\overline{S}_n = \overline{S}_s + \frac{\sigma_s[(l_s^e/l_n^e)^{1/m} - 1]}{\sqrt{\frac{(2/m)!}{[(1/m)!]^2} - 1}}. \quad (18.8)$$

The nominal stress of the notched specimen is less than the local stress by the stress concentration factor:

$$\overline{S}_{n \text{ nom}} = \overline{S}_n/K_t. \quad (18.9)$$

The fatigue strength reduction factor is the ratio of the mean nominal strength of the smooth specimen to that of the notched specimen:

$$K_f = \overline{S}_s/\overline{S}_{n \text{ nom}}. \quad (18.10)$$

Combining Eqs. 18.8, 18.9, and 18.10 and rearranging to obtain the notch insensitivity, we have

$$\frac{K_t}{K_f} - 1 = \frac{\sigma_s}{\overline{S}_s} \frac{[(l_s^e/l_n^e)^{1/m} - 1]}{\sqrt{\frac{(2/m)!}{[(1/m)!]^2} - 1}}. \tag{18.11}$$

For typical values of $m = 3$ and $l_s^e = 100 l_n^e$, evaluating the radical from Fig. 15.15 gives a numerical factor of 10 in Eq. 18.6. With large variability in the strength, σ_s, this can be an important effect. It is important that σ_s be due to the material and not, for example, to extraneous sources such as variability in the tests. McClintock (1955) has shown how this can be assured from studying the variation in position of failure along a specimen with a parabolic distribution of stress. ▶

E. Stress History. Practically all fatigue testing is done at a constant stress amplitude. Most applications involve operations at different stress levels at different times, or, as in auto springs, airplane wings, etc., a complicated stress history at most times. The simplest hypothesis for prediction of fatigue life for parts with complex histories is one proposed by Palmgren (1924) and used by Miner (1945), in which the fraction of fatigue life "used up" at each stress level is calculated. When the accumulated fraction totals unity, failure is assumed:

$$\sum \frac{n_i}{N_i} = 1, \tag{18.12}$$

where n_i is the number of cycles run at stress level i, and N_i is the number of cycles to fracture if stressed only at stress level i. Because of the strong dependence of life on the applied stress level, this equation is quite satisfactory for estimating allowable stress levels for a given life, but may lead to large errors in estimating the life for a given stress.

Starkey and Marco (1957) have applied the Miner hypothesis as well as other theories to tests involving a few types of complex stress cycles and find fairly good agreement for aluminum and steel. Their results were limited to situations where the number of cycles at each stress level were approximately equal. Newmark (1952) gives a good review of the earlier work on the problem.

An interesting practice which is somewhat at variance with the Miner concept is "coaxing." This involves running a test specimen for a considerable time at a stress slightly below its endurance limit. In materials that can strain-age, but are still unaged, Sinclair (1952) finds that the subsequent fatigue limit is raised (Prob. 18.15).

A rather dramatic exception to the Miner equation (18.12) is that while it predicts that occasional overloads should decrease the life, Heywood has found that such occasional overloads will prolong the life, as shown in Table 18.2. These data are consistent with an expected increase in life due to the imposition of favorable residual (mean) stresses at the stress concentrations. Similar tests on smooth specimens showed a small decrease in life, as would be expected. Since the residual stresses gradually disappear, a greater increase in life is noted if the over-

TABLE 18.2*

FATIGUE TESTS ON STRUCTURAL ELEMENTS
(MATERIAL: DTD 364B ALUMINUM ALLOY.
ALL TESTS IN FLUCTUATING TENSION)

Type of specimen	Nominal fatigue stress, psi	Type of high load	Magnitude of high load, psi	Life, 10^6 cycles	Ratio of life, with high load/no high load
Simple lugs	$13{,}000 \pm 4930$	None	0	0.124	1
		10 preloads	39,600	0.155	1.25
			49,500	>10.6	>85
			22,400	0.738	6.3
		Periodic loads			
Transverse hole in specimen	$20{,}200 \pm 7850$	None	0	7.7	1
		Periodic loads	$-22{,}400$	0.494	0.064
			$-44{,}800$	0.130	0.017

* From Heywood (1956).

loading is repeated. Overloading in the negative direction greatly lowers the fatigue performance. In addition to these changes in mean stress during the history, there may be changes in shape due to the creep during fatigue (Coffin, 1960).

One way of determining the fatigue limit involves introduction of history effects through a steadily increasing stress amplitude. Before considering it, recall that the usual determination of a fatigue limit requires a large number of specimens, because if a specimen does not fail, one knows only that the fatigue limit of that specimen lies above that value. Prot (1952) suggested running tests with a continually increasing stress amplitude, so that every specimen fails sooner or later. Specimens run at a greater stress increase per cycle will tend to fail at a higher stress level than those whose stresses increased more gradually. By plotting stress amplitude at failure versus rate of stress increase per cycle and extrapolating the data back to a zero rate of stress increase per cycle, one obtains an estimate of the endurance limit. While this method of testing brings in the effect of stress history, the effect is usually small and the procedure is a recommended one for any program where the specimens are especially expensive. After all, the constant stress amplitude imposed in conventional testing does not simulate service conditions either.

F. Speed of Testing. Since fatigue cracking is a consequence of plastic deformation, one would expect that higher rates would give somewhat less tendency for plastic deformation unless these higher rates, because of hysteresis, led to higher temperatures, which in turn decreased the resistance to plastic flow. The interaction of frequency and temperature makes it difficult to study each separately, but in any event, as with plastic deformation itself, the rate of loading has a small effect on the stress (for a given life), although it may have a large effect on the life for a given stress, a 20% increase in speed producing a 10% increase in the cycles to failure of steel (Fluck, 1951).

18.7 EFFECTS OF MATERIAL VARIABLES

A. Hardness. Because fatigue is associated with plastic deformation, one would first of all expect that the resistance to fatigue would be related to the resistance to plastic deformation. Because of the presence of some strain-hardening, this correlation seems to be better when related to the tensile strength, as indicated in Fig. 18.21. It is clear from this figure that to a first approximation, the fatigue limit is approximately half the tensile strength for a wide variety of ferrous alloys and titanium, and that the fatigue strength at 10^8 cycles is only a third to one quarter the tensile strength for aluminum, copper, and their alloys. Exceptions have been reported by George and Forsyth (1962), who obtained ratios of 0.4 to 0.45 in hard Al-Mn-Zn alloys with the addition of Ni or Mn. The general correlations with tensile strength seem to hold whether increased hardness has been obtained by cold-working, by solution-hardening, or by heat treatment. In spite of all the work that has been done, however, the basic reason for the two classes of fatigue strength ratios, and indeed for the existence of a fatigue limit at all, is not known.

B. Metallurgical Effects. In spite of the general correlation with hardness mentioned above, there are metallurgical details which do produce significant variations in the fatigue limit and hence in life near the fatigue limit.

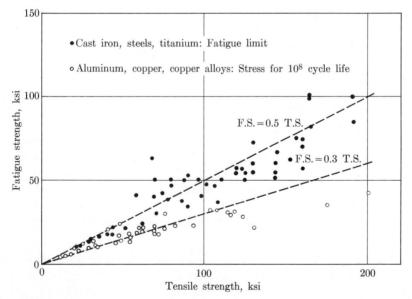

FIG. 18.21. Correlation of endurance limit with tensile strength. (Data from *Metals Handbook*, ASM, 1961.)

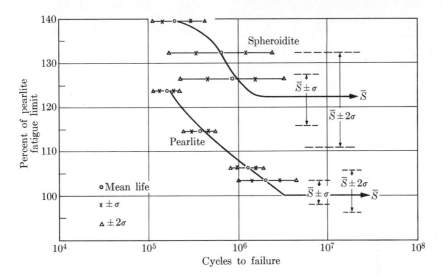

Fig. 18.22. Effect of metallurgical structure on fatigue life. Eutectoid steel in two structures of equal tensile strength 95,000 psi. (Dieter and Mehl, 1953.)

Working, such as rolling or forging, may produce anisotropic effects which give different fatigue properties in different directions relative to the flow (Ransom and Mehl, 1952). Heat treatments which result in quasi-stable metallurgical states may lead to trouble if later mechanical stresses cause transformations. A well-known example is the problem of retained austenite in high-strength steel alloys. This austenite can be transformed to untempered martensite by later stressing. The increase in volume associated with this change leads to serious stresses and possible lowering of fatigue life (Bush et al., 1961).

An example of the effects of different metallurgical structures is given by the comparison of steels with the same tensile strength, one having a pearlitic structure, and the other one a spheroidized structure. The properties of the steels are given in Table 18.3 and fatigue data are plotted in Fig. 18.22. As is so often the case in

TABLE 18.3*

Behavior of Two Structures of Eutectoid Steel

Steel	Endurance limit, psi		Tensile strength, psi	Yield strength, psi	Elongation, %	Reduction of area, %	Hardness
	Mean	Standard deviation					
Pearlite	34,000	1000	98,000	35,000	18	26	$R_B 89$
Spheroidite	41,500	2000	93,000	71,000	29	58	$R_B 92$

* Dieter and Mehl (1953).

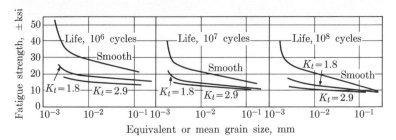

Fig. 18.23. Effect of metallurgical structure on fatigue life. Fatigue life of 70-30 brass versus grain size. (Karry and Dolan, 1953. Courtesy of ASTM.)

fatigue literature, the scale is distorted by suppression of the zero, so the effects are much smaller than might appear at first glance. It is still true, however, that the pearlitic steel which has the more irregular microstructure is likely to have a shorter life by a factor of ten, even in the region in which both steels are above their endurance limits. It should also be noted that this difference in structure affects the reduction of area, and therefore, one would expect this difference to persist even in the high-strain, low-cycle fatigue regime. Finally there is the possibility that the effect is more closely related to whatever is causing the difference in yield strength, rather than the observable differences in the structure.

TABLE 18.4*

THE EFFECTS OF SURFACE CONDITIONS ON FATIGUE LIFE
FOR CHROMIUM PLATING ON SAE X4130 STEEL

Plating conditions	Condition	Hardness	Fatigue limit, psi		% reduction due to plating
			No plating	0.009 in. plating	
Plated at 158°F, current density 1000 amp/ft^2	Normalized	$R_B 89$	44,000	40,000	9
	Quenched and tempered	$R_C 39$	93,000	45,000	52
Plated at 131°F, current density 200 amp/ft^2	Normalized	$R_B 89$	44,000	40,000	9
	Quenched and tempered	$R_C 40$	93,000	49,000	47
Plated at 131°F, current density 350 amp/ft^2	Quenched and tempered	$R_C 40$	93,000	64,000	31
Plated at 185°F, current density 700 amp/ft^2	Quenched and tempered	$R_C 40$	93,000	10,000	89

* From Grover, Gordon, and Jackson (1956).

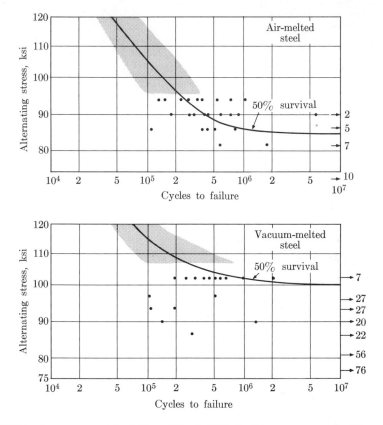

FIG. 18.24. Improvement in fatigue life of 4340 steel from vacuum melting. Tensile strength for both samples 190,000 psi. Note the 15% increase in endurance limit. (Cummings et al., 1957. Courtesy of ASM.)

The effect of grain size is shown in Fig. 18.23. It is likely that this also has an effect on the tensile strength, and so it is not clear from these data whether there is a grain size effect *per se* (Prob. 18.16).

The best performance is also obtained when the inclusions can be reduced to a minimum, preferably in a dispersed form. Figure 18.24 represents the improvements possible by vacuum melting as opposed to air melting steel.

C. Surface Conditions. In addition to the surface effects discussed in Section 18.6, associated with surface irregularities and resulting stress concentrations, and with the residual stresses which are often present, there may be important changes in composition at the surface, either facilitating or preventing plastic flow and hence fatigue. Weak surface layers caused by decarburization (Fig. 18.25), burning during grinding (Fig. 18.26), by soft plating (Fig. 18.27), or by crack-filled platings (such as the chrome plating of Table 18.4) all result in a lower fatigue performance. In chromium plating it is not only the actual existence of the plate,

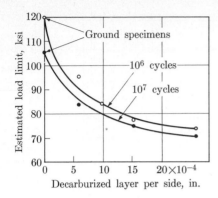

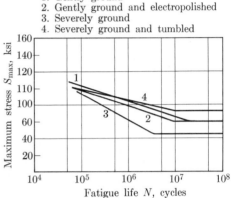

Fig. 18.25. The effects of surface conditions on fatigue life. Decarburized surface layer on 5% Cr-Mo-V sheet specimens 0.042 in. thick. (Simkovich and Loria, 1961. Courtesy of ASM.)

1. Gently ground
2. Gently ground and electropolished
3. Severely ground
4. Severely ground and tumbled

Fig. 18.26. The effects of surface conditions on fatigue life. Steel bars with various ground surfaces (Steel hardness $RC45$). (Grover et al., 1956. Courtesy of Thames and Hudson.)

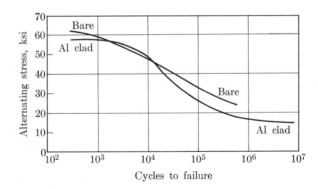

Fig. 18.27. The effects of surface conditions on fatigue life. "Alclad" surface on 2024-T4 aluminum. (Smith et al., 1950.)

but also the method of application which may be of importance in determining the life.

On the other hand, carburizing, nitriding, and flame-hardening can increase the fatigue strength by as much as 25%. Alden and Backofen (1961) showed that the presence of an anodized skin on aluminum crystals extends the fatigue life indefinitely, so long as the skin is intact, whereas fatigue cracking soon develops when the skin is broken or removed.

18.8 EFFECTS OF ENVIRONMENT

A. Effect of Temperature. Most reports of tests ignore any influence of temperature on fatigue life. Since much testing is done in high-speed machines in which the specimen gets quite hot, and since many applications are at slower speeds where heating does not occur, appropriate consideration must be given to temperature effects. The effect of temperature on fatigue roughly parallels the effect of temperature on plastic deformation. Although it generally decreases with increasing temperature, the fatigue limit of steels rises at about 400°F, corresponding to the strain-aging range discussed in Section 5.9.

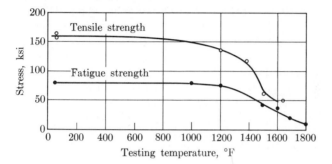

FIG. 18.28. Effect of testing temperature on fatigue strength at 10^8 cycles and tensile strength. Inconel 700 (45 Ni, 30 Co, 15 Cr, 3 Mo, 2.2 Ti, 3.2 Al). (Data from *Metals Handbook*, ASM, 1961.)

TABLE 18.5*

CORROSION FATIGUE STRENGTH OF VARIOUS ALLOYS

Material	Tensile strength, lb/in²	Endurance limit, (in air) lb/in²	Corrosion-fatigue strength, (in salt spray) lb/in²	Cycles endured
Mild steel	—	38,000	2500	100×10^6
0.5% C steel	142,000	56,000	8750	50
15% Cr steel	97,000	55,000	20,500	50
17% Cr, 1% Ni steel	122,000	73,500	27,500	50
18% Cr, 8% Ni steel	148,000	53,500	35,500	50
Beryllium bronze	94,000	36,500	38,800	50
Phosphor bronze	62,000	22,000	26,000	50
Aluminum bronze	80,000	32,000	22,000	50
Duralumin	63,000	20,500	7600	50
Magnesium alloy ($2\frac{1}{2}$% Al)	37,000	15,200	Negligible	50

* From Lessells (1954). Courtesy of Wiley.

The correspondence between the fatigue limit and the tensile strength persists to much higher temperatures, as shown for the nickel-base alloy of Fig. 18.28. The rate of testing now becomes especially important, for with a mean stress, creep fracture may occur. The interaction between creep and fatigue can be represented on a diagram of mean stress versus stress amplitude, much like the Soderberg diagram used at room temperature.

B. Corrosion Fatigue. The influence of the atmosphere on fatigue life is considerable and is often neglected. For an obviously corrosive atmosphere, repeated cycling and corrosion seem to interact and the damage is greater than the simple summing of the two operating independently (McAdam, 1926). Salt and acid solutions are especially detrimental to iron and steel, while caustic solutions improve performance. Figure 18.29 shows typical data for different pH values in salt water. In corrosive atmospheres a fatigue limit does not exist, as shown by the acidic solutions. The basic solutions show no corrosion and do have a fatigue limit. Table 18.5 gives the corrosion fatigue performance of various other metals.

The normal atmosphere has lately been recognized to influence fatigue life. If air is excluded by a coating of oil, or even better, by use of a vacuum, some investigators have reported a life increase of as much as twentyfold at stresses slightly above the normal fatigue limit. These changes have been noticed on such

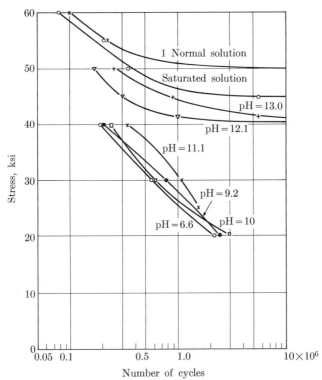

FIG. 18.29. Corrosion fatigue. Effect of pH on fatigue life AISI 1036 steel in 3% NaCl-water solution. (Radd et al., 1959.)

widely different materials as steel, lead, and aluminum. There is a growing feeling among many investigators that the action of the normal atmosphere is far more important in fatigue failures than heretofore thought (Wadsworth and Hutchings, 1958; Wadsworth, 1961).

Even seasonal changes in humidity affect the allowable stress amplitude for a given life by five to ten percent. This would make a major change in fatigue life at a given stress. A plastic sleeve fitted over a fatigue machine is a handy way of carrying out tests on this effect (Bennett, 1963). At high humidity there is much greater likelihood of producing fracture by *fretting corrosion*, which is a form of fatigue in which rubbing at a press fit produces rapid oxidation, corrosion products, and a crack. This process of fretting corrosion is also important in steel.

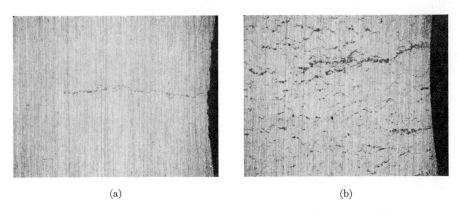

(a)   (b)

FIG. 18.30. (a) 1100-H14 aluminum fatigue specimen fractured in air in 45 minutes. (b) Identical specimen requiring 5 hours to fatigue at 10^{-8} torr.

C. Effect of a Vacuum. Recent interest in space exploration has focused attention on fatigue in a vacuum. The effects are varied. On a carefully cleaned fresh surface, there is less oxide skin to prevent plastic deformation from breaking through the surface and fatigue may actually be hastened. On the other hand, the absence of an oxide skin may prevent fine cracks which could form in it and propagate into the material underneath. Finally, during crack propagation there is more chance of the crack surfaces welding together, thus retarding crack growth. The results of these and probably other phenomena have shown a variety of results depending on the material and conditions of testing. In general, it appears that small increases in fatigue strength are observed in vacuum, and the character of the fatigue fracture changes in that there seem to be many more fine cracks developing and much surface roughening, as shown in Fig. 18.30.

18.9 ROLLING CONTACT FATIGUE

◄Rolling contact devices such as ball and roller bearings, friction transmissions, mechanical ball-disk integrators, cams, gearing, and many other devices in which curved surfaces are repeatedly brought together at high stress levels are commonly subject to fatigue failure. These applications differ in several important respects from most other applications. The stress field is usually triaxial with a very high compressive hydrostatic component. The stressed volume is also very small. There is often a thin lubricant film between the contacting surfaces at very high pressures.

◄First let us examine the state of stress. From the Hertz contact equations discussed in Section 11.7, the maximum shear stress beneath the surface is about three times that on the surface (Prob. 18.17). Thus, if there are inclusions beneath the surface, they can act as nuclei, from which the cracks will grow and break out to the surface taking an entire chip of material away. Alternatively, fatigue failures may begin at the surface of the specimen and work their way in, finally producing a chip of similar shape. In this case, there is direct influence of the lubricant not only through corrosion fatigue but also through the effect of trapped lubricant changing the stress distribution as the crack edge is sealed and rolled over.
◄An empirical relation between load and life, which also has some theoretical foundation is (Prob. 18.18)

$$P^3N = \text{constant.} \quad (18.13) \blacktriangleright$$

Rolling-contact applications are all very sensitive to lubricant changes. Normal mineral oil is almost always the best choice. Other fluids, as Table 18.6 shows, can have enormous effects on fatigue life. This fact supports the supposition that

TABLE 18.6

SUMMARY OF FATIGUE-LIFE REDUCTION IN ROLLING
CONTACT DUE TO LUBRICANT

$$r = \frac{\text{Average life with test lubricant}}{\text{Average life with standard petroleum oil}}$$

r	Fluid	Ref.
0.0154	Hydraulic fluid A	Morrison, 1955
0.0242	Gasoline (nonleaded)	Morrison, 1955
0.0515	Hydraulic fluid B	Morrison, 1955
0.0515	Water-soluble grinding compound	Morrison, 1955
0.0572	Water glycol base	Cordiano et al., 1956
0.0745	Glycerin	Morrison, 1955
0.101	7808B Diester	Moore, 1956
0.129	Hydraulic fluid C	Morrison, 1955
0.165	Sulphonated hydrocarbon cutting oil	Morrison, 1955
0.257	Hydraulic fluid D	Morrison, 1955
0.296	Phosphate ester base	Cordiano et al., 1956
0.385	Grease, petroleum oil, sodium-calcium soap	Morrison, 1955
0.58	Phosphate ester	Cordiano et al., 1956
0.76	Petroleum (1005)	Moore, 1956

this is a corrosion-fatigue problem. However, due to the enormous complexity of the fluid-flow problem under elements which are twisting and sliding slightly as well as rolling, no one has yet been able to untangle all the variables and spell out the exact role of the lubricant.

◀ The scatter in rolling-contact life is usually worse than in other applications, with lives varying by a factor of up to 200. This is probably due to the small stressed volume and the complex, hard, steel alloys used. The probability of hitting a serious nonmetallic inclusion or carbide is subject to considerable scatter. The designer must be very conservative to avoid early failures, although an elaborate statistical correlation of bearing failures by Palmgren (1924) permits some prediction of bearing performance. The statistical distribution may make it possible to "run in" bearings to eliminate very early failures. ▶

18.10 ENGINEERING CONSIDERATIONS IN DESIGN AND SERVICE

Almost all machine elements are subject to cyclic loading and fatigue failure. Rotating shafts are subjected to reversed bending whenever they carry unidirectional loads. The periodic power impulses of a reciprocating engine can cause torsion and tension fatigue. Turbine blades are cyclically loaded as they pass the stator blades. The high noise levels associated with jets and rockets can cause fatigue problems in neighboring parts that vibrate. All types of springs and bellows are frequent victims of fatigue failures. Ball and roller bearings usually fail by "spalling" or "flaking," which is a type of fatigue failure. Even journal bearings are limited in load by the fatigue strength of the babbit liner.

When faced with these complex problems, the designer must make extrapolations from his empirical data. He will usually have a fair idea of the basic fatigue strength of the material he hopes to use.

However, in complex structures, the stresses, stress gradients, residual stresses, and even the loading are not at all well known. For static loading, this imperfect knowledge of stress is not too serious, since local yielding at the points of high stress helps to redistribute the load and there is sufficient ductility to avoid fracture. However, in fatigue applications, the reversed local yielding soon leads to failure. The practical result of the above problem is that many devices are given service testing to discover and eliminate the sources of fatigue failure. This type of testing on full-scale components is expensive and time-consuming, but it is essential where parts of minimimum weight are required.

The use of accelerated service testing can often be dangerously misleading. The first temptation is to test at high stress levels in order to save time. Life at lower stress levels is then predicted by extrapolation. This may cause error when residual stresses are present. The high stress may wipe out the residual stresses, while at the low stress levels encountered in service, the residual stresses would be very important. Accelerated testing in rate of cycling is widely used by the automobile industry on their proving grounds. This unfortunately minimizes any corrosion fatigue problem and may predict longer lives than are later realized.

Service testing is often not feasible or else too much time is consumed so that many items are put in service without these checks. For dangerous applications, the engineer would like techniques to uncover fatigue problems before catastropic failure occurs. Since a fatigue crack grows rather slowly, and for many applications only during periods of high loading, cracks may be present during a fairly long period of the service life. Detection of such cracks can be very important, and, being relatively easy, this should be done. A few of the common techniques are (Lipson, 1950):

1. Listening: Uncracked parts ring clearly when struck, while cracked parts are dull or dissonant. Ultrasonic crack-detection devices can locate smaller or internal cracks (Rasmussen, 1962).
2. Looking: Simple visual inspection will often show some cracks; magnification, of course, helps. Fluorescent dyes which flow into the cracks are a sensitive method of detection. Since magnetic fields are warped by cracks, iron powder sprinkled on a magnetized specimen is also a sensitive method of detection (see McMaster, 1959).
3. Radiographing: x-ray inspection is useful if one can be reasonably sure of hitting the crack edgewise. The very thin fatigue crack does not show up when viewed broadside.

For critical parts such as wing spars, turbine shafts, etc., regular inspection for cracks is often specified. A recent development is automatic inspection by using fine wires bonded to a surface. Fatigue cracks break these wires and give a signal.

In conclusion, one might say that fatigue problems in engineering applications, while difficult to solve, are not complex enough to explain the large number of fatigue failures encountered in everyday life. Most of these failures can be attributed to ignorance of well-known information on material properties, to design features, to manufacturing problems, or to unexpected service loads.

If in spite of preliminary theoretical estimates of the stress and the resulting behavior of the material, and in spite of component testing indicating a reasonable margin of safety, service failures are reported, the question arises as to what the cause of the failure was and how it could be prevented. Examination of a fatigue fracture usually allows its identification through a number of features as well as a determination of the stress history which led up to it. For example, consider the fatigue failure in an engine crankshaft shown in Fig. 18.31. There is no evidence of gross plastic flow. The device has been operated intermittently or at various stress levels so that many "beach marks" are visible. These are simply regions of corrosion or differing rates of growth caused by the changes in the operating conditions. Since there are numerous beach marks, the fatigue crack obviously did not grow all at once but grew gradually, spreading out from the starting point. These marks point out the place at which the fatigue crack started. When the stresses are low, usually only one point of crack initiation is observed. However, for highly stressed parts, several cracks will often be observed all growing simultaneously.

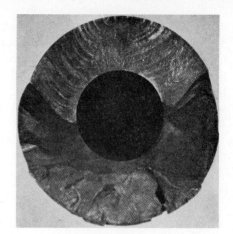

FIG. 18.31. Aircraft engine crankshaft failed in flight from clamping stresses aggravated by roughening caused by rubbing of shaft against propeller hub. Note typical fatigue crack growth originating at a small localized surface imperfection. (Battelle, 1941. Courtesy of Wiley.)

In the range of large-strain short-life fatigue, the fracture will resemble more a ductile fracture in uniaxial tension, i.e., although some beach marks may still be observable, the appearance of the surface will be rough and fibrous. For a more detailed discussion of fatigue fracture surface markings the reader is referred to Peterson (1950), or Beachem and Pelloux (1965).

In metals which can undergo cleavage, such as steel, the final fracture in the last cycle is often transcrystalline, producing characteristic cleavage facets. This is why the early investigators of the 19th century were tempted to attribute fatigue failure to a "crystallization" of the metal resulting from vibration, an explanation still heard from the "garage mechanic" of our day (Peterson, 1950).

18.11 SUMMARY

Fatigue is progressive fracture caused by repeated loading. At low strain amplitudes, reversed plastic flow produces little or no strain-hardening, so that flow continues on a few slip bands rather than shifting to others. At low enough strain levels, this flow finally becomes reversible and no serious damage occurs. At slightly greater strain amplitudes, however, the repeated slip finally produces cracks, usually in slip bands at the surface or at an interface with an inclusion, within 1 to 30% of the fatigue life. Such cracks then grow to fracture, either through metallurgical damage or progressive deformation at the crack tip. Fatigue therefore tends to be controlled by the local rather than by the nominal stress. Stress concentrations are important, although modified by a size effect.

The results of experiments are usually presented on a semilog plot of stress versus number of cycles to failure, called an S-N curve, such as Fig. 18.2. Two characteristics are the rapid increase in life with reduction in stress and the possible existence of a fatigue limit. First, the equation of the sloping part of the curve is

$$\sigma^a N = \text{constant}, \tag{18.2}$$

where a varies from 8 to 15. Thus, a 2% reduction in stress can lead to a 20 to 30% increase in life. Conversely, if one is willing to sacrifice 30% in life, he may obtain only a 2% increase in load-carrying capacity. Secondly, for steel and titanium, there is a stress below which no failures are obtained, called the fatigue limit, which occurs at 10^6 to 10^8 cycles of a stress of roughly half the tensile strength. For other materials, such a limit does not exist, and the life at 10^8 cycles may be as low as 15% of the tensile strength.

Where the stress amplitude is of the order of the tensile strength, fatigue is governed by the strain amplitude, approximately as

$$\epsilon_a = \frac{\epsilon_f}{2(N)^{1/2}} + \epsilon_l, \tag{18.3}$$

where ϵ_f is fracture strain in uniaxial tension and ϵ_l is the elastic strain at the fatigue limit.

For complex stress histories, where the amplitude fluctuates randomly, fatigue failure can be expected to occur when the sum of the ratios of the number of cycles at any particular stress level to the fatigue life at that stress reaches unity, i.e.,

$$\sum_i (n_i/N_i) = 1. \tag{18.12}$$

The presence of a tensile time-mean stress accelerates crack propagation more than initiation. Although the relation between the allowable mean stress and alternating stress varies between materials and even at different stress levels with one material, an approximate linear relation can be used (Eq. 18.4), leading normally to conservative results. Effects of combined stress on initiation can probably be correlated primarily with a maximum shear stress amplitude criterion, but the mean normal stress will have more effect on propagation.

Being very sensitive to local strain, fatigue failure is strongly influenced by the irregularities of various surface finishes and the residual stresses which they may leave behind. Hence, fatigue data obtained with poorly controlled geometry and finishes on the specimen can be very misleading. By the same token, real applications, with their lack of laboratory control, can exhibit very wide variations in life. Since fatigue arises from plastic flow, it depends on hardness, surface effects, temperature, and rate in much the same way that plastic flow does, although it has a greater dependence on surfaces, inclusions, and structural details.

REFERENCES

For general engineering background on fatigue data, Battelle (1941), Lessells (1954), Cazaud (1953), Sines and Waisman (1959), and Grover et al. (1956) furnish a wealth of pictures of failures and data. Kennedy (1963) provides more physical background. The conference of the Institution of Mechanical Engineers and the ASME (1956) is a collection of papers assessing the effect of different variables. Weibull and Odqvist (1956) is a similar engineering conference. Studies of the physical nature of fatigue are covered in Freudenthal (1956) and in a survey of the basic work by Thompson and Wadsworth (1958). A more recent conference reported in *Acta Metallurgica*, July 1963, gives many experimental observations on the micromechanisms of fatigue.

ALDEN, T. H. BACKOFEN, W. A.	1961	"The Formation of Fatigue Cracks in Aluminum Single Crystals," *Acta Met.* **9**, 352–366.
ASM	1961	"Properties and Selection of Metals," *Metals Handbook*, 8th ed., T. Lyman, ed., ASM, Novelty, Ohio.
ASME & IME	1956	*Proc. Int. Conf. Fatigue of Metals*, ASME and IME, London and New York.
ASTM	1949	*Manual on Fatigue Testing*, Committee E9, ASTM. Special Tech. Publ. No. 91, Philadelphia.
BATTELLE MEMORIAL INSTITUTE	1941	*Prevention of the Failure of Metals Under Repeated Stress*, Wiley, New York.
BEACHEM, C. D. PELLOUX, R. M. N.	1965	"Electron Fractography—A Tool for the Study of the Micro-Mechanics of Fracturing Processes," *Fracture Toughness Testing and Its Applications*, ASTM Spec. Tech. Pub. 381.
BENNETT, J. A.	1963	"A Simple Environmental Chamber for Rotating Beam Fatigue Testing Machines," *Mat. Res. Stand.* **3**, 480–482.
BROWN, A. F.	1952	"Surface Effects in Plastic Deformation of Metals," *Advances in Physics* **1**, 427–477 (see p. 469).
BUSH, J. J. GRUBE, W. L. ROBINSON, G. H.	1961	"Microstructure and Residual Stress Changes in Hardened Steel Due to Rolling Contact," *Trans. ASM* **54**, 370–412.
CASWELL, J. S.	1947	"Design of Parts for Conditions of Variable Stress," *Product Eng.* **18**, 118–119.
CAZAUD, R.	1953	*Fatigue of Metals*, Translated by A. J. Fenner, Chapman-Hall, London.
CHRISTENSEN, R. H. DENKE, P. H.	1961	"Crack Strength and Crack Propagation Charactersitics of High-Strength Metals," *ASD Tech. Rept.* 61-207, Aeronautical Systems Division, U.S. Air Force, Wright-Patterson Air Force Base, Ohio.

CLAYTON-CAVE, J. TAYLOR, R. J. INESON, E.	1955	"Reproducibility of Wöhler Type Fatigue Tests," *J. Iron Steel Inst.* **180**, 161–169.
COFFIN, L. R., JR.	1960	"The Stability of Metals under Cyclic Plastic Strain," *Trans. ASME, J. Basic Eng.* **82D**, 671–682.
CORDIANO, H. V. COCHRAN, E. P. WOLFE, R. J.	1956	"A Study of Combustion Resistant Hydraulic Fluids as Ball Bearing Lubricants," *Lubrication Eng.* **12**, 261–266.
COTTRELL, A. H. HULL, D.	1957	"Extrusion and Intrusion by Cyclic Slip in Copper," *Proc. Roy. Soc. (London)* **A242**, 211–213.
CUMMINGS, H. N. STULEN, F. B. SCHULTE, W. C.	1957	"Relation of Inclusions to the Fatigue Properties of 4340 Steel," *Trans, ASM* **49**, 482–516.
DIETER, G. E. MEHL, R. F.	1953	"Investigation of the Statistical Nature of the Fatigue of Metals," *NACA Tech. Note 3019*.
DONALDSON, D. R. ANDERSON, W. E.	1961	"Crack Propagation Behavior of Some Air Frame Materials," *Proc. Crack Propagation Symp.* Vol. 2, The College of Aeronautics, Cranfield, 375–441.
EBNER, M. L. BACKOFEN W. A.	1959	"Fatigue in Single Crystals of Copper," *Trans. AIME* **215**, 510–520.
EWING, J. A. HUMPHREY, J. C. W.	1903	"The Fracture of Metals Under Repeated Alternations of Stress," *Phil. Trans. Roy. Soc. (London)* **A200**, 241–250.
FELTNER, C. E.	1963	"Dislocation Arrangements in Aluminum Deformed by Repeated Tensile Stresses," *Acta Met.* **11**, 817–828.
FINDLEY, W. N.	1957	"Fatigue of Metals Under Combinations of Stresses," *Trans. ASME*, **79**, 1337–1348.
FINDLEY, W. N. MATHUR, P. N. SZCZEPANSKI, E. OITEMEL, A.	1961	"Energy Versus Stress Theories for Combined Stress—A Fatigue Experiment Using a Rotating Disk," *Trans. ASME, J. Basic Eng.* **83D**, 10–14.
FLUCK, P. G.	1951	"The Influence of Surface Roughness on the Fatigue Life and Scatter of Test Results of Two Steels," *Proc. ASTM* **51**, 584–592.
FORSYTH, P. J. E.	1959	"A Study of the Damage Caused by Torsional Fatigue in an Aluminum Alloy," *Tech. Note Met. 310*, Royal Aircraft Establishment, Farnborough, England.
FORSYTH, P. J. E.	1963	"Fatigue Damage and Crack Growth in Aluminum Alloys," *Acta Met.* **11**, 703–715.
FORSYTH, P. J. E. STUBBINGTON, C. A.	1955	"The Slip Band Extrusion Effect Observed in Some Aluminum Alloys Subjected to Cyclic Stress," *J. Inst. Metals* **83**, 395–399.

FRANKEL, A. G. BENNETT, J. A. CARMAN, C. M.	1960	"Fatigue Properties of Some High Strength Steels," *Proc. ASTM* **60,** 501–511.
FREUDENTHAL, A. M. (ed.)	1956	*Fatigue in Aircraft Structures*, Academic Press, New York.
FROST, N. E. PHILLIPS, C. E.	1957	"Some Observations on the Spread of Fatigue Cracks," *Proc. Roy. Soc. (London)* **A242,** 216–222.
GEORGE, R. W. FORSYTH, P. J. E.	1962	"The Influence of Nickel and Manganese on the Fatigue Properties and Microstructures of Three Al-Mg-Zn Alloys," *Tech. Note Met. Phys. 352*, Royal Aircraft Establishment, Farnborough, England.
GOUGH, H. J.	1928	"The Behavior of a Single Crystal of Iron Subject to Alternating Torsional Stresses," *Proc. Roy. Soc. (London)* **A118,** 498–534.
GOUGH, H. J.	1933	"Crystalline Structure in Relation to Failure of Metals, Especially by Fatigue," *Proc. ASTM* **33,** 3–114.
GROSSKREUTZ, J. C.	1962	"Fatigue Crack Propagation in Aluminum Single Crystals," *J. Appl. Phys.* **33,** 1787–1792.
GROVER, H. J. GORDON, A. JACKSON, L. R.	1956	*Fatigue of Metals and Structures*, Thames and Hudson, London.
HEMPEL, M.	1956	"Metallographic Observations on the Fatigue of Steels," *Proc. Int. Conf. Fatigue of Metals*, ASME and IME, London and New York, pp. 543–547.
HEYWOOD, R. B.	1956	"The Effects of High Loads on Fatigue," *Proc. Colloq. Fatigue, Stockholm*, Springer, Berlin, 92–102.
HOLDEN, J.	1961	"The Formation of Sub-Grain Structure by Alternating Plastic Strain," *Phil. Mag.* **6,** 547–558.
HOWELL, F. M. MILLER, J. L.	1955	"Axial-Stress Fatigue Strengths of Several Structural Aluminum Alloys," *Proc. ASTM* **55,** 955–968.
KARRY, R. W. DOLAN, T. J.	1953	"Influence of Grain Size on Fatigue Notch-Sensitivity," *Proc. ASTM* **53,** 789–804.
KENNEDY, A. J.	1963	*Processes of Creep and Fatigue in Metals*, Wiley, New York.
LAIRD, C. SMITH, G. S.	1962	"Crack Propagation in High Stress Fatigue," *Phil. Mag.* **7,** 847–858.
LAZAN, B. J. BLATHERWICK, A. A.	1952	"Fatigue Properties of Aluminum Alloys at Various Direct Stress Ratios, Part 1," *Tech. Rept.* 52-307, Wright Air Development Command, Wright-Patterson Air Force Base, Ohio.

Lessells, J. M.	1954	*Strength and Resistance of Metals*, Wiley, New York.
Lipson, C.	1950	"Methods of Crack Detection," *Handbook of Experimental Stress Analysis*, M. Hetényi, ed., Wiley, New York, pp. 579–592.
MacGregor, C. W. Grossman, N.	1952	"The Effect of Cyclic Loading on the Mechanical Behavior of 24S-T4 and 75S-T6 Aluminum Alloys and 4130 Steel," *NACA Tech. Note 2812*.
McAdam, D. J., Jr.	1926	"Stress-Strain-Cycle Relationship and Corrosion Fatigue of Metals, Part II," *Proc. ASTM* **26**, 224–254.
McClintock, F. A.	1955	"A Criterion for Minimum Scatter in Fatigue Testing," *J. Appl. Mech.* **22**, 427–431.
McClintock, F. A.	1956	Discussion in *Fatigue in Aircraft Structures*, A. M. Freudenthal, ed., Academic Press, New York, pp. 78–79.
McClintock, F. A.	1963	"On the Plasticity of the Growth of Fatigue Cracks," *Fracture in Solids, Met. Soc. AIME Conf. 20*, Interscience, New York, pp. 65–102.
McMaster, R. C. (ed.)	1959	*Nondestructive Testing Handbook*, Vols. I and II, Ronald Press, New York.
Miner, M. A.	1945	"Cumulative Damage in Fatigue," *J. Appl. Mech., ASME*, **67**, A159–A164.
Moore, C. C.	1956	"A Study of Premature Fatigue Failures in 150 mm Ball Bearings," *T.I.S., R. 56, GL 120*, General Electric Co.
Morrison, T. W.	1955	"Some Unusual Conditions Encountered in Lubrication of Rolling Contact Bearings," *Lubrication Eng.* **11**, 405–411.
Mott, N. F.	1958	"A Theory of the Origin of Fatigue Cracks," *Acta Met.* **6**, 195–197.
Newmark, N. H.	1952	"A Review of the Cumulative Damage in Fatigue Fracture of Metals," *Symp. Fatigue and Fracture of Metals*, M.I.T. Press, Cambridge, Mass., and Wiley, New York, pp. 197–228.
Nielsen, L. E.	1962	*Mechanical Properties of Polymers*, Reinhold, New York.
Orowan, E.	1939	"Theory of the Fatigue of Metals," *Proc. Roy. Soc. (London)*, **A171**, 79–106.
Palmgren, A.	1924	"Die Lebensdauer von Kugellagern," *Z. Verein Deutscher Ingenieure* **68**, 339–347.
Paris, P. C. Gomez, M. P. Anderson, W. E.	1961	"A Rational Analytic Theory of Fatigue," *The Trend in Engineering*, University of Washington, pp. 9–14.

Peterson, R. E.	1950	"Interpretation of Service Fractures," *Handbook of Experimental Stress Analysis*, Hetényi, ed., Wiley, New York, pp. 593–635.
Peterson, R. E.	1963	"Fatigue of Metals, Part 3—Engineering and Design Aspects," *Mat. Res. Stand.* **3,** 122–139.
Peterson, R. E. Wahl, A. M.	1936	"Two and Three Dimensional Cases of Stress Concentration, and Comparison with Fatigue Test," *J. Appl. Mech.* **3,** A15–A22.
Prot, M.	1948	"Un nouvelle technique d'essais de fatigue sous charge progressive," *Rev. de Metallurgie* **45,** 481–489; see also **48,** 822–824.
Prot, M.	1952	"Fatigue Testing Under Progressive Loading," translated by E. J. Ward, *Tech. Rept.* 52–148, Wright Air Development Command, Wright-Patterson Air Force Base, Ohio.
Radd, F. J. Crowder, L. H. Wolfe, L. H.	1959	"Relationship of pH to Aerobic Corrosion and Fatigue Life of Steel," *Nature* **184,** 2008–2009.
Ransom, J. T. Mehl, R. R.	1952	"The Anisotropy of the Fatigue Properties of SAE 4340 Steel Forgings," *Proc. ASTM* **62,** 777–790.
Raymond, M. H. Coffin, L. F., Jr.	1963	"Geometric and Hysteresis Effects in Strain-Cycled Aluminum," *Acta Met.* **11,** 801–807.
Rasmussen, J. G.	1962	"Prediction of Fatigue Failure Using Ultrasonic Surface Waves," *J. Soc. Nondestructive Test.* **20,** 103–110.
Segall, R. L.	1963	"Lattice Defects in Fatigued Metals," *Electron Microscopy and Strength of Crystals*, G. Thomas and J. Washburn, eds., Interscience, New York, pp. 515–534.
Simkovich, E. A. Loria, E. A.	1961	"Effect of Decarburization and Grinding Conditions on Fatigue Strength of 5% Cr-Mo-V Sheet Steel," *Trans. ASM* **53,** 109–122, 889–896.
Sinclair, G. M.	1952	"An Investigation of the Coaxing Effect in the Fatigue of Metals," *Proc. ASTM* **52,** 743–758.
Sinclair, G. M. Corten, H. T. Dolan, T. J.	1957	"Effect of Surface Finish on the Fatigue Strength of Titanium Alloys RC 130B and Ti 140A," *Trans. ASME* **79,** 89–96.
Sines, G. Waisman, J. L.	1959	*Metal Fatigue*, McGraw-Hill, New York.
Smith, F. C. Brueggeman, W. C. Harwell, R. H.	1950	"Comparison of Fatigue Strengths of Base and Alclad 24S-T3 Aluminum," *NACA Tech. Note 2231*.
Smith, G. C.	1957	"The Initial Fatigue Crack," *Proc. Roy. Soc. (London)* **A242,** 189–197.
Soderberg, C. R.	1930	"Factor of Safety and Working Stresses," *Trans. ASME J. Appl. Mech.* **52,** 13–28.

STARKEY, W. L. MARCO, S. M.	1957	"Effect of Complex Stress-Time Cycles on the Fatigue Properties of Metals," *Trans. ASME* **79,** 1329–1336.
TARASOV, L. P. HYLER, W. S.	1957	"Effects of Grinding Conditions and Resultant Residual Stresses on the Fatigue Strength of Hardened Steel," *Proc. ASTM* **57,** 601–622.
TAVERNELLI, J. F. COFFIN, L. F., JR.	1959	"A Compilation and Interpretation of Cyclic Strain Fatigue Tests on Metals," *Trans. ASM* **51,** 438–453.
TAVERNELLI, J. F. COFFIN, L. F., JR.	1962	"Experimental Support for Generalized Equation Predicting Low-Cycle Fatigue," *Trans. ASME, J. Basic Eng.* **84D,** 533–541.
THOMAS, W. N.	1923	"Effect of Scratches and Various Workshop Finishes upon the Fatigue Strength of Steel," *Engineering* **116,** 449–454, 483–485.
THOMPSON, N. WADSWORTH, N. J.	1958	"Metal Fatigue," *Advances in Physics* **7,** 72–169.
THOMPSON, N. WADSWORTH, N. J. LOUAT, N.	1956	"The Origin of Fatigue Fracture in Copper," *Phil. Mag.* **1,** 113–126.
WADSWORTH, N. J.	1961	"The Influence of Atmospheric Corrosion on the Fatigue Life of Fe-0.5% C," *Phil. Mag.* **6,** 397–408.
WADSWORTH, N. J. HUTCHINGS, J.	1958	"The Effect of Atmospheric Corrosion on Metal Fatigue," *Phil. Mag.* **8,** 1154–1166.
WEIBULL, W. ODQVIST, K. G. (eds.)	1956	*Colloquium on Fatigue,* Stockholm, 1955, Springer, Berlin.
WILSON, R. N. FORSYTH, P. J. E.	1959	"Some Thin Foil Observations on the Fatigue Process in Pure Aluminum," *Tech. Note Met. 311,* Royal Aircraft Establishment, Farnborough, England.
WÖHLER, A.	1860–1870	"Versuche über die Festigkeit der Eisenbahnwagen-Achsen," *Z. Bauwesen,* 10, 13, 16, 20.
WOOD, W. A.	1956	"Mechanism of Fatigue," *Fatigue in Aircraft Structures,* A. M. Freudenthal, ed., Academic Press, New York.
WOOD, W. A.	1959	"Some Basic Studies of Fatigue in Metals," *Fracture,* B. L. Averbach et al., eds., M.I.T. Press, Cambridge, Mass., and Wiley, New York, pp. 412–434.
WOOD, W. A. COUSLAND, S. M. SARGANT, K. R.	1963	"Systematic Microstructural Changes Peculiar to Fatigue Deformation," *Acta Met.* **11,** 643–652.
WOOD, W. A. DAVIES, R. B.	1953	"Effects of Alternating Strain on the Structure of a Metal," *Proc. Roy. Soc. (London)* **A220,** 255–266.

PROBLEMS

◀ 18.1. Brown (1952) proposed the dislocation model shown in Fig. 18.32 to explain the existence of extrusions or intrusions.

(a) Sketch several stages in the dislocation motion leading to this configuration.

(b) Determine the stress distribution around a dislocation dipole such as that shown in Fig. 18.32(d), neglecting the presence of the free boundary.

(c) From the above result, what stresses would have to be superimposed on the crack surface in order to offset the effect of the dislocation dipole. From this, determine the approximate character and magnitude of the stress concentration at the tip of the intrusion. Compare this with the stress distribution you would estimate from imagining the slipped planes to be removed from the specimen and slit so that they could take a stress-free state as shown in Fig. 18.32(e).

(d) What governs whether a stress concentration at the tip of the inclusion will result in cleavage or plastic flow?

(e) Discuss the significance and limitations of this model.

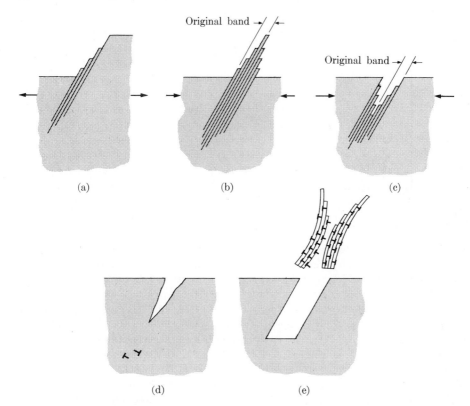

FIG. 18.32. (a) Original slip band. (b) Reversed slip to right of original band. (c) Reversed slip to left of original band. (d) Dislocation dipole near an intersection. (e) Intrusion core imagined as being removed from surface (Prob. 18.1).

620 FATIGUE

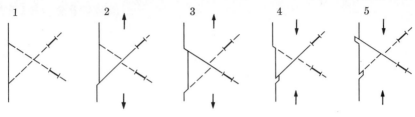

FIGURE 18.33.

◀ 18.2. Cottrell and Hull (1957) proposed the dislocation model shown in Fig. 18.33 for the development of intrusions and extrusions, as a result of the alternating operation of two Frank-Read sources. Discuss the significance and limitations of this model.

◀ 18.3. Mott (1958) presented the dislocation model of extrusion formation shown in Fig. 18.34, in which a screw dislocation makes successive circuits around a subsurface crack by cross slip and gradually raises an extrusion, leaving a void inside the material.

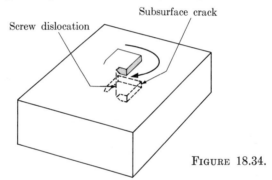

FIGURE 18.34.

(a) The model as originally presented cannot have the proper stress to drive the screw dislocation if the extrusion is perpendicular to the surface. Show, however, that by simply tipping the screw dislocation by 45° to the surface, the necessary shear stress can occur.

(b) Discuss other aspects of the significance and limitations of this model.

◀ 18.4. Ebner and Backofen (1959) suggested that slip starts on the primary slip planes, but cracking does not occur until cross slip starts on secondary intersecting planes. Discuss the significance and limitations of these ideas in terms of a dislocation model. What other effects might be expected to occur as a result of the beginning of cross slip?

E 18.5. Form a roughly cylindrical specimen of "Plasticine" (modeling clay) with your hands and manually perform a torsion fatigue test. Note that the fingerprints on the surface become accentuated and that surface cracks form. Discuss this failure, remembering that this material is amorphous (without slip planes). Predict the life from a monotonic test.

◀ 18.6. Orowan (1939) proposed a phenomenological model of fatigue based on the configuration shown in Fig. 18.35. This model led to a reasonably realistic S-N curve and a fatigue limit above the yield strength of the frictional element C_f, which is assumed to strain-harden linearly, and to fracture at some particular value of the stress or accumulated strain. Discuss how this model should be modified to give a picture of deformation within a grain in the fatigue process as we now understand it.

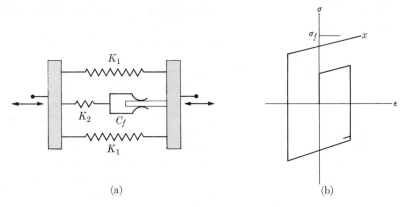

FIG. 18.35. (a) Model. (b) Stress-strain characteristics of element C_f.

18.7. (a) Calculate the exponent n in Eq. 18.2 for the 4340 steel shown in Fig. 18.13, with unnotched specimens.

(b) Calculate the exponent n in Eq. 18.2, assuming that the S-N curve remains constant until 10^2 to 10^3 cycles and then falls off, as for a ferrous alloy, until 10^6 to 10^7 cycles are reached.

18.8. (a) Replot the data for 4340 steel given in Fig. 18.13 as nearly as you can in terms of an ϵ-N diagram. Superimpose the curve corresponding to Eq. 18.3.

(b) Discuss the relative difficulty of determining ϵ-N and S-N curves from experimental data, assuming appropriate observations are made in each case. Pay special attention to the high-stress, short-life region of the curve.

18.9. Sketch on Fig. 18.16 the locus of fracture under the first application of a load, and note how the data tend in this direction for shorter lives.

18.10. Draw a Soderberg diagram for the 7075-T6 aluminum alloy of Fig. 18.16.

18.11. Show that the approximate condition for fatigue failure under combined stress, Eq. 18.5b, reduces to Eq. 18.4b for uniaxial tests.

18.12. From Eq. 18.5b, derive the relation between bending and torsional stress amplitude for fatigue failure, Eq. 18.5c, and show that $\sigma_l/\tau_l \approx 2$.

18.13. Gray cast iron is known to have a low value of the fatigue strength reduction factor K_f (i.e., it is insensitive to notches). Explain how this behavior is consistent with the known fact that cast iron contains many inclusions of thin plates of weak graphite in its matrix.

18.14. Using the data in Fig. 18.23 calculate values of K_f for the curves given. Are these values consistent with the values of K_t shown on the curves? Why is K_f usually smaller? Why is the agreement better for the small grain sizes?

18.15. If strain-aging is important in determining whether or not a metal can be coaxed to a higher fatigue limit, discuss how this phenomenon will vary with temperature and speed of cycling.

18.16. Judging from what you know about the microscopic mechanisms of fatigue, what would you expect for the effect of grain size on the life of parts stressed to the same fraction of their tensile strength?

◀18.17. Show that for spherical bodies in contact, the maximum shear stress beneath the surface is about three times that on the surface.

◀18.18. As given in Eq. 18.13, rolling-contact bearing fatigue life is based on an empirical equation relating life and load in the equation P^3N = constant. Is this variation within the usual power relationship between life and stress? [*Hint:* Consult the Hertz equations relating contact stress to load.]

18.19. Bent shafts and axles are often straightened by cold-bending in an arbor press. What effect is this likely to have on future fatigue life? Can you suggest any way to alleviate these effects?

18.20. Propose and justify a method for designing shafts subject to a residual bending stress σ_{mb}, a rotating bending σ_{ab}, an alternating shear stress τ_{as} at n times the rotating frequency ν_a, and a mean shear stress τ_{ms}. First consider the special case of $n = 1$.

18.21. In view of the possible strong effect of environment on fatigue, discuss the appropriateness of the rather common practice of testing automobiles by driving them 50,000 to 100,000 miles in a short time period. For what components is this most appropriate, and for what is it least appropriate?

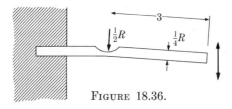

FIGURE 18.36.

18.22. Specimens of cold-rolled steel were prepared for reversed bending fatigue tests by milling a $\frac{1}{2}$-in. radius fillet in one side as shown in Fig. 18.36. When these were tested at a stress range of ± 70,000 psi, it was found that life was shortened equally by either a mean compressive stress or a mean tensile stress on the machined surface. When the specimens were tested at ± 50,000 psi, however, it was found that the reduction in life due to mean stress was less when the machined surface was in tension than when it was in compression. Discuss these results. How can your hypothesis be verified by visual examination of the as-machined specimens?

18.23. (a) An automobile part made of hot-rolled SAE 1020 steel is failing after 20,000 miles. What decrease in stress would be required to extend its life to 200,000 miles? 8000 psi; 16,000 psi; 32,000 psi.

(b) Mention three ways in which the life of the part might be increased the desired amount without changing the nominal applied stress.

18.24. It has been stated that 80% of the service failures in machine parts are due to fatigue, rather than to wear, creep, impact, buckling, or failures as observed in the tensile test. (a) Explain why this should be so. (b) Discuss what steps you can take as an engineer to reduce this figure in equipment you design and supervise.

18.25. Residual stress has been blamed in part for failures in engineering structures. In repeated, cyclic, loading, however, residual stress is often desirable. Why is this so? Describe *briefly* how such desirable residual stress may be introduced.

18.26. What effect is a hardness indentation with a spherical indenter likely to have on a fatigue specimen? What factors will tend to prolong and what factors will tend to shorten its life? If the mark does nucleate fracture, where is it most likely to occur? Why?

18.27. One 1961 American automobile used a flexible drive shaft in place of the conventional rigid drive shaft plus universal joints. Taking typical automobile dimensions, make an estimate of the probable dimensions for such a shaft. State clearly the factors considered and their relative importance.

18.28. Wood's concept of fatigue proposes the possibility of non-workhardening plastic strain. Based on your knowledge of dislocation theory, estimate the order of plastic strain which is possible without work-hardening.

18.29. Why does the spheroidite steel of Fig. 18.22 show greater scatter than the pearlite. Which will show greater notch sensitivity? [*Hint:* Compare the sizes of the two structures.]

18.30. Discuss the logic and validity of the following statement: "For crack growth, the plastic zone at the tip of a fatigue crack is usually quite small, so relatively little time is spent while the crack goes through this region. The history effect is therefore a small one and the Miner equation should hold for crack propagation."

18.31. Discuss the effect of grain size on fatigue crack propagation, especially considering its effect on the mean length of slip bands and the applicability of continuum plasticity.

◀18.32. Show that it is or is not possible to obtain the shear stress-strain curves from the torque-twist curves given in Fig. 18.7 for hardening under alternating torsion.

18.33. According to Battelle (1941), nitriding can reduce the notch sensitivity of specimens to zero. Explain how this is possible.

◀18.34. Estimate how much of the notch insensitivity observed for the 4340 steel reported in Fig. 18.13 is due to the variability shown in those diagrams.

◀18.35. Would the size effect in ball bearings tend to make them more or less sensitive to load than calculated from Eq. 18.2 for normal fatigue behavior and the Hertz equations for contact stress, Eqs. 11.32 and 11.34?

◀18.36. Assuming the existence of an endurance limit, the empirical stress-life relation of Eq. 18.2, and the Miner equation (18.8), calculate how the stress at fracture should depend on the stress increase per cycle when using the Prot scheme for determining the endurance limit.

◀18.37. Discuss the prospects for the ratio of fatigue limit to tensile strength for the dispersion-hardened alloys discussed in Chapter 19 in connection with their creep resistance.

18.38. A string saw for cutting magnesium single crystals consists of a 0.012-in. diameter, stranded, endless, stainless-steel wire, made by brazing together the ends of a length of wire of 88 in. The wire, which was kept wet with concentrated HCl, initially ran over

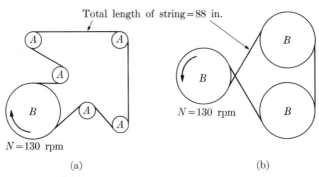

Fig. 18.37. All A pulleys have 2-in. diameter and all B pulleys have 6-in. diameter.

a set of pulleys in an arrangement as shown in Fig. 18.37(a), cutting the crystal by localized chemical dissolution. It was found that although immersion of the brazed joint into HCl for several hours would not reduce its strength significantly, if run over the pulleys, the wire would break repeatedly at the brazed joint after a period of about 15 minutes. Corrosion fatigue was suspected, and the arrangement was changed to the one shown in Fig. 18.37(b), without altering the continuous length of the string. It was also observed that: (1) although the bulk of the wire was stranded and flexible, the brazed joint, being solid, behaved like a wire of 0.012-in. diameter; (2) the breaking strength of the string at the joint was about 15 lb.

(a) Calculate the strain amplitude and the total number of cycles to failure and determine whether in the original setup, corrosion fatigue was important.

(b) Furnish an estimate for the new life of the string in the modified setup.

18.39. For a steel of Rockwell C-50 hardness, estimate the stress level to give a life of 10^5 cycles. Sketch the S-N curve. Compare your sketch with the data of Frankel et al. (1960).

CHAPTER 19

CREEP

19.1 SYNOPSIS

Time-dependent inelastic deformation is known as creep. The time-dependent deformation of polymers and glasses was discussed in Chapter 6. Their extreme temperature and strain-rate sensitivity, coupled with their generally low melting or softening temperatures, make them unattractive for engineering applications at elevated temperatures. For these reasons the discussion in this chapter will be confined to time-dependent deformation of fully crystalline materials.

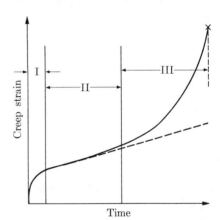

FIG. 19.1. Schematic diagram showing the three possible stages of creep at constant temperature: I, transient; II, steady state; III, tertiary.

Creep is traditionally divided into three stages, though not all are always present. Figure 19.1 illustrates schematically and disproportionately the three stages under a constant uniaxial load. The first stage is called *transient* or primary, the second is called *steady-state* or secondary, and the third is called *tertiary* creep. Usually the increase in the creep rate in the tertiary stage is due to an increase in stress as the area is reduced either by thinning down of the specimen or by internal fracture or void formation. It may also be due to metallurgical changes, such as recrystallization or overaging.

We shall consider first the mechanisms of creep in single crystals and polycrystals as a guide to the interpretation of creep data for design purposes. Data for various engineering materials will be used to illustrate the ways of presenting creep behavior.

Fracture is likely to intervene earlier in the creep process than with plastic deformation, depending on the temperature and stress. Accordingly we shall consider some of the variables which govern fracture in creep.

19.2 MECHANISMS OF CREEP IN SINGLE CRYSTALS

Creep in crystalline materials can be due in part to the time-dependent plastic* deformation of grains and in part to relative sliding at grain boundaries. We shall investigate first the mechanisms of time-dependent plastic deformation of the grains themselves.

A. Transient Creep. The transient part of the creep process results from the temperature sensitivity of usual forms of plastic deformation. In Section 5.5 we discussed mechanisms which could account for the temperature and rate sensitivity of plastic deformation and how a logarithmic transient-creep law could arise. A logarithmic time law for creep is observed only at low temperatures (below room temperature for copper and aluminum) where recovery effects are negligible. At somewhat higher temperatures, but still far below any recrystallization temperature, some recovery effects become observable through the appearance of a second and more rapid component of transient creep, the rate of which is proportional to $t^{-2/3}$, leading to an additional creep strain proportional to $t^{1/3}$. This creep component, which is known as Andrade's β creep, is generally associated with rapid recovery effects which set in at high plastic strains or high temperatures. Mott (1953) has proposed a theory for β creep that is based on a recovery process in which nearly random changes in configuration occur in the neighborhood of any point. A sufficient number of the resulting stress changes then promote further creep at that point.

B. Steady-State Creep. At high temperatures, where diffusion processes become possible, the rate of creep strain becomes constant. In this steady-state creep region the strain-hardening rate is balanced by the rate of recovery. This type of creep is of engineering significance since it limits the use of parts at elevated temperature; most of our considerations will be based on it.

1. Climb of Edge Dislocations. Several mechanisms have been proposed for the steady-state creep component. One theory (Weertman, 1955) assumes the mechanism to be a climb of edge dislocations over obstacles which may be either precipitates or dislocation entanglements, as shown in Fig. 19.2. When the dislocation has climbed out of its slip plane and has cleared its obstacle, it can sweep out a new area and produce creep strain. The rate of creep strain is then determined by the plastic strain resulting from the glide of a dislocation from one obstacle to another,

---

* Note that in this context "plastic" refers to deformation of a crystal by slip, twinning, or kinking, rather than to time-independent, inelastic deformation, as the term is used in continuum mechanics.

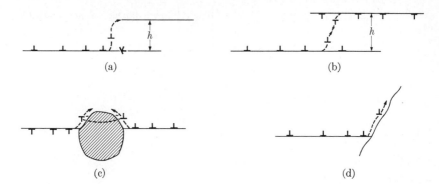

FIG. 19.2. Diagram of recovery mechanisms due to climb. (a) Climbing over the sessile dislocation. (b) Annihilation of dislocations of opposite sign on parallel planes. (c) Disappearance of a loop by climbing over a precipitate. (d) Climbing along a grain boundary. (Schoeck, 1957. Courtesy of ASM.)

by the number of dislocations undergoing the climb and glide sequence, and by the mean time of climb over obstacles. Since the climb of edge dislocations requires diffusion of vacancies, the mean time of climb will be controlled by the rate of arrival of vacancies from vacancy sources, or vice versa. Because the climb of an edge dislocation will become easier in the presence of a local normal stress component σ_N across the extra half-plane of the dislocation, the activation enthalpy of climb would be the difference between the activation enthalpy of self-diffusion and the work done by the normal stress component, i.e.,

$$u_a = h_D - \sigma_N v^*,$$

where v^* is an activation volume of the order of b^3, similar to that discussed in Section 5.5. The climb process can at each stage be undone by a reverse process which, however, has to do work against the local normal stress. Thus the net rate of climb will be proportional to the difference between the forward climb probability and the backward climb probability (similar to the case of viscosity discussed in Section 6.4):

$$\nu \propto \nu_0(e^{-(h_D - \sigma_N v^*)/kT} - e^{-(h_D + \sigma_N v^*)/kT}, \tag{19.1}$$

where ν_0 is a frequency factor dependent on the exact details of the local climb mechanism. The mean time t of climb therefore will be proportional to

$$\tau = \frac{\tau_0 e^{h_D/kT}}{2\sinh(\sigma_N v^*/kT)}$$

Because of the large activation enthalpies involved, it is difficult for any length of edge dislocation greater than an interatomic distance to climb at once.

This does, however, allow the climbing of edge dislocations by the motion of jogs along the length, as shown in Fig. 19.3. The activation volume normally

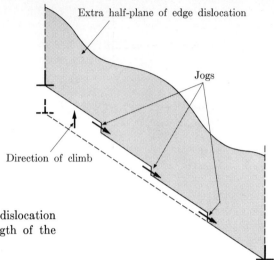

FIG. 19.3. The climb of an edge dislocation by the climb of jogs along the length of the dislocation. (After Friedel, 1956.)

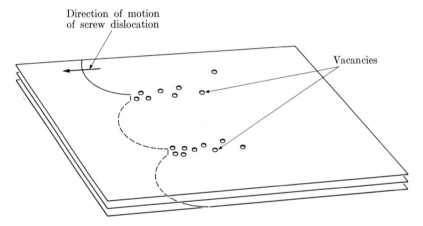

FIG. 19.4. Motion of screw dislocation hindered by drag of vacancy-forming sessile jogs. (After Hirsch and Warrington, 1961.)

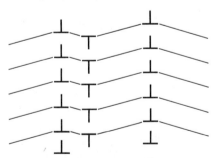

FIG. 19.5. Formation of low-angle tilt boundaries by climb and subsequent glide, known as polygonization.

being of the order of b^3 and the stress being small, say of the order of 20,000 psi, at elevated temperatures only the first term of the expansion of a hyperbolic sine term need be taken, giving

$$\tau = \frac{\tau_0 kT}{2\sigma_N v^*} e^{h_D/kT}.$$

Once climb has occurred, the motion of the dislocation, the area swept out by the freed dislocation, and the number of sites where such climb may occur cannot be determined without a detailed study. Here we simply assume such effects to be describable in terms of a power function of the applied stress, thus obtaining an equation for second-stage creep of the form

$$\dot{\epsilon} = A\sigma^m e^{-h_D/kT}, \qquad (19.2)$$

where A is a constant having the dimensions (stress)$^{-m}$ (time)$^{-1}$. For a more detailed discussion of the climb mechanism, see Friedel (1956).

2. Motion of Sessile Jogs. The climb of edge dislocations is not the only model which leads to a second-stage creep rate of the type given in Eq. 19.2. Another model (Hirsch and Warrington, 1961) attributes steady-state creep to the drag which sessile jogs can create on screw dislocations (see Figs. 19.4 and 4.25). Here again the motion of a jog on a screw dislocation is possible only by diffusion of vacancies to or from the jogs. The analysis can lead to an expression identical with Eq. 19.2.

3. Lattice Friction Stress. In materials such as silicon, aluminum oxide, and the like, which normally show little if any plastic deformation at room temperature, the high shear strength is probably a result of a high lattice friction stress for the motion of dislocations (see Section 4.7). Some of these materials, for instance Al_2O_3, will deform plastically at elevated temperatures (Conrad, 1961a; Kronberg, 1962) by slip on basal planes of the complex hexagonal lattice. Because of the extreme temperature and strain-rate sensitivity that is observed in these materials, bordering on the behavior of glasses, they will be difficult to use in practical structures.

C. Experimental Observations. At elevated temperatures and low stresses the diffusion-controlled climb and subsequent glide of edge dislocations will often produce parallel low-angle boundaries as shown in Fig. 19.5. This is called *polygonization*. Figure 19.6 shows metallographic evidence of such polygonization in coarse-grained aluminum. Note from the as-tested surface that deformation occurs in slip bands, and that kink bands (Fig. 5.15) also form. The polygonization boundaries and kink bands, as would be expected, tend to be normal to the slip bands.

Further evidence of the importance of polygonization in aiding creep has been obtained by Friedel, Boulanger, and Crussard (1955), who studied the change in apparent modulus and damping by cycling polygonized aluminum at temperatures as high as 90% of the melting temperature. They used monocrystalline aluminum

(a) (b)

Fig. 19.6. Polygonization in coarse-grained aluminum after 47% elongation under a stress of 1100 psi at 200°C for 853 hours. (a) Surface appearance showing slip and deformation bands. (b) Similar area after polishing and anodic oxidation, viewed under polarized light. (McLean, 1952. Courtesy of the Institute of Metals.)

to avoid effects due to grain boundaries, and cycled through strains of no more than 10^{-5} to minimize structural changes. The results are shown in Fig. 19.7. The fact that the loss in modulus was primarily due to polygonization was demonstrated by the relative stiffness of the recrystallized specimen. The loss in apparent modulus indicates the ease of plastic deformation after polygonization.

In measuring the activation energy of steady-state creep it is essential that rate measurements at various temperatures be made with the same dislocation structure. This is done by measuring the creep rates before and after a small change in temperature. As shown by Dorn (1956) in Fig. 19.8, for pure metals at temperatures over half their melting point, the activation enthalpy found from creep tests is indeed close to that found for self-diffusion from studies of electrical resistivity and from experiments with radioactive tracer elements. These activation enthalpies are also closely related to the melting-point temperature. From an empirical relation drawn from this plot, Shoeck (1961, p. 87) has explained the beginning of diffusion-controlled creep at half the absolute melting-point temperature (Prob. 19.1). Friedel et al. (1955), from their experiments mentioned above, concluded that the enthalpy of activation varied from 40 kcal/mole at lower temperatures, corresponding to vacancy diffusion, to 55 to 60 kcal/mole at the higher temperatures, corresponding to the sum of the activation enthalpies for vacancy formation and diffusion, i.e., the activation enthalpy of self-diffusion (Probs. 19.2, 19.3).

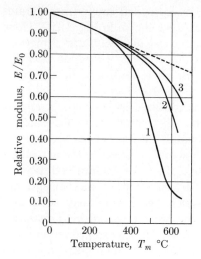

Fig. 19.7. Loss of modulus in 99.96% pure aluminum. (1) Polygonized, (2) partly polygonized, (3) recrystallized. (Friedel et al., 1955. Courtesy of Pergamon Press.)

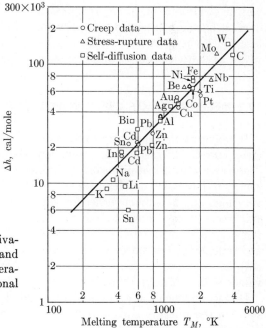

Fig. 19.8. Correlations of the activation energies for creep, stress rupture, and self-diffusion with the melting temperature. (Dorn, 1956. Courtesy of National Physical Laboratory.)

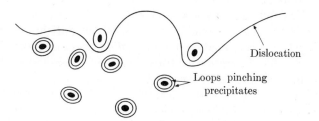

Fig. 19.9. Closed dislocation loops around precipitates left behind by dislocations which have extruded between precipitates.

In precipitation-hardened alloys, dislocation loops are left around precipitates from dislocation extrusion between the precipitates, as shown in Fig. 19.9. These loops, which normally serve to increase the effective size of precipitates somewhat as shown in Fig. 5.43 and block further dislocation motion, can climb over the precipitates, as shown in Fig. 19.2, and collapse by glide, leading to further creep strain by the hitherto blocked dislocations.

The actual mechanisms of creep are rarely as idealized as they would appear from our brief discussion. Other such problems have been studied (see for example, Thornton and Hirsch, 1958, or the review by Conrad, 1961c), but are too detailed to be considered here. It is good to remember that small changes in the activation energy may cause large differences in creep rates. Therefore precise estimates of the activation energies may be necessary to determine what mechanisms are making the current contribution, and which other mechanisms may come into play if conditions of stress, temperature, or structure are changed. This introduction has been intended primarily to sort out the most extreme conditions and mechanisms as a frame of reference to which to refer the more detailed studies which have been made. For a review, see Conrad (1961c).

19.3 MECHANISMS OF CREEP IN POLYCRYSTALS

In polycrystals, there are several additional mechanisms that can contribute to creep: direct diffusion, grain-boundary sliding, and grain-boundary migration. The reason that direct diffusion of vacancies is included in this category is that the energy of formation of vacancies and interstitials simultaneously inside a lattice is very high and it is much easier for vacancies to migrate in from grain boundaries and free surfaces. For an analysis of this mode of creep, see Herring (1950) and the summary by Cottrell (1953, p. 213). This mode of creep has been observed experimentally, but since the rate varies only linearly with stress, it is important only at very low stresses, along with the high temperatures needed to get high values of the diffusion coefficient.

Direct evidence for grain-boundary sliding is shown in Fig. 19.10, where the steps in the initially square grid across the boundary show the relative motion. Measurements of the contribution of grain-boundary sliding to the total strain show fractions as high as 80% (Conrad, 1961b), although values of 30% are more common. Relative sliding along grain boundaries does not alone produce compatible displacements. Various possibilities of accommodation include: migration of grain boundaries by a combination of glide and diffusion, as shown in Fig. 19.11; kinking or so-called fold-formation at triple junctions of grains, as shown in Fig. 19.12; opening up of a crack as shown in Fig. 19.13; and general deformation of the grains themselves. The formation of cracks as a mechanism of accommodation produces subsequent creep fracture, and is evidently not a desirable alternative.

The role of grain-boundary sliding in deformation is dramatically illustrated by Kê's measurements of internal friction peaks in fine wires of aluminum under

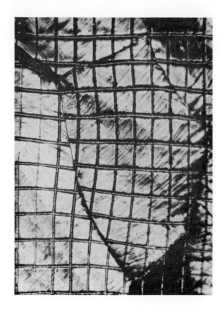

Fig. 19.10. Relative sliding between two grains in a high-purity aluminum specimen. (Brunner, 1957.)

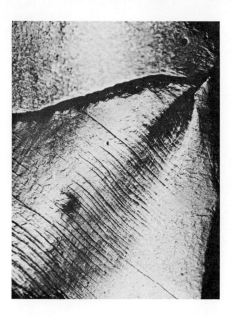

Fig. 19.12. Grain-boundary sliding and fold-formation. Al-2% Cu alloy at 500°F, 3700 psi to 34% elongation. Field of view 0.06 cm high. (Pelloux, 1956.)

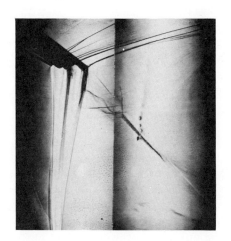

Fig. 19.11. Grain-boundary migration at a triple point. Aluminum at 1000°F, 50 psi to 4.6% elongation in 7.5 hours. Field of view 0.1 cm. (Grant and Chaudhuri, 1957. Courtesy of ASM.)

Fig. 19.13. Initiation of grain-boundary fractures as viewed under polarized light. Field of view 0.1 cm high. (Chang and Grant, 1956. Courtesy of Am. Inst. Mining and Metallurgical Engineers.)

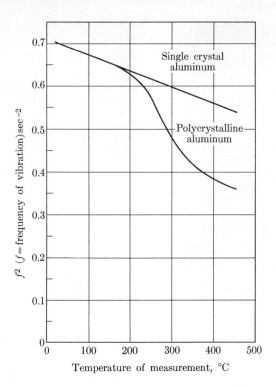

FIG. 19.14. Loss in modulus (proportional to loss in f^2) due to grain-boundary relaxation. (Kê, 1947. Courtesy of the Am. Inst. Phys.)

alternating torsion. Zener (1941) estimated the loss of modulus that would result from grain boundaries which carry no shear stress, by applying the theory of elasticity to find the reduction in strain energy associated with deforming a sphere in the absence of any surface shear stresses, giving

$$\frac{E_{\text{relaxed}}}{E_{\text{unrelaxed}}} = \frac{7 + 5\nu}{2(7 + \nu - 5\nu^2)}. \tag{19.3}$$

Kê's (1947) results are given in Fig. 19.14. It will be noted that the ratio of relaxed to unrelaxed modulus is very near the value of 0.62, given by Eq. 19.3, assuming a Poisson's ratio of $\frac{1}{3}$. The absence of the loss in modulus for single crystals does not entirely prove that the proposed mechanism is the correct one, since there could be other mechanisms triggered by the presence of a grain boundary, that would explain these results. It would be unlikely, however, that the same loss in modulus would result. Another check is to compare the measured coefficient of viscosity from such tests with the viscosity for the liquid, assuming that a layer three atoms wide is contributing to the viscosity. This was done by McLean (1957), and leads to the correct order of magnitude.

Another estimate of the contribution of grain-boundary sliding could be made by considering an elastic solid filled with randomly distributed cracks, the faces of

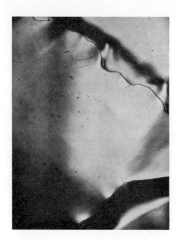

Fig. 19.15. Grain-boundary migration and waviness in Al at 1150°F and 90 psi to 4.2% elongation. Repolished and etched. Field of view 1.5 mm high. (Gervais et al., 1953. Courtesy of Am. Inst. of Mining, Metallurgical, and Petroleum Engineers.)

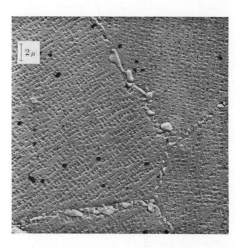

Fig. 19.16. Grain-boundary precipitation in Udimet 700. Composition: 4% Al, 3% Ti, 5% Mo, 15% Cr, 17% Co, 4% max. Fe, balance Ni. Heat-treatment: 2150°F, 4 hr, air cool; 1975°F, 4 hr, air cool; 1550°F, 24 hr, air cool; and 1400°F, 16 hr, air cool. (Courtesy of Special Metals, Inc.)

which support no shear stress. The effect of a single crack has been calculated by Brunner and Grant (1956). Alternatively, one might consider the grain-boundary sliding at the boundary of a possibly plastic cylinder which deformed in such a way as to have no shear strain at its surface (Prob. 19.4). Usual contributions to grain-boundary sliding range from nothing, up to the 30 to 40% expected from these models, leaving the 80% figure reported above unexplained (see also Ishida et al., 1965).

Grain-boundary migration often makes it difficult to determine strain from measurements of grain dimensions, since grains can change shape by boundary migration as well as by deformation. Even away from the triple points, difficulties arise from instability. After small amounts of sliding, a wavy pattern develops, and migration also occurs, as illustrated in Fig. 19.15. This waviness reduces the contribution of grain-boundary sliding at high temperatures.

Grain-boundary sliding can be markedly reduced by introducing alloying elements which form concentrated solutions or precipitates at the grain boundaries. A striking example of this is shown in the high-temperature nickel alloy of Fig. 19.16. For a more detailed discussion of this mechanism of strengthening, see Guard (1961).

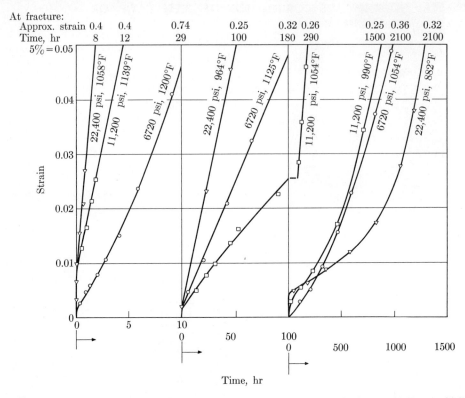

Fig. 19.17. Creep curves for medium carbon forged steel: 0.39% C, 0.24% Si, 0.03% S, 0.03% P, 0.8% Mn. (Data from Tapsell and Johnson, 1931.)

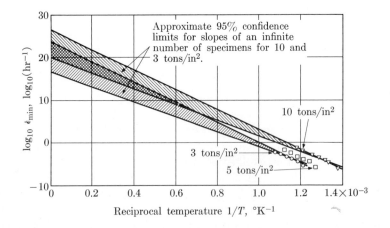

Fig. 19.18. Plot of minimum creep-rate data for medium carbon forged steel of Fig. 19.17.

19.4 PRESENTATION AND CORRELATION OF CREEP DATA FOR ENGINEERING MATERIALS

Although a great wealth of creep information exists, the numerous possible materials and conditions make it unlikely to find exactly the required information. This, and the need to extrapolate fairly short-term creep data for long-term applications, requires at least an empirical correlation technique. A number of methods of correlating the data have been proposed. We shall limit our discussion here to the more fundamental ones, although strictly empirical ones are often as good or better in particular situations.

For low temperatures and for stresses below the yield strength, a type of delayed elastic creep is often observed which more often than not obeys a logarithmic time law resembling that derived in Section 5.6. The exact mechanism of the type of anelastic creep is obscure and data are scarce. Outside of damping and applications in precision apparatus, it is usually negligible. For instance, it amounted to 0.2% of the elastic deflection in 6 hours for a particular household spring scale. At high temperatures or high stress (high in comparison with the yield strength at the temperature in question) a certain portion of the creep strain will be made up of a transient component. If the total strain is limited to one or two percent, this transient creep cannot be ignored. As mentioned in Section 19.2, transient creep usually obeys the relation

$$\epsilon = \beta t^n, \qquad (19.4)$$

where β is a constant that depends on plastic strain, stress, and temperature, and n is usually $\frac{1}{3}$. Data on the stress and temperature sensitivity of β is scarce and is rarely presented, but for copper and aluminum see Wyatt (1953). This type of creep is also often not the controlling mode. For long-term applications, normally the stress has to be kept low to have a low steady-state creep rate, and the transient creep component becomes small in comparison with the steady-state creep component.

The most pressing need for a correlation is in predicting long-time service from data obtained in a reasonably short time. Even if the data from twenty-year tests were available, they would be useless because of the development of obviously superior alloys in the meantime. Of the many different correlations, some discussed by Conrad (1959) and Goldhoff (1959), we shall adopt Eq. 19.2, generalized to allow for a possible dependence of activation enthalpy on stress:

$$\dot{\epsilon} = A\sigma^m e^{-h(\sigma)/kT}. \qquad (19.5)$$

To determine the "constants," Eq. 19.5 is rewritten in the form

$$\ln \dot{\epsilon} = \ln A + m \ln \sigma - h(\sigma)/kT. \qquad (19.6)$$

As an example, consider creep of the carbon steel shown in Fig. 19.17. Suppose it is desired to fit the equation to the steady-state creep rate. More complete data on the minimum creep rates are shown in Fig. 19.18, where the logarithm of the

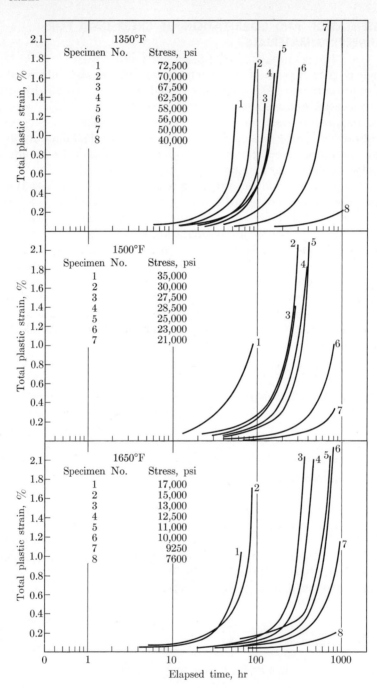

Fig. 19.19. Creep curves for Hastelloy R-235 at three temperatures and various stresses. (McBride et al., 1962. Courtesy of Wright Air Development Division.)

creep rate is plotted versus the reciprocal of the absolute temperature for a series of stress levels. Experience shows that this gives results which are often fairly straight lines. The stress can appear in either of two ways. If the lines have a common intercept, the activation enthalpy varies with stress, and we have the so-called Larson-Miller (1952) correlation (Prob. 19.5) proposed empirically by them, although it had been earlier developed theoretically and applied to strain-rate tests by MacGregor and Fisher (1946). If the straight lines are parallel, the activation enthalpy is independent of stress, as suggested by Dorn (1956). In either event, the slope of the lines gives the activation enthalpy directly. The stress exponent m must be chosen from an appropriate log-log plot. As can be seen from Fig. 19.18, the experimental data can be fitted equally well by either of these assumptions. Taking $h(\sigma)$ independent of stress, the exponent m giving the stress dependence is typically 4 to 8, often dropping at higher temperatures. Note that for Fig. 19.18, the activation enthalpy is $97{,}000 \pm 12{,}000$ cal/mole and the stress exponent about 6 (Prob. 19.6). For a further review of these and other correlations, see Conrad (1959) and Goldhoff (1959), as well as the discussions of Basinski (1957), and references.

Raw creep data are found in various different forms; they are often plotted as strain at constant temperature and stress (usually engineering stress) versus log (time), to shrink the abscissa, such as the curve for Hastelloy R-235 in Fig. 19.19 at 1350, 1500, and 1650°F under various stresses. Note that there are barely sufficient data to determine activation enthalpy, since for only one stress are there data at two temperatures. A log-log plot would make it easier to observe the three stages, if present. For long-term application at low stresses, the second-stage creep rate is often the controlling parameter. Creep data are thus sometimes found in a more compact form as second-stage creep rate plotted against stress for various temperatures, such as Fig. 19.20 for type 316 stainless steel. Another form of presentation of creep data is the relation between stress and time for

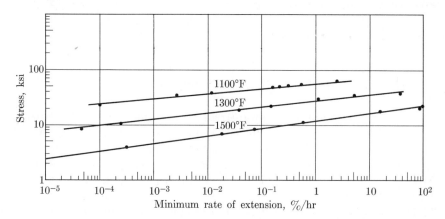

FIG. 19.20. Plot of creep data for type 316 stainless steel. (Shank, 1958. Courtesy of M.I.T.)

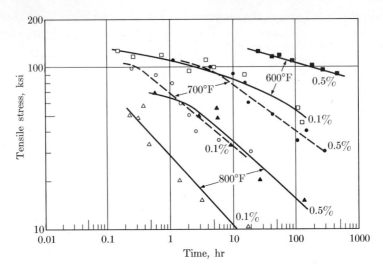

FIG. 19.21. Tensile creep data for titanium alloy sheet (2.5 Al, 16% V, balance Ti), solution-treated and aged. (Kattus and Lessley, 1961. Courtesy of ASTM.)

constant total strain at a given temperature. One such case is the creep curves for solution-treated and aged titanium alloy sheet shown in Fig. 19.21.

Creep data are sometimes cross-plotted as the total strain at a constant time such as 1 hour, 10 hours, etc., versus stress for a certain temperature. This gives a so-called "isochronous stress-strain curve," as shown in Fig. 19.22. Such curves are very convenient for constant-temperature applications where a total strain is the limiting condition.

From the more regular of these representations one can make reasonably good extrapolations by eye, but greater confidence is obtained by cross-plotting to obtain an activation enthalpy as in Fig. 19.18. One must also verify that creep fracture does not intervene, as discussed in Section 19.6.

19.5 EFFECTS OF COMBINED STRESS AND HISTORY

Although most data are for uniaxial stress, more complicated states often occur in practice. Limited experiments such as those by Heimerl and Farquhar (1959) and Kennedy et al. (1959) show that as one would expect from the physical mechanisms underlying creep, creep rates in compression are similar to those in tension, and other states of combined stress can be correlated reasonably well on the basis of the maximum shear stress or the Mises yield criterion.

The history of stress in practice is often complex. This affects creep in a way similar to what might be expected from a knowledge of the Bauschinger effect, fatigue, and also recovery, which is the softening of metal where dislocation climb plays an important role at high temperatures. The effect of recovery is illustrated

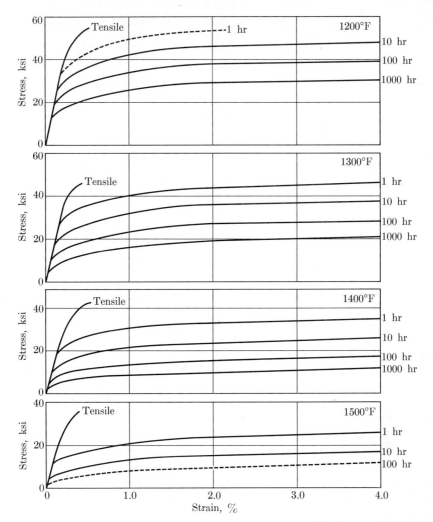

FIG. 19.22. Isochronous creep curves for Allegheny Ludlum Alloy G-192, 0.6C, 8.5Mn, 22.0Cr, balance Fe; 2150°F, 1 hr, water-quench; 1400°F, 16 hr, air cool. (Courtesy of Allegheny Ludlum Steel Corp., 1958.)

for lead in Fig. 19.23. The effect is present, although to a lesser degree, when the stress is simply reduced.

If the unloading is repeated often enough, the process is called fatigue. The acceleration due to the Bauschinger effect becomes continuous, and as Coffin (1960) has shown, a form of creep can occur under cyclic loading even at room temperature.

Another form of varying stress history is encountered in relaxation, which is the decrease in stress at constant total elongation. Since the total strain is the sum of the elastic and creep parts, the occurrence of creep will decrease the elastic

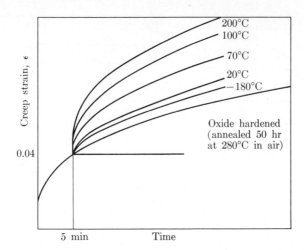

Fig. 19.23. Creep of lead at room temperature and 280 kg/cm². Effect of unloading for 30 min to produce recovery. (Kennedy, 1953. Courtesy of Inst. of Phys.)

part and thus reduce the stress. A rough estimate of relaxation can be obtained from strain rates during creep tests at a series of stress levels. If recovery plays an important part, however, this procedure will not be correct in view of the different stress-strain history in creep and relaxation. Similar difficulties arise in any stress analysis involving creep, as pointed out by Mendelson et al. (1959).

Effects of varying the temperature are summarized by Miller (1954). Although the life may be shortened by a factor of up to 5 compared with that expected from steady-state tests, the required change in stress to compensate for the loss in life is small. There are several reasons for the acceleration of creep under varying temperature. One cause of a temperature effect may be the solution and recondensation of matter from second-phase particles. Another cause can be anisotropic coefficients of thermal expansion in noncubic metals, causing plastic deformation. A third cause, which can be very important, is the presence of thermal stresses. Brophy and Furman (1942) found that a sevenfold increase in creep rate was greatly diminished when a hollow specimen was used to eliminate thermal stresses. Such stresses resulting from cyclic heating and cooling can give rise to thermal fatigue. Coffin (1954) has successfully correlated many examples of thermal strain fatigue on the basis of the plastic strain amplitude per cycle.

The effect of surface conditions on creep is similar to that of surface conditions on plastic deformation in general, and has been reviewed by Kramer and Demer (1961). The primary effect is one of the inhibition of plastic flow by an external film, which is of primary importance only in thin specimens.

Nuclear radiation can affect the creep rate. At low temperatures, where radiation damage cannot be annealed out, radiation causes hardening in a manner discussed in Section 5.2 and a reduction in creep rate results. At temperatures where the radiation damage is annealed out rapidly, the increased concentration of point defects make climb of dislocations easier, with a resulting increase in the creep rate. (See for example, Wilson and Billington 1955.)

19.6 CREEP FRACTURE

For historical reasons, fracture in creep is usually termed "rupture," although to be in keeping with the definition of rupture as reduction of area to the vanishing point, we shall continue to use the word "fracture" to indicate separation with lesser amounts of deformation.

At relatively low temperatures or high strain rates, creep fracture is similar to ordinary ductile fracture in that it occurs in a transgranular fashion. There is a transition, however, at higher temperatures and lower strain rates to intergranular fractures, shown in Fig. 19.24. This typical creep fracture begins at the grain boundaries, as voids at a number of points along a grain boundary or at a triple points, as shown in Fig. 19.13.

Once the cavities have formed, they can grow by sliding of ill-fitting grain boundaries, by general condensation of vacancies into the cavity, and by direct fracture, as discussed by Conrad (1961b). The process of hole growth may in general be a complex phenomenon with diffusion of gas molecules often playing a supporting role (Girifalco and Beeston, 1963).

The time to fracture is a commonly reported variable in creep testing that can be correlated on the basis of an activation energy much as the creep rate data discussed in Section 19.4, except the logarithm of time to fracture is plotted instead of the logarithm of the strain rate. This is somewhat surprising in view of the fundamental differences between fracture and deformation. If intergranular fracture and recovery by overaging do not occur, so that third-stage creep is due to necking, then the equivalence of activation energy is easier to understand. This point was brought out by Hoff (1953), who showed that if at a given temperature, the strain rate could be approximated by an equation of the form

$$\dot{\epsilon} = A\sigma^m, \tag{19.7}$$

(a)

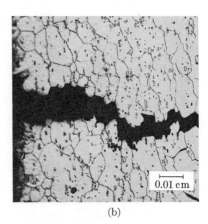
(b)

FIG. 19.24. Creep fracture in Alloy S-590, 20% Cr, 20% Ni, 20% Co, 4% Mo, 4% W, 4% Cb, 0.40% C, 1.2% Mn, 0.4% Si, remainder Fe. (a) Transcrystalline fracture after 0.016 hr at 1350°F and 75,000 psi. (b) Intercrystalline fracture after 30 hr at 1200°F and 60,000 psi. (Courtesy of N. J. Grant.)

the time to rupture is given by (Prob. 19.7)

$$t = 1/(m\dot\epsilon_0), \qquad (19.8)$$

where $\dot\epsilon_0$ is the initial (secondary) creep rate. While Hoff found that this analysis may not give a good estimate of the elongation to fracture, it does give a good estimate of the time to fracture for pure materials that neck substantially before fracture. In most alloys, however, it overestimates life by a factor of 10 or more (Prob. 19.8). The reason for the lower observed life is acceleration of creep rate for reasons other than reduction of area, such as overaging or grain-boundary cracking. The fracture strain may or may not be reduced.

Complex service conditions influence the fracture process in creep; for example, Guarnieri (1954) found that thermal cycling accelerated crack growth under oxidizing conditions, offsetting the oxide strengthening of the type observed by Kennedy (1953) (see Fig. 19.23) and Widmer and Grant (1960).

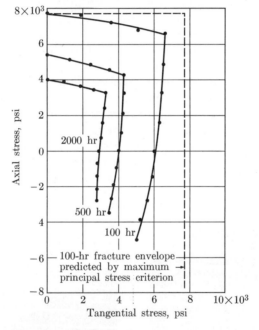

Fig. 19.25. Creep-fracture locus for 76% nickel, 15% chromium, 8% iron alloy in an argon atmosphere, as annealed, 1500°F. (Kennedy et al., 1959. Courtesy of ASME.)

Under multiaxial stress conditions, as shown in Fig. 19.25, a maximum tensile stress criterion appears to be more appropriate for fracture, which is reasonable in view of the importance of opening up cracks. It is interesting to note that anisotropic effects are also of considerable importance, as might be expected from the preferred orientation of inclusions and grains developed during processing of the stock, and of the corresponding effect on nucleating cracks. For further data see Johnson and Frost (1955) and Brown et al. (1959).

19.7 CHOICE OF MATERIALS

The choice of materials for creep resistance is, in general, based on the same principles as the choice of materials for hardness, with the added consideration that the processes must be stable at high temperatures. At high temperatures, interstitial carbon, order-disorder, and strain-hardening forms of hardening are not likely to be of much help. This leaves solute, particle precipitation, and dispersion-strengthening.

Some data on the improvement in creep resistance are presented in Fig. 19.26. By going to a nickel-chromium, solution-hardening alloy, the temperature for 5% strain in 10 hours is increased by almost 400°F. When this alloy is further strengthened by 19% cobalt, along with a few percent of molybdenum, titanium, and aluminum to give the γ' phase Ni_3 (Ti, Al), another 300° is available, although with a loss of some ductility. This illustrates a general problem: hardening by introducing phases which do not form large volumes of brittle intermetallic compounds. One solution is a multistep aging process to nucleate *all* phases without excessive growth of *any*. An example is the nickel alloy of Fig. 19.16. If the precipitate is not very stable, it is sometimes possible to start with an underaged material and let the aging proceed during service so that overaging will not set in too early. The properties of some high-temperature alloys are illustrated for comparative purposes in Figs. 19.27 and 19.28.

Finally, in spite of what can be done, the resistance of solute and precipitation-hardened alloys drops off rapidly at temperatures of about half the melting tem-

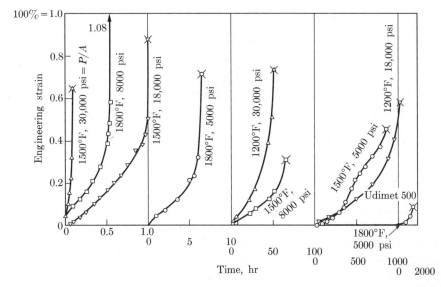

FIG. 19.26. Creep of nickel-chrome alloys. First eight curves for 20% Cr, 1.3% C, 0.1% Fe, balance Ni, water-quenched from 2050°F, aged 24 hr at 1500°F. Udimet 500 is 19% Cr, 0.1% C, 19% Co, 4% Mo, 3% Al, 3% Ti, 4% max. Fe. balance Ni; 1975°F, air cooled; 1550°F, air cooled; and 1400°F, air cooled. (Courtesy of Widmer and Grant, 1961.)

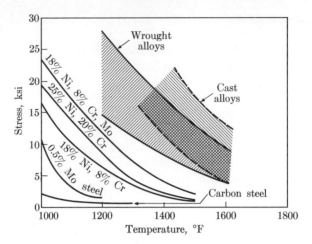

FIG. 19.27. Creep stress for a secondary creep rate of 10^{-6} per hour in wrought and cast high-temperature alloys containing substantial Mo, Co, W, or Cb, in comparison with stresses for alloy steels. (Smith, 1950. Courtesy of McGraw-Hill.)

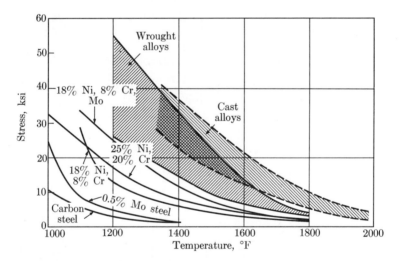

FIG. 19.28. Rupture strength for a life of 1000 hr of wrought and cast alloys containing substantial Mo, Co, W, or Cb, in comparison with stresses for steels. (Smith, 1950. Courtesy of McGraw-Hill.)

perature, as shown in Figs. 19.29 and 19.30. This limitation can be postponed by using metals with still higher melting points. For instance, a molybdenum-base alloy, although plagued by oxidation, has a 100-hour rupture strength of 20,000 psi at 2400°F (Peckner, 1962).

The cost factor must not be ignored, however; e.g., a 6-lb connecting rod costing $10 if made from free-machining B1112 steel would cost $150 to $250 if made from a high-temperature nickel-base alloy (Metals Handbook, ASM, 1961, p. 488).

19.7   CHOICE OF MATERIALS   647

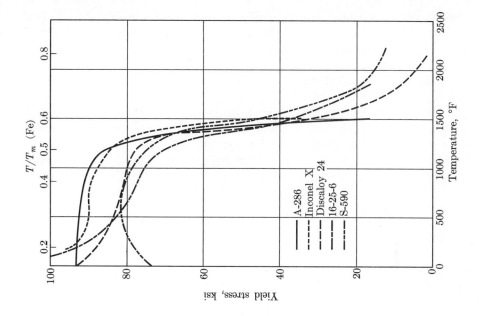

Fig. 19.29. Effect of test temperature on the flow stress of aluminum-magnesium alloys for a true strain of 0.05. (Conrad, 1959, from Sherby et al., 1951. Courtesy of ASME.)

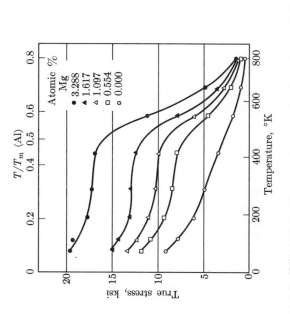

Fig. 19.30. Effect of temperature on the yield strength of some super alloys (containing Mo, Co, W, or Cb). (Conrad, 1959, from Simmons and Cross, 1954. Courtesy of ASME.)

Aside from using materials with higher melting points, the only way of obtaining more stable hardening elements is by *dispersion-strengthening*, in which oxides or insoluble compounds are introduced into metal powder, which is then sintered and worked. This procedure, while expensive, has shown great promise in aluminum and in nickel, where it gives a 1000-hour life to rupture at 7000 psi at 2200°F (Anders et al., 1962).

Another alternative is to turn to ceramics, in which the crystal structure makes dislocation motion much more difficult. Plastic deformation in ceramics, however, shows such extreme temperature and strain-rate sensitivity that these materials are either brittle or flow like a liquid (Kronberg, 1962). This limits their use to a very narrow temperature range if strength and ductility are required simultaneously.

19.8 SUMMARY

Creep is the time-dependent deformation of a solid under applied stress. Three stages can often be distinguished: primary creep at a decreasing rate under constant stress, secondary creep at a more or less constant rate, and tertiary creep at an accelerating rate followed by fracture. When the applied stress is high enough to cause plastic flow, the immediate plastic flow and the rapid primary creep can hardly be distinguished. At very low stress, primary creep, though always present, may be negligible. At moderate temperatures, the amount of primary creep can be greater than the elastic strain, indicating that as some of the sites where thermal activation can produce creep are used up, the stress on other sites increases until they in turn are ready to flow.

Secondary creep is reached when there is an equilibrium between hardening due to plastic flow or precipitation and softening by recovery or overaging. Under these conditions, if a single process with an activation enthalpy of h governs the recovery, the creep rate is given by the following equation, partly empirical and partly justified by statistical mechanics:

$$\dot{\epsilon} = A\sigma^m e^{-h(\sigma)/kT}. \tag{19.5}$$

The activation enthalpy is often nearly independent of stress, and the strain rate then varies as the fourth to eighth power of the stress in structural alloys. In most cases there may be several mechanisms with different activation energies. Since, however, one mechanism may be limiting the entire process, Eq. 19.5 may still give a useful approximation.

In metals, some of the diffusion-controlled processes leading to a steady-state creep rate may be (a) climb of edge dislocations out of slip planes to overcome obstacles, (b) drag of sessile jogs on screw dislocations, (c) elimination of dislocation loops around precipitates, (d) grain-boundary sliding, (e) accommodation mechanisms, such as grain-boundary migration, which are required for strain compatability in grain-boundary sliding, and (f) migration of vacancies between grain boundaries.

At the same time, in many of the nonequilibrium structures which are used, especially in high-temperature materials, there may be gradual metallurgical changes and overaging, resulting in a change of the entire structure. Because many of these processes are associated with vacancies on which self-diffusion also depends, it is often found that the activation energy for creep is approximately that for self-diffusion.

The accelerating creep in the tertiary stage is sometimes due to the decrease in cross-sectional area of a tensile specimen and the resulting increase in true stress at constant load. More often the decrease in area is internal, as a result of fracture beginning either at grain boundaries or due to the growth of voids. The occasional presence of tertiary creep even in compression tests, however, indicates that other mechanisms, such as recrystallization and overaging, are sometimes responsible. Oxidation usually accelerates creep through loss of metal, but it can retard it through oxide hardening. Final fracture is transcrystalline at relatively low temperatures and high strain rates, and intercrystalline at higher temperatures and lower strain rates.

Creep becomes rapidly more important at temperatures of the order of half the melting-point temperature. The plotting of creep data for time to fracture, as though it were a process covered by a single activation energy, is often helpful in correlating data; but other methods of extrapolation and interpolation are frequently as successful. As can be seen from the curves of Fig. 19.17, the creep rate depends markedly on the stress level and the temperature. Furthermore, the allowable creep in an engineering structure is much different for a steam turbine in a central-station power plant from what it is in the nozzle of a rocket, so that it is not possible to set further general rules for allowable stress levels and temperatures for various materials.

There are a number of variables which affect the creep behavior of metals. Changes in temperature and alternating stresses may produce even more creep than would be expected under constant conditions, often as a result of "shaking up" the dislocation structure. Nuclear radiation upsets the structure by introducing interstitials, vacancies which collapse to form dislocation loops, and foreign elements. Radiation thus tends to accelerate the rate of creep at elevated temperatures. Thermal shocking does by macroscopic stress what the fluctuating temperature and stress do on a microscopic scale; namely, it increases the stress levels in a part and shakes up the structure so that it remains in a state of primary creep. Creep under combined stress can be correlated with an equivalent stress, as in the case of plastic deformation. Again, triaxial tension has little effect on deformation but does promote fracture.

Relaxation is the decrease in stress with time in a part subjected to a given initial deformation. Since stresses can be relaxed by strains of the order of magnitude of the elastic strains, a prediction of relaxation from creep data often rests on an understanding of primary creep.

In designing materials for resistance to creep at high temperatures, the same problem arises as in designing materials for high strength at ordinary temperatures; namely, how to get high strength without, at the same time, producing susceptibility to fracture, especially at low temperatures and in the presence of notches.

REFERENCES

An early summary of the information on creep is the book by Smith (1950). Although this is not up to date on current theories of creep, it does contain extensive data and information on testing techniques. An introduction to the theory of creep is found in Cottrell (1953). A recent book containing summaries of theoretical work, experimental data, and extensive references to literature is the collection of papers edited by Dorn (1961). The book *Creep and Recovery* by the American Society for Metals (ASM, 1957) and the National Physical Laboratory symposium (1956) are valuable collections of research papers, many of which have been quoted in this chapter. A recent monograph by Sully (1956) contains basic information. The review article by Finnie (1960) and work by Johnson et al. (1958) are valuable references on stress effects.

Complete stress-strain temperature creep curves are hard to find for many materials, but many characteristics of such curves are summarized in such books as Freeman and Voorhees (1956), the *ASME-ASTM Compilation of Creep Data* (ASME & ASTM, 1938), and the *Metals Handbook* (ASM, 1961, pp. 443–542). For more recent creep data, refer to the technical reports of the Aeronautical Systems Division, U. S. Air Force, manufacturer's literature, and trade magazines.

ALLEGHENY LUDLUM STEEL CORP.	1958	"Total Strain and Isochronous Curves For A-286, AF-71, G-192, S-590, S-816, and V-36 Alloys," *Bulletin No. 358-4*, Allegheny Ludlum Steel Corp., Watervliet, New York.
ANDERS, F. J. ALEXANDER, G. B. WARTEL, W. S.	1962	"A Dispersion-Strengthened Nickel Alloy," *Metal Prog.* **82**, 88–91, 120, 122.
ASM	1957	*Creep and Recovery*, ASM, Novelty, Ohio.
ASM	1961	"Properties and Selection of Metals," *Metals Handbook*, 8th ed., Vol. 1, T. Lyman, ed., ASM, Novelty, Ohio.
ASME & ASTM	1938	*Compilation of Available High-Temperature Creep Characteristics of Metals and Alloys.*
ASTM	1954	*Symposium of Effect of Cyclic Heating and Stressing on Metals at Elevated Temperatures*, ASTM Special Tech. Publ. No. 165.
BASINSKI, Z. S.	1957	"Activation Energy for Creep of Aluminum at Sub-Atmospheric Temperatures," *Acta Met.* **5**, 684–686.
BROPHY, G. R. FURMAN, D. E.	1942	"The Cyclic Temperature Acceleration of Strain in Heat Resisting Alloys," *Trans. ASM* **30**, 1115–1138.
BROWN, W. F., JR. MANSON, S. S. SACHS, G. SESSLER, J. E.	1959	*Literature Surveys on Influence of Stress Concentrations at Elevated Temperatures and the Effects of Non-Steady Load and Temperature Conditions on the Creep of Metals*, ASTM, Special Tech. Publ. No. 260.

BRUNNER, H.	1957	"Effects of Grain Boundary Sliding," Sc.D. Thesis, M.I.T., Cambridge, Mass.
BRUNNER, H. GRANT, N. J.	1956	"Calculation of the Contribution Made by Grain Boundary Sliding to Total Tensile Elongation," *J. Inst. Met.* **85,** 77–80.
CHANG, H. C. GRANT, N. J.	1956	"Mechanism of Intercrystalline Fracture," *Trans. AIME* **206,** 544–551.
COFFIN, L. F.	1954	"The Problem of Thermal Stress Fatigue in Austenitic Steels at Elevated Temperatures," *Symposium on Effect of Cyclic Heating and Stressing on Metals at Elevated Temperatures,* ASTM, Special Tech. Pub. No. 165, pp. 31–52.
COFFIN, L. F.	1960	"The Stability of Metals under Cyclic Plastic Strain," *Trans. ASME, J. Basic Eng.* **D82,** 671–682.
CONRAD, H.	1959	"Correlation of High-Temperature Creep and Rupture Data," *Trans. ASME* **D81,** 617–628.
CONRAD, H.	1961(a)	"Yielding and Flow of Sapphire (Al_2O_3) Crystals," *Rept. NAA-SR-6543,* Atomics International, Canoga Park, California.
CONRAD, H.	1961(b)	"The Role of Grain Boundaries in Creep and Stress Rupture," *Mechanical Behavior of Materials at Elevated Temperatures,* J. E. Dorn, ed., McGraw-Hill, New York, pp. 218–296.
CONRAD, H.	1961(c)	"Experimental Evaluation of Creep and Stress Rupture," *Mechanical Behavior of Materials at Elevated Temperatures,* J. E. Dorn, ed., McGraw-Hill, New York, pp. 149–217.
COTTRELL, A. H.	1953	*Dislocations and Plastic Flow in Crystals,* Oxford University Press, London.
DORN, J. E.	1956	"Some Fundamental Experiments on High Temperature Creep," *Creep and Fracture of Metals at High Temperatures,* National Physical Laboratory, Her Majesty's Stationery Office, London, pp. 89–134.
DORN, J. E. (ed.)	1961	*Mechanical Behavior of Materials at Elevated Temperatures,* McGraw-Hill, New York.
FINNIE, I.	1960	"Stress Analysis in the Presence of Creep," *Appl. Mech. Rev.* **13,** 705–712.
FREEMAN, J. W. VOORHEES, H. R.	1956	*Relaxation Properties of Steels and Super Strong Alloys at Elevated Temperatures,* ASTM Special Tech. Publ. No. 187.
FRIEDEL, J. BOULANGER, C. CRUSSARD, C.	1955	"Constantes élastiques et frottement intérieur de l'aluminum polygonise," *Acta Met.* **3,** 380–391.

FRIEDEL, J.	1956	*Les Dislocations*, Gauthier, Villars, Paris. English trans., Addison-Wesley, Reading, Mass. (1964).
GERVAIS, A. M. NORTON, J. T. GRANT, N. J.	1953	"Subgrain Formation in High Purity Aluminum During Creep at High Temperatures," *Creep and Recovery*, ASM, Novelty, Ohio, pp. 53, 341
GIRIFALCO, L. A. BEESTON, B. E. P.	1963	"Impurity Effects in Strain Induced Void Formation," *Acta Met.* **11**, 161–163.
GOLDHOFF, R. M.	1959	"Comparison of Parameter Methods for Extrapolating High-Temperature Data," *Trans. ASME* **D81**, 629–644.
GRANT, N. J. CHAUDHURI, A.	1957	"Creep and Fracture," *Creep and Recovery*, ASM, Novelty, Ohio, pp. 284–343.
GUARD, R. W.	1961	"Alloying for Creep Resistance," *Mechanical Behavior of Materials at Elevated Temperatures*, J. E. Dorn, ed., McGraw-Hill, New York, pp. 270–287.
GUARNIERI, G. J.	1954	"The Creep Rupture Properties of Aircraft Sheet Alloys Subject to Intermittent Load and Temperature," *Symposium on Effect of Cyclic Heating and Stressing on Metals at Elevated Temperatures*, ASTM, Special Tech. Publ. No. 165, pp. 105–148.
HEIMERL, G. J. FARQUHAR, J.	1959	"Compressive and Tensile Creep of 7075-T6 and 2024-T3 Aluminum Alloy Sheet," *NASA Tech. Note D-160*.
HERRING, C.	1950	"Diffusional Viscosity of a Polycrystalline Solid," *J. Appl. Phys.* **21**, 437–445.
HIRSCH, P. B. WARRINGTON, D. H.	1961	"The Flow Stress of Aluminum and Copper at High Temperatures," *Phil. Mag.* **6**, 735–768.
HOFF, N. J.	1953	"The Necking and the Rupture of Rods Subjected to Constant Tensile Loads," *J. Appl. Mech.* **20**, 105–108.
ISHIDA, Y. MULLENDORE, A. W. GRANT, N. J.	1965	"Internal Grain Boundary Sliding during Creep," *Trans. Met. Soc. AIME* **233**, 204–212.
JOHNSON, A. E. FROST, N. E.	1955	"Note on the Fracture under Complex Stress Creep Conditions of an 0.5% Molybdenum Steel at 550°C and a Commercially Pure Copper at 250°C," *Creep and Fracture of Metals at High Temperatures*, National Physical Laboratory, Her Majesty's Stationery Office, London, pp. 363–378.
JOHNSON, A. E. HENDERSON, J. MATHUR, V. D.	1958	"Creep Under Changing Complex Stress Systems," *Engineer* **206**, 209, 251, 287.

Kattus, J. R. Lessley, H. L.	1961	"Determination of Compressive Bearing and Shear Creep of Sheet Metals," *Proc. ASTM* **61,** 920–929.
Kê, T. S.	1947	"Experimental Evidence of the Viscous Behavior of Grain Boundaries in Metals," *Phys. Rev.* **71,** 533–546.
Kennedy, A. J.	1953	"Creep and Recovery in Metals," *British J. Appl. Phys.* **4,** 225–233.
Kennedy, C. R. Harms, W. O. Douglas D. A.	1959	"Multiaxial Creep Studies on Inconel at 1500°F," *Trans. ASME, J. Basic Eng.*, **D81,** 599–609.
Kramer, I. Demer, L. J.	1961	"Effects of Environment on Mechanical Properties of Metals," *Prog. Mat. Sci.* **9,** 3.
Kronberg, M. L.	1962	"Dynamical Flow Properties of Single Crystals of Sapphire, I" *J. Am. Ceramic Soc.* **45,** 274–279.
Larson, R. F. Miller, J.	1952	"A Time-Temperature Relationship for Rupture and Creep Stresses," *Trans. ASME* **74,** 765–775.
MacGregor, C. W. Fisher, J. C.	1946	"A Velocity-Modified Temperature for the Plastic Flow of Metals," *J. Appl. Mech.* **13,** A11–A16.
McBride, J. G. Mulhern, B. Widmer, R.	1962	*Creep-Rupture Properties of Six Elevated Temperature Alloys*, Report WADC-TR-61-199 Wright Air Development Command, Wright-Patterson Air Force Base, Ohio.
McLean, D.	1952	"Creep Processes in Coarse-Grained Aluminum," *J. Inst. Met.* **80,** 507.
McLean, D.	1957	*Grain Boundaries in Metals*, Oxford University Press, London.
Mendelson, A. Hirschberg, M. H. Manson, S. S.	1959	"The Stability of Metals under Cyclic Plastic Strain," *Trans. ASME, J. Basic Eng.* **D82,** 671–682.
Miller, J.	1954	"Effect of Temperature Cycling on the Rupture Strength of Some High-Temperature Alloys," *Symposium on Effect of Cyclic Heating and Stressing on Metals at Elevated Temperatures*, ASTM Special Tech. Publ. No. 175, 53–66.
Mott, N. F.	1953	"A Theory of Work-Hardening of Metals, II: Flow without Slip Lines, Recovery and Creep," *Phil. Mag.*, **44,** 742–765.
National Physical Laboratory	1956	*Creep and Fracture of Metals at High Temperatures*, Her Majesty's Stationery Office, London.
Peckner, D.	1962	"Refractory Metals Roundup," *Mat. Design Eng.* **56,** 4, 132–146.

Pelloux, R. M. N.	1956	"Creep and Structure Studies of Two-Phase Aluminum Copper Alloys," *M. S. Thesis*, M.I.T., Cambridge, Mass.
Schoeck, G.	1957	"Theory of Creep," *Creep and Recovery*, ASM, Novelty, Ohio, pp. 199–226.
Schoeck, G.	1961	"Thermodynamic Principles in High-Temperature Materials," *Mechanical Behavior of Materials at Elevated Temperatures*, McGraw-Hill, New York, pp. 57–78.
Shank, M. E.	1958	"Some Aspects of the Utilization of Creep Data for Engineering Design Purposes," *Behavior of Metals and Design Requirements for Elevated Temperatures*, Notes for a special summer course at M.I.T., Cambridge, Mass.
Sherby, O. D. Anderson, R. A. Dorn, J. E.	1951	"Effect of Alloying Elements on the Elevated Temperature Plastic Properties of Alpha Solid Solutions of Aluminum," *Trans. AIME* **191**, 643–652.
Simmons, W. F. Cross, H. C.	1954	*Report on the Elevated Temperature Properties of Selected Super-Strength Alloys*," ASTM Special Tech. Publ. No. 160.
Smith, G. V.	1950	*Properties of Metals at Elevated Temperatures*, McGraw-Hill, New York.
Sully, A. H.	1956	"Recent Advances in Knowledge Concerning the Process of Creep in Metals," *Prog. Met. Phys.* **6**, 135–180.
Tapsell, H. J. Johnson, A. E.	1931	"The Strength at High Temperatures of a Cast and a Forged Steel as Used for Turbine Constructions," *Special Rept. No. 17*, Department of Scientific and Industrial Research, Engineering Research, His Majesty's Stationery Office, London.
Thornton, R. P. Hirsch, P. B.	1958	"The Effect of Stacking Fault Energy on Low Temperature Creep in Pure Metals," *Phil. Mag.* **3**, Series 8, 738–761.
Weertman, J.	1955	"Theory of Steady State Creep Based on Dislocation Climb," *J. Appl. Phys.* **26**, 1213–1217.
Widmer, R. Grant, N. J.	1960	"The Creep Rupture Properties of 80 Ni-20 Cr Alloys," *Trans. ASME, J. Basic Eng.* **D82**, pp. 829–838. See also pp. 882–885.
Widmer, R. Grant, N. J.	1961	Personal communication.
Wilson, J. C. Billington, D. S.	1955	"Effect of Nuclear Radiation on Structural Materials," *Problems in Nuclear Engineering*, Pergamon Press, London, pp. 97–106.

| WYATT, O. H. | 1953 | "Transient Creep in Pure Metals," *Proc. Phys. Soc. (London)* **B66**, 459–480. |
| ZENER, C. | 1941 | "Theory of Elasticity of Polycrystals with Viscous Grain Boundaries," *Phys. Rev.* **60**, 906–908. |

PROBLEMS

19.1. From an empirical relation between melting-point temperature and activation energy for diffusion, and assuming that a vacancy diffusion rate of 10 to 10^3 jumps per second is required for climb to give a strain rate of 0.001 per sec $^{-1}$, calculate the corresponding ratio of temperature for creep to melting-point temperature.

19.2. Referring to the statement in connection with the experiments of Friedel et al. mentioned in Section 19.2, why is the higher activation enthalpy found at the higher temperature?

19.3. In connection with the statement referred to in Problem 19.2, why should the activation enthalpy be the sum of two enthalpies?

◀ **19.4.** Propose a mode of deformation of a cylinder which would give zero shear strain everywhere on its surface, referred to coordinates locally parallel to the surface. Estimate the average contribution of the grain-boundary sliding in such a case, and compare it with the total strain. [*Hint:* consider an array of hexagonal cylinders.]

19.5. Larson and Miller (1952) correlated time to rupture, t_r in hours, by plotting stress versus a parameter involving both time and temperature in °F abs:

$$T(C + \log_{10} t_r),$$

where the constant C is often 20. One could also use a similar parameter, with the reciprocal of the strain rate substituted for time, in correlating creep rates. Relate such a procedure to the activation enthalpy ideas of Eq. 19.5, showing the relations between the various terms.

19.6. Show that the activation enthalpy for the creep curves of Fig. 19.17 is about 97 kcal/mole and that the stress exponent m is about 6.

19.7. Derive Eq. 19.8 for the time of rupture in creep.

19.8. Show that a life estimate based on rupture will give an overestimate by about a factor of 10, for the data of Fig. 19.17. See also Problem 19.20.

19.9. It has been proposed that creep is a balance between recovery or loss of flow stress and strain-hardening. Outline experiments by which this hypothesis could be checked and formulate an equation of state based on these ideas.

19.10. Compare the temperatures for the creep data of Figs. 19.17 and 19.26 with the melting-point temperatures of the elements of which the alloys are made.

19.11. If, in a correlation such as shown in Fig. 19.18, curves for constant rates of stress radiate from a common intercept, show that the creep is given by an expression of the form

$$\dot{\epsilon} = c\frac{\sigma^m}{T} e^{-H/RT},$$

656 CREEP

under the assumption that changes in slope due to different stress levels are small compared with the activation enthalpy itself.

19.12. Plot creep rates for the nickel-chromium alloys of Fig. 19.26 on Fig. 19.18. How do the activation enthalpies compare?

19.13. Determine the activation enthalpy for rupture for the forged carbon steel of Fig. 19.17 and compare it with that for minimum creep rate. How would you expect these two quantities to differ and why?

19.14. Compare the intercept creep rate at zero reciprocal temperature for Fig. 19.18 with the factors discussed before Eq. 19.2. Is reasonable agreement obtained in view of the mechanisms involved?

19.15. In Fig. 19.6, why are the polygonization boundaries perpendicular to the slip lines?

19.16. Determine the effect of high hydrostatic pressures on vacancy concentration and, hence, the approximate effect on creep rates.

19.17. Suppose that the creep of a metal can occur by two different mechanisms having different activation enthalpies and different constants multiplying the exponential term. If the correlation of Fig. 19.18 were applied, show what kind of results would be obtained, by taking a specific numerical example.

19.18. The activation enthalpy found from Fig. 19.18 was greater than that given by Dorn in Fig. 19.8. A possible explanation is that no account was taken of changes in structure or composition. Explain why such changes might or might not lead to the difference in activation enthalpy.

19.19. Determine the activation enthalpy for creep of the nickel-chromium alloy of Fig. 19.26.

19.20. Compare the applicability of Hoff's equation for predicting rupture time from initial strain rate to the data for the nickel-chromium alloy of Fig. 19.26 and the forged carbon steel of Fig. 19.17. Explain what conclusions you can draw about the mechanisms leading to tertiary creep in these two cases from this result.

19.21. The medium-carbon steel, shown in Fig. 19.17 is intended for an application at 900°F where only 10,000 hours of service is demanded, and the total strain is not to exceed 2%. Find the maximum allowable stress. From the data determine whether or not fracture is likely to intervene.

19.22. From the creep curves given in Fig. 19.21, estimate the allowable stress for 0.1% creep in 10,000 hours at 500°F. Explain the theoretical validity and limitations of your analysis.

19.23. A bolt of Type 316 stainless steel is pre-tensioned to a stress of 10,000 psi for service at 1500°F. The second-stage creep rate of this material as a function of stress is given in Fig. 19.20. Assume that: (a) the pre-tensioning can be done at 1500°F and that the bolt will plastically deform during the pre-tensioning; (b) the total length of the bolt remains the same in service, i.e., as the stresses are relaxed in the bolt, initial elastic strains are converted into creep strain. The Young's modulus of Type 316 stainless steel at this temperature is about 26×10^6 psi. Calculate the tension in the bolt after 1000 hours.

CHAPTER 20

FRICTION AND WEAR

20.1 SYNOPSIS

In this chapter we shall consider some mechanical interactions between the surfaces of two materials. These surface interactions give rise to a number of important macroscopic phenomena, the main ones being friction, wear, adhesion, and electrical and thermal resistance, although the latter two are beyond the scope of our discussion.

The surface interactions of two materials are governed by those small regions called junctions, where atom-to-atom contact is made. We shall see how elastic and plastic deformation under the high normal and shear stress at the junctions determine the extent of these regions, which constitute the real area of contact. The laws of friction will be derived from the laws of plastic shear of the junctions. These show why very high friction coefficients result from very clean surfaces and how lubricants reduce friction by interposing a low shear-strength layer at the junctions.

Surface interactions can also lead to wear, namely the removal of material from a surface as a result of mechanical action. There are four main forms of wear: adhesive wear, whereby a particle is pulled off one surface and adheres to the other; abrasive wear, whereby a hard rough surface plows grooves in a softer one; corrosive wear, in which mechanical action removes a protective layer from the surface and exposes it to corrosive attack; and surface fatigue, in which spalling occurs after the formation of surface or subsurface cracks. The laws of wear have only been partially worked out as yet, but we shall see that to a first approximation, the volume of wear of a material is proportional to the load and the sliding distance and inversely proportional to the hardness.

Surfaces also interact by adhesion, the process whereby materials become joined together as a result of the attractions of their surfaces. Only soft clean materials show marked adhesional effects. With harder materials, elastic strains on unloading tend to break the adhesional bonds.

20.2 ORIGIN OF SURFACE INTERACTIONS

Suppose we place two solid materials in contact. Some regions on their surfaces will be so close together that the surface atoms of one material "touch" the surface atoms of the other material, (i.e., the distance separating adjacent surface atoms is the sum of their atomic radii). At other regions, the surface atoms are separated by larger distances, ranging from a few to several hundred thousand

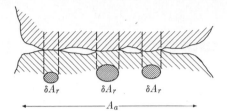

FIG. 20.1. Schematic illustration of the contact between two rough surfaces showing the formation of three junctions.

angstroms. Now, it is known that the strong interaction fields of atoms operate only over very small distances, of the order of a few angstroms, so that essentially all the interaction between contacting materials takes place at regions where atom-to-atom contact is made (Fig. 20.1). These regions will be referred to as *junctions*, and the sum of all the junctions will be called the real area of contact A_r. The rest of the apparent area of contact A_a, although it may be much larger than the real area of contact, plays essentially no part in determining the overall interaction. [*Note:* There do exist very weak long-range forces which operate at those points separated by distances exceeding 10 A; but, as Abrikosova and Deryagin (1956) have shown, it is quite difficult to measure these long-range forces, and they are negligible in comparison with the powerful short-range forces.]

20.3 THE SIZE OF THE REAL AREA OF CONTACT

At first sight, it would seem to be impossible to make any kind of quantitative statement about the real area of contact of two materials pressed together by a force P_n normal to their interface of contact, without having available a great deal of information about the circumstances of the contact (i.e., information about the size and shape of the apparent area of contact, the surface roughness of the two materials, and the way they are placed together). Fortunately, we can make an approximate plastic analysis, and calculate a minimum value for A_r, assuming fully plastic deformation. This *minimum* value turns out in most cases to be the *actual* value of the real area of contact.

The reader might well ask how we can possibly know that the deformation at contact is plastic, when we do not know the shape of the contacting surfaces. Calculations show that the deformation of surfaces may be completely elastic provided that the surface roughness does not exceed a few degrees (Halliday, 1957). However, the examination of surfaces by chemical techniques, in which the amount of a gas required to adsorb one monolayer on to a surface is measured, show that most types of surfaces, in particular those prepared by cutting or lapping techniques, have real areas which are greater than the projected areas by factors of 2 or 3 (O'Connor and Uhlig, 1957), so that a surface roughness of 45° and above is likely to be present. This high surface roughness may take the form of hills and valleys only a few hundred angstroms high, but very close together. On a coarser scale, most smoothly finished surfaces may have roughnesses of only a few degrees. Surfaces produced by wear under boundary lubrication may have much steeper slopes.

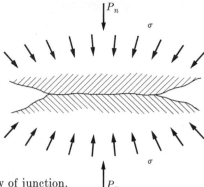

Fig. 20.2. Close-up view of junction.

To calculate the minimum value of A_r under plastic deformation, we note that a typical junction will look as shown in Fig. 20.2, and the interface is seen to be in a state of triaxial constraint. The largest compressive stress that such a region of material can carry without plastic yielding is known as its penetration hardness H, which for an ideally elastic-plastic material is about three times as great as its yield strength in compression Y_c, as discussed in Section 10.7. Then, the least value of the real area of contact is given by

$$A_r = P_n/H. \tag{20.1}$$

This assumes that all the junctions are in a state of incipient plastic flow.

There is evidence that in most cases A_r is indeed equal to P_n/H. It has been shown that most surfaces as prepared technically have ridges and valleys. When one such surface is pressed into another, the geometry is very similar to that prevailing for a hardness test, but on a smaller scale. Hence, as the surfaces come together, plastic deformation will occur so that the initial contact of three points becomes the contact of numerous, sizeable areas, and deformation will continue until A_r becomes equal to P_n/H, at which point deformation will cease. It is not easy to confirm this general picture experimentally, because both the number and the size of contacts change with load. The data of Bowden and Tabor (1939), who measured the electrical resistance of interfaces, were thought to be confirming but can be accounted for without assuming that Eq. 20.1 is correct (Holm, 1958; Archard, 1961).

It is only in exceptional cases that the deformation of surfaces is elastic rather than plastic. In these cases, A_r is greater than P_n/H, sometimes by a considerable margin. Examples are materials with a high maximum elastic strain, such as rubber and diamond, and exceptionally smooth materials, such as highly polished metals used in ball bearings, cleaved surfaces of crystals like mica (Bailey and Courtney-Pratt, 1955), and materials with rounded asperities (Archard, 1961).

Large values of A_r may also be produced when the two contacting materials show a marked tendency toward adhesion. In these cases adhesive forces, similar to the surface-tension forces that operate in liquids, attract the materials closer to each other, thus giving a larger real area of contact (Rabinowicz, 1961).

20.4 THE ADHESIONAL THEORY OF FRICTION

If we try to slide one of the contacting surfaces over the other, we must break all the junctions in shear. If the average shear strength of a junction is τ_{av}, then the tangential friction force P_t will be given by

$$P_t = \tau_{av} A_r. \tag{20.2}$$

By finding an upper-bound to the limit load, it is easy to show that τ_{av} cannot be appreciably larger than the bulk shear strength k of the weaker contacting material, or else the junction would be sheared through along some surface within that material, rather than at the original interface (Fig. 20.3). This would produce a particle of the weaker material adhering to the surface of the stronger material. In fact, the rate of production of such particles is about 1 to 10% of the rate of formation and breaking up of junctions, when uncontaminated metals slide over each other in dry air. From this, we deduce that most junctions are about as strong as, or nearly as strong as, the weaker contacting material. Hence, we write the force in terms of the yield strength in shear, k:

$$P_t = kA_r, \tag{20.3}$$

and combining Eqs. 20.1 and 20.2, we have for the friction coefficient f

$$f = P_t/P_n = kA_r/HA_r = k/H. \tag{20.4}$$

The friction coefficient is merely the quotient of two plastic strength parameters of the weaker contacting material, independent of load, speed, surface geometry, and roughness. Furthermore, since k and H are similarly related for various materials, we would expect f to be almost the same for all materials. These various relationships are quite frequently obeyed. Figure 20.4 shows data by Whitehead (1950) for the friction of steel on aluminum, covering a load range of a factor of 10^6, while Fig. 20.5 gives data by Rabinowicz (1958) for titanium on titanium, covering a speed range of a factor of 10^9. In both cases, the friction is almost

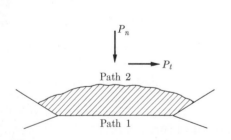

Fig. 20.3. A junction being sheared. If the shear strength of the junction is much greater than the bulk strength of the top material, shear will take place along path 2, producing the shaded fragment.

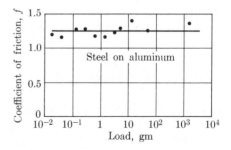

Fig. 20.4. The friction of unlubricated steel on aluminum over a wide load range. The friction coefficient is almost independent of the load. (Data from Whitehead, 1950.)

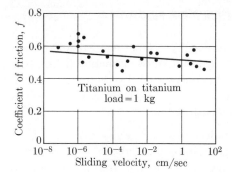

FIG. 20.5. The friction of unlubricated titanium on titanium over a wide speed range. The friction varies but little with the velocity. (Data from Rabinowicz, 1958.)

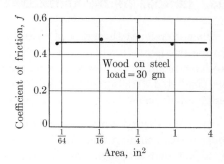

FIG. 20.6. The effect of changes in contact area on the friction of wood on steel. No significant variation is found.

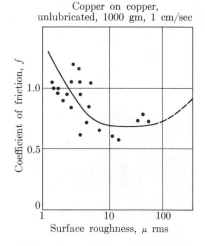

FIG. 20.7. The effect of changes of roughness on the friction of unlubricated copper on copper. Very smooth surfaces give high friction because of adhesional effects, while very rough surfaces give high friction because of asperity interlocking.

independent of the respective variables. Figures 20.6 and 20.7 show that the effect of varying the apparent area of contact and the roughness is relatively small.

It must be stated, however, that the above data are exceptionally good. Because of second-order effects (such as those due to contaminant surface films), variations of the friction coefficient f by up to 10% for every factor of 10 change of load, speed, and roughness are quite common. These may be in the direction of higher friction, or lower friction, or both in turn. Contaminants may also be important in determining the difference between static and dynamic coefficients of friction.

The above presentation constitutes the adhesional theory of friction, first published (but in an obscure German journal) by Holm in 1938, derived independently a little later by Ernst and Merchant in the United States (but they were not allowed to publish it for some years, because their company hoped to obtain some commercial advantage from it; their paper finally appeared in 1940) and, a little later, again independently, the theory was rediscovered by Bowden and Tabor (1942) in England.

20.5 CRITIQUE OF THE ADHESIONAL THEORY OF FRICTION

The adhesional theory is at present supported by a large majority of workers in the friction field. People working in the field of applied mechanics do not seem to have heard of it, however, since quite erroneously they use the terms "smooth surface" and "frictionless surface" interchangeably. The main criticisms of the adhesional theory have centered on the following points.

(1) The theory states that friction is independent of roughness, but this is opposed to common sense and to experience. However, common sense or not, it is a fact that although grossly rough surfaces do show high friction (because of the need during sliding to lift one surface over the humps on the other), atomically smooth surfaces show even higher friction (because of the increase in the real area of contact (Section 20.3)). For normal engineering surfaces, friction and roughness are almost independent. For extremely smooth and clean surfaces, friction will be greater the smoother the surfaces are, while for extremely rough surfaces, the friction will be greater, the rougher the surfaces are (see Fig. 20.7).

Another situation in which a rough surface gives higher friction than a smooth one is that in which a rough hard body slides on a much softer one. In this case, the asperities of the rough surface enable it to dig into the softer material, and, when sliding occurs, a much larger area of the soft surface must be sheared than is represented by A_r.

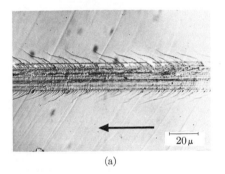
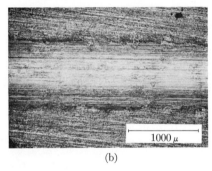

(a) (b)

FIG. 20.8. (a) Wear track on quartz made by a tungsten-carbide rider. Direction of sliding shown by the arrow. (Courtesy of W. F. Brace.) (b) Wear track on lead babbitt made by a steel rider.

(2) The theory applies only to those materials which deform plastically, and yet, brittle solids, which do not deform plastically, generally show very similar frictional behavior. This point loses its validity when it is remembered that under the high compressive stress present at a junction, very few, if any, materials can be expected to show brittle behavior. Figure 20.8(a) shows a scratch on quartz. The appearance is similar to that of scratches on metals (Fig. 20.8b), suggesting that the mode of deformation has been similar. See also Fig. 15.6, showing deformation at a scratch in glass.

20.5 CRITIQUE OF THE ADHESIONAL THEORY OF FRICTION

(3) The theory treats the hardness H and the shear flow stress k as independent. Yet, in fact, they are related by some yield criterion (cf. Chapter 7). This point is treated more fully below.

It is now generally agreed that Eqs. 20.1, 20.2, and 20.3 are only approximations to the truth; however, these equations have proved so useful that few workers in this field would like to abandon them. The best comprehensive explanation of friction is then as follows. When surfaces are pressed together under load, a real area of contact whose magnitude is governed by Eq. 20.1 is produced. When a friction force is applied (even a very small one), then some tangential movement takes place, and, as a result A_r becomes rather larger (say up to 200% larger than the value indicated by Eq. 20.1). During subsequent sliding, a shear force must then be applied to shear this enlarged real area of contact, and this value will be more or less that given by Eq. 20.2.

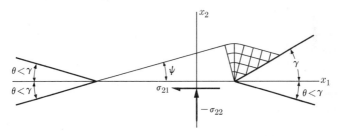

FIG. 20.9. Slip line field for a junction under combined shear and normal forces.

◀This picture has been developed in more detail by Green (1954). Of the many shapes of junctions he treated, we consider the simplest one, a junction under plane strain of the type shown in Fig. 20.9, valid when the maximum flank angle γ is within the range $0 < \gamma < \pi/4$. The indicated slip line field satisfies the conditions imposed by equilibrium, and a compatible set of displacements can be found (Prob. 20.1). Under combined shear and normal stress, the yield locus is as shown in Fig. 20.10. For relatively high values of the normal stress, that is, for

$$k(1 + \pi/2 - 2\gamma) < |\sigma_{22}| < k(2 + \pi - 2\gamma),$$

the locus is given implicitly by (Prob. 20.2)

$$|\sigma_{22}| = k(1 + 2\psi + \pi/2 - 2\gamma + \sin 2\psi),$$
$$|\sigma_{12}| = k \cos 2\psi. \qquad (20.5)$$

For lower values of σ_{22}, that is,

$$|\sigma_{22}| < k(1 + \pi/2 - 2\gamma),$$

the locus is simply

$$|\sigma_{12}| = k. \qquad (20.6)$$

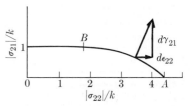

FIG. 20.10. Yield locus for combined and normal stress on a junction with maximum flank angle $\gamma = \pi/8$.

TABLE 20.1

Typical Friction Coefficients

Conditions	Examples	Values
For clean surfaces		
General		
Clean, unlubricated surfaces	Steel on silver Leather on wood Nylon on steel	0.5 to 0.3
Exceptions to above		
Clean, similar metals other than those with close-packed hexagonal structure	Copper on copper Brass on brass Chromium on chromium	1.5 to 0.8
Clean, similar metals with close-packed hexagonal structure	Titanium on titanium Zinc on zinc	0.65 to 0.45
Clean, duplex-structure alloys with a soft constituent, sliding against either a hard metal or hard nonmetal	Copper-lead alloy on steel Babbitt on steel	0.3 to 0.15
Unusual nonmetals		
Rubber on other materials		0.9 to 0.6
Teflon on other materials		0.12 to 0.04
Graphite or carbon on other materials		0.16 to 0.08
For boundary-lubricated surfaces (covered by liquid lubricant)		
Ineffective lubricant	Water, gasoline, nonwetting liquid metals	Same as for clean surfaces
Fairly effective lubricant	Refined mineral oils, wetting liquid-metals, also metal surfaces nominally unlubricated but untreated to remove contaminants	0.3 to 0.15 or unlubricated value, whichever lower
Highly effective lubricant	Mineral oils with "lubricity" additives, fatty oils, good synthetic lubricants	
Metal on metal or metal on nonmetal	Steel on steel Nylon on steel	0.10 to 0.05
Nonmetal on nonmetal	Nylon on nylon	0.20 to 0.10
For solid-film lubricated surfaces		
Hard metals covered by a thin layer of soft metal.	Thin lead film on steel	0.20 to 0.08
Materials lubricated by a layer of graphite or molybdenum disulphide, either alone or compounded by a binder		0.12 to 0.06

TABLE 20.1 (*Cont.*)

	For hydrodynamically lubricated surfaces	
A complete fluid film, produced by the sliding action, separates the surfaces. (This mode of lubrication generally applies only at speeds much in excess of 10 ft/min)		0.001 to 0.01
	For hydrostatically lubricated surfaces	
A complete fluid film, produced by external pressurization, separates the surfaces		0.001 to 0.000001 depending on the design parameters
	For rolling-contact systems	
Pure rolling contact. Geometry carefully arranged so that pure rolling motion occurs over the contacting region	Cylinder rolling over a plane	0.001 to 0.00001
Normal rolling contact. Some shear occurs at the contacting region	Commercial ball bearings	0.01 to 0.001
Arbitrary geometry	Boulder rolling down a hillside	0.2 to 0.05
	For naked surfaces	
Clean metals operated in a good vacuum (10^{-6} mm of mercury or better)		3.0, to adhering, lower for harder metals
Nonmetals (same conditions)		1.0 to 0.4

Under pure normal load, the state of stress is at the point A of Fig. 20.10. As a shear stress is applied, plastic flow occurs. Since the strain increments are normal to the yield locus, the resulting flow will have both normal and shear components of strain. Further shear stress increases the area more. The area stabilizes when the point B is reached (Prob. 20.3). The resulting coefficient of friction is

$$f = 1/(1 + \pi/2 - 2\gamma). \tag{20.7}$$

For values of γ in the range $0 < \gamma < \pi/4$, values of the coefficient of friction range from 0.4 to 1, in reasonable agreement with the values shown in Table 20.1.
◄In summary, Green showed that junctions grow during sliding as a result of the applied shear force. At the same time, the load carried by the junction does not increase correspondingly. In fact, it usually becomes tensile during separation.
◄The above model ignores the surface energy of adhesion α_a of the contacting materials. Recent work (Rabinowicz, 1961) shows that contacting materials with a high ratio of α_a/H compared with the radius of an average junction will give

larger values of A_r and, hence, higher friction than will materials with a low ratio. For example, indium and Teflon have comparable hardnesses, but α_a for indium on indium is about 700 ergs/cm^2, while α_a for Teflon on Teflon is about 40 ergs/cm^2. The coefficient of friction of indium on indium is 2.0, while for Teflon on Teflon, the coefficient is but 0.05.

◀ As the above account will suggest, the final model for the description of friction has not yet been found. ▶

20.6 THE LAWS OF FRICTION

Since friction arises primarily from the plastic shear of the junctions formed between contacting materials, it is clear that the laws of friction must be the laws of plastic shear. The main characteristics of plastic-deformation processes are that the shear force is proportional to the area being sheared but independent of the shape of the area, and but little dependent on the shear rate. When a force insufficient to cause shearing is applied, there is essentially no deformation, so that an equal and opposite resisting force must be set up at the regions of contact. This resisting force has a maximum value and, when the force applied exceeds this value, large-scale shearing commences. In line with these features of plastic deformation processes, the laws of friction are as follows:

(1) When sliding occurs, the friction force P_{kt} has a magnitude characteristic of the nature of the region of contact, and a direction opposite to that of the velocity of sliding. When sliding does not occur, the friction force P_{st} is equal and opposite to the sums of the other applied tangential forces.

(2) The friction force is proportional to the normal force P_n, the constant of proportionality $f(=P_t/P_n)$ being defined as the coefficient of friction.

(3) The friction force is almost independent of the apparent area of contact and of the surface roughness.

(4) The kinetic friction coefficient $f_k(=P_{kt}/P_n)$ is nearly independent of the sliding velocity, and the static friction coefficient $f_s(=P_{st}/P_n)$ is nearly independent of the time the surfaces have been in contact. While f_k and f_s do not differ greatly, f_s is often greater by up to 50%.

Typical coefficients of friction are given in Table 20.1.

20.7 LUBRICATION

The term *lubrication* applies to two different situations. *Fluid lubrication* occurs when a film of some liquid or gas completely separates two solids. Since the phenomena of fluid lubrication are governed by the mechanical properties of the fluid, they fall outside the scope of this book.

Boundary lubrication is a process whereby a soft material is introduced as a surface film at the contact of two harder surfaces. If the lubricant has a shear strength s_l, and covers all the real area of contact except for a portion α, we have the lubricated friction coefficient f_l given in terms of f by the linear relationship

$$f_l = \alpha f + (1 - \alpha)s_l/H. \qquad (20.8)$$

In order for f_l to be low, (a) the lubricating substance must be soft (to minimize the ratio s_l/H), and (b) the lubricating substance must adhere well to the surfaces (to minimize α).

Common boundary lubricants are organic soaps, inorganic layer lattice materials such as graphite and molybdenum disulphide, and soft metals such as lead.

20.8 WEAR

In general, wear may be defined as the removal or relocation of material, arising from the contacting of two solids. Four distinct types of wear are known, and also a few composite types.

FIG. 20.11. Idealized representation of a wear fragment (cf. Fig. 20.3).

A. Adhesive Wear. This is a process whereby, during sliding, a particle is pulled off one surface and adheres to the other. Later, it may work itself loose. The generally accepted quantitative treatment is that of Archard (1953), who considers a sliding process in which n circular junctions each of area $\pi d^2/4$ are present at any instant, and there is a probability K that a junction will give rise to an adhering fragment. If it does, it is assumed that the fragment is hemispherical, of volume $\pi d^3/12$ (Fig. 20.11). It is further assumed that each junction is in existence for a sliding distance d.

First, we need to calculate the number of junctions n formed per sliding distance Δl. At any instant, the number of junctions n_i is given by

$$n_i = A_r/(\pi d^2/4) = 4P_n/\pi d^2 H. \qquad (20.9)$$

Each junction lasts for a distance d, and hence in a distance of Δl, we need a total of

$$N_j = n_i \, \Delta l/d = 4P_n \, \Delta l/\pi d^3 H. \qquad (20.10)$$

Since each junction has a probability K of forming a fragment, the number N_f of fragments found during a sliding distance Δl is given by

$$N_f = KN_j = 4KP_n \, \Delta l/\pi d^3 H. \qquad (20.11)$$

Since each fragment has a volume of $\pi d^3/12$, the volume of wear per unit sliding distance, $\Delta V/\Delta l$, is

$$\frac{\Delta V}{\Delta l} = \frac{N_f}{\Delta l} \frac{\pi d^3}{12} = \frac{KP_n}{3H}. \qquad (20.12)$$

TABLE 20.2

Wear Constant K of Various Sliding Combinations

Combination	Wear constant $K = (\Delta V/\Delta l)/(P_n/3H)$ of first-named material
Zinc on zinc	160×10^{-3}
Copper on copper	32×10^{-3}
Stainless steel on stainless steel	21×10^{-3}
Copper on low-carbon steel	1.5×10^{-3}
Low-carbon steel on copper	0.5×10^{-3}
Bakelite on Bakelite	0.02×10^{-3}

The wear rate $\Delta V/\Delta l$ is seen to be proportional to the load, and inversely proportional to the hardness, and the constant of proportionality consists of a (nondimensional) probability K of forming a wear fragment, and the numerical constant of 3, which is a shape factor for hemispherical fragments, and could be larger for flatter fragments. Values of K, taken from the data of Rabinowicz and Tabor (1951) and of Archard and Hirst (1956) are given in Table 20.2. Note that K never approaches unity, so that the probability of forming a wear particle is small, sometimes very small.

When two materials of different hardness slide over each other, the harder one wears less than the softer one. An empirical relationship is that in terms of the ratio of the hardness of the harder to the softer material, H_H/H_S, the ratio of the wear of the harder to that of the softer material is $(H_S/H_H)^2$.

In many practical situations, we are not concerned with the formation of transferred fragments but only of loose particles. The way in which transferred fragments are transformed into loose particles is still being studied, but a number of mechanisms are known to be applicable in individual cases, among them oxidation (which converts an adherent steel fragment into loose iron oxide powder) and aggregation, which allows small particles to build up until a loose superparticle

TABLE 20.3

Product $K\gamma = (\Delta V/\Delta l)/(P_n/3H)$ for Production of Loose Particles under Adhesive Wear Conditions

Condition	Metal on metal		Nonmetal on metal or nonmetal
	Like	Unlike	
Clean	5×10^{-3}	2×10^{-4}	5×10^{-6}
Poor lubrication	2×10^{-4}	10^{-4}	5×10^{-6}
Average lubrication	10^{-5}	10^{-5}	5×10^{-6}
Excellent lubrication	$10^{-6} - 10^{-7}$	$10^{-6} - 10^{-7}$	10^{-6}

is formed. We may define a conversion ratio γ as the proportion of transferred particles transformed into loose particles. It is found that γ is generally low (i.e., varying from 1/3 to 1/1000). The final wear equation for loose wear formation is then given in the form

$$\Delta V/\Delta l = K\gamma P_n/3H. \tag{20.13}$$

Typical values for $K\gamma$ are given in Table 20.3.

B. Adhesive Wear of Lubricated Surfaces. Lubricants have a very great influence on reducing wear, out of all proportion to their influence in reducing the friction (Rabinowicz, 1951). Thus, let us consider a good lubricant like the organic soap copper palmitate, which is applied as a thin film to a steel surface. This has the effect that on the average junction, all the contact is through the soap film except for a small patch, of diameter say about $\frac{1}{30}$ the junction diameter, at which point metal-to-metal contact occurs (Fig. 20.12). Looking at the boundary lubrication equation (20.8), we see that the fraction of A_r at which metal-to-metal contact occurs is 1/900, while f for steel is 0.6 and s_l/H is 0.08. Then

$$f_l = (1/900)(0.6) + (899/900)(0.08) = 0.0806 \approx 0.08. \tag{20.14}$$

Considering the wear picture, it is reasonable to assume that the metal-to-metal patch on the junction has the same probability K of forming a wear particle as does any junction of an unlubricated surface, and that the size of the particle is proportional to the third power of the patch diameter, i.e., 1/27,000 of the volume of fragments formed for unlubricated surfaces. Hence, because of the lubricant, the wear rate will be decreased by a factor of 27,000, while the friction is reduced by less than a factor of 8 (from 0.6 to 0.08). This typical calculation shows the great importance of good lubrication in reducing adhesive wear rates.

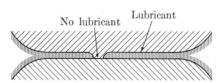

FIG. 20.12. Schematic representation of a lubricated system in which the lubricant separates the two surfaces except at one small patch.

It is only fair to point out that unlubricated surfaces as normally obtained have some contaminant already on their surface, so that the initial effect of adding a lubricant may be slight. However, the contamination accidentally applied to the surface during handling is quite rapidly removed during continued sliding, while the action of a good lubricant is more permanent.

C. Abrasive Wear. Abrasive wear is the process whereby a hard rough surface plows grooves in a softer one, and the material from the grooves later comes off in loose form (two-body wear). Another form of abrasive wear occurs when a hard particle is trapped between two sliding surfaces and abrades material from each (three-body wear).

We may derive a quantitative abrasive wear equation by considering a cone of hard material pressed into a flat softer material, and moving a distance Δl in it, as a result of which process it plows out the disturbed material (Fig. 20.13).

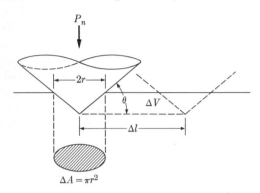

Fig. 20.13. Schematic representation of a conical indenter moving a distance through a softer flat material, and plowing out a groove of volume $r^2 \tan \theta \, \Delta l$.

We have for the real area of contact,

$$P_n = H \, \Delta A = H \pi r^2. \tag{20.15}$$

The volume ΔV swept out in sliding a distance Δl is given by

$$\Delta V = r^2 \tan \theta \, \Delta l, \tag{20.16}$$

$$\Delta V / \Delta l = r^2 \tan \theta = (\tan \theta) P_n / \pi H. \tag{20.17}$$

This is of the same form as the adhesive wear equation (20.12). However, the dimensionless constant is now the tangent of the effective roughness angle, which may vary from 10^{-1} to 10^{-2}, i.e., rather higher than typical adhesive wear constants. This effective angle may be less than the actual angle because some of the material is pushed aside rather than plowed out. Moreover, it is relatively unaffected by lubrication, so that for lubricated surfaces, abrasive wear rates tend to be much greater than adhesive wear rates.

Abrasive wear may be prevented if smooth surfaces are used and hard particles kept out of the sliding system.

D. Corrosive Wear. This occurs when chemical action leads to attack of one of the sliding materials, and the sliding action continually wipes off the products of corrosion, which otherwise would protect the surfaces from further attack. Corrosive wear cannot as yet readily be expressed in quantitative terms.

E. Surface Fatigue Wear. This form of wear is most common in rolling contacts, and is the formation of surface or subsurface cracks, followed by the breaking off of large chunks (spalling), as discussed in Section 18.8. Usually, once a surface has started to spall, complete failure by further spalling is rapid.

Since the stress is applied in a very small region and depends strongly on the local contact geometry according to the Hertz equations (11.32), there is considerable scatter in the life.

F. Fretting. Besides these forms of wear, there is one important mixed form of wear, fretting, produced when two surfaces in contact oscillate slightly. Fretting is generally a combination of adhesive wear and two- or three-body abrasive wear, but corrosive wear also occurs if one of the fretting surfaces is steel. Surface fatigue may also be involved.

20.9 ADHESION OF SOLIDS

Adhesion, that is, the ability of contacting solids to resist tensile forces, is the process by which materials became joined together as a result of the attraction of their surfaces. Since all solids attract each other, and since the tensile strength of their interface is of the same order of magnitude as the bulk tensile strength of the weaker material, one must first explain why most solids placed in contact do not adhere and then discuss those few solids that do.

The reason solids adhere so poorly is that if they are pressed together by a normal load and the load is removed, the elastic deformation present in the region of contact tends to break the junctions one by one. However, in materials which have small elastic limits (i.e., soft materials) and high energies of adhesion (e.g., clean metals), adhesive effects are readily observed. Indium, the softest metal stable at room temperature, shows adhesion readily and, to a lower extent, so do somewhat harder metals like lead.

Since metals are covered by an oxide film in air, it is necessary to press the surfaces together so severely that the oxide layer is largely ruptured. If this is done, even metals as hard as aluminum will show marked adhesion effects (cf. the commercial process of pressure welding).

Soft nonmetals also show adhesion effects, as is well known. Viscous liquids with negligible elastic behavior (e.g., pitch, chewing gum, Plasticine) are the most prominent of these. To be useful as an adhesive between two solids, a material must first wet the surfaces and be soft enough to spread, but must later harden enough to prevent creep, without at the same time becoming brittle enough to crack under the deflections of the neighboring material. For further information see Eley (1961) and Weiss (1962).

20.10 SUMMARY

By considering the plastic deformation of the junctions formed between contacting materials, it is possible to show, in accordance with observations, that when sliding occurs the friction force is in a direction opposite to the velocity of sliding, is relatively independent of the velocity, is proportional to the normal force, and is almost independent of the apparent area of contact and the surface roughness.

Wear rates are proportional to the load, the distance slid, and inversely proportional to the hardness. While lubrication does reduce friction, even when partial metal-to-metal contact occurs, its effect in reducing wear is greater by several orders of magnitude. In addition to adhesive and abrasive wear, corrosive and surface fatigue wear may occur, as well as fretting, which is a combination of these.

For soft metals or in a vacuum, adhesion becomes important.

REFERENCES

Of the few monographs that exist on the subject of surface interaction phenomena, Bowden and Tabor (1954), Holm (1958), and Rabinowicz (1965) are suggested.

Progress in this field is best followed by studying the publications of the conferences and symposia, at which latest developments are presented, and which seem to take place every year or two. A representative selection is given below.

Conferences and Symposia

1937 *Lubrication and Lubricants*, ASME, New York (1938).

1940 *Friction and Surface Finish* (Ernst and Merchant, eds.), M.I.T. Press, Cambridge, Mass. (1940).

1948 *Mechanical Wear*. J. T. Burwell, ed., ASM, Novelty, Ohio (1950).

1950 "The Fundamental Aspects of Lubrication," *Ann. New York Acad. Sci.* **53**, Art. 4, 753–993 (1951).

1950 "Physics of Lubrication," *J. Appl. Phys.* Suppl. No. 1 (1951).

1951 "A Discussion on Friction," *Proc. Roy. Soc. (London)* **A212**, 439–520 (1952).

1952 *Fundamentals of Friction and Lubrication in Engineering*, American Society of Lubrication Engineers, Chicago (1954).

1957 *Friction and Wear*, R. Davies ed., Elsevier, Amsterdam (1959).

1957 *Lubrication and Wear*, IME, London (1957).

1960 "The Nature of Solid Friction," *J. Appl. Phys.* **32**, 1407–1458 (1961).

Recently, the number of special conferences has somewhat diminished because of the holding of regular semiannual conferences by the American Society of Lubrication Engineers. Since 1958, most of these papers have been published in the *ASLE Transactions*, Academic Press, New York. Another journal that is devoted solely to publishing papers in the field of surface interactions (since 1957) is: *Wear*, Elsevier, Amsterdam.

ABRIKOSOVA, I. I. DERYAGIN, B. V.	1956	"Direct Measurement of Molecular Attraction between Solids in Vacuum," *Doklady Akad. Nauk S.S.S.R.* **108**, 214–217. See also *Soviet Physics JETP* **4**, 2–10, 1957.
ARCHARD, J. F.	1953	"Contact and Rubbing of Flat Surfaces," *J. Appl. Phys.* **24**, 981–988.
ARCHARD, J. F.	1961	"Single Contacts and Multiple Encounters," *J. Appl. Phys.* **32**, 1420–1425.
ARCHARD, J. F. HIRST, W.	1956	"The Wear of Metals Under Unlubricated Conditions," *Proc. Roy. Soc. (London)* **A236**, 397–410.

REFERENCES

BAILEY, A. I. COURTNEY-PRATT, J. F.	1955	"The Area of Real Contact and the Shear Strength of Mono-Molecular Layers of a Boundary Lubricant," *Proc. Roy. Soc. (London)* **A227**, 500–515.
BOWDEN, F. P. TABOR, D.	1939	"The Area of Contact Between Stationary and Between Moving Surfaces," *Proc. Roy. Soc. (London)* **A169**, 391–402.
BOWDEN, F. P. TABOR, D.	1942	"The Theory of Metallic Friction and the Role of Shearing and Ploughing," *Bulletin 145*, Commonwealth of Australia, Council for Scientific and Industrial Research.
BOWDEN, F. P. TABOR, D.	1954	*The Friction and Lubrication of Solids*, Oxford University Press, London.
ELEY, D. D., ed.	1961	*Adhesion*, Oxford University Press, London.
ERNST, H. MERCHANT, M. E.	1940	"Surface Friction Between Metals—A Basic Factor in the Metal-Cutting Process," *Proc. Special Summer Conferences on Friction and Surface Finish*, M.I.T. Press, Cambridge, Mass., pp. 76–101.
GREEN, A. P.	1954	"The Plastic Yielding of Metal Junctions due to Combined Shear and Pressure," *J. Mech. Phys. Solids* **2**, 197–211.
HALLIDAY, J. S.	1957	"Application of Reflection Electron Microscopy to the Study of Wear," *Proc. Conf. Lubrication and Wear*, IME, London, pp. 647–651.
HOLM, R.	1938	"The Friction Force over the Real Area of Contact," *Wissenschaftliche Veröffentlichungen Siemens-Werken* **17**, 4, 38–42.
HOLM, R.	1958	*Electric Contacts Handbook*, Springer, Berlin.
O'CONNOR, T. L. UHLIG, H. H.	1957	"Absolute Areas of Some Metallic Surfaces," *J. Phys. Chem.* **61**, 402–410.
RABINOWICZ, E.	1951	"An Investigation of Surface Damage, Using Radioactive Metals," *British J. Appl. Phys. Suppl.* No. 1, 82–85.
RABINOWICZ, E.	1958	"Boundary Lubrication of Titanium," *Proc. 5th World Petroleum Cong.*, Vol. 6, pp. 319–330.
RABINOWICZ, E.	1961	"Influence of Surface Energy on Friction and Wear Phenomena," *J. Appl. Phys.* **32**, 1440–1444.
RABINOWICZ, E.	1965	*Friction and Wear of Materials*, Wiley, New York.
RABINOWICZ, E. TABOR, D.	1951	"Metallic Transfer Between Sliding Metals: An Autoradiographic Study," *Proc. Roy. Soc. (London)* **A208**, 455–475.
WEISS, P., ed.	1962	*Adhesion and Cohesion*, Elsevier, Amsterdam.
WHITEHEAD, J. R.	1950	"Surface Deformation and Friction of Metals at Light Loads," *Proc. Roy. Soc. (London)* **A201**, 109–123.

PROBLEMS

◀ 20.1. (a) Show that the slip line field shown in Fig. 20.9 for a junction under combined shear and normal force satisfies the equilibrium and compatibility equations within the deforming region.

(b) How would you show that this stress field is exact for a non-strainhardening material?

◀ 20.2. Derive the yield locus of Fig. 20.10 for combined shear and normal stress on a junction.

◀ 20.3. Why will no more increase in area of the junction occur when the point B in Fig. 20.10 is reached? [*Hint:* Sketch the slip line field as that point is approached.]

20.4. Using chromium-plated test gauges, it is found that after 20,000 measurements of the bore of steel cylinders, the wear of the gauges is about 10^{-5} in. If the force with which the gauge is pressed into the cylinder is 2 lb, what is the value of the wear coefficient K? Owing to the construction of the gauges, wear is confined to a small raised region of area 2×10^{-2} in.^{2}, and a similar area on the opposite end of the gauge.

20.5. In drawing annealed copper wire through a tungsten carbide die, the die diameter increased by 10^{-4} in. per 15 miles of wire. Assuming that the tungsten carbide is 30 times as hard as the copper, estimate K.

20.6. The piston rings in an automobile wear by 0.002 in. for every 10,000 miles of travel of the automobile. Making plausible assumptions, calculate K. Is the value you obtain a reasonable one compared with values given in Table 20.3?

◀ 20.7. As a tangential load is applied to the junction shown in Fig. 20.9, how, if at all, will the flank angle γ change? How does this affect the estimate of the coefficient of friction for surfaces of given initial roughness?

E 20.8. Note that the coefficient of friction of a pencil eraser on paper can be determined from the angle at which the pencil will just slide over the paper, as indicated in the Fig. 20.14. Determine this angle θ. For a slightly larger angle, the pencil locks. Now increase the load and note that sliding again becomes possible. Explain this in view of the change in real area of contact which rubber is likely to undergo in view of its large elastic deformation. [*Hint:* See Section 11.7.] Repeat the experiment with the other end of the pencil (on a surface which won't be marred). Will increased load cause sliding?

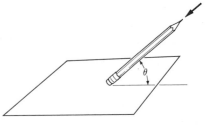

FIGURE 20.14

E 20.9. Study the mechanism by which pieces of modeling clay wear when lightly rubbed together. Do such variables as load and speed affect the size of the wear particles? How does wear in silicone putty compare with that in Plasticine?

20.10. Propose a hardness test based on wear rates and give an approximate equation relating it to the tensile strength.

CHAPTER 21

FIBROUS MATERIALS

21.1 SYNOPSIS

Fibers, both natural and man-made, are either organic polymers or inorganic substances (crystalline or polymeric in nature). Examples are cotton, wool, rayon, nylon, metal, glass, and asbestos. The high surface-to-volume ratio in fibers allows for easy dyeing and chemical modification and greatly increases the magnitude and rapidity of effects due to temperature, humidity, and electrostatics. We shall review the structures of fibrous polymers, ranging from the molecular chain orientation, through the fibrillar structure, to the structure of the fibers, yarns, and entire fabric. Fibrillar stress-strain behavior is due to a combination of the secondary bonds between molecular chains, with the entropy effect tending to force these chains into a random contracted configuration giving rise to a viscoelastic mechanical behavior. The fiber behavior in turn depends on that of its fibrils.

Fibers may be laid out in a mat and bonded by an adhesive or by heat setting; the resultant product is called a nonwoven, or bonded, fabric. Short fibers (1 to 5 inches) may also be gathered and twisted together in the form of yarns in which axial strength is developed through interfiber friction. The yarns in turn may be woven or knitted into cloth. Finally, either yarns or fabric may be embedded in a plastic material to impart stiffness, strength, and tear resistance above those of the unreinforced plastic.

In determining the behavior of a structure from that of its fibrous components, friction and both geometrical and statistical effects (as they interact with the stress-strain behavior of the fibers) must be included in engineering analysis. For certain nonwoven fabrics it is possible to predict the properties of the fabric directly from those of the fibers. Various approximations are available for predicting the modulus and strength of twisted yarn from fiber properties. Expressions derived for the strain in a compound helix are useful in determining the behavior of twisted and bent structures such as ropes or yarns in fabrics. The elastic behavior of a simple woven fabric is developed in terms of the crimp interchange and local shear of its two yarn systems.

21.2 INTRODUCTION

By definition, fibrous materials are composed of fibers, which are in turn characterized by a high slenderness ratio, small diameter, and a high surface to volume ratio. These permit the building of structures with high local curvatures without incurring high levels of strain. Thus fiber fineness leads to high flexibility, but at the same time to a susceptibility to changes in temperature, humidity, and to chemical treatments. Some widely used fibers come from natural sources, while some are manmade. Many fibers used in commerce at present are organic polymers, but inorganic fibers, such as glass, asbestos, and metal, are also used in great quantity. Polymeric fibers are characterized by a high degree of orientation of their molecular chains, and in many cases by a crystallike association between chains. The mechanical properties of polymers have been considered in Chapter 6.

For a fiber to be used commercially, it must have a melting temperature certainly above 212°F and preferably above the temperature of a hot clothes iron. If the melt viscosity of the polymer is too high, it will be difficult to convert it into a fiber by conventional spinning, that is, extrusion through a fine orifice. This is one reason for the difficulty of forming Teflon fibers.

Fibers are spun in three different ways, depending on the melting point of the polymer and its solubility in various types of organic solvents. These methods are (1) wet spinning into a coagulating bath, as in producing viscose rayon fibers; (2) melt spinning, such as is used for nylon and the polyesters due to their thermal stability as they are processed from the melt; (3) dry spinning, such as is used in the acrylics, with solvent evaporation following spinning. The different spinning techniques result in fiber structures and fiber properties which differ both with the material and with the type of spinning. Many of the synthetic fibers are heat-stabilized after drawing to prevent excessive shrinkage in subsequent processes and service to be encountered by the fiber. Frequently a heat treatment is provided for the fiber at the finished fabric stage to relieve recoverable residual strains resulting from drawing, twisting, and weaving. During this heat treatment the crystallinity of the fiber is generally increased and it becomes dimensionally stable at temperatures below the heat-setting temperature.

A *yarn* is a more or less parallel grouping of fibers which serves as a textile structural component. In staple yarns, short fibers (1 to 5 inches) are twisted together to induce lateral pressures and thus provide for force transfer from fiber to fiber through friction. Adhesives can be used to bond zero twist yarns. Filament yarns are made of continuous filaments with twist or random interlacing, or with neither, depending on the severity of handling of the yarn during subsequent processing, and on the yarn roundness desired in the finished fabric.

Textile fabrics, paper, and leather all represent sheet materials made up of fibrous elements. Paper and leather are in effect bonded fibrous materials. Textile fabrics may be bonded with an adhesive or self-bonded; they may be woven, knitted, braided, laced, or knotted. Except for the nonwoven or bonded fabrics, textiles are fabricated from yarn, an intermediate structure.

At present the consumption of fibers in the United States is over 6 billion pounds per year, used roughly as follows: over 3 billion in apparel fabrics, $1\frac{1}{2}$ billion in household and decorative materials, and 1 billion in industrial materials, such as tire cords, conveyor belts, agricultural bagging, etc. Of this total quantity, the cotton fiber counts for over 60%, the wool fiber for less than 10%, and 30% is split between regenerated cellulosics (rayons) and all the synthetics. The hun-

TABLE 21.1
Classification of Textile Fibers

Natural fibers: Cellulosic
  Seed hair: cotton
  Bast
  Leaf
 	
Natural fibers: Protein
  Wool
  Hair
  Silk
 	
Natural fibers: Inorganic
  Asbestos
 	
Man-made fibers: Regenerated organic
  Rayon (regenerated cellulose)
    Viscose
    Cuprammonium rayon
    "Fortisan"
    Cross-linked cellulosic
  Cellulose acetate
    Acetate
    Triacetate
 	
Man-made fibers: Synthetic organic
  Acrylic
    "Acrilan"
    "Creslan"
    "Orlon"
    "Zefran"
	Man-made fibers: Synthetic organic
	  Modacrylic
	    "Verel"
	    "Dynel"
	  Polyester
	    "Dacron"
	    "Fortrel"
	    "Kodel"
	    "Vycron"
	  Polyvinylidene chloride
	    (Saran)
	  Nytril
	  Nylon
	  Spandex
	    "Lycra"
	    "Vyrene"
	  Vinal
	  Olefin
	    Polypropylene
	    Polyethylene
	  Vinyon
	
	Man-made fibers: Inorganic
	  Metal
	  Ceramic
	  Other inorganic polymers: glass, quartz
	
	Rubber fibers
	  Natural
	  Synthetic

dreds of other natural fibers in commercial use together have only a negligible share of the U. S. fiber market. The principal fiber types and generic names to be found on the present market are listed in Table 21.1. Some common trade names are also included.

21.3 STRUCTURE OF POLYMERIC FIBERS

In Chapter 6 an account has been presented of the molecular structure of bulk polymers, both amorphous and partially crystalline in the form of spherulites. In fibers, a useful model of molecular configuration follows from visualizing an aggregate of polymer chains in a rough parallelism, which defines a microfibril. Morton and Hearle (1962) suggest that this fibril is in fact fringed with molecules diverging at different positions along its length. There may be some branching of the crystalline fibers, and many molecules will diverge from the fiber at different points and pass through noncrystalline regions before being incorporated in another crystalline microfibril.

The extent of order and motion of molecules in fibers is not known. Fiber physicists are at work in judging new experimental evidence derived from x-ray diffraction, density measurements, dielectric relaxation, damping, and nuclear magnetic resonance as it relates to polymer structures under varying conditions of stress and temperature.

Most fibers have an additional component of structure superimposed on their molecular configurations. This component, called "fine structure," is denoted by such terms as *fibrils* and *fibril layers* (occurring frequently in native cellulose fibers), *medullae, cortical cells,* and *scales* (occurring in hair fibers), *cells* and *lumen channels* (of the flax fiber), and *skin* and *core structure* (of regenerated cellulose). The details relating to the fine structure of most commercial fibers are given at length by Mauersberger (1954). Only a brief discussion will be provided below on fibers of major commercial importance.

A. Cotton. In view of cotton's major share of today's fiber consumption, it is worth considering its structure in some detail. Cotton is a cellular outgrowth of the cotton seed. The commercial cotton fiber, upon drying and collapsing, is roughly 3000 times as long as it is wide. Viewed in an optical microscope, it looks like a twisted ribbon with a variable cross section, sometimes oval in shape, sometimes kidney shaped. Starting from the outside, the fine structure consists of a primary wall less than 0.5 micron thick with a moderately ordered fibrillar structure, in effect a criss-crossed pattern of about 70° helix angle, shown in Figs. 1.37 and 1.38(a). The secondary wall, laid down after the fiber skin or primary wall has grown, forms the major part of the fiber with a thickness of about 3 microns (Fig. 1.38b). It consists of some 25 to 50 concentric layers of cellulose. The outermost layer of the secondary wall is called the winding layer and is made up of bands of fibril bundles which spiral around the axis of the fiber, shown in Fig. 1.38(b). These fibrils range in diameter from 50 A up to 1 or 2 microns. Meredith (1956) points out that these fibrils do not represent structural units with definite dimensions. They are rather fragments of a submicroscopical network structure. The first layer of fibrils in the secondary wall has a helix angle of about 30° which alternates from a right- to a left-hand helix as many as 20 times in a centimeter of fiber length. The reversal of fiber helix direction influences fiber strength, since rupture tends to originate at points of reversal. In addition, the angle of the fibril spiral affects the tensile modulus of the fiber: the smaller the helix angle, the higher the modulus (Rebenfeld and Virgin, 1957). Meredith describes the system of submicroscopic voids between the fibrils, studied by introducing heavy metals between the fibrils and then examining thin cross sections with the aid of x-rays. The cavities reported varied from 10 A to over 100 A in diameter. The large central opening in the cotton fiber, the lumen, contains the solid residue (in mature dry cotton) from the protoplasm of the single cell organism. The lumen dimension varies with the conditions of growth and with the species of cotton.

It is clear that any attempt to relate the molecular behavior of natural cellulose with the mechanical properties of a cellulose fiber, such as cotton or ramie, must take into account the intermediate level of structure present in the fiber—structure observable by means of both optical and electron microscopes. With the widespread use of "wash and wear" treatments on cotton, it is also important that chemical modification and resin treatments of cotton fiber allow for the interaction between fibrillar bundles in the total complex fiber architecture. There is a tendency for the chemist to worry only about such things as the attachment of side groups along a molecular chain and its effect on the mechanical properties of that chain and its immediate neighbors. It is quite likely that many treatments in use today alter the internal structural rearrangement of the fiber under applied stress and thus mask the direct effect of a change in properties of the individual molecular chain. Such a structural effect, suggested by the data of Rebenfeld (1962), gives considerably different strength and elongation in a cotton fiber which has been resin treated under no stress, as compared with the fiber which has been axially tensioned during treatment. As Rebenfeld points out, fibrillar orientation, as estimated by x-rays, is an extremely important structural parameter in relation to mechanical properties such as elastic modulus, extensibility, and resilience. It is therefore to be expected that changes in these properties resulting from chemical treatment are also functions of fibrillar orientation, as shown in Fig. 21.1 for the effect of angle on modulus.

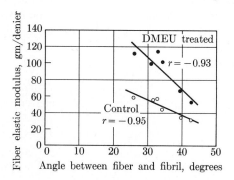

Fig. 21.1. Relationship between x-ray angle and elastic modulus of untreated and dimethylol ethylene urea treated cotton fibers. Here r is the regression coefficient, unity for perfect correlation. See Section 21.4 for a definition of denier. (After Rebenfeld, 1962.)

B. Wool. The wool fiber during growth consists of a *root* (the living region situated beneath the surface of the skin) and the *shaft* (the nonliving portion that extends above the skin surface). Increase in length of the fiber is brought about by the proliferation of new cells in the root and the subsequent emergence of these cells into the shaft. The shaft is composed of cellular units which usually are arranged in three layers: an outer layer of *scales*, a middle region called the *cortex*, and a central core or *medulla*. The scales in the outer layer overlap each other but have free edges pointing toward the fiber tip. The cortex is comprised of spindle-shaped cells 100 to 200 microns in length and 2 to 5 microns wide. Coarse fibers sometimes have a central channel or medulla which may be hollow or filled with a loose honeycomblike network of cells. A cementlike matrix holds the structure together. Electron microscope studies have shown that the cortical cells are made up of fibrillar elements which are in effect bundles of still smaller "protofibrils." At extremely high magnifications in the electron microscope, these protofibrils are shown to consist of globular particles about 110 A in diameter. One unique property resulting from the structure of wool is its directional coefficient of friction, which encourages unidirectional fiber migration and entanglement when fiber assemblies are worked in a wet environment. This is called felting. An up-to-date discussion of the wool fiber, its properties and its applications, is presented by von Bergen (1963).

An additional structural component in the wool fiber is that of asymmetry from side to side. This slight difference in structure is difficult to observe directly, but it is evidenced in a differential dyeing (one side light, the other side dark) with selected dyestuffs. The presence of this slight asymmetry in structure causes the wool fiber to act like a bimetallic strip subjected to varying temperature. In a fabric it is partially restrained from assuming the equilibrium curvature which is dictated by differences in expansion from one side of the fiber to the other. Under changing temperature and humidity, the wool fiber buckles into a helical configuration with occasional pitch reversals, thus reducing its strain energy. It approaches a springlike configuration as it lies in a yarn structure. The net effect, in terms of gross behavior of the wool fabric, is that the contribution of the individual fiber to fabric stiffness is radically reduced as a function of its helical geometry. At the same time the effective extensibility of the coiled fiber is markedly increased.

C. Man-made fibers. Recent "two-component" acrylic fibers permit synthetic reproduction of the natural crimping of wool. The mechanics of this development are discussed by Brand and Backer (1962). Such a two-component man-made fiber represents a deliberate attempt to introduce a substructure in a synthetic fiber. Substructure exists in man-made fibers even when no specific effort is made to introduce or control this intermediate structure. For example Ribi (1951) showed that in viscose rayons and polyamide fibers there is a system of fine fibrils even before extensive orientation of the fiber.

21.4 MECHANICAL BEHAVIOR OF FIBERS

The mechanical behavior of textile fibers reflects the interaction between the viscoelastic phenomena (discussed in Chapters 6 and 7) and the fine structure of the fiber (considered in the preceding section). In addition, the high surface-to-volume ratio, characteristic of fibrous materials, tends to accentuate the speed and magnitude of the effects of ambient temperature and moisture on their mechanical properties. Finally, the geometry of the fiber itself (its crimp and convolutions) greatly influences its measurable behavior. In discussing fiber properties one must be realistic about what constitutes a valid and accurate measurement of mechanical behavior. And one must apply data reported in the fiber literature with full recognition of the restrictions of testing conditions and of fiber identification which should (but does not in many cases) accompany such data.

TABLE 21.2

STRENGTHS OF FIBROUS MATERIALS

Material	Specific gravity	Ultimate tensile strength, ksi		Young's modulus 10^6 psi	
		Direct	Specific*	Direct	Specific*
Steel (piano wire)	7.8	350	45	32	4
Duralumin	2.8	60–90	21–32	10.5	3.75
Flax	1.5	50–160	34–105	12.5–15.6	8.3–10.4
Nylon fiber	1.07	72	67	0.7	0.7
Silica fiber	2.65	2000	750	10	3.8
Commercial glass fiber	2.4	500	210	10.8	4.2
Asbestos fiber[1]	2.4	330	137	26.5	11
Asbestos fiber[2]	3.0	450	150	32.7	11

[1] Syromjatnikoff (1927).
[2] Orowan (1933).
* Value divided by specific gravity.

The load-extension behavior of single fibers which have breaking strengths of less than 10 grams has been commonly obtained only in the last twenty years, with the development of the bonded-strain-gauge load-cell tensile-test instrument. Limited data are available on the torsional behavior of individual fibers. See for example Morton and Permanyer (1949). Data on bending or compressive properties of fibers are scarce. In short, the easiest test of fibers is in uniaxial tension. And so the literature of the last twenty years contains a large body of data on slow-speed tensile stress-strain behavior of fibers, on tensile creep, on stress relaxation (at a constant deformation), and more recently, on dynamic tensile behavior. Typical data for selected fiber and wires are included in Table 21.2 where exceptionally high strengths will be noted. It is the purpose of this section to give the reader some idea of the stress-strain characteristics of fibers in tension and to demonstrate the agreement between this experimental evidence and the behavior of polymeric materials in general.

The stress-strain curves* of most fibers are characterized by an initial relatively stiff region, the slope of which is represented by the initial modulus (see Fig. 21.2). There follows a region of decreasing slope which is termed a yield point. Many fibers manifest a continuous flattening of the stress curve following this yield point, as in the case of wool or cellulose acetate. On the other hand, fibers such as nylon or Dacron appear to develop a stiffening region (concave upward) after the initial yield, and following this stiffening, there is still another yielding region (concave downward).

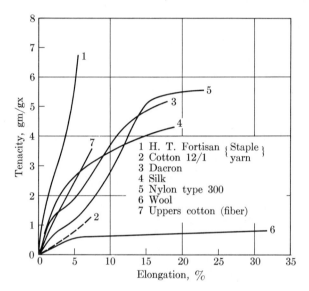

FIG. 21.2. Stress-strain curves of various fibers. (After Susich and Backer, 1951.)

This stress-strain behavior can be partially interpreted in terms of the molecular structure of the fiber, in which the long-chain molecules form regions of a high degree of order (crystalline) and regions of less perfect order. The latter regions are considered to contribute the major component of strain under a given fiber stress. But at the same time, the degree of crystallinity and resultant rigidity in a fiber serves to increase the local strain required in its less ordered regions for

---

* For fibers, strain is expressed in the conventional manner as percent elongation and stress is expressed as tensile "tenacity" given in grams per initial denier, or grams per initial grex. Denier is a long-established measure of fiber or yarn size expressed as grams per 9000 meters. This system is used, since the irregularity of many fiber cross sections and the nonuniform packing of fiber in yarn make it difficult to determine easily the "solid" cross-sectional area of fiber or yarn. Grex (gx) is a measure of grams per 10,000 meters of fiber or yarn. To obtain stress in kg/mm^2, multiply tenacity in gm/gx by 10 times the density, d, in gm/cm^3. To obtain the stress in psi, multiply by $14,223d$. Thus the dimension of tenacity (gm/gx or gpd) is length. The breaking tenacity represents the breaking length of the fiber, i.e., the number of 10,000 (or 9000) meter lengths of itself which the fiber will support.

a given total fiber strain. This is a geometric effect such as one would expect in a series loading of very stiff and relatively soft springs.

In Section 6.5 a discussion was presented of the tendency of the relatively mobile molecular chain network of rubber to contract to the random configuration corresponding to its unstretched state. This configuration represents a condition of maximum entropy and can occur as a result of thermal vibration. In oriented textile fibers, an equilibrium state is achieved in which only part of the noncrystalline structure responds to entropy effects. And even here the resistance to extension is masked by hindrances to motion of the chain segments by bulky side groups or by the presence of crystalline regions.

Further, in drawn man-made fibers, the random chain orientation is not restored after the drawing process due to van der Waals forces and hydrogen bonds which hold the fiber in an extended (or drawn) configuration, in opposition to the retractive entropy effects. (A more extended discussion of these phenomena will be found in Meredith, 1956.)

In a partially drawn fiber, the secondary bonds, which prevent contraction after complete drawing, may aid in extending the fiber, thus showing a reduction in internal energy. This phenomenon was reported by Bryant (1953), who calculated a negative internal energy force for extensions of nylon (in water) up to 10% for drawn and 30% for undrawn fiber. Beyond the stated extensions the increase in tension due to internal energy changes was negligible and entropy changes caused changes in tension with temperature.

Meredith suggests that the short-range elastic forces, resulting from the secondary bonds among molecular chain segments, account for resistance to extensions up to 2%. These forces determine the initial Young's modulus. The first yield point represents a rupture of secondary bonds and slippage of molecular segments over potential barriers to configurations involving new bond locations. The yield region represents this localized flow, and the segments which have been moved to the new position of equilibrium in this manner are often unable to return upon release of load. However, as Meredith points out, application of moisture or heat may reduce the secondary forces between chain segments in the strained state to the point where the contractive force due to the entropy effect may cause a strain recovery.

Following the initiation of yield, the upward curvature of the stress-strain curve of many fibers reflects a more dominant role at this point for chain straightening, influenced, however, by the presence of crystallites and by the continued attraction of secondary bonds between chains. There is much less known about the second yield point, such as occurs in nylon or in Dacron. This yield leads to what Meredith calls the prefracture flow and gives rise to vastly increased strength efficiencies of textile structures, since fibers strained to different levels all carry a high load at the moment of structural disintegration. Fibers which do not possess this prefracture flow exhibit a lower "cooperative" strength in textile structures.

Prefracture flow in the second yield region is highly sensitive to strain rates. The impact tensile behavior of nylon and Dacron (Fig. 21.3) versus slow-speed behavior shows a major reduction in prefracture flow. On the other hand, fibers

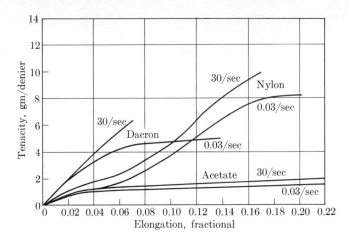

Fig. 21.3. Stress-strain behavior as affected by strain rate.

such as cellulose acetate fiber (and Nomex) exhibit only a single yield point and flow region. The fact that this region remains relatively unaffected by changes in strain rate contributes to textile structural efficiency over a wide range of test speeds for the cellulose acetate specimens. As expected, from the viscoelastic mechanisms discussed in Section 6.4, the stress-strain curves of all polymeric fibers are raised by increasing the rate of strain or lowering the temperature.

The resistance to fracture in fibrous materials is strongly influenced by the high degree of orientation of the fibers. The strength of the interatomic bonds along the chain is not higher than in chemically similar non-fibrous materials; rather, the high strength of the fiber is due to its relative weakness in planes parallel to the fiber axis. In an isotropic material a crack would propagate directly across the specimen. In a strongly anisotropic fiber, however, shear failure occurs easily in a plane parallel to the fiber axis at the tip of the crack AB; as shown in Fig. 21.4, the crack is deflected into BC and made harmless. This is the fundamental cause of the strength of fibrous and lamellar materials (including, prehistorically, asbestos, mica, and later wood). The strength of composite materials such as resin-bonded fiberglass is based on the same principle: when

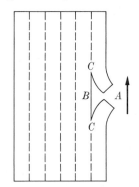

Fig. 21.4. Crack arresting tendency of laminar material.

a fiber breaks, the crack does not propagate across the resin matrix into the next fiber but is deflected into a shear crack at the boundary of the fiber and the matrix.

Relative humidity influences stress-strain curves as does temperature: the higher the humidity, the "softer" the curve. This softening and corresponding increase in extension at fracture reflects the presence of absorbed water, which reduces cohesion of chain molecules, primarily in the noncrystalline region of the fiber. The changes in fiber stress-strain behavior over the range of relative

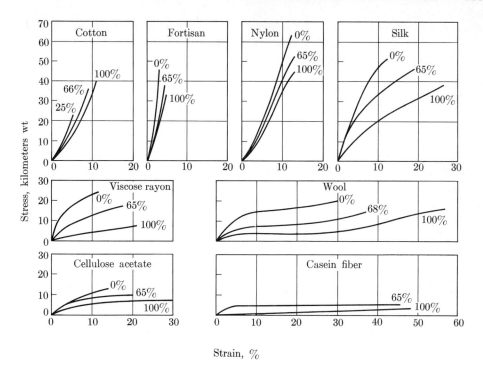

FIG. 21.5. Effect of relative humidity on tensile strength of yarns. (Meredith, 1960. Courtesy of Butterworths.)

humidity 0 to 100% are shown in Fig. 21.5. The magnitude of the effect is related to the amount of moisture absorbed by the fiber and the amount of swelling it undergoes. Hydrophobic fibers such as nylon, Dacron, and the acrylics, in general, show little moisture effect. Hydrophilic and highly swelling fibers such as viscose rayon, wool, and cotton show a large moisture effect.

Note that only cotton (and flax), of all the fibers represented in Fig. 21.5, have a higher breaking strength at higher moisture contents. The reduction in strength is attributed by Meredith (1960) to the reduction in the number of points of adhesion of the chain molecules and concentration of the stress on relatively few molecules which fail by rupture. In the natural cellulose fibers such as cotton or flax, the average chain length is over five times that of the man-made (regenerated cellulose) fibers. In cotton, the presence of moisture is considered to release some of the internal strain between long-chain molecules, leading to a more uniform distribution of internal stress and a higher breaking strength. This hypothesis is in agreement with the behavior of degraded (chain-shortened) cotton, which becomes weaker at higher moisture contents. It would be interesting to relate the strengthening effect of moisture in (undegraded) cotton to the helical fine structure discussed in Section 21.3 (Prob. 21.1).

For small strains, linear viscoelastic theory, discussed for polymers in Section 6.6, will apply. This may be verified not only by the creep and relaxation tests discussed there, but also from tensile-test data obtained at different strain rates. The data are cross-plotted as stress versus time required to reach various strain levels. Here again the ratio of stress to strain will be observed to be a constant at any particular time, if the fiber is acting as a linear viscoelastic material (Prob. 21.2). Data of this sort are seen in Fig. 21.6(a) for fiberglass. Note that at any time value on this graph, applied tensile load, or stress, in grams per denier for the 1% strain is about twice that for the 0.5% strain. However, above the 1% strain, the linearity appears to break down, and the ratio of the stress at 1.5% strain is not consistently 1.5 times that of the stress at 1% strain. In Fig. 21.6(b) stress relaxation data obtained from constant strain-rate tests are shown for drawn nylon. Here linearity does not occur in the range of strains studied, that is, from 2 to 14% extension.

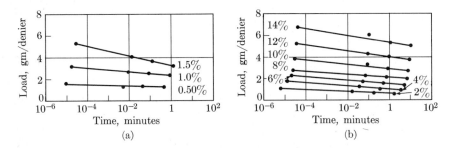

Fig. 21.6. Cross plots from constant strain-rate data of Smith et al. (1957). A test at a given strain rate followed each of the concave-upward series of points. (a) Fiberglass. (b) Drawn nylon.

In short, the examples given indicate nonlinear viscoelastic behavior in fiberglass at 1 to 1.5% strain and in nylon at something below 2%.

The recovery behavior of fibers following the release of stress can be studied in terms of the molecular mechanisms of deformation discussed above. Space limitations preclude discussion of effects of test conditions (time, temperature, humidity) on strain recovery of polymeric fibers, but it should be pointed out that strain recovery can be separated into two categories: (1) immediate elastic recovery, which in practice occurs in the first seconds after stress removal; and (2) delayed recovery, which occurs in an arbitrary period, up to say 5 or 10 minutes after stress removal. That portion of the fiber deformation remaining 10 or more minutes after stress release is called *permanent set*, although if the ambient conditions under which the tension test is conducted are changed during strain recovery, the magnitude of permanent set can be significantly reduced, or eliminated entirely. For example, under standard testing conditions of 65% R.H. and 70°F, wool will display permanent set when subjected to a 15% strain. When tested wet, wool displays *no* permanent set after being subjected to strains as high as 30%. Visualize what this means in running a wool fabric in a continuous belt subjected to tension and strain cycles. This is what occurs in the woolen fabric of the papermakers' felt. Fortunately the belt runs wet.

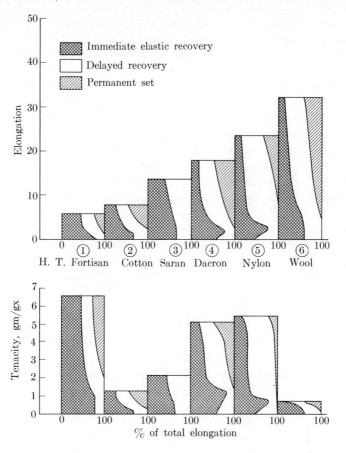

Fig. 21.7. Recovery behavior of fibers. (After Susich and Backer, 1951.)

The important point to remember is that the conditions of use must be known for the proper design of a fibrous system. Likewise knowledge must be available as to the percent of total elongation that is immediately recoverable, recovered in a short time (of say 5 minutes) and remains irrecoverable in a short time for the particular conditions. This information should be available at all levels of initial elongation or tenacity to which the fiber may be subjected. Data of this type have been reported by Susich and Backer (1951) and are shown for selected fibers in Fig. 21.7.

21.5 MECHANICS OF TEXTILE STRUCTURES

Textile structures are made in a variety of ways: by entanglement (felting), adhesive bonding, twisting, weaving, knitting, knotting, or embedding in a bulk elastomeric or plastic matrix. Several examples of the manufacture and behavior of these structures will be discussed.

The conventional textile manufacturing process has long included a fiber-opening and spreading operation called *carding*. The state of the product following the carding operation is a loose two-dimensional web of short fibers with a moderate degree of fiber alignment in the direction of movement of the web through the machine. This alignment can be improved by extending or drafting the web. The appearance of the web is quite similar to that of the fibril structure of the primary wall of the cotton fiber.

If the carded web is sprayed with an adhesive or dipped into an emulsion containing a resin (which is subsequently cured) a fiber-to-fiber bonding occurs. If the web contains thermoplastic fibers, a simple hot-pressing can likewise develop adhesive bonds between fibers. The web in this bonded state is called a *nonwoven* or bonded fabric, usable by itself in a single layer, or in a laminate structure with multiple layers. Paper is a self-bonded web of short wood fibers. Some excellent electron micrographs of paper can be seen in the book edited by Bolam (1962).

The mechanical behavior of a bonded sheet shows many of the mechanisms of deformation and many of the interactions between fiber properties and fiber configurations which are found in fibrous structures. If uniaxial tension is applied to a bonded fabric, its mechanisms of deformation will include: fiber rotation, fiber straightening, bond rotation, bond extension, and fiber extension. It is expected that the mechanisms of rotation and straightening are active before the fibers and the bonds connecting fibers are subjected to any substantial load. This series of deformation mechanisms is similar to those occurring in a single fiber or in a woven fabric subjected to uniaxial tension at a slight angle to warp or filling yarn (Prob. 21.3).

A. Orientation Effects in a Bonded Web.

Some understanding of the anisotropic elastic behavior of nonwoven sheet structures can be obtained from the transformation laws for the elastic constants as given, for example, by Eqs. 3.5 and 3.16. While these equations are useful in the design of engineering structures, they require knowledge of the elastic constants corresponding to the principal directions of orthotropy. They cannot help the producer of nonwoven fabrics until specific products have been made and their elastic constants evaluated. For prediction one needs an analysis based on fiber properties, the geometry, and bond properties.

Such a fiber-web theory has been developed and experimentally confirmed for selected nonwoven materials (Petterson, 1958). It is assumed that (1) the bond strength exceeds the fiber strength, (2) the fibers are straight, (3) the fibers are long enough and frequently enough bonded so that the displacements in the fibers are identical with the average displacements in the fabric as a whole, (4) the bonds are frequent enough so that there is no buckling, (5) there is no variability in properties of the fiber or in the distribution of fiber orientation angles $\phi(\beta)$ from point to point in the web. Let the x_1 and x_2 directions be the machine and transverse or cross-grain directions, respectively; β is the angle between fiber and machine direction. Then if the fabric is subjected to strains ϵ_{11}, ϵ_{22}, and γ_{12}, each fiber lying at the angle β to the direction of applied tension will experience the strain $\epsilon_{\beta\beta}$ given by (Prob. 21.4)

$$\epsilon_{\beta\beta} = \epsilon_{11} \cos^2\beta + \epsilon_{22} \sin^2\beta + \gamma_{12} \sin\beta \cos\beta. \tag{21.1}$$

For the elastic region, the fiber stress will be given by

$$\sigma_{\beta\beta} = E_f \epsilon_{\beta\beta}. \tag{21.2}$$

The contribution of the fiber stresses in the β direction to the stress in the x_1 and x_2 directions is found from the transformation law for stress components and the fraction $\phi(\beta)\,d\beta$ of fibers with orientation between β and $\beta + d\beta$:

$$d\sigma_{11} = E_f(\epsilon_{11} \cos^2\beta + \epsilon_{22} \sin^2\beta + \gamma_{12} \sin\beta \cos\beta) \cos\beta \cos\beta\, \phi(\beta)\, d\beta. \tag{21.3}$$

The contribution of all fibers to the two stress components σ_{11} and σ_{22} will then be (Prob. 21.5)

$$\sigma_{11} = E_f \int_{-\pi/2}^{\pi/2} (\epsilon_{11} \cos^2\beta + \epsilon_{22} \sin^2\beta) \cos^2\beta\, \phi(\beta)\, d\beta,$$

$$\sigma_{22} = E_f \int_{-\pi/2}^{\pi/2} (\epsilon_{11} \cos^2\beta + \epsilon_{22} \sin^2\beta) \sin^2\beta\, \phi(\beta)\, d\beta. \tag{21.4}$$

The coefficients of elastic stiffness can be found by comparing these equations to Eq. 3.3:

$$\sigma_{11} = \epsilon_{11} C_{11} + \epsilon_{22} C_{12}, \qquad \sigma_{22} = \epsilon_{11} C_{12} + \epsilon_{22} C_{22}. \tag{21.5}$$

Similarly the shear component of stiffness is given by

$$\sigma_{12} = E_f \int_{-\pi/2}^{\pi/2} \epsilon_{12} \sin^2\beta \cos^2\beta\, \phi(\beta)\, d\beta. \tag{21.6}$$

Using Eqs. 21.4, 21.5, and 21.6, the web designer can now alter fiber properties and web geometry to control principal direction properties. To predict properties at angles other than the principal directions, he can use these calculated principal direction constants in the theory of orthotropic materials or proceed to recalculate the elastic modulus, shear modulus, and contraction ratio at any angle α (between direction of pull and the machine direction of the web) by substituting $\phi(\alpha + \beta)$ for $\phi(\beta)$ in Eqs. 21.4 and 21.6.

Comparison of the variation of a nonwoven material modulus with angle of test, α, as predicted by orthotropic theory and the fiber-web theory is seen in Fig. 21.8, where the curves represent theory and the circles represent measured moduli.

B. Statistical Effects on Strength of Yarn. Of course, in many fibrous systems, the assumptions made above regarding elastic behavior of the fibers and uniformity in their properties and placement in the web are not valid. Techniques have been successfully applied to such cases, taking into account fiber variability and the shape of the stress-strain curve, particularly in the post-yield region. Platt et al. (1952) analyzed the mechanical behavior of parallel bundles of fibers

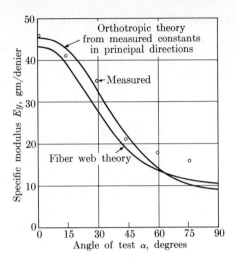

Fig. 21.8. Initial specific moduli, measured, orthotropic, and fiber web theory. (Petterson, 1958.)

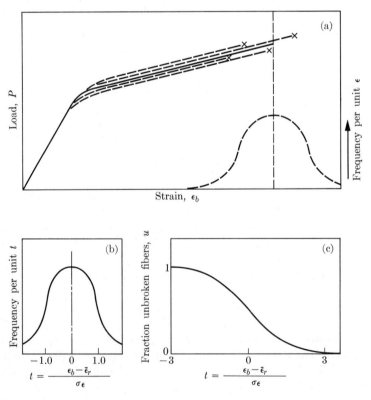

Fig. 21.9. Elongation as number of standard deviations from mean rupture elongation. (After Platt et al., 1952.)

and of twisted yarns. They calculated bundle efficiency in which actual parallel bundle strength is expressed as a ratio of the sum of the individual breaking strengths of N components comprising the bundle. In general, the bundle resistance at any uniaxial strain is simply the sum of the loads on the unbroken fibers, i.e.,

$$P_b = uNP_a, \tag{21.7}$$

where u is the fraction of unbroken fibers, and P_a is the average load on the individual remaining fibers. Note that because of the sharing of the load, the extreme value statistics of Section 15.9 do not apply. The variability in fiber elongation to rupture determines the number of unbroken fibers at any given bundle strain. One first plots the frequency of fiber rupture elongations ϵ_r as a function of fiber elongation ϵ_b (Fig. 21.9a), then normalizes the curve (Fig. 21.9b) by expressing the elongation scale in units t of standard deviations σ_ϵ of the rupture elongation from the mean rupture elongation $\bar{\epsilon}_r$ of the population of fibers. This latter curve is then drawn as a curve (Fig. 21.9c) of cumulative values, specifically a curve of u, the fraction of unbroken fibers versus t. For a typical distribution of rupture elongations, it has been found possible to express u as a simple power series of t. Meanwhile the average load-elongation curve in the region of fiber rupture is simply taken to be linear:

$$P_a = a + b\epsilon_b = b(a/b + \epsilon_b). \tag{21.8}$$

Now

$$t = \frac{(\epsilon_b - \bar{\epsilon}_r)}{\sigma_\epsilon} = \frac{1}{v}\left(\frac{\epsilon_b}{\bar{\epsilon}_r} - 1\right), \tag{21.9}$$

where v is the coefficient of variation of the population ($\sigma_\epsilon/\bar{\epsilon}_r$). It follows that the bundle load is given by

$$P_b = Nu(v, \epsilon_b, \bar{\epsilon}_r)P_a(a, b, \epsilon_b). \tag{21.10}$$

As a test of the given bundle progresses, some fibers will rupture even as the bundle load continues to rise. By setting the derivative of P_b with respect to ϵ_b equal to zero one can predict the breaking load of the assemblage as a function of the average fiber stress-strain curve and the variability in elongation to rupture. This maximum breaking load, expressed as a ratio of the summation of breaking strengths of the individual fibers, is plotted in Fig. 21.10 as a function of mean elongation to rupture $\bar{\epsilon}_r$, coefficient of variation of elongation to rupture v, and the ratio of a/b (stress-strain curve intercept over slope). It is seen that maximum efficiency occurs in bundles of fibers with zero coefficient of variation of rupture elongation, that is, in perfectly uniform material. As fiber variability increases, efficiency in the parallel bundle decreases. For a given mean fiber elongation at rupture $\bar{\epsilon}_r$, and given fiber variability v, the bundle efficiency goes down as the ratio a/b decreases. A high value of a/b suggests the stress-strain curve in the region of rupture is a high flat plateau, and one can reason physically that the bundle in such a case should have a high break efficiency, as Fig. 21.10 shows.

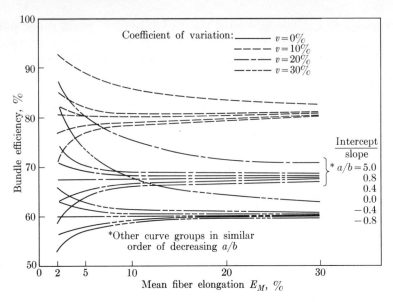

FIG. 21.10. Breaking strength efficiency of parallel fibers. (Platt et al., 1952. Courtesy of *Textile Research Journal*.)

Note that the statistics of bundle fracture are essentially different from those of the brittle fracture of solids, in that failure of the gross structure is not governed by the weakest element alone.

To provide a complete basis for predicting strength efficiencies of assemblages of fibers, one must take into account both parallel- and series-connected components, introduce the complex geometries of textile systems, and allow for variations in rupture strains and stresses as well as in shapes of the stress-strain curves. One must also take into account the fact that failure of a fiber at one point in a twisted structure will not prevent its shifting the load elsewhere.* Such a complete analysis would be a formidable task, but several workers have reported on different combinations of material conditions and have shown success in predicting the mechanical behavior of selected structures. The example given above will suffice to demonstrate one method of approach in the applied mechanics of fibrous structures. However, to permit the reader to go on to problems involving more complex fibrous systems further attention should now be given to the nature of yarn and fabric geometry as it relates to mechanical behavior.

C. Geometrical Effects on Modulus of Yarn. As an intermediate product, card web has little strength. It is usually gathered together into a ropelike form (sliver) and then elongated, allowing slippage of fibers over each other so as to form a continuous strand of substantially aligned fibers. Following final drafting of this strand, it is twisted, fixing the fibers in an essentially helical geometry in the yarn

---

* These statistico-mechanical problems are also encountered in certain other cases of fracture of inhomogeneous materials, such as compression of rocks and tension of solid rocket fuel, which consists of particles of oxidizer in a rubbery matrix.

structure. Only at this point does the material in process resist further drafting. It now becomes a structure usable as a one-dimensional thread, or with subsequent processing such as weaving or knitting, it is converted into a two-dimensional fabric.

It is in the twisting operation that fibers, emerging from the "front roll" of the final drafting operation on staple stock, are turned up into the third dimension to form essentially right circular helices. Early attempts to predict the mechanical behavior of twisted yarns were based on a number of simplifying assumptions: (1) that the individual fibers (or filaments in the case of a continuous filament yarn of silk or of nylon) lie in a perfect helix whose center is the yarn axis, (2) that both the fibers and the yarns are uniform along their entire length and are circular in cross section, (3) that the fibers fall in a rotationally symmetric array in the yarn cross section, (4) that the diameter of the yarn is large compared with that of the fibers, and (5) fibers lying at varying distances from the yarn axis are arranged in concentric helices, all with the same pitch (Platt, 1950).

◀These were simple helices characterized by the geometric parameters derived in the early chapters of most textbooks on differential geometry: the helix angle q, the helix radius r, the helix pitch $1/T$ (where T is the number of turns per inch), the helix torsion τ, and the helix curvature k. To express the torsion and the curvature of a helix in terms of its helix angle and the radius, one starts with the Frenet equations (Struik, 1950)

$$\frac{d\mathbf{x}}{ds} = \mathbf{t}, \qquad \frac{d\mathbf{t}}{ds} = k\mathbf{n}, \qquad \frac{d\mathbf{b}}{ds} = -\tau\mathbf{n}, \qquad (21.11)$$

where $\mathbf{x}$ is the position vector of the curve in cartesian coordinates, s is the distance along the helical curve, $\mathbf{t}$ is the unit tangent vector, $\mathbf{n}$ is the principal normal to the helical curve, a unit vector, and $\mathbf{b}$ is the binormal to the helix, also a unit vector (see Fig. 21.11a). The equations for the primary helix shown are

$$x_1 = r \cos \theta, \qquad x_2 = r \sin \theta, \qquad x_3 = r\theta \cot q, \qquad (21.12)$$

in effect the 1, 2, and 3 components of $\mathbf{x}$. Combining (21.11) and (21.12) it can be shown that (Struik, 1950)

$$k = \frac{\sin^2 q}{r} \quad \text{and} \quad \tau = \frac{\sin 2q}{2r}. \qquad (21.13)$$

These two constants of the helix are of major importance in treating the mechanics of individual fibers as they lie in a twisted yarn, for they determine the bending moment and the local torque required to keep a given fiber in its twisted position. ◀If a fiber lying on a helix of radius r is developed onto a plane surface as shown in Fig. 21.11(b), it can be shown that a relationship similar to Eq. (21.1) exists between any strain which may take place along the yarn axis and the strain which occurs along the fiber axis. Platt assumed that a solidly packed filamentous yarn did not contract radially when pulled axially and therefore

$$\epsilon_{fq} = \epsilon_{33} \cos^2 q, \qquad (21.14)$$

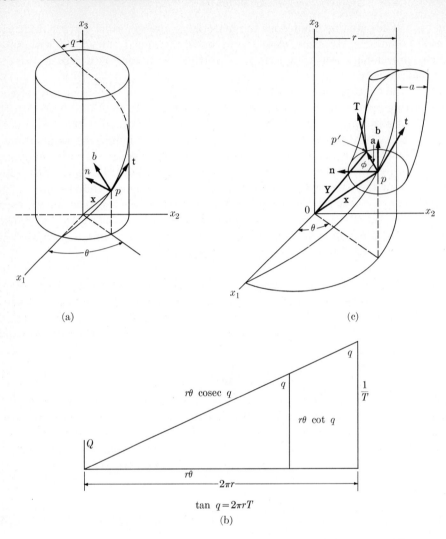

FIG. 21.11. Geometry of a compound helix. (After Schwarz, 1950.)

where q is the local helix angle for the helix of radius r. From Fig. 21.11(c) it is seen that

$$\tan q = 2\pi rT. \qquad (21.15)$$

Hence, by trigonometric identity, we have

$$\cos^2 q = \frac{1}{1 + 4\pi^2 r^2 T^2}. \qquad (21.16)$$

Following the summation procedure illustrated in Eq. (21.4), we see that the axial stress set up in the yarn upon application of an axial strain ϵ_y is, in terms of the slope b and intercept a of the fiber stress-strain curve (Prob. 21.6),

$$\sigma_y = \int_o^R \left[a + \frac{b\epsilon_y}{(1 + 4\pi^2 r^2 T^2)} \right] \left[\frac{2\pi r}{(1 + 4\pi^2 r^2 T^2)} \right] \frac{dr}{\pi R^2}. \qquad (21.17)$$

For the case where the fiber is stressed only in the linear region of the stress-strain curve, the yarn behavior is likewise linear, i.e.,

$$P_y = E_y \epsilon_y, \qquad (21.18)$$

where the effective modulus E_y of the yarn is

$$E_y = \frac{E_f}{1 + 4\pi^2 R^2 T^2}, \qquad (21.19)$$

a function of yarn geometry in turns per inch T, and radius R, and of fiber modulus E_f.

◀Obviously the analysis of yarn behavior is considerably more complicated if any of the five assumptions cited above is eliminated, and several workers in the field of textile mechanics have treated special cases with fewer assumptions than the five listed. Znoek (1961) has studied the behavior of a yarn whose fibers twist about in helices of varying radii. This is a more realistic picture of a twisted yarn in which the fiber or filament wanders from a central position to a peripheral position in order to equalize fiber length distribution in a given segment of yarn. Hearle (1958) and Hearle et al. (1961) have studied the mechanics of yarn deformation, taking transverse stresses into account. Still another assumption to be eliminated is that which fixes the fiber geometry in a straight circular helix. The concentric helices which form a yarn may all be bent or twisted around another "singles" yarn to form a ply twist, or they may be bent around another yarn during weaving or in knitting. In any case, the development of a secondary helix (or helix around a helix) is quite common on the textile scene and has called for an extension of the differential geometry of the primary helix shown in Fig. 21.11(a) to the case shown in Fig. 21.11(c).

◀The path of the secondary helix is determined by movement of the position vector $\mathbf{Y}$ and it can be shown that the components of $\mathbf{Y}$ are

$$\begin{aligned} y_1 &= r \cos \theta - a \cos \theta \cos \lambda\theta + a \cos q \sin \theta \sin \lambda\theta, \\ y_2 &= r \sin \theta - a \sin \theta \cos \lambda\theta - a \cos q \cos \theta \sin \lambda\theta, \qquad (21.20) \\ y_3 &= r\theta \cot q + a \sin q \sin \lambda\theta, \end{aligned}$$

where θ is the rotation angle in the primary helix, ϕ that of the secondary helix, and $\phi = \lambda\theta$. Now using the Frenet equations, we obtain an expression for the unit tangent vector $\mathbf{T}$. The scalar product of $\mathbf{t}$, the unit vector of the primary helix, and $\mathbf{T}$, the unit vector of the secondary helix, leads to an expression for $\cos \psi$, where ψ is the angle between the two unit vectors, i.e., the local helix angle of the fiber relative to the single yarn axis. Expressed in terms of the tan ψ,

$$\tan \psi = \frac{a\lambda + a \cos q}{r \operatorname{cosec} q - a \sin q \cos \lambda\theta}, \qquad (21.21)$$

where a is the radius of the secondary helix (see Schwarz, 1950). Now if the angle q becomes 90°, the yarn will simply be bent, rather than bent and twisted. The expression for tan ψ now simplifies to

$$\tan \psi = \frac{a\lambda}{r - a \cos \lambda\theta}, \qquad (21.22)$$

an equation useful in calculating the change in helix angle which takes place at the top of a yarn bend or crown in a fabric (an important variable affecting abrasion resistance) or at the inside of the bend (which in a fabric determines the extent of embedding of orthogonal yarn systems in a weave structure, and therefore affects cloth density and compactness) (Backer, 1952; Backer et al., 1956).

◀When the deformation of the twisted singles (primary helix) is one of simple bending, the length of the expressions reported by Schwarz (1950) is reduced and it is practical to carry them further via the Frenet relationships. For example, an exact expression for curvature k' of the fiber, in the secondary helix was derived (Backer, 1952) permitting a determination of bending moments required for holding the bent yarn in place. This expression simplified for the fibers at the top (+) and at the bottom (−) of the bend reads

$$k' = \frac{(1 \pm g) + \lambda^2}{a[(1 \pm g)^2 + \lambda^2]}, \qquad (21.23)$$

where g is the ratio of r to a.

◀It was also possible to derive an exact expression for local strain in any fiber in a bent yarn, where the fiber was not free to slip:

$$1 + \epsilon_f = \sqrt{\frac{(g - \cos \phi)^2 + \lambda^2}{g^2 + \lambda^2}}. \qquad (21.24)$$

◀Where complete freedom of slippage occurred, the magnitude of slippage expressed as the length of fiber which passes the central plane of the yarn in the redistribution of material from lower to upper parts of the helix loop is shown to be

$$\text{slippage} = \frac{r}{\lambda\sqrt{\lambda^2 + g^2}}. \qquad (21.25)$$

◀The simple process of twisting a yarn introduces both curvature and torsion to the fiber path. These are geometric parameters, and they must be added to the physical effects of actual fiber twisting or of actual unbending (if the fiber were originally crimped to facilitate processing). Of course, the local angle of fiber inclination must be taken into account in determining the components of fiber torque and bending moment that contribute to the torsional unbalance of the yarn (see Platt et al., 1958). And finally one must take into account the changes in local fiber inclinations and fiber torsion and curvature to derive an expression for yarn-bending rigidity (see Platt et al., 1959). ▶

D. Geometrical Effects on Moduli of Woven Fabrics. During the weaving operation, the warp (longitudinal) yarns are bent about the filling yarns which are inserted by means of a flying shuttle and are initially held in a straight configuration. During this interlacing period, the warp moves up and down in the "crimp" plane above and below successive filling yarns. Upon release of the fabric from the width-restraining templates of the loom, the filling takes on a crimp and a bending of its own and the fabric width accordingly decreases. The warp now becomes a little less crimped, depending on the amount of tension applied to it during subsequent processing. At the same time, the warp may buckle out of the crimp plane and assume a helical configuration within the weave structure. This buckling can be traced to the unbalance in shear stresses set up on the yarn cross section during bending. As a result of the buckling, the yarn will generally reduce its local twist to a degree calculable from the expression for helix torsion (Eq. 21.13). This change in local twist will alter many of the properties of the fabric. It has been observed that the degree to which the fabric weave, particularly its twill line direction (see Fig. 1.39), supports or opposes this buckling tendency also has a significant effect on fabric behavior (see Backer et al., 1956).

There are innumerable combinations and permutations of cloth weave patterns. Rather than try to develop a complete theory, we shall here simply list, in order of ease of occurrence, some of the contributing mechanisms in the plain weave, or simple one-up, one-down pattern of interlacing. We shall then analyze a few of the examples to give an insight into the important variables and the way in which they affect the overall behavior. For a plain weave, the modes of deformation are of a multiple nature and include: *yarn shear* (if the principal stress directions in a plane stress condition are not aligned with the yarns) (see Haas, 1912; Topping, 1961); *crimp interchange* (straightening of one set of yarns and increasing the bend of the other set) (see Peirce, 1947); *yarn flattening* (reduction in vertical diameter); *yarn compaction* (increase of packing factor of the fibers) (Hearle et al., 1961); *fiber straightening and/or rotation* in the yarn structure (Platt, 1950); *fiber extension* (Platt, 1950; Petterson, 1958). See also Weissenberg (1949).

It is not feasible to combine the analyses of each of these modes of deformation into a single gross theory for the plain woven structure. A more practical approach is to treat the deformations in yarn shear and in crimp interchange separately, and then use the new configuration, resulting from one or both of these mechanisms, as the starting point in the more complex analysis of yarn extension and flattening in the distorted fabric structure. As will be shown below, fabric deformation in yarn shear and in crimp interchange is dependent on the *ratio* between the principal stresses acting on the fabric rather than on the *magnitude* of these stresses. Thus it can be reasoned that the first two deformation actions take place at extremely low stresses (in a given ratio) before the remaining modes become active.

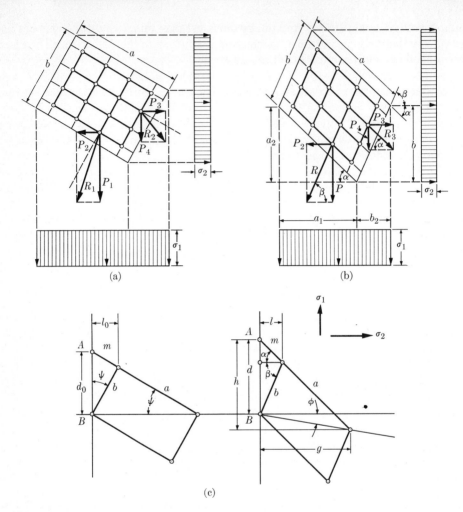

Fig. 21.12. (a) Element of fabric before thread shear. (b) The same element after thread shear. (c) Element of fabric at position with vertical axis irrotational. (After Haas, 1912.)

◀The behavior of a woven fabric in the presence of shear stresses uniquely marks its difference from continuous film or sheet materials. Cloth resistance to shear is of concern to the designers of airships, inflatable shelters, life rafts, radomes, and even of inflated dams, airplanes, and radar antennas, for in these applications a high shear modulus is usually sought. Shear modulus is likewise important in the shaping of a two-dimensional fabric into the double curvature of a garment and in the freedom of movement demanded by the wearer of such a garment; here a low modulus is required at certain fabric orientations. Shear in woven fabric occurs by rotation of the two sets of yarns relative to each other. In tightly woven fabrics or cloths which have been coated with, or laminated to, a plastic film, the textile structure has built-in resistance to this relative rotation. But in more open, unsupported fabric structures, the warp is relatively free to pivot around its point of contact with the filling. Such a fabric will shear when placed in a biaxial stress field in which the principal stress directions do not coincide with the original yarn direction, until the external stresses are balanced solely by tensile forces in the yarns.

◀ Haas (1912) has shown that if a fabric is placed in the stress field with its yarns lying at an angle ψ to the principal stresses σ_{11}, σ_{22}, then the yarns will rotate to new alignments α and β relative to directions 1 and 2 in accordance with the relationship (Fig. 21.12)

$$\tan \alpha \tan \beta = \sigma_{11}/\sigma_{22}. \tag{21.26}$$

He likewise showed that if one of the principal stress axes is parallel to an irrotational axis, as determined from the deformation, then the new yarn orientation angles are governed by a second relationship

$$\cos \alpha / \cos \beta = \cot \psi. \tag{21.27}$$

This condition is observed in torsion of an inflated cylinder, a shape met in many of the applications cited above. The circumferential line of the cylinder is irrotational. Topping (1961) showed that shear modulus G_{xy} (where x, y are the yarn directions), which relates to resistance to yarn rotation, is approximately equal to the stress in the x direction yarns (i.e., those yarns lying farthest from the irrotational axis). A graph of G_{xy} versus σ_{xx} should be a straight line passing through the origin. This G_{xy} is in effect a geometric and stress-induced rigidity. If a material component of rigidity is now added to the fabric in the form of a continuous coating or a "between-pore" matrix which restricts yarn rotation, it should displace the linear relationship between G_{xy} and σ_{xx}. This is what Topping found to be the case in measurements on a single-ply fabric in a pressurized cylinder (Fig. 21.13). In short, the major part of this fabric shear *strain* was determined by the *ratio* between stresses, not by the magnitudes of the principal stresses. It follows

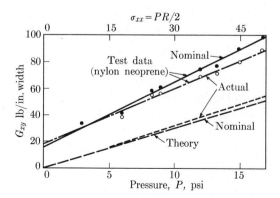

FIG. 21.13. Variation of shear modulus with normal stress in fabric plane. "Nominal" and "actual" refer to dimensions on which curves were based. (After Topping, 1961.)

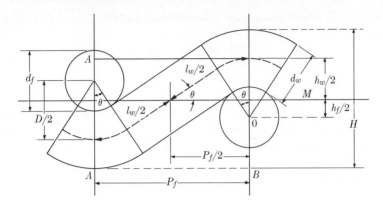

FIG. 21.14. Geometry of a plain-weave cloth structure. (After Peirce, 1947.)

that the ratio between the shear stress and the shear strain, G_{xy}, will vary linearly with the level of the principal stresses.

◀A corresponding stress-ratio effect is observed in the crimp interchange mechanism. Figure 21.14 shows a section of the unit cell of a plane weave. Assumptions applied by Haas (1912) and Peirce (1947) were: (1) The yarns are highly flexible to bending but essentially rigid to tensile extension. (2) Yarns are essentially round (i.e., not flattened more than 20%). (3) Yarn paths are comprised of straight sections and circular arcs. From the geometry of Fig. 21.14 it can be shown that

$$h_w = (l_w - D\theta_w) \sin \theta_w + D(1 - \cos \theta_w), \quad (21.28)$$

$$P_f = (l_w - D\theta_w) \cos \theta_w + D \sin \theta_w, \quad (21.29)$$

$$C_w = \frac{l_w - P_f}{P_f}, \quad (21.30)$$

$$h_w + h_f = D. \quad (21.31)$$

An additional set of three equations can be written by reversing the subscripts in Eqs. 21.28 to 21.30. Thus seven equations are needed to relate the geometric parameters describing the unit cell of a plain weave, including: the yarn spacings P_w, P_f; the yarn wave heights h_w, h_f; the yarn lengths l_w, l_f; the yarn inclination angles θ_w, θ_f; the yarn crimps C_w, C_f; and the sum D of the vertical yarn diameters d_w and d_f. If any four parameters are given for a fabric, it is possible, using the above equations, to completely characterize the geometry of the cloth. More important, it is possible to use them in predicting the changes in fabric geometry resulting from processing strains or from deformations occurring in product utilization, by carrying over from the characterization of the original parameters those values which remain constant. Usually l_w and l_f remain constant as does D, unless the changes involve swelling of the yarn diameters or flattening under stress.

◀The ratio between principal stresses applied to a specimen of woven fabric will cause the geometry of the unit cell to change (through crimp interchange) until force equilibrium is reached at the intersections between warp and filling; this dictates that

$$\frac{F_w}{F_f} = \frac{\tan \theta_f}{\tan \theta_w},\qquad(21.32)$$

where F represents the force component on the yarn in the plane of the fabric, or $\sigma_w P_w = F_w$. Now with Eq. 21.32 and the seven equations of cloth geometry, it is possible to analyze a number of situations which, at first, appear hopelessly complex. For example, it can be shown that the Poisson ratio of woven fabrics is not a constant, a feature which is troublesome in the design of laminated materials. The extensibility of a fabric can be determined in orthogonal uniaxial stress, and in biaxial stress (assuming crimp interchange mechanism alone). The shrinkage of a fabric can be calculated if it occurs primarily as a result of yarn swelling. The limiting fabric thickness in a swollen fabric can be determined and area changes resulting from pressing actions on a fabric can be calculated. The geometric relationships can be used in the design of fabrics with equal crown heights in warp and filling with a view to maximizing abrasion resistance. They can be used to characterize the structural tightness of the fabric, a parameter on which the shear rigidity, the tear resistance, the conformability, and the sewability of the fabric depends.

◀The manipulation of the eight relationships is, of course, too tedious for design purposes and it will be found useful to utilize graphical solutions to aid in computations. Adams et al. (1956) provide a nomographic solution for the seven equations and it is a simple matter to combine Eq. 21.32 with the geometric solutions of this nomograph to depict the mechanical behavior of woven fabrics as it depends on crimp interchange. Numerous examples are given by Adams et al. of solutions to many problems incurred in processing and utilization of fabrics.▶

It is clear that the discussion presented above deals primarily with the reaction of textile structures to either uniaxial or biaxial stress. In a few cases the treatment extends to the general condition of plane stress. It has been shown that the reactions of textile materials to these extremely simple conditions of stress call for analytical treatments which introduce some new concepts and methods not required in the case of isotropic, uniform, elastic materials. Geometric parameters, variability in material properties and structural configurations, viscoelastic fiber behavior, and frictional interaction between components enter the scene to varying degrees, depending on the nature of the material and the conditions of test or usage. The materials engineer will often have to rely on physical intuition as to the importance of these parameters in any given situation. But he should also recognize that the practicing textile engineer must cope with a wide range of textile mechanical behavior in processing or use under other conditions than appear above. A list of mechanical properties of importance to manufacturer, designer, and user of textile materials is furnished in Table 21.3 to give the reader an idea of the range of problems that lie in the special field of textile mechanics. Numerous papers have been written on most of the subjects listed in the table, but our understanding of the various mechanisms involved is still far from complete.

TABLE 21.3

List of Mechanical Properties of Textile Materials

Stress-Strain Behavior in Tension
  Uniaxial tensile behavior
    Breaking strength
    Elongation
    Yield point
    Modulus of elasticity
    Recovery behavior
  Biaxial tensile behavior
  Creep behavior in tension
  Tensile impact behavior
  Fatigue in tension
 	
Stress-Strain Behavior in Compression
(details as listed above)
 	
Torsional and Shear Behavior
 	
Frictional Properties
 	
Abrasion Resistance
	Flexibility and Bending
	  Simple bending
	  Drape (multidirectional)
	  Crease acceptance and retention
	  Flex fatigue resistance
	
	Resistance to Stress Concentrations
	  Tear resistance
	  Snag resistance
	  Puncture resistance
	  Cutting resistance
	  Knot and loop efficiency
	
	Dimensional Stability
	
	Resistance to Delamination
	
	Aesthetic or Subjective Mechanical Properties
	  "Hand"
	  Softness
	  Resilience

Suffice it to say that the purpose of this chapter will be served if the reader can identify one or more of these properties with a specific textile behavior problem in his personal experience and can visualize how a systematic mechanical treatment might aid in its solution.

In general the properties cited in Table 21.3 refer to the behavior of textiles as thin sheet structures. Only the property of resistance to delamination refers specifically to composite, laminated, and fiber-reinforced systems. Such materials often involve bonding of layers which have widely differing properties. More often than not, each individual layer is anisotropic, and so the mechanical considerations involve both the theory of anisotropic elasticity (e.g. Hearmon, 1960) and that of laminated structures (e.g. Hoff, 1949). For a general introduction to such systems, see Riley, et al., (1960), and for examples of the analysis of such structures see Gordon (1952) and Hashin and Rosen (1964).

REFERENCES

ADAMS, D. P. SCHWARZ, E. R. BACKER, S.	1956	"Nomographic Solution of Relationships in Cloth Geometry," *Text. Res. J.* **26**, 653–665.
BACKER, S.	1952	"The Mechanics of Bent Yarns," *Text. Res. J.* **22**, 668–681.
BACKER, S. BEST-GORDON, H. W. ZIMMERMAN, J.	1956	"The Interaction of Twist and Twill Directions as Related to Fabric Structure," *Text. Res. J.* **26**, 87–107.
BOLAM, F. (ed.)	1962	*The Formation and Structure of Paper*, Trans. symposium held at Oxford, Sept. 1961, British Paper and Board Makers Ass'n., London.
BRAND, R. BACKER, S.	1962	"Mechanical Principles of Natural Crimp of Fiber," *Text. Res. J.* **32**, 39–49.
BRYANT, G. M.	1953	"Force-Temperature Behavior of Nylon Filaments at Fixed Extensions," *Text. Res. J.* **23**, 788.
GORDON, J. E.	1952	"On the Present and Potential Efficiency of Structural Plastics," *J. Roy. Aero. Soc.* **56**, 704–728.
HAAS, R. DIETZIUS, H.	1912	"Stretching of the Fabric and Deformation of Envelope in Non-Rigid Balloons," *N.A.C.A. Rept. No. 16, 3rd Annual Rept.* pp. 144–271, 1917, translated by K. K. Darrow (original publication in German by Springer, Berlin, 1912).
HASHIN, Z. ROSEN, B. W.	1964	"The Elastic Moduli of Fiber-Reinforced Materials," *J. Appl. Mech.* **31**, 223–232.
HEARLE, J. W. S.	1958	"The Mechanics of Twisted Yarns: The Influence of Transverse Forces on Tensile Behavior," *J. Text. Inst.* **49**, T389–T408.
HEARLE, J. W. S. El-BEHERY, H. M. A. E. THAKUR, V. M.	1961	"The Mechanics of Twisted Yarns: Theoretical Developments," *J. Text. Inst.* **52**, T197–T220.
HEARMON, R. F. S.	1961	*An Introduction to Applied Anisotropic Elasticity*, Oxford University Press, London.
HOFF, N. J.	1949	"The Strength of Laminates and Sandwich Structural Elements," *Engineering Laminates*, A. G. H. Dietz, ed., Wiley, New York, pp. 6–88.
MAUERSBERGER, H. R.	1954	"*Matthews' Textile Fibers*," Wiley, New York.
MEREDITH, R.	1945	"A Comparison of the Tensile Elasticity of Some Textile Fibers," *J. Text. Inst.* **36**, T147–T164.
MEREDITH, R.	1956	*The Mechanical Properties of Textile Fibers*, Interscience, New York.
MEREDITH, R.	1960	"Effect of Moisture on Mechanical Properties," *Moisture in Textiles*, J. W. S. Hearle and R. H.

		Peters, eds., Textile Institute, Manchester, and Butterworths, London.
Morton, W. E. Hearle, J. W. S.	1962	*Physical Properties of Textile Fibers*, Textile Institute, Manchester, and Butterworths, London.
Morton, W. E. Permanyer, F.	1949	"Twist Relationships in Single and Multiple Rayon Filaments," *J. Text. Inst.* **40,** T371.
Orowan, E.	1933	"The Tensile Strength of Mica and the Problem of the Technical Strength," *Z. Physik* **82,** 235–266.
Peirce, F. T.	1947	"Geometrical Principles Applicable to the Design of Functional Fabrics," *Text. Res. J.* **17,** 123–147.
Petterson, D. R.	1958	"On the Mechanics of Non-Woven Fabrics," Sc.D. Thesis, M.I.T., Cambridge, Mass. See also S. Backer and D. R. Petterson, "Some Principles of Non-Woven Fabrics," *Text. Res. J.* **30,** 704–711 (1960).
Platt, M. M.	1950	"Some Aspects of Stress Analysis of Textile Structures—Continuous Filament Yarns," *Text. Res. J.* **20,** 1–15.
Platt, M. M. Klein, W. G. Hamburger, W. J.	1952	"Factors Affecting the Translation of Certain Mechanical Properties of Cordage Fibers into Cordage Yarns," *Text. Res. J.* **22,** 641–667.
Platt, M. M. Klein, W. G. Hamburger, W. J.	1958	"Torque Development in Yarn Systems, Single Yarn," *Text. Res. J.* **28,** 1–14.
Platt, M. M. Klein, W. G. Hamburger, W. J.	1959	"Some Aspects of Bending Rigidity of Single Yarn," *Text. Res. J.* **31,** 611–627.
Rebenfeld, L.	1962	"Response of Cotton to Chemical Treatments," *Text. Res. J.* **32,** 154–157.
Rebenfeld, L. Virgin, W. P.	1957	"Relationship Between the X-Ray Angle of Cottons and Their Fiber Mechanical Properties," *Text. Res. J.* **27,** 286–289.
Ribi, E.	1951	"Submicroscopic Structure of Fibers and their Formation," *Nature* **168,** 1082–1083.
Riley, M. W. Lieb, J. H. Jaffe, E. H. Miller, S. A. Carter, J. F.	1960	"Filament-wound Reinforced Plastics: State of the Art," *Materials in Design Eng.*, Aug., 128–146. See also "The Promise of Composites," *Materials in Design Eng.*, Sept. 1963, 79–126.
Schwarz, E. R.	1950	"Twist Structure of Plied Yarns," *Text. Res. J.* **20,** 175–179.
Smith, J. C. McCrackin, F. L. Schiefer, H. F.	1957	"The Impact Absorbing Capacity of Textile Yarns," *Bull. ASTM* No. 220, 52–56.
Struik, D. J.	1950	*Differential Geometry*, Addison-Wesley, Reading, Mass.
Susich, G. Backer, S.	1951	"Tensile Recovery Behavior of Textile Fibers," *Text. Res. J.* **21,** 482–509.

Syromjatnikoff, F.	1927	"On an Experiment for the Determination of the Tensile Strength of Asbestos," *Z. Kristallographie* **66,** 191–194.
Topping, A. D.	1961	"An Introduction to Biaxial Stress Problems in Fabric Structures," *Aerospace Eng.* **20,** 18–19, 53–57.
von Bergen, W.	1963	*Wool Handbook*, Interscience, New York.
Weissenberg, K.	1949	"The Use of Trellis Model in the Mechanics of Homogeneous Materials," *J. Text. Inst.* **40,** T89–T110.
Znoek, W.	1961	"Some Properties of Continuous Filament Yarns," *Text. Res. J.* **31,** 504–514.

PROBLEMS

21.1. To what extent is the rotational freedom of the helical windings in the secondary wall of cotton influenced by moisture content?

21.2. Prove that for a linear viscoelastic material, plotting constant strain-rate data as stress versus time to reach a given strain should lead to curves in which the stress at a given time varies linearly with the strain.

21.3. Compare the series of deformation mechanisms in a bonded web with those occurring at the molecular chain level in a single fiber and with those occurring in a woven fabric subjected to uniaxial tension at a slight angle to warp or filling yarn.

21.4. Derive the equation for the strain in a fiber in a bonded web, Eq. 21.1.

21.5. Why does not the shear strain term in Eq. 21.3 contribute to the stress in a bonded web structure as given by Eq. 21.4?

21.6. Derive the equations for the yarn stress in terms of fiber behavior, Eqs. 21.17 through 21.19, and explain them in terms of the units used in the textile industry.

21.7. (a) From data of the type plotted in Fig. 21.6, how is it possible to obtain conventional stress-relaxation or creep data? (b) From such data how can the deformation under an arbitrary service load be determined?

21.8. Conventional textile yarns are twisted to an outside helix angle which is between 15° and 25°. When these yarns are bent around each other (as when a warp bends over filling yarns) the contact fibers of top and of bottom yarns may be aligned and the two yarns may "nest." Under what circumstances is this nesting possible?

21.9. Given a yarn of radius R, with T turns per inch twist. The fibers in the yarn section are round (of radius r) and are closely packed. The fiber modulus is E_f. Estimate the bending rigidity of the yarn. Estimate its torque buildup when stretched axially (without twisting) to a strain ϵ_y.

21.10. A bonded web constructed of straight long filaments is subjected to a small uniaxial load in its principal direction of fiber orientation. The fiber has a tensile modulus E_f. If the distribution of fiber orientation angle β about the principal axis is $\phi(\beta) \propto \cos^2 \beta$, what is the principal modulus of the web relative to the fiber modulus? Assume no buckling.

21.11. As a result of processing inhomogeneities (such as the drafting wave) the volumetric packing factor of fibers in a yarn cross section is found to vary sinusoidally along the yarn length. The wavelength of the variation is l and its amplitude is A, the average packing factor of the yarn is $\bar{\phi}$, where

$$\phi = \frac{\text{solid fiber cross-sectional area}}{\text{total yarn cross-sectional area}}.$$

The yarn radius, which is constant, is R; the yarn twist is T turns per inch, also constant along the yarn length. The fiber modulus of elasticity is E_f. If a length of yarn L (where L is an integral multiple of l) is loaded with a weight P_y, what is the total extension that occurs in L?

◀21.12. One end of a circular horizontal loop of a yarn is kept fixed while a vertical force is applied to the other end, distorting the yarn from the flat loop to a circular helical geometry. Find bending-strain-energy and twisting-strain-energy equations in terms of helix radius r_h or the height h of the moving end. Given E, G of yarn (taken to behave as if it were a monofilament), r_0 = initial radius of the loop, a = radius of the yarn, L = length of yarn (in a loop).

◀21.13. (a) For small yarn inclination angles, show that Eqs. 21.28 through 21.31 for the geometry of a plain weave can be combined and simplified to give

$$h_w = l_w\theta_w - D\theta_w^2/2, \quad l_w = P_f(1+C_w), \quad \theta_w = \sqrt{2C_w(1+C_w)},$$

$$D = h_w + h_f = \left[P_w\sqrt{2C_f(1+C_f)} + P_f\sqrt{2C_w(1+C_w)}\right] \bigg/ \left(1 + \frac{C_f}{1+C_f} + \frac{C_w}{1+C_w}\right).$$

(b) For tightly woven or externally stressed fabrics there is flattening at points of contact between yarns. A useful approximation is that the geometry is unaffected except as a flattening factor differs from unity. This factor, assumed the same for each yarn, is defined as

$$F_w = F_f = D/(d_w + d_f).$$

For the cotton system, the nominal yarn diameters d_w and d_f are found from the yarn numbers N_w and N_f (the number of 840 yard lengths per pound) by the approximate equation (see, for example, Peirce, 1947)

$$d \text{ (in inches)} = 1/(28\sqrt{N}).$$

List the above-mentioned variables for a plain weave, counting both warp and fill. How many of these can be independent? (c) Describe how each geometrical variable could (if possible) be measured directly from a sample of the fabric of a dress shirt. (d) Calculate the flattening factor for a plain-weave cotton cloth with $N_w = 6.7$, $N_f = 6.3$, $C_w = 0.087$, $C_f = 0.062$, $P_w = 0.0309$ in., $P_f = 0.0304$ in. (e) Calculate the fabric shrinkage, assuming swelling until both the straight section of the yarn and the flattening at the contact points disappear.

◀21.14. Given a plain-weave cotton fabric with $N_w = 12.8$, $N_f = 10.3$, $C_w = 0.125$, $C_f = 0.065$, and a count per inch of the warp and filler threads of 81 and 38, respectively. (a) Compute the flattening factor for the yarns of this fabric. (b) If the fabric is pulled fillingwise, what is its effective Poisson's ratio? (c) What is the new thickness after (b)? (d) Starting from the original fabric we extend the warp 5% but hold the width fixed. How much yarn flattening takes place as a result (to the nearest mil)?

21.15. Describe how fiber friction influences the properties of textile products.

CHAPTER 22

CASE STUDIES

22.1 SYNOPSIS

The successful application of book knowledge and laboratory experience to actual engineering problems involving the proper utilization of materials cannot be assured in a university education. A large part of this must still depend on the intelligence of the young engineer, on his capacity to evaluate a myriad of conflicting and often obscuring details, on his ability to discard irrelevant side effects by reasoning and simple calculation, and consequently his ability to recognize the first-order effects. It does, however, help to present a number of representative cases which received proper or practical solutions from engineers in the past. Although better examples of such solutions must abound, not all of them get into print from fear that publicity of the difficulties encountered prior to the proper solution may be damaging to the reputation of the organization concerned.

Several such solutions of different engineering problems will be discussed below. The first one concerns the selection of the pre-tension in connecting rod cap bolts so that the joint will not open under operating loads and so that static yielding in assembly or fatigue in operation is avoided. The second one is on a larger scale and is based on the difficulties encountered in the selection and evaluation of materials for ultra-high-strength fuel casings for solid-fueled rockets. The third problem is the decision of whether or not a pressure vessel shipped to the installation site without a required drain hole should be accepted if the hole is drilled and tapped at the site where no subsequent proof test can be performed. The fourth problem pertains to the winding and unwinding of coiled springs upon stress-relieving and the complexities that enter into the proper prediction of the magnitude of this effect. The fifth problem is based on a turbine wheel fracture during operation and the following detailed investigation undertaken to identify the cause of the failure. Other problems are left to the student.

22.2 DETERMINATION OF OPTIMUM PRELOAD IN CONNECTING ROD CAP BOLTS IN INTERNAL COMBUSTION ENGINES

In internal combustion engines the caps of connecting rods are subjected to high inertia loads during each cycle. In a properly designed assembly, the cap bolts must be pre-tensioned to prevent vibration and excessive wear. The pre-tensioning must be at least enough to assure that even under the largest bearing loads in operation no clearance develops between the cap and the connecting rod. In some cases, the bolts are designed so closely that over-tensioning cannot be permitted if the bolts are not to yield during assembly or operation. The loading of the bolts is further complicated by the cyclic nature of the bearing loads and by nonreproducible friction torques developed during assembly at the bolt threads and bolt abutment interfaces. Such friction torques can be substantial, and influence the pre-tension load in a manner difficult to predict.

This problem, which appears to be straightforward, actually involves a good degree of complexity and is therefore typical. In what follows, a design of the connecting rod cap bolts for a Rover Mark II Diesel engine will be presented. The discussion is based on an analysis by Essex (1960) performed in cooperation with the Rover Company Ltd. of Birmingham, England, and is part of a diploma thesis presented at the University of Birmingham.

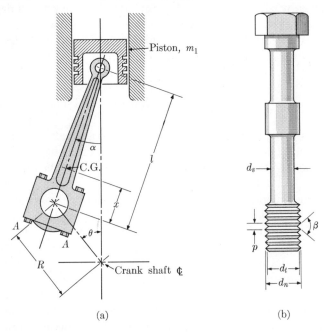

FIG. 22.1. (a) Sketch of connecting rod showing the cap bolts A and characteristic dimensions. (b) Detail of a reduced-shank cap bolt.

A. Design Requirements. In Fig. 22.1 a sketch of a piston, connecting rod, and crank assembly is shown, along with a separate cap bolt A. The requirements for the bolts are that the joint must not open in operation, the bolts must not fracture by fatigue, and the performance must be reproducibly obtainable in actual assembly conditions. Two styles of bolts will be considered, ones with a slender shank to minimize cracking from the root of the threads, and standard bolts, which are less expensive.

B. Analysis of Forces Acting on the Connecting Rod Cap. The first step in a solution is the determination of the loads acting on the caps of the connecting rod under operating conditions. These loads are made up of an inertia component due to the mass of the rotating and reciprocating parts and a pressure component due to the burned gases inside the cylinder. The pressure forces always act between the connecting rod and the crank shaft, while the inertia forces may act on the connecting rod caps as well. It can be seen by inspection that the maximum force on the connecting rod cap occurs at the start of the intake stroke, when the cylinder pressure is only slightly below atmospheric pressure. Thus the inertia forces are the only ones to be taken into account.

If the mass of the reciprocating parts, exclusive of the connecting rod, is m_1, and the center of gravity of the connecting rod of length l and mass m_2 is at a distance x from the crank bearing, the maximum force on the connecting rod cap can be calculated as a function of the angular velocity $\dot{\theta}$ of the crank and the length R of the crank arm, and is (Prob. 22.1)

$$2F = R\dot{\theta}^2[(m_1 + m_2 x/l)(1 + R/l) + m_2(1 - x/l)]. \qquad (22.1)$$

It must be noted here that not all the load calculated in Eq. 22.1 goes into the bolt, but instead a significant part of it releases the clamping pressure on the abutment. To determine the distribution of loads on the bolt and on the abutment as a result of the force F it is necessary to consider the relative stiffnesses of the bolt and the abutment. The stiffness of the bolt must be nearly $A_1 E_1/L_1$, where A_1, E_1, and L_1 are respectively the cross-sectional area, the elastic modulus, and active length (from bolt head to somewhere between the first and second engaged thread between the bolt and nut). The stiffness of the abutment can be determined either experimentally, or by considering an equivalent active cross section as shown in Fig. 22.2. Let these stiffnesses of the bolt and abutment be k_1 and k_2. Under a preload P_L, then, the bolt will extend a certain amount while the abutment will be compressed a smaller amount. If the fluctuating load $2F$ is now applied to the connecting rod cap, the compressive force on the abutment is decreased and the tensile force on the bolts increased, by extending the abutment and the bolts by identical amounts. Thus, the increase in the bolt load is only a certain amount F' of the force F, and depends on the relative stiffnesses of the bolt and the abutment:

$$F'/k_1 = (F - F')/k_2. \qquad (22.2)$$

$$k_2 \approx \frac{E(A_1 + A_2)}{2l_a}$$

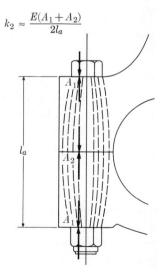

The specified conditions of assembly must be such that even with the lightest reasonable torque and highest friction, the joint will remain in compression:

$$P_L > F - F' = F/(1 + k_1/k_2). \qquad (22.3)$$

FIG. 22.2. Sketch of lines of force in abutment. (After Essex, 1960.)

C. Effect of Thread and Bolt Head Friction. If a torque wrench is used for assembly, the torque will be known to perhaps 10%. This torque not only has to overcome the geometrical frictionless torque of the screw threads in developing the preload, but also the frictional torque in the engaged threads and the torque between the face of the nut and the abutment. For a nut with pitch p, diameter d_t, thread angle β, thread and face coefficients of friction f_t and f_f, and mean face diameter d_f, the applied torque is (Prob. 22.2)

$$T = P_L \frac{d_t}{2} \frac{(p/\pi d_t) + f_t \sec \beta}{1 - (p/\pi d_t) f_t \sec \beta} + P_L \frac{d_f}{2} f_f$$
$$= P_L f(f_t)\, d_t/2 + P_L f_f\, d_f/2. \tag{22.4}$$

From Table 20.1, Section 20.6, coefficients of friction may range from 0.15 to 0.30 for boundary lubrication of normally contaminated surfaces. If the tightening torque is set high enough so as to provide the preload with the highest coefficients of friction, there will then be an overload when the coefficient of friction is low.

D. Strength. First consider requirements on the shank. For trouble-free operation, the bolts must yield little enough under the added dynamic load so that the joint remains tight. While this requirement might allow some plastic deformation on the first loading, a simpler *sufficient* condition for the joint remaining tight is that yielding does not occur under the total load arising from the tightening loads and torques discussed above (Prob. 22.3):

$$\frac{P_L}{A_s Y} \sqrt{\left(1 + \frac{F'}{P_L}\right)^2 + 12 \left(\frac{d_t}{d_s}\right)^2 f^2(f)} \leq 1. \tag{22.5}$$

Finally, fatigue failure should not result from the alternating part $F'/2$ of the pulsating load, in conjunction with the mean load resulting from the preload, the tightening torque, and the mean load $F'/2$ contributed by the pulsating load. Again we take a simple sufficient condition for safety which can be re-examined if the part proves to be critical. From Section 18.6, one such condition is given in terms of the maximum stress amplitude on any slip plane, $(\tau_a)_{max}$, the yield criterion based on the maximum shear stress on any plane at any part of the cycle, τ_{max}, and the fatigue limit, σ_l, by Eq. 18.5b (Prob. 22.4):

$$\frac{F'}{2 A_s \sigma_l} + \frac{P_L}{A_s (\text{T.S.})} \sqrt{(1 + F'/PL)^2 + 16(d_t/d_s)^2 f^2(f)} - \frac{F'}{2 A_s (\text{T.S.})} \leq 1. \tag{22.6}$$

Equations 22.5 and 22.6 are normalized in terms of the ratios of the relevant loads to the strengths of the bolt for convenience in recognizing the importance of the various terms.

E. Numerical Solution. The following information is available for the engine under consideration:

$$\left(m_1 + m_2 \frac{x}{l}\right) = 2.993 \text{ lb}, \quad \left(\frac{l-x}{l}\right) m_2 = 1.855 \text{ lb}, \quad \frac{\dot{\theta}_{max}}{2\pi} = 3650 \text{ rpm},$$

$$R = 1.75 \text{ in.}, \quad l = 7 \text{ in.},$$

SAE fine threads, for which $p/\pi d_t \approx 0.04$, $d_t/d_n \approx 0.92$, $\beta = 30°$,
Shank to nominal diameter ratio (from a reasonable sketch, to be checked later) $d_s/d_n = 0.7$,
Nut face to thread diameter ratio, $d_f/d_t = 2$,
Bolt to abutment area ratio (from a reasonable sketch) $= 0.4$,
Shank to abutment area ratio (from above data) $= 0.2$,
Material: T. S. $= 145{,}000$ psi, $Y = 129{,}000$ psi, $\sigma_l = 59{,}000$ psi.
From these data, $f(f) = 0.21$ to 0.39 for $f = 0.15$ to 0.3, and the pulsating force per bolt is 1850 lb.

The tightening torque must be specified high enough to maintain closure with the highest coefficient of friction, from Eqs. 22.3 and 22.4. If the torque is specified as 10% above this value, applied torques may be 10% higher yet, due to the uncertainties in assembly. If the friction on the face of the nut is less than this high friction, the preload will be higher. The next question is whether conditions will be more critical with low thread friction, producing the highest preload, or high thread friction, producing a higher torque. Likewise it is difficult to tell whether yielding (Eq. 22.5) or fatigue (Eq. 22.6) is more critical. Numerical evaluation shows that fatigue with high thread friction is slightly more critical, requiring a shank area of 0.044 in² (Prob. 22.5).

At the thread root, since fatigue fracture may take place, a strength reduction factor should be used on the alternating load. Taking a thread root radius of 0.002 in., and using a relatively low estimate from Section 11.8 to account for the large flank angle and the notch insensitivity, gives a factor of about 2. A strength reduction factor should not be used in the part of the equation corresponding to general yielding because local yielding will shift the mean load to other parts of the cross section. As a matter of fact, it is conceivable that the creep at the root of the threads under alternating loading would drop the mean load even below the average value in the bolt. For simplicity, however, we shall neglect this effect.

Under the loads required for tightness with the range of possible values of the coefficient of friction, the yielding and fatigue equations indicate no danger in the thread roots. In this case, then, the shank could have been stiffer, requiring less preload and subject to less shear stress, even at the expense of some increase in alternating load. In fact, the necessary diameter of a standard bolt turns out to be less than that for the bolt with the turned-down shank calculated above, for which $d_s/d_n = 0.7$ (Prob. 22.6).

In reviewing the problem, it is clear that a large amount of the stress required comes from the high and uncertain values of the coefficient of friction. Careful lubrication during assembly, and direct measurement of the bolt length to determine preload will markedly reduce the required dimensions. At the same time, fatigue effects will become more important. If weight is extremely critical, a more exact analysis should be undertaken, finally supplemented by tests.

22.3 DEVELOPMENT OF ULTRA-STRONG SOLID-FUEL ROCKET CASINGS

It is well known that in rocketry, excess structural weight limits the range of the missile seriously. Thus, for instance, the fuel casings of solid-fueled rockets must be strong enough to confine the pressures of the burning fuel and yet be no heavier than absolutely necessary. Such casings may be made of ultra-high-strength steel by rather specialized manufacturing methods to operate at design stresses above 200,000 psi. In the following section a discussion will be given of the design considerations and special problems encountered in the testing of full-scale casings built by the Pratt and Whitney Aircraft Division of the United Aircraft Corporation. The work has been performed by Shank et al. (1959).

A. Description of Fuel Casing and the Method of Manufacture. In Fig. 22.3 the rocket chamber is shown in parts before its final assembly. The particular casing discussed here is nearly 6 feet long and 40 inches in diameter, the nominal wall thickness of the cylindrical portion being about 0.070 inch. As can be seen from Fig. 22.3, the design is kept as simple as possible to avoid stress concentrations and possible manufacturing defects. The main cylindrical portion is seamless, flow-turned from a forged ring of the same inside diameter. The ellipsoidal head on the right and the flanged head on the left-hand side are forged and machined from integral pieces. In this fashion (a) the tangential stresses in the cylindrical

FIG. 22.3. Exploded assembly of full-size vessel. Central cylindrical section is flow-turned with no longitudinal seams. Piece at left is a closure for test purposes. Dome-shaped head at left has closure flange integral with shell. Skirt on ellipsoidal head at right is integral with shell portion. Extended skirt section at far right is then welded to skirt on the ellipsoidal head. (Shank et al., 1959. Courtesy of *Metal Progress*.)

section are always made to act across solid material, (b) the shape is kept simple enough to ensure a reasonably exact determination of the stress state in the chamber, and (c) complicated welds leading to residual stresses and embryo cracks are avoided.

The material used in making the casings was a vacuum melted special tool steel of the following composition: 0.42%C, 0.29% Mn, 0.96% Si, 5.17% Cr, 1.35% Mo, 0.52% V with <0.021% P, and <0.006% S.

Material performance tests were carried out on smaller cylindrical vessels in which consistently high strength values were obtained after some initial corrections were made in the design. Some of these values were in excess of 300,000 psi. The full-scale cases were manufactured with great care, the flow-turning being done in three stages with intermediate annealing cycles in a "special" atmosphere. The cylindrical as well as the machined dome sections were stress-relieved before being fusion welded together by an inert-gas, shielded-arc method with parent metal filler rods used as electrodes. Tack welds were avoided by very careful prefitting of parts to be welded. Subsequent to welding the casing was stress-relieved in an atmosphere* and air cooled. After a careful inspection for defects with radiographic, magnetic particle, and fluorescent penetrant methods, the casings were hardened by heating in a reducing gas atmosphere to 1850°F for 30 minutes and air cooled, followed by a double tempering operation in air at 1050°F for 2 hours with subsequent air cooling. In the process, the surfaces of the vessel were deliberately decarburized to a depth of 0.003 to 0.005 inch to provide some surface toughness. Final inspection of the interiors of the vessel uncovered four protrusions, the cause of which could be traced back to the imperfections on the flow-turning mandrel. Some of these surface protrusions were removed by grinding off a thickness of about 0.005 inch from the surfaces.

B. Fracture under Full-Scale Test and Investigation of Its Causes. One of the full-scale vessels was then subjected to the test schedule shown in Fig. 22.4. The casing failed in a hydrostatic test at a stress level of 185,000 psi after having sustained a stress of 216,000 psi shortly before. The catastrophic nature of the failure is seen in Fig. 22.5.

Such a disappointing premature failure received immediate attention. Inspection and tracing of the chevron marks on the fracture surfaces revealed the origin to be one of the protrusions which was ground down. This ground-down region, where the thin oxide layer was broken, had developed corrosion pits, which evidently resulted from the hydrostatic tests. Figure 22.6 shows the fracture and the chevron marks on the fracture surface radiating out of the inner surface of the vessel. Metallographic examination of this region disclosed the presence of fine cracks around some of the corrosion pits, as shown in Fig. 22.7. The question of whether or not such cracks might have existed prior to testing was settled by examination of some of the other protrusions which were not removed by grinding. Figure 22.8 shows a photomicrograph of a section through such a

---

* Not further specified by the workers.

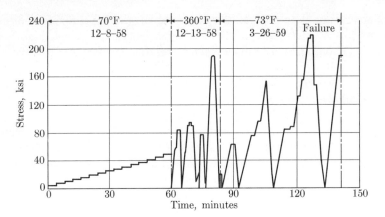

Fig. 22.4. Testing schedule of full-scale rocket casing which failed. (Shank et al., 1959. Courtesy of *Metal Progress*.)

Fig. 22.5. Remains of full-scale missile casing after failure. Conical objects resting on the central portion of the fractured pieces are weights used to hold them in position for photograph. (Shank et al., 1959. Courtesy of *Metal Progress*.)

defect which appeared to be a lap formed in the flow-turning. Etching with 1% nital to distinguish the decarburized surface layer established conclusively that such flaws (with their decarburized surfaces) as shown in Fig. 22.8 were present before testing while the cracks around the corrosion pits of Fig. 22.7 must have resulted after decarburization, i.e., during testing. Since these fine cracks could not be found adjacent to the pits but rather away from them, stress corrosion was ruled out. This, then, suggested to the investigators that a static fatigue phenomenon must have been operative during the testing. The testing medium and the extended test schedule made hydrogen embrittlement a suspect. The chemical

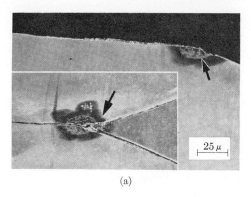
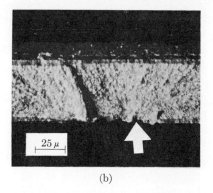

FIG. 22.6. (a) Inner surface of full-scale missile casing after fracture. Arrow points to origin of failure. Inset shows assembled fragments at the origin. (b) Surface of fracture at origin in the failed rocket casing. (Shank et al., 1959. Courtesy of *Metal Progress*.)

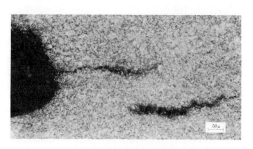

FIG. 22.7. Photomicrograph of area next to fracture origin, showing the edge of a pit and two fine cracks. Etched with 1% Nital. (Shank et al., 1959. Courtesy of *Metal Progress*.)

FIG. 22.8. Transverse section of an "unrepaired" protrusion on the inside surface of the fractured casing. Etched with 1% Nital. Note the decarburized boundaries. (Shank et al., 1959. Courtesy of *Metal Progress*.)

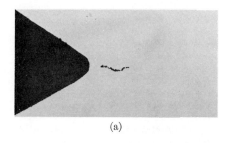
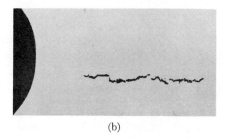

FIG. 22.9. Cracking at regions of high triaxial tension due to hydrogen embrittlement. Note that the localization of the hydrostatic tensile stress in a sharp notch also causes the crack to form there. (Troiano et al., 1958. Courtesy of Am. Inst. Mining and Metallurgical Engineers.)

analysis of the material revealed the hydrogen content to be only a few parts per million, which would not be sufficient on the whole to explain such catastrophic weakening. This overall absence of hydrogen from the analysis is not, however, a strong enough reason to rule out this mechanism. It could still be possible that the local hydrogen concentration was high enough to cause crack growth with eventual embrittling effects. According to Troiano (1958) hydrogen diffusion in steel is strongly enhanced by the presence of a hydrostatic tensile stress, such as the kind that exists at the root of sharp notches. The high hydrogen concentration can then form cracks as shown in Fig. 22.9. Like all static fatigue effects, the production of a crack is associated with a delay time during which diffusion (in this case hydrogen to the crack) takes place. This diffusion time is governed by the absolute temperature and is proportional to the reciprocal of the Boltzmann factor, $e^{u_a/kT}$ (cf. Chapter 1) where u_a is the activation energy for diffusion of hydrogen, which, according to Troiano (1958), can be significantly reduced by the presence of a hydrostatic stress. Such delay times can vary between 10 to 30 minutes at 80°F to 1 to 5 hours at −25°F. The reduction in strength resulting from the production of a crack can be as much as 70%, i.e., a steel with a nominal instantaneous fracture stress of 225,000 psi could have a strength as low as 75,000 psi due to the embrittling effect of hydrogen.

The fine cracks around the corrosion pits, coupled with the fact that in the earlier design evaluation tests fracture always started from the inside surface exposed to water, were strong symptoms pointing toward hydrogen embrittlement. It was reasoned by the investigators that in the full-scale test, in particular, the galvanic action between the ground-down region and its surroundings, could have resulted in enough hydrogen ion concentration, which might have diffused into the steel under the action of the local stress concentration due to the pits. Thus, locally where the effect would be most detrimental, a higher-than-normal hydrogen concentration may have been present in the material to form the cracks revealed in the photomicrograph of Fig. 22.7.

C. Critical Tests to Determine Causes of Premature Catastrophic Failure. The above theory was put to test by constructing and testing a number of small-scale vessels. The results of these tests are summarized below.

(a) Vessels with no protective coating burst at tangential stresses ranging from 224,000 to 256,000 with delay times ranging from 113 minutes to 1046 minutes, their fracture origins being traceable to the inside wall.

(b) Protection of the inside surface with a coat of Ochre Heresite primer raised the bursting stress to 262,000 psi, with fracture origin still traceable to the inside surface of the vessel.

(c) Protection of the inside surface with a primer and exposure of the outside surface to water produced a fracture at 240,000 psi after a delay time of 945 minutes, with the fracture origin now significantly traceable to the outside surface of the vessel.

(d) A vessel with inside and outside protection with a primer sustained hydrostatic loading with water in four stress cycles of 275,000 psi, each lasting 9 hours on the average, with intermediate rest periods of approximately twice that length. The vessel was finally pressurized with oil in which the stress was increased gradually from 275,000 psi in hourly steps until eventual fracture took place at 318,000 psi after a delay time of 45 minutes at this stress.

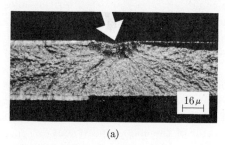
(a)

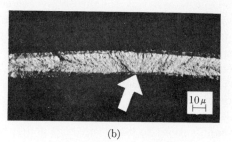
(b)

FIG. 22.10. (a) Typical discolored fracture origin of a vessel exposed to water. (b) Typical fracture in origin without discoloration, in vessels where the surface was protected from the pressurizing water by a primer. (Shank et al., 1959. Courtesy of *Metal Progress*.)

The fracture origins of the unprotected cylinders generally showed a discolored area such as the one shown in Fig. 22.10(a), while those protected with the primer did not show such discoloration as is evident from Fig. 22.10(b).

With the original hypothesis of hydrogen embrittlement thus satisfactorily supported by experiment, full-scale tests were performed on the remaining rocket casings with oil as the pressurizing medium. These casings were cycled four times, first to 220,000 psi and then to 240,000 psi, each time being held at these stress levels for three minutes. Examination of the vessels showed no damage (Prob. 22.7).

Tensile tests carried out on strips cut from the catastrophically failed vessel gave strength results ranging from 240,000 psi to 274,000 psi, with a strip containing an "unrepaired" flow mark being the strongest of all at 280,000 psi. This indicates the irony in the drive toward perfection in that what appeared to be a defective area was stronger than the parent material and that "repairing" actually introduced the weakness.

An epilogue to this successful engineering accomplishment is the fact that as work on the perfection of these vessels was in progress, another solution utilizing a less expensive glass-fiber reinforced plastic construction was being developed by other interested groups, and has proven to be competitive, not only in this field but also in the construction of more conventional pressure vessels (Modern Plastics, 1961).

22.4 ADVISABILITY OF "ON THE SPOT" MODIFICATION OF PRESSURE VESSELS FOR GAS STORAGE

As is not uncommon in engineering practice, a mistake was discovered at installation. In this case, involving 8-in. diameter pressure vessels of 20 foot length made from $\frac{3}{4}$ in. thick 1041 steel to contain gas at 3000 psi, drain holes in the ends, required by the pressure-vessel code, had not been provided. The fact that the installation site was on top of a reinforced concrete building would make the return of the cylinders to the manufacturer costly. It would be more practical to drill and tap the $\frac{3}{4}$ inch holes at the installation site. In this event, however, the proof testing required by the pressure-vessel code could not be performed after the holes were drilled and tapped. The question therefore arose, "Should the installation proceed with the understanding that the holes will be drilled and tapped at the site? If so, what precautions should be taken and what tests should be run to ensure the safety of the pressure vessels?"

The first question is whether or not serious hazard to life exists. If so, there is no alternative but to take all reasonable precautions to eliminate it. In the legal sense, "all reasonable precautions" would include returning the vessel to the manufacturer to make sure that it met the letter of the code requirement.

Assuming that the hazard to life can be eliminated by an appropriate protective shielding, the question arises as to whether or not there is any reasonable probability of a fracture, particularly a brittle fracture. The final question, therefore, is whether or not the modification of drilling and tapping after proof testing will significantly increase the probability of fracture.

It is possible that a fatigue crack will develop under repeated cycling, and that such a crack will grow until brittle fracture occurs. The possibility of fatigue crack growth must be evaluated in terms of the nominal stress near the hole, the stress concentration due to the hole, and the further stress concentration due to checking or sharp corners arising in tapping the hole. In any event, the growth of a crack should not be allowed to lead to brittle fracture.

In order to determine the probability of brittle fracture, it would be desirable to have, for the material involved, a plot of the radius of the plastic zone as a function of plate thickness at the lowest expected operating temperature, such as that shown in Fig. 16.16. If the required radius of the plastic zone were large compared with the thickness of the plate, one could be sure of extensive plastic deformation which would allow leakage and loss of pressure before catastrophic fracture. Such data are not as yet available for this steel, however. In their absence, and in the light of past experience, it has turned out to be reasonable to design on the basis of at least 15 ft-lb of energy absorbed in a V-notched Charpy test. Note that this does not bring in the important factor of plate thickness. It was, however, developed empirically in connection with the fracture of heavy engineering structures, such as ships and storage tanks, which have comparable thicknesses.

The Metals Handbook (ASM, 1961) provides information on the composition of steels, the Charpy energy values from certain steels, and the effects of variations in composition on the transition temperature (pp. 62, 225–243). From these the transition temperature is estimated to be about $-30°F$, which is below the expected service temperature, although not by a very wide margin. Specimens of similar forgings should be tested, taking care that the least favorable orientation, probably circumferential, is studied.

The increase in hazard due to drilling and tapping the hole after proof testing is probably rather small. While less favorable machining conditions may make the initiation of a crack somewhat more likely, it would have been a mistake to have designed the vessel on the assumption of ideal surface finish. The only other factor is that in proof testing, compressive residual stresses might develop around the hole, which would retard the growth of fatigue cracks. Again the effect is relatively slight, and should not have been depended upon in the original design.

In conclusion, then, while the change in procedure is not likely to increase the hazard for a properly designed tank, the consequences of failure may well be such that to avoid legal responsibility in the event of failure, the code should be strictly adhered to. Moral responsibility requires a more careful check of the material and design, plus the continued effort to put our understanding of brittle fracture on a more quantitative basis.

In this case, after weighing the above considerations, the decision was made to perform the alterations at the site.

22.5 DISTORTION OF SPRINGS DUE TO STRESS-RELIEF

Springs are stress-relieved to reduce the residual bending stresses due to forming and hence increase the elastic limit in torsion. If they change shape in stress-relief, one must heat-treat them before final inspection. Final adjustments to the automatic spring-making machines cannot be made until after test batches have been heat-treated. Since the set-up time is often large compared with the running time of the machine, the added delay in waiting for the results of heat-treatment can be expensive. Often with known materials and shapes, spring makers have learned by experience about how much to allow for such changes in shape. With the coming of new materials, however, a method of predicting the behavior of new materials is desirable. An example to test one's understanding of the process is the report by Rimmer (1958) that on stress-relief, springs made of music wire tend to coil more tightly (become smaller in diameter), whereas springs made of stainless steel tend to uncoil.

Music wire, according to ASTM specification A228-51, has 0.7 to 1% carbon, 0.2 to 0.6% manganese, and is drawn to a tensile strength of 250,000 to 500,000 psi with an elastic limit of 150,000 to 350,000 psi. The type of stainless steel referred to was not mentioned, but 18-8 type stainless has a tensile strength of 160,000 to 330,000 psi and an elastic limit of 60,000 to 260,000 psi. Type 316 stainless has a tensile strength of 170,000 to 250,000 psi and an elastic limit of 130,000 to 200,000 psi.

The coiling or uncoiling of a spring on stress relief is due to residual stresses. In principle, the different behavior of the two materials could be due either to a difference in the distribution of residual stress or to a difference in the response to the residual stress on stress-relief. It does not seem that a difference in response is the important variable, however, since in both cases, the rate of relaxation of residual stress due to primary creep will be higher the greater the stress. It therefore seems reasonable to expect that in one case the maximum residual stress is of one sign and in the other case, the maximum residual stress is of the other sign.

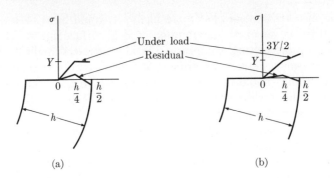

Fig. 22.11. Residual stresses due to coiling in a nonhardening and a strain-hardening material.

As a preliminary hypothesis we assume that the difference is due to the different rates of strain-hardening of the two materials, particularly if the stainless is of the 18-8 type.

As a first approach to the problem, it was assumed that the springs could be considered to be made of square wire, that they were plastic halfway through to the neutral axis, and that strain-hardening was present in one case and absent in the other. This is shown in Fig. 22.11. Somewhat surprisingly, it was found that the ratio of residual stress at the outside to the maximum residual stress of the opposite sign at the quarter-point was exactly the same in these two cases (Prob. 22.8). That this is more than a coincidence can be shown by performing the same analysis for springs plastic all the way through (Prob. 22.9). Again the ratios of the residual stresses of opposite sign are equal for the nonhardening and the strain-hardening materials. It is noted however, that the ratios of the residual stresses do change with change in the depth of the plastically deformed region. This leads to the investigation of possible differences in depth of the plastically deformed region.

If two materials having the same modulus of elasticity are both wrapped around arbors of similar size using the same size wire, then the material with the lower elastic limit will have deformed to a greater depth. Furthermore, the material which is deformed to a greater depth will have the greater residual stress at the inside, where it is of the same sign as the originally applied load, as indicated in Fig. 22.12. Where the plastic zone is shallow, however, with the harder material, the residual stress will be of sign opposite to the originally applied stress. In the first case, relief of the residual stress will have the same effect as applying a reversed stress and the spring will tend to unwind. In the second case, the spring will tend to coil more tightly. Since the elastic limit and tensile strength of the stainless steel springs tend to be below those for music wire, one would expect that for certain combinations of wire and spring diameter, the stainless steel wire would tend to open up and the music wire would tend to close down, as actually observed.

The above analysis indicates that the kind of material is not as important as the ratio of wire diameter to spring diameter, and that either material could be

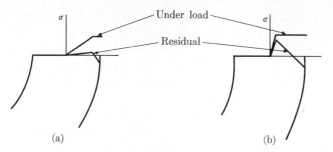

Fig. 22.12. Effect of depth of yielding on sign of maximum residual stress.

made to either open or close, depending on the amount of pretwist. In rechecking with men who make springs, this is found to be the case. It therefore, seems that a reasonable explanation for the behavior of the two materials has been found. It still remains, however, to predict the rate of and amount of opening up or closing down in terms of the relaxation or primary creep rates of the material during stress relief and the actual residual stresses expected in round wire. Sufficient data on primary creep rate as a function of temperature and stress level are not available for quantitative analysis, although it should be possible to work out the form of such relations, in terms of an equation of state (Eqs. 5.22, 5.24, or 19.5). Once this is done, there will be more incentive to obtain the required data on primary creep to evaluate the constants in the equation.

The above analysis was based on the assumption of plastic isotropy. It is, however, to be expected that heavily drawn wire may have a pronounced Bauschinger effect and therefore in coiling, yielding will occur on the compression side of the coil almost immediately on bending. This would result in a highly distorted residual stress distribution, which may also contribute to the above phenomenon (Prob. 22.10).

22.6 INVESTIGATION OF THE CAUSE OF SERVICE FRACTURE OF A TURBINE WHEEL

The required reliability of steam turbines for central power stations is normally assured by extensive tests on the soundness of parts and material before, during, and after manufacturing. Because of this, failure of such turbines is quite rare. In spite of all the care, however, service failure is occasionally still encountered. Thus, for instance in January 1953, after two years of service, the low-pressure turbine element of the No. 1 unit at the Tanners Creek station of the Indiana and Michigan Electric Company suddenly went into violent vibration. The front standard, which contains the No. 1 bearing, the thrust bearing, a hydraulic cylinder, and the governing mechanism, cracked. The resulting oil leakage caused a fire within the station. The turbine which failed was part of the low-pressure turbine running at 1800 rpm, which in turn was the third stage of a 125,000-kw overall unit. The fact that the turbine was designed very conservatively and all features were based on tried and proven techniques, coupled with the fact that many seemingly identical units were operating satisfactorily, all made this failure a very puzzling one. We shall now discuss the report of Rankin and Seguin (1956) of the General Electric Company describing the investigation undertaken to identify the causes of this failure.

When the turbine casing was removed, it was found (as shown in Fig. 22.13) that part of the first-stage wheel of the low-pressure turbine was missing. It had apparently broken loose and was melted by friction while the turbine came to rest. The molten metal was partly scattered and partly solidified on the remnants of the broken wheel.

FIG. 22.13. Photograph of broken wheel after machining from shaft. (Rankin and Seguin, 1956. Courtesy of ASME.)

FIG. 22.14. Oblique sketch of section through center of notch opening in turbine wheel rim, where fracture originated. (Rankin and Seguin, 1956. Courtesy of ASME.)

A. Initial Investigation. The initial investigation revealed that fracture initiated at the two pin holes of the turbine wheel notch opening where the turbine buckets are inserted (see sketch of Fig. 22.14). From the chevron marks on the initial leg of the fracture from A to B (Fig. 22.15), shown in Fig. 22.16, it was clear that this portion of the fracture must have occurred rapidly in a "brittle" manner. The last part of the fracture from C to D appeared to have sheared off on a plane at 45° to the wheel surface in a ductile manner. Metallographic investigation revealed a rather heavy oxide layer on the brittle fracture portion A-B and disclosed it to be primarily intergranular. An oxide skin was absent along the ductile fracture part C-D. This led to the conclusion that the fracture A-B must have happened at an earlier stage and the turbine wheel ran for a considerable time

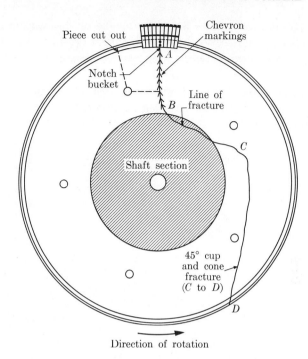

Fig. 22.15. Sketch of first-stage wheel showing line of fracture. (Rankin and Seguin, 1956. Courtesy of ASME.)

Fig. 22.16. Portion of radial fracture face of Fig. 22.15 showing chevron markings. (Rankin and Seguin, 1956. Courtesy of ASME.)

with a radial crack in it until, under the increased stresses, a portion of the wheel separated first by creep fracture and eventually by rapid ductile fracture.

The records of operation, in particular those pertaining to the eccentricity of the rotor, were checked, but nothing unusual could be found until the very violent final vibrations which started, no doubt, after the section of the turbine wheel broke off. The natural frequencies of vibration of an identical spare rotor were rechecked only to reconfirm that the running speed was very much below any natural frequency of vibration.

Some evidence of substantial residual stresses was obtained somewhat unintentionally in removing a rectangular piece of the wheel, shown by dotted lines in Fig. 22.15, containing the initial portion of the fracture. Before the first radial cut could reach the steam balance hole, the tongue broke away from the rest of the wheel, completing the sectioning and leaving a $\frac{1}{16}$-in. circumferential gap. From this experiment a residual tensile tangential stress between 10,500 and 25,500 psi was estimated.

B. Detailed Investigations. To establish the real cause or combination of causes of failure, a series of checks and controlled experiments were performed to duplicate the conditions of failure.

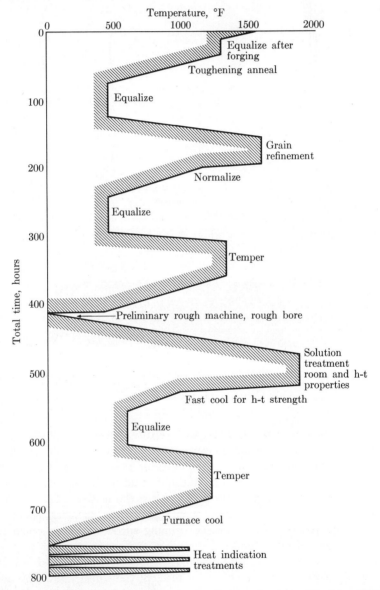

FIG. 22.17. Heat-treating cycle for Tanners Creek rotor forging. (Rankin and Seguin, 1956. Courtesy of ASME.)

1. History of Turbine Rotor Forging. The turbine rotor was made of a low alloy, chrome, molybdenum, and vanadium steel and was given the heat-treatment schedule shown in Fig. 22.17. It had passed all metallurgical tests and mechanical inspections for soundness. The absorbed gases, hydrogen, nitrogen, and oxygen, were about 0.5, 50, and 60 parts per million, respectively. Creep-fracture tests performed on specimens of the forging before and after the failure gave nearly identical results and showed no serious weakening or aging effects, as illustrated in Fig. 22.18.* The total strain to fracture of the rotor material under creep conditions at 1000°F has been, however, rather small at around 1 to 2%, as shown in Fig. 22.19. This, as will become clear later, was one of the contributing causes of the failure.

Metallographic observations at high magnification of the steel of the fractured wheel showed a rather well-defined grain-boundary film, as shown in Fig. 22.20,

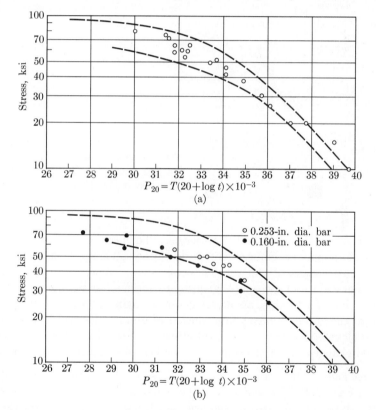

FIG. 22.18. (a) Smooth-bar "rupture" data on Tanners Creek rotor forging in the before-service condition. (b) Smooth-bar "rupture" data on Tanners Creek rotor forging in the after-service condition. (Rankin and Seguin, 1956. Courtesy of ASME.)

---

* The creep-fracture plots of Fig. 22.18 are based on the Larson-Miller parameter, which is similar to the velocity modified temperature of MacGregor and Fisher discussed in Section 5.10 in connection with the mechanical equation of state. (See also Problem 19.5.) Temperature, T, is in °R and time, t, is in hours.

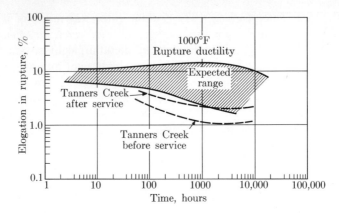

Fig. 22.19. 1000°F "rupture" elongation of Tanners Creek forging with expected range for this alloy. (Rankin and Seguin, 1956. Courtesy of ASME.)

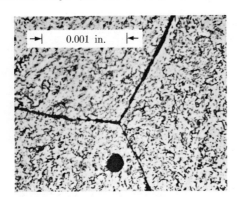

Fig. 22.20. Photomicrograph showing grain-boundary films in Tanners Creek turbine wheel. (Rankin and Seguin, 1956. Courtesy of ASME.)

which was absent in the material before service. To check whether or not a mechanism of grain-boundary embrittlement was active, Charpy experiments were performed on specimens of the turbine wheel steel. When compared with the Charpy impact values of the material prior to service, no significant differences were found. Parallel to this, an investigation of aging at various temperatures disclosed that the grain-boundary film consists of iron carbide and forms gradually with time at elevated temperatures. From this it was concluded that no significant metallurgical changes occurred which could be held responsible for the failure.

2. **Stress in the Rotor.** The stresses in the rotor resulting from the centrifugal forces are shown in Fig. 22.21. Evidently both the radial and tangential nominal stresses are moderate (around 5500 psi) at the origin of fracture. The effect of the stress concentrations due to the two holes at the notch opening is shown in Fig. 22.22. Under creep conditions, since the effect of notches can usually be leveled out, the important effect of the holes is not the maximum stress but the average local stress. This is found to be about 14,000 psi between the two holes, and was confirmed by photoelastic studies. Reference to Fig. 22.18 will show that at this stress level, the life of the piece to fracture is too long (about 700 years) to be a problem.

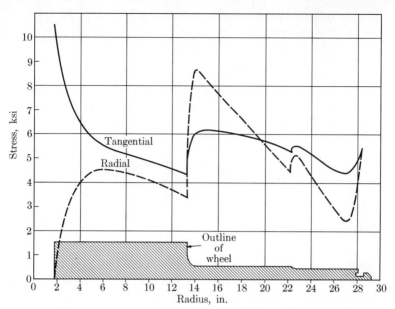

Fig. 22.21. Calculated centrifugal stresses in first-stage wheel. (Rankin and Seguin, 1956. Courtesy of ASME.)

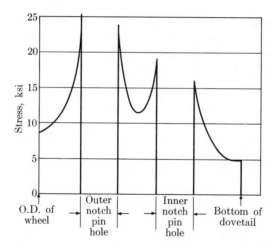

Fig. 22.22. Tangential stresses through center of notch opening as determined by photoelastic tests. (Rankin and Seguin, 1956. Courtesy of ASME.)

From this and from the fact that no difficulty had been encountered with many other identical turbine units (including one which was accidentally oversped by about 50%), it was concluded that the failure could not be due to the level of stress during the normal operation. From the nature of the fracture surface, from the vibrational studies, and from experience with other turbine units it was concluded also that fatigue was not the cause.

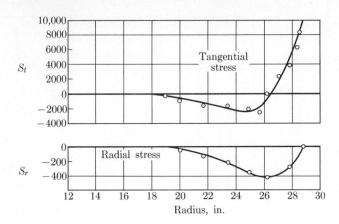

FIG. 22.23. Measured residual stresses in Tanners Creek second-stage wheel. (Rankin and Seguin, 1956. Courtesy of ASME.)

3. Residual Stresses. In view of the negative results of the metallurgical and stress investigations, the evidence of residual stress in the initial sectioning demanded a careful study of residual stresses. A second-stage wheel from the turbine was used for the investigation. Twenty-six strain gauges were mounted on the wheel and the residual stresses were measured by cutting the wheel into nine concentric rings. From the changes in strain, the distribution of residual stresses were measured as shown in Fig. 22.23, giving a result of about 10,000 psi tangential stress at the origin of fracture. Coupled with the 5500 psi due to centrifugal forces and multiplied by the average stress concentration factor of 2.54, a stress of nearly 40,000 psi is obtained. From Fig. 22.18 the life of a part under this stress at a temperature of 960°F is found to be 16,000 hours which is the two-year service life of the machine.

4. Simulation of Conditions of Fracture. To confirm the hypothesis, it was desired to produce a brittle fracture under the estimated service conditions. Toward this end, tensile specimens were prepared with edge holes identical with those existing in the turbine wheel. These were tested in tension at the service temperature, but failed by ductile fracture. Severly notched sheet specimens in tension also did not give a brittle fracture. Additional experiments in slow bending at the service temperature and with specimens subjected to repeated thermal shock also did not produce a brittle fracture. At this time it became evident that the time factor would also have to be simulated. Specimens were then prepared which reproduced the details of the turbine wheel faithfully down to the simulation of the centrifugal forces caused by the turbine bucket. The specimen was then pre-tensioned to a stress somewhat higher than 40,000 psi and inserted into a creep furnace. This time brittle intergranular fracture was observed to start from the outermost hole, within 3000 hours at about 43,000 psi, as shown in Fig. 22.24. The fracture surfaces were covered with an oxide layer and had every bit the appearance of the brittle fracture observed in service.

FIG. 22.24. (a) Photograph showing crack in a wheel-rim test specimen after 3000 hr at 960°F. (b) Photomicrograph of termination of crack of (a). (Rankin and Seguin, 1956. Courtesy of ASME.)

C. Conclusions. From the investigations it was concluded that the presence of an unusually high residual stress, not relieved by the normal stress-relief operations, superimposed on the normal service stresses, was responsible in hastening an intergranular creep fracture sometime before the final fracture. Some time passed between the completion of the radial fracture (enough time to form a heavy oxide layer) and final ductile propagation.

The real cause of the residual stress could not be determined with any certainty. The most likely possibility was that the initial stress-relieving treatment was not effective. Another possibility for production of residual stresses was, however, also considered. Certain difficulties with the condensate drainage from the steam-seal regulator (drainage of the leaked steam from the shaft packing) indicated the possibility that during load dumps, when the steam is shut off, some of this condensate may have crept back into the turbine and chilled the first rotor. The fact that such difficulties were encountered and remedied in other similar turbines without any similar failure left some doubt that this was a likely cause. As a result of this failure much more rigid inspection was undertaken to determine the level of residual stress in turbine rotors. It should be noted here that the particularly low strain to fracture under creep conditions aggravated matters, and a more ductile steel would surely have helped.

One matter remains not well explained in the report of the original investigation. It is apparently assumed that the residual stresses will continue to act even under creep conditions. In reality, however, even a strain to fracture of about 1% is amply sufficient to relax the concentrated residual stresses of about 30,000 psi between the two holes. Thus the agreement of the estimated creep life of two years with reality appears to be not well founded. It seems more likely that the stress in the ligament between the holes indeed relaxed rapidly, increasing the stress concentration at the outer boundaries of the holes. In any event, it is evident that the presence of the initial residual stress was crucial and the basic conclusions of the investigators remain unaltered.

REFERENCES

Reports of similar investigations can occasionally be found in engineering journals. For additional reading, for instance, see the report of two generator rotor fractures by Schabtach et al. (1956), the report of a rotor fracture by De Forest et al. (1957), and the report on the cause of fracture in ends of freight car wheel frames by Tack (1963).

ASM	1961	*Metals Handbook*, 8th ed., T. Lyman, ed., ASM Novelty, Ohio.
DE FOREST, D. R. SCHABTACH, C. GROBEL, L. P. SEGUIN, B. R.	1957	"Investigation of the Generator Rotor Burst at the Pittsburgh Station of the Pacific Gas and Electric Company," *ASME Preprint No.* 57-A-280.
ESSEX, G. S.	1960	*An Investigation of the Forces in Conn-Rod Cap Bolts and Their Members*, Diploma Thesis, Birmingham University, England.
Modern Plastics	1961	"Filament Winding Goes Commercial," *Modern Plastics* **39,** 94–96, 194–200.
RANKIN, A. W. SEGUIN, B. R.	1956	"Report of the Investigation of the Turbine Wheel Fracture at Tanners Creek," *Trans. ASME* **78,** 1527–1546.
RIMMER, S. M.	1958	"The Springback of Coiled Springs," *Mech. Eng.* **80,** April, 74–76, Nov., 140–142.
SCHABTACH, C. FOGLEMAN, E. L. RANKIN, A. W. WINNE, D. H.	1956	"Report of the Investigation of Two Generator Rotor Fractures," *Trans. ASME* **78,** 1567–1584.
SHANK, M. E. SPAETH, C. E. COOKE, V. W. COYNE, J. E.	1959	"Solid Fuel Rocket Chambers for Operation at 240,000 psi and Above," *Metal Prog.* **76,** Nov., 74–81, Dec., 84–92.
SMITHELLS, C. J.	1962	*Metals Reference Book*, 3rd ed., Vol. 2, Interscience, New York, p. 556.
TACK, C. E.	1963	"Fracture," *Mech. Eng.* **85,** 43–46.
TROIANO, A. R.	1958	"Hydrogen Crack Initiation and Delayed Failure in Steel," *Trans. AIME* **212,** 528–536.

PROBLEMS

22.1. Referring to Fig. 22.1, derive Eq. 22.1, for the maximum force on the connecting rod cap.

22.2. Derive Eq. 22.4, for the tightening torque necessary to develop a pre-tension P_L in a bolt.

22.3. Calculate the equivalent stress in the bolt of Fig. 22.1 subjected to a pre-load P_L, a tightening torque T, and a fluctuating load F', and write it in the normalized form of Eq. 22.5.

22.4. Derive the equation relating the allowable fluctuating stress amplitude and the mean stress, along the lines discussed in Section 18.6, and write the relation in the normalized form of Eq. 22.6.

22.5. Show that in the numerical evaluation, the case of fatigue coupled with high thread friction leads to the worst condition, requiring a shank area of 0.044 in^2.

22.6. Show that for the bolts with full shank thickness, equal to the nominal diameter of the threads, the smaller shear stress due to the torque allows a smaller nominal diameter than in the reduced-shank bolt.

22.7. In connection with the final full-scale tests of the rocket casings, is the success of these tests sufficient to establish a stress of 240,000 psi as a reliable working stress for long-term operation?

22.8. In connection with Fig. 22.11, show that the ratios of the residual stresses on the outside layer to the peak residual stresses of the opposite sign in the interior are the same for two square wires in bending, having the same Young's modulus and yield strength but different strain-hardening characteristics, if the plastic zone travels halfway through to the neutral axis.

22.9. Show that the conclusion of Prob. 22.8 is not altered when the plastic zones travel all the way to the neutral axis.

22.10. Consider the alterations in the residual stress distribution due to the presence of a Bauschinger effect, and how this could affect the results of the discussion on coiling or uncoiling of springs on stress relief.

22.11. It is proposed to use foamed polystyrene as a material for crash padding for skull protection in an automobile. The following data are available.

Compression Test
  Modulus of elasticity, 800 psi
  Proportional limit, 20 psi
  Yield strength, 25 psi
  Strain-hardens to 35 psi at $\frac{1}{6}$ original height
  Poisson's ratio during plastic flow ≈ 0
  Fracture, none

Tension Test
  Modulus of elasticity, 2500 psi
  Proportional limit, 40 psi
  Tensile strength, 75 psi
  Elongation at fracture, 1.3 times linear elastic value
  Fracture, brittle

Torsion Test
Brittle fracture perpendicular to axis of twist

Hardness Test
Static, 25 psi
Drop test from 10 in.: 125 in-lb/in^3 of indentation volume.

(a) What characteristics are desirable in crash padding?

(b) How would you design padding made of this or, if desired, of another more appropriate material?

(c) Comment on the relations between the various characteristics that are reported. How do these relations compare with those for metals?

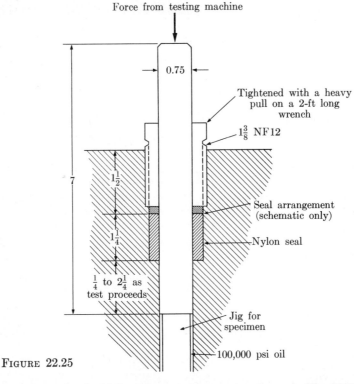

FIGURE 22.25

22.12. A piston for the high-pressure test apparatus shown in Fig. 22.25 was designed out of 420 stainless steel, and had a hardness of $R_C 28$. As designed, the plunger was supposed to handle a pressure of 100,000 psi, but in this service it was also desired that it exert a compressive force on the testing fixture inside the machine which would require another 4000 lb. After a few tests, the piston buckled. The failure was thought to be due to a misalignment of the piston and cylinder in the testing machine supplying the force. A new piston was purchased at a cost of $75. After a few more tests, the piston again began to bind and comparison with a straight edge showed that it was bent.

(a) What is happening? Justify your hypothesis and exclude alternative hypotheses as quantitatively as you can.

(b) How do you suggest obtaining satisfactory performance?

22.13. High-strength steel machine screws fracture before the limit load is reached, apparently due to strain concentrations at the roots of the threads. In order to alleviate this situation, some machine screws of 0.25 in. diameter were decarburized 2 hours in 1 cc/sec of wet hydrogen at 705°C, giving the decarburized layer shown in Fig. 10.11. They were then oil-quenched and tempered to give a core hardness of $R_C 40$. As a control, other machine screws were given the same heat treatment without decarburization. In tension, two decarburized screws failed at 7000 to 7200 lb, whereas four control specimens failed at 6000 to 6200 lb.

(a) Predict the depth of decarburization from a diffusion coefficient which may be obtained from Smithells (1962), the transient diffusion equation, and the assumption that the carbon concentration is zero on the surface.

(b) Was the desired objective of maximizing the fracture load obtained? If not, should further decarburization be deeper or shallower?

(c) Discuss any other advantages and disadvantages of this procedure for machine screws.

AUTHOR INDEX

Numbers in boldface refer to references at the end of chapters.

Abrikosova, I. I., 658, **672**
Adams, D. P., 701, **703**
Adams, R. G., 449, 466, 528, **541**
Adcock, F., 553, **572**
Akita, Y., 500, **510**
Alcoa, 195, 200, **206**
Alexander, G. B., 648, **650**
Alfrey, T., Jr., 249, 262, **263**, 292, 295, **302**
Allen, D. N. deG., 410, **415**
Almen, J. O., 407, **434**
Amelinckx, S., 124, 125, 132, 133, **145**
American Institute of Mining, Metallurgical, and Petroleum Engineers (AIME), 423, **435**
American Society for Metals, 86, 195, **206**, 425, 428, 429, 431, 434, **434**, 444, 446, 449, 460, 466, **466**, 588, 600, 605, **613**, 646, 650, **650**, 718, **730**
American Society of Mechanical Engineers, 613, **613**, 650, **650**
American Society for Testing and Materials, 260, 263, 310, **334**, 406, **415**, 423, **435**, 444, 447, 449, 464, 466, **466**, 538, 539, 541, **541**, 587, **613**, 650, **650**
Anderegg, F. O., 496, 504, **510**
Anders, F. J., 648, **650**
Anderson, O. L., 497, 510, **510**
Anderson, R. A., 647, **654**
Anderson, W. E., 584, 585, **614, 616**
Andrade, E. N. daC., 496, **510**
Ang, A., 407, 410, **415, 416**
Arbogast, C. L., 83, **93**
Archard, J. F., 659, 667, 668, **672**
Argon, A. S., 131, **147**, 162, 163, 170, 171, **206**, 480, **485**, 493, 496, 497, 505, 506, **510**
Ashwell, D. G., 350, **353**
Averbach, B. L., 310, 311, **335**, 557, 558, 572, **572**

Bachman, G. S., 479, **485**
Backer, S., 680, 682, 687, 696, 697, 701, **703, 704**
Backofen, W. A., 327, **335**, 344, **353**, 530, **541**, 614, 620
Bailey, A. I., 659, **673**

Bailey, J. E., 170, **206**
Baker, T. C., 499, **511**
Baldwin, W. M., Jr., 423, 434, **435**
Barnes, R. S., 159, **206**
Barrer, R. M., 17, **43**
Barrett, C. S., 20, 43, **43**, 160, **206**, 403, **415**
Basinski, Z. S., 173, 190, **206**, 639, **650**
Bates, L. F., 91, **92**
Bateson, S., 497, **511**
Battelle Memorial Institute, 594, 595, 611, 613, **613**, 623
Bauschinger, J., 184, **206**
Beachem, C. D., 611, **613**
Bear, I. J., 161, 186, 187, **210**
Becker, G. W., 232, **263**
Beeston, B. E. P., 643, **652**
Bennett, J. A., 607, **613, 615**, 624
Bennewitz, K., 477, **485**
Berg, C. A., 526, 527, **541, 542**
Berkey, D. C., 413, **415**
Bernauer, F., 255, **263**
Best-Gordon, H. W., 696, 697, **703**
Bethlehem Steel Company, 197, **206**
Bierbaum, C. H., 451, **466**
Bilby, B. A., 134, **146**, 176, 187, **207**, 409, **415**
Billington, D. S., 563, **574**, 642, **654**
Billmeyer, F. W., Jr., 262, **263**
Biscoe, J., 229, **265**
Bishop, J. F. W., 285, **302**, 372, 377, 379, **386**, 519, **541**
Blatherwick, A. A., 591, **615**
Blewitt, T. H., 134, **145**, 170, **206**
Boas, W., 78, 83, **92**, 94, 130, 154, **147**, 171, 182, **206, 210**
Bohn, L., 484, **485**
Bolam, F., 688, **703**
Boley, B. A., 352, **353**
Bollmann, W., 140, **147**
Born, J. W., 260, **263**, 268
Born, M., 235, **263**
Boulanger, C., 629, 630, 631, **651**
Bowden, F. P., 659, 661, 672, **673**
Brace, W. F., 662
Brady, G. S., 195, **207**

Bragg, W. L., 102, 114, **145**
Brand, R., 680, **703**
Bredthauer, R. D., 494, **511**
Brenner, S. S., 118, **145**
Bridgman, P. W., 46, **68**, 323, 324, **334**, 464, **466**, 493, **511**, 530, **541**
Brinell, J. A., 444, **466**
Bristow, C. A., 553, **572**
Brock, G. W., 313, **336**
Broderick, R. F., 423, **435**, 437
Broodberg, A., 568, **572**
Broom, T., 176, **207**
Brophy, G. R., 642, **650**
Brown, A. F., **613**, 619
Brown, J. M., 412, **415**
Brown, W. F., Jr., 644, **650**
Brueggeman, W. C., 604, **617**
Brunner, H., 633, 635, **651**
Bryant, G. M., 683, **703**
Buchholtz, H., 423, 426, 433, **435**
Budiansky, B., 289, **302**, 400, **415**
Buerger, M. J., 17, 43, **43**
Bühler, H., 423, 426, 433, **435**
Bullen, F. P., 189, **207**
Bullough, R., 156, **207**
Burgers, J. M., 106, **145**
Bush, J. J., 601, **613**
Byerly, W. E., 289, **302**

Calladine, C. R., 368, **387**
Carman, C. M., **615**, 624
Carothers, W. A., 218
Carrington, W., 554, **572**
Carter, J. F., 702, **704**
Caswell, J. S., 595, **613**
Cazaud, R., **613**, **613**
Chambers, R. H., 172, **207**
Chang, H. C., 633, **651**
Charles, R. J., 498, 499, **511**
Chaudhuri, A., 633, **652**
Chemical Rubber Co., 21
Chi, S. H., 483, **486**
Christensen, R. H., 410, **415**, 586, **613**
Churchill, R. V., 292, **302**
Churchman, A. T., 190, **207**
Civil Aeronautics Authority, 87
Clarebrough, L. M., 108, **145**, 150, 169, 179, **207**
Clark, D. S., 296, **303**, 410, **416**
Classen-Nekludowa, M., 189, **207**
Clauser, H. R., 260, **263**
Clayton-Cave, J., 588, **614**
Cochran, E. P., 608, **614**

Coffin, L. F., 328, **334**, 582, 590, 599, **614**, **617**, **618**, 641, 642, **651**
Cohen, M., 145, **145**, 180, **211**, 310, **335**, 427, **435**, 557, 558, **572**
Coltman, R. R., 134, **145**, 170, **206**
Conrad, H., 116, **145**, 160, 172, **207**, 629, 632, 637, 639, 643, 647, **651**
Cook, N. H., 64, **68**
Cooke, V. W., 712, 714, 715, 717, **730**
Cordiano, H. V., 608, **614**
Corning Glass Works, 260, **263**
Corteu, H. T., 577, **617**
Courtney-Pratt, J. F., 659, **673**
Cousland, S. M., 578, 585, **618**
Cosserat, E., 54, **68**
Cosserat, F., 54, **68**
Cotterell, B., 500, **511**
Cotterill, P., 500, **511**
Cottrell, A. H., 109, 132, 134, 138, 139, 145, **146,** 156, 173, 176, 187, 189, 190, **207**, 409, **415**, 541, **541**, 554, 561, 562, **572**, **573**, 581, **614**, 620, 632, 650, **651**
Cowdrey, I. H., 449, **466**, 528, **541**
Cowper, G. R., 325, **334**
Coyne, J. E., 712, 714, 715, 717, **730**
Craggs, J. W., 298, **302**, 500, **511**
Crandall, S. H., 49, 54, 60, 68, **68,** 77, 86, 87, 88, **92**, 277, 278, **302**, 318, 327, 328, 331, **334**, 339, 346, **353**, 354, 356, 432, **435**, 483, **485**, 487
Cristescu, N., 298, **302**
Cross, H. C., 647, **654**
Crowder, L. H., 606, **617**
Crussard, C., 629, 630, 631, **651**
Cummings, H. N., 589, 603, **614**

Dahl, N. C., 49, 54, 60, 68, **68**, 77, 86, 87, 88, **92,** 277, 278, **302**, 318, 327, 328, 331, **334**, 339, 346, **353**, 354, 356, 432, **435**
Dash, W. C., 120, 121, 128, **146**
Davidenkov, N. N., 564, **572**
Davies, R. B., 582, **618**
Davis, E. A., 281, 288, **302**, 400, **415**, 531, **541**
Davis, H. E., 325, **335**, 568, **572**
Davis, R. S., 532, **542**
Deak, G., 184, 185, **207**
DeForest, D. R., 730, **730**
Dehlinger, U., 282, **304**
delaMacorra, F., 449, **466**
Delavignette, P., 124, **145**
Demer, L. J., 642, **653**
Den Hartog, J. P., 331, 332, **334**

Denke, P. H., 410, **415,** 586, **613**
Deruyttere, A., 555, **572**
Deryagin, B. U., 658, **672**
DeSisto, T. S., 326, **334**
Devenpeck. L. M., 384, **387**
Dewhirst, D. L., **334,** 338
Dickie, H. A., 428, **435**
Diehl, J., 161, 167, **207,** 210
Dienes, G. J., **303,** 308
Dieter, G. E., 601, **614**
Dolan, T. J., 577, 602, **615, 617**
Donaldson, D. R., 585, **614**
Donnell, L. H., 397, **415**
Dorn, J. E., 630, 631, 639, 647, 650, **651, 654**
Douglas, D. A., 640, 644, **653**
Drucker, D. C., 286, 288, **303,** 365, 368, 374, 381, 386, **387,** 572, **572**
Durelli, A. J., 64, **68**
Dutton, V. L., 322, **335**
Duwez, P., 296, **303**
Dyachenko, P. E., 425, **435,** 436

Eaton, D., 410, **417**
Ebner, M. L., **614,** 620
Edwardson, S., 157, **209**
Ehrenfest, P., 189, **207**
Eirich, F. R., 68, **68,** 262, **263**
El-Behery, H. M. A., 695, 697, **703**
Eley, D. D., 671, **673**
Ellington, J. P., **303,** 305, 321, **334**
Ely, R. E., **387,** 389
Emerson. W. B., 446, **466**
Entwistle, K. M., 485, **485**
Erard, H. R., 433, **435**
Erdoğan, F., 405, 407, **416,** 495, **511**
Ernsberger, F. M., 496, 497, 501, **511**
Ernst, H., 661, **673**
Eshelby, J. D., 106, **146,** 402, 403, 405, **416,** 551, **572**
Essex, G. S., 708, **730**
Ewing, J. A., 577, **614**
Eyring, H., 17, **43,** 235, 237, 253, **264**

Farquhar, J., 640, **652**
Farren, W. S., 179, **207**
Faupel, J. H., 322, **335**
Felbeck, D. K., 560, 561, 572, **572**
Felgar, R. P., 193, **209**
Ferry, J. D., 248, 253, 260, 262, **263,** 295, **303**
Feltner, C. E., 581, **614**
Fields, D. S., Jr., 344, **353**
Findley, W. N., 230, 233, 248, **263, 265,** 593, 594, **614**

Finnie, I., 290, **303,** 650, **651**
Fisher, J. C., 189, 191, 192, 193, **209,** 329, **335,** 510, **511,** 639, **653**
Fitzgerald, J. V., 479, **485**
Flanagan, A. E., 325, **335**
Fleischer, R. L., 155, 156, 176, **207, 208**
Flory, P. J., 260, 262, **263**
Fluck, P. G., 600, **614**
Fogleman, E. L., 730, **730**
Fomina, L. H., 298, **304**
Folias, E. S., 407, **415**
Föppl, O., 482, **485,** 582
Ford, H., 520, **541**
Forsbergh, P. W., 91, **92**
Forsyth, P. J. E., 578, 581, 583, 600, **614, 615, 618**
Fourie, J. T., 31, **43**
Frank, F. C., 110, 126, 135, **146,** 551, **572**
Frankel, A. G., **615,** 624
Franklin, P., 65, **68**
Freeman, J. W., 650, **651**
Freudenthal, A. M., 613, **615**
Friedel, J., 119, 141, 145, **146,** 176, **208,** 402, 403, **416,** 628, 629, 630, 631, **651, 652**
Frisch, J., 425, **435**
Frocht, M. M., 64, **68**
Frost, N. E., 597, **615,** 644, **652**
Fuchs, H. O., 433, **435,** 438, 439
Fujiwara, Y., 255, **263**
Furman, D. E., 642, **650**

Garstone, J., 180, 181, **208**
Geil, P. H., 262, **264**
George, R. W., 600, **615**
Gere, J. M., 332, **336**
Gervais, A. M., 635, **652**
Gibson, J. B., 85, **92**
Gilman, J. J., 33, 116, 117, 118, 129, 130, 132, 135, **146,** 160, 161, 162, 163, **208,** 481, **485,** 555, 563, 572, **572**
Girifalco, L. A., 643, **652**
Glasstone, F. W., 17, **43,** 235, 237, 253, **264**
Goens, E., 87
Goland, A. M., 85, **92**
Goldhoff, R. M., 637, 639, **652**
Goldsmith, W., 298, **303,** 452, **466**
Gomez, M. P., 584, 585, **616**
Goodier, J. N., 49, 60, 68, **69,** 82, **93,** 343, **354,** 395, 398, 401, 410, **417**
Gordon, A., 602, 604, 613, **615**
Gordon, J. E., 702, **703**
Gough, H. J., 578, **615**

Grant, N. J., 633, 635, 643, 644, 645, **651, 652, 654**
Graube, E., 256, **265**
Gray, G. A., 288, **304**
Green, A. E., 68, **68**
Green, A. P., 312, **334,** 379, **386,** 520, 541, 663, **673**
Green, H. S., 235, **263**
Greenberg, H. J., 365, **387**
Greenough, G. B., 555, **572**
Greetham, G., 180, 181, **208**
Griffith, A. A., 489, 490, 493, 494, 495, 497, 504, 509, **511**
Grobel, L. P., 730, **730**
Grosskreutz, J. C., 581, **615**
Grossman, N., 577, **616**
Grover, H. J., 602, 604, 613, **615**
Groves, G. W., 285, **303**
Grube, W. L., 601, **613**
Guard, R. W., 635, **652**
Guarnieri, G. J., 644, **652**
Gumbel, E. J., 510, **511**
Guy, A. G., 23, **43**
Gyulai, Z. Z., 188, **146**

Haas, R., 697, 698, 699, 700, **703**
Hager, R. V., 494, **511**
Hahn, G. T., 188, **208,** 557, 558, **572**
Hale, F. K., 554, **572**
Halford, G. R., 531, **542**
Hall, A. S., 412, **415**
Hall, E. O., 188, **211**
Halliday, J. S., 658, **673**
Ham, R. K., 176, **207**
Hamaker, J. C., 198, **211**
Hamburger, W. J., 689, 690, 692, 697, **704**
Handbook of Chemistry and Physics, 82, 234, 260, **264**
Handin, J., 494, **511**
Hargreaves, M. E., 108, **145,** 150, 169, 179, **207**
Harms, W. O., 640, 644, **653**
Harper, G. N., 410, **416**
Harris, W. J., 561, 562, **573**
Hart, E. W., 316, **334**
Harwell, R. H., 604, **617**
Hashin, Z., 702, **703**
Hayes, W., 116, **145,** 160, 172, **207**
Head, A. K., 113, **146,** 169, 179, **207**
Hearle, J. W. S., 678, 695, 697, **703, 704**
Hearmon, R. F. S., 702, **703**
Heimerl, G. J., 640, **652**
Heindlhofer, K., 431, 434, **435**
Heller, W. R., 290, **303**

Hempel, M., 580, 581, 584, **615**
Henderson, J., 650, **652**
Hendrickson, J. A., 312, 316, **334,** 410, **416**
Henriksen, E. K., 425, **435**
Herbert, E. G., 452, **466**
Herring, C., 632, **652**
Hetenyi, M., 64, **69**
Heywood, R. B., 599, **615**
Hill, R., 278, 285, 289, **302, 303,** 321, 322, 325, **335,** 343, 350, **353,** 365, 372, 376, 377, 378, 379, 382, 386, **386,** 387, 397, 400, **416,** 456, **542**
Hill, T. L., 43, **43,** 45
Hirsch, P. B., 140, **147,** 170, **206,** 628, 629, 632, **652, 654**
Hirschberg, M. H., 642, **653**
Hirst, W., 668, **672**
Hodge, P. G., Jr., 277, **304,** 343, **354,** 497, **417,** 422, **435**
Hoel, P. G., 510, **511**
Hoff, N. J., 332, 333, **335,** 643, **652,** 702, **703**
Hoffman, O., **387,** 391
Holden, A. N., 190, **208**
Holden, J., 581, 585, **615**
Holland, A. J., 499, **511**
Hollomon, J. H., 190, **208,** 510, **511**
Holloway, D. G., 497, **511**
Holm, R., 659, 661, 672, **673**
Honeycombe, R. W. K., 180, 181, **208**
Horger, O. J., 422, 424, 426, 429, 434, **435**
Hori, Y., 493, 497, **510**
Horne, R. W., 140, **147**
Horvay, G., 397, **416**
Howell, F. M., 592, **615**
Howie, A., 166, **208**
Huang, T. C., 425, 434, **435**
Hughes, D. H., 563, **574**
Hull, D., 555, 556, **572,** 581, **614,** 620
Hult, J. A. H., 290, **304,** 407, **416**
Hume-Rothery, W., 43, **43**
Humphrey, J. C. W., 577, **614**
Hundy, B. B., 312, **334,** 530, **541**
Huntington, H. B., 139, **146**
Hutchings, J., 607, **618**
Hyler, W. S., 425, **436,** 591, **618**

Ikeda, K., 500, **510**
Ineson, E., 588, **614**
Ingberg, S. H., 328, **335**
Inglis, C. E., 404, **416,** 489, **511**
Institution of Mechanical Engineers, 613, **613**

AUTHOR INDEX 739

International Critical Tables, 234, **264**
International Nickel Co., 199, 201, 202, 203, **208**
Irwin, G. R., 406, 407, **416**, 418, 539, 541, **542**
Ishida, Y., 635, **652**

Jackson, L. R., 602, 604, 613, **615**
Jacobs, J. A., 410, **416**
Jaeger, J. C., 63, 68, **69**
Jaffe, E. H., 702, **704**
Joffe, A., 189, **207**
Johnson, A. E., 636, 644, 650, **652, 654**
Johnston, T. L., 552, 553, **573**
Johnston. W. G., 116, 117, 129, 130, 132 **146**, 160, 161, 162, 163, 175, **208**
Jominy, W. E., 529, **543**
Jones, H., 15, 43, **43**, 83, 87, **93**
Joos, J., 12, 13, **43**, 136, **147**
Joos, P., 495, **512**

Kalpakcioğlu, S., 384, **387**
Kaplan, S. M., 527, **542**
Karry, R. W., 602, **615**
Kattus, J. R., 640, **653**
Kê, T. S., 480, **485**, 634, **653**
Keh, A. S., 162, 163, 168, 170, **208**, 213
Kelly, A., 285, **303**
Kennedy, A. J., 613, **615,** 640, 642, 644, **653**
Kies, J. A., 407, **416**, 539, **542**
Kingery, W. D., 230, **264**
Kittel, C., 16, 43, **43**
Klein, W. G., 689, 690, 692, 697, **704**
Knoop, F., 446, **466**
Knott, J. F., 561, 562, **573**
Kochendörfer, A., 282, **304**
Kocks, U. F., 183, **208**
Koehler, J. S., 113, **147**
Kohlrausch, F., **264**, 270
Kolsky, H., 296, 297, 298, **303**
Koskinen, M. F., 409, **416**, 418
Kramer, C. R., 642, **653**
Krishnan, R. S., 89, **92**
Kronberg, M. L., 629, 648, **653**
Kröner, E., 55, **69**, 279, **303**
Kronmüller, H., 162, **210**
Kropschot, R. H., 499, **512**
Kuhlmann-Wilsdorf, D., 164, 169, 171, **208**
Kuhn, W., 239, **264**
Kunz, F. W., 190, **208**

Laidler, K. J., 17, **43**, 235, 237, 253, **264**
Laing, K. M., 479, **485**

Laird, C., 585, **615**
Lally, J. S., 164, 195, **209**
Landau, H. G., 352, **354**
Lang, A. R., 167, **209**
Lange, H., 165, **209**
Langmuir, I., 138, **147**
Lapsley, J. T., 431, **436**, 464, **467**
Larson, F. R., 326, **335**
Larson, R. F., 639, **653**, 655
Lazan, B. J., 474, 476, 482, 483, **485, 486**, 591, **615**
LeChatelier, F., 189, **210**
Lee, E. H., 292, 297, 298, **303**, 308, 454, **466**
Lehmann, O., 254, **264**
Lensky, V. S., 288, 298, **303, 304**
Leslie, W. C., 159, **209**
Lessells, J. M., **335**, 337, 424, **435**, 437 605, 613, **616**
Lessley, H. L., 640, **653**
Letner, H. R., 425, **436**
Leurgans, P. J., 480, **486**
Levin, F., 372, 387, 454, **466**, 520, **542**
Li, C. H., 552, 553, **573**
Lianis, G. 520, **541**
Leib, J. L., 702, **704**
Linde, J. O., 156, 157, **209**
Lindell, B. O., 156, 157, **209**
Linhart, F., 484, **485**
Lipson, C., 610, **616**
Lode, W., 277, **304**
Lomer, W. M., 102, **145**
Loretto, M. H., 169, 179, **207**
Loria, E. A., 587, 604, **617**
Louat, N., 579, **618**
Love, A. E. H., 49, 68, **69**, 82, 84, **92**
Low, J. R., 117, **147**, 559, 560, 566, 572, **573**
Lubahn, J. D., 193, **209**
Lubars, W., 196, 197, 198, **209**
Lücke, K., 165, **209**
Ludley, J. H., 564, **573**
Ludwik, P., 190, **209**, 462, 463, **467**

MacGregor, C. W., 189, 191, 192, 193, **209**, 329, **335**, 577, **616**, 639, **653**
Mackenzie, A. C., 520, **543**
Mackenzie, J. K., 78, 83, **92**, 94
Mader, S., 162, 165, 166, 167, 177, **209, 210**
Maloof, S. R., 131, **147**, 170, 171, **206**, 433, **435**
Mangasarian, O. L., 400, **415**
Mann, J., 255, **264**

AUTHOR INDEX

Manson, S. S., 352, **353**, 642, 644, **650**, **653**
Mantell, C. L., 86, 195, **209**
Manufacturing Chemists Association, 234, 256, 260, **264**
Marcinkowski, M. J., 190, **209**
Marco, S. M., 598, **618**
Marin, J., 332, **335**, 343, **353**
Marks, L.S., 86
Marlies, C. A., 262, **265**
Marshall, E. R., 324, 327, **335**
Mason, W. P., 76, 89, **92**, 483, **486**
Mathur, P. N., 594, **614**
Mathur, V. D., 650, **652**
Mattson, R. L., 433, **435**, 438, 439
Mauersberger, H. R., 678, **703**
Mazey, D. J., 159, **206**
McAdam, D. J., 606, **616**
McBride, J. G., 638, **653**
McCann, H., 260, **264**
McClaren, S. W., 531, 532, **543**
McClintock, F. A., 78, **92**, 400, 407, **416, 417,** 458, **467,** 493, 494, 500, **512,** 519, 520, 523, 526, 527, 533, 536, 537, 538, 539, **542,** 544, 557, **573,** 578, 582, 586, 588, 598, **616**
McCrackin, F. L., 686, **704**
McLean, D., 554, **572,** 630, 634, **653**
McLoughlin, J. R., 250, **264**
McMahon, C., 552, 553, **573**
McMaster, R. C., 610, **616**
McMillan, C., 565, **573**
Mehl, R. F., 601, **614**
Mehl, R. R., 593, 601, **617**
Mendelson, A., 642, **653**
Merchant, M. E., 661, **673**
Meredith, R., 678, 683, 685, **703**
Merriam, J. C., 196, 197, 198, **209**
Meyer, E., 445, 457, **467**
Meyer, J. A., 282, **304**
Meyer, K. H., 238, **264**
Mikesell, R. P., 499, **512**
Miklowitz, J., 316, **335**
Milgram, M., 85, **92**
Miller, J., 639, 642, **653,** 655
Miller, J. L., 592, **615**
Miller, S. A., 702, **704**
Miner, D. F., 195, **209**
Miner, M. A., 598, **616**
Modern Plastics, 86, 717, **730**
Moffatt, W. G., 532, **542**
Mohs, F., 450, **467**
Moore, A. T., 35, 36
Moore, C. C., 608, **616**

Mordike, B., 172, 173, **209**
Morey, G. W., 86, 230, **264,** 328, **335**
Morrison, T. W., 608, **616**
Morrow, J., 433, **436,** 531, **542**
Morton, W. E., 678, 681, **704**
Mott, B. W., 446, 466, **467**
Mott, N. F., 15, 43, **43,** 83, 87, **93,** 178, **209,** 500, **512, 616,** 620, 626, **653**
Mulhearn, B., 638, **653**
Mulhearn, T. O., 464, **467**
Mullendore, A. W., 635, **652**
Murgatroyd, J. B., 497, **512**
Murray, W. M., 64, **69**
Muskhelishvili, N. I., 68, **69**
Mylonas, C., 564, **573**

Nabarro, F. R. N., 115, 141, 145, **147,** 179, **209,** 551, **572**
Nachtman, E. S., 433, **436**
Nadai, A., 312, **335,** 343, 347, **353,** 354
National Physical Laboratory, 650, **653**
Nebel, R. W., 256, **264**
Neimark, J. E., 379, **387,** 524, 532, 533, **542**
Neuber, H., 409, 411, **417**
Newhouse, D. L., 570, **573**
Newman, R. C., 156, **207**
Newmark, N. H., 598, **616**
Niblett, D. H., 481, **486**
Niegisch, W. D., 254, **264**
Nielsen, L. E., 262, **264,** 584, **616**
Noble, H. J., 428, **436**
Noller, C. R., 224, 262, **264**
Norton, J. T., 635, **652**
Novoshilov, V. V., 277, **304**
Nowick, A. S., 485, **486**
Nunes, J., 326, **335**
Nye, J. F., 90, **93,** 102, **145**
Nyquist, H. L., 459, **468**

Oberst, H., 484, **485**
O'Brien, J. L., 532, **542**
O'Connor, D. G., 233, 248, **265**
O'Connor, T. L., 658, **673**
O'Day, W. R., Jr., 523, **542,** 544
Odqvist, F. K. G., 290, **304,** 613, **618**
Ogilvie, J., 182, **206**
Oitemel, A., 594, **614**
Onat, E. T., 325, **334, 335**
O'Neill, H., 457, 466, **467**
Orowan, E., 101, 118, **147,** 162, 163, 185, 191, **206, 210, 265,** 446, **467,** 490, 493, 495, 497, 510, **510, 512,** 539, 541,

542, 556, 560, 561, **572, 573**, 584, **616,** 620, 681, **704**
Osgood, W. R., 431, 432, 434, **436,** 439
Otto, W. H., 504, **512**
Owen, W. S., 310, **335,** 557, 558, **572**

Padawer, G., 163, **210**
Palmgren, A., 598, 609, **616**
Paris, P. C., **406,** 407, **416,** 584, 585, **616**
Parker, E. R., 325, **335,** 524, **542,** 546, 568, 572, **572, 573**
Pauling, L., 43, **43**
Paxton, H. W., 161, 186, 187, **210**
Payne, A. R., 260, **265,** 269
Peckner, D., 646, **653**
Peierls, R., 115, **147**
Peirce, F. T., 697, 700, **704**
Pelloux, R. M. N., 565, **573,** 611, **613,** 633, **654**
Permanyer, F., 681, **704**
Petch, N. J., 559, 572, **573**
Peters, G., 466, **466**
Peterson, R. E., 411, **417,** 596, 611, **617**
Petterson, D. R., **93,** 95, 688, 690, 697, **704**
Phillips, A., 288, **304**
Phillips, C. E., 597, **615**
Phillips, E. A., 64, **68**
Phillips, P., 179, **210**
Plass, H. J., 298, **304**
Platt, M. M., 689, 690, 692, 693, 697, **704**
Podnieks, E. R., 482, **486**
Podosenova, N. A., 425, **435,** 436
Polanyi, M., 118, **147**
Portevin, A., 189, **210**
Post, D., 64, **69**
Powell, G. W., 327, **335**
Prager, W., 277, 298, **303, 304,** 325, **335,** 343, **354,** 365, 366, 381, 386, **387,** 397, **417**
Prandtl, L., 379, **387**
Preston, F. W., 499, **511**
Preston, G. D., 157, **210**
Prot, M., 599, **617**
Puttick, K. E., 521, 522, 524, **542,** 543

Quinney, H., 179, **211**

Rabinowicz, E., 64, **68,** 659, 660, 661, 665, 668, 669, 672, **673**
Radd, F. J., 606, **617**
Rademacher, H. J., 232, **263**
Randall, R. H., 478, **486**
Randolph, A. F., 263, **265**
Rankin, A. W., 722, 723, 724, 725, 726, 727, 728, 729, 730, **730**

Ransom, J. T., 593, 601, **617**
Rasmussen, J. G., 610, **617**
Raumann, G., 256, **265**
Raymond, M. H., 582, **617**
Read, W. T., 87, 106, 119, 126, 133, 145, **146, 147**
Rebenfeld, L., 678, 679, **704**
Rebstock, H., 167, **210**
Reding, F. P., 255, **265**
Redman, J. K., 134, **145,** 170, **206**
Reinkober, O., 504, **512**
Rhee, S. S., 400, **417,** 458, **467**
Ribi, E., 680, **704**
Richard, K., 256, **265**
Richmond, O., 384, **387**
Riley, M. W., 702, **704**
Rimmer, S. M., 719, **730**
Rinebolt, J. A., 561, 562, **573**
Riparabelli, C., 298, **304**
Roark, R. J., 411, **417**
Robertson, E. C., 494, **512**
Robinson, G. H., 601, **613**
Rockey, K. C., 564, **573**
Rockwell, S. P., 447, **467**
Rogers, H. C., 529, **543**
Roldan-Gonzales, L., 255, **264**
Rollins, M. L., 35, 36
Rose, F. C., 478, **486**
Rosen, B. W., 702, **703**
Rosenfield, A. R., 311, **335**
Ross, A. S., 433, **436**
Ross, S. T., 529, **543**
Rötger, H., 477, **485**
Rutherford, J. L., 561, **573**
Rychkewitsch, E., 495, **512**

Sachs, G., 185, **210,** 387, 391, 644, **650**
Sack, R. A., 490, **512**
Sadowsky, M. A., 366, **387**
Saimoto, S., 169, **210**
Sale, P. D., 328, **335**
Sandland, G., 445, **467**
Sargant, K. R., 578, 585, **618**
Saunders, D. W., 256, **265**
Sautter, W., 282, **304**
Savin, G. N., 411, **417**
Schabtach, C., 730, **730**
Schardin, H., 500, 501, **512**
Schiefer, H. F., 686, **704**
Schmid, E., 87, 130, 145, **147,** 171, 182, **206, 210**
Schmidt, A. X., 262, **265**
Schoeck, G., 139, **147,** 627, 630, **654**
Schulte, W. C., 589, 603, **614**

Schultz, 172, **207**
Schwarz, E. R., 694, 669, 701, **703, 704**
Schwarzl, F., 230, **265**
Scott, J. R., 260, **265**, 269
Seastone, J. B., 195, **209**
Seeger, A., 162, 166, 167, 168, 177, **209, 210**, 481, **486**
Seely, F. B., **387**, 390
Segall, R. L., 581, **617**
Seguin, B. R., 722, 723, 724, 725, 726, 727, 728, 729, 730, **730**
Seitz, F., 43, **43**, 44, 84, 87, **93**
Sernka, R. F., 529, **543**
Sessler, J. E., 644, **650**
Setty, S. K., 431, **436**, 464, **467**
Shaler, A. J., 530, **541**
Shank, M. E., 566, **573**, 639, **654**, 712, 714, 715, 717, **730**
Shanley, F. R., 332, **336**
Shaw, M. C., 324, **335**, 452, 459, 461, 462, **467**
Sherby, O. D., 647, **654**
Shield, R. T., 372, **387**, 454, **467**, 520, **543**
Shockley, W., 106, **146**
Shoji, H., 185, **210**
Sidebottom, O. M., **334**, 338
Sih, G. C., 405, 406, **406**, **416**, 495, **511**
Silcox, J., **147**, 151, 160, **210**
Simkovich, E. A., 587, 604, **617**
Simmons, W. F., 647, **654**
Simonds, H. R., 260, 262, **265**
Sinclair, G. M., 577, 598, **617**
Sines, G., 613, **617**
Slater, J. C., 43, **43**
Small, L., 466, **467**
Smekal, A. G., 502, 503, **512**
Smith, F. C., 604, **617**
Smith, G. C., 579, 581, **617**
Smith, G. S., 585, **615**
Smith, G. V., 646, 650, **654**
Smith, H. L., 407, **416**, 539, **542**
Smith, J. C., 686, **704**
Smith, J. F., 83, **93**
Smith, R., 445, **467**
Smith, R. L., 561, **573**
Smithells, C. J., 86, 195, 196, **210**, **730**, 733
Sneddon, I. N., 454, **467**, 490, **512**
Snoek, J., 479, **486**
Society of Automotive Engineers, 425, 432, **436**, 438
Soderberg, C. R., 591, **617**
Sokolnikoff, I. S., 49, 60, 68, **69**, 360, **387**
Southwell, R. V., 331, **336**, 410, **415**

Spaeth, C. E., 712, 714, 715, 717, **730**
Stade, C. H., 156, 157, **209**
Stanworth, J. E., 230, **265**, 510, **512**
Starkey, W. L., 598, **618**
Staverman, A. H., 230, **265**
Steele, R. K., 433, **435**
Stein, D. L., 117, **147**
Stein, P. K., 64, **69**
Sternglass, E. J., 298, **304**
Stewart, D. A., 298, **304**
Stimson, L. D., 410, **417**
Stokes, R. J., 190, **207**, 552, 553, **573**
Stookey, S. D., 230, 260, **265**
Stroh, A. N., 106, 135, **146**, **147**, 549, 550, 554, 555, 572, **573, 574**
Struik, D. J., 693, **704**
Stuart, H. A., 28, **43**, 262, **265**
Stubbington, C. A., 578, **614**
Stulen, F. B., 589, 603, **614**
Sucov, E. W., 505, **513**
Sukhatme, S. P., 500, **512**
Suhre, J. R., 313, **336**
Sully, A. H., 650, **654**
Susich, G. V., 238, **264**, 682, 687, **704**
Suzuki, H., 189, **211**
Swalin, R. A., 17, **43**
Swann, P. R., 168, **211**
Swinden, K. H., 409, **415**
Sylvestrowicz, W. D., 188, **211**
Syromjatnikoff, F., 681, **705**
Szcepenski, E., 594, **614**
Szczepinski, W., 288, **304**

Tabor, D., 449, 450, 454, 464, 466, **467, 468**, 659, 661, 668, 672, **673**
Tack, C. E., 730, **730**
Tapsell, H. J., 636, **654**
Tarasov, L. P., 425, **436**, 591, **618**
Tavernelli, J. F., 590, **618**
Taylor, A., 172, **211**
Taylor, E. W., 450, **468**
Taylor, G. I., 118, **147**, 169, 179, **202, 207, 211**, 374, **387**, 388
Taylor, R. J., 588, **614**
Terry, E. L., 531, 532, **543**
Tetelman, A. S., 500, **513**
Thakur, V. M., 695, 697, **703**
Thibault, N. W., 459, **468**
Thomas, D. A., 572, **572**
Thomas, G. B., 51, **69**
Thomas, W. N., 595, **618**
Thompson, N., 579, 613, **618**
Thomsen, E. G., 425, 431, **435**, **436**, 464, **467**

AUTHOR INDEX 743

Thornton, R. P., 632, **654**
Timoshenko, S., 49, 60, 68, **69**, 82, **93**, 332, **336**, 343, **354**, 395, 398, 401, 410, **417**, 421, **436**
Tipper, C. F., 521, **543**, 557, **574**
Tobolsky, A. V., 230, 231, 249, 250, 253, 255, 260, 262, **264, 265**
Topping, A. D., 697, 699, **705**
Torvik, P. J., 483, **486**
Träuble, H., 162, **210**
Treloar, R. L. G., 239, 241, 243, **265**
Tripp, V. W., 35, 36
Troiano, A. R., 715, 716, **730**
Troxell, G. E., 568, **572**
Tsao, C. H., 64, **68**
Tsien, L. C., 496, **510**
Tuncel, O., 407, **416**
Tupper, S. J., **303**, 308
Turner, W. E. S., 499, **511**

Uhlig, H. H., 184, **211**, 658, **673**
U. S. Department of Agriculture, 88
U. S. Rubber Company, 86
Uzhik, G. V., 529, **543**

Valko, E., 238, **264**
van Bueren, H. G., 177, **211**
Van Vlack, L. H., 24, **43**
Vineyard, G. H., 85, **92**
Virgin, W. P., 678, **704**
Voigt, W., 78, **93**, 309, **336, 354**, 355
von Bergen, W., 680, **705**
von Weingraber, H., 466, **468**
Voorhees, H. R., 650, **651**

Wadsworth, N. J., 579, 607, 613, **618**
Wahl, A. M., 596, **617**
Waisman, J. L., 613, **617**
Walsh, J. B., 493, 494, **512**, 520, **543**
Walter, E. R., 255, **265**
Wang, N. M., 298, **304**
Warren, B. E., 229, **265**
Warrington, D. H., 628, 629, **652**
Warshaw, I., 497, 498, **513**
Wartel, W. S., 648, **650**
Webber, A. C., 343, **353**
Weertman, J., 626, **654**
Weerts, J., 87
Weibull, W., 510, **513**, 613, **618**
Weiner, J. H., 352, **353, 354**
Weiss, P., 671, **673**

Weissenberg, K., 238, **265**, 697, **705**
Weissmann, G. F., 343, **353**
Weissmann, S., 162, 163, 168, 170, **208**, 213
West, G. W., 108, **145**, 150
Westbrook, J. H., 462, 463, **468**
Westwood, A. R. C., 181, **211**
Whelan, M. J., 125, 140, **147**, 151, 160, **210**
Whitby, G. S., 260, **265**
Whitehead, J. R., 660, **673**
Whitney, W., 257, **266**
Widmer, R., 638, 644, 645, **653, 654**
Wigglesworth, L. A., 406, **417**
Wilks, J., 481, **486**
Williams, M. L., 405, 407, 408, **415, 417**, 539, **543**
Williams, S. R., 466, **468**
Williams, T. R. G., 563, **574**
Wilsdorf, H. G. F., 31, **43**, 164, **208**
Wilson, J. C., 563, **574**, 642, **654**
Wilson, R. N., 583, **618**
Winchell, H., 450, **468**
Winchell, P. G., 180, **211**
Winne, D. H., 406, **417**
Wöhler, A., 576, **618**
Woldman, N. E., 195, **211**
Wolfe, L. H., 606, **617**
Wolfe, R. J., 608, **614**
Wood, D. S., 312, 316, **334**, 410, **416**
Wood, W. A., 582, 585, **618**, 623
Wooley, R. L., 184, **211**
Wulff, J., 532, **542**
Wundt, B. M., 406, **417**, 570, **573**
Wyatt, O. H., 179, **211**, 637, **655**

Yates, D. H., 198, **211**
Yew, C. H., 452, **466**
Yoffe, E. H., 500, **513**
Young, F. W., Jr., 117, 130, **148**, 153, 160, 161, 162, 163, 168, **211**

Zachariasen, W. H., 229, **266**
Zandman, F., 64, **69**
Zaukelies, D. A., 254, **266**
Zener, C., 17, **43**, 137, **148**, 190, **208**, 476, 477, 478, **486**, 549, 550, **574**, 634, **655**
Zerna, W., 68, **68**
Zimmerman, J., 696, 697, **703**
Znoek, W., 695, **705**
Zwicky, F., 489, **513**

SUBJECT INDEX

Abbreviations of polymers, 216 (table)
Abrasive wear, 669
Activation energy, 17
   creep, 627, 631 (data), 639
   dislocation loops, 139f
   dislocation mechanisms, 176 (table)
   formation and motion of vacancies, 137, 138 and 176 (tables)
   self diffusion, 138 (table), 631
   sodium diffusion in glass, 499
   stress dependence in creep, 639
   viscoelasticity of PMMA, 252
   viscosity, 235, 236, 237
Activation enthalpy (*see* Activation energy)
Activation entropy, 137
Adhesion of solids, 671f
Adhesional theory of friction, 660f, 662f
Adhesive wear, 667
   of lubricated surface, 669
Aging, 158, 159 (*see also* Hardening)
   creep, 645
   fatigue, 585
Alloying (*see also* Hardening)
Alloying, against cleavage fracture, 562, 568
   against creep, 645
   for strength, 194
Almen strip, 438
Amine polymerization, 218
Anelasticity, 476 (*see also* Delayed elasticity)
Anisotropy in, bending, 355
   elastic stress-strain relations, 73f, 86–8 (table)
   fibers, 35, 678, 679, 683, 684
   plastic stress-strain relations, 278, 283f, 307
   tensile test, 309f
   textiles, 688
   torsion, 355
   viscous flow, 238
Annealing of metals, 196 (table)
   stress relief, 433f
   recovery, 177, 640
   recrystallization, 177
Antisymmetric tensor, 62
Approximate stress analysis, 359ff

Associated flow rule, 288
   sliding junctions, 663
   structures, 289
Atactic polymers, 225, 230, 232
Atomic bonding, 9f

Ball bearing, contact stress, 410f
   fatigue, 607f
Bands, deformation, 32, 256, 257
   kink, 32, 166, 255, 629
   Lüders, 34, 187, 312
   slip, 32, 177
      fatigue, 577
   twin, 34
Bar, bending (*see* Beams)
   tension, anisotropic, 309
      circumferentially notched, 372, 519 (table)
      isotropic, 309ff, 643
   torsion, anisotropic, 355
      circumferentially notched, 519 (table)
      isotropic, 339f
      transverse hole, 363
Bauschinger effect, 165, 184f, 283, 582
Beam, bending, anisotropic, 355
   isotropic, 344f, 412
   damping, 477, 483
   deflection at base, 412
Bending, 339ff
   beams, anisotropic, 355
      isotropic, 344f, 412
   notched beams and plates, plastic, 519 (table)
   plates, 349f
Bimetallic strips, 351
Binding energy, 9, 83 (data)
Bolt head, limit load in torsion, 392
Bolt, optimum pre-load in, 707f
Boltzmann, constant, 13, 40 (value)
   distribution, 13
   superposition principle, 247
Bonded fabric, 94, 688
Bonding, 9f
   atomic, 9
   covalent, 11
   hydrogen, 11

intermolecular, 11
ionic, 10
metallic, 11
primary, 10
secondary, 11
van der Waals, 11
Born-Mayer law, 85
Bragg angles, 431
Branched polymer, 217
Brittle fracture, 8, 488ff
    combined stress, 490f
    crack formation, 497f
    crack growth, 500f
    data, 496
    energy condition, 494
    evidence of cracks, 496
    experiments on, 495f
    glassy solids, 488ff
    markings, 502f
    necessary condition, 494
    origin of cracks, 497f
    size effect, 504f
    static fatigue, 499f
    statistics, 504f
    stress conditions for, 494, 489
    sufficient condition, 494
    tendencies, in metals, 567f
    uniaxial stress, 488f
Bubble model, crack initiation, 523
    dislocation, 102, 114
    grain boundary, 181
    impurities in, 188
Buckling, 330f
Bulk modulus, 29, 79, 82–83 (data)
    conversion to other elastic contants, 80 (table)
    relation to binding energy, 10, 83
Burgers circuit, 99
Burgers vector, 99

Case studies, 707ff
Cauchy relations, 84
Central forces, 84
Centroid of a section, 346
Characteristics of differential equations, 322
Charpy V-notched impact test, 391, 568, 718, 726
Chevron markings, 565, 722
cis (structure of polymers), 225, 226
Cleavage steps, 567
Climb force, 112
Climb of a dislocation, 105, 112, 177, 626

Coaxing in fatigue, 598
Coefficient of thermal expansion (see Thermal expansion)
Coefficient of variation, 691
Co-energy, 361
Cohesive strength, 489, 495, 523
Cold drawing, of fibers, 256
    of wire, 391
Combined stress, effect on, creep, 640f
    damping, 483
    fatigue, 591
    fracture, 490f, 529, 552
    yielding, 276f
Compatibility equations, 60
Complementary energy, 361
Complex compliance and modulus, 294, 306, 472
Compliance, creep, 247, 294, 306
    elastic (see Elastic compliance)
Composite beams, 347
Compressibility, 29, 79, 82–83 (data)
    relation to binding energy, 10, 83
Compressive deformation, 309ff
    instability, 330f
Compressive strength of, brittle materials, 491 (see also Buckling)
    polymers, 259 (table)
    rocks, 494 (data)
Configurational entropy, 136, 238
Conservative motion, 104
Constants, elastic (see Elastic compliance and stiffness for anisotropic, Modulus of elasticity for isotropic, Bulk modulus, Poisson's ratio)
Constants, physical, 40 (table)
Constitutive relations, 72ff, 273ff
    creep, 290f
    elastic, 72ff
    elastic, anisotropic, 73f
    elastic, isotropic, 79f, 274
    elastic rubber, 243
    plastic anisotropic, 278, 283f, 307
    plastic isotropic, 276f, 283f
    symmetry, 76
    viscoelastic, 290f
    viscous, 290
Contact, area, 657f, 658f
    fatigue (rolling), 607f
Contact stress, elastic, 410f
    plastic, 663f
Conversion, factors, 40 (table)
    between isotropic elastic constants, 80 (table)
    ton, British, 588

Copolymers, 225, 227
Corrosion, stress, 9, 184
  rocket casings, 716
Corrosion fatigue, 606
Corrosive wear, 670
Cost, creep-resistant alloys, 646
  desk-top materials, viii
  manufactured part, 646, 732
  metals, 196–203 (table)
  polymers, 258–260 (table)
Cotton structure, 35, 36, 678
Cottrell dislocation reaction, 124
Couple stresses, 54, 85
Covalent bond, 11
Crack, plastic deformation and limit
    loads, 519 (table)
  stress-strain distributions, elastic, 405f, 411
    elastic-plastic, 409, 535, 537, 544
    plastic, 370f, 378f, 408f
Crack extension force, 539
Crack formation (cleavage cracks) by
    plasticity, 549f
    dislocation pile-up, 549, 554
    twin intersections, 555
  ductile fracture, 521f
  fatigue, 578f, 581
  glassy materials, 497f
Crack growth,
  brittle fracture in glass, 500f
  brittle fracture in steel, 556f
  ductile fracture, 536f
  fatigue, 584f
Crack velocity, 500f
Creep, 6, 625ff
  activation energy, 631 (data)
  Andrade's β, 275, 626, 637
  buckling, 332
  choice of materials, 645f
  compliance, 247, 294, 306
  constitutive relations, 290f
  correlation of, 637f (data)
  cyclic, 599
  dislocation climb in, 626f
  effect of combined stress, 640f
  effect of history, 640f
  fracture, 643f, 725, 728
  interaction with fatigue, 606, 641
  lattice friction stress, 629
  logarithmic, 626
  Newtonian [see Viscosity (linear), 236]
  polycrystals, 632f
  primary (see Creep, transient)
  rupture (see Creep, fracture)
  secondary (see Creep, steady state)
  sessile jog motion, 629
  single crystals, 626f
  steady state, 245, 625, 626, 639
  tertiary, 625
  transient, 5, 178, 245, 625, 626, 637
    (see also Viscoelasticity)
Critical resolved shear stress, 114f, 130
    (table) (see also Dislocation kinetics
    and stress, friction stress)
Crossglide, 129, 175, 176 (table)
Cross linking of polymers, 230, 232
Cross slip, 129, 177
Crystal defects, 23f
  (see also Dislocations, Grain boundaries
    Impurities, Interstitials, Vacancies)
Crystal structure, 17f, 20, 21 (tables)
Crystal symmetry, 17, 78
Crystallinity, in fibers, 682
  in polymers, 27, 232, 253f
Crystallization of, glass, 230
  polymers, 27
  rubber, 225, 243
Crystallographic notation, 21
Cumulative damage in fatigue, 598
Cup and cone fracture, 528, 529
Cylinders, thick-walled, 393
Cylindrical coordinates, equilibrium, 54
  strain, 60
Cylindrical holes, biaxial stress, 394f, 525
  uniaxial stress, 363, 398f

Damage, in ductile fracture, 526
  fatigue, 585
  heat, in polymers, 227, 233, 260
  radiation, in creep, 642
    in fracture of steels, 563
Damping, 6, 471ff
  air, 483
  in a beam, 477, 483
  Bordoni, 481
  coatings for, 483
  combined stress, 483
  dipole flipping, 481
  in fatigue, 578
  grain-boundary, 480
  interstitial, 478
  kink motion, 481
  models of, 471f
  plastic deformation, 480
  point defects, 479
  polycrystalline, 477, 480
  relation to endurance limit, 483
  slip, 483

SUBJECT INDEX 747

Snoek, 478
sources of, 475f
structural, 483f
thermal currents, 475
Davidenkov diagram, 564f
Debye frequency, 15, 235
Debye temperature, 15
Decarburization, of a bolt, 373, 733
  effect on fatigue, 603
  pressure vessel, 713
Defects in crystals, 17f, 23
  (*see also* Dislocations, Grain boundaries, Impurities, Interstitials, Vacancies)
Deformation (*see* Stress-strain distributions or specific shape or item of interest)
Deformation, kinds of, 5f, 29f, 275
  polymers, 230f
  (*see also* Anelasticity, Creep, Elasticity, Plasticity, Viscosity, Viscoelasticity)
Deformation bands, 32, 256, 257, 522
Deformation theory of plasticity, 289
Deformation twin band, 34
Delayed elasticity, 6, 244, 245, 637
Denier, 682
Density, effect of dislocations on, 108
  liquids, 82 (table)
  polymers, 258 (table)
  solids, 83 (figure), 86 (table)
Devitrification, 230
Dielectric constant, 89
Diene polymerization, 218
Diffusion, self-, 138, 631
  vacancies, 136f, 630, 632
Diffusionless transformations (*see also* Twinning), effect on strain hardening, 180f
Dilatational misfit, 153
Dilatational strain, 59, 291
Dilatational (uniaxial strain) wave velocity, 297, 301
Dipole, dislocation, 120, 121, 162, 163, 168, 481
Dipole flipping, 481
Dipole trails, 121, 168
Direction cosines, 51
Dislocation, 23, 30, 96ff, 152ff, 255
  effects, in crack formation, 523, 549f
    creep, 626f
    damping, 480
    fatigue, 578, 619, 620
    polymers, (*see* Crystallinity)
  geometry, of individual, 97f
    bubble model, 102, 114, 181, 188, 523

Burgers vector, 90
  edge, 97, 107
  effect on density, 108
  extended, 123, 166
  imperfect, 122
  jog, 118, 119, 629
  kink, 118, 481
  loop, 159, 194
  partial, 122
  perfect, 100
  screw, 97, 106
  sessile, 105, 159, 167, 554
  stress field, 106
  unit, 100
  width, 114
interactions and arrays of, cells, 168, 583
  density, 106, 150, 169, 170, 176, 212, 213
  density in polymers, 255
  dipoles, 120, 121, 162, 168, 481
  entanglement, 164, 167, 168, 583
  intersections of, 118f, 173
  loop, 139, 151, 159, 194
  networks, 159, 165, 169
  nodes, 126
  pile-ups, 134f, 549f
  point obstacle interaction, 153f, 175
  precipitate interaction, 157f, 194, 631
  surface interaction, 113, 165
motion, 102f
  climb, 105, 112, 177, 626
  cross glide, 129, 175, 176 (table)
  double cross glide, 168, 307
  glide, 103
  mean free path, 162, 163 (table), 165
  mill, 126
  multiplication, 126f, 150, 162, 213
  non-local effect on plasticity, 165, 283
  reactions, 121f, 124
  size effects in single slip, 165
  strain due to, 105, 311
  trails, 167
kinetics and stress, energy, 108f
  energy of high velocity, 110
  "force" on, 111f
  friction stress, 114f, 117 (table), 153f, 160, 552, 629
  "line tension," 109f
  "mass," 110f
  role of thermal activation, 30, 139f
  stress field, 106f, 169
Distortion energy criterion, 277

Double cross-glide, 129, 168
Doubly grooved plate, 371, 378
Drawing, of fibers, 256, 680f
  polyethylene, 256
  wire, 383, 391
Drilled rod, tensile limit load of, 363
Ductile fracture, 8, 518ff
  appearance, 528–530
  crack formation, 521f
    effect of combined stress, 526, 529
    effect of history, 524, 526, 530
    by growth of holes, 524f
    load and deformation factors for, 533
    mechanisms of, 521f
  crack growth, 523
    elastic-plastic mechanics, 534f
    instability, 536, 539 (table)
    load factors for, 534
  elongation in, 326, 327, 526
    notch sensitivity, 533f (see also Ductile fracture, Crack formation or growth)
  rupture, 8, 518f
  unnotched parts, 528f (see also Ductile fracture, Crack formation)
Ductility, 325f
Dulong and Petit's law, 14
Dynamic stress analysis, 295f

Easy glide, 161, 162, 165
Edge dislocation, 97
  stress field, 107 (see also Dislocation)
Elastic after effect, 6, 245, 269, 686
Elastic compliance and stiffness, calculation of polycrystalline, 361, 362
  components of, 74, 75
  components, transformation, 74, 78
  compressibility, 10, 29, 79, 82–83 (data)
  for cubic and orthotropic materials, 87 (table)
  isotropic (see Compressibility, Modulus of elasticity, Modulus of rigidity, Poisson's ratio), conversion table, 80
  magnitude of, 82f
  for wood, 88 (table)
Elastic contact, stress distribution, 410f
Elastic idealization, 275
Elastic limit, 5, 310f
Elastic modulus (see Modulus of elasticity)
Elastic-plastic idealization, 275

Elastic stiffness (see Elastic compliance and stiffness)
Elastic stress-strain distribution (see also Stress-strain distribution), equivalence to viscous, 291
  uniqueness, 81f
Elasticity, 5, 29, 275
  anisotropic, 73f
  constitutive equations, 73f, 79f, 243
  delayed, 6, 244, 245, 637
  energy methods, 360
  in fibers, 683
  isotropic, 79f, 274
  in polymers, 230
  rubber, 4, 29, 230, 238f
Elastomers, 4, 232
Electroplating, effect on fatigue, 602
  embrittlement, 500
Elliptical holes, 404f
Elongation, 326, 327, 526
  for polymers, 259 (table)
End-quench hardenability, 438
Endurance limit, 576
Energy (see also Activation energy),
  binding, 9, 83 (data)
  dislocation, 108f, 110
  jog, 121
  of vacancy, 121
Energy methods, elasticity, 360f
Energy transition temperature, 559, 568 (data)
Engineering strain and stress, 318
Entropy, 13
  configurational, 136, 238
Environment, effect of humidity on textiles, 685
  effect on brittle fracture, 499f
  effect on creep, 642, 644
  effect on fatigue, 605f
  hydrogen embrittlement, 716
  stress corrosion, 9, 184
Equilibrium equations, 52, 274
  cylindrical coordinates, 54
  plane-strain plasticity, 375
  spherical coordinates, 54
Equivalence, elastic and viscous stress distributions, 291
Equivalent flow stress, effect of plastic deformation (strain hardening), 279f
Equivalent plastic strain, 279
Error, propagation of, 240
Ester polymerization, 218
Extended dislocation, 123

Extreme-value statistics, 507, 597, 691, 692
Extrusion, 383
Extrusions in fatigue, 578, 581, 619, 620

Fabric (*see* Textiles)
Factorial function, 507
Fatigue, 8, 576ff
  component testing, 609
  correlation with tensile strength, 600, 606
  crack detection methods, 610
  crack formation, 578f
  crack growth, 584f
  creep in, 599
  corrosion, 606
  damage, 585
  damping in, 578
  effects of variables on, environment, 605, 608
    hardness, 600
    material variables and processing, 600–604
    rate of testing, 599
    residual stress, 591, 598
    size, 594
    strain amplitude, 590
    stress, 588–600
    surface treatment, 595, 602–604
    temperature, 605
  engineering considerations, 576f, 609f
  flaking, 609
  fracture appearance, 585, 610
  fretting corrosion, 483, 607
  history of, 576f
  limit, 576
  method of study, 576f
  Prot testing, 599
  rolling contact, 607f
  scatter, 587, 588, 601, 609
  service testing, 609
  size effect, 594
  spalling, 609
  static, 499f
  statistics, 609
  strain aging in, 605
  strain hardening in, 582
  strength reduction factor, K_f, 595, 597, 711
  surface, 670
  testing machines, 587
  thermal, 642
Felting, 680
Fibers, 675ff
  anisotropy, 35, 678, 679, 683, 684

  classification, 677 (table)
  cold drawing of, 256
  cotton structure, 35, 678
  effect of humidity, 684
  effect of temperature, 684
  mechanical behavior of, 680f
  polymeric structure, 677f
  recovery, 687 (data)
  spinning, 676
  tensile behavior, 680
  use, 676
  wool structure, 679
Fibrous materials, 675ff
Finite strain, plastic, 316f
  rubbery, 241f
Flow lines, in worked metal, 35
Flow localization, in tension, 311f, 320f
  in Lüders bands, 34, 187, 312
Flow stress, 7, (*see also* Hardening, strain)
  relation to hardness, 456 (table)
  strain hardening, 279
  strain-rate dependence, 173, 190f
  temperature dependence, 172f
"Force" on a dislocation, 111
Forging, 380
Formaldehydes, 222, 227
Fracture, appearance, brittle, 502f, 565f
  creep, 722
  ductile, 528, 530
  fatigue, 585, 610
  brittle, 8, 488ff (*see also* Brittle fracture)
  creep, 643f, 725, 728
  ductile, 8, 518ff (*see also* Ductile fracture)
  elongation, 326, 526
  fatigue, 8, 576ff (*see also* Fatigue)
  hydrogen embrittlement, 714
  markings (*see* Fracture appearance)
Fracture transition, 546ff (*see also* Brittle fracture, 488ff, and Ductile fracture, 518ff)
  crystalline materials, 548f
  Davidenkov diagram, 564f
  effect of, combined stress, 552
    composition of steel, 562 (data), 568
    grain size, 559
    impurities, 563
    irradiation, 563
    notches, 548, 560
    pre-strain, 563
    strain rate, 548, 561
    triaxial tension, 548, 560
  engineering tests, 567f

glassy solids, 547f
temperature, 548
  appearance (FATT), 569
  energy, 559, 569
Fracture work, 539, 556 (see also Ductile fracture, Crack growth, Stress singularity factor)
Frank-Read dislocation mill, 126, 140
Free energy of crystal with vacancies, 136
Free volume, 253
Frequency, Debye (atomic), 15, 176 (table), 235
  dislocation, 176 (table)
Fretting, in wear, 671
Fretting corrosion, 483, 607
Friction, 9, 657ff
  adhesional theory, 660f, 662f
  coefficient of, 644 (data), 665
    theory of, 660
  contact area, 657f, 658f
  elastic theory, 658, 659
  laws of, 666f
  of screw threads, 710
Friction stress, of dislocation, 114, 115, 116, 117 (table)
  velocity effect on dislocation, 116
Functionality, 227

Gamma function, 507
Geiringer equations, 377
Generalized loads, 289
Generation of vacancies, 136
Glass, plastic flow, 495, 501
  structure, 26, 228
  surface structure, 498
  transition temperature, 249, 250, 251
Glide force on dislocation, 111
Glide of dislocation, 103
Goodman diagram, 591
Grain boundary, bubble model, 181
  damping, 480
  effect on strain hardening, 181f
  migration, 632
  precipitation at, 553, 635
  sliding, 632
  in steel, 24
Green-Gauss theorem, 65
Grex, 682
Groove (see also Notch, if not long compared with depth)
  elastic, 405, 406 (table), 411
  elastic-plastic shear, 409, 535f
  plastic, 519 (table)
    circumferential, 372

plane strain, double, 371, 379, 408f
plane strain, single, 368f, 372, 378
shear and tension of strips, 305
sliding of junctions, 663
Grooved strips, limit load, 305
Guest yield criterion, 278

Hackle marks, 502
Hardenability, 438
Hardening (see also affected phenomena),
  diffusionless transformations, 180f
  dispersed particles, 645
  grain boundaries, 180, 181f
  interstitial atoms, 121, 139
  interstitial impurity, 153f, 186f, 645
  order-disorder transformations, 645
  oxide particle, 645
  phase transformations, 180
  precipitation, 157, 645
  quench, 159
  radiation, 159
  solution, 153f, 186f, 645
  strain (or work), 160f
    easy glide (Stage I), 160–166
    effect of combined strain, 279f
    effect of surfaces, 180f
    effect on extension to rupture, 519
    effect on necking, 325
    effect on strain concentrations, 400
    effect on tensile strength, 319
    in fatigue, 582
    in hardness test, 457
    rapid hardening (Stage II), 167
    rate of, 319, 327, 457 (table)
    stored energy in, 179f
  structural, 153f, 180
  surface films, 180
  Suzuki, 189
Hardness, 8, 443ff, 458f (data)
  of carbides, constituents of steel, oxides, and various materials, 460–461 (tables)
  conversion, 449, 451
  effects of variables, 458f
    pressure, 464
    rate of testing, 464
    residual stress, 464
    size, 458
    temperature, 461
  effects on (see affected phenomena)
  relation to flow stress, 453f, 455
  rubber, 464f
  tests, indentation, 443f
    Brinell, 444

Knoop, 446
  microhardness, 446
  Monotron, 445
  Rockwell, 447, 448
  Vickers, 445
 tests, other, abrasion, 453
  Bierbaum, 451
  cutting, 451
  damping, 452
  erosion, 453
  file, 450
  Herbert, 452
  Mohs, 450
  pendulum, 452
  rebound, 452
  scratch, 450
  Shore, 452
 work per unit volume, 456
Hardness indentation, limit load of, 373
Heat distortion temperature, for
    polymers, 259 (table), 260
Heat of vaporization, for liquids, 82
    (table)
  for solids, 83 (data)
Heat resistance temperature,
  for polymers, 259 (table), 260
Heat setting of fibers, 256, 676
Heat treatment (*see also* Hardening),
    glass, 499
  metals, 196 (table)
Hencky's first theorem, 377
Hencky plasticity equations, 289
Hencky yield criterion, 276
Hertz equation, 410, 457
Hexagonal symmetry, 78
Hole,
  cylindrical, under biaxial stress,
    394f, 525
  cylindrical, under uniaxial stress,
    363, 398f
  elliptical, 403f
  growth in ductile fracture, 524
  growth in fatigue, 584
  spherical, 401f
Homologous temperature, 462
Huber yield criterion, 276
Humidity, effect on brittle fracture, 499f
  effect on fatigue, 607
  effect on textiles, 685
Hydrogen bond, 11
Hydrogen embrittlement, 500, 714
Hydrostatic pressure, effect on, brittle
    fracture, 490
     ductile fracture, 530

   hardness, 464
   yielding, 530
Hysteresis, 6, 471ff (*see also* Damping)
Hysteresis loop in fatigue, 582, 584

Ideal strength, cohesive, 489, 495, 523
  shear, 118
Idealization, of mechanical behavior, 274f
  of viscoelastic behavior, 243f
Impact test, Charpy V-notch, 561 and
    568 (data), 718
  mechanics, 295, 391
  use in case studies, 718, 726
Imperfect dislocation, 122
Impurity atom (*see also* Hardening), 23
  role in damping, 478
Incompressibility, in plastic plane strain,
    377
  in plasticity, 108, 288
Indentation hardness, 443f
Independent slip systems, 285
Inelastic deformation, 5, 29
Inertia, moment and product of, 346
Instability, compression, 330
  ductile fracture, 536
  material, 186f
    kinking, 190
    Lüders bands, 34, 187, 312
    misfit interactions, 186
    order-disorder, 190
    Portevin-LeChatelier effect (jerky
      glide), 189
    due to pre-strain, 190
    Suzuki effect (chemical interactions),
      189
    twinning, 190
  mechanical, in tension, 319f, 320f, 330f
    in compression, 330f
    low temperature thermo-mechanical,
      190
  machine, 330f
Internal friction (*see* Damping)
Interstitial atom (*see also* Hardening),
    23, 121
  activation enthalpy, 139
  formation, in plasticity, 121
Interstitial impurities (*see also*
    Hardening), 23
  role in damping, 478
Intrusions, in fatigue, 578, 581, 619, 620
Ionic bond, 10
Isentropic compliance, 76
Isochronous stress-strain curve, 640
Isomers, 226

Isotactic polymers, 225, 230, 232
Isothermal compliance, 76
Isotropic elastic constants (*see also* Constitutive relations), 78f
  relations between, 80 (table)
Izod test, 568

Jog, on a dislocation, 118
  energy, 119, 176 (table)
  sessile, in creep, 629
Jominy hardenability, 438
Junctions, between surfaces, 658
  of dislocations (nodes), 126

Kinematically irreversible deformation, 585
Kink bands, 32, 166, 255, 629
Kinking, 135, 632
Kinks on a dislocation, 118, 481
  role in damping, 481

Lamé constants, 81
Laplace transform, 247, 292
Larson-Miller parameter, 639, 725
Lattice friction stress (*see also* Dislocation friction stress, 114f, 153f), 115
  in creep, 629
Limit analysis, 363f
  bounds for limit load, 364f
  less exact approximations, 380f
  plane strain, 375f
Limit load, 363f, 519 (table)
  bar in torsion, 341
  bolt head, 392
  doubly grooved plate, 371, 379
  drawing, 383
  drilled rod, 363
  extrusion, 383
  forging, 380
  grooved strips, 305
  hardness indentation, 373
  machining, 389, 452 (data)
  notched plate, 389
  punching, 391
  ring, 389
  screw threads, 373, 733
  shearing (cutting) of sheet, 391
  singly grooved plate, 368, 378
  sliding junctions, 663
  spinning, 384
  turbine blade root, 390
  wire drawing, 383, 391
Line defects in crystals, 23

Line tension of dislocation, 109
Linear polymer, 25, 217
Linearity, elastic, 74
  viscoelastic, 232, 233, 236, 247f, 275, 686
Localization of flow, 311
Logarithmic decrement, 473
Logarithmic strain, 318
Lorentz factor, 110
Low pressure polyethylene, 254
Lower bounds, limit load, 365 (*see also* Limit analysis)
  to stiffness, 362
Lower yield point, 312
  in fibers, 255, 682
Lubrication, 666f
  effect on friction, 666f, 669
  effect on wear, 669
Lüders bands, 34, 187, 312 (*see also* Instability, material)

Machining of metal, 389
  specific work, 452 (data)
Macroscopic residual stresses, 420
Macrostructure, 36 (*see also* under desired elements of structure)
Madelung constant, 44
Magnetic hysteresis, in fatigue, 578
Magnetostriction, 91f, 90 (data)
Martensite transformation, 180
"Mass" of a dislocation, 111
Maximum load in tension, 319
Maximum plastic resistance, principle of, 366
Maximum pressure in spheres and cylinders, 320
Maximum shear stress criterion, 278
Maxwell-Boltzmann distribution function, 13
Maxwell material (element), 243, 547
Mean free path, dislocation, 162
Mean normal stress, 67
Mechanical behavior, idealization, 274f
  kinds of, 3f (*see also* specific kind)
Mechanical equation of state, 190f
Mechanical instability (*see* Instability)
Mechanics, fundamental equations, 273ff
Melting temperature, of crystalline polymers, 255
  of metals, 631 (data)
Metal cutting (*see* Machining)
Metal forming (*see also* Bending, Limit loads)
Metallic bond, 11
Meyer index, 457

Microhardness, 446
Microstructure, 31 (*see also* under desired elements of structure)
Miller notation in crystallography, 20
Miner criterion, 598
Minimum complementary energy theorem (co-energy), 361
Minimum potential energy theorem, 360
Mises stress-strain relations, 289
Mises yield criterion, 276
Mobility of vacancies, 139
Modulus, bulk, 29, 79, 82–83 (data)
    conversion to other constants, 80 (table)
  complex, 294, 306, 472
Modulus of elasticity (*see also* Elastic compliance for anisotropic), 79, 86 (table)
  calculation of polycrystalline, 361, 362
  conversion to other constants, 80 (table)
  isentropic vs. isothermal, 76
  magnitude from bulk modulus, 82f
  for polymers, 258 (table)
  relaxed and unrelaxed, 472
  for rubber, 242, 268 (table)
Modulus of rigidity, 79, 86 (table)
  conversion to other constants, 80 (table)
  for MgO, 6 (data)
Mohr's circle, 62, 375
Molecular structure, 25f, 215f
Moment of inertia, 346
Momentum equation, 54
Motion, conservative of dislocation, 104
Multiple slip (polyslip), 167f

Necking, 320f
  bars, 320, 322f
  in creep, 322, 643
  spheres, 322
  strips, 321
Network polymer, 25, 227
Neumann bands, 134
Neutral axis, 344
Newtonian viscosity [*see* Viscosity (linear), 236]
Nitriding, 428, 570, 604
Nodes, dislocation, 126
Nominal, strain, stress, "engineering," 318
  smooth specimen, 394
Nonconservative motion (climb), 105
Nonwoven fabric, 94, 688

Notch, elastic (*see also* Groove, if not short and deep), 405f, 406 (table), 411
  elastic-plastic shear, 409, 535f
  plastic, circumferential, 372
  deformation and limit loads, 389, 519 (table)
  plane stress, 389
Notch sensitivity, ductile fracture, 533f
  in fatigue, 595f
Notched plate, limit load of, 389
Nuclear radiation (*see* Radiation)

Octahedral shear stress criterion, 277
Orthotropic materials, 77, 87 and 88 (tables), 688
Overaging, 158
  creep, 645
  fatigue, 585

Partial dislocations, 122
Peierls-Nabarro stress, 115, 116
Pencil glide, 104
Peptide polymerization, 220, 221
Perfect dislocation, 100
Permanent set, 6, 29, 310
  in fibers, 686
Permutation tensor, 63
Phase angle, in complex modulus, 294
  in damping, 473
Phase space, 12
Photoelasticity, 89f
Physical constants, 40 (table)
Piezoelectricity, 89f
Planck's constant, 15, 40
Plane strain, plastic equations of, 375f
Plane strain vs. plane stress, elasticity, 395
  plasticity, 371, 398, 519 (table)
Plastic constraint, 373
Plastic contact in friction, 663f
Plastic limit load (*see* Limit load, 363f)
Plastic strain concentration at notches, 408
Plastic zone, 408f, 534
  at fracture, 539, 538–539 (data)
  related to shear lip, 540
Plasticity, 6, 30 (*see also* Limit load, Stress-strain distribution)
  constitutive relations, 276f, 278, 283f, 307
  in crystalline materials, 152ff
  damping due to, 480f
  by dislocations, 96ff, 152ff

SUBJECT INDEX

effect on brittle fracture, 563
effect on equivalent flow stress, 279f
effect on fatigue, 600
effect on residual stress, 421f
effect on stress strain concentrations, 414
by kinking, 135, 632
plane strain equations, 375f
in polymers, 253
role of thermal activation in, 30, 139f
by twinning, 24, 34, 134f, 180f, 555
Plate, bending, 349f
  elastic, grooved or notched, 406f, 406 (table), 411
  elastic plastic, notched in shear, 409, 535f
  plastic, doubly grooved, 371, 379, 408f, 519 (table)
  plastic, singly grooved, 368f, 372, 378, 519 (table)
  thermal stress, 351f
Point defects, 23, 136f
  role in damping, 479
  stored energy, 179
Poisson's ratio, 79, 86 (data), 701
  conversion to other constants, 80 (table)
  due to dislocations, 108, 288
Polygonization, 629
Polymers, 25, 215ff
  composition, 217 (table)
  crystallinity, 232, 253f
  crystallization, 224, 225
  mechanical behavior, 30, 230f, 257f, 258 (table), (see also Elasticity, Viscoelasticity, Viscosity)
    plasticity, 253f
    stages of deformation, 230f
    thermoplastic vs. thermosetting, 26, 227, 233
  polymerization, 216
    addition, 218
    amine, 220
    condensation, 218
    diene, 218
    ester, 218
    formaldehydes, 222
    silicones, 223
    vinyl, 217
  structure, 215f
    atactic, 225, 230, 232
    branched, 217
    crystalline, 27, 230
    of fiber, 677f

isotactic, 225
linear, 26, 217
space network, 25
syndiotactic, 225
transparency, 253
Portevin-LeChatelier effect, 189
Precipitation hardening, 157, 194, 631
Preload in bolts, 707f
Pressure vessels, 712f, 717f
Primary creep (see Creep, transient)
Primitive cell, 17
Principal axes, 63
Principal slip component, 280
Principle of maximum plastic resistance, 285, 366
Principle of superposition, 82, 247, 294
Principle of virtual work, 364f
Product of inertia, 346
Propagation of error, 240
Proportional limit, 310
Punching sheet, 391

Quality factor, 473
Quantum mechanics, 15, 235
Quench hardening, 159
Quenching, of metals, 196 (table), (see also Hardening, modes of)
  residual stresses, 351, 426 (table), 427f, 499

Radiation damage, 159
  effect on creep, 642
  effect on fracture transition, 563
  effect on yield strength, 159, 563
Radius of gyration, 332
Rate effect (see Strain rate effect)
Rate process, 16f (see also Activation energy)
Rate of strain hardening (see also Hardening, strain), 171, 327
  effect on maximum load, 319f
  effect on roughening in fatigue, 582
Rayleigh wave velocity, 110, 297, 500
Recovery, annealing, 177
  in creep, 640
  elastic after-effect, 6, 245, 269
  for fibers, 686, 687 (data)
Recrystallization, 177
Reduction of area, 8, 196 (table), 325
Reinforced plastic, 77, 87 (data), 701, 717
Relative damping, 473
Relativistic energy of dislocation, 110
Relaxation, 6, 245, 641
  of polyethylene, 232

Relaxation modulus, 230, 247, 294
Relaxation time, 245, 548
Repeat units of polymers, 215
Residual stress, 342, 420ff, 719, 728
  bending, due to, 347, 351, 357, 422
  classification of sources, 420f
    chemical, 427, 428f, 499
    composite, 429f
    mechanical, 421f, 719f
    metallurgical, 427f
    thermal, 421f
  effect on fatigue, 591, 598
  effect on yielding, 432f
  measurement, 429f
  relief, 433f, 719f
    in springs, 719f
  torsion, due to, 342, 357, 438
Resolved shear stress, 131
Rib marks, 503
Rigid idealization of behavior, 275
Rigid-plastic idealization of behavior, 275
Ring, limit load, 389
River marks, 567
Rocket casings, 712f
Rolling contact fatigue, 607f
Rosette fracture, 529
Rotation, 57
  of holes in fracture, 527
Roughening, in fatigue, 582
Rubber, elasticity, 4, 29, 231, 238f, 683
  free energy, 238, 241
  stress-stretch relations, 243
  vulcanization, 227
Rupture, 8, 518f
  creep (*see* Creep fracture, 643f)
  elongation, 519

St. Venant's principle, 397
  breakdown in plasticity, 399
Sandwich construction, 37
Screw dislocation, 97
  stress field, 106 (*see also* Dislocation)
Screw threads, 707, 711
  friction, 711
  limit load, 373
Secondary creep, 245, 625, 626, 639
Self-diffusion, 138, 138 (table), 631 (data)
Sessile dislocations, 105, 159, 167, 554
  jogs on, 167, 629
Shakedown, 422
Shear of, grooved plate, 405, 409
  grooved strip, 305
  junctions, 663
  notched sheet, 389

  rubber, 242
  sheet by cutting, 391
Shear components, of strain, 58
  of stress, 49
Shear fracture, apparent, 529
  torsion, 529, 531 (table), 533 (data)
Shear modulus, 79, 86 (table)
  conversion to other constants, 80 (table)
  of MgO, 6 (data)
Shear spinning, 384
Shear strength, ideal, 118
Shear wave velocity, 297
Shearing (cutting) of sheet, 391
Sheet, notched elastic, 405f, 406 (table), 411
  notched plastic, 389
Shrink fits, 395
Silicones, 223
Single slip, 160f, 165
  orientation effect in, 165
Singly grooved plate elastic, 405, 406 (table)
  plastic, 368f, 372, 408, 519 (table)
Size, structural, for fracture, 536
Size effect, brittle fracture, 504f, 496 (data)
  ductile fracture, 534
  easy glide, 165
  fatigue, 594
  hardness, 458
  statistics, 504f
Sliding junction, limit load of, 663
Slip (*see also* Dislocation), 102, 162f
Slip bands, 32, 162, 177
  in fatigue, 577
Slip domains, 169, 182
Slip line, 32
Slip systems, 104 (table)
Soderberg diagram, 591
Softening temperature of polymers, 253
Solution hardening, dilatational misfit, 153
  distortional misfit, 155
  effect on fracture, 645
  stiffness misfit, 155
Specific damping energy, 473
Specific gravity (*see* Density, 82, 86, 258)
Specific heat, 14
  for elements, 16 (table)
  for polymers, 234 (table)
Spherical coordinates, equilibrium in, 54
  strain in, 61
Spherical holes, 401f
Spheroidizing, effect on fatigue, 601

Spherulites, 27, 28, 254
Spinning, sheet metal, 384
Springs, elastic plastic deformation in, 720
   geometry of helical, 693, 706
   stress relief in, 439, 719f
Stacking-fault energy, 123
Stage I, II hardening (see Hardening, Strain 160f)
Standard deviation, 240, 507
   in fatigue, 588, 595f
Standard linear solid, 471
State of knowledge of mechanical behavior, 37f
Static fatigue, 499
Statistics, brittle fracture, 504f
   coefficient of variation, 691
   extreme-value distribution, 507, 597, 691, 692
   fatigue and notch sensitivity, 595
   regression coefficient, 679
   standard deviation, 240, 507
   strength of yarn, 689
Statistical mechanics, 12f, 239f
Steady-state creep, 245, 625, 626, 639
Stiffness, elastic (see Elastic compliance and stiffness)
Stirling's formula, 136
Stored energy, 150
   of strain hardening, 179f
Strain, 55f
   components, cylindrical coordinates, 60
      normal, 58
      shear, 58
      shear, tangential, 58
      shear, tensor, 58
      spherical coordinates, 61
      transformation, 61f
   deviatoric, 81, 291
   dilatational, 59, 291
   engineering, 318
   finite, plastic, 316
   finite, rubbery, 240
   logarithmic, 318
   measurements, 64f
   nominal, "engineering," 318
      smooth specimen, 394
   true, 318
   uniform, 320
Strain aging, 188
   in fatigue, 605
Strain concentration (see Stress-strain concentration, or Stress-strain distributions, specific shapes)
Strain-displacement equations, 58, 274

Strain-hardening (see Hardening, strain)
Strain rate, effect on damping (see also Creep, Viscoelasticity, Viscosity), 471ff
   fatigue, 599
   fibers, 684
   flow stress, 173, 190f
   fracture transition, 547, 560, 564
   hardness, 464
   wave propagation, 297
Strain-softening, 523
   in fatigue, 585
Strain-strengthening against cleavage, 563
Strength, ideal cohesive, 489, 495, 523
   ideal shear, 118
   metals and alloys, 194f, 196 (table)
   polymers, 259 (table)
   tensile, 8, 320
   yield, 7, 310, 311, 326 (data)
   yield in fibers, 255, 682
Stress, 47ff
   components, deviatoric, 81, 291
      hydrostatic, 67
      mean normal, 67
      normal, 49
      principal, 63
      shear, 49
      transformation, 50f
   engineering, 318
   flow (see Flow stress)
   nominal, "engineering," 318
      smooth specimen, 394
   residual (see Residual stress)
   true, 318
Stress analysis, approximate, 359ff
   experimental, 64
Stress and strain distribution (see Stress-strain distribution)
Stress concentration (see Stress-strain concentrations or specific shapes under Stress-strain distributions)
   factor, 394
   in fatigue, 594, 595, 598, 711
Stress corrosion, 9, 184
Stress distribution (see also Stress-strain distribution), equivalance of viscoelastic and elastic, 291
Stress field (see also Stress-strain distribution)
   of dislocation, 106
Stress relief,
   in bolt, 656
   in springs, 719f

Stress singularity (or intensity) factor, 405f, 407 (table)
  for fracture, 539 (table)
Stress-strain concentrations, 393ff
  amelioration of, 413f
  concept of, 393
  effects of, 412f
  estimation of, 411f
Stress-strain distributions, plane-strain plasticity, 375f
  plane strain vs. plane stress, elastic, 395
  plane strain vs. plane stress, plastic, 371, 398, 519 (table)
  St. Venant's principle, 397, 399
  uniqueness of elastic, 81f
  uniqueness of plastic, 378f
Stress-strain distributions, specific shapes
  bar in tension, 309ff
    circumferentially notched, 372, 519 (table)
    necking, 320f, 322, 643
    transverse hole, 363
  bar in torsion, 339f, 355, 519 (table)
  beams, 344f, 355
    at base, 412
  bimetallic strips, 351
  bolt head, 392
  contact, elastic, 410f
    plastic, 663f
  cracks, elastic, 403f, 405, 406 (table)
    elastic-plastic shear, 409, 535, 537, 544
    plastic, 370f, 378f, 408f, 519 (table)
  dislocation, 106f, 169
    high velocity (see p. 110)
  hardness indentation, 373, 454f
  holes, cylindrical under biaxial stress, 394f, 525
    cylindrical under uniaxial stress, 363, 398f
    elliptical, 404f
    spherical, 401f
  metal forming (see also Bending, 339ff)
    drawing, 383
    extrusion, 383
    forging, 380
    machining, 389, 452 (data)
    punching, 391
    shearing (cutting) of sheet, 391
    spinning, 384
    wire drawing, 383, 391
  plate, bending, 349f
    doubly grooved plastic, 371, 379, 519 (table)
    grooved, elastic-plastic shear, 409, 535f
    notched elastic, 405f, 406 (table), 411
    notched plastic, 389, 519 (table)
    singly grooved plastic, 368, 372, 378, 408, 519 (table)
    thermal stress, 351f
  ring, 389
  screw threads, 373, 707, 711, 733
  sheet, notched elastic, 405f, 406 (table), 411
    notched plastic, 389, 519 (table)
  shrink fits, 395
  sliding junctions 663f
  strip, grooved plastic, 305
    notched elastic, 405f, 406 (table), 411
    notched plastic, 389, 519 (table)
  turbine blade root, 390
  thermal shock, 351
Stress-strain relations (see Constitutive relations)
Stress waves, 295f
  in crystals, 15
  Rayleigh (surface) wave, 297
  shear, 297
  uniaxial strain, 297
  uniaxial stress, 296
Stretch ratios, 239
Strip, grooved plastic, 305
  notched elastic, 405f, 406 (table), 411
  notched plastic, 389, 519 (table)
Structural damping, 483f
Structural hardening, 153f
Structural size for fracture, 536
Structure of, crystals, 17f
  dislocation arrays (see Dislocation)
  glass surfaces, 498
  polymeric fibers, 35, 677f
  polymers, 25f, 215f
  solids, 3ff
    macroscale, 36f
    microscale, 31f
Structures, associated flow rule for, 289
  mechanics of textile, 687f
Subgrain, in fatigue, 581, 583
Substitional impurities (see also Hardening), 23
Summation convention, 63
Superposition principle, 82, 247, 294
Surface defects (see also Twin, Grain boundary), 24
  polygonization, 629
  subgrain, in fatigue, 581, 583

tilt boundaries, 131f
twist boundaries, 132f, 567
Surface effects on, dislocations, 113, 165
  flow stress, 180f
Surface electrical resistance, 659
Surface energy, 490, 539, 556
Surface fatigue, wear, 670
Surface wave velocity, 110, 297, 500
Suzuki hardening, 189
Symmetric tensor, 62
Symmetry, in crystal structure, 17, 78
  in elastic stress-strain relations, 76f
Syndiotactic polymers, 225

T_g, T_i, T_m (see also Transition temperature), 231
  for polymers, 259 (table)
Tangential components of shear strain, 58
Temperature, effect on, creep, 627f, 637f, 645f
  damping, 475f
  fatigue, 605
  flow stress, 172f, 190f
  fracture transition, 547, 560, 564
  hardness, 459f
  rubber elasticity, 242
  viscoelasticity, 230f, 249f
  viscosity, 237f
Tempering (see also Hardening), glass, 499
  metals, 196 (table)
Tenacity, 682
Tensile deformation (see also Stress-strain distributions), 309ff
  anisotropic elastic, 309f
  ductility, 325f
  finite plastic strain, 316f
  initiation of plastic, 310f
  instability, 330f
  localization of plastic, 311f
  maximum load in, 319f
  necking, 320f, 322f
  rubber, 4, 242
  stress-strain curves, 326f
  types of data, 326f
Tensile strength, 8, 320
  for metals, 196 (table)
  for polymers, 259 (table)
Tensor, 62f
  components of stress and strain, 62
  permutation, 63
  symmetry of, 62
  transformation, 62
  unit, 63

Tertiary creep, 625
Textiles (see also Yarn)
  bonded (nonwoven), 95, 688f
  classification of fibers, 677 (table)
  felt, 680
  mechanical properties, 702 (list)
  mechanics, 688f, 697
  structure, 37, 676, 687
  woven, 697f
    modulus, 699
    Poisson's ratio, 701
Thermal activation, 16
  creep, 627f, 637f
  dislocation motion, 140f, 172f
  plastic deformation, 139f, 172f, 190f
  point defects, 136f
  viscoelasticity, 230f, 249f
  viscosity, 237f
Thermal damping, 475f
Thermal diffusivity, 477
Thermal expansion, 86 (table), 87
  for polymers, 86 and 258 (tables)
  stress dependence, 76
Thermal fatigue, 642
Thermal motion, 9, 12f, 30
Thermal shock, 351
Thermal stress, 351f
Thermoplastic and thermosetting polymers, 26, 227, 233
Thick-walled cylinders, 395
Tilt boundaries, 131f
Time constant, in damping, 471f, 477, 478
  of viscoelasticity, 236, 237
Ton, British or long, 588
Torsion, 339ff
  creep of nylon, 343
  fatigue, 582
  fracture, 531 and 533 (tables)
  limit load of round bars, 342
  round cylindrical bars, 339f
  rubber, 243
  stress-strain from torque-twist, 343f
  torque-twist curve, 6, 343f, 582
  wolf's ear fracture, 530
Trade names, of polymers, 216 (table)
  of textile fibers, 617 (table)
trans (structure of polymers), 225, 226
Transformation of components of,
  elastic compliance, 76
  elastic stiffness, 74
  strain, 61f
  stress, 50f
  tensors, 62f

thermal expansion, 75
vectors, 63
Transformation of partial derivatives, 61
Transient creep, 5, 178, 245, 626, 637
Transition (*see also* Fracture transition, Brittle fracture, 488ff, Ductile fracture, 518ff), fracture mode, 546ff rubbery, 249
Transition temperature (*see also* Fracture transition)
  fracture appearance (FATT), 569
  fracture energy, 559, 569
  glass, 231, 249, 255
    for polymers, 259 (table)
  inflection, 231
    for polymers, 259 (table)
  modulus, 231
    for polymers, 259 (table)
  second-order, 231
Tresca yield criterion, 278
"True" strain, in plasticity, 316
  of rubber, 241
"True" stress, 318
Turbine blade root, limit load, 390
Turbine wheel, service fracture, 721f
Twin bands, 34
Twin boundary, 24
Twinning, 24, 34, 134f, 180f, 555
Twist boundaries, 132f, 567

Ultimate tensile strength (*see also* Tensile strength), 320
Uniaxial strain wave velocity, 297, 301
Uniaxial stress (*see* shape of part)
Uniform strain, 320
Uniqueness, in elasticity, 81f
  in plasticity, 371, 378f
Unit cell, 17
Unit dislocation, 100
Unit tensor, 63
Unrelaxed modulus, 472
Upper bounds, limit load (*see also* Limit analysis), 365
Upper bounds, to stiffness, 360
Upper yield point, 186, 311
  in fibers, 255, 682

Vacancies, 23, 121
Vacancy clusters, 25, 151, 159
Vacancy concentration, 137
Vacancy diffusion, 137f
  activation energy for, 138 and 176 (tables), 631 (data)
  in creep, 628f, 632

Vacancy disks, 159
Vacancy formation, 136f
  energy, 121, 138 (table)
Vacuum melting, effect on fatigue, 603
van der Waals bonds, 11
Vegard's law, 403
Velocity modified temperature, 191, 725
Vinyl polymerization, 217
Virtual work, principle of, 364f
Viscoelastic functions, 293
Viscoelastic idealization, 243f
Viscoelasticity, 30, 233ff
  constitutive relations, 290f
  equivalent models, 246, 249, 253
  in fibers, 685
  linearity, 232, 233, 236, 243f, 247f, 275, 686
  nonlinearity, 232, 233
  small strain, 243f
  spring-dashpot models, 243, 244 (table)
Viscosity, 30, 233f
  anisotropic, 238
  coefficient of, 237
  effect of temperature, 237, 249
  liquids, 30, 237
  Newtonian (*see* linear, 236)
Viscous deformation, 6, 30, 275
Voigt material (element), 243, 244
Volume change around an edge dislocation, 108

Wallner lines, 503
Wave mechanics (*see also* Quantum mechanics, 15, 235), 38
Wave motion (*see also* Stress waves), 295f
Wear, 9, 657f
  abrasive, 699
  adhesive, 667
    lubricated surfaces, 669
  coefficient (constant) K, 667, 668 (data)
  corrosive, 670
  fretting, 671
  surface fatigue, 670
Weissenberg effect, 238
Wire drawing, 383, 391
Wolf's ear fracture, 530
Work-equivalent shear strain, 280
Work, in hardness test, 451, 452 (table), 454
  per unit volume, 65f
  virtual, 364f
Work hardening (*see* Hardening, strain)

X-ray line broadening, fatigue, 578

X-ray, stress measurement, 431

Yarn, 676
  embedding, 696
  geometry, 692
  helix, 695
  mechanics of, 692f
  modulus, 695
  statistics of strength, 689
  twisting and bending, 696
Yield criteria (functions or loci),
      anisotropic metals, 278
  combined stress, 276f
  effect of strain hardening, 279f
  structures, 279
Yield in crystals, 186–190
  chemical interaction (Suzuki effect), 189
  dislocation multiplication, 188
  impurity misfit, 186
  order-disorder, 190
  Portevin-LeChatelier effect, 189
  prestrain, 190
  thermomechanical instability, 190
  twinning and kinking, 190
Yield point, in fibers, 255, 682
  vs. yield strength, 311
Yield strength (*see also* Hardening),
      7, 310, 326 (data)
  offset, 311
  set strength, 311
Young's modulus (*see* Modulus of
      elasticity)

Ziegler process, 217, 254

MATERIALS INDEX

For data, micrographs, and references for specific materials

Polymers are indexed by chemical name, with cross references from the trade names. Alloys are indexed under their principal elements and subindexed in alphabetic order of the principal secondary element and its amount, followed by treatments.

Systematically numbered alloys "Aluminum," "Steel," and "Steel, stainless" are at the ends of the lists under those headings. Page entries without a specific property listing denote information on composition or structure. The following abbreviations are used for heat treatment and processing (except for aluminum):

age hard.	age hardened (often in repeated steps)
ann.	annealed
CD	cold drawn
CR	cold rolled
HR	hot rolled
norm.	normalized
Q and T	quenched and tempered
PH	precipitation hardened

for aluminum and its alloys:

—O	annealed wrought materials
—F	as fabricated
—H18	cold worked to commercial hardness
—H19	cold worked to extra hardness
—T3	solution treated and strain hardened
—T4	solution treated and naturally aged
—T6	solution treated and artificially aged
—T8	solution treated, strain hardened, and artificially aged

For each material and treatment the following data are subindexed where available, along with structural characteristics shown in micrographs:

creep
damping
elongation
fatigue
fracture (under increasing load)
hardness
modulus (of elasticity or of shear)
reduction of area
stress-strain curve
tensile strength
yield (point or strength)

MATERIALS INDEX

For other characteristics, see the appropriate entries in the Subject Index. For example, there are tables or figures containing the following data for a variety of materials:

Activation energy,
    for vacancies and self diffusion, 138
    for creep and self diffusion, 631
Bulk modulus, 82
Cost of metals, 196
Cost of polymers, 258
Critical resolved shear stress, 130
Density of metals, 86
Density of polymers, 258
Friction coefficients, 664
Heat resistance of polymers, 258
Machining work, 452
Magnetostriction, 91
Melting point of metals, 462, 631
Poisson's ratio, 86
Shear modulus, 86
Slip systems, 104
Specific heat, 234
Strain hardening exponent, 457
Textile fiber classification, 677
Thermal coefficient of expansion,
    for metals, 86; for polymers, 258
Transition temperature for polymers, 258
Wear constants, 668

ABS (acrylonitrile-butadiene-styrene), 225
Acetate (*see* Cellulose acetate)
d-Alanine, 220
Aluminum (*see also* 1100)
  ann. or CR, hardness, 460
  quench hardened, dislocation loops in, 160
  single crystal, modulus, 87
    stress-strain curve, 165, 171, 183
  unspecified condition, creep, 631, 634
    deformation bands, 33
    deformation in creep, 633, 635
    dislocation cell structure, 168
    dislocation structures in fatigue, 583
    hardness, 455, 460, 462, 463
    modulus, 86, 631, 634
    polygonization, 630
    slip bands, 32
    slip traces, 182
    stress-strain curve, 183
    yield, 455

Aluminum alloy, modulus, 86
  2 Cu, deformation in creep, 633
  5 Cu, single crystal, modulus, 87
  DTD 364B, fatigue, 599
  Duralumin, fatigue, 605
    fracture appearance, 528
    hardness, 460
    modulus, 681
    tensile strength, 605, 681
  Mg, creep, 647
    hardness, 460
    yield, 647
  7.5 Zn, 2.5 Mg, slip bands in fatigue, 581
  1100-F, elongation, 200
    tensile strength, 200
    yield, 200
  1100—H19, fracture, 538
    uniform strain, 538
    yield, 538
  1100-O, elongation, 200
    fatigue, 590

MATERIALS INDEX 763

fracture, 524, 531
modulus, 531
reduction of area, 531, 590
stress-strain curve, 328
tensile strength, 200, 328
yield, 200, 328
2011-T8, elongation, 200
tensile strength, 200
yield, 200
2014-T6, fatigue, 592
fracture, 532
tensile strength, 532, 592
yield, 592
2024-O, modulus, 328
reduction of area, 328
stress-strain curve, 328
tensile strength, 328
yield, 328
2024-T3, fatigue, 585
fracture, 539
hardness, 389
yield, 539
2024-T4, damping, 590
fatigue, 577, 590, 592, 604
fracture appearance, 529
fracture strain, 590
modulus, 328, 363
reduction of area, 328
stress-strain curve, 328, 363
tensile strength, 328, 577, 592
yield, 328, 363, 592
6061-T6, fatigue, 592
tensile strength, 592
yield, 592
6062-T6, elongation, 200
tensile strength, 200
yield, 200
7075-T6, fatigue, 592
fracture, 531, 533, 538, 539
reduction of area, 531
tensile strength, 531, 539, 592
yield, 592
7178-T6, elongation, 200
tensile strength, 200
yield, 200
Aluminum bronze (see Copper, Al)
Aluminum oxide, hardness, 459, 461
Amides, 220
Aminoacetic acid, 220
Antimony, hardness, 462, 463
Apatite, hardness, 450
Asbestos, fiber, modulus, 681
tensile strength, 681

Bakelite (see Phenol formaldehyde)
Beryllium, hardness, 462
Beryllium bronze or copper (see Copper, Beryllium)
Bismuth, hardness, 463
Boron carbide, hardness, 459, 461
Brass (see Copper, Zn)
Bronze (see Copper, Sn; Copper, Al; Copper, Be)
Buna rubber, 218

Cadmium, hardness, 460, 462, 463
single crystal, stress-strain curve, 171
Calcite, hardness, 450
Calcium, hardness, 462
Carbides, hardness, 459, 461, 462
Casein fiber, stress-strain curve, 685
Cellulose, regenerated, "Fortisan," 677
creep, 687
elongation, 687
stress-strain curve, 682, 685
regenerated, rayon (viscose), 221
stress-strain curve, 685
Cellulose acetate, 221
elongation, 259
hardness, 460
modulus, 258
stress-strain curve, 684, 685
tensile strength, 259
Cellulose nitrate, 221
elongation, 259
modulus, 258
tensile strength, 259
Chemcor, 428, 499
Chromium, hardness, 462
Cobalt, hardness, 462
Cobalt, Cr, W, hardness, 462
Copper, annealed, elongation, 199
fracture, 530, 531
hardness, 456
modulus, 86
plastic waves, 296, 297
stress-strain curve, 5, 281, 456
tensile strength, 199
torque-twist in fatigue, 582
yield, 5, 199, 281, 456
cold rolled, recryst, modulus, 87
cold worked, creep, 179
elongation, 199
hardness, 460
modulus, 86, 87
tensile strength, 199
yield, 199

electrolytic, annealed
  fracture, 521, 522
  hardness, 324
oxygen free, high conductivity
    (OFHC), ann.
  fatigue, 590
  fracture strain, 590
single crystal, cross slip, 177
  modulus, 87
  stress-strain curve, 161, 170, 171,
    173, 181
single crystal, chrome plated,
  stress-strain curve, 181
unspecified condition,
  friction stress (disloc.), 117
  hardness, 455, 462, 463
  mechanical twins, 134
  slip bands in fatigue, 150
  yield, 277, 282, 455
Copper alloys, Al (bronze), fatigue, 605
  tensile strength, 605
7 Al, single crystal
  dislocations in easy glide, 166
Au, yield, 156
Be, fatigue, 605
  tensile strength, 605
Be, ann., CR or age hardened
  elongation, 199
  tensile strength, 199
  yield, 199
Ge, yield, 156
In, yield, 156
Mn, damping, 482
  yield, 156
Ni, yield, 156
Sb, yield, 156
Si, yield, 156
Sn (bronze or phosphor bronze), fatigue,
    605
  tensile strength, 605
  yield, 156
10 Sn, hardness, 460
25 Zn, 15 Ni (german silver), damping,
    477
28 Zn (brass), single crystal, modulus,
    87
30 Zn, ann., damping, 478
  fatigue, 602
  hardness, 460
  modulus, 86
  reduction of area, 329
  slip lines in, 31
  stress-strain curve, 329
  tensile strength, 329

30 Zn, CD, elongation, 199
  fatigue, 602
  modulus, 86
  tensile strength, 199
  yield, 199
40 Zn, ann. or CD, hardness, 460
  modulus, 86
40 Zn, CR, fracture, 531
  reduction of area, 531
  tensile strength, 531
Corundum, hardness, 450
Cotton, creep, 687
  elongation, 687
  modulus, 679
  stress-strain curve, 682, 685
"Cronar" (see Polyethylene terephthalate)

"Dacron" (see Polyethylene
    terephthalate)
"Delrin" (see Polyacetal)
Diamond, hardness, 450, 461

Epoxy, 224
  elongation, 259
  modulus, 258
  tensile strength, 259

Fiberglass, polyester, modulus, 87
  stress-strain curve, 686, 717
Flax, modulus, 681
  tensile strength, 681
Fluorite, hardness, 450
Formaldehyde, phenol, urea or melamine,
    222
  elongation, 259
  modulus, 258
  tensile strength, 259
"Fortisan" (see Cellulose, regenerated)

German silver, 60 Cu, 15 Ni, 25 Zn,
  damping, 477
Glass, 228f
  compressive strength, 493
  damping, 479
  elongation, 259, 328
  fracture, 496, 498, 501, 502
  hardness, 461
  modulus, 86, 258
  stress-strain curve, 328
  tensile strength, 259, 328
Glass, fiber,
  fracture, 496
  modulus, 681
  tensile strength, 681

Glycine, 220
Glyptal, 219, 227
Gold, hardness, 462
  annealed, elongation, 202
    hardness, 460
    tensile strength, 202
  single crystal, stress-strain curve, 171
Graphite, extended dislocations and nodes, 124
  hardness, 461
GRS (Polybutadiene, styrene), 218, 225
  modulus, 268
Gutta percha (Polyisoprene, trans), 218, 225, 226
Gypsum, hardness, 450

Hastelloy, R-235 (see Nickel, 15 Cr, 10 Fe)
Heat resisting alloys (see also Nickel, Cobalt, Steel, Steel, stainless), creep, 646

"Inconel" (see Nickel, 30 Co or 15 Cr)
Indium, hardness, 460
Invar, (see Iron, 36 Ni)
Iridium, hardness, 462
Iron (see also Steel)
  carbide precipitates, 159
  cleavage cracking, 557
  cracks in grain boundary carbides, 553
  dislocation density, 213
  fracture markings, 566
  grain boundaries, 24
  hardness, 461
Iron, ann., stress-strain curve, 184
  CR, hardness, 460
  ——modulus, 87
  ingot, norm., elongation, 196
    fracture, 559
    reduction of area, 196
    tensile strength, 196
    yield, 196
  single crystal, fracture, 560
    modulus, 87
    reduction of area, 560
    stress-strain curve, 186
    yield, 560
  spectrographic, fracture, 559
  wrought, fracture appearance, 528
Iron, cast (>4.3C), compressive strength, 328
  elongation, 328
  fracture appearance, 528

  hardness, 460
  modulus, 86, 328
  reduction of area, 328
  stress-strain curve, 328
  tensile strength, 328
  yield, 328
36 Ni (Invar), ann., elongation, 202
  tensile strength, 202
  yield, 202
silicon, cracks at twin intersections, 555
3 Si, friction stress on dislocations, 117
Iron oxide, hardness, 461
Isoprene (see Polyisoprene)

Kennametal carbide, hardness, 459

Lead, creep, 642
  elongation, 202
  hardness, 460, 462, 463
  modulus, 87
  tensile strength, 202
  yield, 202
Lead alloys, 50 Sn (solder), as cast elongation, 202
  tensile strength, 202
  Te, hardness, 455
    yield, 455
Lithium fluoride, dislocation motion, 116
  dislocation pile-up against surface, 180
  dislocation velocities, 117
  friction stress (disloc.), 117
  low angle boundaries, 132
Lucite (see Polymethyl methacrylate)

Magnesium, dislocation dipoles, 164
  dislocation loops around precipitates, 195
  hardness, 460, 462
  single crystal, stress-strain curve, 171
Magnesium alloy, $2\frac{1}{2}$ Al, fatigue, 605
  tensile strength, 605
Magnesium oxide, cracks at dislocation pileups, 553
  modulus, 6
  slip bands, 162
  stress-strain curve, 6
  yield, 6
Melamine formaldehyde (MF), 222
  elongation, 259
  modulus, 258
  tensile strength, 259
Mica, strength, 495

Molybdenum, hardness, 462
  as rolled, elongation, 203
    tensile strength, 203
    yield, 203
Monel, K, ann. and age hard.
  elongation, 201
  tensile strength, 201
  yield, 201
"Mylar" (see Polyethylene terephthalate)

Neoprene (see Polychloroprene)
Nickel,
  annealed, elongation, 201
    fatigue, 590
    fracture strain, 590
    fracture strength, 201
    yield, 201
  single crystal, stress-strain curve, 171
  unspecified conditions, hardness, 460, 462
    slip bands in fatigue, 580
    yield, 277
Nickel alloys
  17 Co, 15 Cr, 5 Mo, 4 Al, 3 Ti, 4 max. Fe, (Udimet 700), age hard., grain boundary precipitation, 635
  30 Co, 15 Cr, 3 Mo, 2.2 Ti, 3.2 Al, (Inconel 700), fatigue, 605
    tensile strength, 605
  15 Cr, 7 Fe, 2 Ti, 1 Al, (Inconel X), ann. or age hard., elongation, 201
    tensile strength, 201
    yield, 201
  15 Cr, 7 Fe, 2.5 Ti, 1 Cb, 1 Al, (Inconel X) yield, 647
  15 Cr, 8 Fe, creep, 644
  15 Cr, 10 Fe, 5 Mo, 2.5 Ti, 2 Al, (Hastelloy R-235), creep, 638
  19 Cr, 10 Co, 4 Mo, 3 Al, 3 Ti, 4 max. Fe, (Udimet 500), age hard., creep, 649
  19 Cr, 11 Co, 10 Mo, 3 Ti, 1.5 Al, 3 Fe, (René 41), fatigue striations, 586
  20 Cr, 1.3 C, age hard., creep, 645
  unspecified superalloys, creep, 646
"Nivco", damping, 482
Nylon,
  creep, 343
  elongation, 259
  modulus, 258
  stress-strain curve, 684, 685
  tensile strength, 259
Nylon fiber, creep, 687
  elongation, 687

  modulus, 681
  stress-strain curve, 684, 686
  tensile strength, 681
Nylon woven with neoprene filler, shear modulus, 699
Nylon 300 fiber,
    stress-strain curve, 682
  6, 220
  6-6, 220
  610, kink band, 254

"Orlon" (Polyacrylonitrile), 217
Orthoclase, hardness, 450
Osmium, hardness, 462

Palladium, hardness, 462
PAN (polyacrylonitrile), 217, 225
PB (see Polybutadiene)
PE (see Polyethylene)
"Perspex" (see Polymethyl methacrylate)
Phenol formaldehyde (PF), 222
  elongation, 259
  hardness, 460
  modulus, 258
  tensile strength, 259
Phosphor bronze (see Copper, P)
PIB (Polyisobutylene), 217
Platinum, hardness, 462
  annealed, elongation, 202
    hardness, 460
    tensile strength, 202
"Plexiglas" (see Polymethyl methacrylate)
PMMA (see Polymethyl methacrylate)
Polyacetal, 223
  elongation, 259
  modulus, 258
  tensile strength, 259
Polyacrylonitrile, 217, 225
Polybutadiene, 218, 225
  styrene rubber, modulus, 268
  25% styrene rubber, elongation, 259
    modulus, 258
    tensile strength, 259
Polychloroprene, 218
  elongation, 259
  modulus, 258, 268
  tensile strength, 259
Polyethylene, 216, 225
  creep compliance, 248
  drawing, 256
  modulus, 86
  relaxation, 232
  spherulites, 27

Polyethylene film, low or high density,
  elongation, 259
  modulus, 258
  tensile strength, 259
Polyethylene terephthalate, 219
  elongation, 259
  modulus, 258
  tensile strength, 259
Polyethylene terephthalate fiber,
  creep, 687
  elongation, 682, 684
  stress-strain curve, 682, 684
Polyisobutylene, 217
Polyisoprene, cis (natural rubber), 218,
    225, 226
    elongation, 259
    hardness, 460
    modulus, 4, 86, 258, 268
    stress-strain curve, 4, 243
    tensile strength, 259
  trans (gutta percha), 218, 225, 226
Polymethyl methacrylate, 217
  activation energy, 260
  elongation, 259
  fracture, 503, 539
  modulus, 86, 258
  relaxation modulus, 250, 251
  tensile strength, 259
  yield, 539
Polypeptides, 220
  stress-strain curve (silk), 685
Polypropylene, 216
  elongation, 259
  modulus, 258
  tensile strength, 259
Polysaccharides, 221
Polystyrene, 216, 224, 225
  elongation, 259
  hardness, 460
  modulus, 230, 258
  relaxation modulus, 230
  tensile strength, 259
  transition temperature, 231
Polystyrene (atactic),
  deformation band, 257
Polystyrene foam, fracture, 731
  hardness, 732
  modulus, 731
  yield, 731
Polysulfide, 223, 224
  elongation, 259
  modulus, 258
  tensile strength, 259
Polytetrafluoroethylene, 217

  elongation, 259
  modulus, 258
  tensile strength, 259
Polyurethane, 219
  elongation, 259
  modulus, 258
  tensile strength, 259
Polyvinyl acetate, 217
Polyvinyl alcohol, 217
Polyvinyl chloride, 216
  creep, 233, 258
  elongation, 259
  tensile strength, 259
Polyvinylidene chloride, 216
  creep, 687
  elongation, 259, 687
  modulus, 258
  tensile strength, 259
Potassium chloride, cleavage steps, 567
  dislocation net, 125, 133
"Propene" (*see* Polypropylene)
PS (*see* Polystyrene)
PTFE (*see* Polytetrafluoroethylene)
PVA (Polyvinyl alcohol), 217
PVAc (Polyvinyl acetate), 217
PVC (*see* Polyvinyl chloride)

Quartz (*see* Silica)

Rayon, viscose (Cellulose, regenerated),
  221
  stress-strain curve, 685
Rhenium, hardness, 462
Rock, compressive strength, 494
Rubber, natural (*see* Polyisoprene, cis)
  synthetic, (*see* Polybutadiene,
    Polychloroprene, Polyisobutylene,
    Polysulfide)
Ruthenium, hardness, 462

Saran (*see* Polyvinylidene chloride)
d-Serine, 220
Silica, 223, 228, 229
  hardness, 450, 461
Silica, fiber, modulus, 681
  tensile strength, 681
Silicon, dislocation mill, 128
  dislocations, 167
  dislocation trails, 120
Silicon carbide, hardness, 461
Silicone, 223
  elongation, 259
  modulus, 258

tensile strength, 259
Silk (polypeptides), stress-strain curve, 682, 685
Silver, hardness, 462
  annealed, elongation, 202
    hardness, 460
    tensile strength, 202
    yield, 202
  single crystal, stress-strain curve, 171
Sodium, hardness, 460, 462
Sodium chloride, hardness, 460
Solder (see Lead, 50 Sn)
Steel, unspecified, fatigue, 604, 605
  fracture, 562
  hardness, 459, 461, 604
  modulus, 86
  yield, 277
Steel, carbon, (see also Steel, numbered compositions)
  as rolled, norm., or Q and T,
    elongation, 569
    fracture, 568
    hardness, 569
    reduction of area, 569
    tensile strength, 569
    yield, 569
  0.5% carbon, fatigue, 605
    tensile strength, 605
  0.8% C (eutectoid), pearlitic or spheroidized, elongation, 601
    fatigue, 601
    hardness, 601
    reduction of area, 601
    tensile strength, 601
    yield, 601
  music wire (0.7 to 1C), CD, modulus, 681
    tensile strength, 681, 719
    yield, 719
  project E, ann., fracture, 558
    reduction of area, 558
    yield, 558
  unspecified carbon,
    creep, 636, 646
    fatigue, 605
    fracture, 539, 559, 560, 561
    fracture markings, 565, 566
    grain and inclusion elongation, 35
    hardness, 455, 456, 462
    hardness of constituents, 461
    Lüders bands in, 34
    micrograph, 24
    modulus, 86

    reduction of area, 560
    slip bands in fatigue, 580
    stress-strain curve, 456
    tensile strength, 605
    yield, 455, 456, 539, 560
Steel, alloy (see also Steel, stainless and Steel, numbered compositions)
  fracture appearance, 528, 529
  hardness, 462
  modulus, 86
  Cr, Mo, V, air hard. and temp., 724
    creep, 725
    elongation, 726
    fatigue, 725
  2 Cr, 1.5 Si, 0.6 Mo (X200), air hard. and tempered, fracture strain, 532
    tensile strength, 532
  5 Cr, Mo, V, air hard. and temp., fatigue, 587, 604
    fracture, 532
    tensile strength, 532
  5 Cr, 1 Mo, 0.5 V, air hard. and temp., fracture, 712f
    tensile strength, 717
  5 Cr, 1.5 Mo, 0.4 V (H11), air hard. and temp., stress-strain curve, 326
  0.5 Mo, creep, 646
  Ni, Mo, fatigue, 596
  Ni, Mo, V, fracture, 539
    yield, 539
  $1\frac{1}{2}$ Ni, 1 Cr, 0.3 Mo (En-24), fatigue, 588
  $3\frac{1}{2}$ Ni, as rolled, elongation, 569
    fracture, 568, 569
    hardness, 569
    reduction of area, 569
    tensile strength, 569
    yield, 569
  18.5 Ni, 9 Co, 5 Mo, 0.6 Ti (VascoMax 300), ann. or maraged,
    elongation, 198
    reduction of area, 198
    tensile strength, 198
    yield, 198
Steel, numbered compositions
  1008, hardness, 460
  1018, ann., damping, 487
    fatigue, 590
    fracture strain, 590
  1020, hardness, 460
    modulus, 4
  1020 ann., hardness, 460
    stress-strain curve, 189, 192

MATERIALS INDEX 769

1020 CR or HR, elongation, 196
  reduction of area, 196, 327
  stress-strain curve, 327
  tensile strength, 196, 327
  yield, 196, 327
1020 HR, elongation, 196
  reduction of area, 196, 318
  stress-strain curve, 318, 327
  tensile strength, 196, 318
  yield, 196, 318
1020 norm., stress-strain curve, 186
  yield, 186
1034, fatigue, 426
  reduction of area, 426
  tensile strength, 426
  yield, 426
1036, fatigue, 606
1040 CR, HR, or Q and T,
  elongation, 196
  reduction of area, 196
  tensile strength, 196
  yield, 196
1045, fatigue, 596
1045 ann., hardness, 460
  reduction of area, 329
  stress-strain curve, 192, 329
  tensile strength, 329
  yield, 329
1045 CD, hardness, 460
  modulus, 433
  yield, 433
1095 ann., HR, or Q and T
  elongation, 196
  reduction of area, 196
  tensile strength, 196
  yield, 196
1330 ann., CR, or Q and T
  elongation, 197
  reduction of area, 197
  tensile strength, 197
  yield, 197
4130 ann., or Q and T, elongation, 197
  reduction of area, 197
  tensile strength, 197
  yield, 197
X4130 normalized or Q and T, fatigue, 602
  hardness, 602
4130 Q and T, reduction of area, 327
  stress-strain curve, 327
  tensile strength, 327
  yield, 327
4140 HR, hardness, 324

4340 air melted, fatigue, 603
  fracture, 603
4340 ann., CR, or Q and T
  elongation, 197
  reduction of area, 197
  tensile strength, 197
  yield, 197
4340 Q and T, elongation, 197
  fatigue, 589
  fracture, 531, 539, 565
  modulus, 313
  reduction of area, 197, 531
  tensile strength, 197, 531
  yield, 197, 313, 539
4340 vacuum melted, fatigue, 603
  fracture, 603
6150, Q and T, fracture, 539
  yield, 539
9262, fatigue, 438
  hardness, 438
  stress-strain curve, 438
  torsion, 438
  yield, 438
Steel, stainless (12–30% Cr) (see also
  Steel, stainless, numbered)
  dislocations in foil, 125
  14 Cr, 26 Ni, 3 Mo, 2 Ti (Discaloy),
    yield, 647
  15 Cr, fatigue, 605
    tensile strength, 605
  15 Cr, 26 Ni, 2 Ti, 1 Mo (A286),
    yield, 647
  16 Cr, 25 Ni, 6 Mo, yield, 647
  17 Cr, 1 Ni, fatigue, 605
    tensile strength, 605
  18 Cr, 8 Ni, creep, 646
    fatigue, 605
    hardness, 460
    tensile strength, 605, 719
    yield, 719
  20 Cr, 20 Ni, 20 Co, 4 Mo, 4 W, 4 Cb
    (S590), creep 643
    yield, 647
  20 Cr, 25 Ni, creep, 646
  22 Cr, 8½ Mn, Q and T, creep, 641
Steel, stainless, numbered
  302 CR, yield, 389
  303 ann. or CR, elongation, 198
    reduction of area, 198
    tensile strength, 198
    yield, 198
  304 ann. or CR, elongation, 198
    reduction of area, 198